T0342551

Design of Thermal Energy Systems

Design of Thermal Energy Systems

Pradip Majumdar

Registered Offices
John Wiley & Sons, Inc., 111 River Street, Hoboken, NJ 07030, USA
John Wiley & Sons Ltd, The Atrium, Southern Gate, Chichester, West Sussex, PO19 8SQ, UK

Editorial Office
111 River Street, Hoboken, NJ 07030, USA

For details of our global editorial offices, customer services, and more information about Wiley products visit us at www.wiley.com.

Wiley also publishes its books in a variety of electronic formats and by print-on-demand. Some content that appears in standard print versions of this book may not be available in other formats.

Library of Congress Cataloging-in-Publication Data
Names: Majumdar, Pradip, 1954- author.
Title: Design of thermal energy systems / Pradip Majumdar.
Description: First edition. | Hoboken, NJ, USA : Wiley, 2021. | Includes
    bibliographical references and index.
Identifiers: LCCN 2020040595 (print) | LCCN 2020040596 (ebook) | ISBN
    9781118956939 (cloth) | ISBN 9781118956946 (adobe pdf) | ISBN
    9781118956915 (epub)
Subjects: LCSH: Heat engineering. | Renewable energy sources.
Classification: LCC TJ808 .M345 2021 (print) | LCC TJ808 (ebook) | DDC
    621.402–dc23
LC record available at https://lccn.loc.gov/2020040595
LC ebook record available at https://lccn.loc.gov/2020040596

Cover Design: Wiley
Cover Images: Illustration courtesy of Pradip Majumdar;
Background © Xanya69/Getty Images

Set in 9.5/12.5pt STIXTwoText by Straive, Chennai, India
Printed and bound by CPI Group (UK) Ltd, Croydon, CR0 4YY

C9781118956939_100521

In loving memory of my late parents Snehalata and Rati Ranjan Majumdar
and
To my wife: Srabani, and Children: Diya and Ishan

Contents

Preface

This book is intended for undergraduate and first year graduate students in various fields of engineering and science to introduce the design and analysis of thermal energy systems. This book is also intended as a textbook for a required course in the core curriculum and for the capstone design course in thermo-fluid science area of the mechanical engineering degree program.

One of the essential requirements of mechanical engineering curriculum is to provide strong coverage in the areas of thermal energy and fluid systems. Students are expected to analyse and design of thermal systems such as conventional and renewable energy systems, cooling systems and pump-piping systems, and thermo-fluid components such as heat sinks, thermal interface materials, heat exchangers, condensers, solar collectors, wind turbines, heat exchanger, piping systems and networks, and able to select and integrate appropriate heat sinks, pumps, cooling tower, turbine, and compressors in thermal systems for different applications.

This book can also be adapted as textbook for courses in various other fields of engineering such as chemical engineering, nuclear engineering, and civil engineering. Students in these programs are expected to have prerequisite knowledge of thermodynamics, fluid mechanics and heat transfer. Students in other field of studies can also benefit from this book since the book is comprehensive with the inclusion of the reviews of laws and principles of thermodynamics, fluid mechanics and heat transfer. The book will continue to be useful as a reference book for practicing engineers in the field of energy and power industries which experience demands for continuous increase in capacity of conventional power generations as well as demands for developing renewable and alternative energy and power generation systems.

The book contains essential topics in thermal energy systems and components such as conventional power generation and cooling systems, renewable energy systems like solar energy system, heat recovery systems, and thermal heat management. Examples are drawn from solar energy systems, battery thermal heat management, electrical and electronics cooling, engine exhaust heat and emissions and manufacturing processes. Contemporary topics such as steady state simulation and optimization methods are also included. The book includes number of worked out design problems to demonstrate iterative design methodologies.

The book is written with a focus to satisfy following learning objectives:

- Apply thermal analysis techniques to generate design specification and ratings.
- Design thermal systems and components to meet engineering specifications.
- Apply design methodologies to design thermal systems and components.
- Apply iterative methodologies to design thermo-fluid systems and components.
- Develop ability to identify, formulate, and solve design problems.
- Understand the functions of various components and systems requiring thermal-fluid principles.
- Decompose a problem into interdependent sub-problems as appropriate.
- Formulate mathematical problems from the physical/engineering description, and choose physically meaningful boundary conditions and constraints.
- Evaluate economics and costs of thermal energy systems and components.
- Familiarize with various engineering standards and codes for thermal energy system and components.

The book evolved from several years of my teaching a course on Design of Thermal Systems. I believe that the content of the book with comprehensive subject matters will help students build a stronger grip on the subject of design and analysis of thermal and fluid systems. I welcome suggestions from interested readers of the book.

I would like to thank my students for theirs comments, feedbacks, and suggestions over many years. They were the continuous source of my motivation to continue and complete the book.

I thank all reviewers for their constructive comments. I would like to express my sincere appreciation to all editors, managers, designers, and editorial staff members at Wiley for their efforts, supports, understanding and patience during the production of this book. Special thanks to my children Diya and Ishan for helping me select the cover page of the book.

I would like to express my deep appreciation to my wife: ***Srabani*** and my children: ***Diya and Ishan*** for their continuous support, understanding and patience during preparation of the manuscript.

Pradip Majumdar

About the Author

PRADIP MAJUMDAR earned his M.S. and Ph.D. in mechanical engineering from Illinois Institute of Technology. He was a professor and the chair in the Department of Mechanical Engineering at Northern Illinois University. He is an adjunct faculty in Department Mechanical, Materials and Aerospace Engineering at Illinois Institute of Technology. He is recipient of the 2008 Faculty of the Year Award for Excellence in Undergraduate Education. Dr. Majumdar has been the lead investigator for numerous federal and industrial projects. Dr. Majumdar authored numerous papers on fluid dynamics, heat and mass transfer, energy systems, fuel cell, Li-ion battery storage, electronics cooling and electrical devices, engine combustion, nano-structured materials, advanced manufacturing, and transport phenomena in biological systems. Dr. Majumdar is the author of three books including Computational *Methods for Heat and Mass Transfer*; Fuel *Cells- Principles, Design and Analysis*; and *Computational Fluid Dynamics and Heat Transfer* (In Press). Dr. Majumdar is currently serving as an editor of the *International Communications in Heat and Mass Transfer*. He has previously served as the *Associate Editor of ASME Journal of Thermal Science and Engineering*. Dr. Majumdar has been making keynote and plenary presentations on Li-ion Battery storage, fuel cell, electronics cooling, nanostructure materials at national and international conferences and workshops. Dr. Majumdar has participated as an international expert in GIAN lecture series on fuel cell and Li-ion battery storage. Dr. Majumdar is a fellow of the American Society of Mechanical Engineers (ASME).

About the Companion Website

This book is accompanied by a companion website:

https://www.wiley.com/go/majumdar

The website includes:
- Presentation Slides
- Solutions Manual

1

Introduction

Mechanical engineering design involves both mechanical and thermal designs. Mechanical design deals with mechanical strength and structural properties of materials; motion and dynamics; geometrical dimensions and tolerances. Mechanical design requires knowledge of engineering mechanics, materials and strength of materials, vibration, and machine design. Thermal design deals with the thermal aspects of the components, processes, and systems, and requires knowledge of thermal science subjects such as thermodynamics, heat transfer, and fluid mechanics. A design of a product may require thermal design analysis first followed by mechanical design and are often interrelated. A product design may not only require mechanical concepts design but may also require knowledge of thermal science concepts and thermal design analysis techniques. Often, the product design requires additional subject areas such as electrical engineering and biomedical engineering, and multiphysics analysis.

1.1 Thermal Engineering Design

A thermal engineering design process involves the applications of concepts from fundamental engineering science topics such as thermodynamics, heat transfer, and fluid dynamics, following some specified well-defined steps and in an iterative process. A successful design process may involve several steps as shown in Figure 1.1 and is described below:

1. *Conception*: Requires some intuition about the final end-product using one's creative sense.
2. *Synthesis*: Some vision of the way the end results might be achieved. Consideration of multiple options and multiple pathways is given before developing the design.
3. *Analysis*: Ways to realize the design by following well-defined methodologies like thermal analysis, computer simulation analysis, economic analysis, and cost estimation. Such knowledge bases can be learned. The analysis step leads to defining the ratings and specifications of the product.

Design of Thermal Energy Systems, First Edition. Pradip Majumdar.
© 2021 John Wiley & Sons Ltd. Published 2021 by John Wiley & Sons Ltd.
Companion website: www.wiley.com/go/majumdar

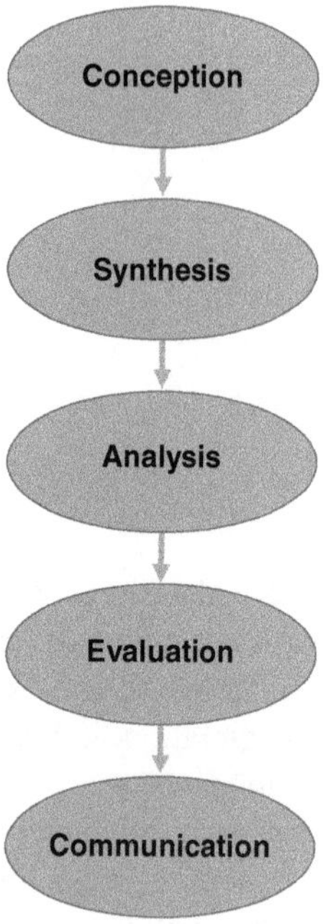

Figure 1.1 Design process.

Conception

Synthesis

Analysis

Evaluation

Communication

4. *Evaluation*: This is the way to prove the functioning of a successful design. This involves testing of a prototype requiring iterations. Use of sophisticated simulation methods and design tools may reduce number prototypes to be made, hence reduce cost in the design, development, and production.
5. *Communication*: Present the design to others in the form of technical reports and oral presentations.

1.2 Elements of Design Analysis of Thermal Systems

Various elements and steps generally used in the design analysis of thermal systems are demonstrated in Figure 1.2. Column one in the figure shows that a design process starts with a conceptual design along with some potential options and alternatives. This is followed by the selection of type of components, selection of ranges for some key variables and parameters, and setting any constraints. Thermodynamic analysis is carried out to establish the initial specification and ratings of the major components.

The component-level analysis and design are next carried out to develop the detail specification of each component in the system. Cost estimation followed by an economic analysis are essential to check the feasibility of the design. Iterative refinement can be carried out by changing the set variables and parameters. A system simulation is required to determine the expected operating conditions and performance at offload or part-load conditions and is generally used in the design stage to provide an improved design. Optimization step is often carried out to ensure the feasibility of the concept based on either the performance or the cost or both.

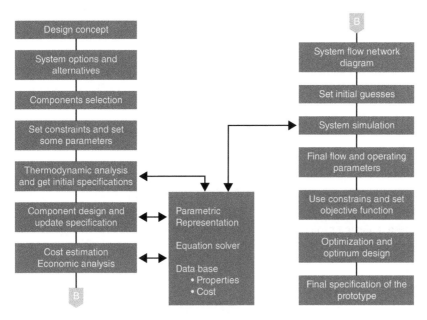

Figure 1.2 Flowchart showing detail elements of thermal design process.

1.2.1 Some Special Aspects of Thermal Design

A thermal system may be very large and have a single application. For example: A utility large thermal power plant that produces 1000 MW of electric power. It could also be some systems that are produced in large numbers. For example, refrigeration units, air-conditioning unit, fuel cells, solar water-heating system, or smaller solar thermal power generation units in the ranges of 1–10 kW. Thermal systems generally involve a large number of components in one design, and often these components can be categorized such as heat exchangers, condensers, boilers, cooling towers heat sinks, pumps, fans, etc. Another important aspect of the thermal system design process is that many parameters must often be set, either arbitrarily or in relation to other aspects of the design. The values of the parameters will, however, affect both **capital** and **operating cost**, including the **energy cost**, and hence will require iterative refinements of the parameter values assumed.

1.2.2 Design Types

Designs can be categorized into different types: nonfunctional, functional, satisfactory, and optimum.

Nonfunctional: The device does not function. For example, the cooling device is designed, but produces no cooling effect, and even produces undesirable effects such as irreversible heating.

Functional: The device performs in the expected manner as it is designed to do so. For example, a designed cooling device is capable of cooling a water stream.

Satisfactory: A functional design that meets some *assigned criteria*. For example, a chiller is designed to transfer 25 kW of heat from an air stream and cool the air stream from 25 to 13 °C.

Optimal: A design that is obtained based on *some specific restrictions*. For example, a solar collector is designed to supply thermal energy to run a domestic solar water heating system.

The collector is designed and fabricated with a restriction of minimum cost and/or minimum weight.

1.3 Examples of Thermal Energy Design Problems

Some typical thermal design projects are discussed in Sections 1.3.1–1.3.7. The objective is to understand some of the basic steps to be followed in the design of thermal systems.

1.3.1 Solar-Heated Swimming Pool

The swimming pools of most hotels in the USA are currently outdoors and heated by gas heaters. It is proposed to use solar energy to heat the pool throughout year as needed. It is also proposed to have flat-plate solar collectors that receive energy from the sun and use the energy to maintain the water at a comfortable temperature range year-round.

A basic solar water-heating system along with two additional optional systems, shown in Figure 1.3, are described here for consideration. The basic water-heating system for swimming pool consists of a solar collector array, a pump, a piping system of valves and fittings.

A few assumptions are made, and a few variables and parameters are set before starting the analysis. A typical design process is described below.

Understand the requirements and set known Data:

1. *Geographical location*: This is important to establish the available solar radiation and to select the design outdoor conditions. For example – Select Santa Barbara, CA.
2. *Pool dimension*: Select the pool dimension to establish the amount of water to be heated and rate at which water will be circulated through the system. For example, select the pool size as 12 m long × 8 m wide with water depth that varies in the lengthwise direction from 0.8 to 3.0 m.
3. *Design operating conditions*:
   - A comfortable water temperature range for the pool
   - Design outdoor air conditions such as the dry-bulb and wet-bulb temperatures.

Select optional systems:

1. Consider system with or without a thermal storage.
2. Consider system with or without an auxiliary gas or electric heater.

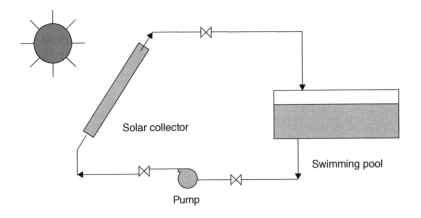

(a) **Option -1:** A basic system: Without an auxiliary heater and without any thermal storage

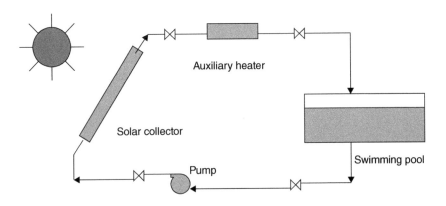

(b) **Option-2:** With an auxiliary heater and without any thermal storage

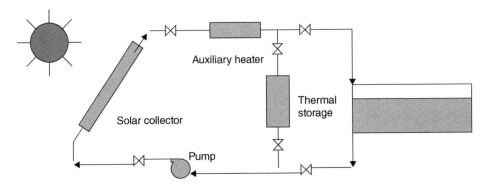

(c) **Option-3:** With an auxiliary heater and a thermal storage

Figure 1.3 Solar-heated swimming pool. (a) Option 1: a basic system: without an auxiliary heater and without any thermal storage. (b) Option 2: with an auxiliary heater and without any thermal storage. (c) Option 3: with an auxiliary heater and a thermal storage.

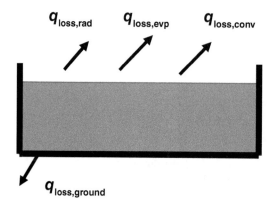

$q_{\text{loss,rad}}$ $q_{\text{loss,evp}}$ $q_{\text{loss,conv}}$

$q_{\text{loss,ground}}$

Figure 1.4   Pool water heat load.

Basic steps for analysis and design:

1. Develop a thermal model of the pool water considering all components of energy transfer or interaction with the surrounding. The model should include all major heat loss
   and heat gain components from/to the pool water body by a number of means such as
   convection heat loss from the water to surface to ambient air ($q_{\text{loss, conv}}$); radiation heat
   loss from the water to ambient air and sky ($q_{\text{loss, rad}}$); evaporation heat loss due to mass
   transfer loss of water to ambient air ($q_{\text{loss, evap}}$); ground heat loss from the water body by
   conduction through pool wall to ground ($q_{\text{loss, ground}}$) Figure 1.4.
   The pool water heat load is the sum of all the heat losses from pool water and given as

$$q_{\text{heat,load}} = q_{\text{loss,conv}} + q_{\text{loss,rad}} + q_{\text{loss,evap}} + q_{\text{loss,ground}}$$

   where

   $q_{\text{loss, conv}}$ = Convection heat loss from pool water surface due to the wind speed flow
   $q_{\text{loss, rad}}$ = Radiation heat loss from water surface to sky
   $q_{\text{loss, evap}}$ = Heat loss due to the evaporation water from pool surface to surrounding
   based on the difference of moisture content of the surrounding air and the
   saturated air at the water surface temperature
   $q_{\text{loss, ground}}$ = Heat loss by conduction through pool wall and ground soil due the
   temperature difference between the water temperature and deep soil
   temperature

   Note here that the solar water-heating system must be designed such that heat collected
   at the solar collector should be ideally equal to the pool water heat load.
2. Perform steady or transient thermodynamic analysis of the solar water heating and
   establish the ratings and specification of the components such as solar thermal collector
   in terms of type, size, and numbers; piping layout, including pipe size and length and
   necessary valves and fitting; and pump type and size.
3. Determine the size and specification of the auxiliary heaters based on available solar
   radiation at off-design hours.
4. Determine the size and specification of energy storage system.
5. Estimate total cost of the system, including initial operating and maintenance.

6. Compare these costs to those associated with the use of a natural gas water-heating system.
7. Perform economic evaluation to establish the feasibility of the system.

1.3.2 A Chilled Water System for Air-Conditioning System

A chilled water system is required to cool 15 kg/s of water from 13 to 8 °C, rejecting the heat to the atmosphere through a cooling tower. It is required to design a system with minimum first cost. The chilled water system, shown in Figure 1.5, consists of vapor compression refrigeration system and a water-cooling tower. The chilled water stream is cooled by the refrigerant, undergoing two-phase flow and boiling heat transfer in the evaporator. The superheated refrigerant vapor from the compressor undergoes two-phase flow and heat transfer in the condensation, transferring heat to the cooling water stream, which is finally cooled using the cooling towers.

Design analysis steps:

1. Select the range of operating conditions such as the evaporation temperature, T_e, and condensation temperature, T_c, and set parameters.
2. Develop a thermodynamic analysis model of the system.
3. Perform thermodynamic analysis over the range of operating conditions and determine the corresponding ratings for key components like the evaporator, condenser, compressor, cooling tower, and the expansion valve.
4. Obtain performance characteristics of some of the components.
5. Obtain parametric representations of the performance characteristics of these components.
6. Develop mathematical models based on fluid flow and heat-transfer concepts for rest of the components.

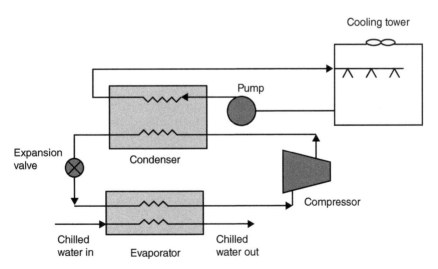

Figure 1.5 Chilled water system.

7. Perform simulation analysis and finalize the design in terms of the size, type, and selection of major components.
8. Perform cost estimation and economic analysis to demonstrate the economic feasibility.
9. Determine cost function for each component as a function of size for all major components.
10. Determine objective function and constraints.
11. Perform optimization analysis to derive a mathematical statement in terms of an objective function that needs to be minimized subject to some constraints.

1.3.2.1 Objective Function

The total cost of the system can be obtained by summing up the cost function for individual components like evaporator, compressor, condenser, cooling tower, and cooling water pump. Cost function of each component can be derived in terms of its critical operating parameters. A typical objective function can then be defined as follows:

$$C = C_{EVP}(Q_{EVP}) + C_{CP}(\dot{m}_r, \Delta P_{CP}) + C_{COND}(Q_{Cond}) + C_{CT}(\dot{m}_W, R, A) + C_P(\dot{m}_w, \Delta H_P)$$
$$(1.1)$$

Constraint #1: Minimum amount of cooling must be produced as defined by the minimum cooling capacity. Required cooling capacity, which is defined as the minimum amount of heat transferred from the chilled water flow to the refrigerant in the evaporator, is given as

$$Q_{EVP} = Q_{cw} = \dot{m}_{cw} C_{p,cw} (T_{cw,i} - T_{cw,o})$$
$$(1.2)$$

The constraints can be stated as

$$Q_{cc} \leq Q_{cw}$$
$$(1.3)$$

Constraint # 2: To avoid freezing of water or formation of frost on evaporator tube surface, the evaporation temperature must be above the freezing temperature of water, which corresponds to the operating pressure of water. This constraint can be stated as follows:

$$T_{evp} \geq 0°C$$
$$(1.4)$$

Constraint # 3: Refrigerant temperature and pressure at compressor outlet is set by the selection of the condenser temperature, and the corresponding saturation pressure represents the exit pressure of the refrigerant vapor out of the compressor. However, to have positive heat transfer, the selection of the condenser temperature is restricted by the cooling water temperature at the outlet of the condenser or at inlet to the cooling tower. The constraint is defined as

$$T_C \geq T_{wo}$$
$$(1.5)$$

1.3.3 Secondary Water System for Heat Rejection

A secondary water system in a power plant is responsible for maintaining the back pressure, also referred to as the condenser pressure, at the exit of the steam turbine in a steam Rankine power plant cycle. Figure 1.6 displays a typical secondary water system for a power plant, showing the major components such as the condenser, cooling tower, cooling water

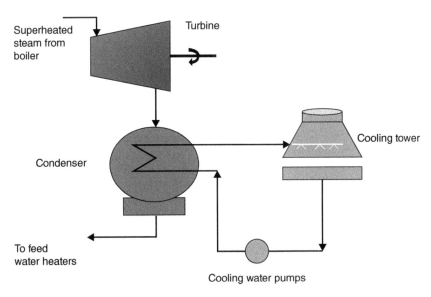

Figure 1.6 Secondary water system for a power plant.

pumps, and piping systems. A secondary water system must be designed for a given set of variables such as the heat rejection rate, Q_c; steam flow rate, $\dot{m}_s$; and thermodynamic states of the steam at inlet to the condenser for a steam power plant. It is necessary to design the system in terms of selection of size and type of major components, e.g. condenser, cooling tower, cooling-water-circulating pumps, and piping system. The major criteria for this design analysis are to maintain the stated heat rejection rate and pressure in the condenser with a minimum first equipment and pumping cost.

Several assumptions need to be made, and several variables and parameters to be set before starting the analysis. A typical design process is described below.

Understand the requirements and set known Data:

1. Mass flow rate and operating state of the steam flow through the steam turbine of the power plant cycle.
2. *Geographical location*: This is important to establish the design outdoor conditions like dry-bulb and wet-bulb temperatures.
3. Physical layout of the power plant is required to determine the elevations and distances between condenser and cooling tower. Based on this information, the piping layout and length can be estimated, and the number and types of pipes, pumps, valves, and fittings can be selected.
4. Design operating conditions
   - Seasonal variation in outdoor conditions can be considered for steady-state and transient analyses.
   - Select design outdoor air conditions such as the dry-bulb and wet-bulb temperatures for steady-state analysis, and consider hourly or daily variation of these date for transient analysis.
5. Select pipe material of construction.

Select optional systems:

1. Consider piping system options with a single pipe or multiple parallel pipes for the delivery of cold water from cooling tower to the condenser.
2. Consider system options with a single large pump or multiple smaller pumps.

Basic steps for analysis and design:

1. Select range of design operating conditions in the condenser and cooling tower based on the design outdoor conditions.
2. Perform thermodynamic analysis to determine the mass flow rate of steam condensed and amount of heat rejected by steam as it condenses or heat-transfer rating of the condenser.
3. Select type and number of condensers, cooling towers, pumps, and valves depending on the size of the power plant.
4. Decide piping-pump network and pipe sizes.
5. Perform piping pressure drop analysis to compute major and minor losses for different options of the piping system.
6. Select cooling tower parameters like approach (A) and range (R); set cooling water inlet and outlet temperature across condenser; and determine the cooling water mass flow rate based on condenser heat-transfer rating.
7. Perform condenser design analysis based on heat-transfer rating, mass flow rates of steam and cooling water, and operating conditions of steam and cooling water.
8. Determine the total pressure drop in the system considering major loss in the pipe, minor loss in the valves and fittings, condenser pressure drop. Vary the volume flow rate (Q) of cooling water and determine system curve given as ΔP_{sys} vs. Q.
9. Use the energy equation for the piping system to determine the required pumping to meet the system pressure drop and maintain cooling water flow through the system and determine the pump head or the pressure rise across the pump.
10. Finalize the pump specification, the required type of pump, and the performance characteristics of the pump to ensure that the desired operating flow rate can be maintained over the design and off-design conditions.
11. Obtain mathematical representation of the heat-transfer model for the condenser and pipe friction/pressure drop model and determine the functional representation of performance characteristics of all other components like pumps, cooling tower, and valves as a function of the operating conditions.
12. Perform system simulation based on solving the system equations.
13. Finalize size of each component over the range of operating conditions.
14. Estimate and express the cost functions for the condenser, cooling tower, pumps, pipes, and fittings in terms of sizes and rated parameters.
15. Determine objective and constraint functions.
16. Perform optimization study to finalize sizes and operating conditions.

1.3.4 Solar Rankine Cycle Power Generation System

Concentrated solar power (CSP) plants are an emerging energy conversion concept of power generation using solar energy. Solar irradiation is concentrated to a focal point or line, and the focused beam of irradiation is converted to heat on an absorbing surface. This heat is then transferred to a *heat-transfer fluid (HTF)* such as steam/water, synthetic oil, or molten salt. A separate thermal power cycle like Rankine power cycle or Brayton cycle then transfers the heat from the HTF to generate electricity by driving a turbine and generator system.

A *thermal energy storage (TES) system* is also used to address the issues such as excess mid-day energy collection and delivery to the electrical grid after sunset to meet the peak demand and to provide extended hours of operations and ensure reliability and profitability.

While central receiver tower-plant technology has been under development for many decades and normally used commercially for large-scale utility applications, the line-focusing collectors like the concentrated parabolic trough (CPT) or concentrated parabolic collector (CPC) are increasingly being considered for relatively lower-power output ratings and operate at relatively lower absorber temperature in the range of 400 °C as compared to the range of 600 °C or higher in a central tower receiver/absorber system.

As an example, let us analyze and design a CPC solar power generation system to produce a rated power of 1 MW. Figure 1.7 shows a solar power generation system that uses CPCs to run a Rankine cycle power generation system considering a number of options: (a) a basic system with direct production of steam to run the turbine, (b) a system with the inclusion of an auxiliary heater to ensure a target steam condition at inlet to the turbine, (c) a system with a separate boiler/heat exchanger to transfer heat for collector HTF to the water and produce steam for the vapor power cycle, and (d) a system with the inclusion of a thermal energy storage (TES) to provide extended period of operation.

TES is used to optimize the use of available solar energy by collecting more energy at peak solar hours than the rated energy needs and use the extra energy during off-peak hours of solar radiations. This requires additional cost for the TES and larger collector area, but this reduces the size of the auxiliary heater and fuel cost. A typical design process is described below.

Understand the requirements and set known Data:

1. Set operating state for the following states: (i) pressure, P_b, and temperature, T_b, of the steam flow at inlet to the steam turbine of the power plant cycle, (ii) pressure, P_c, in the condenser, (iii) temperature, T_c, of the HTF at the outlet of the solar collector, and (iv) select the type of HTF in the CPC solar collector loop.
2. *Geographical location*: This is important to establish the design outdoor conditions such as solar radiation data and ambient temperature. Seasonal and hourly variation in outdoor conditions can also be considered for steady-state and transient analyses.

Basic steps for analysis and design:

1. Perform thermodynamic analysis to determine the mass flow rate of water/steam in the Rankine power cycle based on the rated power output and the given thermodynamics state at inlet to turbine and in the condenser. Ratings and specifications of the heat exchanger/boiler, turbine, condenser, auxiliary heater, and pump can be established.
2. Develop a thermal analysis model of the concentrating solar collector along with setting collector parameters such as solar radiation concentration factor, type of selecting coating of absorber, and transmittance–absorptance optical properties.

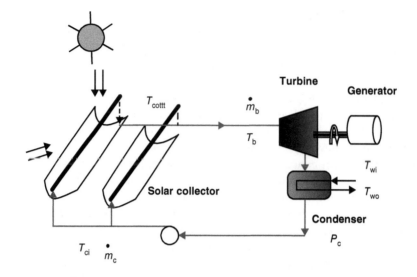

(a) A basic CPC solar powered rankine cycle power generation system with
of direct uses steam

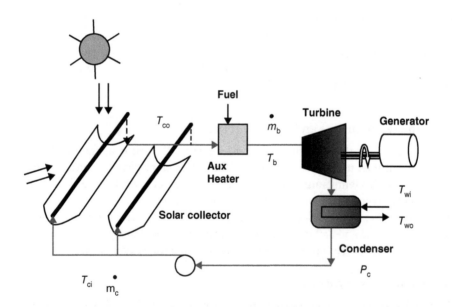

(b) CPC solar powered rankine cycle with an auxiliary heater

Figure 1.7 CPC solar power generation system. (a) Option 1: a basic CPC solar-powered Rankine cycle power generation system with direct uses of steam. (b) Option 2: CPC solar-powered Rankine cycle with an auxiliary heater. (c) Option 3: a CPC solar-powered Rankine cycle with boiler/heat exchanger and auxiliary heater. (d) Option 4: A CPC solar-powered Rankine cycle with thermal energy storage.

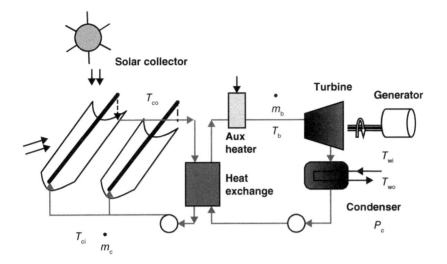

(c) A CPC solar powered rankine cycle with boiler/heat exchanger and
     and auxiliary heater

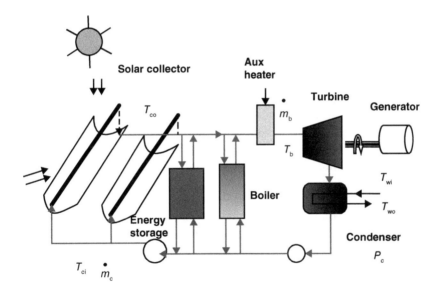

(d) A CPC solar powered rankine cycle with thermal energy storage

Figure 1.7 (*Continued*)

3. Perform solar collector design analysis in terms of solar collector area, efficiency, and number of collectors for the solar radiation date and ambient design conditions for a range of collector parameters. Characterize the collector performance in terms of heat recovery factor, transmittance–absorptance coefficient, and heat loss coefficient.
4. Determine the total pressure drop in the collector system by performing a series-parallel piping network flow analysis and considering major loss in the pipe, minor loss in the valves and fittings. Determine the required pumping head to meet the system pressure drop and maintain HTF flow through the collector system.
5. Finalize the pump specification, the required type of pump, and the performance characteristics of the pump to ensure that the desired operating flow rate can be maintained over the design and off-design conditions.
6. Perform system simulation based on solving the system equations.
7. Finalize size of each component over the range of operating conditions.
8. Estimate and express the cost functions for major components such as concentrating solar collector, TES, heat exchanger/boiler, steam turbine, condensers, and circulating pumps.
9. Determine objective and constraint functions.
10. Perform optimization study to finalize sizes and operating conditions.

1.3.5 Residential Air-Conditioning System

An air-conditioning system is used to maintain the space comfortable for human living by continuously removing excessive heat and moisture from the living space. During summer time, the conditioned space experiences both sensible and latent heat build-up due to both interior and external loads. Similarly, a heating system is used to add heat using a heater and add moisture using a humidifier during the winter to keep the space comfortable during winter time. Most commonly, an integrated cooling and heating system is employed to keep the space comfortable through the year as outside weather conditions varies.

Heating load is the energy transfer from indoor to outdoor during winter. Cooling load is the total heat energy transfer from outdoor to indoor during wintertime. Figure 1.8 shows the major contributing factors that build up the cooling and heating load of a living space.

Cooling load contribution comes from a number of following mechanisms: (i) Heat transfer through the walls, roof, windows by conduction and convection across building elements; (ii) Solar radiation transfer through windows and wall (absorption and transfer); (iii) Infiltration/exfiltration through leaks and cracks; and (iv) Internal heat generations from appliances, equipment, human occupants, cooking, shower, etc.

Heating load contributions come from heat transfer through building elements; transmission of solar radiation; and due to infiltration of warmer air. Moisture build up happens because of internal loads such as cooking and showers, and because of infiltration of humid air through doors, windows, and cracks.

To meet the sensible and latent loads of a space, the warmer and humid air from space is taken out and passed through an air-conditioning system to remove heat and moisture, and then returns the supplied air back to the space at temperature and moisture content (also known as humidity ratio) lower than the space. Figure 1.9 shows a single-zone air-conditioning system with major components such as cooling/dehumidifying coil, heating

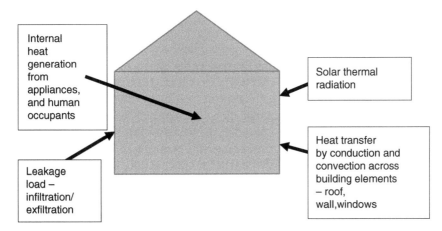

Figure 1.8 Contributing factors for space cooling and heating loads.

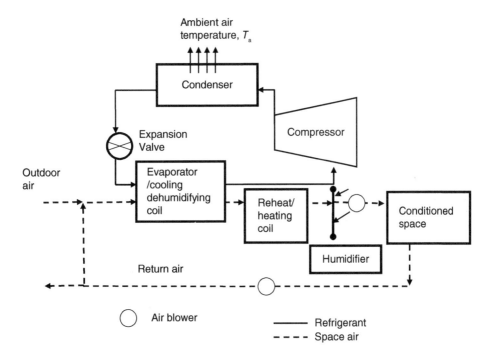

Figure 1.9 Single – zone direct expansion air-conditioning.

coil, humidifier, a direct expansion vapor compression chiller, and fans. A direct expansion chiller is one in which refrigerant flows through the cooling and dehumidifying coil. Another choice is the use of chilled water flow through the air coil.

Based on the thermostat and humidistat setting of the conditioned space, a return air fan/blower takes out the air from the space and supplies to the air-conditioning system. A fraction of the return air is exhausted to outdoor and the rest recirculated air is mixed with a fraction of the fresh outdoor.

The cooling/dehumidifying coil is used to remove heat and moisture from mixed-supply air stream. While trying to meet the latent load and remove enough moisture from the supply air stream, the air may be cooled too much because of the nature of the heat and mass transfer process in the cooling coil and the state of the air moves toward the saturation line at the temperature of the wetted coil surface. A reheating/heating coil is then used in combination with the cooling coil to achieve the desired state of the air stream before supplying to the conditioned space. The combination of cooling and dehumidification and the reheating coil is used to meet both the sensible and latent load of the space. A combination of heating coil and a humidifier is used to meet the heating load of the conditioned space during the winter time. The cooling coil acts as the evaporator for the vapor compression system, which maintains the coil wall temperature below the dew-point temperature of the incoming mixed-supply air stream so that moisture from air condenses.

The air-conditioning system is designed such that a net cooling or the net change in enthalpy of the supply air stream equates with sum of the sensible and latent load of the space. Additionally, the system must remove the sensible heat and the latent heat in an appropriate proportion that matches the ratio of sensible load to total load, i.e. the sum of sensible and latent loads of the space as defined by the following load-ratio line given by Eq. (1.6) below:

$$\frac{C_{pa}\,(T_r - T_s)}{(h_r - h_s)} = \frac{Q_s}{Q_s + Q_l} \tag{1.6}$$

where

Q_s = Sensible cooling load
Q_l = Latent cooling load
C_{pa} = Specific heat of air
T_s = Temperature of the supply air to the space
T_r = Temperature of the return air from space
h_r = Enthalpy of the return air from space
h_s = Enthalpy of the supply air to the space

For the state-s of the air coming out of the cooling and dehumidifying coil, the enthalpy, h_s, is the enthalpy of the saturated air at the temperature of the wetted surface. The thermodynamic saturation data as function of temperature are given for saturated air in moist-air data table. The functional relationship $h_s\,(T_s)$ can be obtained from the thermodynamic property data using curve – fitting techniques described in Appendix A.

The mass flow rate of the supply air is estimated from balancing the sensible cooling load or the total load as follows:

$$\dot{m}_a = \frac{Q_s}{C_{pa}(T_r - T_s)} = \frac{Q_s + Q_l}{h_r - h_s} \tag{1.7}$$

Equations (1.1) and (1.2) can be solved for $\dot{m}_a$, T_s, ω_s, and h_s. Note that the supply point s could be any point on the load-ratio line and satisfy the proportion of sensible load to latent load. The final selection depends on the criteria that enough moisture is removed, and desired level of moisture content is achieved. If the corresponding temperature is too low because of the nature of the load-ratio line, reheating can be used, and mass-flow rate can be recalculated based on this new supply air temperature.

A typical design process is described below.

Understand the requirements and set known Data:

1. Set the geometrical details and orientation of the living space.
2. Set the geographical location.
3. Set design outdoor conditions such as solar radiation data and dry-bulb and wet-bulb temperatures. Seasonal and hourly variation in outdoor conditions can also be considered for steady-state and transient analyses.
4. Designed indoor comfort conditions for dry-bulb temperature and moisture content in terms of humidity ratio or relative humidity are normally set based on ASHRAE standards. Typical design conditions in summer time are considered as the dry-bulb temperature in the range of 24–26 °C and a maximum humidity of 60%. For winter time, the temperature range of 20–22 °C and a minimum relative humidity of 30% are recommended.
5. A layout of ducting system for air distribution can be set.

Basic steps for analysis and design:

1. Perform thermal analysis to determine the sensible and latent cooling loads of the living space considering all major contributing factors such as heat transmission, solar gain, infiltration, and internal generations.
2. Decide the fraction of fresh outdoor air to be mixed with the recirculating return air.
3. Determine the dry bulb temperature, moisture content, and the dew-point temperature of the mixed air state (m) by applying mass and energy balances.
4. Determine mass flow rate, $\dot{m}_a$, and state (T_s, ω_s) of the supply air.
5. Select the type of refrigerant and temperature of evaporation for flow inside the cooling and dehumidifying coil to ensure enough cooling and dehumidification.
6. Perform design analysis of the cooling coil based on heat mass transfer principles to determine needed surface area, length, and diameter of the coil. The option of assuming finned-tube coil can also be considered to make the coil compact in size.
   Alternatively, one can also select a commercially available direct expansion refrigerant chiller based on its performance characteristic date and required cooling needs of the supply air.
7. Determine total pressure drop in the air distribution system by performing a series-parallel piping network flow analysis and considering major loss in the ducts, minor loss in the dampers and fittings such as expansions/contractions, elbows and bends, tee and branch junctions, and diffusers.
8. Determine the required increase in pressure head for the fan to meet the system pressure drop and maintain enough flow and pressures at different parts of the air distribution system.
9. Finalize the fan specification, the required type of fan, and the performance characteristics of the fan to ensure that the desired operating air flow rate can be maintained over the design and off-design conditions.

1.3.6 Heat Recovery from Diesel Engine Exhaust

The product of combustion or the emission gas from an internal combustion engine such as diesel engine exits at high temperature and with higher exergy. The range of exhaust gas temperatures (EGT) varies with type, size, and load of the engines, and amount of intake air mass flow rate. For a typical diesel engine operating at a medium high speed and high load conditions, the EGT can be in the range of 600 °C. There are a number of ways this higher-temperature heat energy and pressure can be recovered for useful purposes, such as using a turbocharger and drive the compressor wheel, using air preheater, and use of an engine gas recirculation (EGR) system for enhanced engine performance and controlling emissions. With such heat recovery techniques, the EGT drops down to the range of 300–400 °C. Organic Rankine power cycle is another alternative approach that has been under extensive investigation for use as high-temperature and pressure exhaust to run an organic Rankine cycle (ORC) power generation system. Figure 1.10 shows a diesel engine heat recovery system that uses exhaust heat to Rankine cycle power generation system. An EGR boiler is used to transfer heat from the exhaust to the circulating Rankine cycle fluid that undergoes to-phase heat transfer, producing high pressure and temperature vapor to run a turbine. The cycle heat rejection is done through a condenser, either air-cooled or water-cooled, depending on the expected power rating of the system. For a vehicle application, the heat rejection can be done through the existing engine radiator and the generated power from ORC can be directly fed to the engine crankshaft. An optional EGR cooler can also be used to preheat the incoming ambient air, and hence recover additional amount of heat before releasing the exhaust gas to atmosphere. Often, a certain amount of EGR is taken back to the cylinder as means to control reaction rates and emission composition, and reduce NO_x formation. A typical design process is described below.

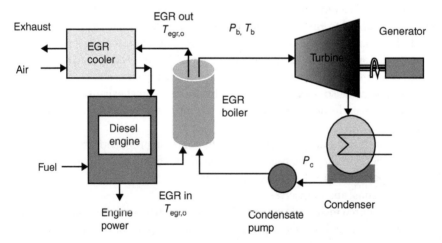

Figure 1.10 Engine exhaust heat recovery using Rankine cycle power generation system.

Understand the requirements and set known Data:

For simplicity, the combustion analysis process and the EGR cooler can be excluded from this design analysis process, and the heat recovery and power generation using the ORC can be evaluated assuming set EGR flow rate and temperature.

1. Decide the mass flow rate, composition, and temperature of the EGR based on engine size and data such as the compression ratio, rpm, and internal volume ratio. The design analysis process can also be extended to include the combustion analysis process based on the set EGR rate and other engine parameter and operating conditions.
2. Set EGR temperature reduction across the EGR boiler.
3. Set efficiencies of the turbine and pump.

Basic steps for analysis and design:

1. Decide a working fluid for the Rankine power cycle. Along with water/steam being the most common Rankine cycle working fluid, other potential organic fluids such as water–ethylene glycol mixture, ethanol, and R245fa (a hydrofluorocarbon) can also be considered as other optional working fluids, and performances can be compared.
2. Perform thermodynamic analysis of the Rankine cycle to determine ratings of all components such as EGR boiler, turbine, condenser, and the pump for the selected working fluid. A parametric study needs to be carried out to see the correlation of the engine ORC performance and variation in major component ratings with engine load output parameters.
3. Perform thermal design analysis for the EGR boiler and the condenser to determine the size and other geometrical details to meet the heat-transfer ratings based on heat-exchangers design principles.
4. Establish appropriate types, specifications, and performance characteristics for the turbine and circulating pump.
5. Estimate and express the cost functions for major components such as EGR boiler, turbine, condensers, and circulating pump.
6. Determine objective and constraint functions.
7. Perform optimization study to finalize sizes and operating condition.

1.3.7 Cooling System for a Li-ion Battery Stack in a Vehicle

The demands for Lithium-ion battery storage are increasing for greater use in several applications such as electric vehicles, integration with renewable power generation systems such as solar and wind farms, and for utility electric grid systems. Major components of a Li-ion battery are shown in Figure 1.11. These components are anode electrode, cathode electrode, and electrolyte/separator.

Batteries are classified based on battery chemistry and active material used. For example, in a Li-ion battery, the active material is lithium. Lithium (active material) is used as an insertion material in base materials like carbon in the negative electrode (anode – undergoes oxidation when discharging). Lithium ions, during the discharging, get

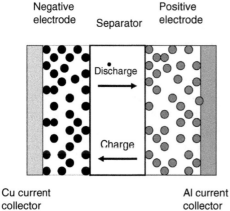

Negative electrode   Separator   Positive electrode

Discharge

Charge

Cu current collector

Al current collector

Figure 1.11 Elements of Li-ion battery cell.

extracted from the base carbon structure and are inserted back during the charging. Most of the cathode electrodes are also insertion compounds made of lithiated oxides of metals. Several different lithium – ion battery types are being developed and available based on the cathode chemistry such as lithium cobalt oxide (LCO) – $LiCoO_2$; lithium manganese oxide (LMO) – $LiMn_2O_4$; lithium nickel manganese cobalt oxide (NMC) – $LiNiMnCoO_2$; lithium iron phosphate (LIP) – $LiFePO_4$; and lithium nickel cobalt aluminum oxide (NCA) – $LiNiCoAlO_2$.

The **separator** is a porous matrix, which keeps the two electrodes separated physically and avoids short circuit and provides mechanical stability to the cell. All the pores or the voids of the negative electrode, separator, and the positive electrode are filled with electrolyte, and forms the solution or the liquid phase of the cell.

At higher discharge and charge rates, the battery performance decreases due to increased polarization losses, which results in increased internal heat generation and temperature rise of the lithium-ion battery. Battery performance is also significantly affected by the ambient environment temperature condition. Temperature variation greatly affects the performance and capacity of the battery. Beyond certain temperature level, thermal runaway will occur and thus increase temperature uncontrollably, causing serious safety problems. Thermal runaway is even more critical for automobile applications, which involve very high discharge and charge rates during driving and fast charging conditions.

Thermodynamic open circuit voltage (OCV) or the maximum possible voltage of an electrochemical battery cell is given by the thermodynamic relation as

$$E^0 = -\frac{\Delta G}{n_e F} \tag{1.8}$$

where

ΔG = Change in Gibbs free energy change between the reactants and the product
n_e = number of electrons transfer in the electrochemical reaction
F = Faraday constant

The voltage obtained from this equation is the *open-circuit voltage*, which is the voltage or potential that is obtained from the battery when it is neither charging nor discharging, i.e. when no current is drawn or given to the cell. So, this voltage is more than the actual voltage of the battery, as it is obtained without considering the losses associated with the process of charging and discharging.

There are three main losses associated with the battery when it is charging and/or discharging, which causes a drop in the voltage obtained from the battery. These losses are also called polarization or overpotential. Polarization is simply a deviation of the potential of the cell from the equilibrium potential when current is drawn from the cell. The effect of all these polarizations is the loss in the voltage and the performance characteristics of a battery cell is called *polarization curve*, which shows the variation in cell voltage with operating current density, i.e. *V* vs. *I* curve as shown in Figure 1.12.

The activation polarization is the voltage loss or overpotential that is expended in overcoming this energy barrier for each of the electrochemical half reaction that takes place at the anode-electrolyte and cathode-electrolyte interfaces. It is the voltage loss to initiate the electrochemical reaction. This polarization is prominent at lower current density. Region I in Figure 1.12 depicts the activation polarization. The ohmic polarization in Region II depicts resistance or ohmic polarization. It is predominant in moderate current densities. This is the overpotential caused by the resistance offered to the movement of the charged species like ions and electrons. The concentration polarization depicted in Region III arises from the difference in the concentration of the reactants and the products at the electrode surface and in the bulk electrolytes. It is predominant at higher current densities.

Drop in battery cell voltage drops causes battery capacity loss, which results in heat generation within the battery cell. In addition, battery capacity and performance get critically affected by surrounding ambient temperature. Under extreme environmental temperature conditions, excessive heat transmission to and from the surrounding significantly affects the cell transport properties, and hence the performance loss and heat generation. Based on the operating charge and discharge rates, one can compute the battery capacity loss and heat generation from the polarization curve or through direct measurement of operating voltage and current density.

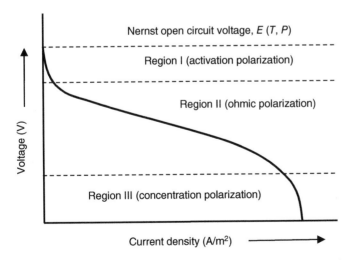

Figure 1.12 Battery performance characteristics showing all major losses.

Heat generation is composed of three major components: (i) reaction heat generation due to deviation of battery potential from equilibrium potential termed as irreversible heat generation, (ii) Reversible heat generation due to the entropic effect, and (iii) Ohmic heating due to charge transport through the electrodes, electrolytes, and current collector plates. A simple model to heat generation in a battery cell is just estimating the deviation of the operating voltage from reversible OCV and the operating current density as follows:

$$Q_{cell} = N_{cell} A_{cell} \, i \, (E_{rev} - V) \tag{1.9}$$

An effective battery-cooling scheme is required to maintain battery cells at a temperature level for acceptable and optimum battery performance. A limited cooling may lead to reduce the battery cell performance and may induce high thermal stresses within stack and cause failure. The thermal management schemes and the cooling techniques vary with rated power output and size of the battery stack. For a smaller fuel cell stack with lower power rating, heat dissipation by conduction through the solid, and natural convection along with use of heat sink may be adequate for maintaining cell operating temperature and performance. However, for a larger stack with high power rating, such a heat management and cooling scheme may not be adequate, and it may demand for alternative advanced cooling schemes involving forced convection using fan-induced air convection or forced liquid cooling and even two-phase forced convection boiling heat transfer. For example, in a battery stack with rated capacity of 20-kW power output and with 25% voltage drop from reversible OCV, it is expected that about 5-kW heat will be generated in the stack. To maintain the battery cell surface temperature at a desired level of 40 °C, one can evaluate and develop a cooling scheme based on either considering air cooling with/without heat sink and forced convection air cooling or liquid cooling using cold plates with integrated cooling channels.

High-performance cooling mechanisms, such as forced convection liquid cooling, two-phase flow boiling heat transfer, have good potential for use in battery cooling. Some of the key criteria for the selection of the cooling fluid are being noncorrosive, nonfreezing, environmentally safe, and electrical nonconductor, i.e. dielectric to avoid any short circuits.

Often, a combined cooling and heating system is considered for a battery stack. The heating system will be used for heating the battery cell during start up under cold environment conditions using the same integrated flow channel flow loop. The heating loop consists of circulating pump and heater. The cooling system will be used to remove the heat generated during the operation of the battery cell to maintain the near-isothermal operating conditions. The cooling loop consisting of a circulating pump and radiator/heat exchanger for dissipating the heat is shown in Figure 1.13. Figure 1.13b shows typical battery stack with cold plates with integrated forced convection cooling channels.

Let us consider a forced convection cooling system for an electric vehicle as shown in Figure 1.13b. Both single-phase and two-phase convection cooling options can be evaluated and compared. The battery-generated heat is removed by circulation liquid coolant through cooling plates with integrated cooling channels. Heat rejection to environment is done through a typical vehicle radiator heat exchanger/condenser. The system includes a coolant reservoir and a circulating pump. A typical battery cooling system design process is described below.

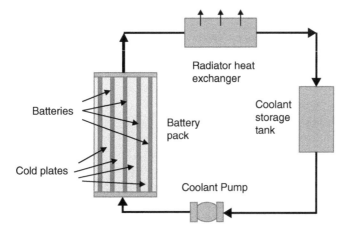

(a) Cooling fluid loop with radiator heat exchanger

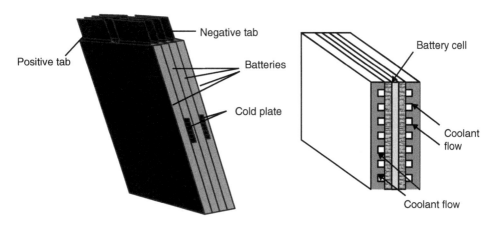

(b) Battery stacks with cold plates with integrated cooling channels

Figure 1.13 Cooling system for a battery stack. (a) Cooling fluid loop with radiator heat exchanger. (b) Battery stacks with cold plates with integrated cooling channels.

Understand the requirements and set known Data:

Information regarding battery energy storage capacity and power rating of the battery storage system must be defined.

1. Decide the peak charge and discharge rates based on the vehicle load cycle.
2. Estimate the battery heat generation rate based on an operating voltage and current density as set by the charge or discharge rates, and the battery performance polarization curve.
3. Set the desired battery surface temperature for efficient battery operation.
4. Determine the geometrical shape, size, and number of battery cells in the battery stack.

Basic steps for analysis and design:

1. Decide a working fluid for the forced convection battery cooling system. Along with the most common coolant choices such as water or water–ethylene glycol mixture single-phase convection, other low-temperature boiling fluids such as FC-72 dielectric fluid may also be considered for two-phase boiling convection.
2. Decide the size and number of the cooling plates.
3. Decide the cold plate material of constructions.
4. Decide the shape and size of the cooling channels.
5. Estimate the required mass flow rate using an iterative process. Selection of the correct coolant mass flow rate will result in a convection heat-transfer coefficient value that will be high enough to sustain a heat-transfer rate and balance with the battery heat generation rate, and hence keep the battery surface temperature at the desired level.
6. Perform thermal design analysis for the cold plate considering conjugate heat transfer considering conduction in solid part of the plate and convection flow and heat transfer in the flow channels of the plate in contact with the battery cell surface. The simplest analysis may involve computation of forced convection heat-transfer coefficients using correlations of convection heat-transfer coefficient and using thermal resistance network analysis. For a more comprehensive and quantitative analysis, a conjugate heat-transfer analysis is based on computational fluid dynamics and heat transfer may be performed.
7. Establish appropriate type, specification, and performance characteristics for the circulating pump using the date for the mass flow rate and pressure drop in the circulating fluid system.

Bibliography

Agarwal, D., Singh, S.K., and Agarwal, A.K. (2011). Effect of exhaust gas recirculation (EGR) on performance, emissions, deposits and durability of a constant speed compression ignition engine. *Appl. Energy* 88 (8): 2900–2907.

Arendas, A., Majumdar, P., Kilaparti, S.R., and Schroeder, D. (2013). Experimental investigation of the thermal characteristics of Li-ion battery for use in hybrid locomotives. *ASME J. Therm. Sci. Eng. Appl.* 6: 3. https://doi.org/10.1115/1.4026987.

Bergman, F.P. and Lavine, A.S.F. (2011). *Fundamentals of Heat and Mass Transfer*, 7e. New York, NY: Wiley.

Bidawai, J., Majumdar, P., Schroeder, D. and Rao, K. (2012). Electrochemical and thermal run-away analysis of lithium-ion battery for hybrid locomotive. Proceedings of the 2012 ASME Summer Heat Transfer Conference, HT-2012-58492, Rio Grande, Puerto Rico (8–12 July 2012).

Boem, R.F. (1987). *Design Analysis of Thermal Systems*. New York, NY: Wiley.

Cipollone, R., Di Battista, D., and Bettoja, F. (2017). Performances of an ORC power unit for waste heat recovery on heavy duty engine, IV international seminar on ORC power systems, ORC 2017, Milano, Italy. *Energy Procedia* 129: 770–777.

Moran, M.J., Shapiro, H.N., Boettner, D.D., and Bailey, M.B. (2018). *Fundamentals of Engineering Thermodynamics*, 9e. Wiley.

Revankar, S. and Majumdar, P. (2014). *Fuel Cells: Principles, Design and Analysis*. New York, NY: CRC Press, Taylor & Francis Group.

Stoecker, W.F. (1971). *Design of Thermal Systems*, 3e. New York, NY: McGraw-Hill.

Stoecker, W.F. and Jones, J.W. (1982). *Refrigeration and Air Conditioning*, 2e. New York, NY: McGraw-Hill.

Wei, H., Zhu, T., Shu, G. et al. (2012). Gasoline engine exhaust gas recirculation – a review. *Appl. Energy* 99 (C): 534–544.

Ziviani, D., Kim, D., Subramanian, S.N., and Braun, J. (2017). Feasibility study of ICE bottoming ORC with water/EG mixture as working fluid, IV international seminar on ORC power systems, ORC 2017, Milano, Italy. *Energy Procedia* 129: 762–769.

2

Thermodynamics Analysis

This chapter presents some of the basic principles, laws, and relations of engineering thermodynamics. This includes thermodynamic properties; energy forms such as heat, work, internal energy, and enthalpy; the first and the second laws of thermodynamics; ideal gas equation of states; and relations of change in enthalpy and entropy; mixtures with applications to moist-air and combustion processes; and applications to thermal power generation and cooling systems.

2.1 Some Basic Concepts of Thermodynamics

2.1.1 Thermodynamic System and Control Volume

The concepts of system and control volume are used for the application of various conservation principles such as mass, momentum, and energy for analyzing thermofluid devices.

A *thermodynamic system*, also referred to as the closed system, is defined as a quantity of fixed mass and identity without the presence of any inlets or outlets as demonstrated in Figure 2.1. Everything outside the system is referred to as the *surrounding*. The system and the surrounding are separated by the system boundary, which could be fixed or movable. While no mass can cross the system boundary, energy in the form of heat and work can cross the boundary and can be transferred to/from the system.

A *control volume*, also referred to as the open system, is defined as the fixed region or volume in space involving a number of inlets and outlets through the boundary surface, also known as the *control surface,* through which mass and energy may flow in and out of the control volume. The control volume is generally selected to enclose the space or region of interest around a device.

The thermodynamic system and the control volume may be finite or infinitesimal and may be fixed or moving.

2.1.2 Thermodynamic Properties, States, and Phases

A *property* can be defined as any quantity that depends only on the state of the system and is independent of the path or process by which the system has arrived at the given state.

Design of Thermal Energy Systems, First Edition. Pradip Majumdar.
© 2021 John Wiley & Sons Ltd. Published 2021 by John Wiley & Sons Ltd.
Companion website: www.wiley.com/go/majumdar

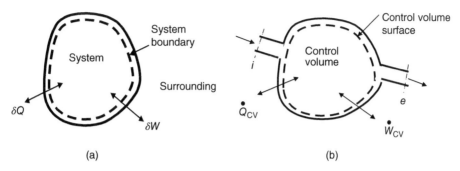

Figure 2.1 Thermodynamic system and control volume (a) thermodynamic system and (b) control volume.

All properties of a system can be divided into two types: **Intensive property** and **Extensive property**. Intensive properties are like pressure, temperature, density, which are independent of mass. Extensive properties are like volume or total energy that varies with mass. Thermodynamic states are defined by thermodynamic intensive properties like pressure, temperature, specific volume, internal energy, enthalpy, and entropy.

2.1.2.1 Pure Substance

A pure substance is one that is homogeneous and has an invariable chemical composition. A pure substance may exist in many phases, but its chemical composition may remain same. For example, water as a pure substance is homogeneous and has invariable chemical composition even when it changes phase from liquid to vapor or to solid ice.

2.1.2.2 Simple Compressible Substance

Simple compressible substance is one that undergoes only one mode of work energy transfer given by the expansion or compression of the control surface, i.e. $\delta W = P dV$. A simple compressible substance requires any two independent intensive properties to define the thermodynamic state. For additional work modes, the number of independent thermodynamic properties needed to define the state is two plus the additional work modes.

2.1.2.3 Phase-Equilibrium Diagram of a Pure Substance

Different phases and states of a pure substance are demonstrated in the phase diagram (Figure 2.2) below:

Some of the important thermodynamic terms associated with the phase diagram are defined as follows:

Saturated liquid: Liquid at a temperature equal to the saturation temperature corresponding to the given pressure

Saturated vapor: Vapor at a temperature equal to the saturation temperature corresponding to the given pressure.

Saturated liquid line: Line connecting all saturated liquid states.

Saturated vapor line: Line connecting all saturated vapor states.

Subcooled liquid: Liquid at a thermodynamic state with temperature lower than the saturation temperature at the corresponding pressure.

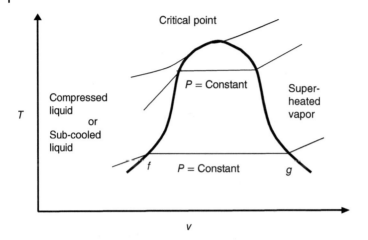

Figure 2.2 Phase diagram of a pure substance.

Compressed liquid: Liquid at a thermodynamic state with pressure higher than the saturation pressure at the corresponding to the temperature.

Mixed phase: Substance exists as a mixture of liquid and vapor at the saturation pressure and temperature.

Critical point: It is a point of inflection at which saturation liquid line and the saturated vapor line meet. At this point, the saturated liquid state and the saturated vapor states are identical.

Since in this mixed phase the temperature and pressure are dependent on each other due to saturation condition, additional independent property is required to define the thermodynamic state of the substance. This property is the quality, **x**, which is defined as the mass of vapor to the total mass and expressed as

$$x = \frac{m_v}{m_v + m_l} \tag{2.1}$$

where m_v = mass of vapor and m_l = mass of vapor.

Quality of a substance in a mixed state varies between 0 and 1, with $x = 0$ for saturated liquid and $x = 1$ for saturated vapor. All other thermodynamic properties of a substance in a mixed state with saturated temperature and pressure conditions are given as a weighted average of the saturated liquid and vapor values. For example:

$$\text{Specific volume, } v = (1 - x) \, v_f + x v_g = v_f + x v_{fg} \tag{2.2}$$

$$\text{Specific enthalpy, } h = (1 - x) \, h_f + x h_g = h_f + x h_{fg} \tag{2.3}$$

and

$$\text{Specific entropy, } s = (1 - x) \, s_f + x v_g = s_f + x s_{fg} \tag{2.4}$$

2.1.3 Thermodynamic Processes and Cycles

When one or more properties of a system change, then we have a change of state. The path of succession of all states through which a system passes through is called the ***thermodynamic process***. For example, as gas expands pushing the piston outward in the cylinder from the front-end dead center (FDC) to the crank-end dead center (CDC), the pressure of the gas decreases and volume increases, and the gas is assumed to pass through an expansion process given by the line connecting all the states the gas passed through as demonstrated in Figure 2.3.

Several processes are generally described by the fact that one of the properties remains constant during the process. For example, **Isothermal process**: Temperature remains constant; **Isobaric process**: Pressure remains constant; **Isochoric process**: Volume remains constant; and **Isentropic process or Reversible adiabatic process**: Entropy remains constant. For an **Adiabatic process,** there is no heat transfer across the system or control volume boundary.

2.1.3.1 Reversible and Irreversible Processes
A reversible process is defined as a process that, on completion, cannot be reversed without making any changes in the surrounding. There are a number of factors that make a process irreversible. These factors are (i) Friction, (ii) Heat transfer through finite temperature difference, (iii) Unrestrained expansion, (iv) Mixing process, (v) Combustion process, (vi) Hysteresis, and many other factors.

2.1.3.2 Thermodynamic Cycle
A system is assumed to undergo a thermodynamic cycle when it starts from an initial state (1), goes through several thermodynamic processes, and finally returns to the same initial state as demonstrated in Figure 2.4.

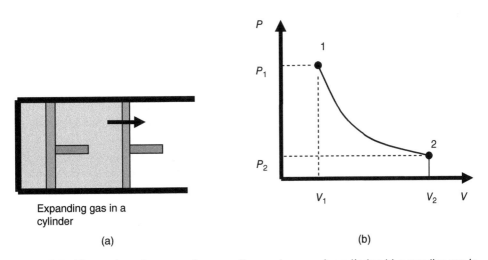

Expanding gas in a cylinder

(a)   (b)

Figure 2.3 Thermodynamic process for expanding gas in an engine cylinder: (a) expanding gas in an engine cylinder and (b) expanding process on P–V diagram.

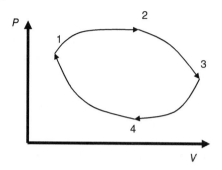

Figure 2.4 Thermodynamic cycle.

2.1.4 Energy and Energy Transfer

Total energy content of a system, E, includes three basic components: (i) the kinetic energy, $KE = \frac{1}{2}mV^2$, associated with the translation velocity of the system, (ii) the potential energy, $PE = mgZ$, associated with the elevation of the system from some reference level, and (iii) the internal energy, U, that includes all energy forms associated with the atomic and molecular structures, orientations, and motion.

The energy content of a system changes due to the transfer of energy in the form of heat and work to or from the surrounding across the system boundary as shown in Figure 2.5.

The rate of energy transfer is called the **power**.

2.1.5 Heat and Work

2.1.5.1 Heat Energy (Q)

Heat is an energy form that is transferred between two parts of a system or between a system and the surrounding due to temperature differences. It is an energy form that is in transit and it can only be identified at the boundary of a system. If there is no difference in temperature between the system and the surrounding, then there is no heat transfer. There are three modes of heat transfer: (i) **conduction** heat transfer through solid or stationary fluid film and the corresponding rate equation as given by Fourier law of conduction; (ii) **convection** heat transfer due to the motion of fluid over the surface and the rate equation is given in Newton's Law of cooling; and (iii) Thermal radiation heat transfers due to the transmission and exchange of electromagnetic waves or Photons, and the rate equation is given by Stefan–Boltzmann law.

For the sign convention, heat transfer to a system is considered as positive ($Q > 0$) and heat transfer from a system is considered as negative ($Q < 0$). Note that heat energy is not a

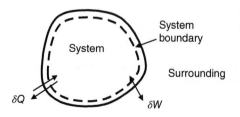

Figure 2.5 Interaction between system and surrounding through energy transfer.

property and it is given as inexact differential as demonstrated below:

$$Q = \int_1^2 \delta Q \qquad (2.5a)$$

and amount of heat transfer over a period is given as

$$Q = \int_{t_1}^{t_2} \dot{Q}\, dt \qquad (2.5b)$$

where $\dot{Q}$ is the rate of heat energy transfer.

2.1.5.2 Work (W)

In thermodynamics, work is expressed in terms of pressure and change in volume for a simple compressible substance as

$$\delta W = P\, dV \qquad (2.6a)$$

and for a process with change of state from 1 to 2 as

$$_1W_2 = \int_1^2 P\, dV \qquad (2.6b)$$

The total work energy transfer over a period is given as

$$W = \int_{t_1}^{t_2} \dot{W}\, dt \qquad (2.6c)$$

where $\dot{W}$ is the rate of work energy transfer.

For the sign convention, work done by a system is considered as positive $(W > 0)$ and work done on a system is considered as negative $(W < 0)$.

2.2 Conservation of Mass

The law of conservation of mass is a statement of the mass balance for flow in and out, and changes of mass storage of a system. For analysis purposes, the conservation of mass law is presented for both the system and the control volume.

2.2.1 System

Since a system is defined as fixed and identifiable quantity of mass, the conservation mass for a system is defined as

$$\left.\frac{dm}{dt}\right)_{system} = 0 \qquad (2.7)$$

2.2.2 Control Volume

A control volume is an open system that involves mass flow in and out. The conservation of mass statement for a control volume considers all mass flow in and mass flow out as well as changes in the mass inside the control volume. The statement is derived as

$$\left.\frac{dm}{dt}\right)_{cv} + \sum \dot{m}_e - \sum \dot{m}_i = 0 \qquad (2.8)$$

where $\dot{m}_i = $ mass flow into the control volume and $\dot{m}_e = $ mass flow out of the control volume.

2.3 The First Law of Thermodynamics

The first law of thermodynamics is a statement of conservation of energy considering all forms of energy transfer, storage, consumption, and generation.

2.3.1 The First Law of Thermodynamics for a System

The first law of thermodynamics for a system or fixed quantity of mass is stated as the balance of total energy content with energy transfer across the system boundary in terms of heat and work. A system interacts with the surrounding through transfer of energy in the form of work energy and heat energy across the system boundary as shown in Figure 2.1.

For a system undergoing a **thermodynamic cycle**, the first law of thermodynamics is given as

$$\oint \delta Q = \oint \delta W \tag{2.9}$$

It physically states that cyclic integral or sum of heat transfer in all processes and equal to cyclic integral or sum of work done in all processes.

The first law of thermodynamics for a **process** states that the change in energy content of system is caused by net transfer of energy in the form of heat and work across the system boundary, and this is stated as

$$\delta Q = dE + \delta W \tag{2.10}$$

and for a process with a change of state from 1 to 2

$$_1Q_2 = E_2 - E_1 + _1W_2 \tag{2.11}$$

where

$E =$ Energy content of the system $= U + \mathrm{KE} + \mathrm{PE}$
$U =$ Internal energy associated with rotational, vibrational, and translational motions, and structures of the atoms and molecules
$\mathrm{KE} =$ Kinetic energy of the system $= \frac{1}{2}mV^2$
$\mathrm{PE} =$ Potential energy of the system $= mg\,Z$.

With the substitution of the expressions for different energy forms, Eq. (2.11) can be written as

$$_1Q_2 = (U_2 - U_1) + \frac{1}{2}m(V_2^2 - V_1^2) + mg(Z_2 - Z_1) + _1W_2 \tag{2.12}$$

In terms of per unit mass of the system, the equation is expressed as

$$_1q_2 = (u_2 - u_1) + \frac{1}{2}(V_2^2 - V_1^2) + g(Z_2 - Z_1) + _1w_2 \tag{2.13}$$

where

$u =$ Specific internal energy $= \frac{U}{m}$
$q =$ Heat transfer per unit mass of the system $= \frac{Q}{m}$
$w =$ Work done per unit mass $= \frac{W}{m}$

2.3.2 The First Law of Thermodynamics for a Control Volume

For a control volume shown in Figure 2.6, *the first law of thermodynamics for a control volume* undergoing a process is derived based on the conservation of energy across a control volume as

$$\dot{Q}_{CV} + \sum \dot{m}_i \left(h_i + \frac{V_i^2}{2} + gZ_i \right) = \frac{dE_{CV}}{dt} + \sum \dot{m}_e \left(h_e + \frac{V_e^2}{2} + gZ_e \right) + \dot{W}_{CV} \quad (2.14)$$

For a stationary control volume with negligible changes in kinetic energy and potential energy, Eq. (2.14) can rewritten as

$$\dot{Q}_{CV} + \sum \dot{m}_i \left(h_i + \frac{V_i^2}{2} + gZ_i \right) = \frac{dU}{dt} + \sum \dot{m}_e \left(h_e + \frac{V_e^2}{2} + gZ_e \right) + \dot{W}_{CV} \quad (2.15)$$

2.3.3 Special Cases

For analysis purposes, the equations for conservation mass and the first law of thermodynamics are simplified with some assumptions. Two such common cases are (i) *the steady-state steady-flow process* and (ii) *the uniform-flow uniform-state process*.

2.3.3.1 Steady-State Steady-Flow (SSSF) Process

A Steady-state steady-flow (SSSF) process is a simplified model that represents a process in which all properties at each point of the system; all properties and flow rates in and out of the control volume; and energy transfer rates across the control volume surface are assumed to be constant and invariable with time. With these assumptions, Eqs. (2.8) and (2.15) reduce to

$$\sum \dot{m}_e = \sum \dot{m}_i \quad (2.16)$$

$$\dot{Q}_{CV} + \sum \dot{m}_i \left(h_i + \frac{V_i^2}{2} + gZ_i \right) = \sum \dot{m}_e \left(h_e + \frac{V_e^2}{2} + gZ_e \right) + \dot{W}_{CV} \quad (2.17)$$

For a single flow in and out, Eqs. (2.16) and (2.17) reduce to

$$\dot{m}_i = \dot{m}_e = \dot{m} \quad (2.18)$$

and

$$\dot{Q}_{CV} + \dot{m} \left(h_i + \frac{V_i^2}{2} + gZ_i \right) = \dot{m} \left(h_e + \frac{V_e^2}{2} + gZ_e \right) + \dot{W}_{CV} \quad (2.19a)$$

Figure 2.6 Flow and energy transfer across a control volume.

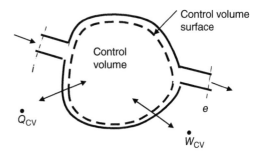

For unit mass flow rate, the first law equation reduces to

$$q_{CV} + \left(h_i + \frac{V_i^2}{2} + gZ_i \right) = \left(h_e + \frac{V_e^2}{2} + gZ_e \right) + w_{CV} \tag{2.19b}$$

2.3.3.2 Uniform-State Uniform-Flow (USUF) Process

A uniform-flow uniform-state process is a simplified model of a transient process. Integrating Eqs. (2.8) and (2.15) over period, conservation mass and energy are derived based on the initial and final states of the control volume as

$$(m_2 - m_1)_{CV} + \sum m_e - \sum m_i = 0 \tag{2.20}$$

$$Q_{CV} + \sum m_i \left(h_i + \frac{V_i^2}{2} + gZ_i \right) = \left[m_2 \left(u_2 + \frac{V_2^2}{2} + gZ_2 \right) - m_1 \left(u_1 + \frac{V_1^2}{2} + gZ_1 \right) \right]_{CV}$$

$$+ \sum m_e \left(h_e + \frac{V_e^2}{2} + gZ_e \right) + W_{CV} \tag{2.21}$$

where states 1 and 2 represent the initial and the final states of the control volume.

2.4 The Second Law of Thermodynamics

The second law of thermodynamics is stated through several statements such as the **Kelvin–Planck statement**, the **Clausius statement**, **Inequality of Clausius**, and **Conservation of Entropy**.

2.4.1 Kelvin–Planck Statement

According to **Kelvin–Planck statement**, it is impossible to make a device that will transfer heat from a single high-temperature reservoir and produce no effect other than producing thermodynamic work. This is demonstrated in Figure 2.7a.

It basically implies that it is impossible to have 100% efficient device that interacts with a single heat reservoir and converts heat into work. What is possible is the transfer of heat from high-temperature reservoir and transform part of the heat energy into thermodynamic work and reject rest of the heat energy into a low-temperature reservoir as shown in Figure 2.7b. This statement eventually leads to the development of the heat engine.

In a **heat engine**, certain amount of heat Q_H is transferred from a high-temperature heat source T_H as shown in Figure 2.7b. The engine converts a part of the heat into work, W, and rejects rest of the amount of heat Q_L into a low-temperature heat sink T_L.

The efficiency of the heat engine is defined as the ratio of the work produced to the heat received from high-temperature source:

$$\eta = \frac{W}{Q_H} \tag{2.22}$$

with $Q_H = W + Q_L$, Eq. (2.22) is written as

$$\eta = \frac{Q_H - Q_L}{Q_H} = 1 - \frac{Q_L}{Q_H} \tag{2.23}$$

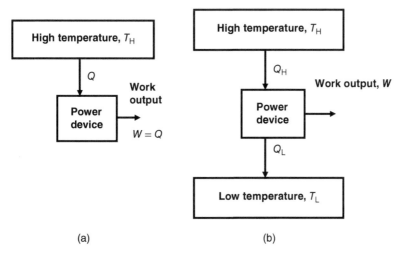

Figure 2.7 Kelvin–Plank statement. (a) Impossible and (b) possible.

It can be noted here that since $Q_H > Q_L$, the thermal efficiency of a heat engine has to be less than 100%, and since $Q_L > 0$ according to Kelvin–Plank's statement, the theoretical upper bound of heat engine efficiency is given as $\eta < 100\,\%$.

2.4.2 Clausius Statement

According to **Clausius statement**, it is **impossible** to have a device that produces no effect other than receiving heat from a low-temperature or cooler reservoir and rejecting it to a high-temperature or hotter reservoir (Figure 2.8a). It can be demonstrated that violating this statement leads to the violation of Kelvin–Planck statement and vice versa. What is possible is that some amount of energy, either in the form of heat energy or work transfer, needs to be added to the cooling device before rejecting heat energy into a high-temperature reservoir as shown in Figure 2.8b.

Clausius statement eventually led to the development of the cooling or refrigeration devices.

In a cooling device, an amount of heat is transferred from a low-temperature heat reservoir, T_L, to the high-temperature reservoir, T_H, but it requires an additional amount of energy either in the form of work or heat as shown in Figure 2.8b. It can be noticed that an increased amount of heat is transferred to the high-temperature reservoir and given as

$$Q_H = Q_L + W \tag{2.24}$$

The performance of the cooling device is given in terms of **Coefficient of Performance (COP)**, which is defined as

$$COP = \frac{Q_L}{W} \tag{2.25}$$

or

$$COP = \frac{Q_L}{Q_H - Q_L} = \frac{1}{\frac{Q_H}{Q_L} - 1} \tag{2.26}$$

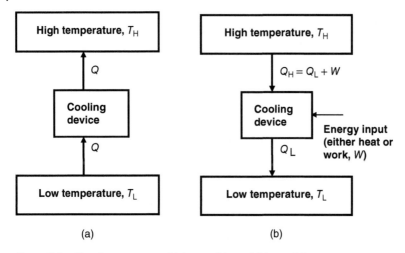

Figure 2.8 Clausius statement. (a) Impossible and (b) possible.

It can noted here that since $Q_H > Q_L$, COP of a cooling system can be greater than one, and since $W > 0$ according to the Clausius statement, the theoretical upper bound is COP < ∞.

2.4.3 Inequality of Clausius

The **Inequality of Clausius** is a consequence of the second law of thermodynamics, and it is stated for a **system** undergoing a **thermodynamic cycle** as

$$\oint \frac{\delta Q}{T} \leq 0 \text{ for a } \textbf{\textit{cycle}} \tag{2.27a}$$

with

$$\oint \frac{\delta Q}{T} = 0 \text{ for a } \textbf{\textit{reversible cycle}} \tag{2.27b}$$

and

$$\oint \frac{\delta Q}{T} < 0 \text{ for an } \textbf{\textit{irreversible cycle}} \tag{2.27c}$$

Application of this statement for a **process** leads to

$$dS \geq \left(\frac{\delta Q}{T} \right) \tag{2.28}$$

where S is the thermodynamic property **entropy**, and this is defined as property because the quantity $\frac{\delta Q}{T}$ is constant for all thermodynamic processes between the same two thermodynamic states.

For a **reversible process**, this leads to

$$dS = \left(\frac{\delta Q}{T} \right)_{\text{rev}} \tag{2.29a}$$

or

$$S_2 - S_1 = \int_1^2 \left(\frac{\delta Q}{T} \right)_{\text{rev}} \tag{2.29b}$$

It states that entropy, S, increases in a process with the addition of heat to the system and decreases for a process with the rejection of heat from the system. For a **reversible and adiabatic process**, also known as **isentropic process**, there is no heat transfer and entropy remain constant and Eq. (2.29b) is reduces to

$$S_2 = S_1 \tag{2.29c}$$

Also, for a reversible process, the heat transfer across a system is given by Eq. (2.29a) as

$$\delta Q = T dS \tag{2.30a}$$

For a **reversible process** between two states 1 and 2 as

$$_1Q_2 = \int_1^2 T dS \tag{2.30b}$$

For an **irreversible process**

$$dS > \left(\frac{\delta Q}{T}\right)_{irrev} \tag{2.31a}$$

or

$$S_2 - S_1 > \int_1^2 \left(\frac{\delta Q}{T}\right)_{irrev} \tag{2.31b}$$

Considering the **entropy generation** or **entropy production** associated with the irreversible process, the Eqs. (2.31a) and (2.31b) are written in the following manner for entropy changes for an irreversible process, respectively

$$dS = \left(\frac{\delta Q}{T}\right)_{irrev} + dS_{gen} \tag{2.32a}$$

or

$$S_2 - S_1 = \int_1^2 \frac{dQ}{T} + {}_1S_{2gen} \tag{2.32b}$$

where dS_{gen} and $_1S_2$ are the entropy generation in the process due to irreversibilities caused by various system factors.

The **second law of thermodynamics for a control volume** is given based on the conservation of entropy as

$$\frac{dS_{CV}}{dt} + \sum \dot{m}_e s_e - \sum \dot{m}_i s_i \geq \sum \frac{\dot{Q}_{CV}}{T} \tag{2.33}$$

Equation (2.33) is written for following cases as follows:

Reversible process:

$$\frac{dS_{CV}}{dt} + \sum \dot{m}_e s_e - \sum \dot{m}_i s_i = \sum \frac{\dot{Q}_{CV}}{T} \tag{2.34}$$

and

Irreversible process:

$$\frac{dS_{CV}}{dt} + \sum \dot{m}_e s_e - \sum \dot{m}_i s_i > \sum \frac{\dot{Q}_{CV}}{T} \tag{2.35a}$$

or

$$\frac{dS_{CV}}{dt} + \sum \dot{m}_e s_e - \sum \dot{m}_i s_i = \sum \frac{\dot{Q}_{CV}}{T} + \dot{S}_{gen} \quad \text{for an irreversible process} \tag{2.35b}$$

where $\dot{S}_{gen}$ is the rate of entropy generation or entropy production in an irreversible process. Equations (2.35a) and (2.35b) is simplified for the SSSF and UFUS processes as follows:

2.4.3.1 Steady-State Steady-Flow (SSSF) Process

$$\sum \dot{m}_e s_e - \sum \dot{m}_i s_i \geq \sum \frac{\dot{Q}_{CV}}{T} \qquad (2.36a)$$

with

$$\sum \dot{m}_e s_e - \sum \dot{m}_i s_i = \sum \frac{\dot{Q}_{CV}}{T} \text{ for a reversible process} \qquad (2.36b)$$

and

$$\sum \dot{m}_e s_e - \sum \dot{m}_i s_i > \sum \frac{\dot{Q}_{CV}}{T} + \dot{S}_{gen} \text{ for an irreversible process} \qquad (2.36c)$$

where $\dot{S}_{gen}$ represents rate of entropy generation in an irreversible process.
For a single flow in and single flow out, Eq. (2.36c) reduces to

$$\dot{m}\left(s_e - s_i\right) = \sum \frac{\dot{Q}_{CV}}{T} + \dot{S}_{gen} \qquad (2.37)$$

For an adiabatic process, $\dot{Q}_{CV} = 0$ and Eq. (2.37) can be written as

$$\dot{m}\left(s_e - s_i\right) = \dot{S}_{gen} \qquad (2.38)$$

For unknown values of entropy generation, we can write Eq. (2.39) in the following form:

$$s_e \geq s_i \qquad (2.39)$$

For a reversible adiabatic or isentropic process:

$$s_e = s_i \qquad (2.40a)$$

and
For an irreversible process:

$$s_e > s_i \qquad (2.40b)$$

2.4.3.2 Uniform-State Uniform-Flow (USUF)

$$\left[m_2 s_2 - m_1 s_1\right]_{CV} + \sum m_e s_e - \sum m_i s_i \geq \int_0^t \frac{\dot{Q}_{CV}}{T} dt \qquad (2.41a)$$

where

$$\left[m_2 s_2 - m_1 s_1\right]_{CV} + \sum m_e s_e - \sum m_i s_i = \int_0^t \frac{\dot{Q}_{CV}}{T} dt \text{ for a reversible process} \qquad (2.41b)$$

and

$$\left[m_2 s_2 - m_1 s_1\right]_{CV} + \sum m_e s_e - \sum m_i s_i = \int_0^t \frac{\dot{Q}_{CV}}{T} dt + {}_1 S_{2gen} \text{ for an irreversible process}$$

$$(2.41c)$$

${}_1 S_{2gen}$ represents the entropy generation during the process from state 1 to 2.

2.4.3.3 Reversible Steady-Flow Work

Energy equation for a reversible steady-flow process in a device is written as

$$\delta q_{rev} = dh + \delta w_{rev} \tag{2.42}$$

Substituting $\delta q_{rev} = Tds$ from Eq. (2.29a) for a reversible process and using the thermo-dynamic relation $Tds = dh - vdP$, Eq. (2.42) can be written as

$$\delta w_{rev} = -vdP \tag{2.43}$$

Reversible work for steady-flow process for a control volume between states is given as

$$w_{rev} = -\int_1^2 vdP \tag{2.44}$$

This is applicable for determining the reversible adiabatic or isentropic work for flow of liquid through a pump. Considering negligible specific volume for liquid flow, the isentropic pump work is given as

$$\delta w_{ps} = -vdP \tag{2.45a}$$

and

$$w_{ps} = -\int_1^2 vdP \tag{2.45b}$$

2.5 Carnot Cycle

The Carnot cycle is an ideal thermodynamic cycle that represents the most efficient thermo-dynamic cycle for a heat engine and refrigeration machine operating between two temper-ature limits. It consists of four reversible processes: (i) reversible isothermal heat addition, Q_H, from high temperature, T_H (ii) reversible adiabatic expansion, (iii) reversible isothermal heat rejection to a low temperature, T_L, and (iv) reversible adiabatic compression. Figure 2.9 shows the Carnot cycle on a T-S diagram.

As we can see from the figure that the heat added at the high temperature, T_H, can be represented by rectangular area 1-2-b-a-1 under the constant temperature line 1-2. Similarly, the heat rejected at the low temperature, T_L, is represented by the area 3-4-a-b-3 under

Figure 2.9 Carnot cycle on temperature–entropy diagram.

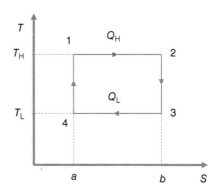

the constant temperature line 3-4. Also, note that the net work, W, is represented by the rectangular area 1-2-3-4-1.

Thermal efficiency of this Carnot cycle is given as

$$\eta = \frac{\text{Work out}}{\text{Heat added}} = \frac{W}{Q_H}$$
$$= \frac{Q_H - Q_L}{Q_H} = 1 - \frac{Q_L}{Q_H} \tag{2.46}$$

Now using Eq. (2.30b) for constant temperature heat addition at high temperature, T_H, and constant temperature heat rejection at low temperature, T_L, we can substitute $Q_H = T_H(S_2 - S_1)$ and $Q_L = T_L(S_3 - S_4)$ for reversible isothermal heat addition and heat rejection processes, respectively, in Eq. (2.46) and, we get

$$\eta = 1 - \frac{T_L (S_2 - S_1)}{T_H (S_3 - S_4)} \tag{2.47}$$

Since $S_2 = S_3$ for reversible adiabatic expansion process and $S_4 = S_1$ for reversible adiabatic compression process, Eq. (2.47) for the Carnot cycle efficiency is given as

$$\eta_{\text{carnot}} = 1 - \frac{T_L}{T_H} \tag{2.48}$$

With the application of second law of thermodynamics for an absolute thermodynamic scale and in the context of Carnot reversible thermodynamic cycle, it has been shown that the ratio of reversible heat addition at the high temperature and heat rejection at low temperature is given as the ratio of absolute temperatures of high-temperature heat source and low-temperature heat sink as

$$\frac{Q_H}{Q_L} = \frac{T_H}{T_L} \tag{2.49}$$

and this shows that high and low temperatures in the Carnot cycle efficiency Eq. (2.49) are given in terms of absolute temperature scale, $^\circ K$ or $^\circ R$

2.6 Machine Efficiencies

Efficiencies of different devices or machines in the power and cooling systems are defined by comparing actual performance of a real process with that achievable in an ideal process or in a reversible process. Examples of some of these devices are given here.

2.6.1 Turbine

The purpose of a steam turbine or a gas turbine is to convert thermal energy into mechanical energy or produce work output. The maximum possible work output is achieved when the working fluid expands in a reversible manner and without losing any heat to surrounding. Such an ideal process is referred to as the reversible adiabatic process or the isentropic process without any changes in entropy. The real process in the turbine involves increase in entropy due to irreversibilities associated with the friction and flow dynamics of the fluid as it passes through the flow passage between rotating blades of the turbine. The turbine efficiency is the defined as

$$\eta_t == \frac{\text{Actual turbine work}}{\text{Isentropic turbine work}} = \frac{w_{ta}}{w_{ts}} \tag{2.50}$$

or

$$\eta_t = \frac{w_{ta}}{w_{ts}} = \frac{h_i - h_e}{h_i - h_{es}} \tag{2.51}$$

2.6.2 Compressor and Pumps

In a compressor or in a pump, energy is added in the form of work to raise the pressure of the working fluid from a low-pressure level to a higher-pressure level. The actual work needed to compress the working fluid for given pressure ratio is greater than the reversible and adiabatic or isentropic compression work due to the presence of irreversibilities associated with the friction and flow dynamics of working fluid through the passages of the rotating blades. Efficiencies for these machines are expressed as follows:

2.6.2.1 Compressor

$$\eta_c = \frac{\text{Isentropic compressor work}}{\text{Actual compressor work}} = \frac{w_{cs}}{w_{ca}}$$

or

$$\eta_c = \frac{w_{cs}}{w_{ca}} = \frac{h_i - h_{es}}{h_i - h_e} \tag{2.52}$$

2.6.2.2 Pump

$$\eta_p = \frac{\text{Isentropic pump work}}{\text{Actual pump work}} = \frac{w_{cs}}{w_{ca}}$$

or

$$\eta_p = \frac{w_{ps}}{w_{pa}} = \frac{h_i - h_{es}}{h_i - h_e} \tag{2.53}$$

2.7 Specific Heat

Specific heat is defined as the energy needed to raise the temperature of a unit mass by a unit degree temperature. For constant volume process, $\delta w = pdv = 0$ and the first law reduces to $\delta Q = dU$. The **constant volume specific heat** is then expressed as

$$C_v = \frac{1}{m}\frac{\delta Q}{\delta T} = \frac{1}{m}\left(\frac{dU}{dT}\right)_v = \left(\frac{\partial u}{\partial T}\right)_v \tag{2.54}$$

For a constant pressure process, $\delta w = pdv$ and the first law reduces to $\delta Q = dU + PdV = dH$. The **constant pressure specific heat** is expressed as

$$C_p = \frac{1}{m}\frac{\delta Q}{\delta T} = \frac{1}{m}\left(\frac{dH}{dT}\right)_p = \left(\frac{\partial h}{\partial T}\right)_p \tag{2.55}$$

2.8 Ideal Gas Equation of State

An *ideal gas equation of state* is the relationship among the three basic measurable intensive properties of gases: temperature, pressure, and volume that is applicable for gases at low densities and defined as

$$P\bar{v} = \bar{R}T \tag{2.56a}$$

where

$$\bar{R} = \text{Universal gas constant} = 8.3145 \text{ kN m/kmol K} = 8.3145 \text{ kJ/kmol K}$$

and on a mass basis as

$$Pv = RT \tag{2.56b}$$

where

$$R = \text{gas constant} = \frac{\bar{R}}{M} \tag{2.57}$$

M = Molecular weight of gas
   In terms of total volume

$$PV = n\bar{R}T \tag{2.58a}$$

and

$$PV = mRT \tag{2.58b}$$

where n and m are the number of moles and mass of the gas, respectively.

   Any gas that satisfies this relationship is defined as the ideal gas. At a higher pressure, all gases deviate from ideal gas behavior and several other equations of state are derived to represent relationship of such real gas behavior. One such relationship that represents the degree of deviation from ideal gas behavior is given as

$$Pv = ZRT \tag{2.59}$$

where Z is defined as the **compressibility factor** and this factor approaches a value of unity as a real gas approaches ideal gas behavior. Compressibility factor is also presented in the form of a *generalized chart* as a function of reduced pressure, $\left(P_r = \frac{P}{P_C}\right)$ and reduced temperature $\left(T_r = \frac{T}{T_C}\right)$.

2.9 Change in Enthalpy, Internal Energy, Entropy, and Gibbs Function for Ideal Gases

2.9.1 Change in Enthalpy and Internal Energy

For ideal gases, enthalpy and internal energy are functions of temperature only. So, change in enthalpy and internal energy for a change of state or process is derived from the definition of specific heats given by Eqs. (2.54) and (2.55) respectively, and expressed as

$$h_2 - h_1 = \int_1^2 C_{po}\,dT \tag{2.60a}$$

and

$$u_2 - u_1 = \int_1^2 C_{vo} dT \tag{2.60b}$$

where C_{po} and C_{vo} represent ideal gas specific heat values. These equations can be evaluated for constant specific heat values and for temperature-dependent specific heat values.

2.9.1.1 Case I: Constant Specific Heat
For constant specific heat values, Eq. (2.60a and b) can be evaluated as

$$h_2 - h_1 = C_{po}(T_2 - T_1) \tag{2.61a}$$

and

$$u_2 - u_1 = C_{vo}(T_2 - T_1) \tag{2.61b}$$

2.9.1.2 Case II: Temperature-Dependent Specific Heat values
For temperature-dependent specific heat functions, equations can be evaluated by simply substituting the functional relations and carrying out the integrations term by term.

$$h_2 - h_1 = \int_{T_1}^{T_2} C_{po}(T) dT \tag{2.62}$$

where $C_{po}(T)$ is a functional relationship of the specific heat as a function of temperature. Table C.5 presents such functional relationship for some of the common ideal gases.

2.9.1.3 Case III
To simplify the computations, the integral Eq. (2.62) is written by computing enthalpy change from a reference temperature as

$$h_2 - h_1 = \int_{T_0}^{T_2} C_{po}(T) dT - \int_{T_0}^{T_1} C_{po}(T) dT \tag{2.63}$$

By defining

$$h_T = \int_{T_0}^{T} C_{po}(T) dT \tag{2.64}$$

the change in enthalpy equation is written as

$$h_2 - h_1 = h_{T_2} - h_{T_1} \tag{2.65}$$

The integral given by Eq. (2.64) is evaluated for different gases over a range of temperatures and assuming a reference temperature of $T_0 = 20 \,°C$ or $298 \,°K$. Table C.7 presents such integral values for some of the common ideal gases.

2.9.2 Entropy Change in a Process

Entropy change of an ideal gas is derived from the thermodynamic relation $TdS = du + pdv$ and expressed as

$$s_2 - s_1 = \int_1^2 C_{vo} \frac{dT}{T} + R \ln \frac{v_2}{v_1} \tag{2.66a}$$

and from the thermodynamic relation $TdS = dh - vdp$ as

$$s_2 - s_1 = \int_1^2 C_{po} \frac{dT}{T} - R \ln \frac{P_2}{P_1} \tag{2.66b}$$

2.9.3 Special Cases

2.9.3.1 Case I: For Constant Specific Heat Values

$$s_2 - s_1 = C_{vo} \ln \frac{T_2}{T_1} + R \ln \frac{v_2}{v_1} \tag{2.67a}$$

and

$$s_2 - s_1 = C_{po} \ln \frac{T_2}{T_1} - R \ln \frac{P_2}{P_1} \tag{2.67b}$$

2.9.3.2 Case II: For Temperature-Dependent Specific Heat Values

$$s_2 - s_1 = \int_1^2 C_{vo}(T) \frac{dT}{T} + R \ln \frac{v_2}{v_1} \tag{2.68a}$$

and

$$s_2 - s_1 = \int_1^2 C_{po}(T) \frac{dT}{T} - R \ln \frac{P_2}{P_1} \tag{2.68b}$$

With the substitution of the functional relations for specific heats, these equations can be evaluated by carrying out the integration term by term.

2.9.3.3 Case III

Using the procedure outlined for enthalpy, the integral Eq. (2.68b) is written by computing entropy change from a reference temperature as follows:

$$s_2 - s_1 = \left(s_{T_2}^0 - s_{T_1}^0\right) - R \ln \frac{P_2}{P_1} \tag{2.69a}$$

where

$$s_T^0 = \int_{T_0}^T \frac{C_{po}}{T} dT \tag{2.69b}$$

The integral given by Eq. (2.69b) is evaluated for different gases over a range of temperatures and assuming a reference temperature of $T_0 = 20$ °C or 298 °K. Table C.7 presents such integral values for some of the common ideal gases.

2.10 Reversible Polytropic Process

For a reversible polytropic process, the pressure-volume relation is given as

$$PV^n = \text{Constant} \tag{2.70}$$

where the superscript index **n** varied depends on type of process as described as: $n = 1$ for Isothermal process (constant temperature process); $n = k$ for isentropic process (constant entropy process); $n = 0$ for isobaric process (constant pressure process); and $n = 0$ for isochoric pressure (constant volume process).

Using the ideal gas relation $PV = mRT$ and for constant specific heat, the following relations are derived for process between two states 1 and 2:

$$\frac{P_2}{P_1} = \left(\frac{V_1}{V_2}\right)^n \tag{2.71a}$$

$$\frac{T_2}{T_1} = \left(\frac{P_2}{P_1}\right)^{\frac{n-1}{n}} \tag{2.71b}$$

$$\frac{T_2}{T_1} = \left(\frac{V_2}{V_1}\right)^{n-1} \tag{2.71c}$$

Work done:

$$_1W_2 = \int_1^2 PdV = \frac{P_2V_2 - P_1V_1}{1-n} = \frac{mR\left(T_2 - T_1\right)}{1-n} \text{ for all values of } \mathbf{n}, \text{ except } \mathbf{n} = 1 \tag{2.72a}$$

and

$$_1W_2 = \int_1^2 PdV = P_1V_1 \ln\frac{V_2}{V_1} = mRT_1 \ln\frac{V_2}{V_1} \text{ for } \mathbf{n} = 1 \tag{2.72b}$$

2.11 Reversible Adiabatic or Isentropic Process

For a reversible adiabatic or isentropic process, the pressure-volume relation is given as

$$PV^k = \text{Constant} \tag{2.73}$$

where the superscript index **k** is defined as

$$k = \frac{C_{po}}{C_{vo}} \tag{2.74}$$

Using the ideal gas relation $PV = mRT$ and for constant specific heat, the following relations are derived for a reversible adiabatic process between two states 1 and 2

$$\frac{P_2}{P_1} = \left(\frac{V_1}{V_2}\right)^k, \frac{T_2}{T_1} = \left(\frac{P_2}{P_1}\right)^{\frac{k-1}{k}} \text{ and } \frac{T_2}{T_1} = \left(\frac{V_2}{V_1}\right)^{k-1} \tag{2.75}$$

Work done:

$$_1W_2 = \int_1^2 PdV = \frac{P_2V_2 - P_1V_1}{1-k} = \frac{mR\left(T_2 - T_1\right)}{1-k} \tag{2.76}$$

2.12 Mixture of Gases

Operations of many power systems involve mixture of gases. The thermodynamic analysis of such systems requires consideration of mixture gas composition and properties. The gas mixture may be a mixture of ideal gases or a mixture of real gases.

2.12.1 Mixture Parameters

Mass fraction and mole fraction are the two most common mixture parameters that are important for the estimation of mixture properties. These parameters are defined as follows:

2.12.1.1 Mass Fraction

The mass fraction of a component in a mixture is defined as the ratio of mass of the component gas to the total mass as

$$\text{Mass fraction, } x_i = \frac{m_i}{\sum_{i=1}^{N} m_i} = \frac{\text{Mass of component-}i}{\text{Total mass of the mixture}} \tag{2.77a}$$

where N is the total number of gas species or components.

2.12.1.2 Mole Fraction

The mole fraction of a component in a mixture is defined as the ratio of number of moles of the component to the total number of moles in the mixture as

Mole fraction,

$$y_i = \frac{n_i}{\sum_{i=1}^{N} n_i} = \frac{\text{Number of moles of component-}i}{\text{Total number of moles in the mixture}} \tag{2.77b}$$

Relation between mass fraction and mole fraction is given as

$$x_i = \frac{y_i M_i}{\sum_{i=1}^{N} y_i M_i} \tag{2.78}$$

Molecular weight of the mixture is given as

$$M_{\text{mix}} = \frac{m_{\text{mix}}}{N} = \sum_{i=1}^{N} y_i M_i \tag{2.79}$$

Application of ideal gas law model leads to the following two important relations for ideal gas mixtures:

Dalton's law assumes each component of the mixture exists at the same temperature and total volume of the mixture and it leads to

$$P = \sum_{i=1}^{N} P_i \tag{2.80}$$

where

$$P_i = \frac{n_i \overline{R} T}{V} = \text{Partial pressure of the component gas} \tag{2.81}$$

Amagat's law assumes that each component of the mixture exists at the same temperature and total pressure of the mixture and it leads to

$$V = \sum_{i=1}^{N} V_i \tag{2.82}$$

where

$$V_i = \frac{n_i \overline{R} T}{P} \tag{2.83}$$

Additionally, it can be shown that the volume fraction, the mole fraction, and the ratio of partial pressure to the total pressure are all equal, i.e.

$$y_i = \frac{n_i}{n} = \frac{V_i}{V} = \frac{P_i}{P} \tag{2.84}$$

Eq. (2.84) is used to determine the partial pressure of gas components based on the mole fraction of those components.

2.12.2 Ideal Gas Mixture Properties

Ideal gas mixture properties are represented based on Dalton's law as the sum of contributions from all components of the mixture. The following is a list of some of the basic mixture properties:

Mixture gas constant:

$$R_{mix} = \frac{\overline{R}}{M_{mix}} = \sum_{i=1}^{N} x_i R_i \tag{2.85}$$

Mixture specific heat:

$$C_p = \sum_{i=1}^{N} x_i C_{p_i}, \overline{C}_p = \sum_{i=1}^{N} y_i \overline{C}_{p_i} \tag{2.86a}$$

$$C_v = \sum_{i=1}^{N} x_i C_{v_i}, \overline{C}_v = \sum_{i=1}^{N} y_i \overline{C}_{v_i} \tag{2.86b}$$

Total properties:

$$U = \sum_{i=1}^{N} U_i, H = \sum_{i=1}^{N} H_i, S = \sum_{i=1}^{N} S_i, \tag{2.87}$$

Specific properties:

$$u = \sum_{i=1}^{N} x_i u_i\,(T), u = \sum_{i=1}^{N} y_i \overline{u}_i\,(T) \tag{2.88a}$$

$$h = \sum_{i=1}^{N} x_i h\,(T), h = \sum_{i=1}^{N} y_i \overline{h}_i\,(T) \tag{2.88b}$$

$$s = \sum_{i=1}^{N} x_i s\,(T, P_i), s = \sum_{i=1}^{N} y_i \overline{s}_i\,(T, P_i) \tag{2.88c}$$

It is important to note that for an ideal gas mixture, enthalpy and internal energy are only a function of temperature, and hence component gas enthalpy and internal energy are estimated as function of mixture temperature, T. All component gas properties are evaluated at the mixture temperature, T, and component partial pressure, P_i. However, entropy of an ideal gas is a function of temperature and pressure, and so the component gas entropy is estimated as function of gas mixture temperature, T, and partial pressure, P_i, of the component in the mixture.

2.12.3 Change of Properties in a Thermodynamic Process

For change of state in a thermodynamic process, the change in thermodynamic properties is given as follows:

Total properties:

$$U_2 - U_1 = \sum_{i=1}^{N} n_i \left[\bar{u}_{i2}(T_2) - \bar{u}_{i1}(T_1) \right] = \sum_{i=1}^{N} m_i \left[u_{i2}(T_2) - u_{i1}(T_1) \right] \tag{2.89a}$$

$$H_2 - H_1 = \sum_{i=1}^{N} n_i \left[\bar{h}_{i2}(T_2) - \bar{h}_{i1}(T_1) \right] = \sum_{i=1}^{N} m_i \left[h_{i2}(T_2) - h_{i1}(T_1) \right] \tag{2.89b}$$

$$S_2 - S_1 = \sum_{i=1}^{N} n_i \left[\bar{s}_{i2}(T_2, P_{i2}) - \bar{s}_{i1}(T_1, P_{i1}) \right] = \sum_{i=1}^{N} m \left[s_{i2}(T_2, P_{i2}) - s_{i1}(T_1, P_{i1}) \right] \tag{2.89c}$$

Specific properties:

$$\bar{u}_2 - \bar{u}_1 = \sum_{i=1}^{N} y_i \left[\bar{u}_{i2}(T_2) - \bar{u}_{i1}(T_1) \right], \ u_2 - u_1 = \sum_{i=1}^{N} x_i \left[u_{i2}(T_2) - u_{i1}(T_1) \right] \tag{2.90a}$$

$$\bar{h}_2 - \bar{h}_1 = \sum_{i=1}^{N} y_i \left[\bar{h}_{i2}(T_2) - \bar{h}_{i1}(T_1) \right], \ h_2 - h_1 = \sum_{i=1}^{N} x_i \left[h_{i2}(T_2) - h_{i1}(T_1) \right] \tag{2.90b}$$

$$\bar{s}_2 - \bar{s}_1 = \sum_{i=1}^{N} y_i \left[\bar{s}_{i2}(T_2, P_{i2}) - \bar{s}_{i1}(T_1, P_{i1}) \right], \ s_2 - s_1 \sum_{i=1}^{N} x_i \left[s_{i2}(T_2, P_{i2}) - s_{i1}(T_1, P_{i1}) \right] \tag{2.90c}$$

For an ideal gas mixture, change of thermodynamic properties can be computed following Eqs. (2.61a and b) as follows:

For constant specific heat

$$\bar{h}_2 - \bar{h}_1 = \sum_{i=1}^{N} y_i \bar{c}_{pi}(T_2 - T_1), \ h_2 - h_1 = \sum_{i=1}^{N} x_i c_{pi}(T_2 - T_1) \tag{2.91a}$$

$$\bar{u}_2 - \bar{u}_1 = \sum_{i=1}^{N} y_i \bar{c}_{vi}(T_2 - T_1), \ u_2 - u_1 = \sum_{i=1}^{N} x_i c_{vi}(T_2 - T_1) \tag{2.91b}$$

For variation in specific heat with temperature, we can use Eq. (2.65) to compute mixture

$$h_2 - h_1 = \sum_{i=1}^{N} x_i \left(h_{iT_2} - h_{iT_1} \right) \tag{2.92}$$

where h_T is computed for different gas species using Eq. (2.64) and given in Table C.7.

Example 2.1 *Ideal Gas Mixture Properties*

The volume composition of a gas mixture is given as $H_2 : 78\%$, $CO_2 : 20\%$ and $H_2O : 2\%$. Determine: (a) mass fraction of the component gasses in the mixture, (b) gas constant of the mixture, (c) constant pressure specific heat of the mixture, and (d) heat transfer to cool the mixture from 500 to $100\,°C$.

Solution

Molar composition is given as

$$y_{H_2} = 0.78, y_{CO_2} = 0.2, \text{ and } y_{H_2O} = 0.02$$

The molecular weight of the mixture or the mass of the mixture per kmol of mixture is

$$M_{mix} = \sum_{i=1}^{N} m_i = m_{H_2} + m_{CO_2} + m_{H_2O}$$

$$= M_{H_2} \times y_{H_2} + M_{CO_2} \times y_{CO_2} + M_{H_2O} \times y_{H_2O}$$

$$= 2.016 \times 0.78 + 44.0 \times 0.2 + 18.016 \times 0.02$$

$$= 1.5724 + 8.8 + 0.3603$$

$$M_{mix} = 10.7327 \text{ kg/kmol of mixture}$$

(a) Mass fractions

$$x_{H_2} = \frac{m_{H_2}}{\sum_{i=1}^{N} m_i} = \frac{1.5724}{10.7327} = 0.1465$$

$$x_{CO_2} = \frac{m_{CO_2}}{\sum_{i=1}^{N} m_i} = \frac{8.8}{10.7327} = 0.8199$$

$$x_{H_2O} = \frac{m_{H_2O}}{\sum_{i=1}^{N} m_i} = \frac{0.3603}{10.7327} = 0.0335$$

(b) Mixture gas constant

The gas constant of the mixture is

$$R_{mix} = \frac{\bar{R}}{M_{mix}} = \frac{8.3144}{10.7327} = 0.7746 \text{ kJ/kg K}$$

(c) Mixture specific heat

The constant pressure specific heat of the mixture is

$$C_{P_{mix}} = \sum x_i C_{P_i}$$

$$= x_{H_2} \times C_{P_{H_2}} + x_{CO_2} \times C_{P_{CO_2}} + x_{H_2O} \times C_{P_{H_2O}}$$

$$= 0.1465 \times 14.209 + 0.8199 \times 0.842 + 0.0335 \times 1.872$$

$$= 2.0816 + 0.69035 + 0.0627$$

$$= 2.835 \text{ kJ/kg K}$$

(d) Heat transfer to cool the mixture

$$_1q_2 = h_2 - h_1 = C_{P_{mix}} (T_2 - T_1) = 2.835 \times (500 - 100)$$

$$_1q_2 = 1134 \text{ kJ/kg}$$

2.12.4 Moist Air: Mixture of Air and Water Vapor

One of the simplified models is developed for dealing with a mixture of ideal gases in contact with liquid or solid phase of one of the components. One such mixture is the moist air,

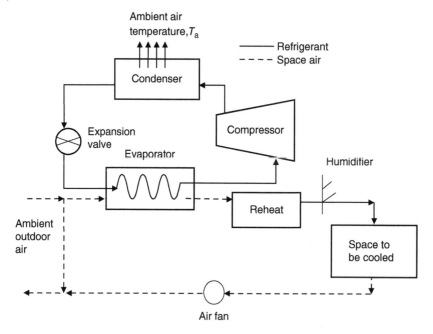

Figure 2.10 Air-conditioning processes dealing with moist-air mixture.

which is a mixture of dry air and water vapor, and may exist in contact with liquid water or ice. Such a moist-air mixture model is used for analyzing and designing air-conditioning processes shown in Figure 2.10.

Figure 2.10 shows some of the common moist-air or psychrometric processes: (i) *Heating*, (ii) *Cooling*, (iii) *Dehumidification*, (iv) *Humidification*, and (v) *mixing of two moist-air streams*. A brief description of this model in terms of psychrometric principles, basic parameters, and governing equations for analyzing and designing air-conditioning processes and systems is given here.

2.12.4.1 Dew-Point Temperature (T_{dp})

It is the temperature at which the vapor condenses or solidifies when the mixture is cooled at a constant pressure corresponding to the partial pressure of the vapor in the mixture.

2.12.4.2 Relative Humidity (RH or ϕ)

Defined as the ratio of mole fraction of the vapor in the mixture to the mole fraction of the vapor in a saturated mixture at the same temperature and the total pressure

$$\phi = \frac{Pv}{Pg \text{ at } T} \tag{2.93}$$

2.12.4.3 Humidity Ratio (ω)

Defined as the ratio of the mass of the vapor (m_v) in the mixture to the mass of the dry air (m_a) and expressed as

$$\omega = \frac{m_v}{m_a} \tag{2.94}$$

Since air–water vapor mixture is considered as the ideal gas mixture, we can compute mass of water vapor and that of dry air using ideal gas equations state

$$m_v = \frac{P_v V}{R_v T} = \frac{P_v V M_v}{\overline{R}_v T} \tag{2.95a}$$

and

$$m_a = \frac{P_a V}{R_a T} = \frac{P_a V M_a}{\overline{R}_v T} \tag{2.95b}$$

Substituting these relations, we can obtain the expression for humidity ratio:

$$\omega = \frac{m_v}{m_a} = \frac{R_a P_v}{R_v P_a} = \frac{M_v P_v}{M_a P_a} \tag{2.96a}$$

Now substituting the molecular weight values of water and air, we get

$$\text{Humidity ratio,}\ \omega = 0.622 \frac{P_v}{P_a} = 0.622 \frac{P_v}{P - P_v} \tag{2.96b}$$

The **degree of saturation** is defined as the ratio of the actual humidity ratio to the humidity ratio of a saturated mixture at the same temperature and pressure. As mixture temperature reaches the dewpoint temperature, moisture starts condensing and the partial pressure of the vapor decreases. But the vapor that remains in mixture is always enough to keep the air saturated, and the liquid or the solid is in equilibrium with the air-water vapor mixture.

2.12.4.4 Dry-Bulb and Wet-Bulb Temperatures

The dry-bulb air temperature (T_{db}) is the temperature measured by a regular thermometer. The wet-bulb temperature (T_{wb}) is the temperature measured by the thermometer with its bulb covered with a wet wick. Depending on the moisture level in air, moisture evaporates from the wick and the thermometer temperature is established by the combined heat and mass transfer process from the moist air. The combination of wet-bulb temperature and dry-bulb temperature measurements is used to measure the moisture or humidity content of air in the instrument called psychrometer. The smaller the difference between dry-bulb temperature and wet-bulb temperature, the higher is the moisture content or humidity level in the air. A state of equal dry-bulb and wet-bulb temperature is an air of 100% relative humidity.

2.12.4.5 Moist-air Enthalpy

Total enthalpy of moist air is given as the sum of the enthalpy of the dry air and that of the moisture in the mixture

$$H = H_a + H_v = m_a h_a + m_v h_v \tag{2.97a}$$

The specific enthalpy of the mixture per unit mass of the dry air is expressed as

$$h = h_a + \omega h_v \tag{2.97b}$$

where $h_a =$ Enthalpy of dry air determined at a temperature with respect to the reference enthalpy of zero at the O $^\circ$C

$$h_a = \int_{O^\circ}^{T} C_{pa} dT = C_{pa} T\ (^\circ C) \tag{2.97c}$$

h_v = Enthalpy of water vapor computed as the saturated water vapor, h_g, at the dry-bulb temperature of the mixture from the steam table.

2.12.4.6 Psychrometric Chart

Psychrometric charts are constructed to display moist-air properties and the process associated with the moist air. All major moist-air properties such as humidity ratio (ω), relative humidity (ϕ), moist-air enthalpy ($h_a + \omega h_v$), dry-bulb temperature (T_{db}) and wet-bulb temperature (T_{wb}), and specific volume of air. Such charts are widely used by the designers of air-conditioning systems. One such psychrometric chart constructed at 1-atm pressure is given in Figure C.1.

2.12.5 Application of Conservation Equations to Air-Conditioning Process

Analysis of processes in air-conditioning systems involves application of conservation mass and first law of thermodynamics along with the moist-air property data using the psychrometric principles. Let us consider the section of the duct from inlet to the outlet without the volume of the heating coil or cooling and dehumidifying for flow of moist air through the duct. Considering SSSF process for a control volume, we can derive the conservation of mass and energy equation for these processes as follows:

2.12.5.1 Conservation of Mass

Setting $\left(\frac{dm}{dt}\right)_{cv} = 0$, we can write the conservation mass equation for SSSF process for control volume from Eq. (2.8) as

$$\sum \dot{m}_e - \sum \dot{m}_i = 0 \tag{2.98a}$$

Now applying this equation individually for each components of moist air, we get

Dry air mass:

$$\dot{m}_{a1} = \dot{m}_{a2} = \dot{m}_a \tag{2.98b}$$

Water mass:

$$\dot{m}_{v1} = \dot{m}_{v2} \pm \dot{m}_{w3} \tag{2.98c}$$

If the mass flow rate of dry air does not vary during the process like in most air-conditioning processes except for a mixing process, we can express water vapor condensed or evaporated in terms of unit flow rate of the dry air. Dividing both sides of Eq. (2.98c) by the mass flow rate of the dry air, $\dot{m}_a$

$$\omega_1 = \omega_2 \pm \frac{\dot{m}_{w3}}{\dot{m}_a} \tag{2.99}$$

where $\dot{m}_{a1}$, and $\dot{m}_{a2}$ =Mass flow rate of dry air at sections 1 and 2, respectively; $\dot{m}_{v1}$, and $\dot{m}_{v2}$ =Mass flow rate of water vapor at sections 1 and 2, respectively; and $\dot{m}_{w3}$ = Mass flow rate of water for condensate in dehumidification or water injected in humidification

2.12.5.2 Conservation of Energy

The conservation energy equation for moist air is derived from the first law of thermodynamics equation for an SSSF process in control volume given by Eq. (2.17) and assuming negligible kinetic and potential energy changes as:

$$\dot{Q}_{cv} + \sum \dot{m}_i h_i = \sum \dot{m}_e h_e + \dot{W}_{cv} \tag{2.100a}$$

Considering single inlet and single outlet for the moist air flow, we can expand the summation terms in the following manner

$$\dot{Q}_{cv} + \dot{m}_{a1} h_{a1} + \dot{m}_{v1} h_{v1} = \dot{m}_{a2} h_{a2} + \dot{m}_{v2} h_{v2} \pm \dot{m}_{w3} h_{w3} \tag{2.100b}$$

Dividing both sides by mass flow rate of the dry air, $\dot{m}_a$.

$$\frac{\dot{Q}_{cv}}{\dot{m}_a} + \dot{m}_a h_{a1} + \frac{\dot{m}_{v1}}{\dot{m}_{v1}} h_{v1} = \dot{m}_a h_{a2} + \frac{\dot{m}_{v2}}{\dot{m}_a} h_{v2} \pm \frac{\dot{m}_{w3}}{\dot{m}_a} h_{w3}$$

or

$$\frac{\dot{Q}_{cv}}{\dot{m}_a} + \left(h_{a1} + \omega_1 h_{v1}\right) = \left(h_{a2} + \omega_2 h_{v2}\right) + \left(\omega_1 - \omega_2\right) h_{w3} \tag{2.100c}$$

where $\dot{Q}_{cv}$ = rate of heat transfer to or from the heating or cooling fluid through the heating coil or the cooling coil, respectively. $\dot{W}_{cv}$ = rate of work energy transfer and usually equal to zero for no shaft work.

We can compute the moist-air properties by using ideal gas properties for the dry air and steam table data for the water vapor or liquid. Also, note that the $(h_{a1} + \omega h_{v1})$ and $(h_{a2} + \omega h_{v2})$ in Eq. (2.100c) represent the moist-air enthalpy at inlet and outlet of the process, respectively. The term $(\omega_1 - \omega_2) h_{w3}$ represents the energy out flow due to the exit flow of the condensate.

2.12.6 Heating of Moist Air

One typical process of heating moist air flowing in an air-conditioning duct is shown in the figure below (Figure 2.11):

In the heating of moist air without any addition or removal of water, the temperature of the moist air increases and the state of the moist air moves to the region of lower percentage of relative humidity as shown in Figure 2.11b for psychrometric chart. The conservation mass and energy equations for the heating process can be derived from Eqs. (2.98b and c) as follows:

2.12.6.1 Conservation Mass
Dry-air mass:

$$\dot{m}_{a1} = \dot{m}_{a2} = \dot{m}_a \tag{2.101a}$$

Water mass:

$$\dot{m}_{v1} = \dot{m}_{v2} \pm \dot{m}_{w3} \tag{2.101b}$$

Since no water is added or removed, $\dot{m}_{w3} = 0$, and water vapor mass balance is given from Eq. (2.101b) as

$$\dot{m}_{v1} = \dot{m}_{v2} \tag{2.102a}$$

or

$$\omega_1 = \omega_2 \tag{2.102b}$$

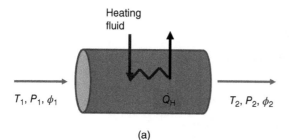

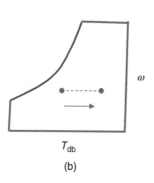

(b)

Figure 2.11 Heating of moist air flowing in a duct: (a) control volume for a heating process and (b) heating process shown on psychrometric chart.

2.12.6.2 Conservation of Energy

From first law or energy balance equation given by Eq. (2.100c), the heating rate is given as

$$\frac{\dot{Q}_{cv}}{\dot{m}_a} + h_{a1} + \omega_1 h_{v1} = h_{a2} + \omega_2 h_{v2} + (\omega_1 - \omega_2) h_{w3} \tag{2.103a}$$

or

$$\frac{\dot{Q}_{cv}}{\dot{m}_a} = (h_{a2} - h_{a1}) + \omega_2 h_{v2} - \omega_1 h_{v1} + (\omega_1 - \omega_2) h_{w3} \tag{2.103b}$$

Example 2.2 *Heating of Moist Air*

Let us consider the heating of moist air for steady-state-steady-flow over a heating coil in an air flow duct. Moist air enters the heating section of the duct at 10 °C and with 80% relative humidity. Air is heated and exits at a temperature of 40 °C. Determine (a) relative humidity of the air at the exit and (b) rate of heat transfer to the moist-air stream from the heating coil.

Solution

From conservation mass equations:

Dry-air mass:

$$\dot{m}_{a1} = \dot{m}_{a2} = \dot{m}_a$$

Water mass:

Since no water is added or removed,

$$\dot{m}_{v1} = \dot{m}_{v2}$$

or

$$\omega_1 = \omega_2$$

Compute partial pressure of water vapor and humidity ratio at inlet and exit:

State 1:

$$T_1 = 10°C, = 80\%, P_{g1} = P_{\text{sat at } T=10°C} = 1.2276 \text{ kPa}$$

$$P_{v1} = \phi_1 P_{g1} = P_{\text{sat at } T=10°C} = 0.80 \times 1.2276 \text{ kPa} = 0.98208 \text{ kPa}$$

$$\omega_1 = 0.622\frac{P_{v1}}{P_1 - P_{v1}} = 0.622\frac{0.98208}{100 - 0.98208 \text{ kPa}} = 0.622\frac{0.98208}{99.1792}$$

$$\omega_1 = 0.00616 \text{ kg H}_2\text{O/kg dry air}$$

$$h_{v1} = h_{g \text{ at } T=10°C} = 2519.74 \text{ kJ/kg from steam table.}$$

State 2:

$$\omega_2 = \omega_1 = \omega = 0.00616 \text{ kg H}_2\text{O/kg dry air}$$

$$T_2 = 40 °C, P_{g2} = P_{\text{sat at } 40°C} = 7.384 \text{ kPa}$$

$$h_{v2} = h_{g \text{ at } T=40°C} = 2574.26 \text{ kJ/kg from steam Table C.1.}$$

Since the water content remains same in the heating process, the mole fraction and partial pressure of the water vapor remain same. So, we can set

$$P_{v2} = P_{v1} = 0.98208 \text{ kPa}$$

and compute the relative humidity of the air at the exit as

$$\phi_2 = \frac{P_{v2}}{P_{g2 \text{ at } 40°C}} = \frac{0.98208}{7.384} = 0.1330$$

or

Relative humidity at the exit state, $\phi_2 = 13.3\%$
From first law or energy balance equation given by Eq. (2.94), the heating rate is given as

$$\frac{\dot{Q}_{cv}}{\dot{m}_a} + h_{a1} + \omega_1 h_{v1} = h_{a2} + \omega_2 h_{v2} + (\omega_1 - \omega_2) h_{w3}$$

or

$$\frac{\dot{Q}_{cv}}{\dot{m}_a} = (h_{a2} - h_{a1}) + \omega (h_{v2} - h_{v1})$$

$$= C_{pa} (T_2 - T_1) + \omega \left(h_{g_2 @ 30°C} - h_{g_1 @ 10°C}\right)$$

$$= 1.004 (40 - 10) + 0.00616 (2574.26 - 2519.74)$$

$$= 30.12 + 0.3358$$

$$\frac{\dot{Q}_{cv}}{\dot{m}_a} = 30.46 \text{ kJ/kg}$$

2.12.6.3 Cooling and Dehumidification Process

In the cooling and dehumidification process, a fraction of water vapor is removed from the moist-air stream by passing it over the cooling and dehumidifying coil. Refrigerant or chilled water flows through the cooling coil and keeps wall temperature below the dewpoint temperature of the moist stream so that water vapor condenses. The figure shows a control volume enclosing a dehumidifier operating at steady state. Moist air enters at state 1 and moves toward the saturation line. Moist air exits at state 2 closer to the saturation state at a lower temperature and humidity. Liquid water condensate exits as saturated liquid at state 2 (Figure 2.12).

The conservation mass and first law of thermodynamics for this process is summarized based on Eqs. (2.98) and (2.100) as follows assuming SSSF process over a control volume:

2.12.6.3.1 Conservation of Mass

Air mass:

$$\dot{m}_{a1} = \dot{m}_{a2} = \dot{m}_a \tag{2.104a}$$

Water mass:

$$\dot{m}_{v1} = \dot{m}_{v2} + \dot{m}_{12} \tag{2.104b}$$

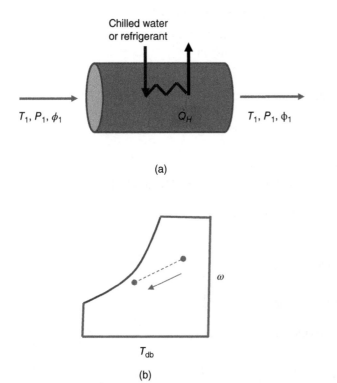

(a)

(b)

Figure 2.12 Cooling and dehumidifying process: (a)control volume for cooling and dehumidifying process and (b) cooling and dehumidifying process on psychrometric chart.

Dividing both sides by mass flow rate of the dry air, $\dot{m}_a$

$$\omega_1 = \omega_2 + \frac{\dot{m}_{12}}{\dot{m}_a} \tag{2.104c}$$

Mass of liquid condensate removed is given as

$$\frac{\dot{m}_{12}}{\dot{m}_a} = \omega_1 - \omega_2 \tag{2.105}$$

2.12.6.3.2 First Law of Thermodynamics

$$\frac{\dot{Q}_{cv}}{\dot{m}_a} + h_{a1} + \omega_1 h_{v1} = h_{a2} + \omega_2 h_{v2} + (\omega_1 - \omega_2) h_{w2} \tag{2.106}$$

Rearranging Eq. (2.106) and solving for the cooling rate, we get

$$\frac{\dot{Q}_c}{\dot{m}_a} == (h_{a2} - h_{a1}) + (\omega_2 h_{v2} - \omega_1 h_{v1}) + (\omega_1 - \omega_2) h_{w2} \tag{2.107a}$$

Assuming ideal gas relation for the change of enthalpy for dry air and using saturated value of water vapor and liquid, we get

$$\frac{\dot{Q}_c}{\dot{m}_a} = C_{pa} (T_2 - T_1) + (\omega_2 h_{g2} - \omega_1 h_{g1}) + (\omega_1 - \omega_2) h_{12} \tag{2.107b}$$

or

$$\frac{\dot{Q}_c}{\dot{m}_a} = C_{pa} (T_2 - T_1) + (\omega_2 h_{g2} - \omega_1 h_{g1}) + \frac{\dot{m}_{12}}{\dot{m}_a} h_{f2} \tag{2.107c}$$

Example 2.3 Cooling and Dehumidifying Coil

For moist-air flow over a cooling and dehumidifying coil, the moist-air state at inlet state 1 is given as $P_1 = 110$ kPa, $T_1 = 35\,°C$, and $Ø_1 = 85\%$. The exit state 2 is $P_2 = 106$ kPa, $T_2 = 10\,°C$, and $Ø_1 = 90\%$. Determine the heat-transfer rate for the cooling of the moist air and the amount of liquid condensate removed.

Solution

Compute partial pressure of water vapor and humidity ratio at inlet and exit:

State 1:

$$P_{v1} = Ø_1 P_{g1} = Ø_1 P_{\text{sat at } T=35°C} = 0.85 \times 5.628 \text{ kPa} = 4.7838 \text{ kPa}$$

$$\omega_1 = 0.622 \frac{P_{v1}}{P_1 - P_{v1}} = 0.622 \frac{4.7838}{110 - 4.7838 \text{ kPa}} = 0.622 \frac{4.7838}{105.2162}$$

$$\omega_1 = 0.02828 \text{ kg H}_2\text{O/kg dry air}$$

$$h_{v1} = h_{g \text{ at } T=35°C} = 2565.3 \text{ kJ/kg H}_2\text{O from steam table.}$$

State 2:

$$P_{v2} = Ø_2 P_{g2} = Ø_2 P_{\text{sat at } T=10°C} = 0.90 = 1.228 \text{ kPa}$$

$$\omega_2 = 0.622 \frac{P_{v2}}{P_2 - P_{v2}} = 0.622 \frac{1.228}{106 - 1.228} = 0.622 \frac{1.228}{104.8948}$$

$$\omega_2 = 0.0655 \text{ kg H}_2\text{O/kg dry air}$$

$$h_{v2} = h_{g \text{ at } T=10^\circ C} = 2519.8 \text{ kJ/kg and } h_{12} = h_{f \text{ at } T=10^\circ C} = 42.01 \text{ kJ/kg H}_2\text{O}$$

From Eq. (2.105), we can compute the mass of liquid water removed as the condensate as

$$\frac{\dot{m}_{12}}{\dot{m}_a} = \omega_1 - \omega_2 = 0.02828 - 0.00\,655 \text{ kg H}_2\text{O/kg dry air}$$

$$\frac{\dot{m}_{12}}{\dot{m}_a} = 0.02173 \text{ kg H}_2\text{O/kg dry air}$$

From Eq. (2.107c), the cooling coil heat-transfer rate is given as

$$\frac{\dot{Q}_c}{\dot{m}_a} = C_{pa}\left(T_2 - T_1\right) + \left(\omega_2 h_{v2} - \omega_1 h_{v1}\right) + h_{12}\frac{\dot{m}_{12}}{\dot{m}_a}$$

$$\frac{\dot{Q}_c}{\dot{m}_a} = 1.004\,(35-10) + (0.00655 \times 2519.8 - 0.02828 \times 2565.3) + 0.02173 \times 42.01$$

$$\frac{\dot{Q}_c}{\dot{m}_a} = -25.1 - 56.0419 + 0.91287$$

$$\frac{\dot{Q}_c}{\dot{m}_a} = -80.229 \text{ kJ/kg dry air}$$

2.12.6.4 Humidification Process

In the humidification process, water is added to increase the amount of moisture or, in other words, to increase the humidity in the flowing moist-air stream. Water vapor can be added by either injecting steam or by spraying liquid water as shown in Figure 2.13a. Both the dry-bulb temperature and humidity ratio of air increase as water is injected as steam as shown in Figure 2.13a. However, temperature of air decreases as humidity ratio increases when water is sprayed as liquid water as shown in Figure 2.13b. Such a humidification process is also referred to as the ***evaporative cooling*** process.

Like in the case of the heating process and in the cooling and dehumidifying process, the analysis of humidification process also starts with application of conservation of mass and energy balance equations. Considering SSSF process for a control volume representing the humidification process, we can state the mass and energy balance equations as follows:

2.12.6.5 Conservation of Mass

Air mass:

$$\dot{m}_{a1} = \dot{m}_{a2} = \dot{m}_a \tag{2.108a}$$

Water mass:

$$\dot{m}_{v1} = \dot{m}_{v2} + \dot{m}_{w3} \tag{2.108b}$$

Expressing in terms of per unit mass flow rate of dry air, we get

$$\omega_1 = \omega_2 + \frac{\dot{m}_{w3}}{\dot{m}_a} \tag{2.108c}$$

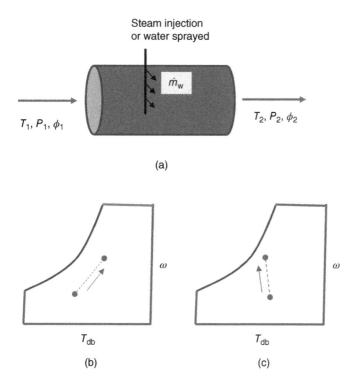

Figure 2.13 Humidification process: (a) control volume for humidification process, (b) process on psychrometric chart for steam, and (c) process on psychrometric chart for water spay in evaporative cooling.

where $\dot{m}_{a1}$, and $\dot{m}_{a2}$ = Mass flow rate of dry air at sections 1 and 2, respectively; $\dot{m}_{v1}$, and $\dot{m}_{v2}$ = Mass flow rate of water vapor at sections 1 and 2, respectively; and $\dot{m}_{w3}$ = Mass flow rate of condensate water in dehumidification or water injected in humidification.

2.12.6.6 Conservation of Energy

$$\dot{Q}_{cv} + \dot{m}_{a1}h_{a1} + \dot{m}_{v1}h_{v1} = \dot{m}_{a2}h_{a2} + \dot{m}_{v2}h_{v2} - \dot{m}_{w3}h_{w3} \qquad (2.109)$$

Dividing both sides by mass flow rate of the dry air, $\dot{m}_a$, and considering no transfer of heat across the control volume surface

$$\dot{m}_a h_{a1} + \frac{\dot{m}_{v1}}{\dot{m}_{v1}}h_{v1} = \dot{m}_a h_{a2} + \frac{\dot{m}_{v2}}{\dot{m}_a}h_{v2} - \frac{\dot{m}_{w3}}{\dot{m}_a}h_{w3}$$

or

$$\left(h_{a1} + \omega_1 h_{v1}\right) = \left(h_{a2} + \omega_2 h_{v2}\right) - \left(\omega_1 - \omega_2\right)h_{w3} \qquad (2.110)$$

where $\dot{Q}_{cv}$ = rate of heat transfer to or from the heating or cooling fluid through the heating coil or the cooling coil, respectively. $\dot{W}_{cv}$ = rate of work energy transfer and usually equal to zero for no shaft work.

We can compute the moist-air properties by using ideal gas properties for the dry-air and steam table data for the water vapor or liquid. Also, note that the terms $\left(h_{a1} + \omega h_{v1}\right)$ and

$(h_{a2} + \omega h_{v2})$ in Eq. (2.110) represent the moist-air enthalpy at inlet and outlet of the process, respectively. The term $(\omega_1 - \omega_2)h_{w3}$ represents the energy inflow due to water spray for humidification.

2.12.6.7 Mixing Process
Figure 2.14 shows the mixing of two different streams at a junction. The state of the mixed state can be determined using conservation of mass and energy equations.

From mass balance equations

Dry-air mass balance:

$$\dot{m}_{a3} = \dot{m}_{a1} + \dot{m}_{a2} \tag{2.111}$$

Water vapor mass balance:

$$\omega_1 \dot{m}_{a1} + \omega_2 \dot{m}_{a2} = \omega_3 \dot{m}_{a3} \tag{2.112}$$

Solving

$$\omega_3 = \frac{\omega_1 \dot{m}_{a1} + \omega_2 \dot{m}_{a2}}{\dot{m}_{a3}} \tag{2.113}$$

From energy balance equation

$$\dot{m}_{a1}\left(h_{a1} + \omega_1 h_{v1}\right) + \dot{m}_{a2}\left(h_{a2} + \omega_2 h_{v2}\right) = \dot{m}_{a3}\left(h_{a3} + \omega_3 h_{v3}\right)$$

or

$$\left(h_{a3} + \omega_3 h_{v3}\right) = \frac{\dot{m}_{a1}\left(h_{a1} + \omega_1 h_{v1}\right) + \dot{m}_{a2}\left(h_{a2} + \omega_2 h_{v2}\right)}{\dot{m}_{a3}} \tag{2.114}$$

Since right-hand-side (RHS) term of the Eq. (2.114) can be computed based on the thermodynamic moist-air properties of the two incoming streams, we can equate this to the enthalpy of the moist-air exit state at 3 given by the left-hand side (LHS). We can solve for the exit-state temperature using an iteration process by trial and error. We can guess T_3 and compute exit-state dry-air and water vapor enthalpies (h_{a3}, h_{v3}), and check if the exit-state enthalpy on the LHS balances with RHS of Eq. (2.114). This iteration process can be repeated until RHS = LHS.

Example 2.4 *Mixing of Moist-Air Streams*
A stream consisting of 35 m³/min of moist air 14 °C, 1-atm, 80% relative humidity mixes adiabatically with a stream consisting of 80 m³/min of moist air at 40 °C, 1-atm, 40% relative

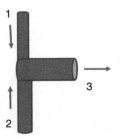

Figure 2.14 Mixing of two moist-air streams.

humidity, giving a mixed stream at 1-atm (see figure below). (a) Determine the specific volume and mass flow rates of the two streams at inlet sections (1 and 2), (b) determine humidity ratios and enthalpy values of the incoming streams, and (c) use mass and energy balance to determine the relative humidity, temperature, and enthalpy of the exiting stream.

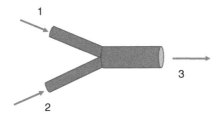

Solution

Compute partial pressure of water vapor and humidity ratio at inlet and exit states:

State 1:

$$T_1 = 14°C \ (287.2 \ K), \phi_1 = 80\%$$

$$P_{v1} = \emptyset_1 P_{g1} = \emptyset_1 P_{\text{sat at } T=14°C} = 0.80 \times 1.598 \ kPa = 1.2784 \ kPa$$

$$\omega_1 = 0.622 \frac{P_{v1}}{P_1 - P_{v1}} = 0.622 \frac{1.2784}{100 - 1.2784 \ kPa} = 0.622 \frac{1.2784}{98.7216}$$

$$\omega_1 = 0.00805 \ kg \ H_2O/kg \ dry \ air$$

$$h_{v1} = h_{g \text{ at } T=14°C} = 2527.1 \ kJ/kg \text{ from steam table.}$$

$$h_{a1} = 1.005 \times 14 = 14.07 \ kJ/kg \ H_2O \text{ at } 14°C \text{ from } (h_a \ (T) = C_{pa} \ T \ (C)$$

$$h_{a1} + \omega_1 h_{v1} = 14.07 + 0.00 \ 805 \times 2527.1 = 34.413 \ kJ/kg \ dry \ air$$

$$\text{Specific volume of dry air, } v_{a1} = \frac{R_a T_1}{P_{a1}} = \frac{0.2869 \times 287.2}{98.7216} = 0.8348 \ m^3/kg$$

$$\text{Mass flow rate of dry air, } \dot{m}_{a1} = \frac{\dot{V}_{a1}}{v_{a1}} = \frac{35}{0.8348} = 41.92 \ kg/ \ min = 0.6986 \ kg/s$$

State 2:

$$T_2 = 40°C \ (313.2 \ K), \phi_1 = 40\%$$

$$P_{v2} = \emptyset_2 P_{g2} = \emptyset_2 P_{\text{sat at } T=40°C} = 0.40 \times 7.384 = 2.9536 \ kPa$$

$$\omega_2 = 0.622 \frac{P_{v2}}{P_2 - P_{v2}} = 0.622 \frac{2.9536}{100 - 2.9536} = 0.622 \frac{2.9536}{97.0464}$$

$$\omega_2 = 0.0189 \ kg \ H_2O/kg \ dry \ air$$

$$h_{v2} = h_{g \text{ at } T=40°C} = 2574.3 \text{ kJ/kg}$$

$$h_{a2} = 1.005 \times 40 = 40.2 \text{ kJ/kg at } T = 40°C$$

$$h_{a2} + \omega_2 h_{v2} = 40.2 + 0.0189 \times 2574.3 = 88.93 \text{ kJ/kg dry air}$$

Specific volume of dry air, $v_{a2} = \dfrac{R_a T_2}{P_{a2}} = \dfrac{0.2869 \times 313.2}{97.0464} = 0.9259 \text{ m}^3/\text{kg}$

Mass flow rate of dry air $\dot{m}_{a2} = \dfrac{\dot{V}_{a2}}{v_{a2}} = \dfrac{80}{0.9259} = 86.40 \text{ kg/ min } = 1.44 \text{ kg/s}$

From dry-air mass balance:

$$\dot{m}_{a3} = \dot{m}_{a1} + \dot{m}_{a2} = 0.6986 + 1.44 = 2.1386 \text{ kg/s}$$

From water vapor mass balance:

$$\omega_1 \dot{m}_{a1} + \omega_2 \dot{m}_{a2} = \omega_3 \dot{m}_{a3}$$

Solving

$$\omega_3 = \dfrac{\omega_1 \dot{m}_{a1} + \omega_2 \dot{m}_{a2}}{\dot{m}_{a3}} = \dfrac{0.00805 \times 0.6986 + 0.0189 \times 1.44}{2.1386} = \dfrac{0.005623 + 0.0272}{2.1386}$$

$$\omega_3 = 0.01534 \text{ kg water/kg air}$$

From energy balance equation

$$\dot{m}_{a1} \left(h_{a1} + \omega_1 h_{v1}\right) + \dot{m}_{a2} \left(h_{a2} + \omega_2 h_{v2}\right) = \dot{m}_{a3} \left(h_{a3} + \omega_3 h_{v3}\right)$$

Substituting known values at sections 1 and 2, we get

$$0.6986 \left(14.07 + 0.0.00805 \times 2527.1\right) + 1.44 \left(40.2 + 0.0189 \times 2574.3\right)$$

$$= 2.1386 \left(h_{a3} + \omega_3 h_{v3}\right)$$

$$= 0.6986 \left(34.41\right) + 1.44 \left(88.854\right) = 2.1386 \left(h_{a3} + \omega_3 h_{v3}\right)$$

$$\left(h_{a3} + \omega_3 h_{v3}\right) = h_{m3} = \dfrac{24.03 + 127.94}{2.1386} = 71.065 \text{ kJ/kg air}$$

Since RHS, i.e. the enthalpy of the moist-air state at 3 is known, we can solve for the exit-state temperature using an iteration process by trial and error. Let us Guess T_3 and compute (h_{a3}, h_{v3}) and check if LHS. Iterate until RHS = LHS.

Trial 1: Guess $T_3 = 30°C$

$$h_{v3} = h_{g \text{ at } T=30°C} = 2556.3 \text{ kJ/kg}$$

$$h_{a3} = 1.005 \times 30 = 30.15 \text{ kJ/kg at } T = 30°C$$

$$\text{LHS} = \left(h_{a3} + \omega_3 h_{v3}\right) = \left(30.15 + 0.01534 \times 2556.3\right)$$

$$\text{LHS} = h_{m3} = 69.363 \text{ kJ/kg}$$

Trial 2 Guess $T_3 = 32\,^\circ C$

$$h_{v3} = h_{g \text{ at } T=32^\circ C} = 2559.9 \text{ kJ/kg}$$

$$h_{a3} = 1.005 \times 32 = 32.16 \text{ kJ/kg at } T = 30\,^\circ C$$

$$\text{LHS} = \left(h_{a3} + \omega_3 h_{v3}\right) = (32.16 + 0.01534 \times 2559.9)$$

$$\text{LHS} = h_{m3} = 71.428 \text{ kJ/kg}$$

Since LHS $\approx$ RHS, we can define the mixed state 3 as

$$T_3 = 32\,^\circ C, \omega_3 = 0.01534 \text{ kg water/kg air}$$

Relative humidity at this mixed state can be computed as follows:

$$\omega_3 = 0.01534 = 0.622\frac{P_{v3}}{P_3 - P_{v3}} = 0.622\frac{P_{v3}}{100 - P_{v3}}$$

Solving for P_{v3}

$$P_{v3} = 2.4068 \text{ kPa}$$

$$\varnothing_3 = \frac{P_{v3}}{P_{g3}} = \frac{P_{v3}}{P_{\text{sat at } T=32^\circ C}} = \frac{2.4068}{4.759} = 0.5057$$

Relative humidity, $\varnothing_3 = 50.57\%$

2.13 Combustion Process

In a chemical reaction, the bonds of the molecules of the reactants are broken, and atoms and electrons regroup to form new products. Chemical reactions can be exothermic or endothermic depending on the type of process.

A combustion process is one kind of a chemical reaction in which a fuel is oxidized, and a large quantity of chemical energy is released. The rapid oxidation of the combustible elements in the fuel results in the energy release as combustion products are formed. The amount of chemical energy converted to thermal energy is the difference between the internal energy content of the original bond structure of the reactants and the internal energy content of the regrouped bond structures of the products.

Some of the combustible elements in many common fuels are Carbon (C), Hydrogen (H), and Sulfur (S). While carbon and hydrogen are the major contributor to the release of thermal energy, sulfur contributes relatively less to energy release. But it is one of the major contributors to pollution.

2.13.1 Combustion Reaction

A general combustion reaction that transforms reactants into products is demonstrated as:

Reactants (R) → Products (P)

or

Fuel (R) + Oxidizer (R) → Products (P)

In a complete combustion, all combustible elements are burned completely, forming compounds as follows:

$$C + O_2 \rightarrow CO_2; H_2 + \frac{1}{2}O_2 \rightarrow H_2O \text{ and } S + O_2 \rightarrow SO_2 \qquad (2.115)$$

In an incomplete combustion, partial combustion takes place for some of the elements.

In a combustion reaction, mass is always conserved through the balance of mass of reactants with the mass of the products. Even though the number of moles of reactants and products may vary, the total mass of each chemical element must be same on both sides of the reaction equation even though the element may exist in different chemical compounds in the reactant and products. Let us consider an example of complete combustion of hydrogen with oxygen and formation of water molecule:

$$H_2 + \frac{1}{2}O_2 \rightarrow H_2O \qquad (2.116)$$

In this complete reaction, hydrogen and oxygen are the two reactants: the fuel and the oxidizer, respectively, and water is the product. We can notice that in this reaction, 1-kmol of H_2 reacts with ½ kmol of O_2 and forms 1-kmol of H_2O. So, the number of moles of the reactants and product is different. However, the mass conserved with equal mass of reactants and products on both sides of the reaction, i.e.

$$2 \text{ kg of } H_2 + 16 \text{ kg of } O_2 = 18 \text{ kg of } H_2O \qquad (2.117)$$

2.13.2 Balanced Reaction Equation

As a first step to any thermodynamic combustion analysis, a mole or mass balanced equations are always written by checking into account of the conservation of mass to all elements. For example, a typical combustion of methane reaction with oxygen can be written as

$$CH_4 + \textbf{(a)} \ O_2 \rightarrow \textbf{(b)} \ CO_2 + \textbf{(c)} \ H_2O \qquad (2.118a)$$

where the constant coefficients **a**, **b**, and **c** are moles of O_2, CO_2, and H_2O per mole of CH_4 in a complete reaction, and these are determined by making balance of mass for each element on both sides of the equation as demonstrated below:

For C: $1 = b, b = 1$

For O: $2a = 2b + c$ or $2a = 2 + c = 2 + 2$

or $a = 2$

For H: $4 = 2c$ or $c = 2$

Substituting coefficient values, we get the balanced reaction as

$$CH_4 + 2O_2 \rightarrow CO_2 + 2H_2O \qquad (2.118b)$$

2.13.3 Hydrocarbon Fuel Types

One of the most commonly available forms of fuel is **hydrocarbon fuels**, which have carbon and hydrogen as the two primary constituents. The fuels are categorized into three basic forms: solid coal, liquid hydrocarbons, and gaseous hydrocarbons. Most liquid and gaseous hydrocarbon fuels are a mixture of many different hydrocarbons and are produced from crude oil using a process of distillation and cracking. Some of the common hydrocarbon fuels that are produced from crude oil are gasoline, diesel, kerosene fuel oil, etc. Other common popular hydrocarbon fuels are **methyl alcohol** or **methanol**, **ethyl alcohol** or **ethanol**, etc. Even though each one of the hydrocarbon fuels is a mixture of a number of hydrocarbons, it is expressed as a single hydrocarbon, C_mH_n, with fixed **m**-number of carbon and **n**-number of hydrogen. For example, **gasoline** is known as **octane**, C_8H_{18}, and diesel fuel is known as **dodecane**, $C_{12}H_{26}$. Common sources of gaseous hydrocarbons fuels are natural gas wells. Other sources of liquid and gaseous hydrocarbon fuel are from coal gasification, shale oil, and shale gas. Solid coal composition varies depending on the geographical locations of the coal mines, and the composition is generally given in terms of relative percentage of carbon, hydrogen, nitrogen, ash, etc.

2.13.4 Combustion Reaction Model

While oxygen in a pure form is used for many special applications, air is generally the source of oxygen in most combustion applications. Even though air is composed of many different elements, for analysis purposes it is modeled as dry air and assumed to be composed of 21% oxygen (O_2) and 79% nitrogen (N_2) on a molar basis, and this equates to a molar ratio of nitrogen (N_2) to oxygen (O_2) as 3.76. So, when air is supplied as a source of oxygen in a combustion reaction, each mole of O_2 is accompanied by 3.76 mol of N_2. Based on this assumption, when humid air is supplied in a combustion reaction, the water vapor composition also needs to be included in writing the balanced combustion reaction.

In combustion of hydrocarbon fuels, carbon, hydrogen, and any other constituents in the fuel are oxidized by reacting with oxygen. For example, a typical combustion reaction of Octane fuel (C_8H_{18}) is shown in Figure 2.15: and is represented in the following manner:

$$C_8H_{18} + 12.5\left(O_2 + 3.76\,N_2\right) \rightarrow 8\,CO_2 + 9\,H_2O + 47N_2 \tag{2.119}$$

where carbon and oxygen on the LHS of the reaction equation are referred to as reactants, and carbon dioxide, water, remaining oxygen and nitrogen on the RHS are referred to as the products. In this reaction, nitrogen is assumed as inert and does not undergo any chemical reactions. Nitrogen thus appears unchanged and influences the product temperature.

2.13.5 Major Combustion Parameters

Several parameters are defined to control and analyze the combustion reaction performance. Two such important parameters are (i) percentage of theoretical air or percentage of excess air and (ii) air-to-fuel ratio.

Figure 2.15 Typical schematic representation of a combustion reaction.

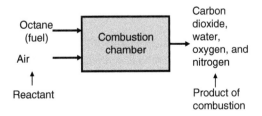

2.13.5.1 Theoretical Air (Stoichiometric) and Excess Air

Theoretical air, also known as *stoichiometric air*, is defined as the minimum amount of air needed to supply enough oxygen for the complete combustion of all combustion elements. For example, in a complete combustion reaction of hydrocarbon fuel with 100% theoretical or stoichiometric air, the product mixture will be composed of CO_2, H_2O, and N_2 as demonstrated in Eq. (2.119).

With 200% stoichiometric air or excess air, the balanced reaction equation is given as

$$C_8H_{18} + (12.5 \times 2.0)\,O_2 + (12.5 \times 2.0 \times 3.76)\,N_2$$
$$\rightarrow \quad 8CO_2 + 9H_2O + (12.5) \times O_2 + (94.0)\,N_2 \tag{2.120}$$

Generally, it is also assumed that in an ideal combustion, N_2 does not react with oxygen and remains inert. In most engine combustion reactions, however, nitrogen undergoes reaction with oxygen and produces air pollutants in the form of oxides of nitrogen (NO_x) such as nitrous oxide (N_2O), nitric oxide (NO), and nitrogen ixide (NO_2) depending on the temperature level of the reaction. In an incomplete combustion reaction, the product may also contain some fuel as unburned fuel, some carbon in the form of CO and even as carbon particles. An incomplete combustion is caused due to insufficient supply of oxygen as well as inadequate mixing of fuel and air in the mixture. For this reason, excess air is supplied to avoid any incomplete reactions caused due to the lack of air or oxygen. Other factors that contribute to incomplete combustions are inadequate mixing and turbulence. In an incomplete combustion reaction, some carbon produces carbon monoxide (CO) in addition to the formation of carbon dioxide (CO_2).

When more than stoichiometric quantity of air or oxidizer or excess is used in a combustion, the air-fuel (AF) mixture is referred to as the *lean* or *fuel lean*. Air-fuel mixture with less than stoichiometric air results in a *fuel-rich* or *rich* mixture.

2.13.5.2 Air-Fuel Ratio (AF)

Air-to-Fuel ratio is defined as the ratio of the amount of air in the reaction to the amount of fuel, and this can be expressed either on a mass basis or on a molar basis as demonstrated below:

$$\text{Mass basis: AF} = \frac{\text{Mass of air}}{\text{Mass of Fuel}} \tag{2.121a}$$

and

$$\text{Molar basis: } \overline{\text{AF}} = \frac{\text{Moles of air}}{\text{Moles of fuel}} \tag{2.121b}$$

The relationship between these two forms of air–fuel ratio is given as

$$\overline{\text{AF}} = \text{AF}\,\frac{\text{Molecular weight of air}}{\text{Molecular weight of fuel}} = \text{AF}\,\frac{M_{air}}{M_{fuel}} \tag{2.121c}$$

2.13.5.3 Equivalence Ratio (Φ)

The equivalence ratio is defined as the ratio of the theoretical Air-Fuel ratio to the Air-Fuel ratio associated with the use of the excess air in a real reaction.

$$\Phi = \frac{\text{AF}_{Stoi}}{\text{AF}} = \frac{\text{FA}}{\text{FA}_{stoi}} \tag{2.122a}$$

The equivalence ratio is related to the percentage of theoretical or percentage of stoichiometric air as

$$\% \text{ of theoretical air} = \frac{100\%}{\Phi} \qquad (2.122b)$$

and to the percentage of excess air as

$$\% \text{ of excess air} = \frac{(1 - \Phi)}{\Phi} 100\% \qquad (2.122c)$$

Example 2.5   *Combustion Reaction Model*

Consider combustion of octane (C_8H_{18}) with 200% theoretical air. Determine (a) The air-fuel ratio, both on a molar basis and on a mass basis, (b) The equivalence ratio, (c) The molar and mass composition of the product for the combustion, (d) The dew-point temperature of the product gas mixture at an operating pressure of 0.1 MPa, and (e) The adiabatic flame temperature of the product gas mixture.

Solution

The balanced reaction with stoichiometric or theoretical air is given as

$$C_8H_{18} + 12.5 \left(O_2 + 3.76 \, N_2 \right) \rightarrow 8 \, CO_2 + 9 \, H_2O + 47 N_2$$

The balanced reaction with 200% theoretical air is given as

$$C_8H_{18} + (12.5 \times 2.0) \, O_2 + (12.5 \times 2.0 \times 3.76) \, N_2$$
$$\rightarrow \quad 8 CO_2 + 9 H_2O + (12.5) \times O_2 + (94.0) \, N_2$$

The total number of moles in the product is

$$\sum_{i=1}^{N} n_i = n_{CO_2} + n_{H_2O} + n_{O_2} + n_{N_2} = 8 + 9 + 12.5 + 94.0 = 123.5$$

(a) Air–fuel ratio for 200% theoretical air is computed as follows:

$$\text{Molar air} - \text{fuel ratio}, \overline{AF} = \frac{n_{air}}{n_{fuel}} = \frac{n_{O_2} + n_{N_2}}{1} = \frac{25 \, (1 + 3.76) \text{ kmol air}}{1 \text{ kmol of } C_8H_{18}}$$

$$\overline{AF} = 119 \text{ kmol of air/kmol of } C_8H_{18}$$

$$\text{Mass air} - \text{fuel ratio}, AF = \overline{AF} \frac{M_{air}}{M_{C_8H_{18}}} = 119 \times \frac{28.97}{114.22} = 30.18 \text{ kg air/kg fuel}$$

Air-fuel ratio for the stoichiometric air

$$AF_{stoi} = \overline{AF}_{stoi} \times \frac{M_{air}}{M_{fuel}} = \frac{12.5 \times 4.76 \text{ kmol air}}{1 \text{ kmol } C_8H_{18}} \times \frac{28.97}{114.22}$$

$$AF_{stoi} = 15.09 \text{ kg air/kg } C_8H_{18}$$

(b) The equivalence ratio is given as

$$\Phi = \frac{AF_{Stoi}}{AF} = \frac{15.09}{30.18}$$

$$\Phi = 0.5$$

(c) Molar composition is given as

$$y_{CO_2} = \frac{n_{CO_2}}{\sum_{i=1}^{N} n_i} = \frac{8}{123.5} = 0.0647; \quad y_{H_2O} = \frac{n_{H_2O}}{\sum_{i=1}^{N} n_i} = \frac{9}{123.5} = 0.0728;$$

$$y_{O_2} = \frac{n_{O_2}}{\sum_{i=1}^{N} n_i} = \frac{12.5}{123.5} = 0.1012; \quad y_{N_2} = \frac{n_{H_2O}}{\sum_{i=1}^{N} n_i} = \frac{94}{123.5} = 0.7611$$

Total mass of the product per kmol of the mixture is

$$\sum_{i=1}^{N} m_i = m_{CO_2} + m_{H_2O} + m_{O_2} + m_{N_2}$$

$$= M_{CO_2} \times y_{CO_2} + M_{H_2O} \times y_{H_2O} + M_{O_2} \times y_{O_2} + M_{N_2} \times y_{N_2}$$

$$= 44 \times 0.0647 + 18.016 \times 0.0728 + 32 \times 0.1012 + 28.01 \times 0.7612$$

$$= 2.8468 + 1.3115 + 3.2384 + 21.3184$$

$$\sum_{i=1}^{N} m_i = 28.7151 \text{ kg/kmol of mixture}$$

Mass fraction is given as

$$x_{CO_2} = \frac{m_{CO_2}}{\sum_{i=1}^{N} m_i} = \frac{2.8468}{28.7151} = 0.0991; \quad x_{H_2O} = \frac{m_{H_2O}}{\sum_{i=1}^{N} m_i} = \frac{1.3115}{28.7151} = 0.0456$$

$$x_{O_2} = \frac{m_{O_2}}{\sum_{i=1}^{N} m_i} = \frac{3.2384}{28.7151} = 0.1127; \quad x_{N_2} = \frac{m_{N_2}}{\sum_{i=1}^{N} m_i} = \frac{21.3184}{28.7151} = 0.7424$$

Partial pressure of water in the mixture product

$$P_{H_2O} = y_{H_2O} \times P = 0.0728 \times 0.1 = 7.28 \text{ kPa}$$

The dew-point temperature of the mixture product is the saturation temperature of water at the partial pressure of water in the mixture and given as

$$T_{dp} = T_{Sat}\left(P_{H_2O}\right) = 39.7 \,^{\circ}\text{C}$$

based on saturation thermodynamic properties of water given in Table C.1.

This indicates that if the mixture temperature is cooled below the dew-point temperature of $T_{dp} = 39.7\,^{\circ}\text{C}$, water vapor in the mixture will condense to liquid water.

2.13.5.4 Evaluation of Enthalpy and Entropy in a Reacting System

Since in a reacting system the composition of elements changes from reactant to products, an arbitrary reference state can no longer be used for evaluation enthalpy, internal energy, and entropy of a specific element. For a reactive system analysis, a common reference state for evaluating enthalpy of basic elements is set by assigning a zero enthalpy value for the elements in their most stable form such as carbon, C; hydrogen, H_2; oxygen, O_2; nitrogen, N_2, and all other stable element forms. This standard reference state is defined by $T_{ref} = 25\,^{\circ}\text{C}$ (298 K) and $P_{ref} = 1 - \text{atm}$ (0.1 MPa). The enthalpy of a **compound** at the standard state is then set equal to the enthalpy **of formation**, denoted by $\overline{h_f}^{\circ}$, which is defined as the energy released or absorbed when the compound is formed from its elements, the compound and elements all being at the reference state. The enthalpy of formation values of common elements and compounds is given in Table C.6. The values of enthalpy of formations are assigned a negative or a positive sign depending on whether the reaction is exothermic or endothermic, respectively, during the formation of the compound.

The enthalpy values of compounds can be written as

$$\overline{h}\,(T,P) = \overline{h}_f^{\circ} + \Delta\overline{h} \tag{2.123}$$

where $\Delta\overline{h} = \left[\overline{h}\,(T,P) - \overline{h}\,(T_{\text{ref}}, P_{\text{ref}})\right]$ is the change in enthalpy of the compound with constant composition at a state from the reference state. The different enthalpy values can be computed from the steam table for water and from ideal gas table for different gas elements.

In a similar manner, a reference state is defined for evaluation entropy in a reacting system. The entropy of the element in a pure crystalline state is assigned a value of zero at a reference state absolute zero (0 K) temperature. Entropy values of all elements with nonpure crystalline structure are assigned nonzero entropy values computed based on experiments. The entropy values relative to this reference state are defined as the absolute entropy, $\overline{s}^{\circ}$. Table C.6 includes the absolute entropy values at reference state of $T_{\text{ref}} = 25\,°C$ (298 K) and $P_{\text{ref}} = 1$ atm (0.1 MPa). The absolute entropy at any other state is then computed based on adding the change of entropy values to the absolute entropy values at the temperature and as

$$\overline{s}\,(T,P) = \overline{s}^{\circ}\,(T,P_{\text{ref}}) + \left[\overline{s}\,(T,P) - \overline{s}\,(T,P_{\text{ref}})\right] \tag{2.124}$$

2.13.6 First Law for Reacting Systems

The first law of thermodynamics for a reacting system under SSSF process with negligible changes in kinetic energy and potential energy is given as

$$\dot{Q}_{\text{CV}} + \sum_R n_R \overline{h}_R = \sum_P \overline{h}_P + \dot{W}_{\text{CV}} \tag{2.125a}$$

or

$$Q_{\text{CV}} + \sum_{R=1} n_R \left(\overline{h}_f^{\circ} + \Delta\overline{h}\right)_R = \sum_{P=1} n_P \left(\overline{h}_f^{\circ} + \Delta\overline{h}\right)_P + W_{\text{CV}} \tag{2.125b}$$

where

$\overline{h}_f^{\circ}$ = enthalpy of formation at the reference state of 25 °C and 0.1 Mpa

$\overline{\Delta h}_{25°\text{C},0.1\,\text{Mpa}\rightarrow T,P}$ = Change in enthalpy between the state of the component and the reference state of 25 °C and atmospheric pressure of 0.1 Mpa.

Change of enthalpy between a state and the reference state of 25 °C or 298 °K temperature and 0.1 MPa pressure

$$= \int_{298,0.1\text{MPa}}^{T,P} C_P\,(T)\,dT \tag{2.126}$$

$\overline{h}_f^{\circ}$ = enthalpy of formation at the reference state of 25 °C and 0.1 Mpa

Another frequently used term relating the enthalpy of combustion and internal energy of combustion is the **heating value,** which is equal to the **negative of enthalpy of combustion** for a **constant pressure process** and **negative of internal energy of combustion** for a **constant volume process**. The heating value for a combustion process also differs depending on whether the product contains liquid water or vapor water. The **higher heating value (hhv)** is referred to the combustion process with liquid water, $H_2O(l)$, in the products and **lower heating value** with vapor water, $H_2O(v)$, in the products. The enthalpy of combustion of some of the common fuels at standard temperature and pressure is given in Table C.8.

2.13.7 Temperature of Product of Combustion

One of the key variables in a combustion process is the temperature of the product of combustion at the exhaust. This temperature depends on the reacting components, heat of combustion of the reaction, any associated heat transfer and work done, and the amount of excess air used. The temperature of the product of combustion can be estimated from the first law equation of the reaction process given by Eq. (2.125) as

$$\sum_{P=1} n_P h_P = +Q_{CV} + \sum_{R=1} n_R h_R - W_{CV} \tag{2.127a}$$

or

$$\sum_{P=1} n_P \left(\overline{h_f^\circ} + \overline{\Delta h} \right)_P = \sum_{R=1} n_R \left(\overline{h_f^\circ} + \overline{\Delta h} \right)_R + Q_{CV} - W_{CV} \tag{2.127b}$$

Considering Q_{CV} as negative for heat loss from the control volume, Eq. (2.127b) can rewritten as

$$\sum_{P=1} n_P \left(\overline{h_f^\circ} + \overline{\Delta h} \right)_P = \sum_{R=1} n_R \left(\overline{h_f^\circ} + \overline{\Delta h} \right)_R - Q_{CV} - W_{CV} \tag{2.127c}$$

For a combustion process with no work involved, the temperature of the product of combustion can be estimated as

$$\sum_{P=1} n_P \left(\overline{h_f^\circ} + \overline{\Delta h} \right)_P = \sum_{R=1} n_R \left(\overline{h_f^\circ} + \overline{\Delta h} \right)_R - Q_{CV} \tag{2.128}$$

The maximum temperature that a mixture can reach is for the case with no heat losses such as in an adiabatic process; involving no work done; and for using 100% theoretical air. Use of any additional excess air and/or heat loss and work results in a lower temperature of the mixture product. This maximum temperature is referred to as **Adiabatic Flame Temperature**, which is determined by substituting $Q_{CV} = 0$ in Eq. (2.116) for negligible heat loss and given by

$$\sum_{P=1}^{N_P} n_P \left(\overline{h_f^\circ} + \overline{\Delta h} \right)_P = \sum_{R=1}^{N_R} n_R \left(\overline{h_f^\circ} + \overline{\Delta h} \right)_R \tag{2.129}$$

RHS of Eq. (2.129) is estimated based on the temperature of the reactants. The mixture product temperature is then estimated using an iterative process until the total enthalpy of the products given by the LHS of the Eq. (2.129) matches with that of the reactants given by the RHS. It can be noted here that adiabatic flame temperature represents the maximum possible temperature of the product of combustion. In reality, there will be positive heat loss from the combustion chamber to outside across the chamber wall geometry and through circulating engine coolant, and hence will result in a lower temperature of the exhaust combustion product given by Eq. (2.129). In a power-producing engine like internal combustion engines, one needs to consider the total work produced as well as heat carried away by cooling fluid while estimating exhaust combustion product using Eq. (2.127c).

Example 2.6 *Temperature of Product of Combustion*

Methane gas (CH_4) at 25 °C, 1 atm, enters an insulated reactor operating at steady state and burns completely 200% theoretical air entering at 25 °C, 1 atm. Determine the adiabatic flame temperature (Show at least two iterations)

Solution

The balanced equation for combustion of methane with 100% theoretical air reaction can be written as

$$CH_4 + (a)\left(O_2 + 3.76\,N_2\right) \rightarrow (b)\,CO_2 + (c)\,H_2O + (d)\,N_2 \qquad (E2.6.1)$$

where the constant coefficients **a, b, c,** and **d** are moles of O_2, CO_2, H_2O, and N_2 per mole of CH_4 in a complete reaction, and these are determined by making balance of mass for each element on both sides of the equation as demonstrated below:

For C: $1 = b$, $b = 1$
For O: $2a = 2b + c$ or $2a = 2 + c = 2 + 2$ Or $a = 2$
For H: $4 = 2c$ or $c = 2$
For N: $3.76\,a = d$ or $d = 7.52$

Substituting coefficient values, we get the balanced reaction 100% theoretical air

$$CH_4 + 2\,O_2 + 7.52\,N_2 \rightarrow CO_2 + 2H_2O + 7.52\,N_2 \qquad (E2.6.2)$$

For combustion reaction with 200% theoretical air, the balanced reaction equation

$$CH_4 + 4\,O_2 + 15.04\,N_2 \rightarrow CO_2 + 2\,H_2O + 2\,O_2 + 15.04\,N_2 \qquad (E2.6.3)$$

The Adiabatic *Flame Temperature is the estimated* maximum temperature of the product of combustion while neglecting any heat loss and workout from the combustion reaction. This is given by the first law of energy balance equation and is derived by Eq. (2.129) as

$$\sum_{P=1}^{N_P} n_P \left(\overline{h_f^o} + \overline{\Delta h}\right)_P = \sum_{R=1}^{N_R} n_R \left(\overline{h_f^o} + \overline{\Delta h}\right)_R$$

Expanding this equation for the combustion reaction of methane with 200% theoretical air, we get

$$\left(h_f^o + \overline{\Delta h}\right)_{CH_4} + 4\left(h_f^o + \overline{\Delta h}\right)_{O_2} + 15.04\left(h_f^o + \overline{\Delta h}\right)_{N_2}$$
$$= 1\left(h_f^o + \overline{\Delta h}\right)_{CO_2} + 2\left(h_f^o + \overline{\Delta h}\right)_{H_2O} + 2\left(h_f^o + \overline{\Delta h}\right)_{O_2} + 15.04\left(h_f^o + \overline{\Delta h}\right)_{N_2} \qquad (E2.6.4)$$

From Table C.6, we get enthalpy of formation data:

$$\left(h_f^o\right)_{CH_4} = -74850,\ \left(h_f^o\right)_{O_2} = 0,\ \left(h_f^o\right)_{N_2} = 0,\ \left(h_f^o\right)_{CO_2} = -393520,\ \left(h_f^o\right)_{H_2O} = -241820$$

At 25 °C, 1 atm: $\left(\overline{\Delta h}\right)_{O_2} = 0,\ \left(\overline{\Delta h}\right)_{N_2} = 0$

The difference in enthalpy, $\overline{\Delta h}$, at any temperature with respect to reference temperature, 25 °C, is given as

$$\left(\overline{\Delta h}\right)_{CO_2} = \left(\overline{h}_{CO_2}(T) - 9364\right), \left(\overline{\Delta h}\right)_{H_2O} = \left(\overline{h}_{H_2O}(T) - 9904\right)$$

$$\left(\overline{\Delta h}\right)_{O_2} = \left(\overline{h}_{O_2}(T) - 8682\right), \left(\overline{\Delta h}\right)_{N_2} = \left(\overline{h}_{N_2}(T) - 8669\right)$$

Substituting values in Eq. (E2.6.4),

$$-74\,850 = \left[-393\,520 - \left(\overline{h}_{CO_2}(T) - 9364\right)\right] + 2\left[-241\,820 - \left(\overline{h}_{H_2O}(T) - 9904\right)\right]$$
$$+ 2\left[\left(\overline{h}_{O_2}(T) - 8682\right)\right] + 15.04\left[\left(\overline{h}_{N_2}(T) - 8669\right)\right]$$

or

$$\overline{h}_{CO_2}(T) + 2\,\overline{h}_{H_2O}(T) + 2\overline{h}_{O_2}(T) + 15.04\,\overline{h}_{N_2}(T) = 979\,227.76 \text{ kJ/kmol} \qquad \text{(E2.6.5)}$$

Equation (E2.6.5) can be iteratively solved for the exhaust temperature by starting with an initial guess and substituting on the LHS to match value on the RHS. The process can be repeated with newer guess value until converge solution is reached. One of the simplest approaches is to compute the LHS with two guess values, and then use either interpolation or extrapolation formulas to compute the solution. Let us demonstrate this process in the following few steps.

Iteration 1: Assume $T = 1500$ K

Substituting the temperature value on LHS of Eq. E2.6.5 and evaluating the corresponding enthalpy values from Table C.7, we get

$$\text{LHS} = \overline{h}_{CO_2}(1500 \text{ K}) + 2\,\overline{h}_{H_2O}(1500 \text{ K}) + 2\overline{h}_{O_2}(1500 \text{ K}) + 15.04\,\overline{h}_{N_2}(1500 \text{ K})$$

$$\text{LHS } (1500 \text{ K}) = 71{,}078 + 2\,(57{,}999) + 2\,(49292) + 15.04\,(47{,}073)$$

$$\text{LHS } (1500 \text{ K}) = 993{,}637.92 \text{ kJ/kmol} > \text{RHS}$$

Iteration # 2: Assume $T = 1400$ K

$$\text{LHS} = \overline{h}_{CO_2}(1400 \text{ K}) + 2\,\overline{h}_{H_2O}(1400 \text{ K}) + 2\overline{h}_{O_2} - (1400 \text{ K}) + 15.04\,\overline{h}_{N_2}(1400 \text{ K})$$

$$\text{LHS } (1400 \text{ K}) = 65\,271 + 2\,(53\,351) + 2\,(45\,648) + 15.04\,(43\,605)$$

$$\text{LHS } (1400 \text{ K}) = 919\,088.2 \text{ kJ/kmol} < \text{RHS}$$

Interpolating between $T = 1400$ and 1500 K, we get the exhaust adiabatic flame temperature as

$$\frac{T - 1400}{1500 - 1400} = \frac{979\,227 - 919\,088.2}{993\,637.92 - 919\,088.2}$$

$$T = 1480.67 \text{ K}$$

2.14 Power-Generating Cycles

As we have discussed, the Carnot cycle represents an ideal cycle for the operation of a heat engine and represents maximum limit of thermal efficiency of heat engine operating between two temperature limits. Real power-generating cycle differs in several ways from the Carnot cycle. Some of the common power-generating cycles are vapor power cycles such as Rankine cycle for vapor power plants, and gas cycles such as Otto and diesel cycles for internal combustion engines, Brayton cycle for rotary gas turbine power generation, and jet propulsion cycle for jet engines.

2.14.1 Vapor Power Cycles

2.14.1.1 Rankine Vapor Power Cycle

A Rankine cycle represents a basic power cycle for all power plants such as fossil power plant, nuclear power plant, and solar thermal power plant. The cycle involves the four processes like those in a Carnot cycle with some deviations. Figure 2.16 shows a simple steam power plant in a schematic diagram and a $T–S$ diagram showing all four processes along with numbers indicating the state points for all processes. In a standard Rankine cycle, the working fluid is at a saturated vapor state 1 or at a superheated state 1 at inlet to turbine and saturated liquid state 3 at exit of condenser.

The **isothermal heat addition process** in a standard Rankine cycle is represented by the heat addition process in the **boiler**. A working fluid enters the boiler in a liquid state 4 and gets heated by the combustion flue gas from the furnace following a constant pressure line as shown in $T–S$ diagram. Heating at a constant pressure is continued until boiling starts at the corresponding saturation temperature, all liquid is transformed into vapor, and reaches the saturated vapor state.

The saturated vapor at high temperature and pressure enters the **turbine** and undergoes expansion process to condenser pressure producing work. The process 1-2s represents the **reversible adiabatic process or the isentropic process** and the process 1-2 represents the real and irreversible process with increase in entropy. The mixture of liquid and vapor at state 2 then enters the condenser and gets cooled by transferring the heat to ambient air in an air-cooled condenser or to circulating cooling water in a water-cooled condenser as shown in the figure.

In the **condenser**, the working fluid undergoes a phase-change process and moves toward the saturated liquid state 3. The **isothermal heat rejection process** in a standard Rankine cycle is represented by the heat rejection process in the condenser. The working fluid in the liquid state is then **pumped** to the higher pressure given by the subcooled or compressed liquid state 4 at inlet to the boiler following the reversible adiabatic or isentropic process 3-4s. The real pumping process is represented by the process 3-4 with increase in entropy. The standard Rankine cycle 1-2-3-4 deviates from the Carnot cycle 1-2-3$^/$-4$^/$ in two ways.

Notice that in the Rankine cycle, the exit state from the condenser could be state 3$^/$ to follow a Carnot cycle 1-2-3$^/$-4$^/$ as shown in Figure 2.12. Instead, steam is condensed all the way to a saturated liquid state 3 and followed by the pumping of the liquid water to a compressed liquid or subcooled liquid state 4. This contrasts with the Carnot cycle where the working fluid exits the condenser at a mixed condensed state 3$^/$ followed by the

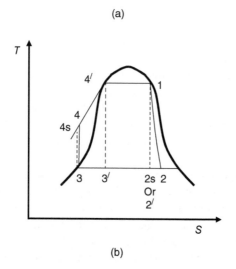

(a)

(b)

Figure 2.16 Standard Rankine vapor power cycle: (a) Rankine cycle power generation system and (b) $T-S$ diagram.

compression of the liquid–vapor mixture to the saturated state $4'$. Since the working fluid enters the boiler at a subcooled state, heat is added at a lower average temperature than that in the Carnot cycle, and this results in a lower thermal efficiency of the Rankine cycle compared to the Carnot cycle.

To improve the thermal efficiency, the working fluid water is heated to a superheated state at inlet to the turbine (see Figure 2.17 along with numbers indicating state points), and hence increased the average temperature of heat addition as shown by the process 4-1. Such superheating is a usual feature in large-size vapor power plants. Superheating also helps to retain quality of the exiting fluid to an acceptable level.

Figure 2.17 Rankine cycle with superheated steam.

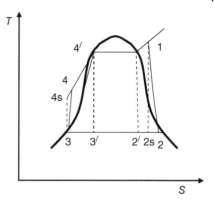

There are additional features such as **reheating** and **regenerative heating** that are also usually employed to improve thermal efficiency of the vapor power cycle.

2.14.1.2   First Law of Thermodynamic Analysis of a Standard Rankine Vapor Power Cycle

A summary of the first law of thermodynamic analysis of a standard Rankine cycle is presented here.

2.14.1.3   Thermodynamic analysis of a standard Rankine cycle:

Requirement: Produce P kW of net thermal power.

Basic assumptions:
1. Use standard Rankine power cycle using water as the working fluid
2. Turbine efficiency, η_t, and pump efficiency, η_p

Known data:

High- and low-pressure levels:

Boiler pressure: P_b

Condenser pressure: P_c

Outside design conditions:

Dry-bulb temperature, T_{DBT}

Wet-bulb temperature, T_{WBT}

State 1: Known $P_1 = P_b$ (kPa), Saturated vapor

Find $T_1 = T_{sat}$, $h_1 = h_g$, $s_1 = s_g$ at $P_1 = P_1$ (kPa) from pressure-based saturation steam Table C.1

State 2: Find $s_f, s_g, s_{fg}, h_f, h_g, h_{fg}$, and v_f at $P_2 = P_3 = P_c$ (kPa) from the pressure-based saturation table

State 3: Saturated liquid state

Set $v_3 = v_f$, $h_3 = h_f$

Control volume: Turbine:

Process 1-2: Reversible adiabatic process

From first law of thermodynamics: $w_{ts} = (h_1 - h_{2s})$

From second law of thermodynamics: $s_{2s} = s_1 = s_f + x_{2s}s_{fg}$

Solve for the isentropic exit state of turbine:

$$x_{2s} = \frac{s_{2s} - s_f}{s_{fg}}$$

Calculate: $h_{2s} = h_f + x_{2s}h_{fg}$

Calculate turbine work:

Isentropic turbine work: $w_{ts} = (h_1 - h_{2s})$

Actual turbine work: $w_t = \eta_t w_{ts}$

Calculate actual turbine exit state 2:

$$h_2 = h_1 - w_t$$

$$x_2 = \frac{h_2 - h_f}{h_{fg}}$$

Total turbine work:

$$W_t = \dot{m}_s \left(w_t \right) = \dot{m}_s \left[(h_1 - h_2) \right]$$

where $\dot{m}_s$ = Mass flow rate of steam. Note that the steam mass flow rate needs to be estimated based on the net thermal power output.

Control volume: condenser:

In the condenser, heat is transferred from the condensing mixture of vapor–liquid working fluid water to circulation cooling water steam from cooling tower in a secondary cooling water system. Figure 2.18 shows a typical temperature variation of both the fluid streams along the length of condenser.

We are assuming here that the cooling water temperatures at the inlet, $T_{w,i}$, and at the outlet, $T_{w,o}$, of the condenser are set based on the selected performance parameters of the cooling tower as discussed in Section 2.14.1.3.1. Temperature drop in the line between the cooling tower and the condenser can be neglected as a first-order approximation and we can set

$$T_{w,i} \approx T_{c,o} \text{ and } T_{w,o} \approx T_{c,i}$$

Calculate heat transfer rejected in the condenser:

From first law of thermodynamics

For steam side:

$$q_c = (h_3 - h_4)$$

and total heat rejected in the condenser

$$\dot{Q}_c = \dot{m}_s (h_3 - h_4)$$

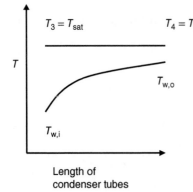

$T_3 = T_{sat}$

$T_4 = T_{sat}$

T

$T_{w,o}$

$T_{w,i}$

Length of
condenser tubes

Figure 2.18 Temperature variation along the length of the condenser.

For cooling water side:

$$\dot{Q}_c = \dot{m}_w \left(h_{w,o} - h_{w,i} \right) \text{ Or } \dot{Q}_c = \dot{m}_w C_{pw} \left(T_{w,o} - T_{w,i} \right)$$

Calculate mass flow rate of cooling water:

$$\dot{m}_w = \frac{\dot{Q}_c}{C_{pw} \left(T_{w,o} - T_{w,i} \right)}$$

Control volume: pump:

$v_3 = v_f$ at p_3 = condenser pressure

Calculate pump work:

Isentropic pump work: $w_{ps} = v_3 (P_4 - P_3)$
Calculate actual pump work:

$$w_p = \frac{w_{ps}}{\eta_p}$$

State of water at the exit of the pump:

$$h_4 = h_3 + w_p$$

Since the process 3-4 in the pump involves only liquid state, we can estimate temperature at the exit state

$$T_4 = T_3 + \frac{w_p}{C_p}$$

Total pump work:

$$W_p = \dot{m}_s w_p$$

Control volume: boiler:

Heat input to the steam:
    From first law of thermodynamics:

$$q_b = h_1 - h_4 \text{ and } Q_b = \dot{m}_s \left(h_1 - h_4 \right)$$

Estimate steam flow rate:
    From total power requirement and using the net power output

$$P = \dot{m}_s \left(W_t \right)$$

Mass flow rate of the steam is computed as

$$\dot{m}_s = \frac{P}{w_t} = \frac{P}{\left(h_1 - h_2 \right)}$$

Thermal efficiency of the power system:

$$\eta_{th} = \frac{W_{net}}{Q_b}$$

where

$$W_{net} = W_t - W_p = \dot{m}_s \left[\left(h_1 - h_2 \right) - \left(h_4 - h_3 \right) \right]$$

2.14.1.3.1 Secondary Water System

The secondary water system in the power cycle consists of the cooling tower, cooling water pump, and associated piping system. While calculating the cooling water flow rate in the condenser side, we have assumed the cooling water temperatures at the inlet and at the outlet of the condenser. These two temperatures are set based on the selected performance parameters of the cooling tower. The cooling tower performance is generally given based on two parameters: **Approach, A**, and **Range, R**.

The parameter **Approach, A,** is defined as the difference in temperature between the cooling tower outlet temperature, $T_{c,o}$, and the outdoor wet-bulb temperature, T_{wht} as

$$A = T_{c,o} - T_{wbt}$$

The cooling tower inlet temperature is then given as

$$T_{c,o} = T_{wbt} + A \tag{2.130a}$$

Note that in a cooling tower, the incoming water stream can only be cooled to temperatures less than the local ambient wet-bulb temperature.

The parameter **Range** is defined as the difference in temperature between the cooling tower inlet and outlet temperature as

$$R = T_{c,i} - T_{c,o}$$

and the cooling tower outlet temperature is then given as

$$T_{c,i} = T_{c,o} + R \tag{2.130b}$$

Depending on the amount of heat loss from the pipeline connecting the cooling tower to the condenser, the cooling water temperature from inlet to the condenser may be few degrees less than cooling tower outlet temperature. As we can see that the cooling tower performance parameters not only directly influence the size and cost of the cooling tower, but also affect the operating conditions, performance, and size of the condenser.

Summary: Ratings and Specification of Major Components:

Boiler:

Net heat addition, Q_b

Pressure, $p_b = p_4 = p_1$

Temperature, T_b

Mass flow rate, $\dot{m}_s$

Determine size, type, cost

Turbine:

Net work output, W_t

Mass flow rate, $\dot{m}_s$

Inlet condition: pressure, P_1, Temperature, T_1

Back pressure, $p_2 = p_3 = p_c$

Efficiency of turbine, η_t

Condenser:

Heat rejected = Heat gained by cooling water = Q_c

Mass flow rate of steam = $\dot{m}_s$

Temperature and pressure of condensing steam,

$$P_c = P_2 = P_3$$
$$T_c = T_2 = T_3$$

Cooling water flow rate = $\dot{m}_w$

Temperature rise for cooling water, $T_{w,o} - T_{w,i}$

Determine size, type, and cost

Pump:

Power: W_p

Pressure rise: $\Delta P = P_4 - P_3$

Mass flow rates: $\dot{m}_w$

Determine size, type, and cost

Example 2.7   Standard Rankine Cycle

A 10-kW power-generating plant operates on standard Rankine vapor power cycle that operates with a 1-MPa pressure in the boiler and 20-kPa pressure in the condenser. Consider turbine and pump efficiency as 90% and 85%, respectively. Perform first law of thermodynamic analysis of the cycle to determine the ratings of the major components of the plant. Consider a secondary cooling water system consisting of cooling tower with operating parameters approach, $A = 4\,°C$, and range, $R = 10\,°C$. Determine cooling water flow rate through the condenser for outdoor dry-bulb temperature and wet-bulb temperature of $25\,°C$ and $23\,°C$, respectively.

Solution

State 1: Known $P_1 = P_b$ (kPa) = 1 MPa, Saturated vapor state

At $P_1 = 1$ MPa (kPa) from pressure-based saturation steam Table C.1

$T_1 = T_{sat} = 179.91\,°C$, $h_1 = h_{g1} = 2728.1$ kJ/kg, $s_1 = s_{g1} = 6.5869$ kJ/kg·K

State 3: Saturated liquid state

At $P_2 = P_3 = P_c$ (kPa) = 20 kPa

At 20 kPa from pressure-based saturation steam table

$$T_4 = T_1 = T_{sat} = 60.06\ °C$$

$$s_f = 0.8319\ \text{kJ/kg}\cdot\text{K}, s_g = 7.9085\ \text{kJ/kg}\cdot\text{K}, s_{fg}, = 7.0765\ \text{kJ/kg}\cdot\text{K},$$

$$= 251.38\ \text{kJ/kg}, h_g = 2609.70\ \text{kJ/kg}, h_{fg} = 2358.3\ \text{kJ/kg}$$

and h_f

$$v_3 = v_f = 0.001\ 017\ \text{m}^3/\text{kg}$$

State 2: $P_2 = P_3 = P_c$ (kPa) = 20 kPa

Control volume: turbine:

Process 1-2: Reversible adiabatic process

From first law of thermodynamics: $w_{ts} = (h_1 - h_{2s})$

From second law of thermodynamics: $s_{2s} = s_1 = 6.5869$ kJ/kg·K

Since $s_{2s} > s_g$ state 2 is mixed state, we can estimate the quality of the isentropic turbine exit state as

$$x_{2s} = \frac{s_{2s} - s_f}{s_{fg}} = \frac{6.5864 - 0.8319}{7.0765} = 0.8132$$

$$h_{2s} = h_f + x_{2s}h_{fg} = 251.38 + 0.8132 \times 2358.3 = 2169.199 \text{ kJ/kg}$$

$$h_{2s} = 2169.199 \text{ kJ/kg}$$

Calculate turbine work:

Isentropic turbine work: $w_{ts} = (h_1 - h_{2s}) = 2778.1 - 2169.199$

$$w_{ts} = 608.95 \text{ kJ/kg}$$

Actual turbine work: $w_t = \eta_T w_{ts} = 0.9 \times 608.95$

$$w_t = 548.055 \text{ kJ/kg}$$

Calculate actual turbine exit state 2:

$$h_2 = h_1 - w_t = 2230.044 \text{ kJ/kg}$$

$$h_2 = 2778.1 - 548.055$$
$$h_2 = 2230.044 \text{ kJ/kg}$$

$$x_2 = \frac{h_2 - h_f}{h_{fg}}$$

$$x_2 = \frac{2230.044 - 251.38}{2358.3} = \mathbf{0.839}$$

Estimate steam flow rate:

From total power requirement:

$$P = W_t = m_s (w_t) = 10 \text{ kW}$$

Compute steam mass flow rate, $\dot{m}_s = \frac{P}{w_t}$

$$\dot{m}_s = \frac{10 \text{ kJ/s}}{548.055 \text{ kJ/kg}}$$

$$\dot{m}_s = 0.01824 \text{ kg/s}$$

Control volume: condenser:

Calculate heat rejected in the condenser:

From first law of thermodynamics
For steam side:

$$q_c = (h_3 - h_2) = 251.38 - 2230.044$$

$$q_c = -1978.664 \text{ kJ/kg}$$

and total heat rejected in the condenser

$$Q_c = \dot{m}_s (h_3 - h_4) = 0.01824 \, (-1978.664)$$

$$Q_c = -36.090 \text{ kW}$$

Note that this is the amount of heat that is transferred to the cooling water side. Application of first law of thermodynamics to cooling water side leads to

$$\dot{Q}_c = \dot{m}_w (h_{w,o} - h_{w,i}) \text{ or } \dot{Q}_c = \dot{m}_w C_{pw} (T_{w,o} - T_{w,i})$$

The required cooling water flow rate through the condenser depends not only on the amount of heat rejected by the condensing steam but also on the cooling tower performance and outdoor weather conditions.

For the given cooling tower parameters of **Range**, $R = 10\,°C$ and **Approach**, $A = 4\,°C$, we can calculate the cooling water inlet and outlet temperatures as follows:

For cooling tower outlet temperature:

$$T_{co} = T_{WBT} + A = 23 + 4 = 27\ °C$$

and for cooling tower inlet temperature:

$$T_{ci} = T_{co} + R = 27 + 10 = 37\ °C$$

Considering no temperature drop in cooling water pipe line from cooling tower to condenser, we can set the cooling water inlet and outlet temperature of the condenser as follows:

$$T_{wi} = T_{co} = 28\ °C \text{ and } T_{wo} = T_{ci} = 38\ °C$$

The mass flow rate of cooling water is calculated as:

$$\dot{m}_w = \frac{\dot{Q}_c}{C_{pw}\left(T_{w,o} - T_{w,i}\right)} = \frac{36.09\ kJ/s}{\left(4.187\ kJ/kg\ K\right)(37 - 27)}$$

$$\dot{m}_w = 0.862\ kg/s$$

Control volume: pump:

$V_3 = v_f$ at p_3 = condenser pressure

Calculate pump work:

Isentropic pump work:

$$w_{ps} = v_3(P_4 - P_3)$$
$$w_{ps} = 0.001017(1000 - 20)$$
$$w_{ps} = 0.9966\ kJ/kg$$
$$v_3 = v_f = 0.001017\ m^3/kg,\ h_3 = h_f = 251.38\ kJ/kg$$

Calculate actual pump work:

$$w_p = \frac{w_{ps}}{\eta_p} = \frac{0.9966}{0.85}$$
$$w_p = 1.1725\ kJ/kg$$

State of water at the exit of the pump or at inlet to boiler:

$$h_4 = h_3 + w_p$$
$$h_4 = 251.38 + 1.1725$$
$$h_4 = 252.552\ kJ/kg$$

Total pump work:

$$W_p = \dot{m}_s w_p$$
$$W_p = 0.01824 \times 1.1725 = 0.02138\ kW$$

Control volume: boiler:

Heat input to the steam:

From first law of thermodynamics:

$$q_b = h_1 - h_4 = 2728.1 - 252.552 = 2475.548\ kJ/kg$$

and

$$\dot{Q}_b = \dot{m}_s \left(h_1 - h_4\right) = 0.01824 \times 2475.548$$
$$\dot{Q}_b = 45.1539 \, \text{kW}$$

Thermal efficiency of the cycle is estimated as

$$\eta_{th} = \frac{\dot{W}_{net}}{\dot{Q}_b} = \frac{W_t - W_p}{\dot{Q}_b} = \frac{10 - 0.0214 \, \text{kW}}{45.1539}$$

$$\eta_{th} = 22.09\%$$

Summary: Ratings and Specification of Major Components:

Boiler:

Net heat addition, Q_b

Pressure, $p_b = P_4 = P_1$

Temperature, T_b

Mass flow rate, $\dot{m}_s$

Determine size, type, cost

Turbine:

Net work output, W_t

Mass flow rate, $\dot{m}_s$

Inlet condition: pressure, P_1, Temperatures, T_1

Back pressure, $p_2 = p_3 = p_c$

Efficiency of turbine, η_t

Condenser:

Heat rejected = heat gained by cooling water = Q_c

Mass flow rate of steam =

Temperature and pressure of condensing steam,

$p_c = p_2 = p_3$

$T_c = T_2 = T_3$

Cooling water flow rate = $\dot{m}_w$

Temperature rise for cooling water, $T_{w,o} - T_{w,i}$

Determine size, type, and cost

Pump:

Power: W_p

Pressure rise: $\Delta P = P_4 - P_3$

Mass flow rates: $\dot{m}_w$

Determine size, type, and cost

2.14.1.4 Effect of Superheating and Reheating

To achieve higher thermal efficiency, the standard Rankine cycle considered in Section 2.14.1.2 is often modified to use superheated steam at inlet to the turbine and/or reheating the steam through multistage expansion in the turbine. Such modification results in an increased average temperature of heat addition.

Superheating and reheating also help in maintaining moisture level at the exit of the turbine to an acceptable level of greater than 10% to avoid any erosion damage to the turbine blades. This is particularly critical while using lower turbine back pressure or condenser pressure to increase thermal efficiency of the cycle.

2.14.1.4.1 Reheat Vapor Power Cycle

The **reheat cycle** is used to provide higher pressure and temperature vapor at inlet to the turbine to achieve higher thermal efficiency. In this cycle, the expansion of the steam in the turbine from the boiler pressure to the condenser pressure takes place in two stages. Steam is first expanded in the first stage of the turbine to some intermediate pressure and then extracted and taken back to the reheat-section of the boiler where it is reheated to the initial high temperature.

Steam is then expanded in the second stage of the turbine to the condenser pressure. Reheating allows use of high temperature and pressure heat addition in the boiler and still maintains reasonably low liquid content, i.e. low-quality steam at the exit section of turbine before entering the condenser as demonstrated in the T–S diagram for a reheat cycle in Figure 2.19 along with numbers indicating state points across the processes.

Reheating the steam results only in a small gain in thermal efficiency. The primary gain in using a reheat cycle is in providing lower moisture content in the exit section of the turbine.

2.14.1.4.2 Regenerative Power Cycle with Feed Water Heaters

As we have noticed, in a Rankine cycle the condensate enters the boiler at a subcooled stage and undergoes significant amount sensible heating in the liquid state before undergoing boiling and phase change to steam. This results in heat addition at a lower average temperature compared to the higher saturation temperature heat addition and significant deviation from the Carnot cycle.

In a regenerative cycle, the condensate from the condenser is preheated in several feed water heaters by using extracted steam fractions from the turbine at different intermediate pressures. This helps in returning the condensate to the boiler at significantly higher subcooled temperature at the boiler pressure. Figure 2.20 shows regenerative cycle with one feed water heater and along with numbers indicating state points across the processes.

Steam at the boiler pressure enters the turbine at state 1 and undergoes expansion to the state 2 at the intermediate boiler feed water pressure. A fraction steam is extracted at this state and supplied to the feed water heater. The rest of the steam is expanded in the turbine to state 3 followed by condensation to the saturated liquid state 4 in the condenser. The condensate then enters pump 1 where it undergoes an increase in pressure from the condenser pressure to the feed water pressure. In the feed water heater, condensate at state 5 mixes with extracted steam at state 2 from the turbine and exits at the saturated liquid state 6. The amount of steam extracted is just enough to ensure a saturated liquid state at the exit of the feed water heater. The saturated liquid at state 6 enters the second pump 2 where it undergoes further increase in pressure from the feed water heater to the boiler pressure. The subcooled liquid water at state 7 enters the boiler where it receives heat at the constant boiler pressure and exits at the saturated or superheated vapor state 1.

The feed water heaters can be of two types: **Open feed water heater** and **Closed feed water heater**.

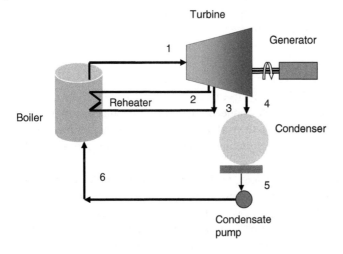

(a)

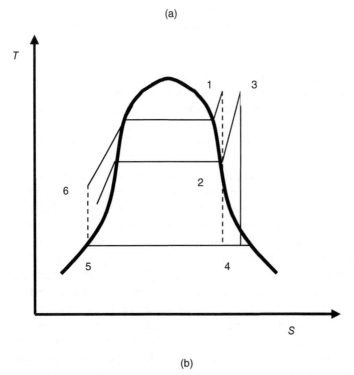

(b)

Figure 2.19 Reheat vapor power cycle: (a) vapor power cycle with reheat and (b) reheat vapor power cycle shown on $T–S$ diagram.

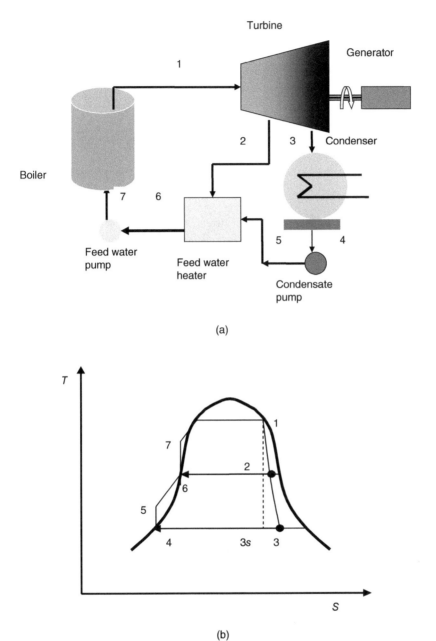

Figure 2.20 Regenerative power cycle with one feed water heater. (a) Regenerative power system with one feed water heater. (b) $T-S$ diagram.

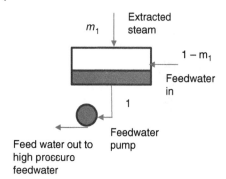

Figure 2.21 Open feedwater heater.

Open Feed Water Heater In an open feed water heater, the extracted steam from turbine mixes directly with the condensate feed water supplied from the condenser through the condensate pump (Figure 2.21).

Based on the application of the first law or energy balance over the feed water heater, the extracted fraction is computed as

Control Volume: Feed water heater

$P_5 = P_6 = P_2 =$ Feed water pressure

From first law:

$$m_1 h_2 + (1 - m_1) h_5 = h_6 \tag{2.131a}$$

where $h_6 = h_f$, enthalpy of saturated liquid at feed water heater pressure, P_6

Solve for mass flow rate steam extracted from turbine at state 2:

$$m_1 = \frac{(h_6 - h_5)}{(h_2 - h_5)} \tag{2.131b}$$

Closed Feed Water Heater In closed feed water heater, there is no direct mixing of the extracted steam from the turbine with condensate from the upstream feed water or condensate pumpspump. It just exchanges heat like in heat exchanger as shown in Figure 2.22.

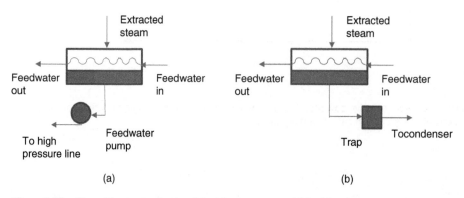

Figure 2.22 Closed feedwater heater: (a) without a trap and (b) with a trap.

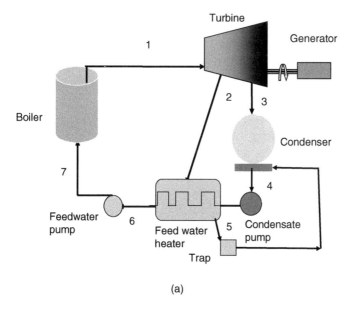

(a)

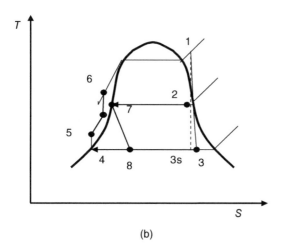

(b)

Figure 2.23 Regenerative cycle with closed feedwater heater and trap. (a) Regenerative system with closed feedwater heater. (b) Regenerative cycle with closed feed water and condensate trap.

The extracted steam condenses, and the condensate can either be pumped to the next feed water heater at a higher-pressure or to a lower-pressure heater or the condenser after passing through a trap (see Figure 2.23 that shows statements across process).

A trap is just like a throttling valve, which allows drop in pressure from feed water heater pressure to condenser pressure and mixes with the condensate in the condenser

CV: Trap:

From first law:

$$m_1 \left(h_2 - h_7 \right) = \left(h_6 - h_5 \right) \tag{2.132a}$$

Solving

$$m_1 = \frac{(h_6 - h_5)}{(h_2 - h_7)}$$ (2.132b)

2.14.1.5 Thermodynamic Analysis of Regenerative Feed Water Power Cycle

Requirement: Produce rated power output.

Basic assumptions:

1. Use standard Rankine power cycle using water as the working fluid
2. One open feed water heater

Known data:

Outside design conditions:

Dry-bulb temperature, T_{DBT}

Wet-bulb temperature, T_{WBT}

High and low temperature and pressure levels:

Boiler pressure: P_b

Feed heater pressure: P_{fwh}

Condenser pressure: P_c

State 1: Known $P_1 = 400 \, \text{kPa}$, Saturated vapor

Find $T_1 = T_{sat}$, $h_1 = h_g$, $s_1 = s_g$ at $P_1 = 400 \, \text{kPa}$.

State 2: Find s_f, s_g, s_{fg}, h_f, h_g, h_{fg}, and v_f at $P_2 = P_{fwh} = 200 \, \text{kPa}$

State 3: Find s_f, s_g, s_{fg}, h_f, h_g, h_{fg}, and v_f based on $P_3 = P_c = 20 \, \text{kPa}$

Control volume: turbine (first-stage turbine):

Process 1-2: Reversible adiabatic process

From first law of thermodynamics: $W_{ts-1} = (h_1 - h_{2s})$

From second law of thermodynamics: $s_{2s} = s_1 = s_f + x_{2s}s_{fg}$

Solve: $x_{2s} = \frac{s_{2s} - s_f}{s_{fg}}$ and $h_{2s} = h_f + x_{2s}h_{fg}$

Calculate turbine work:

Isentropic turbine work: $w_{t-1s} = (h_1 - h_{2s})$

Actual turbine work: $w_{t-1} = \eta_T w_{ts-1}$

Calculate: $h_2 = h_1 - w_{t-1}$

$$x_2 = \frac{h_2 - h_f}{h_{fg}}$$

Similarly, for process 1–3s (Second-stage turbine)

Process 1–3: Reversible adiabatic process

From first law of thermodynamics: $w_{ts-1} = (h_1 - h_{3s})$

From second law of thermodynamics: $s_{3s} = s_1 = s_f + x_{3s}s_{fg}$

Solve $x_{3s} = \frac{s_{3s} - s_f}{s_{fg}}$ and $h_{3s} = h_f + x_{3s}h_{fg}$ for the exit state of the second turbine or inlet to the condenser

Calculate Turbine Work:

Isentropic turbine work: $w_{t-2s} = (1 - m_1)(h_2 - h_{3s})$

where $m_1 =$ fraction steam extracted from turbine to feed water heater

Actual turbine work $w_{t-2} = \eta_t w_{ts-2}$

Calculate: $h_3 = h_2 - w_{t-1}$ and $x_{3s} = \frac{h_3 - h_f}{h_{fg}}$

Total turbine work:
$$W_t = \dot{m}_s \left(w_{t-1} + w_{t-2} \right) = \dot{m}_s \left[(h_1 - h_2) + (1 - m_1)(h_2 - h_3) \right]$$
where $\dot{m}_s =$ Mass flow rate of steam

Control volume: condenser:

From first law of thermodynamics

Steam side:

Heat rejected in the condenser as steam changes phase vapor to liquid state

$$q_c = h_3 - h_4 \text{ and } Qc = \dot{m}_s \left(h_3 - h_4 \right)$$

Cooling water side:

$$\dot{Q}_c = \dot{m}_w \left(h_{w,o} - h_{w,i} \right) \text{ Or } \dot{Q}_c = \dot{m}_w C_{pw} \left(T_{w,o} - T_{w,i} \right)$$

Calculate Mass flow rate of Cooling Water:

$$\dot{m}_w = \frac{\dot{Q}_c}{C_{pw} \left(T_{w,o} - T_{w,i} \right)}$$

CV: Pump-1 (Condensate pump)

State 4: Saturated liquid state, i.e. $v_4 = v_f$ at $p_4 =$ condenser pressure

For exit state of the pump

$P_5 = P_6 = P_2 =$ Feed water pressure

Isentropic pump work: $w_{ps-1} = (1 - \dot{m}_1)v_4 \left(P_5 - P_4 \right)$

Calculate:

Actual pump work: $w_{p-1} = \dfrac{w_{ps-1}}{\eta_p}$

Total pump work: $W_{p-1} = \dot{m}_s w_{p-1}$

Calculate exit state of the pump-1

$$h_5 = h_4 + w_{p-1}$$

Control volume: Feed water heater

$$P_5 = P_6 = P_2 = \text{Feed water pressure}$$
From first law:

$$\dot{m}_1 h_2 + (1 - \dot{m}_1) h_5 = h_6$$

where $h_6 = h_f$ at T_6 or P_6

Solve for mass flow rate steam extracted from turbine at state 2:

$$\dot{m}_1 = \frac{(h_6 - h_5)}{(h_2 - h_5)}$$

Control Volume: Pump-2 (feed water pump):

Calculate:

Isentropic pump work: $w_{ps-2} = v_6 \left(P_7 - P_6 \right)$

where $v_6 = v_f$ at $P_6 = P_2 =$ feed water pressure

Actual pump work: $w_{p-2} = \dfrac{w_{ps-2}}{\eta_p}$

Total pump work: $W_{p-2} = \dot{m}_s w_{p-2}$

For the exit state of the pump, $h_7 = h_6 + w_{p-2}$

Control Volume: *Boiler*

Heat input to the steam

From first law of thermodynamics:

$$q_b = h_1 - h_7 \text{ and } Q_b = \dot{m}_s \left(h_1 - h_7 \right)$$

Total turbine work output

$$w_t = \left[\left(h_1 - h_2 \right) + \left(1 - m_1 \right) \left(h_2 - h_3 \right) \right]$$

$$W_t = \dot{m}_s w_t = \dot{m}_s \left[\left(h_1 - h_2 \right) + \left(1 - m_1 \right) \left(h_2 - h_3 \right) \right]$$

Total pump input power

$$w_p = w_{p-2} + \left(1 - m_1 \right) w_{p-1}$$

$$W_p = \dot{m}_s w_p = \dot{m}_s \left[\left(h_7 - h_6 \right) + \left(1 - m_1 \right) \left(h_5 - h_4 \right) \right]$$

Estimate steam flow rate:

From total power requirement:

$$P = \dot{m}_s \left(w_t \right) \text{ and } \dot{m}_s = P / \left(w_t \right)$$

Summary: Ratings and Specification of Major Components:

Boiler:

Net heat addition, Q_b

Pressure, $p_b = p_8 = p_1$

Temperature, T_b

Mass flow rate, $\dot{m}_s$

Turbine:

Net work out put, W_t

Mass flow rate, $\dot{m}_s$

Mass flow rate of extracted steam, m_1

Inlet condition: pressure, P_1

Temperature, T_1

Back pressure, $p_3 = p_4 = p_c$

Efficiency of turbine, η_t

Condenser:

Heat rejected = Heat gained by cooling water = Q_c

Mass flow rate of steam = $\dot{m}_s$

Temperature and pressure of condensing steam, $p_c = p_3 = p_4$ and $T_c = T_3 = T_4$

Cooling water flow rate = $\dot{m}_w$

Temperature rise for cooling water, $T_{w,o} - T_{w,i}$

Pump 1: Condensate pump

Power: w_{p-1}

Pressure rise: $\Delta P = P_5 - P_4$

Mass flow rates: $\dot{m}_s$

Pump efficiency: η_p

Pump-2: Feed water heater pump

Power: w_{p-2}

Pressure rise: $\Delta P = P_7 - P_6$

Mass flow rates: $(1 - m_1) \dot{m}_s$
Pump efficiency: η_p
Feed water heater: open types
Mixing process.

Example 2.8   *Thermodynamic Analysis of Regenerative Rankine Power Cycle*
A regenerative Rankine regenerative cycle with one open feed water heater is considered to produce 1-MW of power. Steam leaves the boiler and enters the turbine as ***saturated vapor*** at 4 MPa Pressure. Assume feed water heater pressure as 2 MPa and the condenser pressure as 20 kPa. The turbine and pump efficiencies are 90% and 85%, respectively. (a) Perform the first law of thermodynamic analysis to determine (a) net boiler heat transfer, (b) net heat rejected in the condenser, and (c) thermal efficiency of the power cycle.

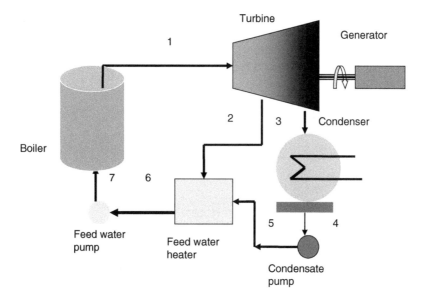

Requirement: Produce 1 MW of power.
Basic assumptions:
   Use standard Rankine power cycle using water as the working fluid
Known data:
   High and low temperature and pressure levels
   Boiler pressure: 4 MPa
   Feed heater pressure: 2 MPa
   Condenser pressure: 20 kPa
State 1: Known $P_1 = 4$ MPa, Saturated vapor
   From Table C.1b from pressure-based saturation data at 4 MPa:
   $T_1 = T_{sat} = 250.4\,°C$, $h_1 = h_g = 2801.38$ kJ/kg, $s_1 = s_g = 6.070$ kg/kg K
State 2: Feed water heater pressure. At $P_2 = P_5 = P_6 = P_{fwh} = 2$ MPa, we can get following data from the pressure-based saturation table:
$$T_2 = T_6 = T_{sat} = 212\,42°C$$

$$h_f = 908.77 \text{ kJ/kg } h_g, h_{fg} = 1890.74 \text{ kJ/kg}, h_g = 2799.51 \text{ kJ/kg}$$

$$s_f = 2.4473 \text{ kJ/kg K}, s_{fg} = 3.8935 \text{ kJ/kg K}, s_g = 3.8935 \text{ kJ/kg K}$$

$$v_f = 0.001177 \text{ m}^3/\text{kg}$$

State 3: Condenser pressure. At $P_3 = P_4 = P_c = 20$ kPa, we can get following data from the pressure-based saturation Table C.1b:

$$T_3 = T_4 = T_{sat} = 60.06°C$$

$$h_f = 251.38 \text{ kJ/kg}, h_{fg} = 2358.33 \text{ kJ/kg}, h_g = 2609.70 \text{ kJ/kg}$$

$$s_f = 0.8319 \text{ kJ/kg K}, s_{fg} = 7.0766 \text{ kJ/kg K}, s_g = 7.90\,085 \text{ kJ/kg K}$$

$$v_f = 0.001017 \text{ m}^3/\text{kg}$$

Control volume: turbine (first-stage turbine)
Process 1-2: From first law of thermodynamics
Reversible adiabatic process: $w_{t-1s} = (h_1 - h_{2s})$
Actual turbine work: $w_{t-1} = (h_1 - h_2)$
From second law of thermodynamics: $s_{2s} = s_1 = 6.07 \text{ kJ/kg K}$
From $s_1 = s_f + x_{2s}s_{fg}$, *Solve*:

$$x_{2s} = \frac{s_{2s} - s_f}{s_{fg}} = x_{2s} = \frac{6.07 - 2.4473}{3.8935} = 0.9304$$

and

$$h_{2s} = h_f + x_{2s}h_{fg} = 908.77 + 0.9304 \times 1890.74$$

$$h_{2s} = 2592.5521 \text{ kJ/kg}$$

Calculate turbine work:
Isentropic turbine work: $w_{t-1s} = (h_1 - h_{2s})$

$$w_{t-1s} = (2801.38 - 2592.5521)$$

$$w_{t-1s} = 208.8279 \text{ kJ/kg}$$

Actual turbine work: $w_{t-1} = \eta_T w_{ts-1}$

$$w_{t-1} = 0.9 \times 208.8279$$

$$w_{t-1} = 187.9451 \text{ kJ/kg}$$

Calculate: $h_2 = h_1 - w_{t-1}$

$$h_2 = 2801.38 - 187.9451 = 2613.4349 \text{ kJ/kg}$$

$$x_2 = \frac{h_2 - h_f}{h_{fg}} = \frac{2613.4349 - 908.77}{1890.74} = 0.9419$$

Similarly for process 1-3s (Second-stage turbine)

Process 1–3: Reversible adiabatic process

From first law of thermodynamics: $w_{ts-1} = (h_1 - h_{3s})$

From second law of thermodynamics: $s_{3s} = s_1 = 6.07 \, \text{kJ/kg K} = s_f + x_{3s}s_{fg}$

Solve $x_{3s} = \frac{s_{3s}-s_f}{s_{fg}}$ *and* $h_{3s} = h_f + x_{3s}h_{fg}$ *for the exit state of the second turbine or inlet to the condenser*

$$X_{3s} = \frac{s_{3s} - s_f}{s_{fg}} = x_{3s} = \frac{6.07 - 0.8319}{7.0766} = 0.7402$$

$$h_{3s} = h_f + x_{3s}h_{fg} = 251.38 + 0.7402 \times 2358.33 = 1997.0158$$

Calculate turbine work: first law of thermodynamics

Isentropic turbine work: $w_{t-2s} = (h_2 - h_{3s}) = 2613.4349 - 1997.0158 = 616.419 \, \text{kJ/kg}$

$$w_{t-2s} = 616.419 \, \text{kJ/kg}$$

$$\textit{Actual turbine work: } w_{t-2} = \eta_t w_{ts-2} = 0.90 \times 616.419 = 554.777 \, \text{kJ/kg}$$

$$w_{t-2} = \eta_t w_{ts-2} = 554.777 \, \text{kJ/kg}$$

Calculate:

$$h_3 = h_2 - w_{t-2} = 2613.4349 - 554.777$$

$$h_3 = 2058.6649 \, \text{kJ/kg}$$

and

$$X_3 = \frac{h_3 - h_f}{h_{fg}} = \frac{2058.6649 - 251.38}{2358.33}$$

$$x_3 = 0.7663 = 76.63\%$$

Total turbine work:

$$W_T = \dot{m}_s \left(w_{t-1} + w_{t-2} \right) = \dot{m}_s \left[(h_1 - h_2) + (1 - \dot{m}_1)(h_2 - h_3) \right]$$

where $\dot{m}_s$ = mass flow rate of steam

Control volume: condenser:

From first law of thermodynamics

Steam side:

Heat rejected in the condenser as steam changes phase vapor to liquid state

$$q_c = (1 - m_1)(h_3 - h_4) = (1 - 0.277)(2058.6649 - 251.38)$$

$$q_c = 0.723 \times 1807.2849 \text{kJ/kg}$$

$$q_c = 1306.667 \text{kJ/kg}$$

Cooling water side:

$$\dot{Q}_C = \dot{m}_w \left(h_{w,o} - h_{w,i}\right)$$

or

$$\dot{Q}_C = \dot{m}_w C_{pw} \left(h_{w,o} - h_{w,i}\right)$$

Calculate: Mass flow rate of cooling water:

$$\dot{m}_w = \frac{\dot{Q}_C}{C_{pw} \left(T_{w,o} - T_{w,i}\right)}$$

CV: pump-1 (condensate pump): process 4–5

State 4: Saturated liquid state

$$v_4 = v_f = 0.001017 \text{ m}^3/\text{kg at } P_4 = P_c = 20 \text{ kPa}$$

Isentropic pump work:

$$w_{ps-1} = v_4 \left(P_5 - P_4\right) = 0.001017 \ (2000 - 20)$$

$$w_{ps-1} = 2.014 \text{ kJ/kg}$$

Calculate:

Actual pump work: $w_{p-1} = \frac{w_{ps-1}}{\eta_p} = \frac{2.014}{0.85}$

$$w_{p-1} = 2.37 \text{ kJ/kg}$$

Total pump work: $W_{p-1} = \dot{m}_s w_{p-1}$
Calculate exit state of the pump:

$$h_5 = h_4 + w_{p-1} = 251.38 + 2.37$$

$$h_5 = 253.75 \text{ kJ/kg}$$

Control volume: feed water heater:

$P_5 = P_6 = P_2 = $ Feed water pressure
From first law:

$$\dot{m}_1 h_2 + \left(1 - \dot{m}_1\right) h_5 = h_6$$

where $h_6 = h_f = 908.77 \text{ kJ/kg at } P_6 = P_{fwh} = 2 \text{ MPa}$
Solve for mass flow rate steam extracted from turbine at state 2:

$$\dot{m}_1 = \frac{\left(h_6 - h_5\right)}{\left(h_2 - h_5\right)} = \frac{908.77 - 253.75}{2613.4349 - 253.75} = \frac{655.02}{2359.6849}$$

$$\dot{m}_1 = 0.277$$

Control volume: pump-2 (feed water pump):

Calculate:

Isentropic pump work: $w_{ps-2} = v_6 \left(P_7 - P_6\right)$

where $v_6 = v_f = 0.001\,177\,\text{m}^3/\text{kg}$ at $P_6 = P_2 = P_{fwf} = 2\,\text{MPa}$

$$w_{ps-2} = v_6\left(P_7 - P_6\right)$$

$$w_{ps-1} = v_6\left(P_7 - P_6\right) = 0.001\,177\,(4000 - 2000)$$

$$w_{ps-2} = 2.358\,\text{kJ/kg}$$

Calculate:

Actual pump work: $w_{p-2} = \dfrac{w_{ps-2}}{\eta_p} = \dfrac{2.358}{0.85}$

$$w_{p-2} = 2.7741\,\text{kJ/kg}$$

Total pump work: $W_{p-2} = \dot{m}_s w_{p-2}$
For exit state of the pump and inlet to the boiler

$$h_7 = 908.77 + 2.7741$$

$$h_7 = 911.5441\,\text{kJ/kg}$$

Control volume: boiler:
Heat input to the steam in boiler per unit mass flow rate
From first law of thermodynamics:

$$q_b = h_1 - h_7 = 2801.38 - 911.544$$

$$q_b = 1889.836\,\text{kJ/kg}$$

and

$$Q_b = \dot{m}_s\left(h_1 - h_7\right)$$

Total turbine work output:

$$w_t = w_{t-1} + w_{t-2} = \left(h_1 - h_2\right) + \left(1 - m_1\right)\left(h_2 - h_3\right)$$

$$= (2801.38 - 2613.4349) + (1 - 0.277)(2613.4349 - 2058.6649)$$

$$= 187.9451 + 401.09871$$

$$w_t = 589.04381\,\text{kJ/kg}$$

Total pump input power:

$$w_p = w_{p-1} + \left(1 - m_1\right)w_{p-2}$$

$$= 2.37 + (1 - 0.277) \times 2.7441 = 2.37 + 1.984$$

$$w_p = 4.354\,\text{kJ/kg}$$

and

$$W_p = w_p \dot{m}_s = \dot{m}_s\left[\left(h_7 - h_6\right) + \left(1 - m_1\right)\left(h_5 - h_4\right)\right]$$

Estimate steam flow rate:

Equating net power requirement to net turbine workout, we get

$$P = \dot{m}_s \, w_T$$

and the steam mass flow rate is given as

$$\dot{m}_s = \frac{P}{w_T} = \frac{1000 \, \text{kJ/s}}{589.04381 \, \text{kJ/kg}}$$

$$\dot{m}_s = 1.698 \, \text{kg/s}$$

Thermal efficiency of the cycle:

$$\eta_{th} = \frac{w_{net}}{q_b} = \frac{w_t - w_p}{q} = \frac{589.04381 - 4.354}{1889.836} = \frac{584.69}{1889.836}$$

$$\eta_{th} = 30.94\%$$

Determine net quantities:

Net boiler heat transfer:

$$\dot{Q}_b = \dot{m}_s q_b = 1.698 \, \text{kg/s} \times 1889.836 \, \text{kJ/kg}$$

$$\dot{Q}_b = 3208.94 \, \text{kW}$$

Net heat rejected in the condenser:

$$\dot{Q}_c = \dot{m}_s q_c = 1.698 \, \text{kg/s} \times 1306.667 \, \text{kJ/kg}$$

$$\dot{Q}_c = 2218.72 \, \text{kW}$$

Net pump work input:

$$\dot{W}_p = \dot{m}_s w_p = 1.698 \, \text{kg/s} \times 4.354 \, \text{kJ/kg}$$

$$\dot{W}_p = 7.393 \, \text{kW}$$

Cooling water side:

$$\dot{Q}_C = \dot{m}_w \left(h_{w,o} - h_{w,i} \right)$$

or

$$\dot{Q}_C = \dot{m}_w C_{pw} \left(h_{w,o} - h_{w,i} \right)$$

Calculate: Mass flow rate of cooling water:

$$\dot{m}_w = \frac{\dot{Q}_C}{C_{pw} \left(T_{w,o} - T_{w,i} \right)}$$

Summary: Ratings and Specification of Major Components:

Boiler:

    Net heat addition, Q_b
    Pressure, $p_b = p_8 = p_1$
    Temperature, T_b
    Mass flow rate, $\dot{m}_s$

Turbine:
   Net work output, W_t
   Mass flow rate, $\dot{m}_s$
   Mass flow rate of extracted steam, m_1
   Inlet condition: pressure, P_1
   Temperature, T_1
   Back pressure, $p_3 = p_4 = p_c$
   Efficiency of turbine, η_t
Condenser:
   Heat rejected = heat gained by cooling water = Q_c
   Mass flow rate of steam = $\dot{m}_s$
   Temperature and pressure of condensing steam, $p_c = p_3 = p_4$ and $T_c = T_3 = T_4$
   Cooling water flow rate = m_w
   Temperature rise for cooling water, $T_{w,\,o} - T_{w,\,i}$
Pump 1: Condensate pump:
   Power: w_{p-1}
   Pressure rise: $\Delta P = P_5 - P_4$
   Mass flow rates: $\dot{m}_s$
   Pump efficiency : η_p
Pump-2: Feed water heater pump:
   Power: w_{p-2}
   Pressure rise: $\Delta P = P_7 - P_6$
   Mass flow rates: $(1 - m_1)\,\dot{m}_s$
   Pump efficiency: η_p
Feed water heater: open types
   Mixing process

Example 2.9   *Thermodynamic Analysis of Regenerative Rankine Power Cycle*
A regenerative Rankine regenerative cycle with one open feed water heater is considered to produce 1-MW of power. Steam leaves the boiler and enters the turbine as ***saturated vapor*** at 4 MPa Pressure. Assume feed water heater pressure as 2 MPa and the condenser pressure as 20 kPa. The turbine and pump efficiencies are 90% and 85%, respectively. (a) Perform the first law of thermodynamic analysis to determine (a) net boiler heat transfer, (b) net heat rejected in the condenser, and (c) the thermal efficiency of the power cycle.

Requirement: Produce 1 MW of power.
Known data:
   High- and low-temperature and pressure levels
      Boiler pressure: 4 MPa
      Feed heater pressure: 2 MPa
      Condenser pressure: 20 kPa
State 1: Known $P_1 = 4$ MPa, Saturated vapor
From pressure-based saturation data at 4 MPa:

$$T_1 = T_{sat} = 250.4°C, h_1 = h_g = 2801.38 \text{ kJ/kg}, s_1 = s_g = 6.070 \text{ kg/kg K}$$

State 2: Feed water heater pressure. At $P_2 = P_5 = P_6 = P_{fwh} = 2$ MPa, we can get following data from the pressure-based saturation Table C.1b:

$$T_2 = T_6 = T_{sat} = 212.42°C$$

$$h_f = 908.77 kJ/kg, h_g, h_{fg} = 1890.74 \text{ kJ/kg}, h_g = 2799.51 kJ/kg$$

$$s_f = 2.4473 \text{ kJ/kg K}, s_{fg} = 3.8935 \text{ kJ/kg K}, s_g = 3.8935 \text{ kJ/kg}$$

$$v_f = 0.001177 \text{ m}^3/\text{kg}$$

State 3: Condenser pressure. At $P_3 = P_4 = P_c = 20$ kPa, we can get following data from the pressure-based saturation Table C.1b:

$$T_3 = T_4 = T_{sat} = 60.06 \text{ C}$$

$$h_f = 251.38 \text{ kJ/kg}, h_{fg} = 2358.33 \text{ kJ/kg}, h_g = 2609.70 \text{ kJ/kg}$$

$$s_f = 0.8319 \text{ kJ/kg K}, s_{fg} = 7.0766 \text{ kJ/kg}, s_g = 7.90085 \text{ kJ/kg K}$$

$$v_f = 0.001\,017 \text{ m}^3/\text{kg}$$

Control volume: turbine (first-stage turbine)

Process 1-2: From first law of thermodynamics
Reversible adiabatic process: $w_{t-1s} = (h_1 - h_{2s})$
Actual turbine work: $w_{t-1} = (h_1 - h_2)$
From second law of thermodynamics: $s_{2s} = s_1 = 6.07$ kJ/kg K
From $s_1 = s_f + x_{2s}s_{fg}$, Solve:

$$x_{2s} = \frac{s_{2s} - s_f}{s_{fg}} = x_{2s} = \frac{6.07 - 2.4473}{3.8935} = 0.9304$$

and

$$h_{2s} = h_f + x_{2s}h_{fg} = 908.77 + 0.9304 \times 1890.74$$

$$h_{2s} = 2667.91 \text{ kJ/kg}$$

Calculate turbine work

Isentropic turbine work: $w_{t-1s} = (h_1 - h_{2s})$

$$w_{t-1s} = (2801.38 - 2667.91)$$

$$w_{t-1s} = 133.47 \text{ kJ/kg}$$

Actual turbine work: $w_{t-1} = \eta_T w_{ts-1}$

$$w_{t-1} = 0.9 \times 133.47$$

$$w_{t-1} = 120.123 \text{ kJ/kg}$$

Calculate: $h_2 = h_1 - w_{t-1}$

$$h_2 = 2801.38 - 120.123 = 2681.257 \text{ kJ/kg}$$

$$x_2 = \frac{h_2 - h_f}{h_{fg}} = \frac{2681.257 - 908.77}{1890.74} = 0.9375$$

Similarly for process 1–3s (second-stage turbine):

Process 1–3: Reversible adiabatic process

From first law of thermodynamics: $w_{ts-1} = (h_1 - h_{3s})$

From second law of thermodynamics: $s_{3s} = s_1 = 6.07 \text{ kJ/kg K} = s_f + x_{3s} s_{fg}$

Solve $x_{3s} = \frac{s_{3s} - s_f}{s_{fg}}$ and $h_{3s} = h_f + x_{3s} h_{fg}$ for the exit state of the second turbine or inlet to the condenser

$$x_{3s} = \frac{s_{3s} - s_f}{s_{fg}} = x_{3s} = \frac{6.07 - 0.8319}{7.0766} = 0.7402$$

$$h_{3s} = h_f + x_{3s} h_{fg} = 251.38 + 0.7402 \times 2358.33 = 1997.0158$$

Calculate turbine work: first law of thermodynamics

Isentropic turbine work:

$$w_{t-2s} = (h_2 - h_{3s}) = 2681.257 - 1997.0158$$

$$= 684.24 \text{ kJ/kg}$$

$$w_{t-2s} = 684.24 \text{ kJ/kg}$$

Actual turbine work:

$$w_{t-2} = \eta_t w_{ts-2} = 0.90 \times 684.24 = 615.82 \text{ kJ/kg}$$

$$w_{t-2} = \eta_t w_{ts-2} = 615.82 \text{ kJ/kg}$$

$$h_3 = h_2 - w_{t-2} = 2681.257 - 615.82$$

$$h_3 = 2065.44 \text{ kJ/kg}$$

and

$$x_3 = \frac{h_3 - h_f}{h_{fg}} = \frac{2065.44 - 251.38}{2358.33}$$

$$x_3 = 0.7692 = 76.92\%$$

Control volume: condenser

From first law of thermodynamics

Steam side:

Heat rejected in the condenser as steam changes from phase vapor to liquid state

$$q_c = (1 - m_1)(h_3 - h_4) = (1 - 0.277)(2065.44 - 251.38)$$

$$q_c = 0.723 \times 1814.06 \text{ kJ/kg}$$

$$q_c = 1311.57 \text{ kJ/kg}$$

CV: pump-1 (condensate pump): process 4–5

State 4: Saturated liquid state

$$v_4 = v_f = 0.001017 \text{ m}^3/\text{kg at } P_4 = P_c = 20 \text{ kPa}$$

Isentropic pump work:

$$w_{ps-1} = v_4 \left(P_5 - P_4 \right) = 0.001\,017 \, (2000 - 20)$$

$$w_{ps-1} = 2.0136 \text{ kJ/kg}$$

Actual pump work: $w_{p-1} = \dfrac{w_{ps-1}}{\eta_p} = \dfrac{2.0136}{0.85}$

$$w_{p-1} = 2.37 \text{ kJ/kg}$$

Total pump work: $W_{p-1} = \dot{m}_s w_{p-1}$

Calculate exit state of the pump

$$h_5 = h_4 + w_{p-1} = 251.38 + 2.37$$

$$h_5 = 253.75 \text{ kJ/kg}$$

Control volume: feed water heater

$$P_5 = P_6 = P_2 = \text{Feed water pressure}$$
From first law:

$$m_1 h_2 + \left(1 - m_1 \right) h_5 = h_6$$

where $h_6 = h_f = 908.77 \text{ kJ/kg at } P_6 = P_{fwh} = 2 \text{ MPa}$
Solve for mass flow rate steam extracted from turbine at state 2:

$$m_1 = \frac{\left(h_6 - h_5 \right)}{\left(h_2 - h_5 \right)} = \frac{908.77 - 253.75}{2681.257 \text{ kJ} - 253.75} = \frac{655.02}{2427.507}$$

$$m_1 = 0.27$$

Control volume: pump-2 (feed water pump)

Isentropic pump work: $w_{ps-2} = v_6 \left(P_7 - P_6 \right)$
where $v_6 = v_f = 0.001\,177 \text{ m}^3/\text{kg at } P_6 = P_2 = P_{fwf} = 2 \text{ MPa}$

$$w_{ps-2} = v_6 \left(P_7 - P_6 \right)$$

$$w_{ps-1} = v_6 \left(P_7 - P_6 \right) = 0.001\,177 \, (4000 - 2000)$$

$$w_{ps-2} = 2.358 \text{ kJ/kg}$$

Actual pump work: $w_{p-2} = \dfrac{w_{ps-2}}{\eta_p} = \dfrac{2.358}{0.85}$

$$w_{p-2} = 2.7741 \text{ kJ/kg}$$

Total pump work: $W_{p-2} = \dot{m}_s w_{p-2}$

For exit state of the pump and inlet to the boiler

$$h_7 = 908.77 + 2.7741$$

$$h_7 = 911.5441 \text{ kJ/kg}$$

Control volume: boiler

Heat input to the steam in boiler per unit mass flow rate
From first law of thermodynamics:

$$q_b = h_1 - h_7 = 2801.38 - 911.544$$

$$q_b = 1889.836 \text{ kJ/kg}$$

Total turbine work output per unit mass flow rate:

$$w_t = w_{t-1} + w_{t-2} = \left(h_1 - h_2\right) + \left(1 - m_1\right)\left(h_2 - h_3\right)$$

$$= (2801.38 - 2681.257 \text{ kJ}) + (1 - 0.27)(2681.257 - 2065.44)$$

$$= 120.123 + 449.55$$

$$w_t = 569.67 \text{ kJ/kg}$$

Total pump input power:

$$w_p = w_{p-1} + \left(1 - m_1\right) w_{p-2}$$

$$= 2.37 + (1 - 0.27) \times 2.7441 = 2.37 + 2.003$$

$$w_p = 4.373 \text{ kJ/kg}$$

Estimate steam flow rate:

Equating net power requirement to net turbine workout, we get

$$P = \dot{m}_s w_T$$

and the steam mass flow rate is given as

$$\dot{m}_s = \frac{P}{w_t} = \frac{1000 \text{ kJ/s}}{569.67 \text{kJ/kg}}$$

$$\dot{m}_s = 1.755 \text{ kg/s}$$

Thermal efficiency of the cycle:

$$\eta_{th} = \frac{w_{net}}{q_b} = \frac{w_t - w_p}{q} = \frac{569.67 - 4.373}{1889.836} = \frac{565.297}{1889.836}$$

$$\eta_{th} = 29.91\%$$

Determine net quantities: 5 points:

Net boiler heat transfer:

$$\dot{Q}_b = \dot{m}_s q_b = 1.755 \text{ kg/s} \times 1889.836 \text{ kJ/kg}$$

$$\dot{Q}_b = 3316.66 \text{ kW}$$

Net heat rejected in the condenser:

$$\dot{Q}_c = \dot{m}_s q_c = 1.755 \text{ kg/s} \times 1311.57 \text{ kJ/kg}$$

$$\dot{Q}_c = 2301.8 \text{ kW}$$

Net pump work input:

$$\dot{W}_c = \dot{m}_s w_p = 1.755 \text{ kg/s} \times 4.373 \text{ kJ/kg}$$

$$\dot{W}_c = 7.675 \text{ kW}$$

2.14.1.5.1 Real Power Plant Cycle with Multiple Feed Water Heaters

Regenerative cycle efficiency increases with the increase in feed water heater, but with additional cost for each feed water heater. Real number of feed water heaters is determined based on an optimization study. Real big power plant uses 5–7 feed water heaters as demonstrated in Figure 2.24 below:

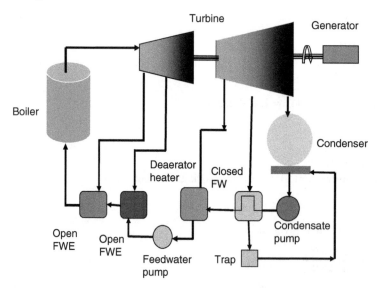

Figure 2.24 Power plant system with multiple feed water heaters.

2.14.2 Gas Power System

As we have noticed, in vapor power systems, working fluid changes between liquid and vapor phases as it is alternately vaporized and condensed. In contrast, the working fluid in gas power cycles remains in gaseous phase all throughout as it undergoes a number of processes. Objective of this section is to study power systems utilizing working fluids that are always a **gas**. Some of the most common gas power systems are **spark ignition engine**, **compression-ignition or diesel engine**, and **gas turbine power generation system**. All these systems are referred to as the internal combustion engine types with combustion taking place inside the system. In contrast, a steam power plant is called an external combustion engine because heat is transferred from the products of combustion to the working fluid of system in the boiler, while combustion takes place in a furnace, which is located outside the thermodynamic cycle.

Another important characteristic of these systems is that in all these systems there is a change in the composition of the working fluid, because during the combustion process it *changes from a mixture of air and fuel to a mixture of products of combustion*.

2.14.2.1 Reciprocating Internal Combustion Systems

There are two types of reciprocating internal combustion engines: the *spark-ignition engine* and the *compression-ignition engine*. In a *spark-ignition engine*, a mixture of fuel and air is ignited by a spark plug. In a **compression-ignition engine**, air is compressed to a high enough pressure and temperature so that combustion occurs spontaneously when fuel is injected.

The *spark-ignition engines* have advantages in applications requiring power up to about 225 kW (300 HP). They are relatively light and lower in cost and suited for use in automobiles. The *compression-ignition engines* are normally preferred for applications when fuel economy and relatively large amount of power are required such as in *heavy* trucks, buses, locomotives, and ships. In the *middle range*, both spark ignition and compression-ignition engines are used.

2.14.2.2 Simplified Model for the Analysis of Internal Combustion Engine: Air Standard Cycles

Air standard cycles are used to analyze performance of different engines that operate using gas as the working fluid such as in internal combustion engine, gas turbine engine, and jet propulsion engine. In these cycles, several assumptions are made to simplify the complex process involved in a real engine to get a first-order approximation and get a qualitative understanding of the engine performance. Working fluid in an internal combustion engine or gas turbine engine does not go through a complete thermodynamic cycle even though the engine goes through a mechanical cycle. Basically, these engines operate on an open cycle. However, for analysis purposes, it is advantageous to devise a closed cycle that closely approximates the open cycles.

Basic assumptions for air standard cycles are as follows: (i) A fixed mass of air is the working fluid throughout the entire cycle, and air is always the ideal gas. (ii) The combustion process is replaced by a heat-transfer process from an external source. (iii) The cycle is completed by a heat-transfer process to surrounding – in contrast to the exhaust and intake process of an actual engine. (iv) All processes are internally reversible. Additionally, specific heat of air is constant over a process for simplicity.

2.14.2.3 Otto Cycle for Spark Ignition Engine

The Otto cycle is an ideal cycle for spark ignition internal combustion. The typical four processes in an Otto cycle are described along with movement of the piston in the cylinder of the spark ignition engine (see Figure 2.25 that shows statements across processes).

Four reversible processes are

1-2: Isentropic compression as the piston moves from the CDC center to head-end dead center (HDC).

2-3: Heat addition at constant volume when piston is momentarily at rest at HDC.

3 4: Isentropic expansion as piston moves from HDC to CDC (work output).

4-1: Rejection of heat when piston is at the CDC.

2.14.2.4 First Law of Thermodynamic Analysis

Process 1-2:

State 1: Ambient states T_1, P_1

For an isentropic compression process 1-2, $s_1 = s_2$ and for a given compression ratio, $r_v = \frac{V_1}{V_2}$

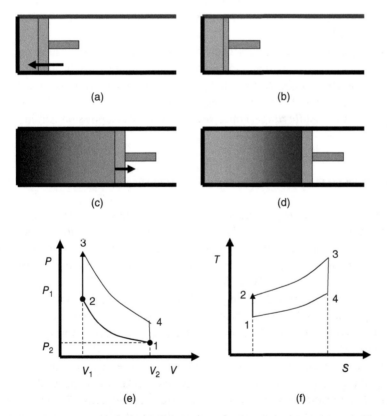

Figure 2.25 Processes in an Otto cycle of a spark ignition internal combustion engine. (a) 1-2: air intake and compression, (b) 2-3: heat addition – combustion, (c) 3-4: expansion – work, (d) 4-1: exhaust – heat rejection, (e) process on a *P-V* diagram and (f) Process on a *T-S* diagram.

The isentropic relationships between states 1 and 2 are given as

$$\frac{T_2}{T_1} = \left(\frac{V_1}{V_2}\right)^{k-1} \text{ and } \frac{P_2}{P_1} = \left(\frac{V_1}{V_2}\right)^k \tag{2.133}$$

State 2

Calculate temperature and pressures values at state 2 as follows:

$$T_2 = T_1 \left(\frac{V_1}{V_2}\right)^{k-1} \tag{2.134a}$$

and

$$P_2 = P_1 \left(\frac{V_1}{V_2}\right)^k \tag{2.134b}$$

The first law of thermodynamics for the air mass as the system that undergoes the process 1-2 is written as

$${}_1Q_2 = U_2 - U_1 + {}_1W_2 \tag{2.135}$$

For the reversible adiabatic or isentropic process 1-2, ${}_1Q_2 = 0$ and the compressive work in the process is given as

$${}_1W_2 = -\left(U_2 - U_1\right) \tag{2.136}$$

Process 2-3:

State 3: Known maximum combustion gas temperature, $T_3 = T_{max}$

In this process, heat is added through the combustion of fuel injected in the compressed air and use of the spark ignition. For this constant volume process, there is no work output and the first law equation reduces to the expression for the amount heat transferred at the high temperature in the process as

$$Q_H = {}_2Q_3 = U_3 - U_2 \tag{2.137a}$$

For ideal gas and constant specific heat, the equation reduces to

$$Q_H = {}_2Q_3 = mC_v \left(T_3 - T_2\right) \tag{2.137b}$$

Process 3-4:

For an isentropic expansion process 3-4, $s_4 = s_3$ and for the given compression ratio, $r_v = \frac{V_1}{V_2} = \frac{V_4}{V_3}$ the thermodynamic property relationships are given as

$$\frac{T_4}{T_3} = \left(\frac{V_3}{V_4}\right)^{k-1} \text{ and } \frac{P_4}{P_3} = \left(\frac{V_3}{V_4}\right)^k \tag{2.138}$$

State 4:

Calculate temperature and pressures values at state 4 using Eq. (2.138) as follows:

$$T_4 = T_3 \left(\frac{V_3}{V_4}\right)^{k-1} \tag{2.139a}$$

and

$$P_4 = P_3 \left(\frac{V_3}{V_4}\right)^k \tag{2.139b}$$

The first law of thermodynamics for the process 3-4 is written as

$$_3Q_4 = U_4 - U_3 + _3W_4 \tag{2.140a}$$

For the reversible adiabatic or isentropic expansion process 3-4, $_3Q_4 = 0$ and the expansion compressive work in the process is given as

$$_3W_4 = (U_3 - U_4) \tag{2.140b}$$

Process 4-1:

In this process, heat is rejected into the environment in a constant volume process as the combustion gas exits the cylinder when the piston stops momentarily at the CDC. For this constant volume process, there is no work output and first law equation reduces to the expression for the amount heat transferred at the low temperature as

$$Q_L = _4Q_1 = U_4 - U_1 \tag{2.141a}$$

For ideal gas and constant specific heat, the equation reduces to

$$Q_L = _3Q_4 = mC_v (T_4 - T_1) \tag{2.141b}$$

The thermal efficiency of the Otto cycle is given as

$$\eta_{th} = \frac{Q_H - Q_L}{Q_H} = 1 - \frac{Q_L}{Q_H}$$

$$= 1 - \frac{mC_v (T_4 - T_1)}{mC_v (T_3 - T_2)}$$

$$\eta_{th} = 1 - \frac{T_1 \left(\frac{T_4}{T_1} - 1\right)}{T_2 \left(\frac{T_3}{T_2} - 1\right)} \tag{2.142}$$

Using the isentropic process relationships, following relations can be derived

$$\frac{T_2}{T_1} = \left(\frac{V_1}{V_2}\right)^{k-1} = \left(\frac{V_4}{V_3}\right)^{k-1} = \frac{T_3}{T_2}$$

or

$$\frac{T_4}{T_1} = \frac{T_3}{T_2} \tag{2.143}$$

Substituting (2.143) into (2.142), we have

$$\eta_{th} = 1 - \frac{T_1}{T_2} = 1 - (r_v)^{1-K}$$

or

$$\eta_{th} = 1 - \frac{T_1}{T_2} = 1 - \frac{1}{(r_v)^{K-1}} \tag{2.144}$$

where

Compression ratio, $r_v = \dfrac{V_1}{V_2} = \dfrac{V_4}{V_3}$

Equation (2.144) indicates that the thermal efficiency of an Otto cycle is only a function of the compression ratio and increases only with the increase in the compression ratio. This is also true for a real spark ignition engine.

The compression ratio of an internal combustion engine represents the maximum volume of the combustion chamber at the CDC to the minimum volume of the combustion chamber at the HDC.

A higher compression ratio or higher expansion ratio allows greater expansion of the combustion gas producing more work and causing lower temperature of the exhaust gas. A higher compression ratio is always desirable as the engine can produce more mechanical power from the fuel-air mixture in the combustion chamber due to the higher thermal efficiency of the engine. Higher engine compression ratio is limited by the fact there is increased potential for detonation of the fuel as the compression ratio is increased, particularly for a low-octane fuel. Detonation is characterized by an extremely rapid burning of the fuel, which causes a stronger pressure wave and knocking of the engine. Compression ratio is an important specification of the internal combustion engine and the typical value varies from vendor to vendor depending on the use of knock sensor to control the engine stroke cycle and timings. The maximum compression ratio is limited by the fact that detonation must be avoided. Typical compression ratio of a spark ignition engine varies in the range of $10:1$ to $12:1$ depending on type of fuel and use of knock sensors.

Example 2.10   Otto Cycle

Consider an Otto cycle with a compression ratio of $r_v = 12$ and air intake condition of $T_1 = 25\,°C$, $P_1 = 0.1\,MPa$. The maximum temperature at the exit of the combustion chamber $T_{max} = 1200\,°C$. Determine (a) the temperature and pressure at the end of each process of the cycle and (b) the thermal efficiency.

Solution

Given compression ratio $r_v = 12$,

Air intake conditions: $T_1 = 25\,°C = 25 + 273 = 298$ K

Maximum temperature or the Temperature at the exit of the combustion chamber:

$$T_3 = T_{max} = 1200°C = 1200 + 273 = 1473 \text{ K}$$

$$P_1 = 0.1 \text{ MPa}$$

Process 1-2: Isentropic compression process

$$S_1 = S_2$$

$$r_v = \frac{v_1}{v_2}$$

(a) *Compute state point:*

State 2:

$$\left(\frac{T_2}{T_1}\right) = \left(\frac{v_1}{v_2}\right)^{k-1}$$

$$\left(\frac{T_2}{298}\right) = (12)^{k-1} \quad \left(\frac{T_2}{298}\right) = (12)^{k-1}$$

for $k_{air} = 1.4$

$$T_2 = 298(12)^{0.4} = 805.172 \text{ K}$$

$$P_2 = P_1 \left(\frac{v_1}{v_2}\right)^k = 0.1 \, (12)^{1.4}$$

$$P_2 = 3.2423 \text{ MPa}$$

State point 4:

Known data – $T_3 = T_{max} = 1473$ K and $r_v = \frac{v_1}{v_2} = 12$

$$T_4 = T_3 \left(\frac{v_3}{v_4}\right)^{k-1}$$

$$= 1473 \left(\frac{1}{12}\right)^{0.4}$$

$$T_4 = 545.1678 \text{ K}$$

$$P_4 = P_3 \left(\frac{v_3}{v_4}\right)^k$$

For $\dfrac{P_3}{P_2} = \dfrac{T_3}{T_2}, P_3 = 5.927$ MPa

$$P_4 = 0.18 \text{ MPa}$$

(b) *Thermal efficiency:*

$$\eta_{th} = 1 - \frac{T_1}{T_2}$$

$$= 1 - \frac{298}{805.172}$$

$$= 1 - 0.30710$$

$$= 0.6298$$

$$\eta_{th} = 63\%$$

2.14.2.5 The Diesel Cycle for Compression–Ignition Engine

The diesel cycle is the air standard cycle for Compression–Ignition Engine, also known as diesel engine. Figure 2.26 shows the processes, state points, and piston position in a design engine cycle.

The four processes in a diesel cycle are like the Otto-cycle, except that heat is added at constant pressure. A relatively higher compression ratio is used, and the air mass is compressed to high enough pressure and temperature so that combustion starts instantaneously as the fuel is injected in the compressed air mass without the need of any spark. This process

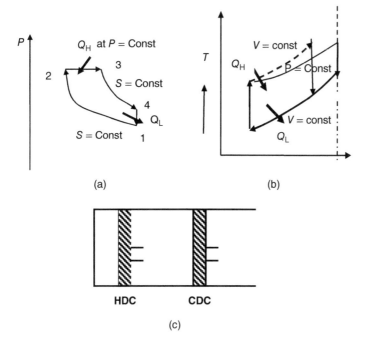

Figure 2.26 Diesel engine cycle representation. (a) $P-V$ diagram, (b) $T-S$ diagram, (c) piston–cylinder representation of an internal combustion engine.

corresponds to injection and burning of fuel in the actual engine. As fuel is injected in the high pressure and temperature air mass, combustions start instantaneously releasing high thermal energy and causing the piston to start expanding. Since the gas is expanding during the heat addition process, the heat transfer must be just enough to maintain constant pressure. The four reversible processes in an ideal diesel cycle are:

1-2: Isentropic compression as the piston moves from the CDC center to HDC.
2-3: Heat addition at constant pressure when fuel start burning and the piston starts expanding and the same time.
3-4: Isentropic expansion as piston moves to CDC producing major part of the work output.
4-1: Rejection of heat when piston is at the CDC.

First law of thermodynamic analysis

Known data:

Compression ratio, $r_{vc} == \dfrac{V_2}{V_1}$, Expansion ratio, $r_{va} == \dfrac{V_3}{V_4}$

Combustion mixture temperature or maximum temperature, $T_3 = T_{max}$

Process 1-2: isentropic compression

State 1: Ambient States T_1, P_1

State 2

Calculate temperature and pressures values at state 2 from isentropic relationships as follows:

$$T_2 = T_1 \left(\frac{V_1}{V_2}\right)^{k-1}$$

and

$$P_2 = P_1 \left(\frac{V_1}{V_2}\right)^{k}$$

The first law of thermodynamics for the air mass as the system that undergoes the process 1-2 is written as

$$_1Q_2 = U_2 - U_1 + _1W_2$$

For the reversible adiabatic or isentropic process 1-2, $_1Q_2 = 0$ and the compressive work in the process is given as

$$_1W_2 = -\left(U_2 - U_1\right) \tag{2.145}$$

Process 2-3:

State 3: Known maximum combustion gas temperature, $T_3 = T_{max}$ and $T_3 = T_{max}$

In this process, heat is added through the combustion of fuel injected in the compressed air in a constant pressure process. The first law equation reduces to the expression for the amount heat transferred at the high temperature in the process as

$$Q_H = _2Q_3 = U_3 - U_2 + _2W_3 \tag{2.146a}$$

The workout output for the constant pressure expansion process is expressed as

$$_2W_3 = P_2 \left(V_3 - V_2\right) \tag{2.146b}$$

Substituting Eq. (2.146b) into (2.146a) and considering constant specific heat, the equation reduces to

$$_2Q_3 = \left(U_3 - U_2\right) + P\left(V_3 - V_2\right) = \left(H_3 - H_2\right)$$

$$Q_H = _2Q_3 = mC_p \left(T_3 - T_2\right) \tag{2.147}$$

Process 3-4: isentropic expansion process

For an isentropic expansion process 3-4, $s_4 = s_3$ and for the expansion ratio, $r_v = \frac{V_3}{V_4}$, the thermodynamic property relationships are given as

$$\frac{T_4}{T_3} = \left(\frac{V_3}{V_4}\right)^{k-1} \quad \text{and} \quad \frac{P_4}{P_3} = \left(\frac{V_3}{V_4}\right)^{k}$$

State 4

Calculate temperature and pressures values at state 4 as follows:

$$T_4 = T_3 \left(\frac{V_3}{V_4}\right)^{k-1} \quad \text{and} \quad P_4 = P_3 \left(\frac{V_3}{V_4}\right)^{k}$$

The first law of thermodynamics for the process 3-4 is written as

$$_3Q_4 = U_4 - U_3 +_3W_4$$

For the reversible adiabatic or isentropic expansion process 3-4, $_3Q_4 = 0$ and the expansion compressive work in the process is given as

$$_3W_4 = (U_3 - U_4) \tag{2.148}$$

Process 4-1:

In this process, heat is rejected into the environment in a constant volume process as the combustion gas exits the cylinder when the piston stops momentarily at the CDC. For this constant volume process, there is no work output and first law equation reduces to the expression for the amount heat transferred at the low temperature as

$$Q_L = \, _4Q_1 = U_4 - U_1 \tag{2.149a}$$

For ideal gas and constant specific heat, the equation reduces to

$$Q_L = \, _4Q_1 = mC_v (T_4 - T_1) \tag{2.149b}$$

The thermal efficiency of the Otto cycle is given as

$$
\begin{aligned}
\eta_{th} &= \frac{Q_H - Q_L}{Q_H} = 1 - \frac{Q_L}{Q_H} \\
&= 1 - \frac{mC_v (T_4 - T_1)}{mC_p (T_3 - T_2)} \\
\eta_{th} &= 1 - \frac{T_1 \left(\frac{T_4}{T_1} - 1 \right)}{k T_2 \left(\frac{T_3}{T_2} - 1 \right)}
\end{aligned}
\tag{2.150}
$$

Note that in a diesel engine cycle, the isentropic compression ratio is greater than the isentropic expansion ratio. This is unlike that in an Otto cycle where compression ratio and expansion ratio are equal. For a given state before compression and a compression ratio (i.e. states 1 and 2 are known), the cycle efficiency decreases with increase in maximum temperature.

The Otto cycle has higher efficiency than the Diesel cycle for a given compression ratio. However, higher compression ratio can be used with the diesel cycle. The Otto cycle has the problem of detonation of fuel due to the compression fuel and air mixture. However, detonation is not an issue in a diesel engine since air is only compressed during the compression stroke of the engine.

Example 2.11   *Diesel Cycle*
Consider a diesel cycle with a compression ratio of $r_v = 12$ and a cutoff ratio $V_3/V_2 = 2$. The air intake condition of $T_1 = 25\,°C$, $P_1 = 0.1\,MPa$. The maximum temperature at the exit of the combustion chamber $T_{max} = 1200\,°C$. Determine (a) the temperature and pressure at the end of each process of the cycle and (b) the thermal efficiency

Solution

Known data:

$$\text{Compression ratio, } r_v = \frac{v_1}{v_2} = 12$$

$$\text{Cutoff ratio, } \frac{v_3}{v_2} = 2$$

(a) *State points:*

State 1: $T_1 = 298$ K, $P_1 = 0.1$ MPa

$$T_3 = T_{max} - 1473 \text{ K}$$

State 2: end of isentropic compression process

$$T_2 = T_1 \left(\frac{v_1}{v_2}\right)^{k-1} \text{ and considering } k_{air} = 1.4$$

$$= 298(12)^{0.4}$$

$$T_2 = 805.172 \text{ K}|$$

$$P_2 = P_1 \left(\frac{v_1}{v_2}\right)^{k} = 0.1 \, (12)^{1.4}$$

$$P_2 = 3.2423 \text{ MPa}$$

State 3:

$$P_3 = P_2 = 3.42 \text{ MPa}$$

State 4: End of isentropic expansion process

$$P_4 = P_3 \left(\frac{v_3}{v_4}\right)^{k} \text{ and } v_1 = v_4$$

Given $\frac{v_1}{v_2} = 12$, $\frac{v_3}{v_2} = 2$

$$\frac{v_4}{v_2} - 12$$

$$\frac{v_3}{v_2} \times \frac{v_2}{v_4} = \frac{2}{12}$$

$$\frac{v_3}{v_4} = 1/6$$

$$P_4 = 3.42 \left(\frac{1}{6}\right)^{1.4}$$

$$P_4 = 0.2783 \text{ MPa}$$

$$T_4 = T_3 \left(\frac{v_3}{v_4}\right)^{k-1} = 1473 \times \left(\frac{1}{6}\right)^{1.4}$$

$$T_4 = 719.35 \text{ K}$$

(b) *Thermal efficiency*:

$$\eta_{th} = 1 - \frac{(T_4 - T_1)}{k(T_3 - T_2)}$$

$$= 1 - \frac{719.35 - 298}{1.4(1473 - 805.172)}$$

$$= 1 - 0.450\,66$$

$$= 0.5493$$

$$\eta_{th} = 54.9\%|$$

2.14.2.6 Brayton Cycle: A Standard Cycle for Gas Turbine Engine

Brayton cycle is an air standard cycle representation of a gas turbine engine. A simple gas turbine operating on a Brayton cycle is shown in Figure 2.27. This cycle also involves four basic thermodynamic processes (i) air intake and compression in a rotary air compressor, (ii) combustion of fuel with the incoming compressed air in a combustion chamber that represents the high-temperature heat addition, (iii) expansion and work in a rotary gas turbine, and (iv) heat rejection in an exhaust process to the environment. The rotary compressor and the turbine are connected in a common shaft so that a fraction of the work produced by the gas turbine is supplied to the air compressor, and the rest results in the engine net work output.

2.14.2.7 Gas Turbine with Regenerative Heat Exchanger for Heat Recovery

To recover the heat from the exhaust gas stream and improve thermal efficiency of the gas turbine engine, a regenerator is generally used in the exhaust gas stream to preheat the incoming compressed airstream. Figure 2.28 shows a gas turbine engine with a regenerator for heat recovery from the exhaust.

As in the case of reciprocating internal combustion engines, several assumptions are made for the air standard gas turbine cycle to perform simplified thermodynamic analysis as a first-order approximation. These are (i) air is the working fluid with fixed mass flow rate, (ii) combustion process is replaced by a constant pressure combustion process, (iii) heat rejection at the exhaust is represented by a low-temperature heat-transfer process, (iv) air is assumed as ideal gas, (v) all irreversible losses in the rotary compressor and in turbine are taken into consideration by assuming efficiency values.

Figure 2.27 Brayton cycle for a standard gas turbine engine.

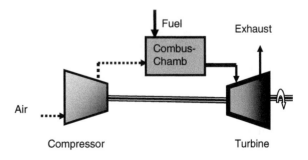

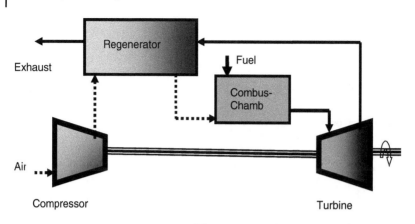

(a)

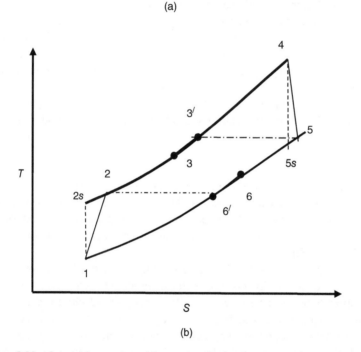

(b)

Figure 2.28 Gas turbine engine with regenerative heat recovery: (a) gas turbine power engine with a regenerator and (b) thermodynamic processes and state points on a T–S diagram.

2.14.2.8 First Law of Thermodynamic Analysis of a Gas Turbine Cycle with Regenerative Heat Recovery

Know Data:

Pressure Ratio, PR

Maximum temperature at the exit of combustion chamber, T_{max}

Compressor efficiency, η_c

Turbine efficiency, η_t

Regenerator efficiency, η_r

Intake air state

Analysis Steps

State 1

Process 1-2

  Control volume: compressor:

  From first law of thermodynamic:

  For isentropic compression work 1-2s

$$w_{cs} = h_{2s} - h_1 = C_p \left(T_{2s} - T_1 \right) \tag{2.151a}$$

  For real compression work 1-2

$$w_{ca} = h_2 - h_1 = C_p \left(T_2 - T_1 \right) \tag{2.151b}$$

From second law:

$$s_{2s} = s_1$$

$$\frac{T_{2s}}{T_1} = \left(\frac{P_2}{P_1} \right)^{k-1/k}$$

Calculate isentropic exit state 2

$$T_{2s} = T_1 \left(\frac{P_2}{P_1} \right)^{k-1/k} \tag{2.152a}$$

Calculate real exit state 2:

  For compressor efficiency defined as

$$\eta_c = \frac{w_{cs}}{w_{ca}} = \frac{h_{2s} - h_1}{h_2 - h_1} = \frac{T_{2s} - T_1}{T_2 - T_1}$$

$$T_2 = T_1 + \frac{T_{2s} - T_1}{\eta_c} \tag{2.152b}$$

  For real compression work

$$w_{ca} = h_2 - h_1 = C_p \left(T_2 - T_1 \right) \tag{2.153}$$

Process 3-4 – Expansion process for work output

Control volume: turbine

  From first law of thermodynamic:

  For isentropic expansion work 3-4s in the turbine

$$w_{ts} = h_3 - h_{4s} = C_p \left(T_3 - T_{4s} \right) \tag{2.154}$$

  From second law:

$$s_3 = s_{4s}$$

$$\frac{T_{4s}}{T_3} = \left(\frac{P_4}{P_3} \right)^{k-1/k}$$

Calculate isentropic exit state 4

$$T_{4s} = T_3 \left(\frac{P_4}{P_3} \right)^{k-1/k} \tag{2.155a}$$

Calculate real exit state 4:

For turbine efficiency defined as

$$\eta_c = \frac{w_{ta}}{w_{ts}} = \frac{h_3 - h_4}{h_3 - h_{4s}} = \frac{T_3 - T_4}{T_3 - T_{4s}}$$

Calculate turbine exit state

$$T_4 = T_3 - \eta_t \left(T_3 - T_{4s} \right) \tag{2.155b}$$

For real expansion work 3-4 in the turbine

$$w_t = h_3 - h_4 = C_p \left(T_3 - T_4 \right) \tag{2.156}$$

Control volume: regenerator:

Regenerator is a heat exchange in which heat is transferred from exiting exhaust stream from the turbine to the incoming air stream. The regenerator efficiency is defined as the ratio of actual heat transfer to air stream to the maximum possible heat transfer in the regenerator.

$$\eta_r = \frac{q_{reg}}{q_{reg,max}} = \frac{h_3 - h_2}{h_{3'} - h_2} = \frac{T_3 - T_2}{T_{3'} - T_2}$$

For maximum heat transfer across the regenerator, $T_{3'} = T_5$ and we can calculate air exit temperature from the regenerator as

$$T_3 = T_2 + \eta_r \left(T_5 - T_2 \right) \tag{2.157}$$

Control volume: combustion chamber:

Application of first law of thermodynamic

$$q_H = h_4 - h_3 = C_p \left(T_4 - T_3 \right) \tag{2.158}$$

Gas turbine thermal efficiency

$$\eta_r = \frac{w_{net}}{q_H} = \frac{w_{ta} - w_{ca}}{q_H} \tag{2.159}$$

Example 2.12   *Gas Turbine Cycle with Regenerative Heater*

The gas turbine cycle shown in the figure below is considered as an automotive engine. In the first-stage turbine, the gas expands to pressure P_5, just low enough for the turbine to drive the compressor. The gas is then expanded through the second turbine to produce a net power of $\dot{W}_{net} = 150\text{kW}$. The data for the engine are given below and shown in the figure. Assume 90% efficiency for the regenerator. For simplicity, assume 100% efficiency for compressor and turbine, and assume ideal gas air with constant specific heat (can assume no variation in temperature). Determine: (i) Intermediate temperature and pressure states at 2, 3, 5, and 6, (ii) mass flow rate through the engine, (iii) thermal efficiency of the engine.

Consider following known data:

$$P_1 = 100\text{kPa}, T_1 = 300\text{K}, \frac{P_2}{P_1} = 8, T_4 = 1600\text{K } P_7 = 100\text{kPa}$$

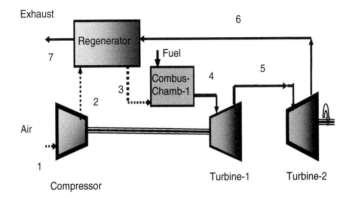

Solution

Process 1-2:

Control volume: compressor:

For isentropic compression process 1-2s:

From second law, $S_{2s} = S_1$

$$T_{2s} = T_1 \left(\frac{P_2}{P_1}\right)^{\frac{k-1}{k}} = T_1 \left(\frac{P_2}{P_1}\right)^{\frac{1.4-1}{1.4}} = 300(8)^{0.286} = 543.757 \text{ K}$$

From first law, the isentropic compressor works:

$$w_{cs} = C_{po} \left(T_{2s} - T_1\right)$$

$$w_{cs} = C_{po} \left(T_{2s} - T_1\right) = 1.005 \,(543.757 - 300.0) = 244.975 \text{ kJ/kg}$$

$$P_2 = 8P_1 = 0.8 \text{ MPa}$$

$T_2 = T_{2s} = 543.757\, K$ and $w_{ca} = w_{cs} = 244.975$ kJ/s for $\eta_c = 100\%$

Determine intermediate pressure:

Equate compressor work to first-stage turbine work

$$w_{ta-1} = w_{ca} = C_{po} \left(T_4 - T_5\right) = 244.975 \text{ kJ/kg}$$

$$T_5 = T_4 - \frac{w_{ca}}{C_{po}} = 1600 - \frac{244.975}{1.005}$$

$$T_5 = 1356.2437 \text{ K}$$

For $\eta_t = 100\%$, $T_{5s} = T_5 = 1356.2437$ K

Now determine the intermediate pressure.

Control volume: turbine stage 1:
Process 4-5 – Expansion process:
For isentropic expansion process

$$P_5 = P_4 \left(\frac{T_{5s}}{T_4} \right)^{\frac{1.4-1}{1.4}} = (800 \text{ kPa}) \left(\frac{1356.2437}{1600} \right)^{0.286} = 763.0627 \text{ kPa}$$

Intermediate pressure: $P_5 = 763.0627$ kPa

Process 5-6: Turbine expansion in second stage for power output

Second-stage turbine
From second law: $s_5 = s_{6s}$

$$\frac{T_{6s}}{T_5} = \left(\frac{P_6}{P_5} \right)^{\frac{k-1}{k}}$$

$$T_6 = T_{6s} = T_5 \left(\frac{P_6}{P_5} \right)^{\frac{k-1}{k}} = 1356.2437 \left(\frac{100}{763.0627} \right)^{0.286} = 758.4474 \text{ K}$$

Second-stage turbine work:

$$w_{ta-2} = C_{po} \left(T_5 - T_6 \right)$$

$$w_{ta-2} = 1.005 \left(1356.2437 - 758.4474 \right) = 600.785 \text{ kJ/kg}$$

Net mass flow rate:
Power output, $P = \dot{m} w_{ta-2} = W_{ta}$

$$\dot{m} = \frac{P}{w_{ta-2}} = \frac{150 \text{ kJ/s}}{600.785 \text{ kJ/kg}} = 0.2496 \text{ kg/s}$$

Control volume: regenerator:
For an ideal regenerator, the exit air temperature is given as

$$T_{3'} = T_6 = 758.4474 \text{ K}$$

The real exit air temperature is computed based on the efficiency of the regenerator. A real regenerator with efficiency is defined as

$$\eta_{reg} = 0.9 = \frac{h_3 - h_2}{h_{3'} - h_2} = \frac{T_3 - T_2}{T_{3'} - T_2} = \frac{T_3 - 543.757}{758.4474 - 543.757}$$

The air exit temperature out of regenerator is given as

$$T_3 = T_2 + \eta_r \left(T_{3'} - T_2 \right)$$

$$T_3 = 543.757 + 0.9 \times \left(758.4474 - 543.757 \right)$$

$$T_3 = 736.978 \text{ K}$$

Control volume: combustion chamber:
Total heat added in the combustion chamber

$$q_H = \left(h_4 - h_3 \right) = C_{po} \left(T_4 - T_3 \right) = 1.005 \left(1600 - 736.978 \right)$$

$$q_H = 867.3367 \text{ kJ/kg}$$

Total heat transfer in the combustion chamber

$$Q_H = \dot{m}q_H = 0.2496 \times 867.3367$$

$$Q_H = 216.487 \text{ kJ/s} = 216.487 \text{ kW}$$

The thermal efficiency of a single-stage gas turbine power system is given as

$$\eta_{Th} = \frac{W_{ta}}{Q_H} = \frac{150 \text{ kW}}{216.487 \text{ kW}} = 69.28\%$$

2.14.2.9 Gas Turbine with Multistage Compressions and Expansions

To achieve higher thermal efficiency, gas turbines are often used with multistage compression with intercooler and multistage expansion with interreheating. Figure 2.29 shows a gas turbine power generation system with two-stage turbine and compressor.

In general, the compression ratio in each stage of an n-stage multistage gas turbine is assumed as constant and given as the n-th square root of the total pressure rise. The pressure

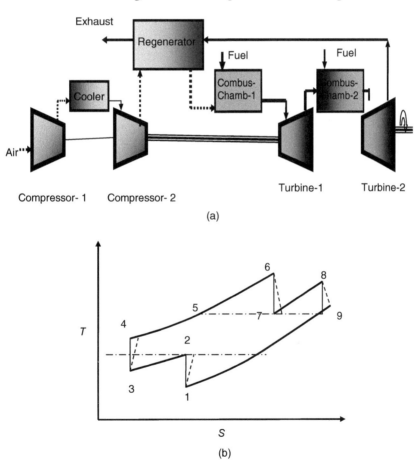

(a)

(b)

Figure 2.29 Multi-stage gas turbine system: (a) schematic of multistage gas turbine power generations system and (b) processes and state points on T–S diagram.

rise in each stage of the two-stage gas turbine, shown in the Figure 2.29, is then given as

$$\frac{P_2}{P_1} = \frac{P_4}{P_3} = \sqrt{PR} \text{ for the compressor stages} \tag{2.160a}$$

and

$$\frac{P_6}{P_7} = \frac{P_8}{P_9} = \sqrt{PR} \text{ for the turbine stages} \tag{2.160b}$$

Following the thermodynamic analysis presented for the single-stage gas turbine, we can express the net work, heat input, and thermal efficiency of the multistage gas turbine as follows:

Net work of compressor:

$$w_{ca} = n \left(w_{ca-1}\right) = n \left(h_2 - h_1\right) = nC_p \left(T_2 - T_1\right) \text{ for } n - \text{stages} \tag{2.161a}$$

and

$$w_{ca} = 2 \left(w_{ca-1}\right) = 2 \left(h_2 - h_1\right) = 2C_p \left(T_2 - T_1\right) \text{ for } 2 - \text{stages} \tag{2.161b}$$

Net work of turbine:

$$w_{ta} = n \left(w_{ta-1}\right) = n \left(h_6 - h_7\right) = nC_p \left(T_6 - T_7\right) \text{ for } n - \text{stages} \tag{2.162a}$$

and

$$w_{ta} = 2 \left(w_{ta-1}\right) = 2 \left(h_6 - h_7\right) = 2C_p \left(T_6 - T_7\right) \text{ for } 2 - \text{stages} \tag{2.162b}$$

Heat input:

$$q_H = q_{Primary} + q_{Reheat} = \left(h_6 - h_5\right) + (n - 1)\left(h_8 - h_7\right) = nC_p \left(T_6 - T_7\right)$$
$$= C_p \left(T_6 - T_5\right) + (n - 1)C_p \left(T_8 - T_7\right) \tag{2.163}$$

Efficiency:

$$\eta_r = \frac{w_{net}}{q_H} = \frac{w_{ta} - w_{ca}}{q_H} \tag{2.164}$$

Example 2.13 *Gas Turbine Cycle with a Regenerator and Multistage Cooling and Heating*

Consider an operating compression ratio and a maximum combustion temperature of the system as 12 and 1400 K, respectively. The state of the air at intake is given as pressure of 0.1 MPa and temperature of 300 K. Assume efficiencies of turbine, compressor, and regenerators as 90%. Perform thermodynamic analysis assuming air standard cycle to determine the operating states at each point in the cycle and the thermal efficiency considering (a) single-stage and four-stage compression and expansion. Considering a net mass flow of air as 6.0 kg/s, determine the net power output and the net heat addition. Give a summary table for the ratings for each component of the gas turbine system.

Solution

Known data:

$$PR = P_2/P_1 = 12$$
Maximum temperature, $T_{max} = 1400$ K

Intake conditions: $P_1 = 0.1$ MPa and $T_1 = 300$ K, $K = 1.4$, $C_p = 1.005$
Components efficiencies: $\eta_{reg} = 90\%$, $\eta_T = 90\%$ and $\eta_c = 100\%$
Net mass flow of air $\dot{m} = 6$ kg/s

Single stage
Process 1-2

Control volume: compressor:
For isentropic compression process 1-2s:
From second law, $S_{2s} = S_1$

$$T_{2s} = T_1 \left(\frac{P_2}{P_1}\right)^{\frac{k-1}{k}} = T_1 \left(\frac{P_2}{P_1}\right)^{\frac{1.4-1}{1.4}} = 300(12)^{0.286} = 610.1706 \text{ K}$$

From first law, the isentropic compressor work:

$$w_{cs} = C_{po} \left(T_{2s} - T_1\right)$$

$$w_{cs} = C_{po} \left(T_{2s} - T_1\right) = 1.005 \,(610.1706 - 300.0) = 311.7309 \text{ kJ/kg}$$

Actual compressor work is given as

$$w_{ca} = \frac{w_{cs}}{\eta_c} = \frac{311.7309}{0.9} = 346.3632 \text{ kJ/kg}$$

Temperature at exit of each compressor stage

$$T_2 = T_1 + \frac{w_{ca}}{C_{po}} = 300 + \frac{346.3632}{1.005} = 644.64 \text{ K}$$

$$P_2 = 12P_1 = 1.2 \text{ MPa}$$

Process 4-5 – Expansion process:

Control volume: turbine:
For isentropic expansion process

$$T_{5s} = T_4 \left(\frac{P_7}{P_6}\right)^{\frac{1.4-1}{1.4}} = (1400) \left(\frac{1}{12}\right)^{0.286} = 688.32 \text{ K}$$

Isentropic turbine work:

$$w_{ts} = C_{po} \left(T_4 - T_{5s}\right)$$

$$w_{ts} = C_{po} \left(T_4 - T_{5s}\right) = 1.005 \,(1400 - 688.32) = 715.2384 \text{ kJ/kg}$$

Actual turbine work:

$$w_{ta} = w_{ts} \cdot \eta_T = 715.2384 \times 0.9 = 643.7145 \text{ kJ/kg}$$

Temperature at exit of each turbine

$$T_5 = T_4 - \frac{w_{ta}}{C_{po}} = 1400 - \frac{643.7145}{1.005} = 759.488 \text{ K}$$

Control volume: regenerator:
For an ideal regenerator, the exit air temperature is given as

$$T_{3'} = T_5 = 759.488 \text{ K}$$

The real exit air temperature is computed based on the efficiency of the regenerator. For a real regenerator with efficiency defined as

$$\eta_{reg} = 0.9 = \frac{h_3 - h_2}{h_{3'} - h_2} = \frac{T_3 - T_2}{T_{3'} - T_2} = \frac{T_3 - 644.64}{759.488 - 644.64}$$

The air exit temperature out of regenerator is given as

$$T_3 = T_2 + \eta_r \left(T_{3'} - T_2\right)$$

$$T_3 - 644.64 + 0.9\,(759.488 - 644.64)$$

$$T_3 = 748 \text{ K}$$

Control volume: combustion chamber

Total heat added in the combustion chamber

$$q_H = \left(h_4 - h_3\right) = C_{po} \left(T_4 - T_3\right) = 1.005\,(1400 - 748)$$

$$q_H = 655.26 \text{ kJ/kg}$$

Total heat transfer in the combustion chamber

$$Q_H = \dot{m}q_H = 6 \times 655.26$$

$$Q_H = 3931.56 \text{ kJ/s}$$

The thermal efficiency of a single-stage gas turbine power system is given as

$$\eta_{Th} = \frac{w_{ta} - w_{ca}}{q_H} = \frac{643.7145 - 346.3632}{655.26} = 45.38\%$$

Consider four stages:

$n = 4$

Pressure rise in single stage $P_2/P_1 = \sqrt[n]{PR}$

Pressure ratio for single stage: $n = 4$

$P_2/P_1 = \sqrt[4]{12} = 1.861$

Control volume: compressor

For isentropic compression process

$$T_{2s} = T_1 \left(\frac{P_2}{P_1}\right)^{\frac{1.4-1}{1.4}} = 300(1.861)^{0.286} = 358.25 \text{ K}$$

From first law, the isentropic compressor work:

$$w_{cs-1} = C_{po} \left(T_{2s} - T_1\right)$$

$$w_{cs-1} = 1.005\,(358.25 - 300) = 58.541 \text{ kJ/kg}$$

$$w_{cs} = nw_{cs-1} = 4 \times 58.541 = 234.164 \text{ kJ/kg}$$

Actual compressor work in each stage is given as

$$w_{ca-1} = \frac{w_{cs-1}}{\eta_c} = \frac{58.541}{0.9} = 65.045 \text{ kJ/kg}$$

Total compressor work:

$$w_{ca} = 4 \times w_{ca-1} = 4 \times 65.045 = 260.18 \text{ kJ/kg}$$

Temperature at exit of each compressor stage

$$T_2 = T_1 + \frac{w_{ca} - 1}{C_{po}} = 300 + \frac{65.045}{1.005} = 364.72 \text{ K}$$

Control volume: turbine

For isentropic expansion process

$$T_{7s} = T_6 \left(\frac{P_7}{P_6} \right)^{\frac{1.4-1}{1.4}} = (1400) \left(\frac{1}{1.861} \right)^{0.286} = 1172.35 \text{ K}$$

Isentropic turbine work in a single stage:

$$w_{ts-1} = C_{po} \left(T_6 - T_{7s} \right)$$

$$w_{ts-1} = C_{po} \left(T_6 - T_{7s} \right) = 1.005 (1400 - 1172.35) = 228.788 \text{ kJ/kg}$$

Actual turbine in single stage:

$$w_{ta-1} = w_{ts-1} \cdot \eta_T = 228.788 \times 0.9 = 205.909 \text{ kJ/kg}$$

Total turbine work:

$$w_{ta} = n\, w_{ta-1} = 4 \times 205.909 \text{ kJ/kg} = 823.636 \text{ kJ/kg}$$

Temperature at exit of each turbine

$$T_7 = T_9 = T_6 - \frac{w_{ta-1}}{C_{po}} = 1400 - \frac{205.909}{1.0035} = 1195.115$$

Control volume: regenerator

For an ideal regenerator, the exit air temperature is given as

$$T_{5'} = T_7 = T_9 = 1195.115 \text{ K}$$

The real exit air temperature is computed based on the efficiency of the regenerator. A real regenerator with efficiency is defined as

$$\eta_{reg} = 0.9 = \frac{h_5 - h_4}{h_{5'} - h_4} = \frac{T_5 - T_4}{T_{5'} - T_4} = \frac{T_5 - 364.72}{1195.115 - 364.72}$$

The air exit temperature out of regenerator is given as

$$T_5 = 1112.0755 \text{ K}$$

Control volume: combustion chamber

Total heat added in the four-stage gas turbine system is

$$q_H = q_{H-1} + (n-1)q_{H-2}$$

$$q_H = C_{po}\left(T_6 - T_5\right) + (n-1)C_{po}\left(T_8 - T_7\right)$$
$$= 1.005\,(1400 - 1112.0755) + (4-1)\,1.0035\,(1400 - 1195.115)$$

$$= 289.364 + 3\,(204.885)$$

$$q_H = 904.019\ \text{kJ/kg}$$

The thermal efficiency of the four-stage gas turbine power system is given as

$$\eta_{Th} = \frac{w_{ta} - w_{ca}}{q_H} = \frac{823.636 - 260.18}{904.019} = 62.327\%$$

2.15 Cooling and Refrigeration System

The ***Clausius statement*** led to the development of all cooling devices like refrigerator or heat pump. For an air-conditioning system or in a refrigerator, a vapor compression system is used to transfer heat from lower space temperature, T_L, to a higher ambient temperature, T_H, using a rotary compressor where work energy is the required input energy to drive the system.

2.15.1 Vapor Compression Refrigeration System

The ideal refrigeration cycle is like an ideal power cycle, i.e. Carnot cycle consisting of four reversible processes, but each process is in reverse of that in the power cycle. A vapor compression refrigeration system that operates on reverse Carnot cycle and within two-phase liquid–vapor region is shown in Figure 2.30a, and the operating states of the cycle are represented by the process lines 1-2'-3-4'-1 in Figure 2.30b. The four processes in the ideal vapor-compression refrigeration cycle or reverse Carnot cycle are: (i) ***reversible isothermal heat addition in the evaporator as low-temperature heat source***, (ii) ***reversible adiabatic compression in the compressor***, (iii). ***reversible isothermal heat rejection in the condenser*** as the high-temperature heat sink, and (iv) ***reversible adiabatic expansion through turbine***.

A basic ideal refrigeration cycle 1-2-3-4-1, as shown in Figure 2.31, differs from the reversed Carnot cycle in several ways:

1. Expansion of the saturated liquid vapor from the higher-pressure condenser exit state, 3 to lower evaporation pressure takes place through a throttling device–like expansion valve without recovering any work in contrast to the turbine in a power cycle. This expansion process is irreversible, causing working fluid to enter the evaporator pressure with increase in entropy and quality at state 4.

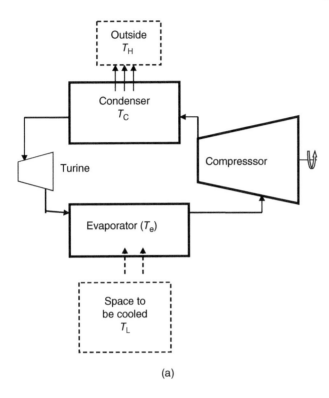

(a)

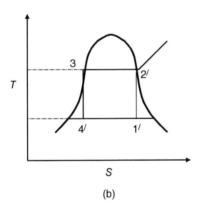

(b)

Figure 2.30 Vapor compression refrigeration system following reversed Carnot cycle: (a) schematic of a vapor compression refrigeration system and (b) processes and state points on a *T–S* diagram.

2. The two-phase working fluid undergoes isothermal and constant pressure evaporation process 4-1 until it reaches the saturated vapor state, 1. Saturated vapor at state 1 enters a compressor and undergoes the reversible adiabatic or isentropic compression process 1-2 to condenser pressure and exits as at a superheated vapor state 2.

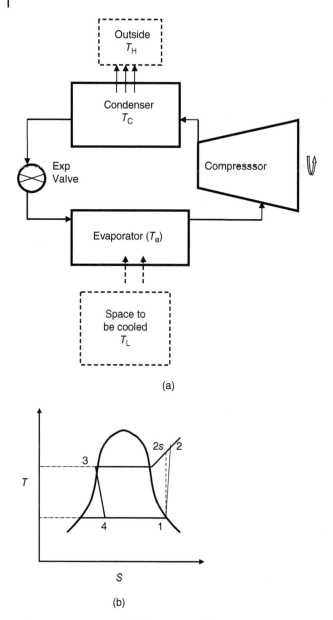

Figure 2.31 Standard vapor compression refrigeration system and cycle: (a) schematic of a standard vapor compression refrigeration system and (b) processes and statements on a $T–S$ diagram.

3. The working fluid at the superheated vapor state 3 enters the condenser and undergoes de-superheating followed by condensation to reject heat at a constant temperature, T_C, and constant condenser pressure, P_C.
4. The heat-transfer processes in the evaporator and in the condenser take place through a fine temperature difference.

A **vapor compression refrigeration cycle** is a standard cycle for cooling devices that receive heat from a low-temperature reservoir represented by the evaporator and rejects heat at high-temperature reservoir represented by the condenser.

A working fluid enters the evaporator at lower pressure, P_e, and the corresponding saturation temperature, which is also defined as the evaporation temperature, T_e, and is lower than the temperature of the low-temperature medium by a temperature band of ΔT_L. Heat is added, during which the working fluid undergoes two-phase boiling at a constant temperature, T_e, until it reaches saturated vapor state 1. At state 1, saturated vapor enters a compressor where the working fluid undergoes reversible adiabatic compression process 1-2s to the higher pressure, P_c, in the condenser. For a real compression process with irreversible loss, the process is represented by the lines 1-2. The vapor at higher pressure and temperature state 2 enters a condenser where it undergoes isothermal heat rejection process 2-3. The working fluid undergoes two-phase condensation at a constant temperature, T_c, which is higher than the high-temperature medium by a temperature band ΔT_H. At state 3, the working fluid at the saturated liquid enters an expansion valve and undergoes expansion process 3-4 to the lower pressure and mixed state 4.

The performance of a refrigeration system is expressed as the COP, which is defined as the ratio of the cooling effect, i.e. the amount heat transferred in the evaporator to cause cooling to the amount energy in the form of work used in the compressor and given as

$$\text{COP} = \frac{\text{Cooling effect}}{\text{Work input}} = \frac{Q_L}{W_c} = \frac{Q_L}{Q_H - Q_L}$$

$$\text{COP} = \frac{1}{\dfrac{Q_H}{Q_L} - 1} \tag{2.165a}$$

Now using Eq. (2.49), COP based on reverse Carnot cycle is given as

$$\text{COP}_{carnot} = \frac{1}{\dfrac{T_H}{T_L} - 1} = \frac{T_L - T_H}{T_L} \tag{2.165b}$$

A vapor compression refrigeration system is used to maintain low temperature in a refrigerator or space air-conditioning is used to transfer heat from a **space** maintained at lower-space temperature to the refrigerant as the working fluid and reject the heat at higher outside ambient temperature by drawing heat from the low temperature, $T_L (= T_e + \Delta T_L)$, and rejecting heat to the high temperature, $T_H (= T_c - \Delta T_H)$. Note that while the space was to be kept cool during the summer time, for example, using the refrigeration chiller system, the space is to be kept warm during the winter time using the same system in the heat pump mode.

A heat pump system is used to maintain a **space** higher than the ambient by drawing heat from the low temperature, $T_L (= T_e + \Delta T_L)$, and rejecting heat to the high temperature, $T_H (= T_c - \Delta T_H)$. Note that while the space was to be kept cool during the summer time, for example, using the refrigeration chiller system, the space is to be kept warm during the winter time using the same system in the heat pump mode.

The COP for a heat pump is system is defined as

$$\text{COP}_{HP} = \frac{\text{Heating Effect}}{\text{Work input}} = \frac{Q_H}{W_c} = \frac{Q_H}{Q_H - Q_L}$$

$$\text{COP}_{HP} = \frac{1}{1 - \dfrac{Q_L}{Q_H}} \tag{2.166a}$$

and using Eq. (2.49), COP based on reverse Carnot cycle is given as

$$\text{COP}_{\text{carnot}} = \frac{1}{1 - \frac{T_L}{T_H}} \tag{2.166b}$$

2.15.1.1 Thermodynamic Analysis of Vapor Compression Refrigeration Cycle

Let us consider a chiller that operates using a vapor compression system with a rated cooling capacity of P-kW, low-temperature space at T_L, and high-temperature ambient at T_H

State 1: Saturated vapor at the temperature and pressure in the evaporator

$$T_1 = T_e = T_L - \Delta T_L, P_1 = P_4 = P_e = P_{\text{sat}} \text{ at } T_e$$

From the saturation Table C.2a for R-134, we get following saturated data at $T_1 = T_e$

$$h_f, h_g, s_f, s_g, h_{fg}, s_{fg,} \text{ and } P_{\text{sat}}$$

Set $h_1 = h_g$, $s_1 = s_g$ and $P_1 = P_4 = P_e = P_{\text{sat}}$
State 3: Saturated liquid state at the condenser temperature $T_3 = T_c$ and pressure $P_3 = P_2 = P_c = P_{\text{sat}}$
From the saturation Table C.2a for R-134, we get following saturated data at the condenser temperature $T_1 = T_e$: h_f and P_{sat}
Set $h_3 = h_f$, and $P_2 = P_3 = P_c = P_{\text{sat}}$ at the condenser temperature.
Control volume: compressor
From second law of thermodynamics: $s_{2s} = s_1$
For $s_{2s} = s_1$ and $P_2 = P_c = P_{\text{sat}}$, we get following data for state 2s from the superheated Table C.2b:

$$h_{2s} \text{ and } T_{2s}$$

Process 1-2: Reversible adiabatic process
From first law of thermodynamics: Isentropic compressor work, $w_{cs} = (h_{2s} - h_1)$
For actual compressor work:

$$w_{cs} = \frac{w_{cs}}{\eta_c}$$

Total compressor work: $W_c = \dot{m}_r w_{cs}$
For actual exit state of the compressor $h_2 = h_1 + w_c$
Control volume: expansion valve
From first law of thermodynamics for the throttling process

$$h_4 = h_3$$

Control volume: evaporator
Application of first law of thermodynamics over a control around the refrigerant fluid of the evaporator leads to the expression for the heat-transfer rate across the evaporator

$$Q_e = \overset{*}{m}_r \left(h_1 - h_4\right)$$

The cooling effect of the refrigeration system is given

$$q_e = \left(h_1 - h_4\right)$$

Noting that the total heat-transfer rate is equal to the rated cooling capacity of Q_e^* associated with the vapor compression system, the mass flow rate of the refrigerant in the refrigeration system is given as

$$\overset{*}{m}_r = \frac{Q_e}{\left(h_1 - h_4\right)}$$

Control volume: condenser

From the first law of thermodynamics, heat transferred from refrigerant to ambient outside is given as

$$q_c = h_2 - h_3$$

and the total heat transfer is

$$Q_c = \dot{m}_r * q_c = q_c = h_2 - h_3$$

The amount of heat released due to the condensation of the steam is rejected to ambient outside either by the circulating cooling water in a water-cooled condenser (Figure 2.32a) in secondary water system or through air-cooled condenser (Figure 2.32b).

Example 2.14   *Vapor-Compression Refrigeration*

A vapor-compression chiller system is used to cool the air stream in a room air-conditioning system with total cooling load of 10-kW as shown in the figure below. The room air stream enters the evaporator at temperature of $T_{ro} = 26$ °Cand cooled to a temperature of $T_{ri} = 12$ °C. The condenser is air-cooled with design outdoor temperature of $T_a = 35$ °C and compressor is assumed to have an efficiency of $\eta_c = 85\%$. Perform first law analysis of the refrigeration system assuming R-134 as the refrigerant and determine: (a) mass flow rate of the refrigerant, (b) COP of the system, and summarize the ratings of all major components of the vapor compression refrigeration system.

Solution

In the first step, we need to select the operating refrigeration temperatures in evaporation and condensation. Considering that the air is to be cooled from 26 to 12 °C in the evaporator, we can assume a temperature band of $\Delta T_e = 5$ °C across the evaporator and that leads to an evaporation temperature of

$$T_e = T_L - \Delta T_L = 12 - 5 = 7 \ °C$$

Similarly, for the condenser, the condensing refrigerant temperature must be higher than the outside the ambient temperature of 30 °C. We can assume a temperature band of $\Delta T_H = 10$ °C and select the condensing refrigerant temperature as

$$T_c = T_a + \Delta T_c = 35 + 10 = 45 \ °C$$

$$T_c = 45 \ °C$$

The vapor-compression cycle with the selection of evaporation and condensation temperature are shown in the figure below.

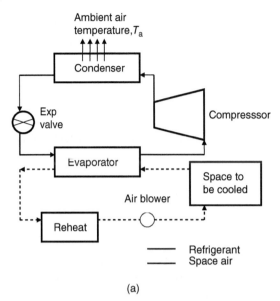

(a)

Figure 2.32 Vapor compression refrigeration system with air-cooled and water-cooled condensers for heat rejection: (a) vapor compression refrigeration with air-cooled condenser and (b) vapor compression refrigeration with water cooled condenser.

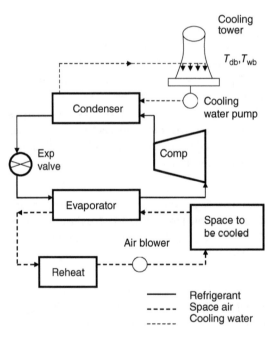

(b)

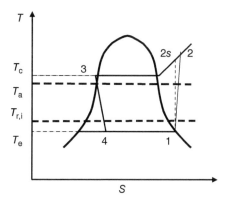

Refrigerant: R-134a

State 1: Saturated vapor at the temperature and pressure in the evaporator

$$T_1 = T_e = 7 \ ^\circ C, P_1 = P_4 = P_e = P_{sat} \text{ at } T_e$$

From the saturation Table C.2a for R-134, get following saturated data at $T_1 = T_e = 7 \ ^\circ C$, $h_f, h_g, s_f, s_g, h_{fg}, s_{fg},$ and P_{sat}

$$h_1 = h_g = 402.775 \text{ kJ/kg and } s_1 = s_g = 1.7209 \text{ kJ/kg K}$$

and

$$P_1 = P_4 = P_e = P_{sat}$$

State point 3: Saturated liquid at

State 3: Saturated liquid state at the condenser temperature $T_3 = T_c = 45\ ^\circ C$ and pressure $P_3 = P_2 = P_c = P_{sat}$
From the saturation Table C.2a for R-134, we get following saturated data at the condenser temperature, $T_3 = T_c = 45\ ^\circ C$

$$h_f = 264.11 \text{ kJ/kg and } p_{sat} = 1160.2 \text{ kPa}$$

Set $h_3 = h_f = 264.11$ kJ/kg and $p_2 = p_3 = p_c = p_{sat} = 1160.2$ kPa

Control volume: compressor

Process 1-2: Reversible adiabatic process
From second law of thermodynamics: $s_{2s} = s_1 = 1.7209$ kJ/kg K
For $s_{2s} = 1.7209$ kJ/kg K and $p_2 = p_c = 1160.2$ kPa, we get following data for state 2s from the superheated Table C.2b $T_{2s} = 40\ ^\circ C$ and $h_{2s} = 420.25$ kJ/kg
From first law of thermodynamics:
Isentropic compressor work,

$$w_{cs} = (h_{2s} - h_1) == 420.25 - 402.775 = 17.475 \text{kJ/kg}$$

For actual compressor work:

$$w_{ca} = \frac{w_{cs}}{\eta_c} = \frac{17.475}{0.85} = 20.5588 \text{ kJ/kg}$$

For actual exit state of the compressor:

$$h_2 = h_1 + w_{ca} = 402.775 + 20.5588 \ 423.33 \ kJ/kg$$

Total compressor work:

$$W_c = \dot{m}_r w_{ca}$$

where $\dot{m}_r$ = Mass flow rate of refrigerant.

Control volume: expansion valve

From the first law of thermodynamics for the throttling process

$$h_4 = h_3 = 264.11 \ kJ/kg$$

Control volume: condenser

According to the first law of thermodynamics, heat transferred from refrigerant to ambient outside is given as

$$q_c = h_2 - h_3 = 423.33 - 264.11 = 159.22 \ kJ/kg$$

and the total heat transfer is

$$Q_c = \dot{m}_f * q_c$$

Control volume: evaporator

Application of the first law of thermodynamics over a control around the refrigerant fluid of the evaporator leads to the expression for the heat-transfer rate per unit mass flow rate across the evaporator or the cooling effect of the refrigeration system is given

$$\text{Cooling effect, } q_e = h_1 - h_4 = 402.775 - 264.11 = 138.665 \ kJ/kg$$

For total heat transfer, the total heat gain by the refrigerant in the evaporator is equal to the total cooling load (P) or the heat lost by the air stream:

$$Q_e = \dot{m}_f * q_e = \dot{m}_a \left(h_{w,i} - h_{w,o} \right) = 10 \ kW$$

Noting that the total heat-transfer rate is equal to the rated cooling capacity of $Q_e = 10 \ kW$ of the vapor compression system, the mass flow rate of the refrigerant in system is given as

$$\overset{*}{m}_r = \frac{Q_e}{(h_1 - h_4)} = \frac{10}{138.665} = 0.0721 \ kg/s$$

Total heat real work in the compressor and total heat transfer in the condenser are now estimated as follows:

$$W_{ca} = \dot{m}_f * w_{ca} = 0.0721 * 20.5588 = 1.4822 \ kW$$

and

$$Q_c = \dot{m}_f * q_c = 0.0721 * 159.22 = 11.47 \ kW$$

Coefficient of performance:

$$COP = \frac{q_e}{w_{ca}} = \frac{h_1 - h_4}{w_{ca}} = \frac{138.665}{20.5588} = 6.744$$

Maximum possible COP based on the Carnot cycle is given as

$$\text{COP}_{\text{carnot}} = \frac{T_L}{T_H - T_L} = \frac{Q_L}{Q_L - Q_H} = \frac{7 + 273}{45 - 7} = 7.3684$$

2.15.1.2 The Absorption Refrigeration System

In contrast to the mechanical work energy input to the compressor in a vapor compression system, the **absorption refrigeration system** uses heat energy input to drive the cooling device as shown in the Figure 2.33:

Note that there is additional equipment such as absorber and generator in comparison to the vapor compression system. In this refrigeration system, ammonia is the refrigerant and water the absorbent. The ammonia refrigerant circulates through the condenser, expansion valve, and the evaporator like in a vapor compression refrigeration system. However, the mechanical compressor is replaced with a compression process driven by thermal heat energy. As we can see, the compression of ammonia vapor from low-pressure evaporator to high-pressure condenser is achieved through use and transfer of an ammonia solution using two components such as the **generator** and the **absorber**. The generator–absorber unit replaces the mechanical compressor used in the vapor compression system. Heat energy is added from the high-temperature heat source to the ammonia solution passing through the generator. Some heat is also rejected at the low-temperature heat sink from the absorber. Operating principles of the absorption refrigeration cycle are described as follows:

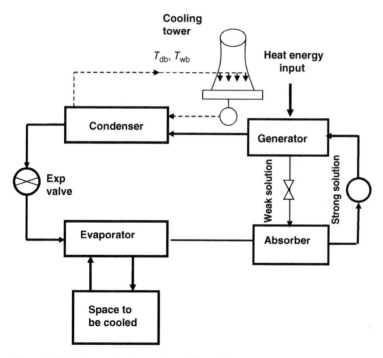

Figure 2.33 Ammonia absorption refrigeration system.

2.15.1.2.1 Low-Temperature Heat Addition Process

The low-pressure ammonia liquid-vapor mixture enters the evaporator where it receives heat from low-temperature space to be cooled and exits as vapor.

2.15.1.2.2 Compression Process

The low-pressure ammonia vapor then enters the **absorber** where it is absorbed in the **weak ammonia solution with water as the absorbent.** The absorption capacity of the ammonia-water solution is function of the temperature. The absorption of the refrigerant in the solution increases with the decrease in the solution temperature. The ammonia is absorbed in incoming weak solution by cooling the solution using a circulation medium.

The **low-pressure strong liquid ammonia solution** is then pumped through to the high-pressure **generator** where heat is added from a high-temperature source. Ammonia vapor is driven off from the solution as heat is added in the generator. In the generator, heat transfer from a high-temperature source increases the solution temperature and releases the ammonia vapor, leaving a strong ammonia-water solution in the generator and flows into the absorber through a valve.

2.15.1.2.3 High-Temperature Heat Rejection Process in Condenser

The high-pressure ammonia vapor from the generator then undergoes phase-change condensation in the condensation and rejects heat to the surrounding ambient temperature.

2.15.1.2.4 Expansion Process in the Expansion Valve

The high-pressure liquid ammonia undergoes expansion process through the expansion valve and enters the evaporator in a mixed state.

Net mechanical work input is considerably less as compression work input is only necessary in a pump with low-specific volume liquid as compared to compression of high-specific volume vapor in a compressor. The simple ammonia refrigeration system described here is normally modified to improve the performance by including additional components such as a heat exchanger to preheat the weak solution from the absorber before entering the generator, and a rectifier to remove any trace of water from ammonia refrigerant vapor before entering the condenser. Absorption system becomes economically more justifiable where a suitable **heat energy source is** available such as the renewable energy source like **solar energy, geothermal energy, or industrial waste heat**.

Another popular absorption refrigeration system is **lithium-bromide absorption system** that uses lithium bromide as the absorbent and water as the refrigerant. In a solar-powered refrigeration absorption system, heat energy input at the generator comes from circulating fluid circulating through the solar collector loop.

2.16 The Second Law or Exergy Analysis

The second law analysis, also known as the **exergy analysis**, is increasingly being used for designing energy systems. Let us consider here a SSSF process for a control volume involving inlet and outlet flows, heat transfer from a heat temperature reservoir and work as shown in Figure 2.34.

Figure 2.34 Control volume for the conservation analysis of mass, energy, and entropy.

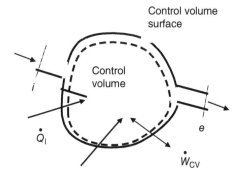

The first and second laws for a real process with negligible kinetic and potential energies are given as

First law:

$$\sum \dot{Q}_i = \sum \dot{m} \, h_e + \sum \dot{m} \, h_i + \dot{W} \tag{2.167}$$

Second law:

$$\dot{m} \, (s_e - s_i) = \sum_{i=1} \frac{\dot{Q}_i}{T_i} + \dot{S}_{gen} \tag{2.168a}$$

$$\left[\sum \dot{m} \, s_e - \sum \dot{m} \, s_i \right] - \sum \frac{\dot{Q}_i}{T_i} = \dot{S}_{gen} > 0 \tag{2.168b}$$

To quantify the amount of entropy generation or the amount of irreversibility, let us consider the same process as reversible operating under the same operating conditions. It is assumed that the control volume involves additional reversible heat-transfer process from the environment at a temperature, T. The second law for this reversible process is given as

$$\left[\sum \dot{m} \, s_e - \sum \dot{m} \, s_i \right] - \sum \frac{\dot{Q}_i}{T_i} - \frac{\dot{Q}_{rev,o}}{T_o} = 0 \tag{2.169}$$

Solving for the reversible heat transfer

$$\dot{Q}_{rev,o} = T_o \left[\sum \dot{m} \, s_e - \sum \dot{m} \, s_i \right] - T_o \sum \frac{\dot{Q}_i}{T_i} \tag{2.170}$$

The first law of thermodynamic for the reversible process is given as

$$\sum \dot{Q}_i + \dot{Q}_{rev,o} = \sum \dot{m} \, h_e - \sum \dot{m} \, h_i + \dot{W}_{rev} \tag{2.171}$$

Substituting for Eq. (2.170) for the reversible heat transfer and rearranging for the reversible work

$$\dot{W}_{rev} = \sum \dot{Q}_i + T_o \left[\sum \dot{m} \, s_e - \sum \dot{m} \, s_i \right] - T_o \sum \frac{\dot{Q}}{T} - \left[\sum \dot{m} \, h_e - \sum \dot{m} \, h_i \right] \tag{2.172}$$

Reversible work for a SSSF for a control volume is derived as

$$\dot{W}_{rev} = T_o \left[\sum \dot{m} \, s_e - \sum \dot{m} \, s_i \right] - \sum \dot{m} \, h_e + \sum \dot{m} \, h_i + \sum \dot{Q}_i \left(1 - \frac{T_o}{T_i} \right) \tag{2.173}$$

For a control volume with single flow in and single flow out, and heat transfer from a single high-temperature reservoir, T_H, reversible work written from Eq. (2.173) as

$$\dot{W}_{rev} = T_o \dot{m} \left(s_e - s_i \right) - \dot{m} \left(h_e - h_i \right) + \dot{Q}_H \left(1 - \frac{T_o}{T_H} \right) \tag{2.174a}$$

Dividing both sides by the mass flow rate, we can express the reversible work per unit mass flow as

$$w_{rev} = T_o \left(s_e - s_i \right) - \left(h_e - h_i \right) + q_H \left(1 - \frac{T_o}{T_H} \right) \tag{2.174b}$$

It can be noted here that the third term of the expression on the right represents the contribution of heat transfer to the reversible work.

2.16.1 Irreversibility

Irreversibility is defined as the difference between the reversible work and the actual work.

$$I = W_{rev} - W \tag{2.175}$$

where actual work for SSSF process for a control volume is given as

$$W = \dot{m} \left(h_i - h_e \right) - Q_H \tag{2.176}$$

Substituting Eq. (2.174a) for the reversible work and Eq. (2.176) for actual work into Eq. (2.175), we can have another expression for the irreversibility as

$$I = T_o \dot{m} \left(s_e - s_i \right) - Q_H \frac{T_o}{T_H} \tag{2.177a}$$

and for multiple flow in and out of the control volume as

$$I = T_o \sum \dot{m}_e s_e - T_o \sum \dot{m}_i s_i - Q_H \frac{T_o}{T_H} \tag{2.177b}$$

Dividing both sides of the equation by the mass flow rate, we can express the irreversibility per unit mass flow as

$$i = T_o \left(s_e - s_i \right) - \frac{T_o}{T_H} \tag{2.178a}$$

or

$$I = T_o s_{gen} \tag{2.178b}$$

where $s_{gen} = \left(s_e - s_i \right) - \frac{q}{T_H}$ is the **net entropy generation** or **production** per unit mass flow rate caused by irreversible losses and by the heat-transfer process.

2.16.2 Availability or Exergy

Availability, also termed as **exergy**, is defined as the potential of any state of a mass to produce maximum possible work. The maximum possible work is achieved when the mass undergoes a complete reversible process and reaches a complete equilibrium state with the surrounding or a dead state. The complete equilibrium state includes thermal, mechanical, and chemical equilibrium so that there is no more potential left for any heat transfer, work, or chemical reactions. So, availability of a mass at any given initial state is defined as the associated potential of that state to perform maximum possible work. Let us summarize here the general expression for the availability for a system or fixed mass and a control volume.

System:

Availability or *exergy of a system* is the *maximum reversible work* that is achievable as the system reaches equilibrium with the surrounding or the dead state.

$$\phi = W_{rev,max} \tag{2.179a}$$

Considering an energy balance over an overall system that includes both the system and surrounding, it can be shown that the maximum reversible work of a system or fixed mass is

$$W_{rev,max} = \left(U + P_o V - T_o S + \frac{1}{2} m V^2 + mgZ \right) - \left(U_o + P_o V_o - T_o S_o + mgZ_o \right) \tag{2.179b}$$

The availability or exergy for a system is written per unit mass as

$$\phi = \left(u + P_o v - T_o s + \frac{V^2}{2} + gZ \right) - \left(u_o + P_o v_o - T_o s_o + gZ_o \right) \tag{2.179c}$$

Control volume:

From Eq. (2.173), the reversible work for a SSSF process along with no additional heat transfer can be given as

$$\dot{W}_{rev} = T_o \left[\sum \dot{m}_e s_e - \sum \dot{m}_i s_i \right] - \sum \dot{m}_e \left(h_e + \frac{V_e^2}{2} + gZ_e \right) + \sum \dot{m}_i \left(h_i + \frac{V_i^2}{2} + gZ_i \right) \tag{2.180}$$

For a single mass flow in and out through a control volume for a SSSF process, the reversible work will be maximum when the mass flow exits the control volume in equilibrium with the surrounding state, i.e. $P_e = P_o$, $T_e = T_o$, $h_e = h_o$, $s_e = s_o$, $V_e = 0$ and $Z_e = Z_o$.

The *availability or exergy for a control volume*, which is the *maximum possible reversible work per unit mass flow rate without any additional heat transfer*, is derived from Eq. (2.180) as

$$\psi = W_{rev,max}$$

or

$$\psi = \left(h - T_o s + \frac{V_i^2}{2} + gZ \right) - \left(h_o - T_o s_o + gZ_o \right) \tag{2.181}$$

In a simple term, *availability* or exergy represents natural resources that need to be used with minimum increase in irreversibility. For a reversible process with no irreversible losses, the decrease in availability is the reversible work. In reality, the real work is less than the reversible or ideal work due to irreversible losses, and hence a decrease in availability.

2.16.3 Second Law Efficiency

The second law efficiency of a working producing machine like turbine is defined as the ratio of actual work with the decrease of availability from the same inlet state to the same exit state, and is expressed as

$$\eta_{\text{second law}} = \frac{w_{\text{ta}}}{\psi_i - \psi_e} \tag{2.182}$$

For devices involving input energy in the form work, the second law efficiency is defined as the ratio of increase of availability to the work input. For compressors and pumps, this is defined as follows:

For a compressor:

$$\eta_{\text{second law,c}} = \frac{\psi_i - \psi_e}{w_{\text{ca}}} \tag{2.183}$$

For a pump:

$$\eta_{\text{second law,d}} = \frac{\psi_i - \psi_e}{w_{\text{pa}}} \tag{2.184}$$

In a heat exchanger as shown in the figure, the heat energy is transferred from a hot fluid stream to a cold fluid stream. The second law efficiency of the heat exchanger is defined as the ratio of the availability increase of the cold fluid stream to that of the availability decrease of the hot fluid stream and is written as

$$\eta_{\text{second law, HE}} = \frac{\text{Increase in availability of cold stream}}{\text{Decrease in availability of the hot stream}} = \frac{\dot{m}_C (\psi_e - \psi_i)_{\text{cold stream}}}{\dot{m}_H (\psi_e - \psi_i)_{\text{hot stream}}} \tag{2.185}$$

Example 2.15 *Second Law Analysis*

Let us consider a boiler/heat exchanger where water is heated to produce steam by recovering heat from the hot exhaust stream of a diesel engine and producing power using a Rankine Power cycle. Engine exhaust is cooled from 950 to 600 °C and boiler feedwater is heated from a subcooled state with temperature of 160 °C and pressure of 5 MPa to a temperature of 450 °C, assuming a constant pressure heat addition process. Assume the specific heat of the engine exhaust gas as $C_p = 1.08$ kJ/kg K. Determine: (a) the irreversibility of the process, (b) the change in availability of water and gas streams, and (c) the second law efficiency of the process.

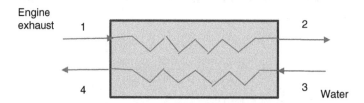

Solution

Determine water properties from thermodynamics table for the inlet and outlet states:

State 3: Subcooled or compressed water state at $T_3 = 150 °C$, $P_3 = 5$ MPa
From thermodynamic table for compressed liquid water:

$$h_3 = 678.10 \text{ kJ/kg and } s_3 = 1.9374 \text{ kJ/kgK}$$

State 4: Superheated steam at $T_4 = 450\,°C$, $P_3 = 5$ MPa
From thermodynamics table for superheated vapor:

$$h_4 = 3316.15 \text{ kJ/kg and } s_4 = 6.8185 \text{ kJ/kg K}$$

The control volume for this process is shown as the dotted control surface enclosing the heat exchangers with flow inlets and outlets for the water and exhaust gas streams. The conservations equations for the control volume are shown as

Conservation of mass:
Water: $\dot{m}_i = \dot{m}_e = \dot{m}_{H_2O}$
Exhaust gas: $\dot{m}_i = \dot{m}_e = \dot{m}_{gas}$

First law of thermodynamics:
Heat lost by the exhaust gas stream: $\dot{Q} = \dot{m}_{gas}\left(h_1 - h_2\right)$
Heat gained by the water stream: $\dot{Q} = \dot{m}_{gas}\left(h_1 - h_2\right)$

Combining above two equations, we can write the alternate first law equation for the process as

$$\dot{m}_{gas}\left(h_1 - h_2\right) = \dot{m}_{H_2O}\left(h_4 - h_3\right)$$

Rearranging this equation, the ratio of mass flow rates is given

$$\frac{\dot{m}_{gas}}{\dot{m}_{H_2O}} = \frac{\left(h_4 - h_3\right)}{\left(h_1 - h_2\right)} = \frac{\left(h_4 - h_3\right)}{C_{po}\left(T_1 - T_2\right)} \qquad \text{(E2.15.1)}$$

Substituting the enthalpy data for water, we get

$$\frac{\dot{m}_{gas}}{\dot{m}_{H_2O}} = \frac{(3316.15 - 678.10)}{1.08\,(950 - 600)} = \frac{2638.05}{378}$$

or

$$\frac{\dot{m}_{gas}}{\dot{m}_{H_2O}} = 6.979$$

Second law of thermodynamics:

$$\dot{m}_{gas}\left(s_2 - s_1\right) + \dot{m}_{H_2O}\left(s_4 - s_3\right) \geq 0$$

(a) The irreversibility of this process can be computed based on Eq. (2.177b) with multiple flows in and out of the control volume as

$$I = T_o \sum \dot{m}_e s_e - T_o \sum \dot{m}_i s_i - Q_H \frac{T_o}{T_H}$$

Expanding the expression for the water and gas flow streams, we get

$$I = T_o \dot{m}_{gas}\left(s_2 - s_1\right)_{gas} + T_o \dot{m}_{H_2O}\left(s_4 - s_3\right)_{H_2O} - Q_{CV} \frac{T_o}{T_H}$$

Assuming negligible heat loss across the control volume, we get

$$I = T_o \dot{m}_{gas}\left(s_2 - s_1\right)_{gas} + T_o \dot{m}_{H_2O}\left(s_4 - s_3\right)_{H_2O} \qquad \text{(E2.15.2)}$$

To determine the irreversibility per unit mass flow of water, divide both sides of the equation by $\dot{m}_{H_2O}$

$$\frac{I}{\dot{m}_{H_2O}} = T_o \frac{\dot{m}_{gas}}{\dot{m}_{H_2O}}\left(s_2 - s_1\right)_{gas} + T_o\left(s_4 - s_3\right)_{H_2O} \qquad \text{(E2.15.3)}$$

The change in entropy of the exhaust gas stream can be computed using Eq. (2.66b), which is derived based on ideal gas assumption and constant specific heat as follows:

$$s_2 - s_1 = \int_1^2 C_{po} \frac{dT}{T} - R \ln \frac{P_2}{P_1}$$

For the negligible pressure drop in the heat-transfer process, the equation simplifies

$$\left(s_2 - s_1\right)_{gas} = C_{po} \ln \frac{T_2}{T_1} = 1.08 \ln \frac{(600 + 273.2)}{(950 + 273.2)}$$

$$\left(s_2 - s_1\right)_{gas} = 1.08 \ln \frac{(873.2)}{(1223.2)}$$

$$\left(s_2 - s_1\right)_{gas} = -0.3640$$

The change in entropy of the water stream can be computed as

$$\left(s_4 - s_3\right)_{H_2O} = (6.8185 - 1.9374)$$

$$\left(s_4 - s_3\right)_{H_2O} = 4.8811 \text{ kJ/kg K}$$

The irreversibility per unit mass flow of water is computed from Eq. (E2.15.3)

$$\frac{I}{\dot{m}_{H_2O}} = T_0 \frac{\dot{m}_{gas}}{\dot{m}_{H_2O}} \left(s_2 - s_1\right)_{gas} + T_0 \left(s_4 - s_3\right)_{H_2O}$$

$$\frac{I}{\dot{m}_{H_2O}} = 298.2 \times 6.979 \ (-0.4225) + 298.2 \times (4.8811)$$

$$\frac{I}{\dot{m}_{H_2O}} = -879.28 + 1{,}455.55$$

Irreversibility per unit mass flow of water: $\frac{I}{\dot{m}_{H_2O}} = 576.27 \text{ kJ/kg H}_2\text{O}$

(b) Determine availability of the exhaust gas and water streams using Eq. (2.179a)
Increase in availability of water:

$$\psi_4 - \psi_3 = \left(h_4 - h_3\right) - T_0 \left(s_4 - s_3\right)$$

$$\psi_4 - \psi_3 = (3316.15 - 678.10) - 273.2 \ (6.8185 - 1.9374)$$

$$\psi_4 - \psi_3 = 2638.05 - 1333.52 = 1304.53 \text{ kJ}$$

Increase in availability of water stream: $\psi_4 - \psi_3 = 1304.53 \text{ kJ/kg H}_2\text{O}$
Decrease in availability of exhaust gas stream

$$\left(\psi_1 - \psi_2\right) = \left(h_1 - h_2\right) - T_0 \left(s_1 - s_2\right) = C_{po} \left(T_1 - T_2\right) - T_0 \left(C_{po} \ln \frac{T_1}{T_2}\right)$$

$$\left(\psi_1 - \psi_2\right) = 1.08 \times (950 - 600) - 273.2 \times \left(1.08 \ln \frac{(950 + 273.2)}{(600 + 273.2)}\right)$$

$$\left(\psi_1 - \psi_2\right) = 378 - 273.2 \times (0.3640) = 378 - 99.45$$

$$\left(\psi_1 - \psi_2\right) = 278.55 \text{ kJ/kg gas}$$

Decrease in availability of exhaust stream per unit mass flow of water:

$$\frac{\dot{m}_{gas}}{\dot{m}_{H_2O}} (\psi_1 - \psi_2) = 6.979 \times 278.55 = 1944 \text{ kJ/kg H}_2\text{O}$$

(c) The second law of efficiency: From Eq. (2.185)

$$\eta_{second\ law,HE} = \frac{\text{Increase in availability of water stream}}{\text{Decrease in availability of gas stream}}$$

$$\eta_{second\ law,HE} = \frac{\dot{m}_{H_2O} (\psi_4 - \psi_3)}{\dot{m}_{gas} (\psi_1 - \psi_2)} = \frac{(\psi_4 - \psi_3)}{\frac{\dot{m}_{gas}}{\dot{m}_{H_2O}} (\psi_1 - \psi_2)}$$

$$\eta_{second\ law,HE} = \frac{1304.53}{1944} = 0.671$$

$$\eta_{second\ law,HE} = 67.1\%$$

2.17 Case Study Problems

2.17.1 Case Study Problem: Analysis and Design of Solar-Driven Irrigation Pump

A solar-energy-driven irrigation pump operating on a solar driven Rankine vapor cycle is to be analyzed and designed. The solar energy striking the solar collector during a daylight hour $0 \le t \le 10$ is given in the table below.

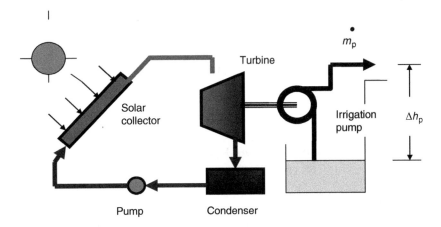

A **standard Rankine cycle power system** is to be analyzed and designed assuming the working fluid at the collector exit as **saturated vapor** at T_H and with a condenser back pressure of P_c. The irrigation pump must lift water to a height of $\Delta h_p = 50$ m and overcome a piping system friction drop of $\Delta h_f = 10$ m. Assume turbine efficiency of 90% and pump efficiencies as 87%. (a) Perform thermodynamic analysis for **two options:** with and without an auxiliary heater. (b) Determine the ratings and initial specifications of all major components, including the collector, the auxiliary heater, the turbine, the condenser, the irrigation pump, and the condenser water pump.

Known data:

Available solar radiation:

The variation of incident solar radiation striking the solar collector during typical daylight hours ($0 < t < 10$) is given in the table below:

Time	Radiation on tilted collector surface, G_t (W/m^2)
7–8	50
8–9	377
9–10	486
10–11	746
11–12	1050
12–13	1075
13–14	948
14–15	720
15–16	352
16–17	51

Solar collector: flat plate

Transmittance–absorptance product: $\tau\alpha = 0.87$

Collector removal factor: $F_R = 0.85$

Overall loss coefficient: $U_L = 10.0\,\text{W/m}^2\,{}^\circ\text{C}$

Collector efficiency: $\eta_c = F_R\left(\tau\alpha - \frac{U_L}{G_t}\left(T_{in} - T_a\right)\right)$

where T_{in} = fluid temperature at inlet to the collector, T_a = ambient air temperature, and G_t = incident solar radiation.

Analysis and design steps:

Find the collector area, A_c, and working fluid flow rate, $\dot{m}_r$, (can be varied) in the Rankine power cycle to provide $\dot{m}_p = 500\,\text{kg/min}$ of water pumped *at midnoon (maximum capacity at t = 6 as design condition)*. Perform the thermodynamic analysis for the base case of $T_H = 250\,^\circ\text{C}$ and $P_c = 20\,\text{kPa}$.

Solution

Perform thermodynamics analysis of the Rankine cycle and consider the base operating case: *T*

State 1: At $T_1 = T_{sat} = 150\,^\circ\text{C}$ from temperature-based saturation steam Table C.1a:

$$P_1 = 476.16\,\text{kPa}, h_1 = h_{g1} = 2745.9\,\text{kJ/kg}, s_1 = s_{g1} = 6.8371\,\text{kJ/kg K}$$

State 3: Saturated liquid state

At $P_2 = P_3 = P_c\,(\text{kPa}) = 20\,\text{kPa}$

At 20 kPa from pressure-based saturation steam Table C.1b

$$T_3 = T_2 = T_{sat} = 60.06 \ °C$$

$$s_f = 0.8319 \ kJ/kg \ K, s_g = 7.9085 \ kJ \ kg \ K, s_{fg} = 7.0765 \ kJ \ kg \ K,$$

$$h_f = 251.38 \ kJ/kg, h_g = 2609.68 \ kJ/kg, h_{fg} = 2358.3 \ kJ/kg \ and$$

$$v_3 = v_f = 0.001017 \ m^3/kg$$

State 2: $P_2 = P_3 = P_c \ (kPa) = 20 \ kPa$ and mixed state
Control volume: turbine
  Process 1-2: Reversible adiabatic process
  From first law of thermodynamics: $w_{ts} = (h_1 - h_{2s})$
  From second law of thermodynamics: $s_{2s} = s_1 = 6.837 \ kJ/kg \ K$
  Since $s_{2s} > s_g$, the state 2 is mixed state. We can estimate the quality of the isentropic
    turbine exit state is given as

$$x_{2s} = \frac{s_{2s} - s_f}{s_{fg}} = \frac{6.8371 - 0.8319}{7.0765} = 0.8486$$

$$h_{2s} = h_f + x_{2s}h_{fg} = 251.38 + 0.8486 \times 2358.3 = 2252.63 \ kJ/kg$$

$$h_{2s} = 2252.63 \ kJ/kg$$

Calculate turbine work:
  Isentropic turbine work:

$$w_{ts} = (h_1 - h_{2s}) = 2745.9 - 2252.63$$
$$w_{ts} = 493.27 \ kJ/kg$$

Actual turbine work:

$$w_t = \eta_T w_{ts} = 0.9 \times 493.27$$

$$w_t = 443.94 \ kJ/kg$$

Enthalpy of steam at turbine exit is given as

$$h_2 = h_1 - w_t = 2745.9 - 443.94$$

$$h_2 = 2302.5 \ kJ/kg$$

Control Volume: Condenser
From first law of thermodynamics
  For steam side

$$q_c = (h_3 - h_2) = 251.38 - 2302.5$$

$$q_c = -2051.12 \ kJ/kg$$

and total heat rejected in the condenser

$$Q_c = \dot{m}_s q_c = 0.01824 \ (-2051.12)$$

$$Q_c = -37.4124 \ kW$$

Note that this is the amount heat that is transferred to the cooling water side. Application of first law of thermodynamics leads to

$$Q_c = \dot{m}_w \left(h_{w,o} - h_{w,i} \right) \text{ Or } Q_c = \dot{m}_w C_{pw} \left(T_{w,o} - T_{w,i} \right)$$

where $\dot{m}_w$ is the mass flow rate of cooling water and can be calculated as:

$$\dot{m}_w = Q_c / \left[c_{pw} \left(T_{w,o} - T_{w,i} \right) \right]$$

The required cooling water flow rate through the condenser depends not only on the amount of heat rejected by the condensing steam but also on the cooling tower performance and outdoor weather conditions.

For the given cooling tower parameters of range, $R = 10\,°C$ and approach, $A = 4\,°C$, we can calculate the cooling water inlet and outlet temperatures as follows:

Control volume: pump
$v_3 = v_f$ at $p_3 =$ condenser pressure
Calculate pump work:
  Isentropic pump work:

$$w_{ps} = v_3 \left(P_4 - P_3 \right)$$

$$w_{ps} = 0.001017 \,(1000 - 20)$$

$$w_{ps} = 0.9966 \text{ kJ/kg}$$

$$v_3 = v_f = 0.001017 \,\mathrm{m^3/kg}, \; h_3 = h_f = 251.38 \,\mathrm{kJ/kg}$$

Calculate actual pump work:

$$w_p = \frac{w_{ps}}{\eta_p} = \frac{0.9966}{0.85}$$

$$w_p = 1.1725 \text{ kJ/kg}$$

State of water at the exit of the pump or at inlet to boiler:

$$h_4 = h_3 + w_p$$

$$h_4 = 251.38 + 1.1725$$

$$h_4 = 252.552 \text{ kJ/kg}$$

Temperature at inlet to the collector is estimated from the following

$$w_p = \left(h_4 - h_3 \right) = C_{pw} \left(T_4 - T_3 \right)$$

$$T_4 = T_3 + \frac{w_p}{C_{pw}}$$

$$T_4 = 60.06 + \frac{1.1725 \text{ kJ/kg}}{1.187 \text{ kJ/kg K}}$$

$$T_4 = 60.06 + \frac{1.1725 \text{ kJ/kg}}{1.187 \text{ kJ/kg K}}$$

$$T_4 = 61.05°C$$

Total pump work:

$$W_p = \dot{m}_s w_p$$

$$W_p = 0.01824 \times 1.1725$$

Control volume: solar collector:

Heat input at the solar collector:

From first law of thermodynamics:

$$Q_{Col} = \dot{m}_s \left(h_1 - h_4 \right)$$

Estimate steam flow rate:

Since the irrigation pump is required to lift water to a height of $\Delta h_p = 50$ m and overcome a piping system friction drop of $\Delta h_f = 10$ m, the total pumping head of the irrigation pump is given

$$H_p = \Delta h_p + \Delta h_f = 60 \text{ m}$$

The required pumping flowing rate is

$$\dot{m}_p = 500\text{kg/ min} = 500/60 = \text{kg/s}$$

The power input to the irrigation pump is estimated as

$$W_{IP} = \dot{m}_p g \, H_p / \eta_p = \frac{500 \times 9.8 \times 60}{60 \times 0.87} = 5.632 \text{ kW}$$

Equating the turbine power output to the required power input to the irrigation pump, we get

$$\dot{m}_s \left(w_t \right) = W_{IP}$$

Compute steam mass flow rate, $\dot{m}_s = \frac{W_{IP}}{w_t}$

$$\dot{m}_s = \frac{5.632 \text{ kJ/s}}{548.055 \text{ kJ/kg}}$$

$$\dot{m}_s = 0.01824 \text{ kg/s}$$

Control volume: solar collector

Solar collector efficiency:

$$\eta_c = F_R \left(\tau\alpha - \frac{U_L}{G_t} \left(T_{in} - T_a \right) \right)$$

and

$$\eta_c = 0.85 \left(0.87 - \frac{10}{2100} (61.04 - 20) \right)$$

$$\eta_c = 0.573 = 57.3\%$$

Estimate the required solar collector area from the energy balance equation as

$$G_{sol} \times A_c \times \eta_{col} = \dot{m}_s \left(h_1 - h_4 \right)$$

$$A_c = \frac{\dot{m}_s \left(h_1 - h_4 \right)}{G_{sol} \times \eta_{col}} = \frac{0.01824 \left(2745.9 - 252.552 \right)}{1.050 \, \text{kW/m}^2 \times 0.573}$$

$$A_c = 75.58 \, \text{m}^2$$

Bibliography

Adrian, B. (1982). *Entropy Generation through Heat and Fluid Flow*. Wiley.

Cengel, Y.A. and Boles, M.A. (2015). *Thermodynamics*, 8e. McGraw Hill.

Howell, J.R. and Buckius, R. (1987). *Fundamentals of Engineering Thermodynamics*. New York, Ny: Mc-Graw-Hill.

Moran, M. (1982). *Availability Analysis, a Guide to Efficient Energy Use*. Prentice-Hall.

Moran, M.J., Shapiro, H.N., Boettner, D.D., and Bailey, M. (2011). *Fundamentals of Engineering Thermodynamics*, 7e. Wiley.

Revankar, S. and Majumdar, P. (2014). *Fuel Cells: Principles, Design and Analysis*. CRC Press/Taylor & Francis Group.

Sonntag, R.E., Borgnake, C., and van Wylen, G.J. (2003). *Fundamentals of Thermodynamics*, 6e. Wiley.

Problems

2.1 A gas turbine burns octane (C_8H_{18}) completely with 400% of theoretical air. Determine (a) the balanced reaction equation and (b) the air-fuel ratio, both on a molar basis and on a mass basis.

2.2 Propane (C_3H_8) at 298 K, 1 atm, enters a combustion chamber operating at steady state with molar flow rate of 0.8 kmols s^{-1} and burns completely with 200% of theoretical air entering at 298 K, 1 atm. Kinetic and potential energy effects are negligible. If the combustion products exit at 800 K, 1 atm, determine the rate of heat transfer from the combustion chamber, in kW.

2.3 Methane gas (CH_4) at 25 C, 1 atm, enters an insulated reactor operating at steady state and burns completely 150% theoretical air entering at 25 C, 1 atm. Determine the adiabatic flame temperature.

2.4 Consider a standard refrigeration cycle (Vapor compression) chiller driven by a solar-powered standard Rankin cycle. The cooling capacity of vapor-compression chiller is of 5 kW. The chiller is used to cool an air stream. The combined cycle is shown below:

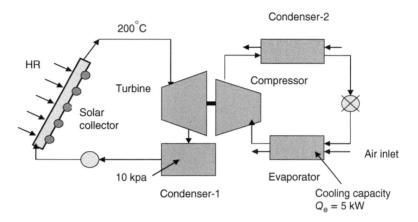

The temperature at the exit of the collector and inlet to turbine is maintained at
200 °C. The pressure at turbine exit and the condenser of the Rankine cycle is at
10 kPa. The evaporator temperature and the condenser temperature are 5 °C and
30 °C for the refrigeration cycle, respectively. Assume working fluid for the Rankine cycle as water and working fluid for the refrigeration cycle as R-134. Assume
efficiencies of the turbine and the compressor as 85% and that of the pump as 80%.
Perform a first law analysis for the performance of the system, and determine:
(i) All necessary properties at all state points. Show the processes for both the cycles
on T-S diagrams, (ii) Heat input at the collector; work output at the turbine; work
input at the compressor; heat rejected at the condensers, (iii) Flow rates in refrigeration cycle, m_r, and in Rankine cycle, m_s; (iv) COP of the combined cycle; and v. Give
a summary of the ratings for the major components: collector, turbine, compressor,
evaporator, condenser, and pump.

2.5 A combined gas turbine and-vapor power generation plant is considered to produce
a net power output of 100 MW. Air enters the compressor at 30 °C and 0.1 MPa and
compressed to pressure of 1.2 MPa. The state of the combustion product at the exit
of the combustion chamber and at inlet to the gas turbine is 1100 °C and 1.2 MPa.
The combustion product mixture expands through the gas turbine to a pressure of
0.1 MPa. Consider isentropic efficiencies of the compressor and the gas turbine as
85% and 90%, respectively. The exhaust combustion gas from the gas turbine then
passes through a heat recovery steam generator of Rankine power generation plant
and discharged to outside at a temperature of 120 °C. Steam exits the steam generator
at 400 °C and 6 MPa for expansion in a steam turbine followed by condensation in
a condenser at a pressure of 10 kPa. Consider isentropic efficiencies of the steam
turbine and pump in the steam power plant as 80% and 90%. Determine: (i) the mass
flow rate of the air and the steam; (ii) the net power output from the gas turbine and
vapor power cycles; (iii) the thermal efficiency of the combined power plant; and
Net exergy analysis of the air/combustion gas in the gas turbine cycle.

2.6 Consider a solar-powered standard Rankine cycle to produce 7 kW of net power for a residential building. The power cycle is shown as below:

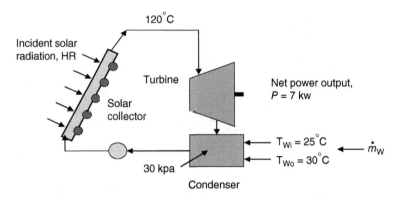

The temperature at exit of the collector and inlet to turbine is maintained at 120 °C. The condenser is water-cooled and maintained at 20 kPa. The cooling water inlet and outlet temperatures are 25 °C and 30 °C, respectively. Assume working fluid for the Rankine cycle as water. Assume efficiencies of the turbine and the pump as 85% and 80%, respectively. Perform a first law analysis for the performance of the system and determine: (i) flow rate of the working fluid in the Rankine cycle, $\dot{m}_s$; (ii) required area of the solar collector assuming the efficiency of the collector as $\eta_{col} = 35\%$ and the peak solar radiation in a typical day as HR = 2000 W/m^2; (iii) determine cooling water flow rate in the condenser; and (iv) give a summary of the sizes and ratings of the major components of the solar Rankine power cycle.

2.7 In an air-standard gas turbine cycle, the air enters the compressor at 0.15 MPa at 20 C and leaving at a pressure 1.0 MPa. The maximum temperature is 1200 C. Assume a compressor efficiency of 80%, a turbine efficiency of 85%, and an ideal regenerator. Determine the compressor work, turbine, heat addition in combustion chamber, and cycle efficiency. If net power output is 25 MW, what are the ratings of the components?

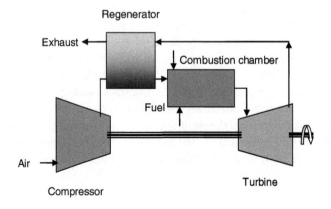

2.8 The gas turbine cycle shown in the figure below is considered as an automotive engine. In the first-stage turbine, the gas expands to pressure P_5, just low enough for the turbine to drive the compressor. The gas is then expanded through the second turbine to produce a net power of $\dot{W}_{net} = 150kW$. The data for the engine are given below and shown in the figure. Assume 90% efficiency for the regenerator. For simplicity, assume 100% efficiency for compressor and turbine, and assume ideal gas air with constant specific heat (can assume no variation in temperature). Determine: (i) Intermediate temperature and pressure states at 2, 3, 5, and 6, (ii) mass flow rate through the engine, (iii) thermal efficiency of the engine.
Use following known data set:

$$P_1 = 100kPa, T_1 = 300K, \frac{P_2}{P_1} = 8, T_4 = 1600K \; P_7 = 100kPa$$

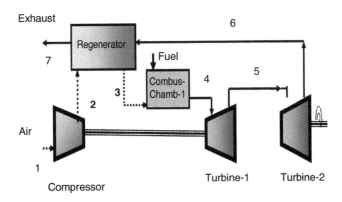

2.9 Consider the case-study problem considered under Section 2.16.1 and Step-I. Step-II. Determine the irrigation pump flow rates at different hours of the day based on the estimated collector area. Plot the irrigation pump flow rate during the daylight hours at off-design hours. (i) Find the total water pumped during the entire day. (*Note:* Flow rate for Rankine cycle must be adjusted.) (ii) If the irrigation pumping rate is to be kept constant at $\dot{m}_p = 500$ kg/min by using an auxiliary after-heater and maintaining the $T_H = 250\,^\circ C$ (saturated vapor) constant at inlet to the turbine for all hours of operation.
Determine the auxiliary energy, Q_{aux}, needed at the after-heater. Plot Q_{aux} vs. hour of the day. Repeat steps I and II with varying T_H. Show the effect of variation in this operating condition on the size of the collector, auxiliary power needed to operate the irrigation pump. Perform the analysis for varying operating conditions given as $T_H = 100\,^\circ C, 150\,^\circ C, 200\,^\circ C$ (*base case*), $250\,^\circ C$, and $300\,^\circ C$, and considering the condenser pressure as $P_c = 20$ kPa.Repeat steps I and II with varying T_H. Show the effect of variation in this operating condition on the size of the collector, auxiliary power needed to operate the irrigation pump. Summarize your results for (a) solar collector area needed and (b) total auxiliary energy needed in tables and plots and (c) present a summary table for the ratings and initial specifications of all major components.

2.10 A Rankine cycle power generation system is considered to recover heat from the exhaust of a diesel engine to improve the overall efficiency of the power generation using the diesel fuel. Water is heated in the boiler/heat exchanger to produce steam using the exhaust of a diesel engine with the purpose of recovering exhaust heat and producing power in steam turbine. Engine exhaust with a mass flow rate of $\dot{m}_g = 0.9\,\text{kg/s}$ and specific heat of $C_{pg} = 0.9$ kJ/kg. C is cooled from 600°C to 400°C. Water in Rankine cycle is heated from condensate pump exit temperature to a saturated steam state of 250°C and run the steam turbine. Assume a condenser pressure of 20 kPa, and turbine and pump efficiencies are 90% and 85%, respectively. (a) Perform the first law of thermodynamic analysis to determine (a) net power output of the turbine, (b) thermal efficiency of the power cycle, and (c) the ratings of the major components of the system.

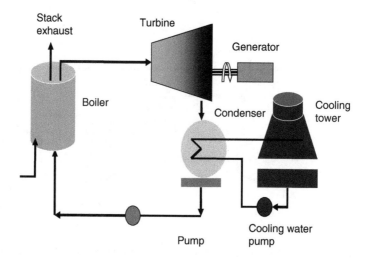

3

Review of Basic Laws and Concepts of Heat Transfer

In this chapter, we will review some of the basic laws and concepts of heat transfer that are essential in the design and analysis of some thermal components. Correlations for convection heat-transfer coefficients in particular are reviewed in a comprehensive manner for different types and modes of convection heat transfer.

3.1 Heat-Transfer Modes and Rate Equations

Heat transfer is defined as the energy transfer due to the presence of spatial temperature variations. There are three basic modes of heat transfer: conduction, convection, and radiation.

3.2 Conduction Heat Transfer

This mode is primarily important for heat transfer in solids and in stationary fluid. Figure 3.1 demonstrates heat transfer by conduction in a plane slab representing a one-dimensional solid or stationary fluid layer.

The conduction rate equation is governed by Fourier's law, which states the heat flow rate per unit area or heat flux as

$$\vec{q} = -k\nabla T \tag{3.1}$$

where

$\vec{q} = \dfrac{q}{A}$ = heat flow per unit area per unit time or heat flux

k = Thermal conductivity of the material

The heat-flux vector in Cartesian coordinate system is written as

$$\vec{q} = -\left(\hat{i} k_x \frac{\partial T}{\partial x} + \hat{j} k_y \frac{\partial T}{\partial y} + \hat{k} k_z \frac{\partial T}{\partial z} \right) \tag{3.2}$$

where the heat-flux components are

$$q_x'' = -k_x \frac{\partial T}{\partial x}, q_y'' = -k_y \frac{\partial T}{\partial y} \text{ and } q_z'' = -k \frac{\partial T}{\partial z} \tag{3.3}$$

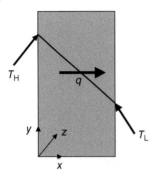

Figure 3.1 Heat transfer by conduction in a solid or stationary fluid layer.

3.2.1 Conduction Heat-Transfer Resistance

The temperature distribution in the plane slab layer is simply obtained by the solution of the one-dimensional heat conduction equation and constant surface temperature boundary conditions as given below:

$$\frac{d}{dx}\left(k\frac{dT}{dx}\right) = 0 \tag{3.4}$$

3.2.1.1 Boundary Conditions

$$\text{BC} - 1: \text{ at } x = 0, T(0) = T_0 \tag{3.5a}$$

$$\text{BC} - 2: \text{ at } x = L, T(L) = T_L \tag{3.5b}$$

Solution to this heat equation leads to the linear temperature distribution in the plane slab as

$$T(x) = T_0 + \frac{T_L - T_0}{L}x \tag{3.6}$$

and the heat-transfer rate based on Fourier law is given as

$$q = -kA\frac{dT}{dx} = \frac{kA}{L}(T_0 - T_L) \tag{3.7}$$

Rearranging,

$$q = -kA\frac{dT}{dx} = \frac{(T_0 - T_L)}{\frac{L}{kA}} \tag{3.8}$$

Using a similarity to the electrical circuit, the conduction heat-transfer resistance is defined as

$$R_{\text{cond}} = \frac{L}{kA} \tag{3.9}$$

Conduction heat-transfer resistance increases with the increase in thickness or physical dimensions of the solid or stationary fluid layers; decreases with the use of higher thermal conductivity solids and fluids; and decreases with increased surface area for heat transfer.

3.2.2 Thermal Resistances and Heat Transfer in Composite Layers

A composite wall includes number of different materials with varying thicknesses and thermal conductivities as shown in Figure 3.2.

Figure 3.2 Composite plane wall.

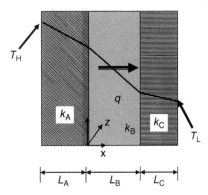

The total thermal resistance for the composite wall is given by taking the sum of all conduction thermal resistances associated with each composite layer and given as

$$R_{cond} = R_A + R_B + R_C \tag{3.10a}$$

or

$$R_{cond} = \frac{L_A}{k_A A} + \frac{L_B}{k_B A} + \frac{L_C}{k_C A} \tag{3.10b}$$

3.3 Convection Heat Transfer

Convection heat transfer is the transfer of heat energy due to the **combined effect** of molecule motion or **diffusion** plus energy transfer by **bulk fluid motion,** which is also referred to as **advection**. The convection heat transfer occurs between a moving fluid and an exposed solid surface. Let us consider the fluid flow over a solid surface at a temperature T_S as shown in Figure 3.3. The fluid upstream temperature and velocity are T_∞ and u_∞, respectively.

Due to the effect of viscosity or **no-slip condition**, there is a development of thin-fluid region, known as the **hydrodynamic boundary layer**, inside which velocity varies from the solid-surface velocity to the outer-stream velocity, u_∞. Similarly, there is development of a **thermal boundary layer** inside which fluid temperature changes from solid-surface temperature T_S to outer fluid temperature T_∞.

Figure 3.3 Hydrodynamic and thermal boundary layers for flow over a solid surface.

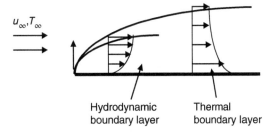

Since the fluid is stationary at the solid surface, the heat is transferred by conduction normal to the surface, and is expressed by the conduction rate equation as

$$q_S'' = k_f \frac{\partial T}{\partial y}\bigg|_{y=0} \tag{3.11}$$

where k_f is the thermal conductivity of the fluid.

To estimate the heat flux by convection, the temperature gradient or the temperature distribution in the thermal boundary layer needs to be known.

3.3.1 Convection Modes

Based on the nature of the flow field, the convection heat transfer is classified as *forced convection*, *free or natural convection*, or *phase-change heat transfer* such as in *condensation* and *boiling*. In *forced convection*, the flow field is induced by some external forces generated by pumps, fans, or winds. In *free or natural convection*, the flow is induced by natural forces such as buoyancy or Marangoni forces. In both forced and free convections, energy being transferred is in the form of sensible energy of the fluid. On the other hand, in *phase-change heat transfer*, the energy transfer is in the form of latent heat of the fluid, and the flow field is created due to the formation of vapor bubbles as in boiling heat transfer or due to the condensation of vapor on a solid surface as in condensation heat transfer.

3.3.2 Convection Heat-Transfer Coefficient

Irrespective of this classification of convection heat-transfer modes, the overall effect is given by a *convection rate equation* governed by *Newton's law of cooling* expressed as

$$q_c'' = h_c(T_S - T_\infty) \tag{3.12}$$

where h_c is the *convection heat-transfer coefficient* or *film coefficient*.

The defining equation for the convection heat-transfer coefficients is given as

$$h_c = \frac{q_c''}{(T_S - T_\infty)} \tag{3.13}$$

Combining Eqs. (3.11) and (3.12), we have the alternate defining equation for convection heat-transfer coefficient

$$h_c = \frac{-k_f \dfrac{\partial T}{\partial y}\bigg|_{y=0}}{(T_S - T_\infty)} \tag{3.14}$$

The convection heat-transfer coefficient encompasses all the effects that influence the convection mode. It depends on conditions in the boundary layer, which are influenced by surface geometry; nature of fluid motion; fluid thermodynamic and transport properties.

To determine the convection heat-transfer coefficient, it is necessary to solve the energy equation for the temperature distribution along with the equation of motion for the velocity field. Typical values of the convection heat-transfer coefficient are given in Table 3.1.

Table 3.1 Typical values of the convection heat-transfer coefficient.

Convection mode	h (W/m² °C)
Free convection	
Gas	5–25
Liquid	50–1000
Forced convection	
Gas	100–500
Liquid	500–20 000
Boiling or condensation	2500–100 000

3.3.2.1 Local Convection

Local convection heat flux and local convection heat-transfer coefficient vary from point to point on the surface as flow conditions vary.

$$q''_x = h_x(T_s - T_\infty) \tag{3.15}$$

where

h_x = Local convection heat-transfer coefficient

Total heat-transfer rate is sum of all incremental heat transfer from the surface and is given as

$$q = \int q'' dA \tag{3.16}$$

3.3.2.2 Average or Mean Heat-Transfer Coefficient

The average or mean convection heat-transfer coefficient is very useful for analysis and design of heat-transfer problems. The average heat-transfer coefficient is computed, based on the total heat transfer and expressed as an integral mean value of local heat-transfer coefficient values as

$$\overline{h} = \frac{1}{A_s} \int_{A_s} h_x dA_s \quad \text{for a 2 – D surface} \tag{3.17a}$$

and

$$\overline{h} = \frac{1}{L} \int_0^L h_x \, dx \quad \text{for a 1D surface} \tag{3.17b}$$

The total heat transfer based on average heat-transfer coefficient is given as

$$q = \overline{h} A (T_s - T_\infty) \tag{3.18}$$

3.3.3 Controlling Forces in Convection

The study of convection heat transfer requires an analysis of energy transport and balance along with an analysis of the fluid flow of the problem. A fluid motion is induced by the application of two types of forces: surface forces and body forces.

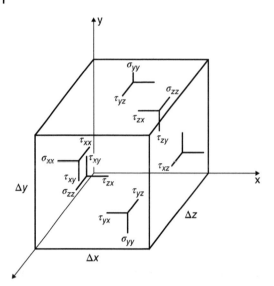

Figure 3.4 Surface stresses over fluid element.

3.3.3.1 Surface Forces

Surface forces act on the boundaries of the media by direct contact in normal as well as tangential to the surface. Two most common surface forces are (i) pressure forces that act normal to a surface (P) and (ii) viscous forces due to normal viscous stresses (σ) that act normal to the surface and viscous shear stresses (τ) that act tangential to the surface. Figure 3.4 shows the three component stresses for each of the six surfaces of rectangular fluid control volume.

3.3.3.2 Body Forces

These are the forces that are developed without any body contact and are distributed over the entire volume of the fluid. Some examples of these types of forces are inertia force; gravitational force that leads to buoyancy force; Marangoni force; electromagnetic force; centrifugal force; Coriolis force, etc.

3.3.4 Major Factors and Parameters in Convection Heat Transfer

3.3.4.1 Thermophysical and Transport properties

Density, ρ; Specific heat, C_p; Thermal conductivity, k; Dynamic, μ; Kinetic viscosity, $v = \frac{\mu}{\rho}$.

These properties along with the flow variables such as the characteristic flow velocity and characteristics dimension of the flow region are combined to define the number of dimensionless parameters to characterize the flow dynamics and heat transfer. A few of them are defined as follows.

3.3.4.1.1 Prandtl Number (Pr)

Prandtl number represents a relative effectiveness of molecular transport of momentum to heat energy transport, and is defined as

$$Pr = \frac{\mu C_p}{k} = \frac{v}{\alpha} \tag{3.19}$$

Table 3.2 Prandtl number values of some typical fluids at 300 K.

Fluid type	Prandtl number (*Pr*)
Air	0.707
Water-saturated liquid	5.83
Water-saturated vapor	0.857
Engine oil	6400
Engine oil at 400 K	152
Ethylene glycol	151
Refrigerant-134a	3.4
Sodium (liquid metal at 366 K)	0.011
Blood	4.62

where

v = Kinematic viscosity
α = Thermal diffusivity

A high Prandtl number value indicates the fluid velocity field develops at a faster rate than thermal field due to the higher kinematic viscosity compared to thermal diffusivity. Prandtl values of some typical heat-transfer fluids are summarized in Table 3.2.

3.3.4.1.2 Reynolds Number (Re)

$$Re_{L_c} = \frac{\rho U_c L_c}{\mu} = \frac{U_c L_c}{v} \tag{3.20}$$

where

L_c = Characteristic length such as D, D_H, x, or L
U_c = Characteristic velocity such as U, U_∞, or u

Reynolds number represents an order of magnitude ratio of inertia force to viscous force and is also used as an index that indicates the nature of the fluid flow as **laminar** or **turbulent**.

3.3.4.1.3 Characteristics Length (L_c)
The characteristics length used in defining Reynolds number and Nusselt number in a problem varies from problem to problem depending on which dimension is critical in characterizing flow dynamics and heat transfer. For example, it is the local position coordinate x or length L for external flow over a flat surface; it is the outer diameter D_o for flow over a cylinder or sphere; it is the **hydraulic diameter** D_H for flow over a noncircular cylinder. For internal flows, it is often the internal diameter D_i of the circular tube or the hydraulic diameter D_H of the nonconduits like rectangular ducts or hexagonal flow tube.

3.3.4.1.4 Hydraulic Diameter (D_H)

This is defined as the ratio of four times the flow area A to the wetted perimeter P_w of the flow geometry as

$$D_H = \frac{4A}{P_w} \tag{3.21}$$

A few examples are given below:

Circular tube – $D_H = 4\frac{\frac{\pi D^2}{4}}{\pi D} = D$

Circular annulus – $D_H = \frac{4\pi(D_o^2-D_i^2)/4}{\pi(D_o+D_i)} = D_o - D_i$

Rectangular duct – $D_H = \frac{4ab}{(2a+2b)} = \frac{2ab}{(a+b)}$

Parallel plates with width w and depth z) – $D_H = 4\frac{wz}{2z} = 2w$

3.3.4.1.5 Characteristic Velocity (U_C)

The characteristic velocity also varies from problem to problem. For example, it could be the upstream velocity u_∞ for external flow over flat plate or a cylinder or a sphere. It is the average velocity u_{av} for flow in an internal flow geometry.

3.3.4.1.6 Critical Reynolds Number (Re_crit)

Even though the transition of a flow changes over a range of Reynolds number, for design purposes the flow is classified based on a critical value of the Reynolds number that is defined specifically for the application. A few examples are given as follows.

Forced Convection Over a Flat Plate   Flow is laminar for $Re_x < 5 \times 10^5$ and flow is turbulent for $Re_x \geq 5 \times 10^5$

Where Reynolds number and critical Reynolds number is defined as

$$Re_x = \frac{\rho u_\infty x}{\mu} \tag{3.22a}$$

and

$$Re_{crit} = \frac{\rho u_\infty x_c}{\mu} = 5 \times 10^5 \tag{3.22b}$$

Forced Convection in a Pipe

Flow is laminar for $Re_D < 2300$

and

Flow is turbulent for $Re_D \geq 2300$

where Reynolds number and critical Reynolds number is defined as

$$Re_D = \frac{\rho u_{av} D}{\mu} \tag{3.23a}$$

and

$$Re_{crit} = 2300 \tag{3.23b}$$

3.3.4.1.7 Grashof Number (Gr)

The Grashof number is an important dimensionless parameter in free convection study and plays an important role in a similar manner as Reynolds number in forced convection study. The Grashof number is a measure of the ratio of the buoyancy force to the viscous force acting on the fluid and defined as

$$Gr_{L_c} = \frac{g\beta(T_s - T_\infty)L_c^{\ 3}}{v^2} \tag{3.24}$$

3.3.4.1.8 Rayleigh Number (Ra)

Often, free convection correlations are given in terms of Rayleigh number, which is defined as the product of *Grashof number* and *Prandtl number* and defined as

$$Ra_{L_c} = Gr_{L_c} = \frac{g\beta(T_s - T_\infty)L_c^{\ 3}}{v\alpha} \tag{3.25}$$

Transition between laminar and turbulent in external free convection flow over a vertical flat-plate surface is defined based on the critical Rayleigh number as $Ra_L \sim 10^9$.

3.3.4.1.9 Nusselt Number (Nu)

Nusselt number represents the ratio of convective heat transfer to conductive heat-transfer rates across the boundary layer. This also can be looked at as a way to express dimensionless heat-transfer coefficient and is defined as

$$Nu = \frac{hL_c}{k_f} \tag{3.26}$$

where

h = Local or average convection heat-transfer coefficient
L_c = Characteristic length (D, D_H, or L)
k_f = Thermal conductivity of the fluid

3.3.4.1.10 Graetz Number (Gz)

The Graetz number is a dimensionless number that is typically used for characterizing thermally developing flow and heat transfer and is defined as

$$Gz = RePr\frac{D_H}{L} \tag{3.27}$$

where

Re = Reynolds number
Pr = Prandtl number
D_H = Hydraulic diameter of tubes or channels
L = Length of tube or channel

Physically, **Graetz number** represents the ratio of time for heat to dissipate by conduction in the radial or normal direction in the fluid to the time taken by the fluid to travel a distance while moving with a certain velocity (Table 3.3).

3.3.4.2 Flow Geometry

3.3.4.2.1 External Flows

These are cases where fluid flows over a completely submerged body such as flow over a flat plate, blocks, flow over a cylinder, or over a bundle of cylinders and sphere.

Table 3.3 Commonly used dimensionless parameters in thermal design.

Dimensionless parameters	Expression	Physical representation
Friction factor	$f = \dfrac{\Delta P}{\rho \dfrac{L}{D} \dfrac{V^2}{2g}}$	Dimensionless pressure drop
Coefficient of friction	$C_f = \dfrac{\tau_w}{\frac{1}{2}\rho u_\infty^2}$	Dimensionless shear stress
Biot number	$Bi = \dfrac{hL_c}{k}$	Ratio of conduction resistance to convection resistance
Fourier number	$Fr = \dfrac{\alpha t}{L_c^2}$	Dimensionless time
Grashof number	$Gr = \dfrac{g\beta(T_s - T_\infty)L_c^3}{v^2}$	Ratio of buoyancy force to viscous force
Colburn j factor	$j = StPr^{2/3}$	Dimensionless heat-transfer coefficient
Nusselt number	$Nu = \dfrac{hL_c}{k_f}$	Dimensionless heat-transfer coefficient representing the ratio of convective heat transfer to conductive heat-transfer rates across the boundary layer
Peclet number	$Pe = RePr$	Ratio of advective transport rate to diffusive transport rate
Prandtl number	$Pr = \dfrac{\mu C_p}{k} = \dfrac{v}{\alpha}$	Ratio of momentum diffusivity to thermal diffusivity
Reynolds number	$Re = \dfrac{\rho V L_c}{\mu}$	Ratio of inertia force to viscous force
Rayleigh number	$Ra_L = Gr_L Pr = \dfrac{g\beta(T_s - T_\infty)L_c^3}{v\alpha}$	Ratio of buoyancy and viscous forces multiplied by the ratio of momentum and thermal diffusivities
Stanton number	$St = \dfrac{h}{\rho V C_p} = \dfrac{Nu}{RePr}$	Dimensionless mass transfer coefficient
Graetz number	$Gz = RePrD/L$	Represents the ratio of time for heat to dissipate by conduction normal direction to the time taken by the fluid to travel in the axial flow

Figure 3.5 shows boundary-layer flow development. As the fluid comes in contact and flows over a solid surface, a thin layer of fluid forms over the solid surface over which fluid velocity varies from zero velocity, maintaining no-slip condition at the solid surface at $y = 0$ to the outer-stream velocity at the edge of the boundary thickness, $y = \delta$. In a similar manner, a thermal boundary layer also forms as the fluid at a temperature, $T = T_\infty$, comes in contact with the solid surface at temperature, $T = T_s$. Temperature of the fluid varies from the solid-surface temperature to the outer-stream fluid temperature at $y = \delta_t$.

3.3.4.2.2 Internal Flows
These are the cases where flows are completely bounded by solid surfaces such as in flows through pipes, tubes, channels, and rectangular cavities.

(a)

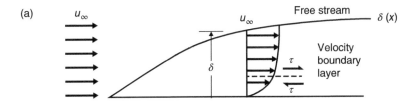

(b)

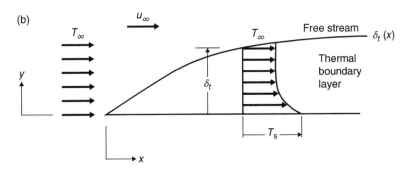

Figure 3.5 Boundary-layer flow over a flat plate. (a) Velocity or hydrodynamic boundary-layer development. (b) Thermal boundary-layer development.

Characteristics of Internal Flows in a Duct As flow enters the pipe, boundary layers develop and grow on both top and bottom surfaces.

At some distance from the entrance, the boundary layers meet and flow is assumed as viscous over the entire cross-section of the pipe. The flow in a pipe and tube is said to be **fully developed** if the velocity and dimensionless temperature profiles show no variation with the longitudinal distance along the length of the pipe (Figure 3.6).

Hydrodynamic-entry length: The length required for the velocity profile to become fully developed, i.e. the profile shows no variation with longitudinal distance along the pipe. For internal flows, **thermal boundary layers** develop from both top and bottom surfaces and develop into two regions: **thermal-entry length** and **thermal fully developed** regions similar to hydrodynamic internal flow as shown in Figure 3.7.

Thermal fully developed region: The region where a dimensionless temperature profile remains invariable along the longitudinal length of the channel. A common dimensionless temperature variable is defined based on the local temperature of fluid and the wall surface temperature and the fluid inlet temperature.

Thermal-entry length ($L_{e, th}$): The length required for the dimensionless temperature profile to become fully developed.

Criterion for Entry Length Criterion for **hydrodynamic-entry length** for laminar flow:

$$\frac{L_{e,h}}{D} \approx 0.06 Re_D \tag{3.28a}$$

For example, for a maximum laminar flow, Reynolds number in a circular channel is given by the critical Reynolds $Re_{crit} = 2300$,

$$\frac{L_{e,h}}{D} \approx 0.06 \times 2300$$

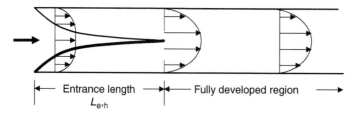

Figure 3.6 Hydrodynamic-entry length and developed flow.

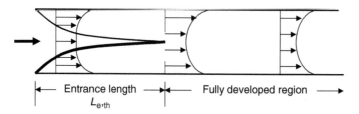

Figure 3.7 Thermal-entry length and thermally fully developed region.

or

$$L_{e,h} = 138D$$

For example, if we consider a circular channel of 1-mm diameter, then the hydrodynamic entrance length is about 13.8 cm.

Criterion for **thermal-entry length** for laminar flow is given as

$$\frac{L_{e,th}}{D} \approx 0.06 Re_D Pr \tag{3.28b}$$

For turbulent flows, both the hydrodynamic- and the thermal-entry lengths are reduced due to increased mixing caused by the turbulence. Criterion for hydrodynamic- and thermal-entry length for **turbulent flow is given as**:

$$20 \approx \frac{L_e}{D} \approx 40 \tag{3.29}$$

3.3.4.2.3 Nature of Fluid Motion
Depending on the nature of the flow dynamics, the fluid flow is classified as laminar or turbulent. A flow may initially be turbulent while entering the flow region over the solid surface but may turn into a laminar flow after passing through a region of transition due to the rapid amplification of small disturbances that originate from the fluid-solid interface.

Figure 3.8 shows flow patterns of laminar flow and turbulent flow and transformation of laminar flow into turbulent flow. Laminar flow is characterized as flow regions where fluid moves in layers in contrast to turbulent-flow regions where fluid exhibits enhanced mixing due to the presence of eddies that carry momentum as well as energy from one layer to another in addition to molecular diffusion motion.

3.3.4.3 Convection Heat-Transfer Correlations
Convection heat-transfer coefficients are derived either analytically by solving the governing equations for flow fields and heat equations for temperature distributions or

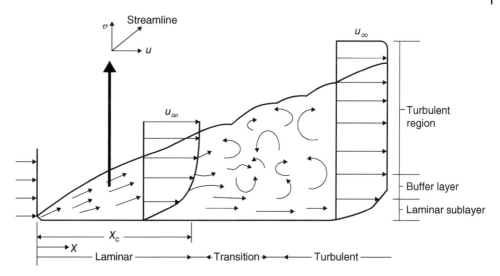

Figure 3.8 Laminar and turbulent-flow characteristics.

by experiments for many flow conditions and geometries. Convections heat-transfer coefficients are computed based on Eq. (3.13) or (3.14) and expressed in the form of a correlation and in terms of dimensionless parameters.

3.3.4.4 Forced Convection Heat Transfer and Correlations

In forced convection heat transfer, the flow field is mainly dominated by inertia and viscous forces. For a particular flow geometry, the convection heat-transfer coefficient (h) or the Nusselt number (Nu) is either derived from the solution of flow field mass and momentum equations and the temperature field equation given by heat equation or energy equation or derived experimentally, and expressed as a function of Reynolds number (Re) and Prandtl number (Pr)

$$Nu_x = f(x, Re_x, Pr)$$
(3.30)

or

$$Nu_L = f(Re_L, Pr)$$
(3.31)

A large body of experimental data and analytical solution led to the following general correlation

$$Nu = cRe^m Pr^n$$
(3.32)

where c, m, and n are constants that vary with problems.

Nu = Nusselt number = $\dfrac{h_c L_c}{k}$

Re = Reynolds number = $\dfrac{\rho U_c L_c}{\mu}$

Pr = Prandtl number = $\dfrac{\mu C_p}{k} = \dfrac{\nu}{\alpha}$

L_c and U_c represent the characteristic length and velocity in the problem.

For **free convection**, the heat-transfer correlations are given either in terms of Grashof number, *Gr*, and Prandtl number or in terms of Rayleigh number as

$$Nu = f(Gr, Pr) = f(Ra) \tag{3.33}$$

where

$$Gr = \text{Grashof number} = Gr_{L_c} = \frac{g\beta(T_s - T_\infty)L_c^{\,3}}{v^2} \tag{3.34a}$$

$$Ra_L = \text{Rayleigh number} = Gr_{L_c} Pr = \frac{g\beta(T_s - T_\infty)L_c^{\,3}}{v\alpha} \tag{3.34b}$$

For simplicity, such correlations are used as **convective boundary conditions** for many heat conduction problems to take into account convection heat transfer from/to solid surfaces instead of solving complete set of differential equations for flow field and convection heat equations.

Let us briefly discuss a few cases where these correlations are derived analytically or semi-analytically for fully developed hydrodynamic and thermal flow conditions. Some of the widely used correlations that are applicable to the flow channels in different heat-transfer applications are outlined here.

3.3.5 Forced Convection Internal Flow and Heat Transfer

3.3.5.1 Laminar Flows
3.3.5.1.1 Fully Developed Flow and Heat Transfer in Pipe or Tubes
During heating and cooling of fluid in channel, the fluid temperature must continue to change with *x*. A fully developed thermal condition is, however, defined in a different manner from fully developed velocity condition of invariable axial component of velocity, i.e. $du/dx = 0$ in the fully developed region. In contrast to heat transfer, dT_m/dx and dt/dx at any *x* location is not zero. The temperature distribution $T(r)$ or $T(y)$ changes continuously with position *x*, and it seems that the fully developed thermal condition can never be reached. Fully developed temperature profile is explained through the concept of dimensionless temperature profile.

The nondimensional temperature is defined in terms of mixed mean temperature, T_m, and the wall temperature, T_s as

$$\theta = \frac{T_s - T}{T_s - T_m} \tag{3.35}$$

The mixed mean fluid temperature, T_m, is also referred to as the bulk fluid temperature, and is derived at any flow cross-section from the net heat energy convected through the flow cross-section as

$$T_m = \frac{1}{A_c V} \int_{A_c} uT \, dA_c \tag{3.36}$$

For internal flows, the heat-transfer coefficient, h, is defined using the temperature difference between mixed mean fluid temperature T_m and solid-surface temperature T_s as

$$\dot{q}_s'' = h(T_s - T_m) \tag{3.37}$$

and also defining the conduction heat flux at the wall surface as

$$\dot{q}_s'' = -k \left(\frac{\partial T}{\partial r} \right)_{r=r_0} \tag{3.38}$$

For invariable dimensional temperature profile in the fully developed flow region, the convection heat-transfer coefficient, h, is also invariant with axial flow direction. This is apparent from the defining equation for convection coefficient. For example, in the fully developed section of a pipe, the radial temperature gradient is invariant as demonstrated in Figure 3.7 and so the convection is constant in this region. For **invariant nondimensional temperature profile** in the flow direction, we can write the following the conditions:

$$\left[\frac{\partial}{\partial x} \left(\frac{T_s - T}{T_s - T_m} \right) \right]_{r=r_0} = \text{const} = -\frac{\left(\frac{\partial T}{\partial r} \right)_{r=r_0}}{T_s - T_m} \tag{3.39}$$

Simplified form of momentum and heat equation are derived for fully developed flow and heat transfer in internal flow conduits. Solutions to these equations are obtained analytically and semianalytically for a number of limiting case problems and results are presented in references (Incropera et al. 2007; Shah and London 1978; Asako et al. 1988; Kays and Crawford 1980). Two such important limiting cases are:

Case-1: Constant surface heat-flux rate and
Case-2: Constant surface temperature.

Results for these two cases are

Constant heat-flux rate: $Nu = \frac{hD}{k} = 4.364$
Constant-surface temperature: $Nu = \frac{hD}{k} = 3.658$

Detailed solution procedure for these two limiting cases for circular pipes, annular flow channels, and rectangular channels are given in the books by Kays and Crawford (1980). As we can see, Nusselt number for fully developed flow is constant with lower limit or value given for the case of constant surface temperature and upper limit or value given for the case of constant surface heat flux. In a similar manner, the fully developed correlations can be derived for other flow geometries. Convection heat-transfer coefficients and friction coefficients for fully developed flows are summarized in Table 3.4 for some common flow geometries.

Technically, the **case of constant heat-flux rate problems** arises in a number of situations such as electric resistance heating like the wall heated electrically; radiant heating like outer surface of wall is uniformly irradiated by solar radiation; nuclear heating, and in

Table 3.4 Fully developed flow correlations for heat-transfer coefficients.

Geometry	b/a	$q_s'' = $ Const	$T_s = $ Const	fRe_D
(circle)		4.36	3.66	64
(square with sides a, b)	1.0	3.61	2.98	57
(rectangle b, a)	2.0	4.12	3.39	62
(rectangle)	4.0	5.33	4.44	73
(rectangle)	8.0	6.49	5.60	82
(parallel plates)	∞	8.23	7.54	96
(parallel plates)	∞	5.38	4.861	96
(triangle)		3.11	2.47	53
(hexagon)		4.02	3.35	60.25
Octagon		4.20	3.46	61.52

Where $q_s'' = $ Const represents the case with constant surface heat flux and $T_s = $ Const represents the case with constant surface heat.
Sources: Incropera et al. (2007), Shah and London (1978), Asako et al. (1988), and Kays and Crawford (1980).

counter flow heat exchangers when the fluid capacity rates are the same. The **case of constant surface temperature** is another very common convection application and occurs in such heat exchangers where phase change occurs at the outer tube surfaces of evaporators, condensers, and in any heat exchanger where one fluid has a very much higher thermal capacitance than the other.

The constant surface-temperature and constant heat-flux boundary conditions do cover the usual extremes seen in the heat-exchanger design. However, it must be remembered that near the tube entrance a fully developed condition does not exist and the fully developed constant heat-transfer coefficient values are only applicable to a long flow conduit.

3.3.5.1.2 Thermal-Entry-Length Solution for a Circular Tube

For thermal-entry-length solution, it is assumed that the velocity field is fully developed and temperature field is developing. The fully developed velocity profile is substituted before solving the energy equation for the temperature.

Case: Uniform Surface Temperature The applicable differential energy equation is equation:

$$\frac{1}{r}\frac{\partial}{\partial r}\left(r\frac{\partial T}{\partial r}\right) + \frac{\partial^2 T}{\partial x^2} = \frac{u}{\alpha}\frac{\partial T}{\partial x} \tag{3.40}$$

It is convenient to put this equation in nondimensional form before discussing its solution. The equation is expressed in terms of dimensionless variables, defined as follows:

$$\theta = \frac{T_s - T}{T_s - T_e}, \quad r^+ = \frac{r}{r_0}, \quad u^+ = \frac{u}{V}, \quad x^+ = \frac{x/r_0}{RePr} \tag{3.41}$$

where x is the axial distance from the point where heat transfer starts, T_s is the constant surface temperature, and T_e is the uniform entering fluid temperature.

The substitution of the dimensionless variables into the equation results

$$\frac{\partial^2 \theta}{\partial r^{+2}} + \frac{1}{r^+}\frac{\partial \theta}{\partial r^+} = \frac{u^+}{2}\frac{\partial \theta}{\partial x^+} - \frac{1}{(RePr)^2}\frac{\partial^2 \theta}{\partial x^{+2}} \tag{3.42}$$

Note that the last term in the equation takes into account heat conduction in the axial direction. The magnitude of this term depends on the magnitude of the Peclet number, i.e. the product of Reynolds number, Re, and Prandtl number, Pr. At large values of the Peclet number ($Pe = RePr$), the axial conduction term becomes negligibly small. Based on experience, it has been indicated that as a general rule this term is negligible for $Re. Pr > 100$. Neglecting this term, the governing equation reduces to

$$\frac{\partial^2 \theta}{\partial r^{+2}} + \frac{1}{r^+}\frac{\partial \theta}{\partial r^+} = \frac{u^+}{2}\frac{\partial \theta}{\partial x^+} \tag{3.43}$$

For hydrodynamically fully developed laminar flow, the parabolic velocity profile is

$$u = 2V_1(1 - r^2/r_0^2) \tag{3.44}$$

and in dimensionless form as

$$u^+ = 2(1 - r^{+2}) \tag{3.45}$$

Substituting the velocity in the dimensionless form of the energy equation in equation, the final mathematical statement of the problem is given in the form of the differential equation

$$\frac{\partial^2 \theta}{\partial r^{+2}} + \frac{1}{r^+}\frac{\partial \theta}{\partial r^+} = (1 - r^{+2})\frac{\partial \theta}{\partial x^+} \tag{3.46}$$

Subject to the boundary conditions: $\theta(0, r^+) = 1$ and $\theta(x^+, 1) = 0$

The solution to this linear homogeneous partial differential equation is obtained using the method of separation of variables. The final solution must take on the form

$$\theta(x^+, r^+) = \sum_{n=0}^{\infty} C_n R_n(r^+) \exp(-\lambda^2{}_n x^+) \tag{3.47}$$

where λ_n are the **eigenvalues**, R_n are the corresponding *eigenfunctions*, and C_n are constants.

Once we have obtained this solution, we can calculate the heat flux at any x^+ from the slope of the temperature profile at the wall.

$$\dot{q}_0''(x^+) = -k\left(\frac{\partial T}{\partial r}\right)_{r=r_o} = k\frac{T_s - T_e}{r_0}\left(\frac{\partial \theta}{\partial r^+}\right)_{r^+=1} \tag{3.48}$$

$$\dot{q}_0''(x^+) = \frac{2k}{r_0} \sum G_n \exp(-\lambda^2_n x^+)(t_o - t_e) \tag{3.49}$$

where the constant $G_n = -\left(\frac{C_n}{2}\right) R_n'(1)$.

Based on this solution, local fluid mean temperature, local and mean Nusselt numbers are computed and are summarized as follows:

Local Mean Temperature:

$$\theta_m = 8 \sum \frac{G_n}{\lambda^2_n} \exp(-\lambda^2_n x^+) \tag{3.50}$$

Local Nusselt Number:

$$Nu_x = \frac{\sum G_n \exp(-\lambda^2_n x^+)}{2 \sum \frac{G_n}{\lambda^2_n} \exp(-\lambda^2_n x^+)} \tag{3.51}$$

Mean Nusselt Number:

$$Nu_m = \frac{1}{2x^+} \ln\left[\frac{1}{8 \sum \frac{G_n}{\lambda^2_n} \exp(-\lambda^2_n x^+)}\right] \tag{3.52}$$

The constants and **eigenvalues** for these infinite series are given in Table 3.5.

Table 3.5 Eigenvalues and constants for thermal-entry-length infinite series solution in circular tube with constant surface temperature.

n	λ_n	λ^2_n	G_n
0	2.704	7.312	0.749
1	6.679	44.62	0.544
2	10.667	113.8	0.463
3	14.666	215.12	0.414
4	18.668	348.5	0.382
5	22.667	513.778	0.358
6	26.667	711.111	0.339
7	30.667	940.445	0.324
8	34.667	1201.778	0.311
9	38.667	1495.111	0.299
10	42.667	1820.444	0.289

Table 3.6 Thermal-entry-length solution for local and mean Nusselt number for laminar flow in a circular tube with constant surface temperature.

x^+	Nu_x	Nu_m
0	∞	∞
0.001	12.80	19.29
0.004	8.03	12.09
0.01	6.00	8.92
0.04	4.17	5.81
0.08	3.77	4.86
0.10	3.71	4.64
0.20	3.66	4.15
∞	3.66	3.66

These eigenvalues were originally derived by Sellars et al. (1956) and are reproduced here from that presented by Kays and Crawford (1980). For higher eigenvalues, i.e. for $n > 2$, the eigenvalues are given by $\lambda_n = 4n + \frac{8}{3}$, $G_n = 1.012\,76\lambda_n^{-1/3}$.

Based on these eigenvalues (λ_n) and constants (G_n), results of constant-surface-temperature solution in terms of mean fluid temperature, local and mean Nusselt number values are computed as function of the axial position using Eqs. (3.50)–(3.52) and are presented in Table 3.6.

We can notice that as x^+ increases, the series becomes increasingly more convergent and approaches the fully developed value of 3.66. It has been demonstrated that the series is dominated by the first few terms. In fact, for $x^+ > 0.1$, only the first term of the infinite series dominates and there are insignificant contributions from rest of the terms. Considering only the first term, Eq. (3.51) can be reduced to the following form

$$Nu_x = \frac{G_o \exp(-\lambda^2{}_o x^+)}{2\frac{G_o}{\lambda^2{}_o} \exp(-\lambda^2{}_o x^+)} \tag{3.53}$$

and for $x^+ = 0.1$ it reduces to

$$Nu_x = \frac{\lambda_o{}^2}{2} = 3.658 \tag{3.54}$$

This one term approximation value is within 0.05% of the fully developed value estimated or fully developed flow in circular pipe with constant surface temperature, and hence the **thermal-entry length** can be approximated as equal to

$$x^+ = 0.1 = \frac{x/r_o}{RePr} = \frac{2(x/D)}{RePr}$$

or

$$\left(\frac{x}{D}\right)_{\text{fully dev}} \sim 0.05(RePr)$$

Table 3.7 Variation in thermal-entry length with fluid types for $Re = 1000$.

Fluid type	Pr	Entry length, $L_{e,th}$ (ft)
Air	0.7	2.04
Refrigerant-R134	3.4	14.166
Water	5.83	24.291
Ethylene glycol	151	629.2
Oil	6400	26 666.6
Sodium (liquid metal) at 266 K	0.011	0.0458
Blood	4.62	19.25

For example, for **air** with $Pr = 0.7$ and flowing with *Reynolds number*, $Re = 700$,

$$\left(\frac{x}{D}\right)_{fully\ dev} \sim 24.5$$

and for a tube diameter of 1-in., the thermal-entry length is about $L_e = 24.5$-in. or 2.04 ft.
For **water** with $Pr = 5.83$ and flowing with Reynolds number, $Re = 1500$

$$\left(\frac{x}{D}\right)_{fully\ dev} \sim 437.25$$

and for a tube diameter of 1-in., the thermal-entry length is about $L_e = 437.25$-in. or 36.4 ft.
For **oil** with $Pr = 152$ at around 400 K and flowing with $Re = 1000$,

$$\left(\frac{x}{D}\right)_{fully\ dev} \sim 7600$$

and for a tube diameter of 1-in., the thermal-entry length is about 633 ft.

Table 3.7 shows the variation in thermal-entry length for different fluids with Prandtl values at 300 K and at $Re = 1000$.

Note that in an oil-flow heat exchanger, the flow in the tube side will be primarily in the developing thermal entry region and requires considerably more length to reach fully developed flow. Considerably large error may be involved in selecting fully developed flow correlations for fluid with large Prandtl numbers. In the design of heat exchanger, one can benefit from a higher convection heat-transfer coefficient associated with developing flow region, particularly dealing with liquids. For example in a 1-in. diameter tube with oil flow at $Re = 500$, $Pr = 150$, and length $x = 3.125$-ft, we can estimate $x^+ = \frac{2\frac{x}{D}}{RePr} = \frac{2\times\frac{37.5}{1}}{500\times150} = 0.001$ and $Nu_m = 12.09$ from Table 3.3. This is about 3.3 times higher than the fully developed flow Nusselt number. As we can see, the use of a fully developed flow correlation will lead to considerable overdesign of the heat exchanger.

Case: Uniform Surface Heat Flux Infinite-series solution for the laminar thermal-entry-length flow in circular tube with constant heat flux is obtained in similar method and written as follows:

$$Nu_x = \left[\frac{1}{Nu_\infty} - \frac{1}{2}\sum\frac{\exp(-\gamma^2_m x^+)}{H_m\gamma^4_m}\right]^{-1} \tag{3.55}$$

The eigenvalues (γ_m) and the functions (H_m) for the first few terms $(m < 6)$ computed and are listed in Table 3.8. For $m > 6$, these values are computed from the simplified equations given as $\gamma_m = 4m + \frac{4}{3}$ and $H_m = 0.358\,\gamma_m^{-2.32}$ to extend the data presented in Table 3.8.

Table 3.8 The eigenvalues and functions for thermal-entry-length solution for uniform surface heat flux.

n	γ_m	γ_m^2	$H_m \times 10^3$
1	5.067	25.68	7.630
2	9.157	83.86	2.058
3	13.198	174.2	0.901
4	17.219	296.5	0.487
5	21.234	450.9	0.297
6	25.333	641.778	0.198
7	29.333	860.444	0.141
8	33.333	1111.111	0.104
9	37.333	1393.778	0.080
10	41.333	1708.444	0.063

Table 3.9 Local thermal-entry-length Nusselt numbers for laminar flow in a circular tube with constant surface heat flux as function of nondimensional position.

x^+	Nu_x
0	∞
0.002	12.00
0.004	9.93
0.010	7.49
0.020	6.14
0.040	5.19
0.100	4.51
∞	4.36

The series solution Eq. (3.55) is used to compute the local thermal-entry-length Nusselt numbers for laminar flow in a circular tube as a function of nondimensional position and listed in Table 3.9.

Figure 3.9 shows the variation of thermal entry-length Nussel number for laminar flow in circular tube for the limiting cases of constant wall surface temperature and constant surface wall heat flux.

3.3.5.1.3 Thermal-Entry Length in Circular Tube: Empirical Correlation
Thermal-entry-length empirical correlation for constant surface temperature condition is given by Housen (Incropera et al. 2006):

$$\overline{Nu}_D = 3.66 + \frac{0.0668\left(\frac{D}{L}\right)Re_D Pr}{1 + 0.04\left[\left(\frac{D}{L}\right)Re_D Pr\right]^{2/3}} \tag{3.56a}$$

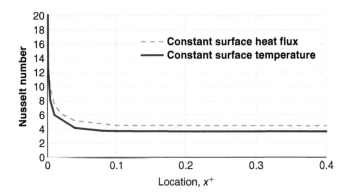

Figure 3.9 Variation of mean thermal-entry-length Nusselt number for laminar flow in a circular tube.

and when expressed in terms of Graetz number

$$\overline{Nu_D} = 3.66 + \frac{0.0668 Gz_D}{1 + 0.04 \, [Gz_D]^{2/3}} \tag{3.56b}$$

This thermal-entry-length correlation is also **valid for combined entry length with** $Pr \geq 5$ for which velocity field develops at a faster rate than thermal field. For higher Prandtl numbers, the velocity profile develops at a faster rate than that for the temperature profile and one can use the thermal-entry-length solution while assuming fully developed velocity profile.

3.3.5.1.4 *Thermal-Entry-Length Solutions for Flow in Rectangular Channels*

The flow in **Thermal-Entry-Length Solutions** for flow between **two parallel planes** is a special case of the rectangle family. A number of such cases are analyzed with a range of surface conditions and presented in the book by Kays and Crawford (1980). Let us present here a brief summary of the analytical solution for heat transfer in a parallel-flow channel with uniform constant surface temperature maintained at both plane surfaces (Figure 3.10).

Mixed mean temperature

$$\theta_m = 8 \sum_n \frac{G_n}{\lambda^2_n} \exp(-\lambda^2_n x^+) \tag{3.57}$$

Local Nusselt numbers

$$Nu_x = \frac{\sum G_n \exp(-\lambda^2_n x^+)}{2 \sum \frac{G_n}{\lambda^2_n} \exp(-\lambda^2_n x^+)} \tag{3.58}$$

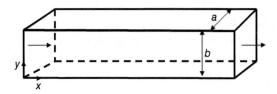

Figure 3.10 Convection heat transfer for flow through a rectangular channel.

Table 3.10 The thermal-entry-length infinite-series-solution functions for the parallel-planes system with both surfaces at constant temperature.

n	λ_n	λ_n^2	G_n
0	3.849	15.09	1.717
1	13.087	171.259	1.139
2	22.324	498.370	0.952
3	31.562	996.148	0.030
4	40.799	1664.592	0.022
5	50.037	2503.704	0.017
6	59.275	3513.481	0.013
7	68.512	4693.926	0.011
8	77.749	6045.037	0.009
9	86.987	7566.815	0.008
10	96.225	9259.259	0.007

Mean Nusselt numbers

$$Nu_m = \frac{1}{2x^+} \ln\left[\frac{1}{8 \sum \frac{G_n}{\lambda_n^2} \exp(-\lambda_n^2 x^+)}\right] \tag{3.59}$$

For the solutions to be presented, define the nondimensional distance as

$$x^+ = \frac{2\left(\frac{x}{D_h}\right)}{RePr} \tag{3.60}$$

where D_h is the hydraulic diameter defined as in Eq. (3.21).

In Table 3.10, the eigenvalues and functions for $n > 2$ are computed based on following formulas: $\lambda_n = \frac{16n}{\sqrt{3}} + \frac{20}{3\sqrt{3}}$ and $G_n = 2.68\lambda_n^{-1/3}$

Thermal-entry-length solutions based on infinite-series function for the rectangular channel with constant surface temperature condition are presented in Table 3.11 and in Figure 3.11 with varying aspect ratios.

Note that as x^+ increases, Nusselt number values approach fully developed values as presented in Table 3.4. While aspect ratio $b/a = 1$ represents a square channel, the aspect ratio $b/a = \infty$ represents the case of the infinite parallel-plane channels.

3.3.5.1.5 Combined Hydrodynamic and Thermal-Entry Length

The applicable energy differential equation for calculation of the **developing temperature field** under conditions of a **developing velocity field**, axisymmetric heating, and negligible axial conduction is given by the following equation:

$$\frac{1}{r}\frac{\partial}{\partial r}\left(r\frac{\partial T}{\partial r}\right) = \frac{u}{\alpha}\frac{\partial T}{\partial x} + \frac{v_r}{\alpha}\frac{\partial T}{\partial r} \tag{3.61}$$

The combined entry-length solution was obtained using a linearized approximation form of this energy equation and by obtaining a generalized entry-region temperature profile.

Table 3.11 Thermal-entry-length mean Nusselt numbers for rectangular tubes with constant surface temperature.

x^+	$\overline{Nu}_m$				
	b/a				
	1	2	4	6	∞
0	∞	∞	∞	∞	∞
0.01	8.63	8.58	9.47	10.01	11.63
0.02	6.48	6.84	7.71	8.17	9.83
0.05	4.83	5.24	6.16	6.70	8.48
0.10	4.04	4.46	5.44	6.04	8.02
0.20	3.53	3.95	5.00	5.66	7.78
∞	2.98	3.39	4.51	5.22	7.55

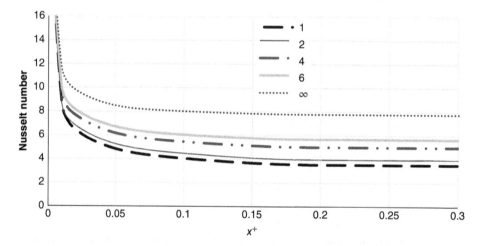

Figure 3.11 Thermal-entry-length mean Nusselt numbers for rectangular tubes with constant surface temperature for different aspect ratios.

The energy integral form equation is then solved using the generalized temperature profile for the case of constant surface temperature and for constant surface heat flux for different Prandtl number.

For combined entry length, the heat-transfer coefficient is strongly dependent on the Prandtl number because of strong dependence of velocity development on viscosity of the fluid. Prandtl number is a parameter in the solution of combined hydrodynamic and thermal-entry-length problems or combined entry-length solution.

The Nusselt number values for combined entry length are presented for the circular tube with constant surface temperature in Figure 3.12 and Table 3.12 for Prandtl number values of $Pr = 0.7$, $Pr = 2.0$, and $Pr = 5.0$. Similarly, Nusselt number values are presented in Table 3.13 for constant heat-flux conditions.

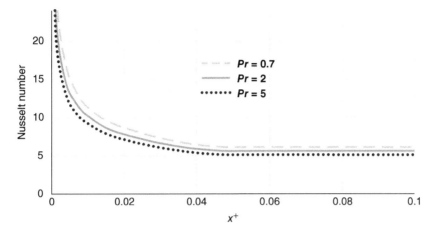

Figure 3.12 Combined entry-length Nusselt numbers in a circular tube with constant surface temperature for different Prandtl number, *Pr*.

Table 3.12 Nusselt numbers for combined thermal- and hydrodynamic-entry-length flow in a circular tube with constant surface temperature.

	Nu_m		
x^+	$Pr = 0.7$	$Pr = 2$	$Pr = 5$
0	∞	∞	∞
0.001	30.6	25.2	22.1
0.002	22.1	19.1	16.8
0.004	16.7	14.4	12.9
0.006	14.1	12.4	11
0.01	11.3	10.2	9.2
0.02	8.7	7.8	7.1
0.05	6.1	5.6	5.1
∞	3.66	3.66	3.66

Table 3.13 Combined thermal- and hydrodynamic-entry-length Nusselt numbers for the circular tube with *constant heat flux*.

	Nu_m		
x^+	$Pr = 0.01$	$Pr = 0.7$	$Pr = 10$
0	∞	∞	∞
0.002	24.2	17.8	14.3
0.010	12.0	9.12	7.87
0.020	9.10	7.14	6.32
0.01	6.08	4.72	4.51
0.20	5.73	4.41	4.38
∞	4.36	4.36	4.36

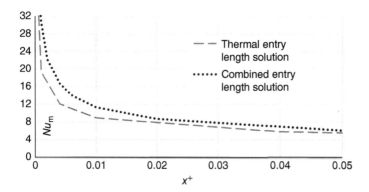

Figure 3.13 Combined entry-length and thermal-entry length solutions for laminar flow in a circular tube with constant surface temperature for $Pr = 0.7$.

For combined entry length, Nusselt number decreases with increase in Prandtl number in the entrance regions and approaches the fully developed values.

Figure 3.13 shows a comparison of combined entry-length solution and thermal entry-length solution for laminar flow in a circular tube with constant surface temperature and for $Pr = 0.7$.

As we can see, the Nusselt number for combined entry-length solution is higher than that for the thermal entry-length solution and approaches thermal entrance solution for larger distances. Both solutions approach the fully developed values at around $x^+ \cong 0.05$.

We can summarize it with the following few important notes, which could be helpful while selecting heat-transfer fluid and performing design analysis for heat exchangers: (i) Lower the Prandtl number, the higher is the combined entry-length Nusselt number values and (ii) The Nusselt number for the combined entry length is always higher than the thermal-entry-length solution.

3.3.5.1.6 Combined Entry Length: Empirical Correlation

For developing both velocity and temperature or combined entry length, the recommended correlation of **Baehr and Stephan correlation** (Bergman et al. 2011) **is** given for **constant surface temperature** and for $Pr \geq 0.1$ as follows:

$$\overline{Nu}_D = \frac{\frac{3.66}{\tanh(2.264Gz_D^{-1/3}+1.7Gz_D^{-2/3})} + 0.0499Gz_D \tanh(Gz_D^{-1})}{\tanh(2432Pr^{1/6} \, Gz_D^{-1/6})} \tag{3.62}$$

where

Gz_D = Graetz number = $Re_D PrD/L$

Another simple form of combined entry-length laminar-flow correlation for low Prandtl number values ($Pr < 5$) and for constant surface temperature is given by **Sieder and Tate correlation** (Incropera et al. 2006) as follows:

$$\overline{Nu}_D = 1.86 \left(\frac{Re_D Pr}{L/D} \right)^{1/3} \left(\frac{\mu}{\mu_s} \right)^{0.14} \tag{3.63a}$$

or

$$\overline{Nu}_D = 1.86(Gz)^{1/3}\left(\frac{\mu}{\mu_s}\right)^{0.14} \tag{3.63b}$$

For $0.60 \leq Pr \leq 5$ and $0.0044 \leq \frac{\mu}{\mu_s} \leq 0.75$

3.3.5.2 Internal Turbulent-Flow Heat-Transfer Correlations

Since turbulent flows are very complex, empirical correlations are mostly used for design purposes. For *fully developed turbulent flow* in a smooth circular tube, the heat-transfer correlation is given by Colburn correlation:

$$Nu_D = 0.023 Re_D^{4/5} Pr^{1/3} \tag{3.64}$$

Another preferred correlation that is applicable for moderate fluid-temperature variations is given by *Dittus–Boelter* correlation as

$$Nu_D = 0.023 Re_D^{4/5} Pr^n \tag{3.65}$$

For $0.6 \leq Pr \leq 160$, $Re_D \geq 10\,000$ and $\left(\frac{L}{D}\right) \geq 10$

where $n = 0.4$ for heating of fluid and $n = 0.3$ for cooling of fluid.

For larger fluid property variations, the recommended equation is by *Sieder and Tate correlation*:

$$Nu_D = 0.027 Re_D^{4/5} Pr^{1/3}\left(\frac{\mu}{\mu_s}\right)^{0.14} \tag{3.66}$$

For $0.7 \leq Pr \leq 16\,700$, $Re_D \geq 10\,000$ and $\left(\frac{L}{D}\right) \geq 10$

While correlations of *Dittus–Boelter* and *Sieder and Tate correlations* are generally used for simplicity in design computations, very large errors may result for some applications. A more accurate correlation applicable to a wider range of Reynolds number is given by *Gnielinski correlation:*

$$Nu_D = \frac{(f/8)(Re_D - 1000)Pr}{1 + 12.7(f/8)^{1/2}(Pr^{2/3} - 1)} \tag{3.67a}$$

For $0.5 \leq Pr \leq 2000$ and $3000 \leq Re_D \leq 5 \times 10^6$

where the friction factor, f, is given by Moody's diagram or by the curve-fit equation of Moody diagram as

$$f = (0.790)\ln(Re_D - 1.64)^{-2} \tag{3.67b}$$

Gnielinski's correlation is *applicable for both constant surface temperature* and *constant heat-flux* conditions.

As we have mentioned before, for turbulent flows, both hydrodynamic and thermal developing regions are significantly reduced (see Eq. (3.29)) and fully-developed turbulent-flow correlation values can be assumed reasonably as the average Nusselt number for the entire tube length.

3.3.5.3 Liquid Metals

Some of major characteristics that make liquid metal attractive coolants are high thermal conductivity, lower heat capacity, and wider working temperature range as single-phase

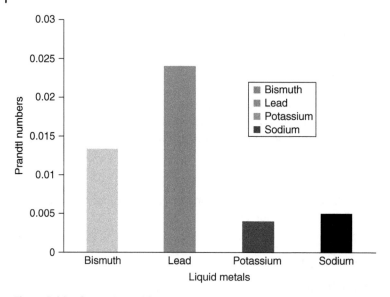

Figure 3.14 Comparison of Prandtl number values for four different liquid metals.

liquid coolant. Liquid metal heat transfer is of considerable interest because of the high heat-transfer rates that can be achieved due to their high thermal conductivity compared to other fluids. Unlike the typical cooling fluids, e.g. water, liquid metals have the advantage of being liquid even at high temperatures without phase transition to gas phase, allowing more compact heat-exchanger design. Some of the applications where liquid metals can be used as coolants are solar concentrator power generation, liquid-metal-fuel reactors, and liquid-metal magneto hydrodynamic power generators.

Liquid metals have high thermal conductivity and low specific heat when compared to ordinary liquids. The Prandtl number, which is a measure of momentum diffusivity over thermal diffusivity, is about a hundred times lower than that of water. Liquid metals have higher convection heat rate compared to water and oil. The Prandtl number for four different liquid metals is shown in Figure 3.14. Some of the most commonly used liquid metal coolant properties are given in Appendix C.

Applicable correlations for liquid metals are:

Skupinski's correlation for ***constant surface heat flux***:

$$Nu_D = 4.82 + 0.0185 Pe_D^{0.827} \tag{3.68}$$

For $3.6 \times 10^3 \leq Re_D \geq 9.05 \times 10^5$ and $10^2 \leq Pe_D \geq 10^4$

Seban and Shimazaki's correlation for ***constant surface temperature***:

$$Nu_D = 5.0 + 0.025 Pe_D^{0.8} \tag{3.69}$$

where

Pe_D = Peclet number = $Re_D Pr$

3.3.6 External Flows

A number of external flow applications are useful for the analysis and design of thermal components and processes such as flow over a flat horizontal or vertical surface, flow of a cylinder, flow over a circular and noncircular cylinder, flow over sphere, and flow over a banks of circular tubes. A summary of applicable correlations is presented here.

3.3.6.1 Laminar Flow Over a Flat Plate

Hydrodynamic boundary-layer thickness:

$$\frac{\delta}{x} = Re_x^{-\frac{1}{2}} \tag{3.70}$$

Local coefficient of friction:

$$\frac{C_{fx}}{2} = 0.332 Re_x^{-\frac{1}{2}} \tag{3.71}$$

Average coefficient of friction:

$$\frac{\overline{C_{fL}}}{2} = 0.664 Re_x^{-\frac{1}{2}} \tag{3.72}$$

3.3.6.1.1 Convection Heat Transfer Correlations: Constant Surface Temperature Condition

Local heat-transfer coefficient:

$$Nu_x = 0.332 Re_x^{\frac{1}{2}} Pr^{\frac{1}{3}} \tag{3.73a}$$

Valid in the range **of** $0.6 \leq Pr \leq 50$

$$Nu_x = 0.332 Re_x^{\frac{1}{2}} Pr^{\frac{1}{3}} \left[1 - \left(\frac{x_0}{x} \right)^{\frac{3}{4}} \right]^{-\frac{1}{3}} \quad \text{with unheated length of } x_0 \tag{3.73b}$$

Mean Nusselt number over plate of length:

$$\overline{Nu_L} = 2 Nu_{x=L} = 0.644 Re_L^{\frac{1}{2}} Pr^{\frac{1}{3}} \tag{3.74}$$

For a wider range of Prandtl number, the following two correlations are also applicable:

$$Nu_x = \frac{0.3387 Re_x^{\frac{1}{2}} Pr^{\frac{1}{3}}}{\left[1 + \left(\frac{0.0468}{Pr} \right)^{\frac{2}{3}} \right]^{\frac{1}{4}}} \quad \text{for a wider range of } Pr : RePr > 100 \tag{3.75}$$

Constant surface heat-flux condition:

$$Nu_x = 0.453 Re_x^{\frac{1}{2}} Pr^{\frac{1}{3}} \text{ valid in the range of } 0.6 \leq Pr \leq 50 \tag{3.76a}$$

$$Nu_x = \frac{0.4637 Re_x^{\frac{1}{2}} Pr^{\frac{1}{3}}}{\left[1 + \left(\frac{0.0207}{Pr} \right)^{\frac{2}{3}} \right]^{\frac{1}{4}}} \quad \text{for a wider range of } Pr : RePr > 100 \tag{3.76b}$$

3.3.6.2 Turbulent Flow Over a Flat Plate

3.3.6.2.1 Local Heat-Transfer Coefficient

$$Nu_x = 0.0296 Re_x^{\frac{4}{5}} Pr^{\frac{1}{3}} \quad \text{for } 0.6 \leq Pr \leq 60 \tag{3.77}$$

Average mixed heat-transfer coefficient with laminar flow till the critical Reynolds, $Re = Re_{crit}$, and turbulent flow thereafter with Reynolds number greater than the critical Reynolds number value is given as

$$\overline{Nu_L} = (0.037Re_L^{4/5} - A)Pr^{1/3} \tag{3.78a}$$

For $0.6 \leq Pr \leq 60$ and $Re_{crit} \leq Re_L \leq 10^8$
where

$$A = 1700 \quad \text{for } Re_{crit} = 5 \times 10^5 \tag{3.78b}$$

and

$$A = 0.037Re_{crit}^{\frac{4}{5}} - 0.664Re_{crit}^{\frac{1}{2}} \text{ for a range of critical Reynolds numbers}$$

for $0.6 \leq Pr \leq 60$ and $Re_{crit} \leq Re_L \leq 10^8$.

3.3.6.3 External Cross Flow Over a Cylinder

As fluid moves in cross flow and progresses along the front side of cylinder, boundary layer develops over the cylinder surface and the velocity parallel to the surface decreases from upstream velocity, u_∞, at the edge of the boundary layer to zero velocity at the cylinder surface. For flow along the front side of the cylinder, the pressures decrease and then increases along the back side as flow moves to the back. This increase in pressure causes decrease in velocity throughout the boundary layer following the velocity–pressure relationship given by the Bernoulli equation, and flow reverses at certain point near the surface. The separation point exists where the normal component of the velocity gradient becomes zero, i.e. $\left.\frac{\partial u}{\partial y}\right]_{y=0}$. Beyond this point of separation, flow reversal takes place and results in additional drag force, called **pressure drag**. The net drag force then has two components: the viscous drag and the pressure drag. For no-flow separations, the drag force is caused by the viscous friction only.

Reynolds number for cross flow over a cylinder (Figure 3.15) is defined as $Re_D = \frac{\rho u_\infty D}{\mu}$. Transition to turbulent flow takes place at an approximate critical Reynolds number of $Re \sim 10^5$.

Regarding the heat transfer, the heat-transfer coefficient decreases over the front side as boundary layer thickens with a minimum taking place at the point of separation. Beyond this point of separation, heat-transfer coefficient increases on the back side of the cylinder surface due to the formation of circulating eddies in the separated lows. As flow changes from laminar to turbulent, a second jump in convection heat-transfer coefficient takes place. Due to the complex nature of the flow, most convection correlations are empirical and derived, based on experimental analysis.

An average heat-transfer coefficient is given as

$$\overline{Nu_D} = \frac{hD}{k} = CRe_D^m Pr^{1/3} \tag{3.79}$$

where C and m are constants and vary with Reynolds number as given in Table 3.14

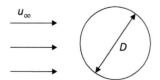

Figure 3.15 Cross-flow over a circular cylinder or tube.

Table 3.14 Constants C and m for heat-transfer correlation for cross-flow over a circular cylinder.

Re_D	C	m
0.4–4	0.989	0.330
4–40	0.911	0.385
40–4000	0.683	0.466
4000–40 000	0.193	0.618
40 000–400 000	0.027	0.805

A more comprehensive relation is given by **Churchill and Bernstein correlation** for a wide range of available data as

$$\overline{Nu}_D = 0.3 + \frac{0.62 Re_D^{1/2} Pr^{1/3}}{\left[1 + \left(0.4/Pr^{2/3}\right)\right]^{1/4}\left[1 + \left(\frac{Re_D}{282\,000}\right)^{5/8}\right]^{4/5}} \quad \text{for } Re_D Pr \geq 0.2 \quad (3.80)$$

where all properties are evaluated at the film temperature.

3.3.6.4 Flow Over a Sphere

A number of correlations are available for gases and liquids with different restrictions. Most popular one is given by **Whitaker correlation** as

$$Nu = 2 + \left(0.4 Re_D^{1/2} + 0.06 Re_D^{2/3}\right) Pr^{0.4}\left(\frac{\mu}{\mu_s}\right)^{1/4} \quad (3.81)$$

For $0.71 \leq Pr \leq 380$, $3.5 \leq Re_D \leq 7.6 \times 10^4$, and $1.0 \leq \left(\mu/\mu_s\right) \leq 3.2$.

3.3.6.5 Flow Over Tube Banks

Heat transfer to or from tube banks in cross flow is relevant to many industrial applications, including heat exchangers such as condensers, boilers, cooling coils, etc. A bundle of tubes may be arranged in inline or in a staggered manner as shown in Figure 3.16.

Heat is transferred between the fluid-1 flowing inside the tubes and fluid-2 flowing over the banks of tubes. For fluid flowing inside the tubes, we will select appropriate correlation for inside tube flow. Here, we will present correlations for external fluid assuming outside wall temperature T_s as known.

For flow across tube banks, the experimental correlation is given by **Zukauskas correlation** as

$$\overline{Nu}_d = \frac{hd}{k} = C_1 Re_{d,max}^m Pr^{0.36}\left(\frac{Pr}{Pr_s}\right)^{\frac{1}{4}} \quad (3.82)$$

for $N_L \geq 20$, $0.7 \leq Pr \leq 500$, and $1000 \leq Re_{d,\,max} \leq 2 \times 10^6$
where constant C_1 and m is given in Table 3.15

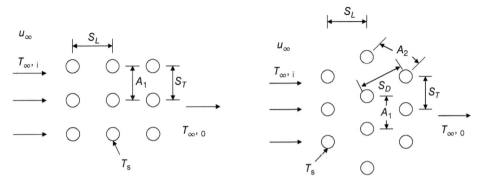

Figure 3.16 Banks of tubes for heat transfer between fluids in cross-flow. (a) Inline tube arrangement. (b) Staggered tube arrangement.

Table 3.15 Values of constants C_1 and m used in *Zukauskas correlation*.

Configuration	$Re_{D,max}$	C_1	m
Aligned	$10-10^2$	0.80	0.40
Staggered	$10-10^2$	0.90	0.40
Aligned	$10^2 - 10^3$	Approximate as a single (isolated) cylinder	
Staggered	$10^2 - 10^3$		
Aligned	10^3 to 2×10^5	0.27	0.63
	$(S_T/S_L < 2)$		
Staggered	10^3 to 2×10^5	0.40	0.60
	$(S_T/S_L > 2)$		
Aligned	2×10^5 to 2×10^6	0.021	0.84
Staggered	2×10^5 to 2×10^6	0.022	0.84

Source: Bergman et al. (2011). © 2011, John Wiley & Sons.

For $N_L < 20$, a correction factor is used as

$$\overline{Nu_d}\Big|_{N_L<20} = C_2 \overline{Nu_d}\Big|_{\geq 20} \tag{3.83}$$

where the constant C_2 is given in Table 3.16.

Reynolds number used in the correlation is based on the maximum fluid velocity occurring within the tube banks as follows:

$$Re_{d,max} = \frac{\rho V_{max} d}{\mu} \tag{3.84}$$

For **aligned arrangement**, the maximum velocity occurs at the transverse plane area A_1 as shown in Figure 3.16 and the maximum velocity is estimated

$$V_{max} = \frac{S_T}{S_T - d} V \tag{3.85a}$$

Table 3.16 Values of constant C_2 used in Eq. (3.83).

N_L	1	2	3	4	5	7	10	13	16
Aligned	0.70	0.80	0.86	0.90	0.92	0.95	0.97	0.98	0.99
Staggered	0.64	0.76	0.84	0.89	0.92	0.95	0.97	0.98	0.99

Source: Bergman et al. (2011). © 2011, John Wiley & Sons.

For **staggered alignment**, the maximum velocity may occur at either the normal area A_1 and or the inclined area A_2 based on following conditions:

V_{max} at area A_2 if $2(S_D - d) \le (S_T - d)$

and have the value given as:

$$V_{max} = \frac{S_T}{2(S_D - d)} V \tag{3.85b}$$

3.3.6.5.1 Energy Balance Over Banks of Tubes

Energy balance for the fluid over the banks of tube is written as

$$q = \dot{m}C_p(T_i - T_o) = h_o A \Delta T_{lmtd} \tag{3.86}$$

where

$$A = N_T \pi d L \tag{3.87a}$$

$$\Delta T_{lmtd} = \frac{(T_s - T_i) - (T_s - T_o)}{\ln\left(\frac{T_s - T_i}{T_s - T_o}\right)} \tag{3.87b}$$

The outlet temperature can be determined from the following relation

$$\frac{T_s - T_o}{T_s - T_i} = \exp\left(-\frac{\pi d N_t h_o}{\rho V N_T S_T C_p}\right) \tag{3.88}$$

A simplified form of the energy equation can be written as

$$q = \dot{m}C_p(T_i - T_o) = h_o A \left(T_s - \frac{T_i + T_o}{2}\right) \tag{3.89}$$

where

$\dot{m} = \rho V S_T N_T$
A = Total area of heat transfer = $N_t \pi d L$
L = Length of tubes
D = Outside diameter of tubes
N_T = Number of rows of tubes in flow direction
N_L = Number of tubes per tube row
N_t = Total number of tubes in the banks = $N_T \times N_L$

3.3.6.5.2 Pressure Drop in Banks of Tubes
Pressure drop for the flow across the banks of tubes is given as

$$\Delta P = N_L \chi \left(\frac{\rho V_{max}^2}{2} \right) f \tag{3.90}$$

where f is the friction factor obtained experimentally for different tube types and tube arrangements shown in Figure 3.17; χ is the correction factor introduced for $S_T \neq S_L$ and is given in subplots for friction factor (Kays and London 1984; Bergman et al. 2011).

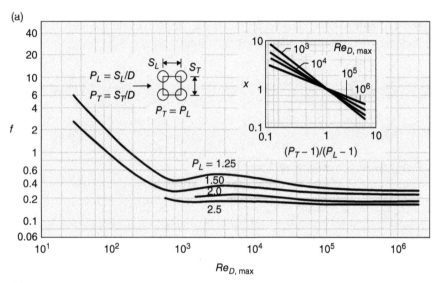

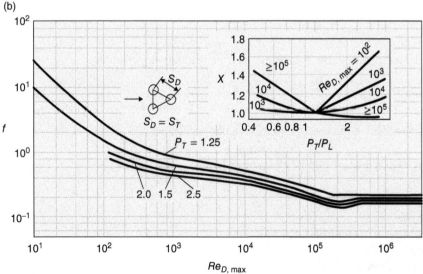

Figure 3.17 Friction factor values for flow over a bank of tubes arrangement. (a) Inline tube arrangement. (b) Staggered tube arrangement. Source: Bergman et al. (2011). © 2011, John Wiley & Sons.

The heat-transfer coefficient and pressure drop given by Eqs. (3.89) and (3.90), respectively, are applied for flow over the banks of tubes in the shell side of a shell and tube heat exchanger.

3.3.6.6 Jet Cooling

Enhanced convective heat transfer is achieved for some heat-transfer applications by using impinging jets of fluid. Due to its capability to extract high heat power, jet impingement is an attractive cooling method for highly concentrated heat-generating areas. The jet is confined or semiconfined to the radial spread area of the narrow channel between the impingement surface and orifice. The jet nozzle may be of different kinds: round jet nozzle, slot jet nozzle, an array of jet nozzles, etc. Figure 3.18 shows impingement of a single nozzle jet over a heated surface.

A recommended correlation for average convection heat-transfer coefficients for a circular jet is given by Martin (1977) as

$$\frac{\overline{Nu}}{Pr^{0.42}} = \left[2A_r^{\frac{1}{2}} \frac{1 - 2.2A_r^{\frac{1}{2}}}{1 + 0.2\left(\frac{H}{D} - 6\right)A_r^{\frac{1}{2}}} \right][2Re^{1/2}(1 + 0.005Re^{0.55})^{1/2}] \tag{3.91}$$

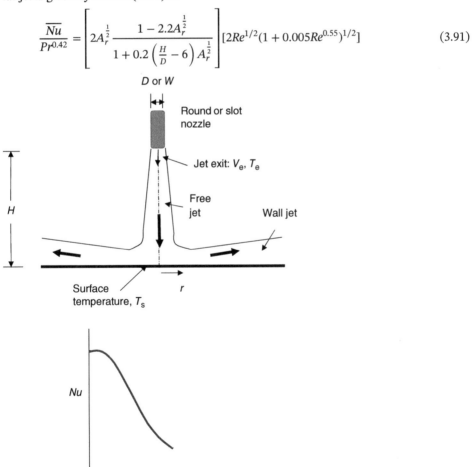

Figure 3.18 (a) Impinging jet from single round or slot nozzle on a heated surface maintained at a constant temperature. (b) Typical variation of local Nusselt number for impinging jet.

For $2000 \leq Re \leq 400\,000$, $2 \leq H/D \leq 12$, and $0.004 \leq A_r \leq 0.04$
where

$$\overline{Nu} = \frac{hD}{k}, Re = \frac{V_{ne}D}{v}, \text{ and } A_r = \frac{D^2}{4r^2}$$

Similarly, correlations for a square slot nozzle and for an array of slot nozzles are available in references (Martin 1977; Bergman et al. 2011).

For relatively large spacing between nozzle and the surface, the Nusselt number varies from a maximum value at the stagnation point ($r = 0$ or $x = 0$) in a bee-shaped manner to lower values as the fluid spreads outward with increasing r or x values. For a relatively smaller spacing, a second maximum peak value of Nusselt number is seen at a distance away from the stagnation. The presence of turbulence and formation of local eddies is observed.

3.3.7 Free or Natural Convection

In natural convection, heat transfer from heated surface to the adjacent fluid medium takes place due to the fluid motion induced by the buoyancy force. Fluid adjacent to the heated surface becomes lighter due to decrease in density as it gets heated and moves upwards. This body of fluid is replaced by the colder and heavier bulk fluid and sets in fluid motion termed as natural or forced convection. In such a free convection, fluid motion is caused by the buoyancy, which is induced due to the presence of the fluid density gradient caused by the temperature gradient and the gravitational body force that is proportional to the density. Other forms of free convection motion are also feasible due to other types of body forces such as Marangoni force caused by surface tension gradient, centrifugal force, or Coriolis force caused by rotation. There are different forms of free convections motions that are classified based on whether the flow is bounded by a surface or not. In the absence of the adjacent surface, the convection motion can be caused by a submerged heated object and such a free boundary flow is referred to as plume or buoyancy jet. Most common form of free convection flow is the boundary-layer flow development adjacent to a surface such as the heated vertical surface shown in Figure 3.19.

The boundary-layer equations for laminar free convection heat transfer from a heated semi-infinite vertical plate maintained at a constant surface temperature are summarized here. For detailed derivation of these equations, one may use reference books such as Kays

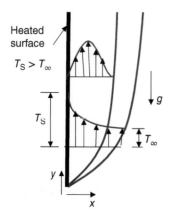

Figure 3.19 Velocity and thermal boundary-layer developments for free convection over a heated vertical surface.

Heated surface

$T_S > T_\infty$

T_S

g

T_∞

y

x

and London (1980) and Bergman et al. (2011). Major assumptions that are made in deriving these equations are laminar flow; free convection is due to gravity-induced buoyancy body force; and Boussinesq *approximation* for density difference $(\rho_\infty - \rho) \approx \rho\beta(T - T_\infty)$.

Governing Equations:

$$\text{Mass conservation:} \frac{\partial u}{\partial x} + \frac{\partial v}{\partial y} = 0 \tag{3.92a}$$

$$x - \text{momentum:} \quad u\frac{\partial u}{\partial x} + v\frac{\partial v}{\partial y} = g\beta(T - T_\infty) + v\frac{\partial^2 u}{\partial y^2} \tag{3.92b}$$

$$\text{Energy:} \quad u\frac{\partial T}{\partial x} + v\frac{\partial T}{\partial y} = \alpha\frac{\partial^2 T}{\partial y^2} \tag{3.92c}$$

where u and v are the fluid velocity components in x and y directions, respectively; β is the fluid volumetric thermal expansion; and g is the acceleration due to gravity.

It can be noted here that the first term on right-hand side of x-momentum equation is the buoyancy body force that drives the flow. The governing flow equations are solved using semianalytical methods such as similarity and integral methods (Kays and London 1980) for different vertical surface heating conditions such as constant surface temperature and constant surface heat flux.

The boundary conditions for constant surface temperature are given as

$$\text{At } y = 0 \quad u, v = 0 \quad T = T_s \tag{3.93a}$$

$$\text{At } x = 0 \quad u = 0 \quad T = T_\infty \tag{3.93b}$$

$$\text{At } y \to \infty \quad u \to 0 \quad T \to T_\infty \tag{3.93c}$$

The local convection heat-transfer coefficient is defined as

$$h_x = \frac{\dot{q}_0''}{T_s - T_\infty} \tag{3.94a}$$

where

$$\dot{q}_0'' = -k\left(\frac{\partial T}{\partial y}\right)_0 \tag{3.94b}$$

A semianalytical solution for the set of Eqs. (3.92a)–(3.92c) along with boundary conditions (3.93a)–(3.93c) is derived for laminar free convection from a vertical plate with constant surface temperature by of Ostrach (1952) using similarity method. Based on this solution, a local Nusselt number is derived as

$$Nu_x = \frac{h_x x}{k} = -\frac{\theta'(0)}{\sqrt{2}} Gr_x^{1/4} \tag{3.95a}$$

or

$$\frac{Nu_x}{Gr_x^{1/4}} = -\frac{\theta'(0)}{\sqrt{2}} \tag{3.95b}$$

where

$$\theta = \text{Dimensionless temperature} = \frac{T - T_\infty}{T_s - T_\infty}$$

$$Gr_x = \text{Local Grashof number} = \frac{g\beta x^3(T_s - T_\infty)}{v^2}$$

Table 3.17 Variation of free convection Nusselt number with Prandtl number.

Pr	Constant surface temperature $\dfrac{Nu_x}{Gr_x^{1/4}}$	Constant surface heat flux $\dfrac{Nu_x}{Gr_x^{1/4}}$
0.01	0.0570	0.0669
0.1	0.164	0.189
0.72	0.357	0.406
1.0	0.401	0.457
10	0.827	0.931
100	1.55	1.74
1000	2.80	3.14

Source: Based on London and Kays (1984).

Since the dimensionless form of the energy equation is found to be a function of Prandtl number, Pr, the local Nusselt number is computed for different Pr values using similar solution and is given in Table 3.17.

Based on this solution, the following correlation for local Nusselt number for natural free convection over a vertical surface is obtained:

For constant surface temperature

$$Nu_x = \frac{3}{4}\left[\frac{2Pr}{5\left(1 + 2Pr^{\frac{1}{2}} + 2Pr\right)}\right]^{1/4} [Gr_x Pr]^{1/4} \tag{3.96a}$$

The average free convection **Nusselt number** over a vertical plate of length, L, is computed based on integral average and is given as

$$\overline{Nu} = \frac{4}{3}Nu_L \tag{3.96b}$$

For constant surface heat flux

$$Nu_x = \left[\frac{Pr}{\left(4 + 9Pr^{\frac{1}{2}} + 10Pr\right)}\right]^{1/5} [Gr_x^* Pr]^{1/5} \tag{3.97}$$

where

$Gr_x^* = $ Modified Grashof number $= \frac{g\beta x^3}{k\nu^2}q_s''$

3.3.7.1 Effects of Turbulence

An initial laminar flow may transform into a turbulent flow depending on hydrodynamic instabilities and rate of amplification of any disturbances present. Transition depends on the relative magnitude of the buoyancy and viscous forces in the fluid, and generally occurs in a Grashof number (Gr_x) range of 10^9–10^{10} depending on the Prandtl number (Pr) values, surface orientation, and heating rates. For computation purposes, the critical **Rayleigh number** for a vertical surface is given as

$$Ra_{x,c} = Gr_{x,c}Pr = \frac{g\beta(T_s - T_\infty)x^3}{\nu\alpha} \approx 10^9 \tag{3.98}$$

Turbulent free convection correlations are developed based on experimental data.

3.3.7.2 Empirical Free Convection Correlations
External free convection flows:

General form of the correlations is given in terms of Rayleigh *Number* (Ra) and in some cases in terms of

$$\overline{Nu}_L = \frac{\overline{h}L}{k} = CRa_L^n \tag{3.99a}$$

where the Rayleigh number,

$$Ra_L = Gr_L Pr = \frac{g\beta(T_s - T_\infty)L^3}{\nu\alpha} \tag{3.99b}$$

3.3.7.3 Free Convection Over a Vertical Plate
For Laminar flow

$$\overline{Nu}_L = \frac{\overline{h}L}{k} = 0.59Ra_L^{\frac{1}{4}} \quad \text{for } 10^4 \le Ra_L \le 10^9 \tag{3.100a}$$

For turbulent flow

$$\overline{Nu}_L = \frac{\overline{h}L}{k} = 0.10Ra_L^{\frac{1}{3}} \quad \text{for } 10^9 \le Ra_L \le 10^{13} \tag{3.100b}$$

For entire range of Ra_L, the heat-transfer correlation is given as

$$\overline{Nu}_L = \left\{ 0.825 + \frac{0.387Ra_L^{\frac{1}{6}}}{\left(1 + \left(\frac{0.492}{Pr}\right)^{\frac{9}{16}}\right)^{\frac{8}{27}}} \right\}^2 \tag{3.101a}$$

and for laminar flow as

$$\overline{Nu}_L = 0.68 + \frac{0.670Ra_L^{\frac{1}{4}}}{\left(1 + \left(\frac{0.492}{Pr}\right)^{\frac{9}{16}}\right)^{\frac{4}{9}}} \quad Ra_L \le 10^9 \tag{3.101b}$$

The free convection correlations for flow over a vertical surface may also be applied to **vertical cylinders** if the cylinder diameter is significantly larger than boundary-layer thickness, satisfying the following condition

$$\frac{D}{L} \ge 35/Gr_L^{1/4} \tag{3.102}$$

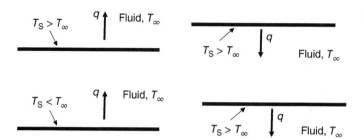

Figure 3.20 Buoyancy-driven flows over a horizontal cold ($T_s < T_\infty$) and hot ($T_s > T_\infty$) surfaces. (a) Hot surface facing up. (b) Hot surface facing down. (c) Cold surface facing up. (d) Cold surface facing down.

3.3.7.4 Free Convection Over a Horizontal Surface (Figure 3.20)
3.3.7.4.1 Uniform Surface Temperature
Hot surface facing up or cold surface facing down:

For laminar flow:

$$\overline{Nu}_L = 0.54 Ra_L^{1/4} \quad 10^5 \le Ra_L \le 2 \times 10^7 \tag{3.103a}$$

For Turbulent flow:

$$\overline{Nu}_L = 0.15 Ra_L^{\frac{1}{3}} \quad 2 \times 10^7 \le Ra_L \le 10^{11} \tag{3.103b}$$

Hot surface facing down or cold surface facing up

$$\overline{Nu}_L = 0.27 Ra_L^{1/4} \quad 3 \times 10^5 \le Ra_L \le 10^{10} \tag{3.104}$$

3.3.7.4.2 Uniform Surface Heat Flux
For the horizontal plate with heated surface facing up:

$$\overline{Nu}_L = 0.13 Ra_L^{1/3} \quad Ra_L \le 2 \times 10^8 \tag{3.105a}$$

and

$$\overline{Nu}_L = 0.16 Ra_L^{1/3} \quad 5 \times 10^8 \le Ra_L \le 10^{11} \tag{3.105b}$$

For the horizontal plate with heated surface facing down:

$$\overline{Nu}_L = 58 Ra_L^{1/5} \quad 10^6 \le Ra_L \le 10^{11} \tag{3.106}$$

Free Convection on Inclined Surface Free convection on an inclined surface can be computed from correlation Eqs. (3.100a)–(3.101b) for vertical surface by replacing the gravitational term in the Grashof number or *Rayleigh number (Ra)* term with inclusion of the effect of angle of inclination by replacing g with $g \cos \theta$. However, this recommendation is only restricted for the top inclined surface of a cold plate and bottom inclined surface of hot plate with a restriction of the angle inclination in the range of $0 \le \theta \le 60°$ (Bergman et al. 1911) (Figure 3.21).

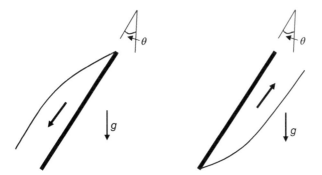

Figure 3.21 Free convection flow over an inclined cold surface. (a) Cold surface up. (b) Hot bottom surface down.

3.3.7.4.3 Uniform Surface Heat Flux
For the horizontal plate with heated surface facing down:

$$\overline{Nu}_L = 0.56(Ra_L \cos \theta)^{\frac{1}{4}} \tag{3.107}$$

For $10^5 < Ra_L \cos \theta \leq 10^{11}$ and $\theta < 88°$

For the horizontal plate with heated surface facing up:

$$\overline{Nu}_L = 0.145 \left[Ra_L^{\frac{1}{3}} - Ra_c^{\frac{1}{3}} \right] + 0.56 \left(Ra_c^{\frac{1}{3}} \cos \theta \right)^{\frac{1}{4}} \tag{3.108}$$

For $Ra_L \leq 10^{11}$ and $-15° < \theta < -75°$.

Free Convection Heat Transfer from Horizontal Heated Cylinder The boundary-layer development for free convection heat transfer is from a horizontal heated horizontal cylinder. The average heat-transfer coefficient over a long horizontal heated cylinder with constant surface temperature is given by the Churchill and Chu correlation:

$$\overline{Nu}_D = \left\{ 0.60 + \frac{0.387 Ra_D^{\frac{1}{6}}}{\left(1 + \left(\frac{0.559}{Pr} \right)^{\frac{9}{16}} \right)^{\frac{8}{27}}} \right\}^2 \quad \text{for } Ra_D \leq 10^{12} \tag{3.109}$$

Free Convection Heat Transfer from Heated Spheres

$$\overline{Nu}_D = 2 + \frac{0.589 Ra_D^{\frac{1}{4}}}{\left(1 + \left(\frac{0.469}{Pr} \right)^{\frac{9}{16}} \right)^{\frac{4}{9}}} \quad \text{for } Pr \geq 0.7 \text{ and } Ra_D \leq 10^{11} \tag{3.110}$$

Free Convection Within Parallel-Plate Channels Free convection through vertical or inclined flow channels, formed by two parallel planes, represents a base case of many common applications such as heat dissipation from electronics circuit boards stacked vertically upward; an array of equally spaced vertical plates with surface-mounted or

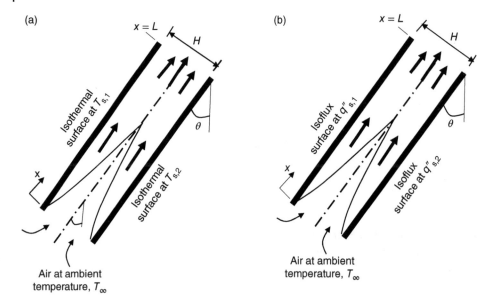

(a)

(b)

Figure 3.22 Free convection flow between parallel heated plates.

surface-embedded electrical heaters; a stack of equally spaced battery packs mounted vertically forming parallel-flow channels; a heat sink that consists of an array of vertical rectangular fins, which are used to cool an electronic device; and an inclined solar collector that consists of a parallel-plate channel through which water or air flows under the influence of the buoyancy force (Figure 3.22).

3.3.7.4.4   Vertical Channels
For symmetrically heated, isothermal plates

$$\overline{Nu}_H = \frac{1}{24} Ra_H \left(\frac{H}{L}\right) \left\{1 - \exp\left[\frac{-35}{Ra_s \left(\frac{H}{L}\right)}\right]\right\}^{\frac{3}{4}} \tag{3.111}$$

For $10^{-1} \le Ra_H \left(\frac{H}{L}\right) \le 10^5$.

where average Nusselt and Rayleigh numbers are defined as

$$\overline{Nu}_H = \left(\frac{\frac{q}{A}}{T_s - T_\infty}\right) \frac{H}{k} \tag{3.112a}$$

and

$$Ra_H = \frac{g\beta(T_s - T_\infty)H^3}{\nu\alpha} \tag{3.112b}$$

3.3.7.4.5   Inclined Vertical Channels
Free convection on an inclined channel can be computed from the correlation Eqs. (3.112a) and (3.112b) for vertical channel by replacing the gravitational term in the *Rayleigh number* (*Ra*) replacing g with g cos θ.

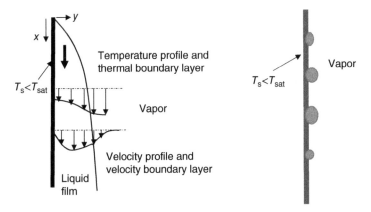

Figure 3.23 Condensation over vertical plate. (a) Film-wise condensation. (b) Drop-wise condensation.

3.3.8 Condensation Heat Transfer

Condensation is a mode of convection heat transfer where fluid changes phase from vapor to liquid, releasing latent heat. Condensation takes place when vapor comes in contact with a cooler surface with temperature below the saturation of the vapor like in a water-cooled tube in a steam condenser. Vapor undergoes change of phase and transfers the released latent heat to the cooled surface and forms the condensate film over the surface. Another variation of condensation is the dropwise condensation where vapor condenses in small droplets over the surface and without covering the entire cooled surface. Such dropwise condensation leads to higher heat-transfer rate, but it is quite difficult to sustain without forming the film (Figure 3.23).

3.3.8.1 Laminar Film Condensation Over a Vertical Plate

One of the simplest analytical models for film condensation is derived by neglect advection in the falling liquid file, and hence assuming linear temperature profile in the film and based on the solution of conduction heat equation in the film. From Fourier law of heat conduction, we can write the surface heat flux as

$$q_s'' = k_1 \frac{(T_{sat} - T_s)}{\delta} \tag{3.113}$$

where the thickness of the film, $\delta(x)$, is derived from the solution of the x-momentum equation for u-component velocity, film mass flow rate, and energy balance equation. The momentum equation is derived through a balance of the weight of a differential film element with the viscous drag and buoyancy force. Equating the heat removal rate over the differential element with heat released by the condensate and integrating the differential equation for the film thickness, we get

$$\delta_x = \left[\frac{4k_1\mu_1(T_{sat} - T_s)x}{g\rho_1(\rho_1 - \rho_v)h_{fg}'} \right]^{\frac{1}{4}} \tag{3.114}$$

The local convection heat-transfer coefficient is defined from Newton's law of cooling as

$$h_x = \frac{q''_s}{(T_{sat} - T_s)} \tag{3.115}$$

Combining Eqs. (3.113) and (3.115), we get the **local convection heat-transfer coefficient** as

$$h_x = \left[\frac{g\rho_1(\rho_1 - \rho_v)k_1^3 h'_{fg}}{4\mu_1(T_{sat} - T_s)x} \right]^{\frac{1}{4}} \tag{3.116}$$

Average heat-transfer coefficient for **laminar film condensation** over vertical plate of length L is computed from

$$\overline{h}_L = \frac{1}{L} \int_0^L h_x dx \tag{3.117}$$

Substituting Eq. (3.116) for local heat-transfer coefficient in Eq. (3.117), we get the **average film heat-transfer coefficient** as

$$\overline{h}_L = 0.943 \left[\frac{g\rho_1(\rho_1 - \rho_v)k_1^3 h'_{fg}}{\mu_1(T_{sat} - T_s)L} \right]^{\frac{1}{4}} \tag{3.118}$$

where

k_1 = Thermal conductivity of liquid
μ_1 = Dynamic viscosity of liquid
ρ_1 = Density of liquid
ρ_v = Density of vapor
$h'_{fg} = h_{fg} + 0.68 C_{p,1}(T_{sat} - T_s)$
h_{fg} = Latent heat of vaporization

Use of h'_{fg} instead of h_{fg} takes into account the nonlinear variation of temperature of the liquid within the film and the thermal advection effects. The vapor density, ρ_v, and h'_{fg} are computed based on saturation temperature, T_{sat}. It is recommended to compute all other properties at the film temperature given as

$$T_f = \frac{T_s + T_{sat}}{2} \tag{3.119}$$

Average Nusselt number:

$$\overline{Nu}_L = \frac{\overline{h}_L L}{k_1} = 0.943 \left[\frac{g\rho_1(\rho_1 - \rho_v)L^3 h'_{fg}}{\mu_1 k_1(T_{sat} - T_s)L} \right]^{\frac{1}{4}} \tag{3.120a}$$

where

$$h'_{fg} = h_{fg}(1 + 0.68 C_{p,1}\Delta T / h_{fg}) \tag{3.120b}$$

All properties are computed at the film temperature

$$T_f = \frac{T_{sat} + T_s}{2} \tag{3.120c}$$

The total heat-transfer rate is given as

$$q = \overline{h}_L A(T_{sat} - T_s) \tag{3.120d}$$

The total condensation rate is given as

$$\dot{m} = \frac{q}{h'_{fg}} = \frac{\overline{h}_L A(T_{sat} - T_s)}{h'_{fg}} \tag{3.120e}$$

3.3.8.2 Turbulent Condensation

A condensation film may initially remain laminar in the initial part of the vertical plate, but as the film thickens, it transitions into turbulent film downstream with the formation of waves and eddies, and finally turns into fully turbulent. The transition criterion is based on the film condensation Reynolds number defined as

$$Re_\delta = \frac{4\dot{m}}{\mu_1 W} \tag{3.121}$$

where mass flow rate in the condensation film is given as

$$\dot{m} = \rho_1 u_m W\delta \tag{3.122}$$

The Reynolds number range for different regimes of film condensations is:

Laminar film condensation: $Re_\delta \le 30$
Transition region or laminar with wavy film: $30 < Re_\delta \le 1800$
Turbulent film condensation: $Re_\delta > 1800$

Nusselt number correlations are available in different forms. Considering the difficulties in computing film thickness and Reynolds number in terms of film thickness, correlations are presented here in an alternate form as convenient form as follows:

Laminar wave free film condensation: $Re_\delta \le 30$ or

$$\overline{Nu}_L = \frac{\overline{h}_L \left(\frac{v_1^2}{g}\right)^{1/3}}{k_1} = 0.943P^{-1/4} \quad \text{for } P \le 15.8 \tag{3.123a}$$

Transition region or laminar wavy region: $30 < Re_\delta \le 1800$

$$\overline{Nu}_L = \frac{\overline{h}_L \left(\frac{v_1^2}{g}\right)^{1/3}}{k_1} = \frac{1}{P}(0.68P + 0.89)^{0.82} \quad \text{for } 15.8 \le P \le 25.30 \tag{3.123b}$$

Fully turbulent region: $Re_\delta > 1800$

$$\overline{Nu}_L = \frac{\overline{h}_L \left(\frac{v_1^2}{g}\right)^{1/3}}{k_1} = \frac{1}{P}[(0.024P - 53)Pr_1^{1/2} + 89]^{4/3} \quad \text{for } P \ge 25.30, Pr_1 \ge 1 \tag{3.123c}$$

where P is a dimensionless parameter given as

$$P = \frac{k_1(T_{sat} - T_s)L}{\mu_1 h'_{fg}(v_1^2/g)^{1/3}} \tag{3.123d}$$

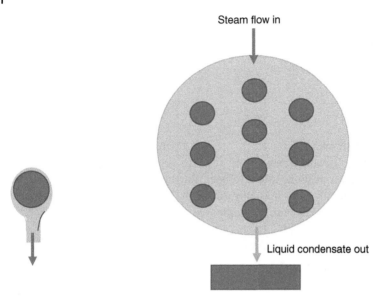

Figure 3.24 Film condensation horizontal tube. (a) Film condensation over a horizontal tube. (b) Film condensation over a bundle of tubes.

It can be noted here that the convection heat-transfer correlation for film condensation over a vertical-plate surface can be applied to a vertical circular tube with diameter of the tube significantly larger than the film thickness.

3.3.8.3 Condensation Over Horizontal Cylindrical Tube

Many of the heat-transfer applications involve film condensation over a horizontal cylindrical tube or over a bundle of horizontal cylindrical tubes such as in a condenser of steam power plant. Steam, exiting out of the turbine, enters the condenser, a type of shell and tube heat exchange, which consists of a bundle of tubes cooled by circulating cooling water through the bundle tubes.

Figure 3.24 shows film condensation over a single horizontal cylindrical and a bundle of tubes with vertically aligned tier of tubes.

Analysis and design of such heat-transfer devices use heat-transfer convection correlations.

Empirical correlation for *film condensation over a single cylindrical tube* is given as

$$\bar{h}_d = 0.729 \left[\frac{g\rho_1(\rho_1 - \rho_v)k_1^3 h'_{fg}}{\mu_1(T_{sat} - T_s)d} \right]^{\frac{1}{4}} \tag{3.124}$$

where d is the outer diameter of the cylindrical tube.

3.3.8.3.1 Condensation Over a Vertical Tier of Horizontal Tubes

In a vertically aligned tier of tubes with continuous film condensation, as shown in Figure 3.25, it is expected that the condensation film thickens as it flows vertically down, and hence it is expected that the condensation coefficients at the lower tubes are less than

Figure 3.25 Film condensation with flow inside a tube. (a) Condensation at low velocity. (b) Condensation at high vapor velocity.

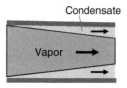

Condensate

Vapor

Condensate

those at the top layer. The average heat-transfer coefficient over circular tube in tier of n number tubes is given as

$$\overline{h}_{D,N} = \frac{\overline{h}_D}{\sqrt[4]{n}} \tag{3.125a}$$

or

$$\overline{h}_{D,N} = 0.729 \left[\frac{g\rho_1(\rho_1 - \rho_v)k_1^3 h'_{fg}}{\mu_1(T_{sat} - T_s)nd} \right]^{\frac{1}{4}} \tag{3.125b}$$

It can be noticed here that this equation is identical to the equation for a single horizontal circular tube with the exception of using n-times tube diameter in place of diameter d.

3.3.8.3.2 Film Condensation with Flow Inside Horizontal Tubes

Condensation for flow inside a horizontal tube has many practical applications such as condensation heat transfer for refrigerant flow in a vapor compression refrigeration and air-conditioning systems. As a fluid with higher vapor quality (x) flows inside a tube and comes in contact with the cooler tube surface of temperature $T_s < T_{sat}$, the vapor condenses over the tube surface like in the case of the film condensation over a vertical plate or over the cylindrical tube surface. However, vapor–liquid flow regime varies depending on the vapor-flowing velocity. Laminar film condensation inside a horizontal tube is given in terms of two correlations for two vapor velocities.

For a lower vapor velocity, condensed liquid tends to flow downward and accumulates at the bottom of the tube forming a liquid pool as shown in Figure 3.25a. For low vapor velocity and hence considering negligible vapor shear, the heat-transfer correlation given is based on falling condensate film with a different correlation constant given by Chato (1962) as

$$\overline{h}_d = 0.555 \left[\frac{g\rho_1(\rho_1 - \rho_v)k_1^3 h'_{fg}}{\mu_1(T_{sat} - T_s)d} \right]^{\frac{1}{4}} \tag{3.126a}$$

With a restricted low vapor Reynolds number given as

$$Re_v = \left(\frac{\rho_v V_v d}{\mu_v} \right) < 35\,000$$

For higher vapor velocities, the two-phase flow becomes annular with vapor flows in the annulus core and liquid film condensate flows along the tube surface under the influence of vapor shear force as shown in Figure 3.25b. The annular two-phase flow heat-transfer correlation is derived based on assumption that the two-phase annular flow Nusselt number is related to the single-phase liquid-flow Nusselt number with a two-phase multiplying factor given as a function of the Martenelli parameter (X_{tt}). One of the most popular empirical correlations for such flow regime and conditions is given by Dobson and Chto (1998) as

$$Nu_d = \frac{hd}{k_1} = 0.023 \, Re_{d,L}^{0.8} Pr_1^{0.4} \left[1 + \frac{2.22}{X_{tt}^{0.89}}\right] \qquad (3.126b)$$

where

$$Re_{d,L} = \frac{4\dot{m}(1-x)}{\equiv \pi d\mu_1}$$

$$x = \frac{\dot{m}_v}{\dot{m}} = \text{quality or the mass fraction of vapor in the liquid}$$

$$X_{tt} = \left(\frac{1-x}{x}\right)^{0.9} \left(\frac{\rho_v}{\rho_1}\right)^{0.5} \left(\frac{\mu_1}{\mu_v}\right)^{0.1} = \text{Martinelli parameter for}$$
$$\text{two-phase multiphase factor}$$

3.3.9 Boiling Heat Transfer

In heat exchangers like boilers or evaporators, fluids undergo phase change from liquid to vapor through formation of bubbles caused by latent heat-transfer process, known as boiling. Because of the phase-change nature involving latent heat transfer, the boiling heat-transfer coefficient is an order of magnitude higher than single-phase convection heat-transfer coefficient. The boiling process takes place when the liquid at a saturation temperature, T_{sat}, corresponding to its pressure, comes in contact with a hotter solid surface of temperature T_s. Following Newton's law of cooling, the boiling heat-transfer coefficient is written as

$$h = \frac{q_s''}{(\Delta T_e)} \qquad (3.127)$$

where

$$\Delta T_e = T_s - T_{sat} = \text{Excess temperature}$$

The boiling heat-transfer coefficient depends on a number of factors like flow field defined by the dynamics of bubble formation; the excess temperature; nature of the heating surface; and thermophysical properties of the liquid, including surface tension.

Boiling heat transfer is categorized into different types: *Pool boiling, Forced convection Boiling, Subcooled boiling,* and *Saturated boiling.* In **pool boiling**, the motion in a liquid body is caused by buoyancy-driven free convection and mixing induced by the formation and detachment of bubbles. In *forced convection*, flow is primarily influenced by the external means that acts on top of the motion caused by the free convection and bubble-induced mixing. In subcooled boiling, the temperature of the fluid in major part of the fluid body is less than the saturation temperature and so the bubble formed at the heating surface eventually condenses after rising to the subcooled region. This is in contrast to the *saturated boiling* where fluid temperature remains higher than the saturation temperature.

3.3.9.1 Pool Boiling

In a typical pool-boiling case, the fluid body comes in contact with a heated surface that maintains a temperature, T_s, higher than the saturation temperature of the fluid as shown in Figure 3.26. Heat transfer to the fluid body takes place with a heat flux, q_s'', due to the difference in temperature of the surface, T_s, and fluid saturation temperature, T_{sat}.

Figure 3.26   Pool-boiling diagram.

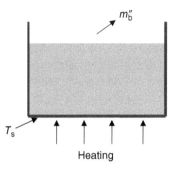

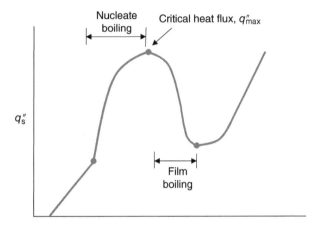

Excess temperature, $\Delta T_e = T_s = T_{sat}$

Figure 3.27   Typical boiling curve: variation of surface heat flux with excess temperature, $\overline{\Delta T}_e = T_s - T_{sat}$.

The heat flux, q_s'', and hence the convection heat-transfer coefficient, h, are strongly influenced by the excess temperature, $\overline{\Delta T}_e = (T_s - T_{sat})$ as demonstrated by the popular Nukiyama's boiling curve shown in Figure 3.27.

Based on the pool-boiling diagram, four boiling modes are identified: **Free Convection Boiling, Nucleate Boiling, Transition Boiling,** and **Film Boiling.**

Free convection boiling takes place at a lower range of excess temperature with the onset of the formation bubbles, but the fluid motion is predominant due to the free convection effect and the free convection correlation can be used to estimate the heat-transfer rate. **Nucleate boiling** is characterized by increased rate bubble formation, coalesces and collapse of bubbles contributing to the convection and jet flow nature of the fluid at higher excess temperature. Heat-transfer rate increases at a varying rate over a range of excess temperatures until it reaches a maximum heat-flux point, referred to as the **critical heat flux,** due to a balance of decreased heat-transfer coefficient and increased excess temperature. Nucleate boiling heat-transfer region is the preferred range of convection heat transfer in many heat-transfer applications due to the high heat-transfer rate and convection heat-transfer coefficient achieved at small excess temperature.

The boiling region beyond the critical point is referred to as the ***film-boiling regime*** where with increase in excess temperature, the bubble formation rate is so rapid that a vapor layer covers the heating surface and prevents the liquid to be in direct contact with the heating surface. Because of this vapor barrier layer, the heat-transfer rate and boiling-heat-transfer rate decrease with increase in excess temperature until it reaches minimum point. With further increase in the excess temperature, radiation becomes increasingly predominant in transferring heat through the vapor film layer and causes higher heat-transfer rate. This phenomenon, however, leads to a condition of uncontrolled increase in heating surface temperature approaching the melting point temperature, and hence needs to be avoided.

The surface heat-flux correlations used for different boiling modes are empirical and summarized here.

3.3.9.1.1 Rohsenow's Nucleating Boiling Correlation

$$q_s'' = \mu_1 h_{fg} \left[\frac{g(\rho_1 - \rho_v)}{\sigma} \right]^{\frac{1}{2}} \left[\frac{C_{p,1} \Delta T_e}{C_{s,f} h_{fg} Pr_1^n} \right]^3 \tag{3.128a}$$

where σ = surface tension. $C_{s,f}$ and n are experimentally determined values for different liquid-heating surface combinations. A few examples of these combination values are given in Table 3.18.

3.3.9.1.2 Correlation for Critical Heat Flux in Nucleate Boiling

$$q_{s,max}'' = 0.31 h_{fg} \rho_v \left[\frac{\sigma g(\rho_1 - \rho_v)}{\rho_v^2} \right]^{\frac{1}{2}} \tag{3.128b}$$

3.3.9.2 Film Pool Boiling

The film-boiling heat-transfer coefficient is derived by drawing a similarity with laminar film condensation. The film heat-transfer coefficient for boiling over a cylinder of diameter, D, as

$$\overline{Nu}_D = \frac{\overline{h}D}{k_v} = C \left[\frac{g(\rho_1 - \rho_v)h_{fg}' D^3}{v_v k_v (T_s - T_{sat})} \right]^{\frac{1}{4}} \tag{3.129}$$

Table 3.18 Nucleate boiling correlation coefficients for some liquid-heating surface combinations.

Fluid – heating surface combinations	$C_{s,f}$	n
Water – copper		
Scored	0.0068	1.0
Polished	0.0128	1.0
Water – stainless Steel		
Chemically etched	0.0133	1.0
Mechanically etched	0.0132	1.0
Ground and polished	0.0080	1.0

where

C = Correlation coefficient = 0.62

3.3.10 Internal Forced Convection Two-phase Flow Boiling

Two-phase flow boiling is important in the design of heat exchanger where liquid flowing inside a heated tube undergoes phase change through formation bubbles at the inner surface. Two-phase correlation was derived and presented (Kandlikar 1990; Bergman et al. 2011) for saturated flow-boiling region based on assuming a sum of contributions from both convective boiling term and nucleate boiling term and assuming as

$$h_{tp} = C_1\, Co^{C_2} h_{sp} + C_3 Bo^{C_4} h_{sp} \tag{3.130a}$$

h_{tp} = two-phase heat-transfer coefficient, $W/m^2\,°C$
h_{sp}= single-phase liquid heat-transfer coefficient given by Eq. (3.67a), $W/m^2\,°C$
Co = convection number = $\left(\frac{1-x}{x}\right)^{0.8}\left(\frac{\rho_v}{\rho_l}\right)^{0.5}$
x = quality or vapor mass fraction
Bo = boiling number = $\dfrac{q_s''}{\dot{m}'' h_{fg}}$
q_s'' = surface heat flux, W/m^2
$\dot{m}'' = \frac{\dot{m}}{A_c}$ = mass flow rate per unit are cross-sectional area, kg/m^2 s
h_{fg} = latent heat of vaporization, J/kg

Equation (3.130a) was further enhanced with the inclusion of some additional factors such *fluid-dependent parameter* for different fluid, F_{fl}, and *effect in horizontal flow* using *Froud number function*, $f(Fr_{lo})$, as follows

$$h_{tp} = C_1 Co^{C_2} h_{sp}(25Fr_{lo})^{C_5} + C_3 Bo^{C_4} h_{sp}(25Fr_{lo})^{C_6} F_{fl} \tag{3.130b}$$

Final form of the best-fitted correlation of wide of experimental data is given using two different sets of constants corresponding to the convective boiling and nucleate boiling regions, respectively:

$$\frac{h_{tp}}{h_{sp}} = 1.1360Co^{-0.9}(25Fr_{lo})^{0.3} + 667.2Bo^{0.7}F_{fl} \quad \text{in convective boiling region} \tag{3.131a}$$

and

$$\frac{h_{tp}}{h_{sp}} = 0.6683Co^{-0.2}(25Fr_{lo})^{0.3} + 1058.0Bo^{0.7}F_{fl} \quad \text{in nucleate boiling region} \tag{3.131b}$$

The proposed fluid-dependent parameter, F_{fl}, for different fluids is suggested as 1.0 for water, 1.30 for R_{11}, 1.50 for R_{12}, 2.20 for R_{22}, 1.31 for R_{138}. All properties of the fluid are estimated at the saturation temperature, T_{sat}.

The heat-transfer correlation at any condition is evaluated based on both these two sets, and the higher of the values is used as the predicted two-phase heat-transfer coefficient. This is to take into account of the fact that the transition from one region to other region takes place at the intersection of the two correlations.

Also, for vertical flows and horizontal flows with $Fr_{lo} > 0.4$, the Froude number factor, Fr in the convective boiling term becomes unity, i.e.

For vertical flows, $f(Fr_{lo}) = 1$

and

For horizontal flow with $Fr_{lo} > 0.4, f(Fr_{lo}) = 1$.

3.3.11 Effect of Temperature

Most of convection correlations are often developed based on assuming constant fluid properties. To include temperature-dependent properties, two following approaches are generally used.

3.3.11.1 Approach – I

Evaluates the properties at some intermediate temperature between solid-surface temperature and fluid temperature such as film temperature defined as

$$T_f = \frac{1}{2}(T_s + T_\infty)$$

This approach is generally followed for external flows.

3.3.11.2 Approach – II

Evaluates the properties for use in correlations at the mean temperature called mixed fluid temperature or mixing cup temperature and then **specifies a correction factor** in the correlated equation. Some of the common correction factors are $\left(\frac{Pr_m}{Pr_s}\right)$ or $\left(\frac{\mu_m}{\mu_s}\right)$.

Where subscript s refers to surface temperature and m refers to mean temperature. This approach is generally used for internal flow. The mean temperature can be computed in the following ways. (i) The mean temperature, T_m, at a given cross-section and (ii) An average mean temperature $\left(\overline{T_m} = \frac{T_{m,i} + T_{m,o}}{2}\right)$ of the inlet mean temperature, $T_{m,i}$, and outlet mean temperature, $T_{m,o}$.

Average heat-transfer coefficients are evaluated based on the properties evaluated as at average mean temperature, $\overline{T_m}$.

3.4 Thermal Radiation Heat Transfer

Radiation energy is considered as being energy emitted and transported in the form of electromagnetic waves or as photons. Radiation energy comprises a wide range of wavelengths covering short-wavelength γ-rays–X-rays to long-wavelength radio waves. Thermal radiation is the radiation energy emitted by a substance entirely due to its temperature. Thermal radiation is the intermediate portion (0.3–$50\,\mu m$) of the electromagnetic radiation, covering a part of ultraviolet radiation, entire range of visible radiation, and a part of infrared radiation. Thermal radiation heat transfer involves transmission and exchange of electromagnetic waves or photon particles as a result of temperature difference.

The thermal radiation emitted by a substance encompasses a range of wavelengths (λ), and it is referred to as spectral or monochromatic distribution ($E_{b\lambda}$) given by Planck's law.

The total black body emissive power is obtained by integrating the spectral emissive power over the entire range of wavelengths and derived as

$$E_b = \sigma T^4 \tag{3.132}$$

where σ = Stefan–Boltzmann constant = 5.6697×10^{-8} W/m^2 K^4 and T is the temperature in absolute scale, K.

Equation (3.132) is referred to as **Stefan–Boltzmann law**, which gives total emissive power of **black** or **ideal body** at a given temperature. For a **non-black body or real body**, the emitted power is less than the black body emissive power, and it is given as

$$E = \varepsilon E_b \tag{3.133a}$$

or

$$E = \varepsilon \sigma T^4 \tag{3.133b}$$

where ε is the **emissivity factor** of a **non-black surface**, and it is defined as the ratio of its emissive power to that of a black body

$$\varepsilon = \frac{E}{E_b} \tag{3.134}$$

Equations (3.133a and b) and (3.134) for the real body radiation and emissivity factor are the total quantities. The corresponding spectral or monochromatic quantities are given as

Real body spectral or monochromatic emissive power:

$$E_\lambda = \varepsilon_\lambda E_{b\lambda} \tag{3.135a}$$

Real body spectral or monochromatic emissivity:

$$\varepsilon_\lambda = \frac{E_\lambda}{E_{b\lambda}} \tag{3.135b}$$

A real surface also possesses three additional optical properties: **absorptivity** (α), **transmissivity** (τ), and **reflectivity** (ρ). All these three properties are associated with characteristics of a real surface as it interacts with incident thermal (G_r) radiation energy. As total radiation energy (G) is incident upon a surface, a fraction of energy is absorbed (G_a), a fraction of energy is transmitted (G_t) through the surface, and rest of the energy (G_r) is reflected back from the surface. This is expressed in a mathematical form as follows:

$$G_a + G_t + G_r = G \tag{3.136}$$

or

$$\alpha + \tau + \rho = 1 \tag{3.137}$$

where

$$\alpha = \frac{G_t}{G} = \text{absorptivity} \tag{3.138a}$$

$$\tau = \frac{G_t}{G} = \text{transmissivity} \tag{3.138b}$$

$$\rho = \frac{G_t}{G} = \text{reflectivity} \tag{3.138c}$$

For an **opaque object**, $\tau = 0$ and $\rho = 1 - \alpha = 1 - \varepsilon$ (3.138d)

Like the optical property emissivity, all these three optical properties are written here as total quantities, which are computed from integral averaging of the corresponding monochromatic or spectral properties: α_λ, τ_λ, and ρ_λ. All such optical properties not only depend on the wavelength but also depend on the direction.

Considering a surface in thermodynamic equilibrium, Kirchhoff's law states the relationship between α_λ and ε_λ as

$$\alpha_\lambda = \varepsilon_\lambda \tag{3.139}$$

Even though Eq. (3.139) is derived based on the assumption of equilibrium state of the system, this equation is also applicable for non-equilibrium state of the system as these optical properties are surface properties and depend on the nature and temperature of the surface.

3.4.1.1 The Gray Body

A gray body is defined with the assumption that the monochromatic emissivity, ε_λ, is independent of wavelength, and so it is stated that

$$\varepsilon = \varepsilon_\lambda \tag{3.140}$$

Emissivity values of substances vary with wavelength, temperature, and surface types. The peak of the variation curve shifts toward the shorter wavelengths for the higher temperatures. Total emissivity of some common substances is listed in Table A.4 .

3.4.1.2 Radiation Heat Exchange

Let us consider two black surfaces exchanging thermal radiation as shown in Figure 3.28.

Based on the Stefan–Boltzmann law, ***the radiation heat transfer between two black surfaces*** through nonparticipating medium is expressed as

$$q_R = A_i F_{ij}(Eb_i - Eb_j) = \sigma A_i F_{ij}(T_i^4 - T_j^4) \tag{3.141a}$$

Using a reciprocity relation $A_i F_{ij} = A_j F_{ji}$, Eq. (3.141a) can also be written as

$$q_R = \sigma A_j F_{ji}(T_i^4 - T_j^4) \tag{3.141b}$$

where A_i, A_j = areas of two surfaces, T_i, T_j = temperatures of two surfaces, and F_{ij} and F_{ji} are the shape factors or view factors.

The ***shape factor*** F_{ij} is defined as the fraction of energy leaving surface – i with a surface area that is intercepted by surface-2 with area A_2. The shape factor depends on the size, shape, and orientations of the two surfaces i and j.

Figure 3.28 Radiation exchange between two black bodies.

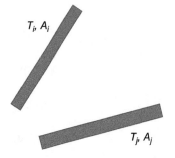

T_i, A_i

T_j, A_j

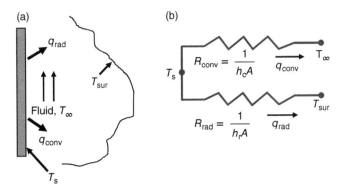

Figure 3.29 (a) Combined convection and radiation heat transfer, (b) resistances for convection and radiation heat transfer.

The shape factors are computed for many common geometries and configurations and presented in the form of formulas and graphs (Sparrow and Cess 1978; Siegal and Howell 1992).

Equation (3.141a) can also be written in an electrical network analogy as

$$q = \frac{E_{bi} - E_{bj}}{\frac{1}{A_i F_{ij}}} = \frac{E_{bi} - E_{bj}}{R_{ij}} \tag{3.142}$$

where $R_{ij} = \frac{1}{A_i F_{ij}}$ is the termed as the **radiation space resistance** between the two surfaces i and j (Figure 3.30a).

For many applications, the expression for radiation heat transfer given by Eq. (3.141b) is expressed in a linearized form as

$$q = h_r A (T_i - T_j) \tag{3.143}$$

where $h_r = \sigma(T_i + T_j)(T_i^2 + T_j^2)/R_{ij}$ and is referred to as the **radiation film heat-transfer coefficient**.

In many applications, a surface may exchange heat by both convective and radiative modes as demonstrated in Figure 3.29.

3.4.1.2.1 Radiation Heat Exchange Between Two Diffuse Gray Surfaces

The radiation exchange between two non-black objects is complicated due to the fact that the net thermal radiation that leaves the non-black surface has two components: (i) the emitted radiation from a real surface, E, and (ii) the reflected energy from the surface. This net energy that leaves a real surface is termed as **radiosity**, J, and expressed as

$$J = \varepsilon E_b + \rho G \tag{3.144}$$

Using Eq. (3.138d)

$$J = \varepsilon E_b + (1 - \varepsilon)G \tag{3.145}$$

The thermal radiation energy that leaves a non-black or gray surface is the difference between the radiosity, J, and the total surface irradiance, G; and expressed as

$$\frac{q_R}{A} = J - G = \varepsilon E_b + (1 - \varepsilon)G - G$$

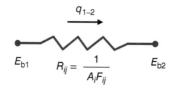

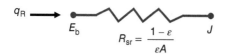

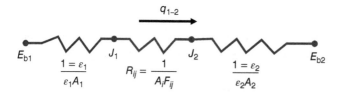

Figure 3.30 Electrical network analogy for thermal radiation transfer. (a) Equivalent electrical circuit for thermal radiation exchange between two black objects. (b) Equivalent electrical circuit for radiation heat transfer with surface resistance. (c) Thermal radiation exchange between two gray bodies.

or

$$q_R = \frac{\varepsilon A}{1 - \varepsilon}(E_b - J) = \frac{E_b - J}{\frac{1-\varepsilon}{\varepsilon A}} \tag{3.146}$$

The quantity in the denominator of the Eq. (3.146) is considered as the **surface resistance** to the radiation heat transfer from the surface subjected to a potential difference, $(E_b - J)$. An equivalent circuit with surface resistance is shown in Figure 3.30b.

The **radiation heat exchange between two diffuse gray surfaces** is derived as

$$q = \frac{E_{bi} - E_{bj}}{\frac{1-\varepsilon_i}{\varepsilon_i A_i} + \frac{1}{A_i F_{ij}} + \frac{1-\varepsilon_j}{\varepsilon_j A_j}} = \frac{\sigma(T_i^4 - T_j^4)}{\frac{1-\varepsilon_i}{\varepsilon_i A_i} + \frac{1}{A_i F_{ij}} + \frac{1-\varepsilon_j}{\varepsilon_j A_j}} \tag{3.147}$$

Equation can also be written in an electrical network analogy, with equivalent electrical circuit depicted in Figure 3.30c as

$$q = \frac{E_{bi} - E_{bj}}{R_{ij}} \tag{3.148}$$

where $R_{ij} = \frac{1-\varepsilon_i}{\varepsilon_i A_i} + \frac{1}{A_i F_{ij}} + \frac{1-\varepsilon_j}{\varepsilon_j A_j}$ = the total radiation resistance between two gray surfaces i and j.

The term $\frac{1-\varepsilon}{\varepsilon A}$ is defined as the **radiation surface resistance**. We can see that for material surfaces with higher emissivity values, the radiation surface resistance values are less, and hence enhance radiation heat transfer.

3.5 Heat-Transfer Resistances

In the transport of heat through a plane slab that is exposed to moving fluids on both sides of the slab, there are number of heat-transfer resistances that influence the heat transport. These resistances are

i. Convective heat-transfer resistance at the inner side of the slab, $R_{conv, i}$
ii. Conduction resistance in the plane slab layer, R_{cond}
iii. Convective heat-transfer resistance at the outer side of the slab, $R_{conv, o}$

Additionally, there could be heat-transfer resistances due to radiation losses from one or both sides of the surfaces as discussed earlier.

Figure 3.31 shows a typical temperature distribution across the plane slab.

While the fluid-temperature distribution in the fluid is obtained from the solution of **Navier–Stokes equations** along with the governing equation for heat transport, the overall resistance for convection heat-transfer rate from bulk fluid stream to the adjacent surface of the solid surfaces is often given by the convection mass transfer coefficients. As we can see higher the convective heat-transfer coefficient, the lower the convective heat-transfer resistance and that leads to smaller temperature drop between the bulk fluid and the solid surface. Conduction resistance in solid part of the slab is primarily controlled by the thermal conductivity of the slab material, slab thickness, and surface area of conduction. Higher the thermal conductivity of the material, the lower is the thermal conduction resistance and enhanced the heat transmission.

For one-dimensional heat-transfer across a plane composite wall, the heat-transfer rate is given as

$$q = \frac{(T_i - T_o)}{\sum R} = \frac{T_i - T_o}{\frac{1}{h_i A} + \frac{L}{kA} + \frac{1}{h_o A}} \tag{3.149}$$

and in terms of overall heat-transfer coefficient

$$q = UA(T_i - T_o) \tag{3.150}$$

Figure 3.31 Temperature distributions across a plan slab with convection on two sides.

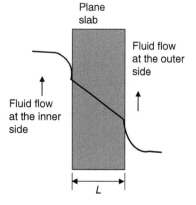

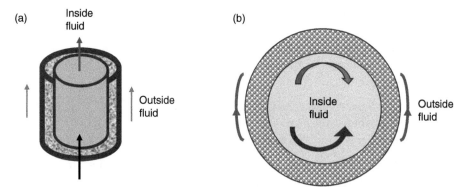

Figure 3.32 Heat transfer in radial system. (a) Heat transfer in a cylindrical wall. (b) Heat transfer through a spherical wall.

where U is the overall heat-transfer coefficient defined as

$$U = \frac{1}{\sum R} = \frac{1}{\frac{1}{h_i A} + \frac{L}{kA} + \frac{1}{h_o A}} \tag{3.151}$$

In a similar manner, conduction and convection resistances for a radial system are derived. For a **cylindrical system**, shown in Figure 3.32a, the resistances are derived as follows

$$R_{cond} = \frac{\ln \frac{r_o}{r_i}}{2\pi L k}, \quad R_{conv,i} = \frac{1}{h_{c,i} A_i} = \frac{1}{2\pi r_i L h_{c,i}} \quad \text{and} \quad R_{conv,o} = \frac{1}{h_{c,o} A_o} = \frac{1}{2\pi r_o L h_{c,o}} \tag{3.152}$$

where U is the overall heat-transfer coefficient defined as

$$U = \frac{1}{\sum R} = \frac{1}{\frac{1}{h_{c,i} A} + \frac{L}{kA} + \frac{1}{h_{c,o} A}} \tag{3.153}$$

For a spherical system (Figure 3.32b), the resistances are derived as follows:

$$R_{cond} = \frac{\frac{1}{r_i} - \frac{1}{r_o}}{4\pi k}, \quad R_{conv,i} = \frac{1}{h_{c,i} A_i} = \frac{1}{4\pi r_i^2 h_{c,i}} \quad \text{and} \quad R_{conv,o} = \frac{1}{h_{c,o} A_o} = \frac{1}{4\pi r_o^2 L h_{c,o}} \tag{3.154}$$

The overall heat-transfer coefficients for a cylindrical tube and spherical storage wall with convections inside and outside surfaces are defined as follows.

3.5.1.1 Overall Heat-Transfer Coefficient in a Heat-Exchanger Circular Tube

For a typical tube wall surface, the equivalent resistances across the tube wall can be shown (Figure 3.33).

The heat-transfer rate at any section is given as

$$q = \frac{T_i - T_o}{\frac{1}{h_{c,i} A_i} + \frac{\ln \frac{r_o}{r_i}}{2\pi L k} + \frac{1}{h_{c,o} A_o}} \tag{3.155}$$

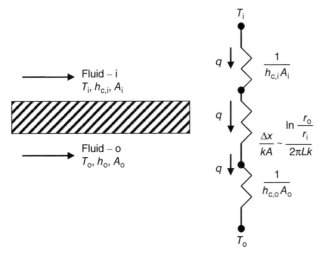

Figure 3.33 Steady-state heat transfer in a cylindrical system.

and overall heat-transfer coefficient is defined as

$$U_i A_i = U_o A_o = \cfrac{1}{\cfrac{1}{h_{c,i} A_i} + \cfrac{\ln \frac{r_o}{r_i}}{2\pi L k} + \cfrac{1}{h_{c,o} A_o}} \tag{3.156a}$$

and

$$U_i = \cfrac{1}{\cfrac{1}{h_i} + \cfrac{\ln \frac{r_o}{r_i}}{2\pi L k} A_i + \cfrac{A_i}{A_o}\cfrac{1}{h_o}} \quad \textbf{\textit{based on inner surface area}} \tag{3.156b}$$

and

$$U_o = \cfrac{1}{\cfrac{A_o}{A_i}\cfrac{1}{h_i} + \cfrac{\ln \frac{r_o}{r_i}}{2\pi L k} A_o + \cfrac{1}{h_o}} \quad \textbf{\textit{based on outer surface area}} \tag{3.156c}$$

where

A_i = Inner surface area of the tube = $2\pi r_i L$
A_o = Outer surface area of the tube = $2\pi r_o L$

3.5.1.2 Overall Heat-Transfer Coefficient in a Spherical Storage wall

For a typical spherical wall surface, the equivalent resistances across the wall can be shown as in Figure 3.34.

The heat-transfer rate at any section is given as

$$q = \cfrac{T_i - T_o}{\cfrac{1}{h_{c,i} A_i} + \cfrac{\frac{1}{r_i} - \frac{1}{r_o}}{4\pi L k} + \cfrac{1}{h_{c,o} A_o}} \tag{3.157}$$

and overall heat-transfer coefficient is defined as

$$U_i A_i = U_o A_o = \cfrac{1}{\cfrac{1}{h_{c,i} A_i} + \cfrac{\frac{1}{r_i} - \frac{1}{r_o}}{4\pi k} + \cfrac{1}{h_{c,o} A_o}} \tag{3.158a}$$

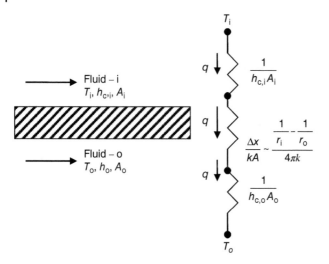

Figure 3.34 Steady-state heat transfer in a spherical system.

Overall heat-transfer coefficient based on inner spherical wall surface is given as

$$U_i = \cfrac{1}{\cfrac{1}{h_{c,i}} + \cfrac{\frac{1}{r_i} - \frac{1}{r_o}}{4\pi k} A_i + \cfrac{A_i}{A_o} \cfrac{1}{h_{c,o}}} \qquad \textbf{\textit{based on inner surface area}} \qquad (3.158b)$$

and overall heat-transfer coefficient based on outer spherical wall surface is given as

$$U_o = \cfrac{1}{\cfrac{A_o}{A_i}\cfrac{1}{h_i} + \cfrac{\frac{1}{r_i} - \frac{1}{r_o}}{4\pi k} A_o + \cfrac{1}{h_o}} \qquad \textbf{\textit{based on outer surface area}} \qquad (3.158c)$$

where

A_i = Inner surface area of the tube = $4\pi r_i^2$
A_o = Outer surface area of the tube = $4\pi r_o^2$

3.6 Contact Resistances and Thermal Interface Materials

In a composite system, heat transfer takes place through the interface of two dissimilar materials of different thermal conductivity values. The interface is composed of direct contact spots and gaps, usually filled with air as shown in Figure 3.35a. Formation of such contact interface is due to the presence of surface roughness and causes a temperature drop across the interface between two materials as shown in Figure 3.35b.

The heat transfer through the interface is due to two major mechanisms in parallel: (i) conduction through the solids at contact spot area and (ii) by conduct, and/or convection and/or radiation through the gaps filled with air or other gases. In metallic or higher conductivity composite materials, the heat transfer is limited by the gap thermal resistance, which is referred to as the **thermal contact resistance** and defined as

$$q_c'' = \frac{T_i - T_j}{R_c''} \qquad (3.159)$$

(a)   (b)

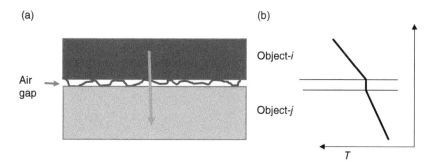

Figure 3.35 (a) Heat transfer through air gap and (b) temperature gradient through dissimilar materials with air gap.

Table 3.19 Typical contact resistance values.

Material surface type (at ambient pressure)	Contact resistance (m² °C/W × 10⁴)
Stainless steel	2.64
Aluminum	1.5–5.0
Magnesium	1.5–3.5
Copper	1–10

where

q_c'' = Conduction heat transfer at contact through the composite system
T_i = Temperature of material-i at the contact interface
T_j = Temperature of material-j at the contact interface
R_c'' = Interface contact resistance

The thermal contact resistance depends on a number of factors such as the mating materials and the corresponding roughness values; the type of gas entrapped and its pressure in the gap; and the pressure applied at the mating interface. Roughness of material surface depends on machining and abrasive processes used and the contact spot area decreases with increase in applied pressure at mating surfaces. Contact resistance values are generally determined experimentally. Some typical values of materials are listed in Table 3.19.

Thermal management is one of the most critical issues in many heat-generating devices such as power electronics, battery storage, and others due to increasing power densities. When a heat sink is placed directly on top of a heat-generating device, the presence of contact resistance between the two mating surfaces reduces heat dissipation rate and raises the device operating temperature as shown in Figure 3.36a.

Thermal interface material (TIM) is used to fill the air gaps and reduce the resistance between two material objects such as the heat-generating module and heat sink as shown in Figure 3.36b. TIMs should have high thermal conductivity, good capability to fill up the micro gaps, and good thermal stability. TIM placed in between the heat source and heat sink significantly increases the heat-transfer capability of the system.

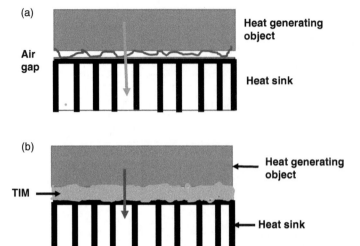

Figure 3.36 Air gap and TIM-filling at the interface of heat-generating module and heat sink. (a) Direct contact. (b) Air gap filled with TIM.

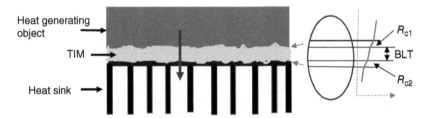

Figure 3.37 Thermal contact resistance and temperature drop across contact interface of dissimilar materials and thermal interface material.

In the development of newer high thermal conductivity composite system of TIMs, additional resistance to heat transfer exists at the contact interface of dissimilar materials and results in temperature drop across the interface as depicted in Figure 3.37.

Improper selection of TIM in terms of **bond line thickness (BLT)**, melting point, reliability, material properties, etc. increases the interfacial thermal resistance and contact resistance. Conventional TIMs such as wax, grease, thermal tapes, gels, and pads face challenges to satisfy the above-mentioned requirements due to decrease in size and increased speed, reliability of the new generation of electronics, and power modules systems. The commonly used TIMs and associated contact resistance values are listed in Table 3.20.

High thermal conductivity and low thermal resistance of TIM are achieved by dispersing high-conductivity material particles such as fillers in a polymeric base or matrix materials. However, adding these high-conductivity particles also affects the viscosity or fluidity to wet the substrate surfaces and the BLT of the TIM as demonstrated in Figure 3.37.

Thermal resistance associated with the bulk TIM is given as

$$R_{\text{TIM,bulk}} = \frac{\text{BLT}}{K_{\text{TIM}}} \tag{3.160}$$

Table 3.20 Common thermal interface materials.

Type of materials	$(m^2 \,°C/W \times 10^4)$
Gap filler pad	1.5–7.0
Thermal gel	0.7–3.5
Thermal tapes	0.4–1.4
Thermal grease	0.5–3.0 (0.3–0.6)
Epoxy	0.8–2.7

Thermal resistance of the TIM depends not only on the thermal conductivity, but also on the thickness of the BLT. The BLT associated with TIM varies with type and volume fraction of the particles and the applied pressure. The effective thermal conductivity of the TIM has two components: one associated with the bulk resistance of the TIM and second associated with the contact resistances of the TIM with the two substrate interfaces, R_{c1} and R_{c2}, at the top and the bottom, respectively. The effective thermal resistance of the TIM is then given as

$$R_{TIM} = R_{TIM,bulk} + R_{c1} + R_{c2} \qquad (3.161)$$

Major factors that contribute to the reduction of the thermal resistance of the TIM are (i) increased thermal conductivity of TIM; (ii) reduced BLT; reduced contact resistances (R_{c1} and R_{c2}).

To evaluate the effect of type and volume fraction of the high-conductivity nanomaterials on the thermal resistance of the polymer-filler TIM, one also needs to understand its effect on the rheology of the polymer mixture. Often, attention is primarily given to the estimation of the thermal conductivity of filler matrix system based on available semiempirical formulae developed based on effective medium theory (EMT).

Bibliography

Adrian, B. (1984). *Convection Heat Transfer*. New York: Wiley.

Asako, Y., Nakamura, M., and Faghri, M. (1988). Three-dimensional laminar heat transfer and fluid flow characteristics in the entrance region of a rhombi duct, ASME. *J. Heat Transfer* 110: 855–861.

Bergman, T.L., Lavine, A.S., Incropera, F.P., and Dewitt, D.P. (2011). *Fundamentals of Heat and Mass Transfer*, 7e. Wiley.

Cengel, A.Y. and Ghajar, A.J. (2015). *Heat and Mass Transfer: Fundamentals & Applications*, 5e. McGraw-Hill.

Chato, J.C. (1962). Laminar condensation inside horizontal and inclined tubes. *ASHRAE (American Society of Heating, Refrigeration and Air-Conditioning Engineers)* 4: 52–60.

Dobson, M.K. and Chto, J.C. (1998). Condensation in smooth horizontal tubes. *ASME J. Heat Transfer* 120: 193.

Holman, J.P. (1990). *Heat Transfer*, 7e. McGraw-Hill.

Incropera, F.P., Dewitt, D.P., Bergman, T.L., and Lavine, A.S. (2007). *Fundamentals of Heat and Mass Transfer*, 6e. Wiley.

Kandlikar, S.G. (1990). A general correlation saturated two phase flow boiling heat transfer inside horizontal and vertical tubes. *J. Heat Transfer* 112: 219–228.

Kays, W.M. and Crawford, M.E. (1980). *Convective Heat and Mass Transfer*. New York: McGraw-Hill.

London, A.L. and Kays, W. (1984). *Compact Heat Exchangers*, 3e. New York: McGraw-Hill.

Martin, H. (1977). Heat and mass transfer between impinging gas jets and solid surfaces. *Adv. Heat Transfer* 13. Academic Press, New York: 1–60.

Shah, R.K. and London, A.L. (1978). *Laminar Flow Forced Convection in Ducts, Advances in Heat Transfer*. New York: Academic Press.

Siegal, R. and Howell, J.R. (1992). *Thermal Radiation Heat Transfer*, 3e. Washington: Hemisphere Publishing Corp.

Sparrow, E.M. and Cess, R.D. (1978). *Radiation Heat Transfer* (Augmented Edition). New York: McGraw-Hill.

4

Design and Selection of Fins and Heat Sinks

A fin is simply an extended surface that is used over the external surface of a heat-dissipating solid to enhance heat transfer by convection. While use of fins is very common in compact heat exchanger surfaces to achieve heat transfer in a reduced space, an array of fins on a base metal surface are the basis for the construction of a heat sink. A heat sink is a thermal device that is formed by using an array of extended surfaces such as fins and used to dissipate heat passively from a number of different heat-generating components such as electronics devices; electric powder modules such as power amplifiers, inductors, transformers, and motors; and power-generating devices such as engine blocks, fuel cells, and electric batteries. Heat sinks are designed to conform to the shape of the heat-producing device, and hence heat sinks are available in different shapes and sizes. Thermal Interface Materials (TIMs) are often used in the interface between the heat-producing devices and the heat spreader or carrier such as the heat sinks or cold plates. In this chapter, we will consider design and thermal performance of different types of fins and heat sinks.

4.1 Design Requirements for Fins and Heat Sinks

All heat-producing devices require some kind of heat management and/or cooling system to maintain the device temperature within an acceptable range and to sustain optimum operation of the device (Figure 4.1).

Let us consider a heat-producing device (shown in Figure 4.2) with external solid surface maintained at a surface temperature of T_0 and that dissipates heat by convection to the surrounding fluid at a temperature T_∞. The convection heat transfer from this solid surface is given by Newton's law of cooling as

$$q = hA\left(T_0 - T_\infty\right) \tag{4.1}$$

It can be noticed from the expression for the convective heat transfer rate that for a fixed target surface temperature T_0, the heat dissipation rate from the surface to the surrounding moving fluid can be increased either by having a convection mode with higher convection heat transfer coefficient or by increasing the surface area using an extended surface, also referred to as the fin. However, once the higher bound for the convection heat transfer coefficient is reached for a selected convection mode such as the free convection or forced

Design of Thermal Energy Systems, First Edition. Pradip Majumdar.
© 2021 John Wiley & Sons Ltd. Published 2021 by John Wiley & Sons Ltd.
Companion website: www.wiley.com/go/majumdar

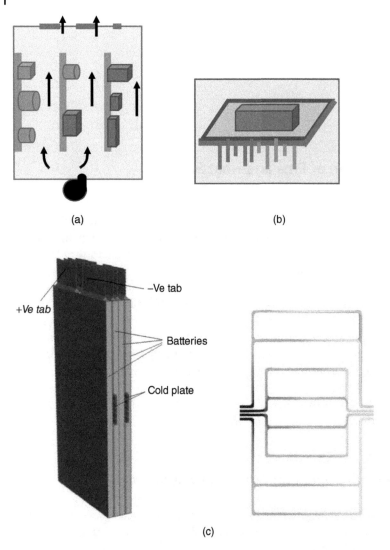

(a)

(b)

(c)

Figure 4.1 Convective heat transfer from a heat generating module. (a) without any heat sink or fins, (b) with heating module mounted on a heat sink and (c) cold plates for a battery pack.

convection, the heat transfer rate can only be increased using extended surfaces or fins as shown in Figure 4.2b.

When using a fin, heat transmits from the base of the fin through fin-projected cross-sectional area, A_c, and then dissipates from the fin surface area, A_s, by convection to the surrounding environment. There is also additional heat loss from the bare area, A_b, of the surface by convection.

The total heat transfer from the finned surface is given as

$$q_{\text{total}} = q_f + q_{\text{wof}} \tag{4.2}$$

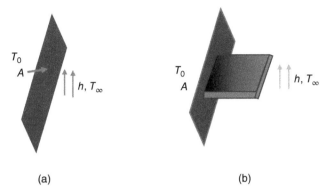

Figure 4.2 Cooling of sold external surface. (a) Convective cooling of solid surface, (b) convective cooling of solid surface with extended fin surface.

where

q_f = Heat transfer from the fin

q_{wof} = Heat transfer from the bare area without the fin

$$q_{wof} = hA_b \left(T_0 - T_\infty\right) \tag{4.3}$$

A_b = Bare area of the surface = $A - A_c$

A_c = Base conduction area of the fin

Primary objective is to have the heat loss from the surface with fin, q_{total}, given by Eq. (4.2), to be significantly higher than the heat loss from the surface without the fin (Eq. (4.1)). Generally, fins are used for cases where convection heat transfer coefficients are low such as in the cases of heat transfer to a gaseous environment. It is, also, however, used in the liquid side of a heat exchange to achieve further reduction in size. Use of fins also promotes additional turbulence by disrupting boundary layer formation over the surface and enhancing convection heat transfer.

4.2 Configurations and Types of Fins

There are different types of fins that are categorized based on geometrical shapes and are designed for different applications. Figure 4.3 shows few examples of different types of fin geometries.

Heat sinks are formed with a collection of number of fins as shown in Figure 4.4a. Often, heat sinks are fabricated by forming flow passages in between two parallel plates and using separating finned walls as shown in Figure 4.4b. Figure 4.4d shows typical direct attachment of a heat sink over the surface of a heat-producing device.

Heat sinks are designed in variety of shapes to meet the cooling needs of wide range of heat-producing devices with adjacent air or gas as a cooling media.

Another popular means of cooling heat-producing devices is the use of **cold plates** that provide customized flow passages formed by finned walls embedded on metallic plates as

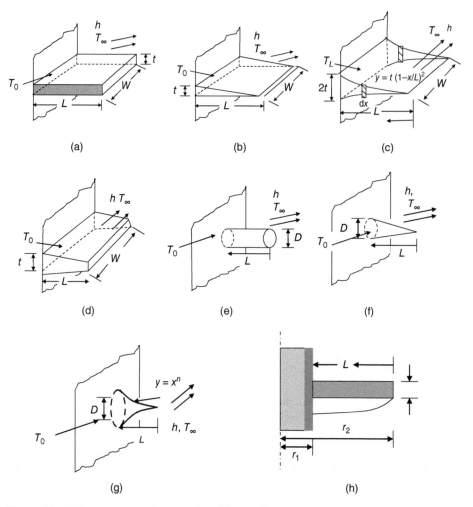

Figure 4.3 Different types and geometries of fins. (a) Straight rectangular fin, (b) straight triangular fin, (c) straight parabolic fin, (d) straight fin trapezoidal fin, (e) cylindrical spine or pin fin, (f) conical spine, (g) straight parabolic spine, and (h) circular or annular fin.

shown in Figure 4.4c. Typically, cold plates are used along with liquid coolants and for very strong heat flux applications. Often, TIM is used in between the heat-producing device and the heat sink to reduce contact resistance (Figure 4.4b,d).

Figure 4.4d shows computer-simulated three-dimensional temperature distribution in a high heat flux-generating Insulated Gate Bipolar Transistor (IGBT) module. A TIM is used in between the generating module and the heat sink to reduce the contact resistance and restrict the junction temperature to the acceptable level for efficient and safe operation of the power-generating modules.

Finned surfaces may be formed as a continuous or an integrated part of the base heat-producing surface (Figure 4.4a) or by directly attaching it to the heat-producing device by pressing or soldering. Some of the complex shapes of heat sinks are formed by

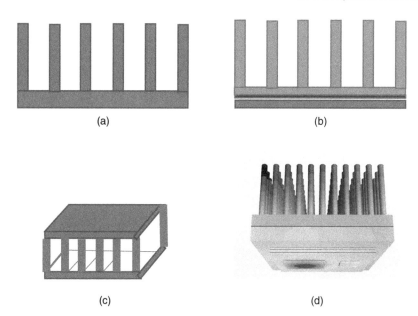

(a)

(b)

(c)

(d)

Figure 4.4 Different heat sink designs. (a) Heat sink as integral part of the base material, (b) heat sink attached to the base material, (c) heat sink with finned flow passages, and (d) heat sink attached over a heat-producing device.

fabrication processes such as extrusion, stamping processes, and more recently by additive manufacturing processes or 3D printing.

4.3 Fin Performance Modeling and Solutions

The fin heat loss can be calculated based on the heat transfer rate at the base of a fin using Fourier law of conduction as

$$q_f = -k_f A_c \frac{dT}{dx}$$
(4.4a)

Alternatively, heat loss is also given as the integral sum of heat loss by convection from the fin surface area

$$q_f = \int_{A_s} h_c dA_s \left(T - T_s \right)$$
(4.4b)

The fin heat loss, given by Eq. (4.4a) or from Eq. (4.4b), can be estimated once the temperature distribution $T(x)$ in the fin is known based on the solution of the fin heat equation.

4.3.1 A General Fin Heat Equation

Let us demonstrate here the procedure for evaluating the performance of a fin based on an analytical solution of the fin heat equation. To determine heat loss from the fin, the temperature distribution within the fin is required as given from the solution of the fin heat equation, which is derived based on energy balance of heat conducted through a differential

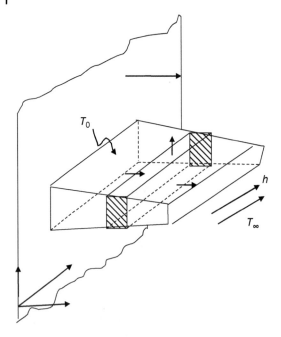

Figure 4.5 Straight fin of variable cross-sectional area.

volume of the solid and heat loss by convection form the extended surface of the differential fin surface as shown in Figure 4.5:

For such fins of variable cross-section, the area of conduction, A, and perimeter, P, are functions of the spatial variables. Some of the major assumptions made in deriving the fin heat equations are (i) The transverse dimension, such as the thickness of the fin, is significantly smaller compared to axial dimensions like the length, and so the temperature distribution along the thickness is assumed to be negligible compared to the temperature variation along the length of the fin, and is small compared to the temperature difference between the fin and the environment; (ii) For a three-dimensional fin in Cartesian co-ordinate, the temperature distribution in the fin could be two-dimensional, i.e. $T(x, z)$, However, for simplicity, thermal resistance in the width-wide or z-direction is assumed large, and the temperature is assumed to be spatially uniform considering large width-wide dimension; (iii) steady-state operation; (iv) constant fin conductivity; (v) convection heat transfer coefficient is uniform over the surface given by the spatially averaged value heat-transfer coefficient value.

A general form of heat equation for the temperature distribution in a straight fin of nonuniform cross-section can be derived based on an energy balance over a differential control volume, shown in Figure 4.5, and is given as

$$\frac{d^2\theta}{dx^2} + \left[\frac{1}{A_c(x)}\frac{dA_c(x)}{dx}\right]\frac{d\theta}{dx} - \left[\frac{1}{A_c(x)}\frac{h}{k}\frac{dA_f(x)}{dx}\right]\theta = 0 \qquad (4.5)$$

where

$A_c(x)$ = Fin variable area of conduction
$A_f(x)$ = Fin variable surface area of convection

Let us now consider a few examples for a straight longitudinal fin of uniform cross-section as well straight fins with variable cross-sections.

4.3.1.1 Straight Longitudinal Fin of Uniform Cross-section

For a straight fin of uniform cross-section, the fin thickness, area of conduction, A_c, and perimeter, P, of the fin are constant. By setting $A_c(x) = A_c$ and $A_f(x) = P\,dx$, the heat equation in straight fin of uniform cross-section is then given as

$$\frac{d^2\theta}{dx^2} - \frac{hP}{kA_c}\theta = 0 \tag{4.6a}$$

or

$$\frac{d^2\theta}{dx^2} - m^2\theta = 0 \tag{4.6b}$$

where

$$\theta = T - T_\infty \tag{4.6c}$$

$$m^2 = \frac{hP}{kA_c} \tag{4.6d}$$

and h = spatially averaged convection heat transfer coefficient, P = perimeter of fin surface, and A_c = cross-sectional area for heat conduction.

One such fin is the straight rectangular fin as shown below:

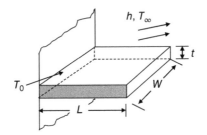

The geometrical parameters for such a straight rectangular fin are summarized below:

$$A_c = Wt, P = (2W + 2t), A_f = PL \text{ and } V_f = A_cL = LWt \tag{4.7}$$

The heat equation for such fin can be solved assuming the following set of boundary conditions:

Boundary conditions:

    1.   At $x = 0, \theta = \theta_0 = T - T_0$ (4.8a)

    2.   At $x = L, -k\left.\dfrac{\partial\theta}{\partial x}\right|_{x=L} = h\theta\,(L)$ (4.8b)

The solution to this fin heat equation is given as

$$\frac{T - T_\infty}{T_0 - T_\infty} = \frac{\theta}{\theta_0} = \frac{\cosh\ m\,(L-x) + (h/mk)\sinh\ m\,(L-x)}{\cosh\ mL + (h/mk)\sinh\ mL} \tag{4.9}$$

The fin heat loss is given either based on the heat transfer rate at the base of the fin as

$$q_f = kA_c \frac{d\theta}{dx}\bigg|_{x=0} \tag{4.10a}$$

or based on total convection heat loss from the extended surface of the fin as

$$q_f = \int_{A_f} h\left(T(x) - T_\infty\right) dA \tag{4.10b}$$

Determining the temperature derivative from Eq. (4.7) and substituting $x = 0$, the fin heat loss from straight longitudinal fin is given as

$$q_f = \sqrt{hPkA_c} \frac{\sinh\ mL + (h/mk)\cosh\ mL}{\cosh\ mL + (h/mk)\sinh\ mL} \tag{4.11}$$

For this straight rectangular fin, the geometrical parameters are summarized as

$$A_c = Wt, P = (2W + 2t), A_f = PL \text{ and } V_f = A_cL = LWt \tag{4.12}$$

Two other limiting case solutions are also obtained based on the boundary condition assumed at the tip of the fin: Case-2 – Insulated tip, i.e. $\frac{\partial T}{\partial x}\big|_{x=L} = 0$ and Case-3 – Long fin with tip temperature assumed to be equal to the surrounding fluid temperature, i.e. $T|_{x=L} = T_\infty$.

The fin temperature distribution and the heat transfer rate for these three cases are summarized in Table 4.1.

Another common example of straight longitudinal fin of uniform cross-section is the **straight cylindrical spine or pin fin** (Figure 4.3e) shown below:

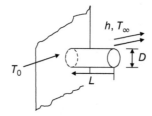

Table 4.1 Solutions for fin temperature and heat loss for straight rectangular fin for different cases with different fin tip boundary conditions.

Case	Fin temperature distribution	Fin heat loss
1. Convective tip	$\dfrac{T - T_\infty}{T_0 - T_\infty} =$ $\dfrac{\cosh\ m(L - x) + (h/mk)\sinh\ m(L - x)}{\cosh\ mL + (h/mk)\sinh\ mL}$	$q_f = \sqrt{hPkA_c}\left(T_0 - T_\infty\right)$ $\dfrac{\sinh\ mL + (h/mk)\cosh\ mL}{\cosh\ mL + (h/mk)\sinh\ mL}$
2. Adiabatic or insulated tip	$\dfrac{T - T_\infty}{T_0 - T_\infty} = \dfrac{\cosh\ (L - x)}{\cosh\ mL}$	$q_f =$ $\sqrt{hPkA_c}\left(T_0 - T_\infty\right)\tanh\ mL$
3. Long fin	$\dfrac{T - T_\infty}{T_0 - T_\infty} = e^{-mx}$	$q_f = \sqrt{hPkA_c}\left(T_0 - T_\infty\right)$

The geometrical parameters for this fin are summarized below:

$$A_c = \frac{\pi}{4}D^2, P = \pi D, A_f = PL, \text{ and } V_f = A_c L = \frac{\pi}{4}D^2 L \qquad (4.13)$$

4.3.1.2 Straight Fin of Variable Cross-section

For a straight fin of variable cross-section, the thickness (t), the cross-section area $A_c(x)$,and the perimeter $P(x)$ vary from the base of the fin to the tip of the fin depending on the fin surface profile $y(x)$. Few such fins are considered in the following in the following Sections 4.3.1.2.1–4.3.1.7.

4.3.1.2.1 Straight Fin of Triangular Profile

Figure 4.6 shows a straight triangular fin whose top and bottom surface varies linearly from the base of the fin located at $x = L$ to the tip of the fin located at $x = 0$.

Appropriate differential heat equation for the straight fin of triangular profile is given as

$$\frac{d}{dx}\left(x\frac{d\theta}{dx}\right) - m^2\theta = 0 \qquad (4.14a)$$

or

$$x\frac{d^2\theta}{dx^2} + \frac{d\theta}{dx} - m^2\theta = 0 \qquad (4.14b)$$

where

$$\theta = T - T_L \text{ and } m^2 = \frac{2h}{kt}\sqrt{L^2 + t^1} \qquad (4.15)$$

The general solution of Eq. (4.12) is given as

$$\theta = C_1 I_0\left(2m\sqrt{x}\right) + C_2 K_0\left(2m\sqrt{x}\right) \qquad (4.16)$$

where I_0 and K_0 are the zero-order modified Bessel functions of the first and second kinds, respectively.

This is subjected to the boundary conditions:
Boundary conditions:

   1. At $x = L, \theta = \theta_L = T - T_L$ (4.17a)

   2. At $x = 0$, the temperature at the tip of the fin is finite   (4.17b)

Figure 4.6 Straight fin of triangular profile.

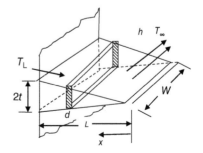

Use of these two boundary conditions in Eq. (4.16) leads to $C_1 = \dfrac{\theta_0}{I_0\left(2m\sqrt{L}\right)}$ and $C_2 = 0$, and the resulting temperature distribution is given as

$$\frac{\theta}{\theta_0} = \frac{I_0\left(2m\sqrt{x}\right)}{I_0\left(2m\sqrt{L}\right)} \tag{4.18}$$

Fin heat loss is computed based on the heat transfer rate at the base of the fin as

$$q_f = kA_{cb}\frac{d\theta}{dx}\bigg|_{x=L} \tag{4.19}$$

where A_{cb} = Conduction area heat flow at the base of the triangular fin = $W \times 2t$

Substituting the expression for the temperature distribution based on Eq. (4.18), we get

$$q_f = k\,2Wt\,\frac{\left(mL^{-\frac{1}{2}}\right)I_1\left(2m\sqrt{L}\right)}{I_0\left(2m\sqrt{L}\right)}\theta_L \tag{4.20a}$$

or

$$q_f = k\,2Wt\,\frac{\left(mL^{-\frac{1}{2}}\right)I_1\left(2m\sqrt{L}\right)}{I_0\left(2m\sqrt{L}\right)}\left(T_L - T_\infty\right) \tag{4.20b}$$

where I_1 and K_1 are the first-order modified Bessel functions of the first and the second kinds, respectively. Modified Bessel function values of first and second kinds are presented in Appendix D: Table D.1. These Bessel function values can also be obtained from the online link www.mhtl.uwaterloo.ca.

4.3.1.3 Spine Fin of Circular Cone Shape

Let us consider the solution for the case of straight spine fin of circular cone shape as shown in Figure 4.7.

Appropriate differential heat equation for the straight fin is given as

$$\frac{d^2\theta}{dx^2} + \frac{2}{x}\frac{d\theta}{dx} - \frac{m^2}{x}\theta = 0 \tag{4.21a}$$

where

$$\theta = T - T_\infty \quad \text{and} \quad m^2 = \frac{2h}{kR}\sqrt{L^2 + R^2} \tag{4.21b}$$

Figure 4.7 Spine fin of circular cone shape.

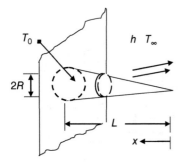

The general solution to this equation is given as

$$\theta = x^{-\frac{1}{2}} \left[C_1 I_1 \left(2m\sqrt{x} \right) \right] + C_2 K_1 \left(2m\sqrt{x} \right) \tag{4.22}$$

Consider the following two boundary conditions

1) At $x = 0$, the temperature at the tip of the fin is finite
2) At $x = L$, $T = T_L$, $\theta = \theta_L$,

Use of boundary conditions 1 and 2 in the general solution given by Eq. (4.19) results in $C_1 = 0$ as $I(0) \rightarrow \infty$ and $C_2 = \frac{\theta_L}{\sqrt{L}I_1(2m\sqrt{L})}$, respectively. Substituting values of constants C_1 and C_2 into general solution Eq. (4.22), we obtain the temperature distribution in conical spine fin as

$$\frac{\theta}{\theta_L} = \frac{\sqrt{L}\, I_1(2m\sqrt{x})}{\sqrt{x}\, I_1(2m\sqrt{L})} \tag{4.23a}$$

or

$$\frac{T - T_\infty}{T_L - T_\infty} = \frac{\sqrt{L}\, I_1(2m\sqrt{x})}{\sqrt{x}\, I_1(2m\sqrt{L})} \tag{4.23b}$$

Fin heat loss is computed based on the heat transfer rate at the base of the fin as

$$q_f = kA_{cb} \frac{d\theta}{dx}\bigg|_{x=L} \tag{4.24}$$

A_{cb} = Conduction area heat flow at the base of the triangular fin = πR^2

Evaluating the temperature derivative at the base of the fin at $x = L$ from Eq. (4.23b) and substituting it in Eq. (4.24), we get the following expression for fin heat loss as

$$q_f = \frac{k\pi R^2}{L} \left[\frac{\left(2m\sqrt{L} \right) I_0 \left(2m\sqrt{L} \right)}{2 I_1 \left(2m\sqrt{L} \right)} - 1 \right] \theta_L \tag{4.25a}$$

or

$$q_f = \frac{k\pi R^2}{L} \left[\frac{\left(2m\sqrt{L} \right) I_0 \left(2m\sqrt{L} \right)}{2 I_1 \left(2m\sqrt{L} \right)} - 1 \right] (T_L - T_\infty) \tag{4.25b}$$

Example 4.1 Consider a spine fin of circular cone shape mounted over a heated surface maintained at a temperature $T_L = 300°F$ as shown in Figure 4.7 with base radius, $R = 0.25$-in. and length, $L = 2.5$-in. Assume the fin material as aluminum with thermal conductivity, $k_f = 120$ Btu/h ft $°$ F and following convection operating conditions: $h = 20$ Btu/h ft^2 $°$ F and $T_\infty = 25°F$. Determine the fin heat loss, q_f. Repeat this computation to show: (i) Effect of decreasing fin base radius for $L = 2.0$ and 2.5-in.: $R = 0.5, 0.1875$, and 0.25-in.; (ii) Effect of decreasing fin length at the base radius of $R = 0.0.25$-in.: $L = 2.5, 2.2, 2.1$, and 2.0-in.; (iii); Effect of fin material conductivity, k_f: =120, 160, and 200 Btu/h ft $°$ F; and (iv) Effect of convection heat transfer coefficient, h: 20, 100, and 100 Btu/h ft^2 $°$ F

Solution

(a) **Effect of base radius**

$R = 0.25\text{-in.} = 0.0208\text{-ft}$ and $L = 2.5\text{-in.} = 0.2083\text{-ft}$

From Eq. (4.21b)

$$m^2 = \frac{2h}{kR}\sqrt{L^2 + R^2} = \frac{2 \times 20}{120 \times (0.25/12)}\sqrt{(0.2083)^2 + (0.0208)^2}$$

$$m^2 = \frac{2 \times 20}{120 \times 0.0208}\sqrt{(0.0434 + 0.00043)}$$

$$m^2 = \frac{2 \times 20}{120 \times 0.0208} \times 0.2094$$

$$= 16.0256 \times 0.2094 = 3.356$$

$$m = 1.8318$$

$$2m\sqrt{L} = 2 \times 1.8318 \times \sqrt{0.2083} = 2 \times 1.8318 \times 0.4563 = 1.672$$

From Eq. (4.25b), heat loss is given as

$$q_f = \frac{k\pi R^2}{L}\left[\frac{\left(2m\sqrt{L}\right) I_0\left(2m\sqrt{L}\right)}{2 I_1\left(2m\sqrt{L}\right)} - 1\right](T_L - T_\infty)$$

$$I_0\left(2m\sqrt{L}\right) = I_0\,(1.672) = 1.8309$$

$$I_1\left(2m\sqrt{L}\right) = I_1\,(1.672) = 1.1642$$

$$q_f = \frac{120 \times \pi \times (0.0208)^2}{0.2083}\left[\frac{1.672\, I_0\,(1.672)}{2 I_1\,(1.672)} - 1\right](300 - 65)$$

$$q_f = \frac{0.1631}{0.2083}\left[\frac{1.672 \times 1.8309}{2 \times 1.1642} - 1\right](300 - 65)$$

or

$$q_f = 0.7830\,[1.3147 - 1] \times 235 = 57.90$$

Fin heat loss, $q_f = 57.90$ Btu/h.

$R = 0.1875\text{-in.} = 0.0156\text{-ft}$ and $L = 2.5\text{-in.} = 0.2083\text{-ft}$

$$m^2 = \frac{2h}{kR}\sqrt{L^2 + R^2} = \frac{2 \times 20}{120 \times (0.25/12)}\sqrt{(0.2083)^2 + (0.0156)^2}$$

$$m^2 = \frac{2 \times 20}{120 \times 0.0208}\sqrt{(0.0434 + 0.000244}$$

$$m^2 = \frac{2 \times 20}{120 \times 0.0208} \times 0.2089$$

$$= 16.0256 \times 0.2089 = 3.348$$

$$m = 1.829$$

$$2m\sqrt{L} = 2 \times 1.829 \times \sqrt{0.2083} = 2 \times 1.829 \times 0.4563 = 1.669$$

From Eq. (4.22b), heat loss is given as

$$q_f = \frac{k\pi R^2}{L}\left[\frac{\left(2m\sqrt{L}\right) I_0\left(2m\sqrt{L}\right)}{2 I_1\left(2m\sqrt{L}\right)} - 1\right](T_L - T_\infty)$$

$$I_0\left(2m\sqrt{L}\right) = I_0\left(1.669 = 1.8274\right)$$

$$I_1\left(2m\sqrt{L}\right) = I_1\left(1.672\right) = 1.1608$$

$$q_f = \frac{120 \times \pi \times (0.0156)^2}{0.2083}\left[\frac{1.669\, I_0\,(1.669)}{2I_1\,(1.669)} - 1\right](300 - 65)$$

$$q_f = \frac{0.0917}{0.2083}\left[\frac{1.669 \times 1.8274}{2 \times 1.1608} - 1\right](300 - 65)$$

or

$$q_f = 0.4404\,[1.313 - 1] \times 235 = 32.39$$

Fin heat loss, $q_f = 32.39$ Btu/h.

R = 0.125-in. = 0.0104-ft and L = 2.5-in. = 0.2083-ft

$$m^2 = \frac{2h}{kR}\sqrt{L^2 + R^2} = \frac{2 \times 20}{120 \times (0.25/12)}\sqrt{(0.2083)^2 + (0.0104)^2}$$

$$m^2 = \frac{2 \times 20}{120 \times 0.0208}\sqrt{(0.0434 + 0.000\,108\,5}$$

$$m^2 = \frac{2 \times 20}{120 \times 0.0208} \times 0.208\,58$$

$$= 16.0256 \times 0.20858 = 3.3427$$

$$m = 1.8283$$

$$2m\sqrt{L} = 2 \times 1.8283 \times \sqrt{0.2083} = 2 \times 1.8283 \times 0.4563 = 1.6685$$

From Eq. (4.25b), heat loss is given as

$$q_f = \frac{k\pi R^2}{L}\left[\frac{\left(2m\sqrt{L}\right) I_0\left(2m\sqrt{L}\right)}{2I_1\left(2m\sqrt{L}\right)} - 1\right](T_L - T_\infty)$$

$$I_0\left(2m\sqrt{L}\right) = I_0\,(1.6685) = 1.8268$$

$$I_1\left(2m\sqrt{L}\right) = I_1\,(1.6685) = 1.160\,25$$

$$q_f = \frac{120 \times \pi \times (0.0104)^2}{0.2083}\left[\frac{1.6685\, I_0\,(1.6685)}{2I_1\,(1.6685)} - 1\right](300 - 65)$$

$$q_f = \frac{0.0409}{0.2083}\left[\frac{1.6685 \times 1.8268}{2 \times 1.16025} - 1\right](300 - 65)$$

or

$$q_f = 0.19638\,[1.3135 - 1] \times 235 = 14.468$$

Fin heat loss, $q_f = 14.468$ Btu/h.

L = 2.0-in. = 0.1667-ft and R = 0.1875-in. = 0.0156-ft

From Eq. (4.21b)

$$m^2 = \frac{2h}{kR}\sqrt{L^2 + R^2} = \frac{2 \times 10}{120 \times (0.1875/12)}\sqrt{(0.1667)^2 + (0.0156)^2}$$

$$m^2 = \frac{2 \times 20}{120 \times 0.0156}\sqrt{(0.0277 + 0.000\,244\,1}$$

$$m^2 = \frac{2 \times 20}{120 \times 0.0156} \times 0.167\,16$$
$$= 21.3675 \times 0.16716 = 3.5717$$

$$m = 1.8899$$

$$2m\sqrt{L} = 2 \times 1.8899 \times \sqrt{0.1667} = 2 \times 1.8899 \times 0.408\,28 = 1.543$$

From Eq. (4.25b), heat loss is given as

$$q_f = \frac{k\pi R^2}{L} \left[\frac{\left(2m\sqrt{L}\right) I_0 \left(2m\sqrt{L}\right)}{2 I_1 \left(2m\sqrt{L}\right)} - 1 \right] (T_L - T_\infty)$$

$$I_0 \left(2m\sqrt{L}\right) = I_0 (1.543) = 1.6898$$

$$I_1 \left(2m\sqrt{L}\right) = I_1 (1.543) = 1.0250$$

$$q_f = \frac{120 \times \pi \times (0.0156)^2}{0.1667} \left[\frac{1.543 \, I_0 \, (1.543)}{2 I_1 \, (1.543)} - 1 \right] (300 - 65)$$

$$q_f = 0.5503 \left[\frac{1.543 \times 1.6898}{2 \times 1.0250} - 1 \right] (300 - 65)$$

or

$$q_f = 0.5503 \, [1.27075 - 1] \times 235 = 35.01 \, \text{Btu/h}.$$

$L = 2.0\text{-in.} = 0.1667\text{-ft}$ and $R = 0.125\text{-in.} = 0.0104\text{-ft}$

From Eq. (4.21b)

$$m^2 = \frac{2h}{kR} \sqrt{L^2 + R^2} = \frac{2 \times 10}{120 \times (0.25/12)} \sqrt{(0.1667)^2 + (0.0104)^2}$$

$$m^2 = \frac{2 \times 20}{120 \times 0.0208} \sqrt{(0.0277 + 0.00010)}$$

$$m^2 = \frac{2 \times 20}{120 \times 0.0208} \times 0.16675$$

$$= 16.0256 \times 0.1845 = 2.6724$$

$$m = 1.6348$$

$$2m\sqrt{L} = 2 \times 1.6348 \times \sqrt{0.1667} = 2 \times 1.6348 \times 0.40\,828 = 1.3349$$

From Eq. (4.25b), heat loss is given as

$$q_f = \frac{k\pi R^2}{L} \left[\frac{\left(2m\sqrt{L}\right) I_0 \left(2m\sqrt{L}\right)}{2 I_1 \left(2m\sqrt{L}\right)} - 1 \right] (T_L - T_\infty)$$

$$I_0 \left(2m\sqrt{L}\right) = I_0 (1.3349) = 1.4976$$

$$I_1 \left(2m\sqrt{L}\right) = I_1 (1.3349) = 0.82\,757$$

$$q_f = \frac{120 \times \pi \times (0.0104)^2}{0.1667} \left[\frac{1.3349 \, I_0 \, (1.3349)}{2 I_1 \, (1.3349)} - 1 \right] (300 - 65)$$

$$q_f = 0.2446 \left[\frac{1.3349 \times 1.4976}{2 \times 0.82757} - 1 \right] (300 - 65)$$

or

$$q_f = 0.2446 \, [1.207 - 1] \times 235 = 11.898 \text{ Btu/h.}$$

Fin heat loss, $q_f = 11.898$ Btu/h.

Effect of base radius on fin heat loss for two fin lengths is given below

L (in.)	R (in.)	h (W/m² K)	k (W/m K)	Fin heat loss, q_f
Effect of fin base radius				
2.0	0.25	20	120	48.104
2.0	0.1875	20	120	35.013
2.0	0.125	20	120	11.898
2.5	0.25	20	120	57.90
2.5	0.1875	20	120	32.39
2.5	0.125	20	120	14.468

Results show decrease in fin heat loss with decrease in fin radius. However, even though heat loss is decreased in a single fin with the use of a smaller-radius fin, a larger number of smaller-radius fins can be attached over the base material surface and are expected to increase total heat loss and effectiveness of using the fin.

(b) Effect of fin length

$L = 2.2$-in. $= 0.1833$-ft and $R = 0.25$-in. $= 0.0208$-ft

From Eq. (4.21b)

$$m^2 = \frac{2h}{kR}\sqrt{L^2 + R^2} = \frac{2 \times 10}{120 \times (0.25/12)}\sqrt{(0.1833)^2 + (0.0208)^2}$$

$$m^2 = \frac{2 \times 20}{120 \times 0.0208}\sqrt{(0.0336 + 0.00043)}$$

$$m^2 = \frac{2 \times 20}{120 \times 0.0208} \times 0.1845$$

$$= 16.0256 \times 0.1845 = 2.9567$$

$$m = 1.7195$$

$$2m\sqrt{L} = 2 \times 1.7195 \times \sqrt{0.1833} = 2 \times 1.7195 \times 0.4281 = 1.4723$$

From Eq. (4.25b), heat loss is given as

$$q_f = \frac{k\pi R^2}{L}\left[\frac{\left(2m\sqrt{L}\right) I_0\left(2m\sqrt{L}\right)}{2I_1\left(2m\sqrt{L}\right)} - 1\right](T_L - T_\infty)$$

$$I_0\left(2m\sqrt{L}\right) = I_0\,(1.4723) = 1.6199$$

$$I_1\left(2m\sqrt{L}\right) = I_1\,(1.4723) = 0.95446$$

$$q_f = \frac{120 \times \pi \times (0.0208)^2}{0.1833}\left[\frac{1.4723\,I_0\,(1.4723)}{2I_1\,(1.4723)} - 1\right](300 - 65)$$

$$q_f = \frac{0.1631}{0.1833} \left[\frac{1.4723 \times 1.6199}{2 \times 0.95446} - 1 \right] (300 - 65)$$

or

$$q_f = 0.8898 \ [1.2493 - 1] \times 235 = 52.147 \ \text{Btu/h}.$$

L = 2.1-in. = 0.175-ft and R = 0.25-in. = 0.0208-ft

$$m^2 = \frac{2h}{kR} \sqrt{L^2 + R^2} = \frac{2 \times 10}{120 \times (0.25/12)} \sqrt{(0.175)^2 + (0.0208)^2}$$

$$m^2 = \frac{2 \times 20}{120 \times 0.0208} \sqrt{(0.0306 + 0.00043)}$$

$$m^2 = \frac{2 \times 20}{120 \times 0.0208} \times 0.17622$$

$$= 16.0256 \times 0.17622 = 2.8240$$

$$m = 1.6805$$

$$2m \sqrt{L} = 2 \times 1.6805 \times \sqrt{0.175} = 2 \times 1.6805 \times 0.4183 = 1.406$$

From Eq. (4.25b) heat loss is given as

$$q_f = \frac{k\pi R^2}{L} \left[\frac{\left(2m\sqrt{L}\right) I_0 \left(2m\sqrt{L}\right)}{2 I_1 \left(2m\sqrt{L}\right)} - 1 \right] (T_L - T_\infty)$$

$$I_0 \left(2m\sqrt{L}\right) = I_0 \ (1.406) = 1.5587$$

$$I_1 \left(2m\sqrt{L}\right) = I_1 \ (1.406) = 0.8916$$

$$q_f = \frac{120 \times \pi \times (0.0208)^2}{0.175} \left[\frac{1.406 \ I_0 \ (1.406)}{2 \ I_1 \ (1.406)} - 1 \right] (300 - 65)$$

$$q_f = \frac{0.1631}{0.175} \left[\frac{1.406 \times 1.5587}{2 \times 0.8916} - 1 \right] (300 - 65)$$

or

$$q_f = 0.932 \ [1.2289 - 1] \times 235 = 50.133 \ \text{Btu/h}.$$

L = 2.0-in. = 0.1667-ft and R = 0.25-in. = 0.0208-ft

From Eq. (4.21b)

$$m^2 = \frac{2h}{kR} \sqrt{L^2 + R^2} = \frac{2 \times 10}{120 \times (0.25/12)} \sqrt{(0.1667)^2 + (0.0208)^2}$$

$$m^2 = \frac{2 \times 20}{120 \times 0.0208} \sqrt{(0.0277 + 0.00043}$$

$$m^2 = \frac{2 \times 20}{120 \times 0.0208} \times 0.16795$$

$$= 16.0256 \times 0.1845 = 2.6915$$

$$m = 1.6406$$

$$2m \sqrt{L} = 2 \times 1.6406 \times \sqrt{0.1667} = 2 \times 1.6406 \times 0.4083 = 1.3396$$

From Eq. (4.25b), heat loss is given as

$$q_f = \frac{k\pi R^2}{L}\left[\frac{\left(2m\sqrt{L}\right) I_0\left(2m\sqrt{L}\right)}{2I_1\left(2m\sqrt{L}\right)} - 1\right](T_L - T_\infty)$$

$$I_0\left(2m\sqrt{L}\right) = I_0\,(1.3396) = 1.5015$$

$$I_1\left(2m\sqrt{L}\right) = I_1\,(1.3396) = 0.8317$$

$$q_f = \frac{120 \times \pi \times (0.0208)^2}{0.1667}\left[\frac{1.3396\,I_0\,(1.3396)}{2\,I_1\,(1.3396)} - 1\right](300 - 65)$$

$$q_f = \frac{0.1631}{0.1667}\left[\frac{1.3396 \times 1.5015}{2 \times 0.8317} - 1\right](300 - 65)$$
$$\frac{}{0.1667}$$

or

$$q_f = 0.9784\,[1.2092 - 1] \times 235 = 48.1037 \text{ Btu/h.}$$

Fin heat loss, $q_f = 48.1037$ Btu/h.

L (in.)	R (in.)	h (W/m² K)	k (W/m K)	Fin heat loss, q_f
Effect of fin length				
2.5	0.25	20	120	57.90
2.2	0.25	20	120	52.147
2.1	0.25	20	120	50.133

Results show that fin heat loss increases with use of a longer fin. However, final selection must be made based on considering fin effectiveness and restrictions such as space limitation, weight restriction, and cost.

(c) Effect of fin material conductivity
k = 160 Btu/h ft F, L = 2.5-in. = 0.2083-ft, and R = 0.25-in. = 0.0208-ft, h = 20 Btu/h ft² F
From Eq. (4.21b)

$$m^2 = \frac{2h}{kR}\sqrt{L^2 + R^2} = \frac{2 \times 20}{120 \times (0.25/12)}\sqrt{(0.2083)^2 + (0.0208)^2}$$

$$m^2 = \frac{2 \times 20}{160 \times 0.0208}\sqrt{(0.0434 + 0.00043)}$$

$$m^2 = \frac{2 \times 20}{160 \times 0.0208} \times 0.2094$$

$$= 12.019 \times 0.2094 = 2.516$$

$$m = 1.5864$$

$$2m\sqrt{L} = 2 \times 1.5864 \times \sqrt{0.2083} = 2 \times 1.5864 \times 0.4563 = 1.4480$$

From Eq. (4.25b), heat loss is given as

$$q_f = \frac{k \pi R^2}{L} \left[\frac{\left(2m\sqrt{L}\right) I_0 \left(2m\sqrt{L}\right)}{2 I_1 \left(2m\sqrt{L}\right)} - 1 \right] (T_L - T_\infty)$$

$$I_0 \left(2m\sqrt{L}\right) = I_0 \ (1.4480 = 1.5970)$$

$$I_1 \left(2m\sqrt{L}\right) = I_1 \ (1.4489) = 0.9310$$

$$q_f = \frac{160 \times \pi \times (0.0208)^2}{0.2083} \left[\frac{1.4480 \, I_0 \ (1.4480)}{2 I_1 \ (1.4480)} - 1 \right] (300 - 65)$$

$$q_f = \frac{0.2174}{0.2083} \left[\frac{1.4489 \times 1.5970}{2 \times 0.9310} - 1 \right] (300 - 65)$$

or

$$q_f = 1.0436 \ [1.2426 - 1] \times 235 = 59.519$$

Fin heat loss, $q_f = 59.519$ Btu/h.
$k = 200$ Btu/h ft F, $L = 2.5$-in. $= 0.2083$-ft, and $R = 0.25$-in. $= 0.0208$-ft, $h = 20$ Btu/h ft² F

$$m^2 = \frac{2h}{kR} \sqrt{L^2 + R^2} = \frac{2 \times 20}{120 \times (0.25/12)} \sqrt{(0.2083)^2 + (0.0208)^2}$$

$$m^2 = \frac{2 \times 20}{200 \times 0.0208} \sqrt{(0.0434 + 0.00043)}$$

$$m^2 = \frac{2 \times 20}{200 \times 0.0208} \times 0.2094$$

$$= 9.6153 \times 0.2094 = 2.0135$$

$$m = 1.4189$$

$$2m\sqrt{L} = 2 \times 1.4189 \times \sqrt{0.2083} = 2 \times 1.4189 \times 0.4563 = 1.2949$$

Heat loss is given as

$$q_f = \frac{k \pi R^2}{L} \left[\frac{\left(2m\sqrt{L}\right) I_0 \left(2m\sqrt{L}\right)}{2 I_1 \left(2m\sqrt{L}\right)} - 1 \right] (T_L - T_\infty)$$

$$I_0 \left(2m\sqrt{L}\right) = I_0 \ (1.2949) = 1.4652$$

$$I_1 \left(2m\sqrt{L}\right) = I_1 \ (1.2949 = 0.79297)$$

$$q_f = \frac{200 \times \pi \times (0.0208)^2}{0.2083} \left[\frac{1.2949 \, I_0 \ (1.2949)}{2 I_1 \ (1.2949)} - 1 \right] (300 - 65)$$

$$q_f = \frac{0.2718}{0.2083} \left[\frac{1.2949 \times 1.4652}{2 \times 0.79297} - 1 \right] (300 - 65)$$

or

$$q_f = 1.3048 \ [1.1963 - 1] \times 235 = 60.196$$

Fin heat loss, $q_f = 60.196$ Btu/h.

L (in.)	R (in.)	h (W/m² K)	k (W/m K)	Fin heat loss, q_f
Effect of fin conductivity				
2.5	0.25	20	120	57.90
2.5	0.25	20	160	59.519
2.5	0.25	20	200	60.191

Results show that the fin heat loss increases with use of a higher conductivity material. However, final selection must be made based on considering restrictions such as weight and cost.

(d) Effect of convection heat transfer coefficient

$h = 100\ Btu/h\,ft^2\ F, L = 2.5$-*in*. $= 0.2083$-*ft, and* $R = 0.25$-*in*. $= 0.0208$-*ft, k = 120 Btu/h ft F*

From Eq. (4.21b)

$$m^2 = \frac{2h}{kR}\sqrt{L^2 + R^2} = \frac{2 \times 20}{120 \times (0.25/12)}\sqrt{(0.2083)^2 + (0.0208)^2}$$

$$m^2 = \frac{2 \times 100}{120 \times 0.0208}\sqrt{0.0434 + 0.00043}$$

$$m^2 = \frac{2 \times 100}{120 \times 0.0208} \times 0.2094$$

$$= 80.1282 \times 0.2094 = 16.7788$$

$$m = 4.0961$$

$$2m\sqrt{L} = 2 \times 4.0961 \times \sqrt{0.2083} = 2 \times 4.0961 \times 0.4563 = 3.381$$

Heat loss is given as

$$q_f = \frac{k\pi R^2}{L}\left[\frac{\left(2m\sqrt{L}\right)I_0\left(2m\sqrt{L}\right)}{2I_1\left(2m\sqrt{L}\right)} - 1\right](T_L - T_\infty)$$

$$I_0\left(2m\sqrt{L}\right) = I_0\,(3.381) = 6.6779$$

$$I_1\left(2m\sqrt{L}\right) = I_1\,(3.381) = 5.5737$$

$$q_f = \frac{120 \times \pi \times (0.0208)^2}{0.2083}\left[\frac{3.381\,I_0\,(3.381)}{2I_1\,(3.381)} - 1\right](300 - 65)$$

$$q_f = \frac{0.1631}{0.2083}\left[\frac{3.381 \times 6.6779}{2 \times 5.5737} - 1\right](300 - 65)$$

$$q_f = 0.7830\,[2.0254 - 1] \times 235 = 188.67$$

Fin heat loss, $q_f = 188.67$ Btu/h.

$h = 500\ Btu/h\,ft^2\ F, L = 2.5$-*in*. $= 0.2083$-*ft, and* $R = 0.25$-*in*. $= 0.0208$-*ft, k = 120 Btu/hr ft F*

From Eq. (4.21b)

$$m^2 = \frac{2h}{kR}\sqrt{L^2 + R^2} = \frac{2 \times 20}{120 \times (0.25/12)}\sqrt{(0.2083)^2 + (0.0208)^2}$$

$$m^2 = \frac{2 \times 500}{120 \times 0.0208} \sqrt{(0.0434 + 0.00043}$$

$$m^2 = \frac{2 \times 500}{120 \times 0.0208} \times 0.2094$$

$$= 400.6410 \times 0.2094 = 83.8942$$

$$m = 9.1593$$

$$2m\sqrt{L} = 2 \times 9.1593 \times \sqrt{0.2083} = 2 \times 9.1593 \times 0.4563 = 8.3588$$

From Eq. (4.25b), heat loss is given as

$$q_f = \frac{k\,\pi\,R^2}{L}\left[\frac{\left(2m\sqrt{L}\right) I_0\left(2m\sqrt{L}\right)}{2I_1\left(2m\sqrt{L}\right)} - 1\right](T_L - T_\infty)$$

$$I_0\left(2m\sqrt{L}\right) = I_0\,(8.3588) = 598.3602$$

$$I_1\left(2m\sqrt{L}\right) = I_1\,(8.3588) = 561.3363$$

$$q_f = \frac{120 \times \pi \times (0.0208)^2}{0.2083}\left[\frac{8.3588 I_0\,(3.381)}{2I_1\,(3.381)} - 1\right](300 - 65)$$

$$q_f = \frac{0.1631}{0.2083}\left[\frac{8.3588 \times 598.3602}{2 \times 561.3363} - 1\right](300 - 65)$$

or

$$q_f = 0.7830\,[4.4550 - 1] \times 235 = 635.748$$

Fin heat loss, $q_f = 635.748$ Btu/h.
Effect of surrounding environment convection heat transfer coefficient on fin heat loss is seen in the table below.

L (in.)	R (in.)	h (W/m² K)	k (W/m K)	Fin heat loss, q_f
Effect of ambient convection heat transfer coefficient				
2.5	0.25	20	120	57.90
2.5	0.25	100	120	188.67
2.5	0.25	500	120	635.748

Results show that fin heat loss increases with higher convection. However, justification for using fins must be made based on considering fin effectiveness and the cost associated with maintaining higher convection rate and/or mode and parasitic cost.

Table below summarizes all results for the spine fin of circular cone shape for a selected range of dimensional parameters, different material conductivities, and convection environmental conditions.

L (in.)	R (in.)	h (W/m² K)	k (W/m K)	Fin heat loss, q_f
Effect of fin length				
2.5	0.25	20	120	57.90
2.2	0.25	20	120	52.147
2.1	0.25	20	120	50.133
Effect of fin base radius				
2.0	0.25	20	120	48.104
2.0	0.1875	20	120	35.013
2.0	0.125	20	120	11.898
2.5	0.25	20	120	57.90
2.5	0.1875	20	120	32.39
2.5	0.125	20	120	14.468
Effect of fin conductivity				
2.5	0.25	20	120	57.90
2.5	0.25	20	160	59.519
2.5	0.25	20	200	60.191
Effect of ambient convection heat transfer coefficient				
2.5	0.25	20	120	57.90
2.5	0.25	100	120	188.67
2.5	0.25	500	120	635.748

In a design optimization study, such computations can be extended to wider range of parameters, including consideration of constraints such as weight, size, and cost.

4.3.1.4 Straight Parabolic Fin with Circular Base

The geometry of a concave parabolic spline or pin fin is shown in Figure 4.8. The concave profile of this fin is given as $y = \frac{D}{2}\left(\frac{x}{L}\right)^2$, while the origin $x = 0$ is considered at the tip of the fin. The diameter of the fin at the base is assumed as D and length as L. Note that at any position x, the local diameter of the circular fin is given as $d = 2y = D\left(\frac{x}{L}\right)^2$.

The local perimeter and the area of conduction heat transfer are given as

$$\text{Perimeter, } P(x) = P(x) = \pi d(x) = \pi D\left(\frac{x}{L}\right)^2$$

Figure 4.8 Straight parabolic concave spline fin with circular base.

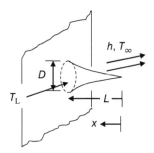

Area of conduction heat transfer, $A_c(x) = \pi(d(x))^2 = \pi\left[D\left(\frac{x}{L}\right)^2\right]^2 = \pi\frac{D^2 x^4}{4L^4}$

The corresponding fin heat equation can be derived from the general Eq. (4.11) as

$$\frac{d}{dx}\left(x^4\frac{d\theta}{dx}\right) - \frac{4hL^2}{kD}\theta = 0 \tag{4.26}$$

Considering $m^2 = \frac{4h}{kD}$, the equation can be written

$$x^4\frac{d^2\theta}{dx^2} + 4x^3\frac{d\theta}{dx} - m^2 L^2\theta = 0 \tag{4.27}$$

Consider two boundary conditions as

At $x = 0$, Temperature is finite, $\theta = \text{Finite}$ \hfill (4.28a)

and

At $x = L$, $T = T_L$, $\theta = \theta_L$ \hfill (4.28b)

4.3.1.5 Straight Concave Parabolic Fin with Rectangular Base

The geometry of concave parabolic fin is shown in Figure 4.9. The concave profile of the fin is given as $y = \frac{t}{2}\left(\frac{x}{L}\right)^2$ while considering origin $x = 0$ is considered at the tip of the fin. The thickness at the base of the fin is assumed as t and the width is assumed as W.

The perimeter and cross-sectional area of conduction heat transfer are given as

$$P(x) = 2\left[t\left(\frac{x}{L}\right)^2 + W \approx 2W\right] \tag{4.29a}$$

and

$$A(x) = t\,W\left(\frac{x}{L}\right)^2 \tag{4.29b}$$

The corresponding fin heat equation can be derived from the general Eq. (4.11) as

$$\frac{d}{dx}\left(x^2\frac{d\theta}{dx}\right) - \frac{2hL^2}{kt}\theta = 0 \tag{4.30}$$

Considering $m^2 = \frac{2h}{kt}$, the equation can be written

$$x^2\frac{d^2\theta}{dx^2} + 2x\frac{d\theta}{dx} - m^2 L^2 = 0 \tag{4.31}$$

Figure 4.9 Straight concave parabolic fin.

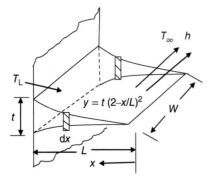

Consider two boundary conditions as

$$\text{At } x = 0, \text{ Temperature is finite and so } \theta = \text{Finite} \tag{4.32a}$$

and

$$\text{At } x = L, T = T_L, \theta = \theta_L \tag{4.32b}$$

The solution to the fin governing Eq. (4.31) and boundary conditions is derived as

$$\frac{\theta}{\theta_L} = \frac{T - T_\infty}{T_L - T_\infty} = \left(\frac{x}{L}\right)^a, \quad \text{where } a = \frac{-1 + \sqrt{1 + 4m^2L^2}}{2} \tag{4.33}$$

The fin heat loss is derived as

$$q_f = \frac{kWt}{2L} \theta_L \left(-1 + \sqrt{1 + 4m^2L^2}\right) \tag{4.34}$$

4.3.1.6 Straight Fin of Trapezoidal Cross-section

The geometry of a straight longitudinal fin of trapezoidal cross-section is shown in Figure 4.10. As demonstrated, this straight fin type has a variable area of conduction and a variable perimeter of surface area for convection due to variable thickness along the longitudinal direction, x.

The length and width of the fin are L and W, respectively. The thickness is shown to vary in linear manner from $2t_0$ at the base ($x = 0$) to $2t_0^*$ at $x = L$. Based on this linear variation of the fin thickness, the local area of conduction can be expressed as follows:

$$A_c(x) = \text{Fin variable area of conduction} = W \times 2\left[t_0 + \left\{\frac{(L-x)(t_0 - t_0^*)}{L}\right\}\right] \tag{4.35a}$$

$$P(x) = \text{Perimeter} = 2\{W + 2\left[t_0^* + \left\{\frac{(L-x)(t_0 - t_0^*)}{L}\right\}\right]\} \tag{4.35b}$$

The corresponding fin heat equation can be derived from the general Eq. (4.11) as (Poulikakos 1994):

$$\frac{\partial^2\theta}{\partial x_*^2} + \frac{1}{x_*}\frac{\partial\theta}{\partial x_*} - B^2\frac{\theta}{x_*} = 0 \tag{4.36a}$$

Figure 4.10 Straight fin of trapezoidal cross-section.

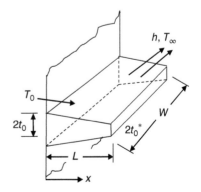

where

$$x_* = t_0 + \left\{ (L - x) \left(t_0 - t_0^* \right) / L \right\} \tag{4.36b}$$

$$B^2 = \frac{h/k}{\left[(t_0 - t_0^*) / L \right]^2} \tag{4.36c}$$

Considering boundary conditions as $\theta(0) = \theta_0$ and $\frac{d\theta}{dx}\big|_{x=L} = 0$, the solution to Eq. (4.36a) is obtained as fin temperature distribution as

$$\frac{\theta}{\theta_0} = \frac{I_0 \left(2B\sqrt{x_*} \right) K_1 \left(2B\sqrt{t_0^*} \right) + K_0 \left(2B\sqrt{x_*} \right) I_1 \left(2B\sqrt{t_0^*} \right)}{K_1 \left(2B\sqrt{t_0^*} \right) I_0 \left(2B\sqrt{t_0} \right) + K_0 \left(2B\sqrt{t_0} \right) I_1 \left(2B\sqrt{t_0^*} \right)} \tag{4.37}$$

Fin heat transfer rate

$$q_f = -k\,A_c\,\frac{\partial\theta}{\partial x}\bigg|_{x=0} = -k\,(2t_0 W)\,\frac{\partial\theta}{\partial x}\bigg|_{x=0} \tag{4.38a}$$

or

$$q_f = 2Bk\,\sqrt{t_0}\frac{W}{L}\left(t_0 - t_0^* \right) \theta_0 \frac{I_1 \left(2B\sqrt{t_0} \right) K_1 \left(2B\sqrt{t_0^*} \right) - K_0 \left(2B\sqrt{t_0} \right) I_1 \left(2B\sqrt{t_0^*} \right)}{K_1 \left(2B\sqrt{t_0^*} \right) I_0 \left(2B\sqrt{t_0} \right) + K_0 \left(2B\sqrt{t_0} \right) I_1 \left(2B\sqrt{t_0^*} \right)} \tag{4.38b}$$

4.3.1.7 Annular or Circular Fin

Annular or circular fins are used over the surfaces of cylindrical tubes and pipes. Annular fins are available in different cross-sectional profiles like rectangular, triangular, or parabolic. Let us briefly discuss the annular fin of rectangular profile as depicted in Figure 4.11.

Appropriate differential heat equation for the annular fin, shown in Figure 4.11, is given as

$$\frac{d^2\theta}{dr^2} + \frac{1}{r}\frac{d\theta}{dr} - m^2\theta = 0 \tag{4.39}$$

where $\theta = T - T_0$ and $m = \sqrt{\frac{2h}{kt}}$.

This is subjected to the boundary conditions:

Boundary condition 1

At the base of the circular fin, $r = r_1, \theta = \theta_0 = T - T_0$ $\tag{4.40a}$

Figure 4.11 Annular fin of rectangular profile.

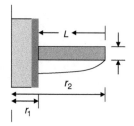

Boundary condition 2

Case I: At the tip of the fin, convection condition exists, and the boundary condition is defined as

$$\text{At } r = r_2, \; -k \frac{\partial \theta}{\partial x}\bigg|_{r=r_2} = h\theta\left(r_2\right) \tag{4.40b}$$

Case II: One limiting case of this boundary condition is an insulated tip and defined as

$$\text{At } r = r_2, \; \frac{\partial \theta}{\partial x}\bigg|_{r=r_2} = 0 \tag{4.40c}$$

The solution to this fin heat equation (4.39) is given as

$$\theta = C_1 I_0\left(mr\right) + C_2 K_0\left(mr\right) \tag{4.41}$$

where I_0 and K_0 are the zero-order modified Bessel functions of the first and second kinds, respectively.

The constants C_1 and C_2 are determined by using **boundary condition 1** (Eq. (4.40a)) and **case II of boundary condition 2** (Eq. (4.40c)),

Substituting BC-1 given by Eq. (4.40a)

$$\theta_0 = C_1 I_0\left(mr_1\right) + C_2 K_0\left(mr_1\right) \tag{4.42a}$$

Substituting BC-2 given by Eq. (4.40c)

$$\frac{d\theta}{dr} = 0$$

or

$$\theta_0 = C_1 I_1\left(mr_2\right) - C_2 K_1\left(mr_2\right) = 0 \tag{4.42b}$$

where I_1 and K_1 are the first-order modified Bessel functions of the first and second kinds, respectively.

Two simultaneous Eqs. (4.42a) and (4.42b) are solved using Cramer's rule for the constants as

$$C_1 = \frac{\begin{vmatrix} \theta_0 & K_0\left(mr_1\right) \\ 0 & -K_1\left(mr_2\right) \end{vmatrix}}{\begin{vmatrix} I_0\left(mr_1\right) & K_0\left(mr_1\right) \\ I_1\left(mr_2\right) & -K_1\left(mr_2\right) \end{vmatrix}} = \frac{-\theta_0 K_1\left(mr_2\right)}{-I_0\left(mr_1\right) K_1\left(mr_2\right) - I_1\left(mr_2\right) K_0\left(mr_1\right)} \tag{4.43a}$$

and

$$C_2 = \frac{\begin{vmatrix} I_0\left(mr_1\right) & \theta_0 \\ I_1\left(mr_2\right) & 0 \end{vmatrix}}{\begin{vmatrix} I_0\left(mr_1\right) & K_0\left(mr_1\right) \\ I_1\left(mr_2\right) & -K_1\left(mr_2\right) \end{vmatrix}} = \frac{-\theta_0 I_1\left(mr_2\right)}{-I_0\left(mr_1\right) K_1\left(mr_2\right) - I_1\left(mr_2\right) K_0\left(mr_1\right)} \tag{4.43b}$$

Now substituting expressions C_1 and C_2 into Eq. (4.41), we get the expression for the temperature distribution in the annular fin as

$$\frac{T - T_\infty}{T_0 - T_\infty} = \frac{\theta}{\theta_0} = \frac{I_0\left(mr\right) K_1\left(mr_2\right) + K_0\left(mr\right) I_1\left(mr_2\right)}{I_0\left(mr_1\right) K_1\left(mr_2\right) + K_0\left(mr_1\right) I_1\left(mr_2\right)} \tag{4.44}$$

Modified Bessel function values of first and second kinds are presented in Appendix D: Table D.1.

Fin heat transfer rate is computed as

$$q_f = -kA_c \frac{dT}{dr}\bigg|_{r=r_1} = -k2\pi r_1 t \frac{d\theta}{dr}\bigg|_{r=r_1} \tag{4.45}$$

Computing the temperature derivative at the base of the fin from the fin temperature distribution Eq. (4.44) and substituting into Eq. (4.45), we get the fin heat loss from the annular fin as

$$q_f = 2\pi r_1 t k m \theta_0 \frac{K_1(mr_1) I_1(mr_2) - I_1(mr_1) K_1(mr_2)}{K_0(mr_1) I_1(mr_2) + I_0(mr_1) K_1(mr_2)} \tag{4.46a}$$

or

$$q_f = 2\pi r_1 t k m \frac{K_1(mr_1) I_1(mr_2) - I_1(mr_1) K_1(mr_2)}{K_0(mr_1) I_1(mr_2) + I_0(mr_1) K_1(mr_2)} (T_0 - T_\infty) \tag{4.46b}$$

Fin heat transfer for **convection tip based** on the **boundary condition 1** given by Eq. (4.40a) and **boundary condition 2** for **Case I** given by Eq. (4.40b) is computed and approximated as

$$q_f = 2\pi r_1 t k m \frac{K_1(mr_1) I_1(mr_{2c}) - I_1(mr_1) K_1(mr_{2c})}{K_0(mr_1) I_1(mr_{2c}) + I_0(mr_1) K_1(mr_{2c})} (T_0 - T_\infty) \tag{4.47}$$

where r_{2c} is the corrected outer radius of the fin given as $r_{2c} = r_2 + t/2$.

Fin temperature distributions and heat losses for different fins are summarized in Table 4.2.

4.4 Parameters for Fin Performance Characterization

Performance of a fin is characterized by number of parameters like fin efficiency, fin surface resistance, and fin effectiveness.

4.4.1 Fin Effectiveness

The fin effectiveness is defined as the ratio heat loss from a surface with fins to the heat transfer from the surface without any fins

$$\varepsilon_f = \frac{q_f}{q_{wof}} \tag{4.48}$$

where q_f is fin heat loss and q_{wof} is heat loss by convection from the surface without the fin given as

$$q_{wof} = hA_c (T_0 - T_\infty) \tag{4.49}$$

Substituting Eq. (4.49) into Eq. (4.48), we have the general expression for the fin efficiency

$$\varepsilon_f = \frac{q_f}{hA_c (T_0 - T_\infty)} \tag{4.50}$$

Table 4.2 Temperature distribution and heat loss for different types of fin.

Type of fin	Temperature distribution	Fin heat loss
Straight rectangular fin	$$\frac{T - T_\infty}{T_0 - T_\infty} = \frac{\cosh m\,(L - x) + (h/mk)\sinh m\,(L - x)}{\cosh mL + (h/mk)\sinh mL}$$	$$q_f = \sqrt{hPkA}\,(T_0 - T_\infty) \times \frac{\sinh mL + (h/mk)\cosh mL}{\cosh mL + (h/mk)\sinh mL}$$
Straight triangular fin	$$\frac{\theta}{\theta_0} = \frac{I_0\left(2m\sqrt{x}\right)}{I_0\left(2m\sqrt{L}\right)}$$	$$q_f = k2Wt\,\frac{\left(mL^{-\frac{1}{2}}\right)I_1\left(2m\sqrt{L}\right)}{I_0\left(2m\sqrt{L}\right)}\theta_L$$
	$$\frac{\theta}{\theta_L} = \frac{T - T_\infty}{T_L - T_\infty} = \left(\frac{x}{L}\right)^a,$$ where $$a = \frac{-1 + \sqrt{1 + 4\,m^2L^2}}{2}$$	$$q_f = \frac{kWt}{2L}\,\theta_L\left(-1 + \sqrt{1 + 4m^2L^2}\right)$$

(continued)

Table 4.2 (Continued)

Type of fin	Temperature distribution	Fin heat loss
Conical spine 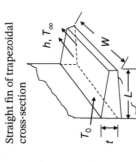	$\dfrac{T-T_\infty}{T_0-T_\infty} = \dfrac{\sqrt{L}\,I_1(2m\sqrt{x})}{\sqrt{x}\,I_1(2m\sqrt{L})}$	$q_f = \dfrac{k\pi R^2}{L}(T_0-T_\infty)\left[\dfrac{\left(2m\sqrt{L}\right)I_0\left(2m\sqrt{L}\right)}{2I_1\left(2m\sqrt{L}\right)} - 1\right]$
Straight fin of trapezoidal cross-section	$\dfrac{\theta}{\theta_*} =$ $\dfrac{I_0\left(2B\sqrt{x_*}\right)K_1\left(2B\sqrt{t_0^*}\right) + K_0\left(2B\sqrt{x_*}\right)I_1\left(2B\sqrt{t_0^*}\right)}{K_1\left(2B\sqrt{t_0^*}\right)I_0\left(2B\sqrt{t_0}\right) + K_0\left(2B\sqrt{t_0}\right)I_1\left(2B\sqrt{t_0^*}\right)}$	$q_f = 2Bk\sqrt{t_0}\,\dfrac{W}{L}\,(t_0-t_0^*)\,\theta_0 \times$ $\dfrac{I_1\left(2B\sqrt{t_0}\right)K_1\left(2B\sqrt{t_0^*}\right) - K_0\left(2B\sqrt{t_0}\right)I_1\left(2B\sqrt{t_0^*}\right)}{K_1\left(2B\sqrt{t_0^*}\right)I_0\left(2B\sqrt{t_0}\right) + K_0\left(2B\sqrt{t_0}\right)I_1\left(2B\sqrt{t_0^*}\right)}$
Circular or annular fin	$\dfrac{T-T_\infty}{T_0-T_\infty} = \dfrac{\theta}{\theta_0} =$ $\dfrac{I_0(mr)K_1(mr_2) + K_0(mr)I_1(mr_2)}{I_0(mr_1)K_1(mr_2) + K_0(mr_1)I_1(mr_2)}$	$q_f = 2\pi r_1 tkm\,\dfrac{K_1(mr_1)I_1(mr_{2c}) - I_1(mr_1)K_1(mr_{2c})}{K_0(mr_1)I_1(mr_{2c}) + I_0(mr_1)K_1(mr_{2c})}$ $(T_0 - T_\infty)$

Fin effectiveness for different fins is evaluated from Eq. (4.50) by substituting expressions for heat loss. For example, for a straight fin of uniform cross-section and rectangular profile, we can obtain the expression for fin effectiveness by substituting fin heat loss Eq. (4.11) for convective tip as

$$\varepsilon_f = \frac{\sqrt{hPkA_c} \; \dfrac{\sinh\; mL + (h/mk)\cosh\; mL}{\cosh\; mL + (h/mk)\sinh\; mL}}{hA_c\left(T_0 - T_\infty\right)} \tag{4.51}$$

While Eq. (4.51) is used for quantitative evaluation and analysis of the straight rectangular fin, a simpler expression derived based on the limiting case 3 solution with long fin is used for qualitative evaluation. Substituting the case 3 heat loss solution from Table 4.1 into Eq. (4.50.), we get the following simplified expression for fin effectiveness:

$$\varepsilon_f = \frac{\sqrt{hPkA}\left(T_0 - T_\infty\right)}{hA_c\left(T_0 - T_\infty\right)} \tag{4.52a}$$

or

$$\varepsilon_f = \sqrt{\frac{kP}{hA_c}} \tag{4.52b}$$

Following qualitative conclusions can be made based on this simplified expression: (i) Fin effectiveness increases with the use of higher thermal conductivity fin materials, (ii) Fin effectiveness increases with increased P/A ratio of the fin geometry. This is achieved by using increased number of thin and closely packed fins. This is, however, limited by the increased pressure drop across the finned surface, (iii) Use of fin can be more justified for the cases with lower convection heat transfer coefficients such as in air flow in free convection heat transfer.

4.4.2 Fin Efficiency

Fin efficiency is defined as the ratio of fin heat transfer to the maximum possible fin heat transfer.

$$\eta_f = \frac{q_f}{q_{max}} \tag{4.53}$$

where q_{max} is the maximum possible heat transfer from the fin surface area.

The maximum possible heat transfer from a fin is the case where the conduction resistance in the fin is negligible and the entire fin is assumed to be at a uniform constant temperature equal to the fin base temperature, T_0. This is expressed as

$$q_{max} = h_c A_f\left(T_0 - T_\infty\right) \tag{4.54}$$

The fin efficiency is then expressed as

$$\eta_f = \frac{q_f}{h_c A_f\left(T_0 - T_\infty\right)} \tag{4.55}$$

For a given fin type, the fin efficiency can be evaluated by substituting the appropriate expression for fin heat loss in Eq. (4.55). As an example, the fin efficiency can be computed for the **circular** or **annular fin** as follows:

$$\eta_f = \frac{q_f}{h A_f (T_0 - T_\infty)} = \frac{q_f}{h \, 2\pi \, (r_2^2 - r_1^2)(T_0 - T_\infty)}$$

Substituting $A_f = 2\pi \, (r_{2c}^2 - r_1^2)$ and Eq. (4.33) for the fin heat loss, we get

$$\eta_f = \frac{2\pi r_1 tkm \dfrac{K_1(mr_1) I_1(mr_{2c}) - I_1(mr_1) K_1(mr_{2c})}{K_0(mr_1) I_1(mr_{2c}) + I_0(mr_1) K_1(mr_2)}(T_0 - T_\infty)}{h 2\pi (r_{2c}^2 - r_1^2)(T_0 - T_\infty)} \qquad (4.56a)$$

or

$$\eta_f = \frac{(2r_1/m)}{(r_{2c}^2 - r_1^2)} \frac{K_1(mr_1) I_1(mr_{2c}) - I_1(mr_1) K_1(mr_{2c})}{K_0(mr_1) I_1(mr_{2c}) + I_0(mr_1) K_1(mr_{2c})} \qquad (4.56b)$$

where

$$A_f = \text{Fins surface area for heat transfer} = 2\pi \, (r_{2c}^2 - r_1^2)$$
$$r_{2c} = r_2 + t/2 = \text{Corrected outer radius of the fin}$$
$$L_{2c} = L_2 + t/2 = \text{Corrected length of the fin}$$

For a **straight fin rectangular profile,** the fins efficiency is computed as follows:

$$\eta_f = \frac{\sqrt{hpKA_c} \, \dfrac{\sinh mL + (h/mk)\cosh mL}{\cosh mL + (h/mK)\sinh mL}(T_0 - T_\infty)}{hPL(T_0 - T_\infty)} \qquad (4.57a)$$

or

$$\eta_f = \frac{\sqrt{hpKA_c}}{hPL}\left(\frac{\sinh mL + (h/mk)\cosh mL}{\cosh mL + (h/mK)\sinh mL}\right) \qquad (4.57b)$$

This expression is approximated to a simplified form as

$$\eta_f = \frac{\tanh mL_c}{mL_c} \qquad (4.58)$$

where

$m = \sqrt{\dfrac{hP}{kA_c}}$, $A_c = Wt$, $A_f = 2WL_c$, $A_p = L_c t$ and L_c is the corrected length of the fin given as

$$L_c = L + \frac{t}{2} \qquad (4.59)$$

The expression for mL_c can be approximated as

$$mL_c = \sqrt{\frac{2h}{kA_p}} L_c^{3/2} \qquad (4.60)$$

Fin efficiency is often estimated from the graphical plot of η_f vs. mL_c as shown in Figure 4.12. Similar plots for other types of fin geometries are also shown in Figure 4.12.

Solution to other fin geometries is obtained in similar manner and presented in many heat transfer books, including those by Jacob (1949), Incropera and Dewitt (1996), Cengel and Ghajar (2015), and Holman (1997). A list of results for few common fins is summarized in Table 4.3.

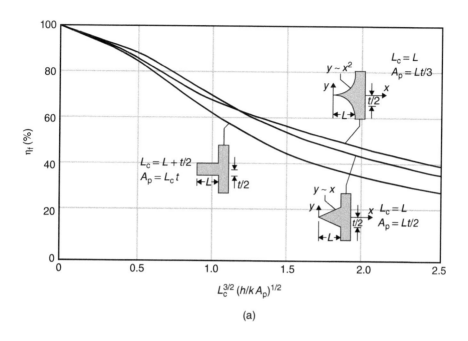

(a)

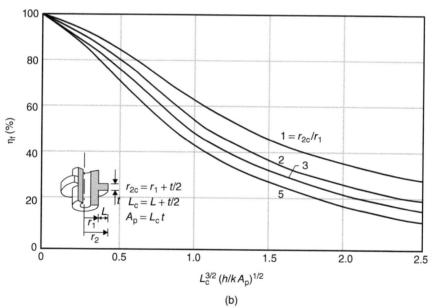

(b)

Figure 4.12 Fin efficiency plots for different fin types. (a) Straight fin and (b) annular fin. Source: Incropera et al. (2007). © 2007, John Wiley & Sons.

Table 4.3 Expressions for heat transfer rate and efficiency for different types of fins.

Fin type	Heat transferrate	Fin efficiency
Straight fin Uniform cross-section of rectangular profile 	$q_{\mathrm{f}} = \dfrac{\sinh mL + (h/mk)\cosh mL}{\sqrt{hPkA_{\mathrm{c}}}\;\cosh mL + (h/mk)\sinh mL}\,(T_0 - T_\infty)$ $L_{\mathrm{c}} = L + \dfrac{t}{2},\; A_{\mathrm{c}} = Wt$ $A_{\mathrm{f}} = 2WL_{\mathrm{c}},\; V_{\mathrm{f}} = WLt$ $m = \sqrt{\dfrac{2h}{kt}}$	$\eta_{\mathrm{f}} = \dfrac{\dfrac{\sinh mL + (h/mk)\cosh mL}{\cosh mL + (h/mk)\sinh mL}}{mL}$ Or $\eta_{\mathrm{f}} = \dfrac{\tanh mL_{\mathrm{c}}}{mL_{\mathrm{c}}}$
Straight triangular fin 	$A_{\mathrm{f}} = 2W[L^2 + (t)^2]^{1/2}$ $A_{\mathrm{p}} = (t)L$ $m = \sqrt{\dfrac{2h}{kt}}$	$\eta_{\mathrm{f}} = \dfrac{2}{mL}\dfrac{I_1\,(2mL)}{I_0\,(2mL)}$

Straight parabolic fin

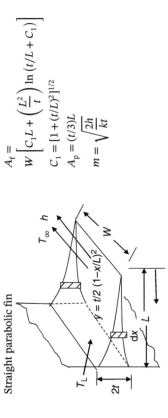

$$A_f = W\left[C_1 L + \left(\frac{L^2}{t}\right)\ln\left(t/L + C_1\right)\right]$$

$$C_1 = [1 + (t/L)^2]^{1/2}$$

$$A_p = (t/3)L$$

$$m = \sqrt{\frac{2h}{kt}}$$

$$\eta_f = \frac{2}{[1 + 4(mL)^2]^{1/2} + 1}$$

Pin fins – cylindrical

$$A_f = \pi D L_c$$

$$L_c = L + D/4$$

$$V_f = \pi(D^2/4)L$$

$$\eta_f = \frac{\tanh mL_c}{mL_c}$$

Pin fin of triangular profile

$$A_f = \pi D/2\, (L^2 + (D/2)^2)^{1/2}$$

$$L_c = L + D/4$$

$$V_f = \pi/3(R^2)L$$

$$m = \sqrt{\frac{4h}{kD}}$$

$$\eta_f = \frac{2}{mL}\frac{I_2(2mL)}{I_1(2mL)}$$

(continued)

Table 4.3 (Continued)

Fin type	Heat transferrate	Fin efficiency
Pin fin – parabolic $y = \left(\dfrac{D}{2}\right)\left(\dfrac{x}{L}\right)^2$ 	$A_f = \dfrac{\pi L^2}{8D}\left[C_1 C_2 - \dfrac{L}{2D}\ln\left(2DC_2/L\right) + C_1\right]$ $C_1 = (1 + 2(D/L)^2)$ $C_2 = (1 + (D/L)^2)^{1/2}$ $V = (\pi/20)D^2 L$ $m = \sqrt{\dfrac{4h}{kD}}$	$\eta_f = \dfrac{2}{\left[1 + 4/9(mL)^2\right]^{1/2} + 1}$
Annular fin – rectangular profile 	$q_f =$ $2\pi r_1 t k m \dfrac{K_1\left(mr_1\right) I_1\left(mr_{2c}\right) - I_1\left(mr_1\right) K_1\left(mr_{2c}\right)}{K_0\left(mr_1\right) I_1\left(mr_{2c}\right) + I_0\left(mr_1\right) K_1\left(mr_2\right)}\left(T_0 - T_\infty\right)$ $A_f = 2\pi\left(r_{2c}^2 - r_1^2\right)$ $r_{2c} = r_2 + t/2$ $L_{2c} = L_2 + t/2$ $A_p = L_c\, t$	$\eta_f = \dfrac{(2r_1/m)}{\left(r_{2c}^2 - r_1^2\right)} \times$ $\dfrac{K_1\left(mr_1\right) I_1\left(mr_{2c}\right) - I_1\left(mr_1\right) K_1\left(mr_{2c}\right)}{K_0\left(mr_1\right) I_1\left(mr_{2c}\right) + I_0\left(mr_1\right) K_1\left(mr_{2c}\right)}$

4.4.3 Fin Thermal Resistance

Fin efficiency is also used to represent fin resistance as follows:

$$R_f = \frac{1}{hA_f\eta_f} \tag{4.61}$$

Example 4.2   Consider a cylindrical heat-generating power module of outer diameter $D = 60$ mm, height $H = 200$ mm, and a base surface temperature $85\,^\circ$ C. Consider 10 equally spaced annular fins of thickness of $t = 3$ mm and length of 20 mm mounted outer surface of the circular module to enhance heat loss. The temperature and the convection coefficient associated with the adjoining fluid are 25 $^\circ$ C and 100 W/m^2 K Use thermal conductivity of aluminum as $k = 230$ W/m K Determine fin efficiency, total heat loss, and effectiveness while assuming a target outer surface temperature of $T_0 = 85\,^\circ$C and considering 10 equally spaced fins.

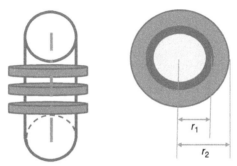

Solution
Annular fin geometrical parameters:

Inner radius, $r_1 = 30$ mm, Length, $L = 20$ mm, Thickness, $t = 3$ mm
Outer radius, $r_2 = r_1 + L = 30 + 20 = 50$ mm

Annular fin parameters:

Corrected outer radius of the fin, $r_{2c} = r_2 + t/2 = 50 + 3/2 = 51.5$ mm $= 0.0515$ m
Corrected length of the fin, $L_c = L + t/2 = 20 + 3/2 = 21.5$ mm $= 0.0215$ m

Let us first show the computation by using analytical solution for fin heat loss given by Eq. (4.47)

$$m = \sqrt{\frac{2h}{kt}} = \sqrt{\frac{2 \times 100}{230 \times 0.003}} = 17.025$$

For $mr_1 = 17.025 \times 0.030 = 0.51075$ and $mr_{2c} = 17.025 \times 0.0515 = 0.8767$, we can compute the following set of Bessel functions from Table D.1

$$I_0\left(mr_1\right) = I_0\left(0.510\,75\right) = 1.066\,287, k_0\left(mr_1\right) = k_0\left(0.51\right) = 0.906\,853\,7$$

$$I_1\left(mr_1\right) = I_1\left(0.510\,75\right) = 0.263\,793, k_1\left(mr_1\right) = k_1\left(0.510\,75\right) = 1.611\,838$$

$$I_1\left(mr_{2c}\right) = I_1\left(0.8767\right) = 0.481\,835, k_1\left(mr_{2c}\right) = k_1\left(0.8767\right) = 0.747\,267$$

Substituting in Eq. (4.47), the fin heat loss is given as

$$q_f = 2\pi r_1 tkm$$

$$\frac{K_1(mr_1)I_1(mr_{2c}) - I_1(mr_1)K_1(mr_{2c})}{K_0(mr_1)I_1(mr_{2c}) + I_0(mr_1)K_1(mr_{2c})}(T_0 - T_\infty)$$

$$q_f = 2\pi \times 0.030 \times 0.003 \times 230 \times 17.025$$

$$\frac{K_1(0.510\,75)I_1(0.8767) - I_1(0.510\,75)K_1(0.8767)}{K_0(0.510\,75)I_1(0.8767) + I_0(0.510\,75)K_1(0.8767)}(85 - 25)$$

$$q_f = 2.2143\,\frac{1.611\,838 \times 0.481\,835 - 0.263\,793 \times 0.747\,267}{0.906\,853 \times 0.481\,835 + 1.066\,287 \times 0.747267} \times 60$$

$$q_f = 2.2143\,\frac{0.7766 - 0.1971}{0.4369 + 0.7968} \times 60 = 2.2143\,\frac{0.5795}{1.2337} \times 60$$

$$q_f = 62.407\,W$$

Heat loss from $N_f = 10$ fins, $Q_f = N_f\,q_f = 10 \times 62.407\,W = 624.07\,W$
Heat loss from bare surface, $Q_b = h\,2\pi\,r_1(H - N_f t)(T_0 - T_\infty)$

$$Q_b = 100 \times 2\pi \times 0.030\,(0.2 - 10 \times 0.003)\,(85 - 25) = 192.26\,W$$

Total heat loss from the heat-generating module

$$Q_t = Q_f + Q_b = 624.07 + 192.26 = W = 816.33\,W$$

Fin Effectiveness, $\varepsilon_f = \dfrac{q_f}{q_{wof}} = \dfrac{q_f}{hA_{cb}(T_0 - T_\infty)}$

$$\varepsilon_f = \frac{59.37\,W}{h\,2\pi\,r_1 t\,(T_0 - T_\infty)} = \frac{62.407\,W}{100 \times 2\pi \times 0.030 \times 0.003 \times (85 - 25)} = \frac{62.407}{3.3929}$$

$$\varepsilon_f = 18.39$$

Alternate solution:

Alternately, we can compute the fin heat loss based on estimating the fin efficiency using the data presented in Figure 4.12b.

$$\frac{r_{2c}}{r_1} = \frac{0.0515}{0.030} = 1.766,\ A_p = L_c t = 0.0215 \times 0.003 = 6.45 \times 10^{-5}\,m^2$$

$$L_c^{3/2}\left(\frac{h}{kA_p}\right)^{1/2} = (0.0215)^{3/2}\left(\frac{100}{230 \times 6.45 \times 10^{-5}}\right)^{1/2}$$

$$= 0.003\,152\,5 \times 82.1024 = 0.2588$$

From Figure 4.12, for $\frac{r_{2c}}{r_1} = 1.766$ and $L_c^{3/2}\left(\frac{h}{kA_p}\right)^{1/2} = 0.2588$, we can approximately estimate fin efficiency, $\eta_f = 0.90$

Let us compute fin heat loss based on fin heat loss, q_f and maximum possible heat loss from the annular fin using

$$q_f = \eta_f\,q_{max}$$

or

$$q_f = \eta_f\,h\,A_f\,(T_0 - T_\infty) = \eta_f\,h\,\left[2\pi\,(r_{2c}^2 - r_1^2)\right](T_0 - T_\infty)$$

$$q_f = \eta_f\,h\,A_f\,(T_0 - T_\infty) = 0.9 \times 100\,\left[2\pi\,(0.0515^2 - 0.030^2)\right](85 - 25)$$

$$q_f = 0.9 \times 100 \, [2\pi \, (0.00265 - 0.0009] \, (85 - 25)$$

$$q_f = 59.37 \text{ W}$$

Heat loss from $N_f = 10$ fins, $Q_f = N_f \, q_f = 10 \times W = 590.37$ W
Heat loss from bare surface, $Q_b = h \, 2\pi \, r_1 (H - N_f t) \, (T_0 - T_\infty)$

$$Q_b = 2\pi \times 0.030 \, (0.2 - 10 \times 0.003) \, 100 \, (85 - 25) = 192.26 \text{ W}$$

Total heat loss: $Q_t = Q_f + Q_b = 590.37 + 192.26 = W = 782.635$ W

Fin Effectiveness, $\varepsilon_f = \dfrac{q_f}{q_{wof}} = \dfrac{q_f}{hA_{cb}(T_0 - T_\infty)}$

$$\varepsilon_f = \frac{59.37 \text{ W}}{h 2\pi \, r_1 t \, (T_0 - T_\infty)} = \frac{59.37 \text{ W}}{100 \times 2\pi \times 0.030 \times 0.003 \times (85 - 25)} = \frac{59.37}{3.3929}$$

$$\varepsilon_f = 17.5$$

4.5 Multiple Fin Arrays and Overall Surface

It can be noticed from the expression for the overall heat transfer coefficient that the heat transfer dissipation rate can be increased by reducing the convection surface resistance by adding extended surface or fins to the convection surface as shown in Figure 4.13.

The convection heat loss from the solid surface maintained at a temperature T_0 without any fin is given as

$$q = h_c A \, (T_0 - T_\infty) \tag{4.62}$$

Let us consider an array of fins mounted on this solid surface at an equal spacing **s** and providing a total surface area of A_t. The total heat transfer from the finned surface is given as

$$q_{total} = N_f q_f + q_{wof} \tag{4.63a}$$

where

$$q_f = \eta_f h_c A_f \, (T_0 - T_\infty) = \text{Heat transfer from a single fin} \tag{4.63b}$$

$q_{wof} = $ Heat transfer from the bare area of the surface without fins

$$q_{wof} = h_c \, (A_t - N_f A_{fl}) \, (T_0 - T_\infty) \tag{4.63c}$$

Figure 4.13 Array of fins.

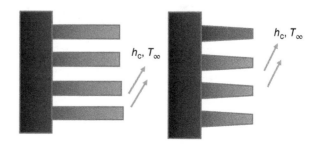

Substituting Eq. (4.63b) and Eq. (4.63c) into Eq. (4.63a), we get

$$q_{total} = N_f \eta_f h_c A_{f1} \left(T_0 - T_\infty\right) + h_c \left(A_t - N_f A_{f1}\right) \left(T_0 - T_\infty\right)$$

With further simplification, we have

$$q_{total} = hA_t \left[1 - \frac{N_f A_{f1}}{A_t} \left(1 - \eta_f\right)\right] \left(T_0 - T_\infty\right) \tag{4.64}$$

or

$$q_{total} = hA_t \eta_0 \left(T_0 - T_\infty\right)$$

where η_0 is the fin surface area effectiveness derived as follows:

$$\eta_0 = 1 - \frac{N_f A_{f1}}{A_t} \left(1 - \eta_f\right) \tag{4.65a}$$

or

$$\eta_0 = 1 - \frac{A_f}{A_t} \left(1 - \eta_f\right) \tag{4.65b}$$

4.5.1 Finned-Surface Convection Thermal Resistance

Using the basic concept of thermal resistance, Eq. (4.66) can be recast as follows:

$$q_{total} = \frac{\left(T_0 - T_\infty\right)}{\eta_0 hA_t} \tag{4.66}$$

and the finned-surface convection resistance can be expressed as

$$R_{conv} = \frac{1}{\eta_0 hA_t} \tag{4.67}$$

4.5.2 Overall Heat Transfer Coefficient for a Finned Surface

The overall heat transfer coefficient for a surface with finned convection on both sides is written as follows:

4.5.2.1 Plane Wall
For a typical plane wall surface, the equivalent resistances across the wall can be shown as:

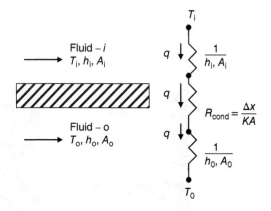

The overall heat transfer coefficient is defined as

$$U_i = \frac{1}{\frac{1}{\eta_{oi}h_i} + \frac{\Delta x}{k} + \frac{1}{\eta_{oo}h_o}} \quad \textbf{\textit{based on inner surface area}} \tag{4.68}$$

where A = Surface area of the plane wall

4.5.2.2 Cylindrical Surface

For a typical tube wall surface, the equivalent resistances across the tube wall can be shown as:

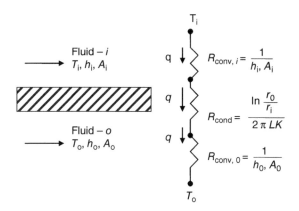

The overall heat transfer coefficient is defined as

$$U_i = \frac{1}{\frac{1}{\eta_{oi}h_i} + A_i \frac{\ln \frac{r_o}{r_i}}{2\pi Lk} + \frac{A_i}{A_o} \frac{1}{\eta_{oo}h_o}} \quad \textbf{\textit{based on inner surface area}} \tag{4.69a}$$

and

$$U_o = \frac{1}{\frac{A_o}{A_i} \frac{1}{\eta_{oi}h_i} + A_o \frac{\ln \frac{r_o}{r_i}}{2\pi Lk} + \frac{1}{\eta_{oo}h_o}} \quad \textbf{\textit{based on outer surface area}} \tag{4.69b}$$

where

A_i = inner surface area of the tube = $2\pi r_i L$
A_o = outer surface area of the tube = $2\pi r_o L$

Example 4.3 Consider annular fin of rectangular profile over the outer surface of a circular pipe as shown in the figure below.

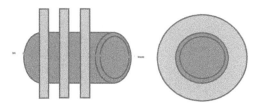

The pipe and fin dimensions are given as

Pipe: Inner diameter, $D_i = 2.0$-in. and outer diameter, $D_o = 2.5$ in.
Fin: Inner diameter, $d_{fi} = 2.5$-in., outer diameter, $d_{fo} = 6$-in. , Thickness, $t = 0.1$-in., and fin spacing, $s = 0.2$-in. $= 0.0166$-ft
Fin material: Copper with $k = 230$ Btu/h ft F
Outside fluid convection coefficient, $h_o = 40$ Btu/h ft F and
Inside fluid convection coefficient, $h_i = 3000$ Btu/h ft F

Determine the fin efficiency, the fin area effectiveness, and the overall heat transfer coefficient.

Solution
Pipe dimensions:

Inner diameter, $D_i = 2.0$ - in. $= 0.1666$ - ft
Outer diameter, $D_o = 2.5$ - in. $= 0.2083$ - ft

Fin dimensional parameters can be computed as

Fin inner radius, $r_1 = 1.25$ - in. $= 0.1042$ - ft, Fin outer radius, $r_2 = 3.0$-in. $= 0.25$-ft
Fin corrected outer radius, $r_{2c} = r_2 + t/2 = 3.0 + 0.1/2 = 3.05$-in. $= 0.254$-ft
Fin length, $L = r_2 - r_1 = 3.0 - 1.25 = 1.75$ - in.
Corrected fin length, $L_c = L + \frac{t}{2} 1.75 + \frac{0.1}{2} = 1.8$-in. $= 0.15$-ft
Fin profile area, $A_P = L_c \times t = 1.8 \times 0.1$ - in. $= 0.18$ - in.^{2}

$$A_P = 1.25 \times 10^{-3} \text{ ft}^2$$

Fin efficiency:

Fin parameters:

$$L_c^{3/2} \sqrt{h/kA_P} = (0.15)^{3/2} \sqrt{10/(230*1.25*10^{-3})} = 0.058\,09 \sqrt{34.78} = 0.3426$$

and

$$r_{2c}/r_1 = 3.05/1.25 = 2.44$$

For the values of $r_{2c}/r_1 = 2.44$ and $L_c^{3/2*} \sqrt{h/kA_P} = 0.3426$, we can estimate from Figure 4.12. The fin efficiency is

$$\eta_f = 0.88 = 88\%$$

Outside fin effectiveness

$$\eta_0 = 1 - \frac{A_{fi}}{A_{total}} \left(1 - \eta_f\right)$$

Let us now compute the fin area and the total area for heat transfer as follows:

Area of a fin, $A_f = 2\pi \left(r_{2c}^2 - r_1^2\right)$

$$A_f = 2\pi \left[(0.254)^2 - (0.1042)^2\right] = 2\pi \left[0.0645 - 0.01086\right]$$

$$A_f = 0.337 - ft^2$$

Area of bare surface of the pipe in between two fins,

$$A_b = \pi D_o s = \pi (0.2083)(0.01666)$$

$$A_b = 0.001\ 090 - ft^2$$

Total area for heat transfer:

$$A_t = A_f + A_b = 0.337 + 0.0109$$

$$A_t = 0.347 - ft^2$$

Area ratio,

$$\frac{A_f}{A_t} = \frac{0.337}{0.3479} = 0.968$$

Fin area effectiveness,

$$\eta_{0o} = 1 - \frac{A_f}{A_t}\left(1 - \eta_f\right) = 1 - 0.968\ (1 - 0.88)$$

$$\eta_{0o} = 0.8838 = 88.38\%$$

We can write overall heat transfer coefficient based on inside area of the tube as

$$U_i = \cfrac{1}{\cfrac{1}{\eta_{oi} h_i} + \cfrac{r_i \ln(r_o/r_i)}{k} + r_i/r_o * \cfrac{1}{\eta_{oo} h_o}}$$

Since no fins are used inside the tubes, we can set $\eta_{oi} = 1.0$. Substituting we get,

$$U_i = \cfrac{1}{\cfrac{1}{3000} + \cfrac{0.0833\ \ln\left(0.1041/0.0833\right)}{230} + \cfrac{0.0833}{0.1041}\ \cfrac{1}{0.8838 \times 40.0}}$$

or

$$U_i = \frac{1}{0.000\ 33 + 0.000\ 080\ 73 + 0.022\ 634\ 9} = \frac{1}{0.0230}$$

or

$$U_i = 43.392 \text{Btu/h ft F}$$

Bibliography

Abramowitz, M. and Stegun, I.A. (1964). *Handbook of Mathematical Functions, National Bureau of Standards, Applied Mathematic Series 55*. Washington, DC: U.S. Government Printing Office.

Arpaci, V.S. (1966). *Conduction Heat Transfer*. San Francisco, CA: Addison-Wesley.

Bergman, T.L., Lavine, A.S., Incropera, F.P., and Dewitt, D.P. (2011). *Fundamentals of Heat and Mass Transfer*, 7e. Wiley.

Boem, R.F. (1987). *Design Analysis of Thermal Systems*. New York: Wiley.

Carslaw, H.S. and Jaeger, J.C. (1959). *Conduction Heat in Solids*. Oxford.

Cengel, A.Y. and Ghajar, A.J. (2015). *Heat and Mass Transfer: Fundamentals & Applications*, 5e. McGraw-Hill.

Hodge, B.K. and Taylor, R.P. (1999). *Analysis and Design of Energy Systems*, 3e. Prentice-Hall.

Holman, J.P. (1990). *Heat Transfer*, 7e. McGraw-Hill.

Incropera, F.P., and Dewitt, D.P. (1996). *Fundamentals of Heat and Mass Transfer*, 4e. John Wiley and Sons.

Incropera, F.P., Dewitt, D.P., Bergman, T.L., and Lavine, A.S. (2007). *Fundamentals of Heat and Mass Transfer*, 6e. Wiley.

Jacob, M. (1949). *Heat Transfer*, vol. I. Wiley.

London, A.L. and Kays, W. (1984). *Compact Heat Exchangers*, 3e. New York: McGraw-Hill.

Majumdar, P. (2005). *Computational Methods for Heat and Mass Transfer*. New York: Taylor & Francis.

Myers, G.E. (1998). *Analytical Methods in Conduction Heat Transfer*, 2e. Madison: AMCH Publications.

Poulikakos, D. (1994). *Conduction Heat Transfer*. Hoboken, NJ: Prentice Hall.

Stoecker, W.F. (1971). *Design of Thermal Systems*, 3e. New York: McGraw-Hill.

Websites

http://keisan.casio.com
www.mhtl.uwaterloo.ca

Problems

4.1 Consider a straight triangular fin mounted on a hot surface a temperature of $100\,°C$. In the base case, the length and the thickness of the fin at the base are $L = 3\,cm$ and $2t = 6\,mm$, respectively, and the fin is constructed of copper. The surrounding free convection environment is given as $h_c = 15\,W/m^2\,°C$ and $T_\infty = 25\,°C$. Determine: (a) fin heat loss per unit width of the fin, (b) fin efficiency, (c) fin effectiveness, (d) weight of the fin, and (e) repeat steps (a) through (d) by reducing the fin length from 3.0 to 2.0-in. using a step size of 0.2-in. Presents results in a summary table.

4.2 Repeat Problem 4.1 and compare results for two other fin materials: (i) aluminum and (ii) magnesium.

4.3 Consider a spine fin of circular cone shape mounted over a heated surface maintained at a temperature $T_L = 300\,°F$ with base radius, $R = 0.25$-in. and length, $L = 2.5$-in. Assume the fin material as aluminum with thermal conductivity, $k_f = 120\,Btu/h\,ft\,F$

and following convection operating conditions: $h = 20$ Btu/h ft² ° F and $T_\infty = 65$ °F. Determine: the fin heat loss, q_f, and fin efficiency.

4.4 Consider a spine fin of parabolic profile mounted over a heated surface maintained at a temperature $T_L = 300°$ F with base radius, $R = 0.25$-in. and length, $L = 2.5$-in. Assume the fin material as aluminum with thermal conductivity, $k_f = 110$ Btu/h ft² ° F and following convection operating conditions: $h = 20$ Btu/h ft² ° F and $T_\infty = 70°$ F. Determine the fin heat loss, q_f, and fin efficiency.

4.5 A commercial steel pipe with pin fins of circular profile has the following characteristics:

Pipe Dimension: Inner Diameter = 1.75-in. and Outer Diameter = 2-in.

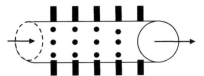

Fins: Consider circular pin fins on outside surface with following dimensions: $L_f = 2.0 - $ in., $d_f = 0.1 - $ in., and $S = 0.2$-in.
Fin material thermal conductivity, $k = 60$ Btu/h ft² ° F
Pipe and fins are exposed gas with gas side convection heat transfer coefficient: $h_o = 100$ Btu/h ft F
Determine fin efficiency η_f and fin surface area effectiveness η_o.

4.6 Consider a cylindrical tube of outer diameter $D = 60$ mm and height $H = 200$ mm. The module is cooled in ambient air at temperature $T_\infty = 25°$C and convection coefficient $h_c = 70$ W/m² K. Ten annular fins of thickness of $t = 10$ mm and fin-to-fin spacing of $s = 20$ mm are used on the outer surface of the circular tube to enhance heat loss. (a) Determine fin efficiency, total heat loss, and effectiveness of the fin while assuming a target outer surface temperature of $T_0 = 50°$C. (b) Show the effect of increased number of fins from 10 to 20 on the total heat loss, fin efficiency, and fin effectiveness.

4.7 Both rectangular and triangular straight fins, and parabolic fins are considered for dissipation from a power-generating module to be maintained at a temperature of $100°$ C. The base dimensions of the fin are: length, $L = 4$-cm, $t = 2$-mm and width, $W = 100$-cm. The surrounding free convection environment is given as $h_c = 15$ W/m² C and $T_\infty = 25°$ C. The objective of the design analysis study is to vary the dimensional parameters such as length and thickness of the fin to finalize the fin design considering high thermal performance while keeping minimum weight as the primary restriction. In addition to the consideration of three different fin shapes, two different materials such as aluminum and magnesium are also taken into account. This analysis will be carried out using a computer program to perform the repetitive computations. Follow the steps outlined below:
(a) Determine: fin heat loss of the fin, fin efficiency, and weight of the fin considering the fin shape as rectangular or two different materials: Aluminum and Magnesium.

(b) Increase the thickness of the magnesium fin while keeping the base-case length ($L = 4$ cm) fixed until the fin heat loss is same as in the case of aluminum fin. Summarize the results for heat loss, efficiency, and weight in a table.

(c) Increase the length of the magnesium fin while keeping the base-case fin thick ($t = 2$-mm) fixed until the fin heat loss is same as in the case of aluminum fin. Summarize the results for heat loss, efficiency, and weight in a table.

(d) Summarize the results obtained in steps (a)–(c) in a table. Discuss results in finalizing the fin design considering fin efficiency and fin weight.

(e) Compare the performance of the final two rectangular fin designs made of aluminum and magnesium in terms of fin heat loss, fin efficiency, and weight considering the convection heat transfer coefficient values as 30, 60, 90, 180, and 360.

(f) Repeat steps (a)–(d) considering two fin types: triangular and parabolic.

(g) Present your final conclusions for the fin design considering minimum weight while keeping the fin thermal performance as high as possible.

5

Analysis and Design of Heat Exchangers

Heat exchangers play a critical role in the design and operation of all thermal energy systems. Heat exchangers may be of single-phase liquid–liquid, gas–liquid, and gas–gas types as well as those involving phase change such as condensers and boilers or evaporators. In this chapter, heat-exchanger types and classifications; geometrical details and standards; and analysis and design algorithms are presented.

5.1 Heat-exchanger Types and Classifications

Heat exchangers are classified into four major categories based on the flow arrangements and amount of surface area for heat transfer. These are *1. Double-Pipe or Concentric Tube Heat Exchangers; 2. Shell-and-Tube Heat Exchangers; 3. Cross Flow; and 4. Compact Heat Exchangers*.

5.1.1 Double-Pipe Heat Exchanger

Double-pipe heat exchanger consists of two concentric tubes with one fluid moving through the inner tube and other flowing through the annular space between the two tubes as shown in Figure 5.1.

5.1.2 Shell-and-Tube Heat Exchangers

A shell-and-tube heat exchanger consists of a shell and a bundle of tubes as shown in Figure 5.2. This is just a variation from the double-pipe heat exchanger to create more surface area for heat transfer. The single inner pipe of the double-pipe heat exchanger is replaced by several smaller tubes to increase the heat-transfer area.

Fluid-1 (Tube-side fluid): enters one side of the tube box and passes through the bundle of tubes and exits through the tube box at other side.

Fluid-2 (Shell-side fluid): enters one side of the shell and flows over the bundle of tubes and exits at other end of the shell. Flow partitions, named ***baffles***, are installed in the shell side to control flow pattern and create enhanced heat-transfer coefficients.

Design of Thermal Energy Systems, First Edition. Pradip Majumdar.
© 2021 John Wiley & Sons Ltd. Published 2021 by John Wiley & Sons Ltd.
Companion website: www.wiley.com/go/majumdar

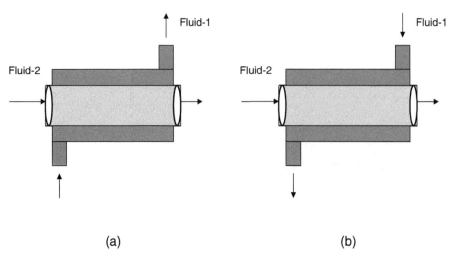

Figure 5.1 Double-pipe heat exchanger. (a) Parallel flow and (b) counter flow.

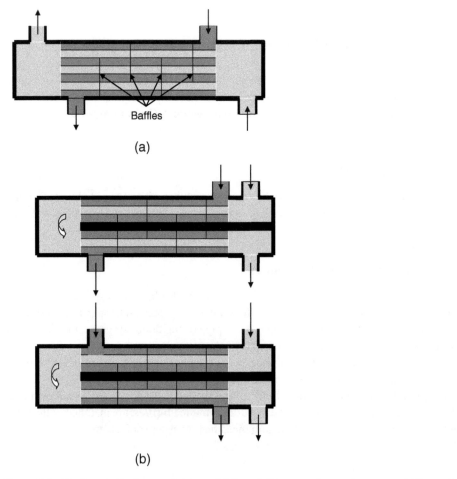

Figure 5.2 Shell-and-tube heat exchangers. (a) One shell pass and one tube pass and (b) two shell passes and four tube passes.

5.1.3 Cross Flow Heat Exchangers

In this category of heat exchangers, the two fluids flow perpendicular or flow across over each other. The flow passages for the fluids could be rectangular parallel finned or unfinned passages as shown in Figure 5.3a or through finned or unfinned tubes as shown in Figure 5.3b. Another important aspect of this type of heat exchanger is that one fluid can freely move inside the flow passage as shown in Figure 5.3a. Based on this flow aspect, crossflow heat exchangers are categorized as (i) **both fluids unmixed** and (ii) **one fluid mixed and other unmixed**.

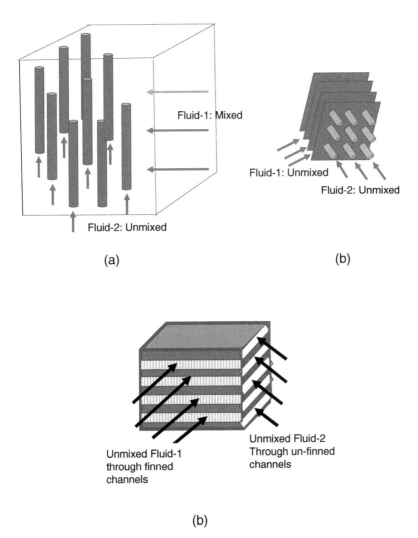

(a)

(b)

(b)

Figure 5.3 Crossflow heat exchangers. (a) One fluid mixed and other unmixed and (b) both fluid unmixed.

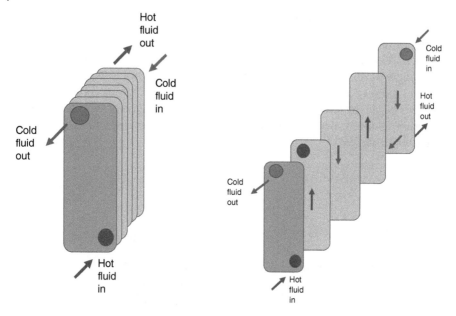

Figure 5.4 Compact heat exchangers.

5.1.4 Compact Heat Exchangers

Compact heat exchangers provide large heat-transfer surface area per unit volume by using dense array of finned tubes or lower flow passage area and/or plates of different flow geometries.

The flow passages are typically very small of the ***order of millimeters*** and ***the surface area per unit volume is of the order of*** $1000\ m^2/m^3$ ***or more***. Configurations of many compact heat-exchanger types are just similar in shape like the box-shaped cross-flow heat exchangers shown in Figure 5.3, except that the flow passages are significantly smaller in thickness formed by stacking thin plates with surface-mounted fins of all different shapes and configurations as shown in Figure 5.4a. Often, the flow passages are formed with different patterns on plate surface by using stamping process. Fluids flow through the passages formed in between the two plates. As shown in Figure 5.4b with an enlarged view, the hot fluid enters the front endplate at the bottom and flows up the channel 1 and skips the next flow channel 2 through which cold fluid flows and enters the channel 3 and flows down the channel 3. The hot fluid flow patterns continue in this manner until the liquid exits through the other side of the back endplate. Similarly, the cold fluid enters through backend plates and flows through the alternate flow channels and exits from the front endplate as shown. At any time, heat is transferred between the hot and cold fluids across the heat-exchanger plates.

5.2 Heat-exchanger Codes and Standards

Several standards are available for shell-and-tube heat exchangers. These are

1. Tubular Exchanger Manufacturers Associations (***TEMA***) standard.
2. American Petroleum Institute (**API**) Standard 600 for Shell and Tube Heat Exchanger.

3. ASME Section VIII, Div. 2: Boiler and Pressure Vessel Code.
4. Heat Exchanger Institute (HEI) standard.
5. ASME B31.1 for Power Piping.
6. API Standard 662 for Plate-Heat Exchangers for General Refinery Services.
7. HEI 3092 stand for Gasketed Plate-Heat Exchangers.

5.2.1 TEMA Standard

TEMA standard is prepared by the technical committee of the Tubular Manufacturers Association comprising a large number of different heat-exchanger manufacturers. The main purpose is to assist heat-exchanger designers in designing, preparing specifications, and installing tubular exchangers. The standard is prepared based on proven engineering principles, research findings, and field experiences. The standard is also subjected to continuous revisions based on new information obtained from state-of-the-art research and experiences.

TEMA standard is the most popular and widely used by manufacturers, designers, and practitioners. It deals with primarily some basic areas such as (i) **Heat-exchanger nomenclature**, (ii) Heat-exchanger **fabrication tolerances**, (iii) General fabrication and performance information, (iv) Installation, operation, and maintenance (v) **Mechanical standard**, (vi) Flow-induced vibration, (vii) Thermal relations, (viii) Physical properties of fluids, (ix) General information, and (x) Recommended good practice.

According to this standard, the shell type, front-end, and rear-end design are designated by combination of letters as shown in TEMA standard in Figure 5.5.

For example, the designation given as **AES** stands for a heat exchanger that includes a *front-end stationary head* **(A)**, *a one-pass shell* **(E)**, and *a rear-end head type with floating head and backing device* **(S)**.

Typical components, parts, and connectors of a heat exchanger are described using a standardized terminology in TEMA standard. An example of the list of nomenclature of the heat-exchanger components and parts for the one with AEP designation is given in Figure 5.6.

TEMA-recommended fabrication tolerances to maintain process flow nozzle and support locations are shown in Figure 5.7.

5.2.2 API Standard 600 (2015)

This standard was developed for the use of shell-and-tube heat exchangers in the petroleum, petrochemical, and natural gas industries. This, however, excludes the use in vacuum steam condenser and feedwater heaters. It specifies requirements and recommendations for the mechanical designs, materials selection, fabrication, testing and inspection, and for the shipment.

5.2.3 ASME Boiler and Pressure Vessel Code (BVPC) (2017)

ASME–BVPC provides rules for the construction of stationary pressure vessels and for the allowable working pressures for pressure vessels covering industrial and residential boilers as well as nuclear reactor components and other forms of pressure vessel, including transportation tanks. The following is a list of some of the key sections of the BVPC code:

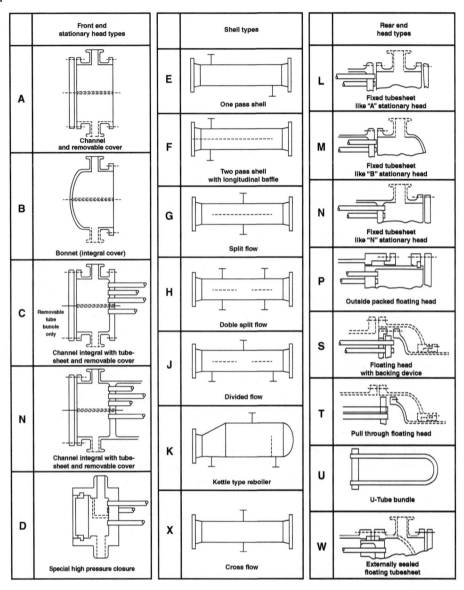

Figure 5.5 TEMA standard designation for the shell type, front-end, and rear-end designs using letters. Source: TEMA Standard (2007). © 2007, Tubular Exchanger Manufacturers Association, Inc..

(i) Section I: provides requirements for all methods of construction of power, electric, and miniature boilers; heat recovery steam generators; solar receiver steam generators boilers; power boilers in locomotive; and some fired pressure vessels; (ii) Section II: Materials; (iii) Section III: Rules for Construction of Nuclear Facility Components; (iv) Section IV: Heating Boilers; (v) Section V: Nondestructive Examination; (vi) Section VI: Care and Operation of Heating Boilers; (vii) Section VII: Care of Power Boilers; (viii) Section VIII, Div. 1, 2, and 3: Rules for Construction of Pressure Vessels; (ix) Section IX: Welding, Brazing, and

1. Stationary head-channel
2. Stationary head-bonnet
3. Stationary head flange-channel or bonnet
4. Channel cover
5. Stationary head nozzle
6. Stationary tubesheet
7. Tubes
8. Shell
9. Shell cover
10. Shell flange-stationary head end
11. Shell flange-rear head end
12. Shell nozzle
13. Shell cover flange
14. Expansion joint
15. Floating tubesheet
16. Floating head cover
17. Floating head cover flange
18. Floating head backing device
19. Split shear ring
20. Slip-on backing flange

21. Floating head cover-external
22. Floating tubesheet skirt
23. Packing box
24. Packing
25. Packing gland
26. Lantern ring
27. Tierods and spacers
28. Transverse baffles or support plates
29. Impingement plate
30. Longitudinal baffle
31. Pass partition
32. Vent connection
33. Drain connection
34. Instrument connection
35. Support saddle
36. Lifting lug
37. Support bracket
38. Weir
39. Liquid level connection
40. Floating head support

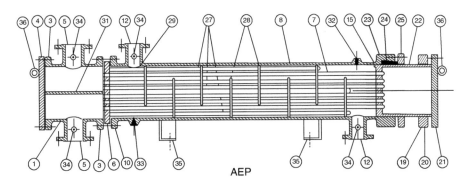

AEP

Figure 5.6 Nomenclature of the components and parts for heat exchanger with AES designation as per TEMA standardized terminology. Source: TEMA Standard (2007). © 2007, Tubular Exchanger Manufacturers Association, Inc.

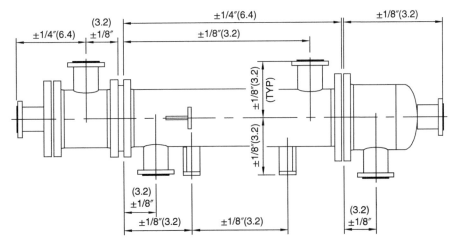

Figure 5.7 Recommended fabrication tolerances as TEMA standard. Source: TEMA Standard (2007). © 2007, Tubular Exchanger Manufacturers Association, Inc.

Fusing Qualification; (x) Section X: Fiber-Reinforced Plastic Vessels; (xii) Section XI: Rules for Inservice Inspection of Nuclear Power Plant Components; (xii) Transport Tanks. This code is being used all over the world, ensuring safety, reliability, and operational efficiency.

5.2.4   Heat Exchanger Institute (HEI) Standard

This HEI standard (2011) was developed by Heat Exchanger Institute for the for shell-and-tube heat exchanger for use in power plants. A recent version (5th Edition, 2011) of this standard also includes a new section on plate-heat exchanger.

5.2.5   API-662 Standard for Plate-heat Exchangers

This international standard outlines requirements and recommendations for the mechanical design, materials selection, fabrication, inspection, testing, and shipment of plate-heat exchangers for use in petroleum and natural gas industries. This standard also covers gasketed, semiwelded, and welded plate-heat exchangers.

5.2.6   HEI 3092 Standard for Gasketed Plate-heat Exchangers

This standard applies to plate-heat exchangers with elastomeric gaskets and carbon steel frames used in power plants.

5.2.7   ASME B31.1 for Power Piping

There are different versions of B31 codes for power piping. Each version is recommended for a specific application. For example, *ASME B31.1* code (1916) was developed for power piping system used in electric power generation systems; industrial plants; geothermal heating; and central and district heating and cooling. The code includes requirements for the design, materials, fabrication, erection test and inspection, operation, and maintenance of piping system.

B31.3 code was developed for power piping in petroleum refineries, chemical, pharmaceutical, and cryogenic plants. B31.4 was developed for piping transportation systems for liquids and slurries. B31.5 was developed for refrigeration piping and heat-transfer components.

5.3   Heat-exchanger Design Options

5.3.1   Categories of Shell-and-Tube Heat Exchanger

The shell-and-tube heat exchanges can be categorized into three basic categories while considering the configuration of the tube bundle and tube sheet as shown in Figure 5.8. These are (i) Fixed tube-sheet shell and tube, (ii) Return-bend or U-tube shell and tube and (iii) Floating tube-sheet or floating head shell and tube.

Some of key the features of these three categories of shell-and-tube heat exchangers are discussed here with the purpose to help making decisions during analysis and design stage.

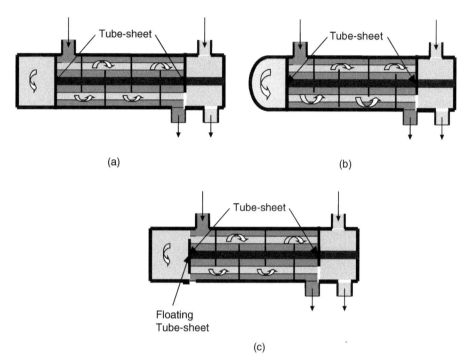

Figure 5.8 Different categories of shell-and-tube heat exchanger. (a) Fixed tube-sheet shell-and-tube, (b) return bend or U-tube shell-and-tube, and (c) floating tube-sheet or floating head shell-and-tube.

5.3.1.1 Fixed Tube Sheet
5.3.1.1.1 Advantages
(i) Allows multiple passes with even–odd number of passes; (ii) Involves fewer number of joints and hence provides higher leakage protection; (iii) Leads to lowest shell diameter for the required heat-transfer surface area with same tube length, tube diameter, and number of passes. Sizes; and (iv) This design can handle fouling fluids on the tube side as it allows easier removal of tube end box and periodic cleaning of inner tube surface using mechanical means.

5.3.1.1.2 Disadvantages
(i) During operations, including start-up and shut-down operations, uneven temperature profiles cause uneven expansions of tubes and the shell, and hence may lead to excessive stresses and may cause fractures and leaks. To encounter this uneven expansion, costlier expansion joints need to be installed.

5.3.1.2 Return Bend or U-Tube
Bend tubes are used to allow tube-side fluid to flow around the hemispherical back-end side of the shell-and-tube heat exchange and return back to the front-end tube box as shown in Figure 4.4b. Longitudinal baffle is used to separate tubes in two sides of the tubes and to create a partition between the inlet and outlet sections at the front-end tube box. Baffle

plates are used to control the flow patterns of the shell-side fluid. These baffle plates also serve as tube-support plates to provide structural support and prevent bending of the long tubes.

5.3.1.2.1 Advantages

(i) Bend sections of tubes can allow easy expansions and contractions due to uneven temperature variations at the back end, and hence can tolerate larger variations in temperature in tubes and avoid cracking. This helps to exclude use of any additional expansion joints; (ii) Higher pressure of tube-side fluid can be allowed; (iii) Lesser number of packings and/or gaskets are required; and (iv) The tube bundle can be removed from the front-end tube box for repair and cleaning purposes.

5.3.1.2.2 Disadvantages

(i) Cleaner fluids are recommended for use in the tube side because services by mechanical means are restricted for tube side and shell side of the tubes; (ii) Many tubes may be difficult to replace if there is failure or leaks due to cracks; and (iii) Use of U-tubes results in only even number of tube passes.

5.3.1.3 Floating Tube Sheet

The design is like the fixed tube-sheet design. Main difference is the replacement of the front-end fixed tube sheet with a floating tube sheet to all movements of the tubes during expansion and contractions.

5.3.1.3.1 Advantages

Because of the floating nature of the front-end tube sheet design, the movement of tubes due to uneven temperature variation can be allowed to some extent. Most of the advantageous features of the fixed tubes-sheet design are applicable to this floating tube-sheet design.

5.3.1.3.2 Disadvantages

Simple modification through use of a floating nature of the tube sheet brings in additional complexity to the design. Extra seals and packings are required to prevent leakages and these results increase in initial cost as well as operation maintenance cost of the heat exchanger.

5.3.2 Recommended Design Assumptions

5.3.2.1 Tube Geometrical Parameters

Tube diameter: Smaller-diameter tubes result in higher heat-transfer surface area per unit volume. It is desirable to make it as small as possible. The lower limit, however, is set by the pressure drop and the ability to clean the inside and outside of the tube.

Tube length: Longer tube cost less. Consider the heat-transfer duty in terms of heat-transfer rate and pressure drop per unit length and see if longer tube is indeed needed to increase performance.

Tube pitch: Triangular tube pitch may result in enhanced heat transfer and in reduced overall size. This will, however, result in higher pressure drop and cause a problem for mechanical cleaning to some extent.

5.3.2.2 Shell Geometrical Parameters

Shell diameter: Shell diameter is restricted by the number of tubes, tube size, and spacing in the tube banks based on heat-transfer duty.

Shell internal design: Internal shell design will have a significant effect on fluid flow pattern and strongly influences the shell-side heat transfer and pressure drop. Often, shell-side fluid flow tends to be laminar. Consideration must be given to ensure that flow does not take a shortcut path or a slower velocity. Desired velocity or Reynolds number and turbulence level are controlled by using properly designed baffles through selection of configuration, height, and spacing.

Multiple shells in series or parallel: Multiple shells in series are preferred to achieve desired shell-side velocity where volume flow rate is low. Multiple shells in parallel are preferred in applications that involve high flow rates.

5.3.2.3 Counter Flow Vs. Parallel Flow

Counter flow is generally preferred as it results in a greater temperature change of a fluid compared to the parallel flow. This greater temperature change also results in a greater mean temperature difference between two fluids from inlet to outlet sections of the heat exchanger. Exceptions are only seen for situations where physical layer prohibits its use as counter flow. However, there is no variation in the mean temperature difference for cases where one of the heat-exchanging fluids undergoes phase change such as in a condenser or in an evaporator, and so either configurations could be selected.

5.3.2.4 Choice of a Fluid in Shell Side Vs. Tube Side

Since turbulence can easily be initiated by controlling the shell-side flow pattern through use of baffles, higher viscosity and lower flow rate fluids are usually used in the shell side. If one of the fluids requires periodic cleaning due to its special characteristics such as fouling or scale formation, corrosivity, and toxicity; it is used in the tube side of the shell-and-tube heat exchanger. Condensing fluids are generally used in shell side like in a steam condenser. Only exceptions are in situations where specialty metals or metal-coating contacts with the fluid are necessary to initiate or enhance condensation.

5.4 Heat-exchanger Design Analysis Methods

Heat-exchanger design analysis method relates to size of the heat exchanger to the required heat-transfer ratings. There are two basic approaches for the analysis and design of heat exchangers. These are 1. Log Mean Temperature Difference (LMTD) and 2. Effectiveness-NTU method, $(\varepsilon\text{-NTU})$.

5.4.1 Log Mean Temperature Difference (LMTD)

This method relates heat-transfer rate to inlet and outlet fluid temperatures, overall heat-transfer coefficients, and to the surface area for heat transfer. Let us demonstrate this method considering both parallel- and counter flow in a double-pipe arrangement. The analysis method will then be extended to the design of shell-and-tube heat exchangers.

5.4.1.1 Parallel-Flow Arrangement

In parallel-flow arrangements, both fluids enter at the same end of the heat exchanger and flow in the same direction as shown in Figure 5.9.

Energy balance over the control volume for each fluid leads to the following equations:

Heat lost by the hot fluid:

$$q = \dot{m}_h c_{p,h}(T_{h,i} - T_{h,o}) = C_h(T_{h,i} - T_{h,o}) \tag{5.1}$$

Heat gained by the cold fluid:

$$q = \dot{m}_c c_{p,c}(T_{c,o} - T_{c,i}) = C_c(T_{c,o} - T_{c,i}) \tag{5.2}$$

For a differential control volume:

$$\text{Heat lost by hot fluid: } dq = \dot{m}_h c_{p,h} \Delta T_h \tag{5.3a}$$

$$\text{Heat gained by the cold fluid: } dq = \dot{m}_c c_{p,c} \Delta T_c \tag{5.3b}$$

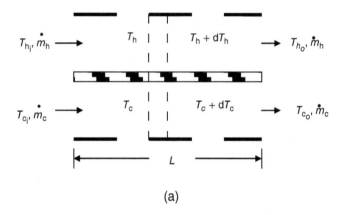

(a)

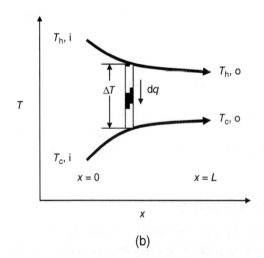

(b)

Figure 5.9 Parallel double-pipe heat exchanger (a) Parallel-flow arrangements and (b) temperature variation in the fluid along the length of the heat.

where C is the capacitance of the fluid defined as $C = \dot{m}c_p$.

In terms of heat-transfer rate across tube wall:

$$dq = UA\Delta T. \tag{5.3c}$$

where $\Delta T = T_h - T_c$ is the local temperature difference between the two fluids and UA is the product overall heat-transfer coefficient across the tube wall and the total surface area. Note that ΔT varies from inlet $\Delta T_{max} = T_{h,i} - T_{c,i}$ to the outlet $\Delta T_{min} = T_{h,o} - T_{c,o}$.

Integrating Eq. (5.3c) over the entire length of the heat exchangers and simplifying, the heat-transfer rate is given as

$$q = UA\Delta T_m \tag{5.4}$$

where

$$\Delta T_m = \text{Log mean temperaure difference} = \frac{\Delta T_{max} - \Delta T_{min}}{\ln \frac{\Delta T_{max}}{\Delta T_{min}}}$$

$$\Delta T_m = \frac{(T_{h,i} - T_{c,i}) - (T_{h,o} - T_{c,o})}{\ln \frac{(T_{h,i} - T_{c,i})}{(T_{h,o} - T_{c,o})}} \tag{5.5}$$

Eqs. (5.1), (5.2), (5.4), and (5.15) are used to perform analysis of a heat exchanger.

5.4.1.2 Counter Flow

In a counter-flow double-pipe heat exchanger, fluids enter at two different ends of the heat exchangers and flow in the opposite directions as shown in Figure 5.10.

In a similar manner, the heat-transfer equations can be derived as

Heat lost by hot fluid:

$$q = \dot{m}_h c_{p,h}(T_{h,i} - T_{h,o}) = C_h(T_{h,i} - T_{h,o}) \tag{5.1}$$

Heat gained by cold fluid:

$$q = \dot{m}_c c_{p,c}(T_{c,o} - T_{c,i}) = C_c(T_{c,o} - T_{c,i}) \tag{5.2}$$

Heat transfer across the tube walls:

$$q = UA\Delta T_m \tag{5.4}$$

where

$$\Delta T_m = \text{Log mean temperaure difference} = \frac{\Delta T_{max} - \Delta T_{min}}{\ln \frac{\Delta T_{max}}{\Delta T_{min}}} \tag{5.5}$$

Special cases:

Case I: $C_h \to \infty$

Capacitance of the hot fluid is very large compared to that of the cold fluid, i.e.

$$C_h = \dot{m}_h c_{p,h} \gg C_c = \dot{m}_c c_{p,c}$$

For this case, the **temperature of the hot fluid remains approximately constant** throughout the heat exchanger, while the temperature of the cold fluid increases.

One example of this special case is the condenser where a vapor stream undergoes condensation at a constant saturation temperature, T_{Sat}, transferring heat to the circulating cooling fluid as shown in Figure 5.11.

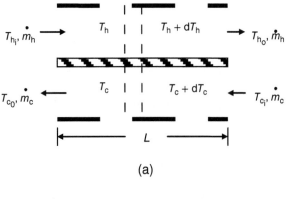

Figure 5.10 Counter flow double-pipe heat exchanger. (a) Counter flow arrangements and (b) temperature variation in the fluid along the length of the heat.

(a)

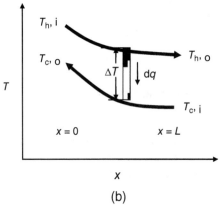

(b)

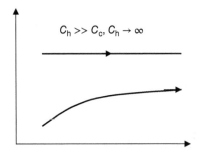

Figure 5.11 Temperature variation in a condenser.

Case II: $C_c \rightarrow \infty$

The capacitance of the cold fluid is very large compared to that of the hot fluid, i.e. $C_c = \dot{m}_c c_{p,c} \gg C_h = \dot{m}_h c_{p,h}$

For this case, the temperature of the cold fluid remains approximately constant throughout the heat exchanger, while that of the hot fluid decreases. An example of this special case is the ***evaporator or boiler*** where the cold fluid undergoes evaporation or boiling at a constant saturation temperature, T_{Sat}, receiving heat from the circulating heating fluid as shown in Figure 5.12.

Case III: **Counter flow,** $C_h = C_c \rightarrow \Delta T = \text{Constant} = \Delta T_m$

For such a case, the temperature difference between two fluid streams remains constant across the length of the heat exchanger as demonstrated in Figure 5.13.

Figure 5.12 Temperature variation in an evaporator.

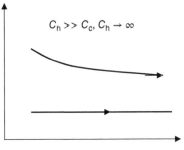

$$C_h >> C_c, \ C_h \rightarrow \infty$$

Figure 5.13 Temperature variation in a limiting case, where $C_h = C_c$.

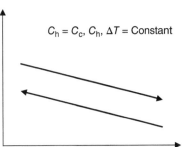

$$C_h = C_c, \ C_h, \ \Delta T = \text{Constant}$$

Few things can be noted from the LMTD analysis equations: 1. The temperature difference, $\Delta T = T_h - T_c$, between two fluids at any sections of the heat exchanger is largest at *the inlet region of the parallel-flow* heat exchanger; 2. The cold fluid exit temperature, $T_{c, o}$, may exceed hot fluid exit temperature, $T_{h, o}$, in the case of a counter-flow heat exchanger; 3. Log mean temperature difference is always less than the arithmetic mean; 4. *LMTD* for *counter flow* exceeds that for *parallel flow,* considering the same inlet and outlet temperatures of the fluids. So, *the surface area required* to effect a prescribed heat-transfer rate q is smaller *for the counter flow* than that for the parallel-flow arrangement, assuming same value for overall heat-transfer coefficient, U; and 5. For a heat exchanger in which one fluid undergoes phase change like in a condenser or evaporator, both parallel flow and counter flow result in same terminal temperature differences.

Example 5.1   A hot fluid enters a double-pipe heat exchange at an inlet temperature of 220 °C and exits at 150 °C while transferring heat to a cold fluid that enters at 40 °C and is heated to 70 °C. Compute the LMTD values considering parallel-flow as well as counterflow heat exchangers.

Solution:
For parallel flow:

$$X = 0: \ (T_{h,i} - T_{c,i}) = 220 - 40 = 180°C = \Delta T_{max}$$

$$X = L: \ (T_{h,o} - T_{c,o}) = 150 - 70 = 80°C = \Delta T_{min} \ =$$

$$\text{LMTD}, \Delta T_m = \frac{\Delta T_{max} - \Delta T_{min}}{\ln \frac{\Delta T_{max}}{\Delta T_{min}}} = \frac{(T_{h,i} - T_{c,i}) - (T_{h,o} - T_{c,o})}{\ln \frac{(T_{h,i}-T_{c,i})}{(T_{h,o}-T_{c,o})}} = \frac{180 - 80}{\ln \left(\frac{180}{80} \right)}$$

$$\Delta T_m = 123°C$$

For counter flow:

$$X = 0: (T_{h,i} - T_{c,o}) = 220 - 80 = 140°C = \Delta T_{max}$$

$$X = L: (T_{h,o} - T_{c,i}) = 150 - 40 = 110°C = \Delta T_{min} =$$

$$\text{LMTD, } \Delta T_m = \frac{\Delta T_{max} - \Delta T_{min}}{\ln \frac{\Delta T_{max}}{\Delta T_{min}}} = \frac{(T_{h,i} - T_{c,o}) - (T_{h,o} - T_{c,i})}{\ln \frac{(T_{h,i} - T_{c,o})}{(T_{h,o} - T_{c,i})}} = \frac{140 - 110}{\ln \left(\frac{140}{110}\right)}$$

$$\Delta T_m = 124.4°C$$

We can see that for the same inlet and exit temperature of the fluids, the LMTD is higher for counter flow than the parallel flow.

5.4.1.3 Multi-Pass Shell-Tube and Cross Flow Heat Exchanger

The LMTD methodology discussed in the previous section on double-pipe heat exchanger is applicable to other types of heat exchangers such as shell-and-tube or Cross Flow types. The heat-transfer rate is calculated based on the same set of equations and by *using a correction factor* applied to the LMTD as listed below:

Heat lost by hot fluid:

$$q = \dot{m}_h c_{p,h}(T_{h,i} - T_{h,o}) = C_h(T_{h,i} - T_{h,o}) \tag{5.1}$$

Heat gained by cold fluid:

$$q = \dot{m}_c c_{p,c}(T_{c,o} - T_{c,i}) = C_c(T_{c,o} - T_{c,i}) \tag{5.2}$$

Heat transfer across the tube walls:

$$q = FUA\Delta T_m \tag{5.6}$$

where

The **correction factor, F,** is estimated as a function of two parameters P and Z as

$$F = f(P, R) \tag{5.7}$$

where

$$P = \frac{T_{to} - T_{ti}}{T_{si} - T_{ti}} \text{ or } \frac{t_2 - t_1}{T_1 - t_1} \text{ and } R = \frac{T_{si} - T_{so}}{T_{to} - T_{ti}} \text{ or } \frac{T_1 - T_2}{t_2 - t_1} \tag{5.8}$$

where T_s is for shell-side temperatures and T_t is for tube-side temperatures. A closed form expression for this correction factor is given as

$$F = \frac{\sqrt{R^2 + 1}\ln(1 - P)/(1 - PR)}{(R - 1)\ln((2 - P(R + 1 - \sqrt{R^2 + 1}))/(2 - P(R + 1 + \sqrt{R^2 + 1})))} \tag{5.9}$$

This is derived for two tube passes. It is also used for multiple tube passes with minor error since correction factor F varies slightly from values for two tube passes. This factor is presented in several charts for different types of heat exchangers as shown in Figures 5.14 and 5.15.

It can be noted here that for heat exchanger that involves one fluid undergoing phase change such as in condensers and evaporators or boilers, the fluid remains at a constant

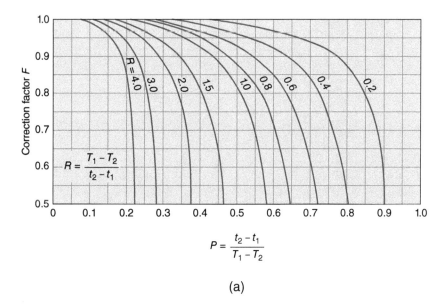

$$P = \frac{t_2 - t_1}{T_1 - T_2}$$

(a)

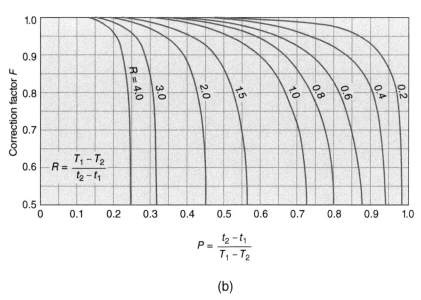

$$P = \frac{t_2 - t_1}{T_1 - T_2}$$

(b)

Figure 5.14 LMTD Correction factors for Shell-and-tube heat exchanger. (a) LMTD correction factor with one shell pass and multiple of 2, 4, ... tube passes and (b) LMTD correction factor for two shell passes and a multiple of 4, 8, ... tube passes. Source: Holman (2010). © 2010, McGraw-Hill Education.

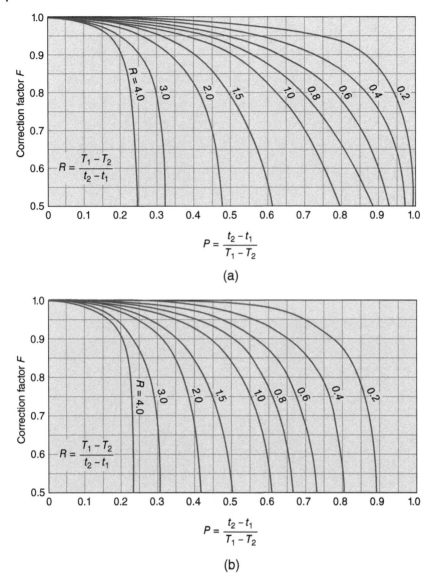

Figure 5.15 LMTD correction factors for crossflow heat exchangers. (a) Both fluids unmixed and (b) one fluid mixed and other unmixed. Source: Holman (2010). © 2010, McGraw-Hill Education.

temperature in most part based on the saturation condition. This condition leads to $P = 0$ or $R = 0$, and hence results in a correction factor of $F = 1.0$.

LMTD method is straightforward when the inlet and outlet temperatures of both hot and cold fluids are known. In the analysis method, LMTD is first calculated and then the heat flow, surface area, or overall heat-transfer coefficient may be determined. The LMTD method, however, involves iterations when the inlet and/or exit temperature is to be evaluated for a given heat exchanger. In such situations, *an alternative method* such as *Effectiveness-NTU* method may be preferred. Figure 5.16 shows the iterative algorithm flowchart for the LMTD method.

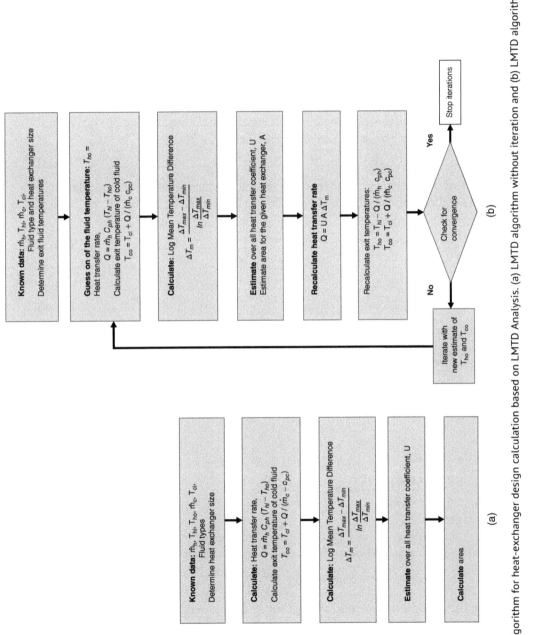

Known data: $\dot{m}_h$, T_{hi}, $\dot{m}_c$, T_{ci},
Fluid type and heat exchanger size
Determine exit fluid temperatures

Guess on of the fluid temperature: T_{ho} =
Heat transfer rate,
$Q = \dot{m}_h\, C_{ph}\, (T_{hi} - T_{ho})$
Calculate exit temperature of cold fluid
$T_{co} = T_{ci} + Q / (\dot{m}_c\, c_{pc})$

Calculate: Log Mean Temperature Difference
$$\Delta T_m = \frac{\Delta T_{max} - \Delta T_{min}}{\ln \frac{\Delta T_{max}}{\Delta T_{min}}}$$

Estimate over all heat transfer coefficient, U
Estimate area for the given heat exchanger, A

Recalculate heat transfer rate
$Q = U\, A\, \Delta T_m$

Recalculate exit temperatures:
$T_{ho} = T_{hi} - Q / (\dot{m}_h\, c_{ph})$
$T_{co} = T_{ci} + Q / (\dot{m}_c\, c_{pc})$

Check for convergence

No → Iterate with new estimate of T_{ho} and T_{co}

Yes → Stop iterations

(b)

Known data: $\dot{m}_h$, T_{hi}, T_{ho}, $\dot{m}_c$, T_{ci},
Fluid types
Determine heat exchanger size

Calculate: Heat transfer rate,
$Q = \dot{m}_h\, C_{ph}\, (T_{hi} - T_{ho})$
Calculate exit temperature of cold fluid
$T_{co} = T_{ci} + Q / (\dot{m}_c - c_{pc})$

Calculate: Log Mean Temperature Difference
$$\Delta T_m = \frac{\Delta T_{max} - \Delta T_{min}}{\ln \frac{\Delta T_{max}}{\Delta T_{min}}}$$

Estimate over all heat transfer coefficient, U

Calculate area

(a)

Figure 5.16 Algorithm for heat-exchanger design calculation based on LMTD Analysis. (a) LMTD algorithm without iteration and (b) LMTD algorithm with iteration.

5.4.2 Effectiveness – NTU Method

In this analysis method, two parameters are introduced: **the heat-exchanger effectiveness, ε,** and the **Net Transfer Unit (NTU)**. The heat-exchanger effectiveness is defined as the ratio of actual heat transferred, q, to the maximum possible heat transfer, q_{max}:

Effectiveness,

$$\varepsilon = \frac{q}{q_{max}} \tag{5.10}$$

The **maximum possible heat transfer,** q_{max}, can be calculated based on heat transfer of fluid with minimum capacitance $\dot{m}c_p = C$ to experience the maximum range of temperature differences that exist in the heat exchanger, i.e. $(T_{h,i} - T_{c,i})$

$$q_{max} = (\dot{m}c_p)_{min}(T_{h,i} - T_{c,i}) \tag{5.11}$$

Note that the minimum capacitance $C_{min} = (\dot{m}c_p)_{min}$ could be either for the cold fluid or for the hot fluid, and can be decided based on following criterion:

If

$$C_h < C_c \quad \text{then} \quad (\dot{m}c_p)_{min} = (\dot{m}c_p)_h \tag{5.12a}$$

and

If

$$C_c < C_h \quad \text{then} \quad (\dot{m}c_p)_{min} = (\dot{m}c_p)_c \tag{5.12b}$$

Effectiveness for parallel-flow or counter-flow heat exchangers can be defined as follows:

Based on hot fluid:

$$\varepsilon_h = \frac{m_h c_h (T_{h_i} - T_{h,o})}{q_{max}} \approx \varepsilon_h = \frac{T_{h,i} - T_{h,o}}{T_{h,i} - T_{c,i}} \tag{5.13a}$$

Or

Based on cold fluid:

$$\varepsilon_c = \frac{m_c c_c (\dot{T}_{c,o} - T_{c,i})}{q_{max}} \approx \varepsilon_c = \frac{T_{c,o} - T_{c,i}}{T_{h,i} - T_{c,i}} \tag{5.12b}$$

Net transfer unit (NTU)

The **Net Transfer Unit** (NTU) is defined as the ratio of the product of overall heat-transfer coefficient and surface area for heat transfer to the minimum capacitance

$$\text{NTU} = \text{Number of transfer units} = \frac{UA}{C_{min}} \tag{5.14}$$

The heat-transfer rates for the two fluid streams are same as stated before for the case of LMTD method:

Heat lost by the hot fluid:

$$q = \dot{m}_h c_{p,h}(T_{h,i} - T_{h,o}) = C_h(T_{h,i} - T_{h,o}) \tag{5.1}$$

Heat gained by the cold fluid:

$$q = \dot{m}_c c_{p,c}(T_{c,o} - T_{c,i}) = C_c(T_{c,o} - T_{c,i}) \tag{5.2}$$

Heat transfer across the tube wall

$$q = \varepsilon \, q_{max} \tag{5.15}$$

Equation (5.15) relates the heat-transfer rates to the size of the heat exchanger through the relation $\varepsilon - $ NTU relationship function:

$$\varepsilon = f\left(NTU, \frac{C_{min}}{C_{max}}\right) \tag{5.16}$$

For a heat-exchanger design problem where heat-exchanger size and type are known, the heat-transfer rates and exit fluid temperatures can be determined by first estimating NTU and then estimating effectiveness from Eq. (5.16).

Alternatively, in problems where heat-transfer rates are known, one can determine the heat-exchanger size and type by first computing the effectiveness and then computing the NTU from the relationship based on effectiveness and capacitance ratio given as

$$NTU = f\left(\varepsilon, \frac{C_{min}}{C_{max}}\right) \tag{5.17}$$

Expressions for effectiveness, ε, and NTU are derived for different heat exchangers and presented in the form of analytical expressions in Table 5.1 and as graphical charts in Figure 5.17.

NTU-Effectiveness method is preferred when not all outlet temperatures are known and in developing a computer code. A typical algorithm for the computations of heat exchangers is demonstrated in Figure 5.18.

Example 5.2 In a shell-and-tube counter-flow heat exchanger with one shell pass and two tube passes, 3800 lbm/h of chilled water ($C_{pc} = 1.00$ Btu/lbm °F) enters at 50 °F and cools 10 000 lbm/h of oil ($C_{ph} = 0.48$ Btu/lbm °F), which enters at 170 °F. Using UA = 6500 Btu/h °F as an estimate for the heat exchanger, determine the effectiveness (ε), heat-transfer rate, exit fluid temperatures using method.

Solution:
For chilled water, $C_c = \dot{m}_c \, C_{pc} = 3800 \, \frac{lbm}{h} \times 1.0 \, Btu/lbm°F = 3800 \frac{Btu}{h-F}$

For oil, $C_h = \dot{m}_h \, C_{ph} = 10\,000 \frac{lbm}{h} \times 0.5 \, Btu/lbm°F = 5000 \frac{Btu}{h-F}$

$$NTU = \frac{UA}{C_{min}} = \frac{6500\frac{Btu}{h-F}}{10\,000} = 0.65$$

$$C_r = \frac{C_{min}}{C_{max}} = \frac{3800}{5000} = 0.76$$

Effectiveness of the counter flow heat exchanger with one shell pass and one tube pass heat exchanger is given as

$$\varepsilon = \frac{1 - \exp[-NTU(1 - C_r)]}{1 - C_r \exp[-NTU(1 - C_r)]}$$

$$\varepsilon = \frac{1 - \exp[-0.65(1 - 0.76))]}{1 - 0.76 \exp[-0.65(1 - 0.76)]} = \frac{1 - \exp[-0.156]}{1 - 0.76 \exp[-0.156]}$$

Effectiveness $\varepsilon = 0.413$

Table 5.1 List of relations for ε-NTU and NTU-ε.

Heat-exchanger type	ε-NTU	NTU-ε
Parallel single pass	$\varepsilon = \dfrac{1 - \exp[-\mathrm{NTU}(1+C)]}{1+C}$	$\mathrm{NTU} = -\dfrac{\ln[1 - \varepsilon(1+C)]}{1+C}$
Counter – Single Pass	$\varepsilon = \dfrac{1 - \exp[-\mathrm{NTU}(1-C)]}{1 - C\exp[-\mathrm{NTU}(1-C)]}$ for $C<1$ $\qquad$ $\varepsilon = \dfrac{\mathrm{NTU}}{\mathrm{NTU}+1}$ for $C=1$	$\mathrm{NTU} = \dfrac{1}{C-1}\ln\left(\dfrac{\varepsilon-1}{\varepsilon C-1}\right)$ for $C<1$ $\qquad$ $\mathrm{NTU} = \dfrac{\varepsilon}{1-\varepsilon}$ for $C=1$
Shell-and-tube – 1 shell pass and 2, 4, 6 tube passes	$\varepsilon_1 = 2\left\{1 + C + (1+C^2)^{1/2} \times \dfrac{1 + \exp[-\mathrm{NTU}(1+C^2)^{1/2}]}{1 - \exp[-\mathrm{NTU}(1+C^2)^{1/2}]}\right\}^{-1}$	$\mathrm{NTU} = -(1+C^2)^{-1/2}\ln\left(\dfrac{E-1}{E+1}\right)$ $\quad E = \dfrac{2/\varepsilon_1 - (1+C)}{(1+C^2)^{1/2}}$
n-Shell pass and $2n$, $4n$ tube passes	$\varepsilon = \left[\left(\left(\dfrac{1-\varepsilon_1 C}{1-\varepsilon_1}\right)^n - 1\right)\right]\left[\left(\left(\dfrac{1-\varepsilon_1 C}{1-\varepsilon_1}\right)^n - C\right)\right]^{-1}$	$\mathrm{NTU} = -(1+C^2)^{-1/2}\ln\left(\dfrac{E-1}{E+1}\right)$ $\quad E = \dfrac{2/\varepsilon_1 - (1+C)}{(1+C^2)^{1/2}}$ $\quad \varepsilon_1 = \dfrac{F-1}{F-C}$ and $F = \left(\dfrac{\varepsilon C-1}{\varepsilon-1}\right)^{1/n}$ $\quad \mathrm{NTU} = n(\mathrm{NTU})_1$
For all heat exchangers with $C=0$	$\varepsilon = 1 - \exp(-\mathrm{NTU})$	$\mathrm{NTU} = 1 - \ln(1-\varepsilon)$
Cross-flow	*Both fluids unmixed:* $\varepsilon = 1 - \exp\left[\left(\left(\dfrac{1}{C}\right)\right)(\mathrm{NTU})^{0.22}\left\{\exp\left[-C(\mathrm{NTU})^{0.78}\right] - 1\right\}\right]$ For $C_{\max} = $ mixed and $C_{\min} = $ unmixed $\varepsilon = \left(\dfrac{1}{C}\right)(1 - \exp\{-C[1 - \exp(-\mathrm{NTU})]\})$ For $C_{\max} = $ unmixed and $C_{\min} = $ mixed $\varepsilon = 1 - \exp\left(1 - \dfrac{1}{C}\{1 - \exp[-C\,\mathrm{NTU}]\}\right)$	For $C_{\max} = $ mixed and $C_{\min} = $ unmixed: $\mathrm{NTU} = -\ln\left[1 + \left(\dfrac{1}{C}\right)\ln(1-\varepsilon C)\right]$ For $C_{\max} = $ unmixed and $C_{\min} = $ mixed $\mathrm{NTU} = -\dfrac{1}{C}\ln[C\ln(1-\varepsilon) + 1]$

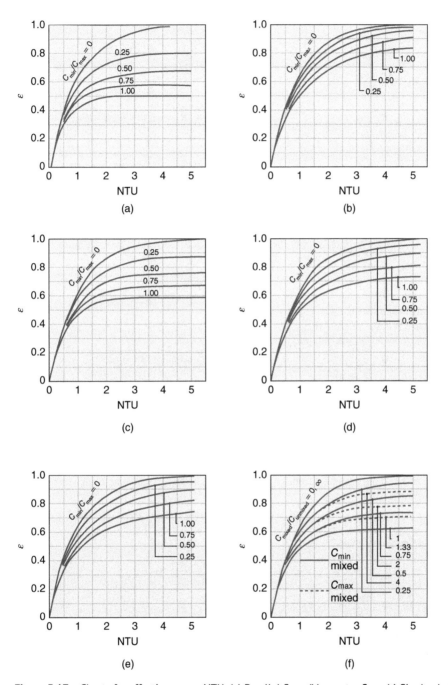

Figure 5.17 Charts for effectiveness – NTU. (a) Parallel flow, (b) counter flow, (c) Single-shell multiple of 2, 4, ... tube passes, (d) two shells and multiples of 4, 8 tube passes, (e) cross flow with both fluid unmixed, and (f) cross flow with one fluid mixed and other unmixed. Source: Incropera et al. (2007). © 2007, John Wiley & Sons.

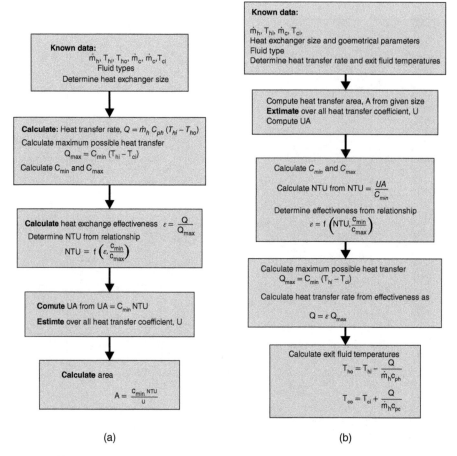

(a)   (b)

Figure 5.18 Algorithm – flow chart for NTU-ε method. (a) Determine size based on required heat transfer rate and (b) determine heat transfer rate and exit fluid temperatures for the given heat exchanger type and size.

Heat-transfer rate is given

$$q = \varepsilon \times C_{\min} \times (T_{h,i} - T_{c,i}) = 0.413 \times 3800 \frac{\text{Btu}}{\text{h} - \text{F}} \times (170 - 50)$$

$$q = 188\,328.0 \frac{\text{Btu}}{\text{h}} \ (55.193 \text{ KW})$$

Exit fluid temperatures:

From the heat-transfer rate equation, the hot fluid exit temperature is given as

$$T_{h,o} = T_{h,i} - \frac{q}{\dot{m}_h c_{ph}}$$

$$= 170 - \frac{188\,328}{10\,000\,(0.5)}$$

$$T_{h,o} = 37.66°\text{F}$$

Similarly, from the heat-transfer rate equation for the cold fluid, the cold fluid exit temperature is given as

$$T_{c,o} = T_{c,i} + \frac{q}{\dot{m}_c c_{pc}}$$

$$= 50 + \frac{188\,328}{3800\,(1.0)}$$

$$T_{c,o} = 99.56°F$$

5.4.3 Overall Heat-transfer Coefficient in Heat Exchanger

For a typical tube wall surface, the heat-transfer rate across the tube wall is governed by the three major resistances across the tube wall. Three major resistances are inner convection resistance, $R_{conv} = \frac{1}{h_i A_i}$; outer convection resistance, $R_{conv} = \frac{1}{h_o A_o}$; and the conduction resistance, $R_{conv} = \frac{\ln \frac{r_o}{r_i}}{2\pi L K}$, as demonstrated in Figure 5.19.

The heat-transfer rate at any section is given as the ratio of the temperature difference between two fluid streams to the net sum of the three resistances and is

$$q = \frac{T_i - T_o}{\frac{1}{h_i A_i} + \frac{\ln \frac{r_o}{r_i}}{2\pi L k} + \frac{1}{h_o A_o}} \tag{5.18}$$

The net sum of all the three resistances is also used to define the product of overall heat-transfer coefficient and heat-transfer surface area as

$$U_i A_i = U_o A_o = \frac{1}{\frac{1}{h_i A_i} + \frac{\ln \frac{r_o}{r_i}}{2\pi L k} + \frac{1}{h_o A_o}} \tag{5.19}$$

And the overall heat-transfer coefficient as

$$U_i = \frac{1}{\frac{1}{h_i} + \frac{\ln \frac{r_o}{r_i}}{2\pi L k} A_i + \frac{A_i}{A_o} \frac{1}{h_o}} \quad \textbf{\textit{based on inner surface area}} \tag{5.20a}$$

and

$$U_o = \frac{1}{\frac{A_o}{A_i} \frac{1}{h_i} + \frac{\ln \frac{r_o}{r_i}}{2\pi L k} A_o + \frac{1}{h_o}} \quad \textbf{\textit{based on outer surface area}} \tag{5.20b}$$

Figure 5.19 Thermal resistances across a heat-exchanger tube wall.

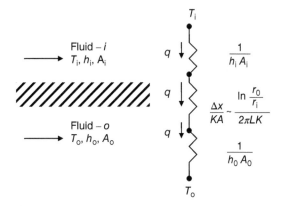

where

$$A_i = \text{inner surface area of the tube} = 2\pi r_i L = \pi d_i L \tag{5.21a}$$
$$A_o = \text{outer surface area of the tube} = 2\pi r_o L = \pi d_o L \tag{5.21b}$$

Following the expressions for heat-transfer rates through the resistances across the tube, we can derive the **tube surface temperature** as given below:

Inner tube wall temperature:

$$T_{wi} = T_i - \frac{U_i A_i}{h_i}(T_i - T_o) \tag{5.22a}$$

and

Outer tube wall temperature:

$$T_{wo} = T_o + \frac{U_i A_i}{h_o A_o}(T_i - T_o) \tag{5.22b}$$

5.4.4 Finned Surface

When fins are used over a surface to increase surface area for heat transfer, the convection resistance is written as

$$R_{conv} = \frac{1}{\eta_0 h A_t} \tag{5.23}$$

where

$\eta_0 =$ Fin surface area effectiveness
$A_t =$ Total surface area for convection heat transfer
$\quad = =$ Bare surface area, $A_b +$ Fin surface area, A_f

We can see that convection resistance decreases with the increase in total surface area with addition of the fin and with increase in the fin area effectiveness.

Figure 5.20 shows a circular tube with fins on the outer surface of the tube.

The convective area for this type of fin is given as

$$A_f = N_f \times [2\pi(r_2^2 - r_1^2) + 2\pi r_2 t] \tag{5.24a}$$

and the bare area of tube is given as

$$A_b = \pi d_o(L - N_f t) \tag{5.24b}$$

The total external surface area of the finned circular tube is then given as

$$A_t = N_f \times [2\pi(r_2^2 - r_1^2) + 2\pi r_2 t] + \pi d_o(L - N_f t) \tag{5.24c}$$

For cases where fins are used on either inside surface or outside surface of the tubes, the overall heat-transfer coefficients are written as

$$U_i A_{i,t} = \frac{1}{\dfrac{1}{\eta_{oi} h_i A_{i,t}} + \dfrac{\ln \frac{r_o}{r_i}}{2\pi L k} + \dfrac{A_{i,t}}{A_{o,t}}\dfrac{1}{\eta_{oo} h_o}} \tag{5.25}$$

$$U_i = \frac{1}{\dfrac{1}{\eta_{oi} h_i} + \dfrac{\ln \frac{r_o}{r_i}}{2\pi L k}A_{i,t} + \dfrac{A_{i,t}}{A_{o,t}}\dfrac{1}{\eta_{oo} h_o}} \quad \textbf{\textit{based on inner area}} \tag{5.26a}$$

and

$$U_o = \frac{1}{\dfrac{A_{o,t}}{A_{i,t}}\dfrac{1}{\eta_{oi} h_i} + \dfrac{\ln \frac{r_o}{r_i}}{2\pi L k}A_{o,t} + \dfrac{1}{\eta_{oo} h_o}} \quad \textbf{\textit{based on outer area}} \tag{5.26b}$$

Figure 5.20 Circular tube with annular fin.

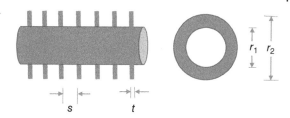

where η_o is the **total surface effectiveness** defined as

$$\eta_o = 1 - \frac{A_f}{A_t}(1 - \eta_f) \tag{5.27}$$

η_f = fin efficiency

A_f = fin surface area

A_t = total surface area that includes both bare surface area without any fin and fin surface area

Example 5.3   A copper pipe with annular fins of circular profile has the following characteristics:

Pipe dimension:

   D_i= 1.75-in. (0.1458-ft) and D_o = 2-in. (0.1666-ft)

Air – outside:

   Air temperature: 32 °F side convection heat-transfer coefficient: $h_o = 120\frac{Btu}{ft^2 hr°F}$

Water – (Inside pipe):

   Liquid inlet temperature: 200 °F, Liquid outlet temperature: 100 °F

   Average velocity: 4.0 ft/s

Fin geometry:

   Consider copper **circular pin fins** with following dimensions: L_f= 2.5-in., $t_f = 0.15 -$ in., and $S = 0.2$-in.

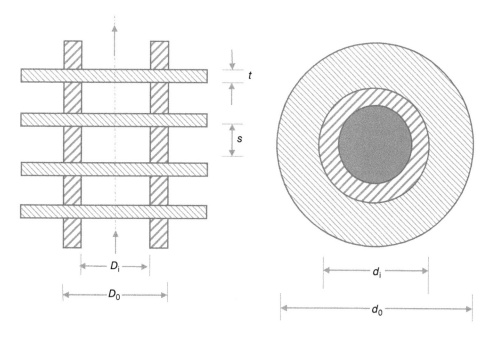

Determine the length of the pipe to cool the liquid from 200 to 100 F. Show clearly all steps for the estimation of following items: Determine the tube-side heat-transfer coefficient, overall heat-transfer coefficient, and total surface area for the heat transfer. If it is required to reduce the length by 20% to maintain the same cooling rate, what design changes can be made to achieve that goal?

Solution:

Air properties: At 32 F

$$\rho = 0.81 \text{ lbm/ft}^3, \mu = 1.165 \times 10^{-5} \text{lbm/ft} - \text{s}, C_p = 0.24 \text{ Btu/lbm}°\text{F},$$

$$k = 0.0140 \text{ Btu/h-ft}°\text{F}, \text{Pr} = 0.72$$

Water properties: At $T = 150°$ F

$$\rho = 61.2 \text{ lbm/ft}^3, \mu = 2.92 \times 10^{-4} \text{lbm/ft-s}, C_p = 1.0 \text{ Btu/lbm}°\text{F},$$

$$k = 0.384 \text{ Btu/hr-ft}°\text{F},$$

$$\text{Pr} = 2.74$$

Tube cross-sectional area for fluid flow

$$A_t = \frac{\pi}{4}D_i^2 = \frac{\pi}{4}\left(\frac{1.75}{12}\right)^2 = 0.0167 - \text{ft}^2$$

Liquid mass flow rate, $\dot{m}_t = \rho_t A_t V_t = 61.2 \times 0.0167 \times 4.0 = 4.088 \text{ lbm/s}$

$$\dot{m}_t = 14716.8 \text{ lbm/hr}$$

Heat-transfer rate is computed from energy balance on tube side as

$$q = \dot{m}_t c_{p,t}(T_{t,i} - T_{t,o}) = 4.088 \times 1.0 \times (200 - 100) = 408.8 \text{ Btu/s}$$

$$q = 408.81\,471\,680 \text{ Btu/hr}$$

The log-mean temperature difference is computed as

$$\Delta T_m = \frac{\Delta T_{max} - \Delta T_{min}}{\ln \frac{\Delta T_{max}}{\Delta T_{min}}}$$

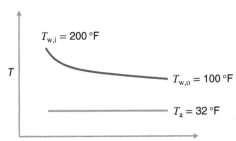

$T_{w,i} = 200 °\text{F}$

T

$T_{w,o} = 100 °\text{F}$

$T_a = 32 °\text{F}$

Axial distance along the tube length, x

$$\Delta T_m = \frac{(T_{t,i} - T_a) - (T_{t,o} - T_a)}{\ln \frac{(T_{t,i}-T_a)}{(T_{t,o}-T_a)}}$$

$$\Delta T_m = \frac{(200 - 32) - (100 - 32)}{\ln \frac{(200-32)}{(100-32)}} = \frac{168 - 68}{\ln \left(\frac{168}{68} \right)}$$

$$\Delta T_m = 110.564°F$$

Computation of fin area effectiveness (η_{oi}):

To determine fin surface area effectiveness, let us compute the fin efficiency, η_f, and and the ratio of fin area to the total surface area, $\frac{A_f}{A_t}$.

Estimation of fin efficiency:

For annular circular fin, the efficiency can be computed either by using the analytical solution given by Eq. (4.56b) or from Figure 4.12b.

Annular fin geometrical parameters:

Inner radius, $r_1 = 2$-in. (0.148-ft), Length, $L_f = 2.5$-in., Thickness, $t = 0.15$-in.

Outer radius, $r_2 = r_1 + L = 2 + 2.5 = 4.5 -$ in.

Annular fin parameters:

Corrected outer radius of the fin, $r_{2c} = r_2 + \frac{t}{2} = 4.5 + \frac{0.15}{2} == 4.575 -$ in.

Corrected length of the fin, $L_{f,c} = L_f + \frac{t}{2} = 2.5 + \frac{0.15}{2} = 2.575 -$ in.

For simplicity, we can compute the fin efficiency using the data presented in Figure 4.12b.

Using fin parameters:

$\frac{r_{2c}}{r_1} = \frac{4.575}{2.0} = 2.2875$, $A_p = L_{f,c} \, t = 2.575 -$ in. $\times 0.15 -$ in. $= 2.682 \times 10^{-3}$ ft^2

$$L_{f,c}^{3/2} \left(\frac{h}{k A_p} \right)^{1/2} = (0.2146)^{3/2} \left(\frac{10}{230 \times 2.682 \times 10^{-3}} \right)^{1/2} = 0.0994 \times 4.026 = 0.4$$

From Figure 4.12b, for $\frac{r_{2c}}{r_1} = 2.2875$ and $L_c^{3/2} \left(\frac{h}{k A_p} \right)^{1/2} = 0.4$, we can approximately estimate fin efficiency, $\eta_f \approx 0.85$ or 85%.

For more accurate computation of the fin efficiency, use analytical solution of fin efficiency for annular fin given by and as demonstrated in Example 4.1

Estimate fin surface area ratio:

Fin surface area for convection heat transfer

$$A_f = \text{Total fin convective area} = 2 \times \frac{\pi}{4} \times (d_o^2 - d_i^2) + \pi d_o t$$

$$= 2 \times \frac{\pi}{4} \times (7^2 - 2.5^2) + \pi \times 7 \times 0.15$$

$$A_f = 67.152 + 3.298 \text{ in.}^2 = 70.45 \text{ in.}^2 = 0.489 \text{ ft}^2$$

Unfinned tube convective surface area

$$A_{unfin} = \pi D_o s = \pi \times 2.5 \times 0.2$$

$$A_{unfin} = 1.570 \text{ in}^2$$

Total surface area

$$A_t = (A_f + A_{unfin})m = 70.45 + 1.570 = 72.02 - \text{in.}^2$$

Fin surface area effectiveness:

$$\eta_0 = 1 - \frac{A_f}{A_t}(1 - \eta_f) = 1 - \frac{70.45}{72.02}(1 - 0.85)$$

$$\eta_0 = 0.853$$

Compute convection heat-transfer coefficient inside the tube:

Sectional area for fluid flow

$$A_t = \frac{\pi}{4}D_i^2 = \frac{\pi}{4}\left(\frac{1.75}{12}\right)^2 = 0.0167 - \text{ft}^2$$

Liquid mass flow rate, $\dot{m}_t = \rho_t A_t V_t = 61.2 \times 0.0167 \times 4.0 = 4.088 \text{ lbm/s}$

Reynolds Number, $\text{Re} = \frac{\rho V_t d_i}{\mu} = \frac{61.2 \times 4.0 \times 0.1458}{2.92 \times 10^{-4}} = 122\,232$

Since Re > 2300, the flow is turbulent. Let us consider the flow as fully developed turbulent flow as a first approximation and use fully developed heat-transfer correlation given by Colburn as

$$\text{Nu}_D = 0.023(\text{Re}_D)^{4/5}(\text{Pr})^{1/3}$$

$$\text{Nu}_D = 0.023(122\,232)^{4/5}(2.74)^{1/3}$$

$$= 0.023 \times 11\,742.158 \times 1.3992$$

$$\text{Nu}_D = 377.87$$

Convection heat-transfer coefficient for inside fluid flow

$$h_i = \frac{\text{Nu} \times k}{d_i} = \frac{377.87 \times 0.384}{0.1458} = 995.21$$

$$h_i = 995.21\frac{\text{Btu}}{\text{ft}^2\text{hr}^\circ\text{F}}$$

Estimate overall heat transfer coefficient:

$$U_i = \cfrac{1}{\cfrac{1}{\eta_{oi}h_i} + R_{fi} + \cfrac{\ln\frac{r_o}{r_i}}{2\pi L k}A_i + \cfrac{R_{fo}A_i}{A_o} + \cfrac{1}{\eta_{oo}h_o}\cfrac{A_i}{A_o}}$$

Use $\eta_{oi} = 1$ for no fins inside and $R_{fi} = R_{fo} = 0$ for negligible fouling.

$$U_i = \cfrac{1}{\cfrac{1}{h_i} + \cfrac{\ln\frac{D_o}{D_i}}{2\pi L k}\pi D_i L + \cfrac{1}{\eta_{oo}\times h_o}\cfrac{D_i}{D_o}} = \cfrac{1}{\cfrac{1}{h_i} + \cfrac{\ln\frac{D_o}{D_i}}{2k}D_i + \cfrac{1}{\eta_{oo}\times h_o}\cfrac{D_i}{D_o}}$$

$$U_i = \cfrac{1}{\cfrac{1}{995.21} + \cfrac{\ln\frac{0.1666}{0.1458}}{2\times230}\times 0.1458 + \cfrac{1}{0.853\times120\times}\cfrac{0.1458}{0.1666}}$$

$$= \frac{1}{0.001 + 0.00029 + 0.00854} = \frac{1}{0.00983}$$

$$U_i = 101.73\frac{\text{Btu}}{\text{ft}^2\text{h}^\circ\text{F}}$$

From heat-transfer rate across the tube wall

$$q = FU_i A_i \Delta T_m$$

We can set correction factor $F = 1$ since this is for single tube pass over an unbounded open ambient condition.

The total surface area based on inside tube diameter and tube length

$$A_i = \frac{q}{FU_i \Delta T_m} = \frac{1\,471\,680}{101.73 \times 110.564} = 130.84 \text{ ft}^2$$

Once the total inside surface area is known, we can estimate the length of the tube based on the selected tube diameter as

$$L = \frac{A_i}{\pi D_i} = \frac{130.84}{\pi \times 0.1458} = 285.65 - \text{ft}$$

Length of the pipe, $L = 285.65 - \text{ft}$

Note that in most applications where space may be restricted, the long straight tube section can be replaced with multipass serpental sections like in coiled tube section.

5.4.5 Fouling Factor

Another important factor that affects the overall heat-transfer coefficient and pressure drop, and hence the design of the heat exchanger, is the *fouling factor*.

A **fouling factor** or **dirt factor** is introduced to consider if a **thin layer of deposits or scales** forms over the tube's inner and outer surfaces over the service period of the heat exchanger. Fouling mechanisms are classified into different types: (i) Precipitation fouling caused by crystallization insoluble salts like calcium chloride; (ii) Particulate fouling due to sedimentation of particles of clay or sand; (iii) Chemical reaction fouling due to chemical reaction of the fluid with the tube surface such as the hard deposits of hydrocarbon due to high surface temperature and presence of oxidation elements; (iv) Corrosion fouling caused by electrochemical reactions and forms over ion-conduction-exposed surfaces of heat exchangers; and (v) Biological fouling due to the growth of organic materials over the heat-transfer surface when in contact with untreated water like river water or sea water. In many power plant applications, cooling water systems water treatment and make-up water subsystems.

This adds two more resistances, R_{fi} and R_{fo}, to the computation of the overall heat-transfer coefficient, U. The overall resistance of heat flow across the tube wall increases and the overall heat-transfer coefficients (decreases) are modified in the following manner

$$U_i = \frac{1}{\frac{1}{\eta_{oi} h_i} + R_{fi} + \frac{\ln \frac{r_o}{r_i}}{2\pi L k} A_i + \frac{R_{fo} A_i}{A_o} + \frac{1}{\eta_{oo} h_o} \frac{A_i}{A_o}} \tag{5.28a}$$

and

$$U_o = \frac{1}{\frac{A_o}{A_i} \frac{1}{\eta_{oi} h_i} + \frac{R_{fi} A_o}{A_i} + \frac{\ln \frac{r_o}{r_i}}{2\pi L k} A_o + R_{fo} + \frac{1}{\eta_{oo} h_o}} \tag{5.28b}$$

As we can notice, the heat-transfer coefficient value of the heat exchanger decreases in value from the beginning of the service or for a new heat exchanger to later years over period of service. This will cause a reduction in the desired heat-transfer rate causing an elevated hot fluid exit temperature and lower cold fluid temperature than the desired values. In anticipation of the effects, it is a standard practice to include fouling factors in designing heat exchangers. In other words, the heat exchanger is overdesigned under clean condition and expected to perform adequately till the next cleaning period.

Some of the physical and operating conditions that influence fouling are type of fluids and their properties; surface and bulk fluid temperatures; local fluid velocities; tube material, surface finish, and configurations; heat-exchanger geometry and orientation; heat-transfer process such as heating, cooling, or phase change; and planned cleaning methods and schedule. As mentioned previously, more fouling fluid is used on the tube side rather than shell side. This helps in avoiding the possibility of low velocity and stagnant-flow regions.

Fouling factors or resistances are obtained experimentally for different types of fluids and a comprehensive list of recommended design fouling resistances is given in TEMA standard (2007). These values are the recommended estimate of fouling taking place annually. It is therefore necessary to have regular schedule for periodic cleaning of the heat exchanger. Recommended design fouling values for some selected fluids are given in Table 5.2 and 5.3.

Table 5.2 presents the fouling factor data for different types of water under different operating conditions. Table 5.3 shows the fouling factor data for different heat-transfer fluids.

Following example demonstrates the change in overall heat-transfer coefficient value from a clean heat exchanger to dirty one as scales build up over a period.

Example 5.4 Consider a double-pipe heat exchanger with convection heat-transfer coefficients of $h_i = 1000$ Btu/h ft^2 ° F and $h_0 = 200$ Btu/h ft^2 ° F in the tube side and in the shell side of the heat exchanger. Determine (i) the overall heat-transfer coefficient for a clean heat exchanger and (ii) the overall heat-transfer coefficient for the dirty heat exchanger considering anticipated fouling resistances as $R_{fi} = 0.001$ h ft^2 ° F/Btu and $R_{fo} = 0.001$ h ft^2 ° F/Btu for inside and outside of the tube, respectively, over a period of one year.

Table 5.2 Fouling factors for water.

Temperature of heating medium	≤240 °F		240–400 °F	
Temperature of water	≤150 °F		>240 °F	
Water	Water velocity		Water velocity	
	≤3 fps	>3 fps	≤3 fps	>3 fps
Sea water	0.0005	0.0005	0.001	0.001
Cooling tower				
Treated make-up	0.001	0.001	0.002	0.002
Untreated	0.003	0.003	0.005	0.004
City water	0.001	0.001	0.002	0.002
Distilled water	0.0005	0.0005	0.0005	0.0005
Treated Boiler feedwater	0.001	0.0005	0.001	0.001

Table 5.3 Fouling factors for some common fluid.

Fluid types	Fouling resistance (ft² h °F/Btu)
Refrigerant liquids	0.001
Hydraulic fluid	0.001
Ammonia liquid	0.001
Methanol solutions	0.002
Ethanol solutions	0.002
Ethylene Glycol solutions	0.002
Fuel Oils	0.002–0.006
Engine lube oil	0.001
Engine exhaust gas	0.010
Steam (non-oil bearing)	0.0005
Exhaust steam (oil bearing)	0.0015–0.002
Refrigerant vapors (oil bearings)	0.002
Compressed air	0.001
Ammonia vapor	0.001
Coal flue gas	0.010
Natural gas flue gas	0.005
Natural gas	0.001–0.002

Solution:

To demonstrate the effect of fouling resistance, let us neglect the conduction resistance across tube wall and write the overall heat-transfer coefficient for the clean tube from Eq. (5.8) as follows

$$U_c = \frac{1}{\frac{1}{h_i} + \frac{1}{h_o}} = \frac{1}{\frac{1}{1000} + \frac{1}{200}} = \frac{1}{0.001 + 0.005}$$

Clean overall heat-transfer coefficient, $U_c = 166.67$ Btu/h ft² ° F

Let us now include the fouling resistance and compute the overall heat-transfer coefficient as

$$U_D = \frac{1}{\frac{1}{h_i} + R_{fi} + R_{fo} + \frac{1}{h_o}} = \frac{1}{\frac{1}{1000} + 0.0015 + 0.002 + \frac{1}{200}}$$

$$= \frac{1}{0.001 + 0.0015 + 0.002 + 0.005}$$

Dirty overall heat-transfer coefficient, $U_D = 105.26$ Btu/h ft².°F

We can see that the overall heat-transfer coefficient has decreased by about 36% because of fouling over a period of one year. To compensate for this degradation, same order of increase in heat-transfer surface area must be considered in the design and selection stage, and yearly cleaning must be scheduled.

5.5 Shell-and-tube Heat Exchanger

Analysis procedure for shell-and-tube heat exchanger is a bit more complex than simple double pipe due to complexities in flow passages involving number of additional variables and parameters. Shell-and-tube heat exchanger includes a shell and a bundle of tubes. Fluid in tube-side passes through a bundle of tubes, and the flow and heat transfer are influenced by the number of tube-side geometrical parameters such as tube diameter and length, tube surface roughness, and presence of any fins. On the shell side, the flow and heat transfer are influenced by the shell-side geometrical parameters such as baffle type, size, and spacing; tube spacing, arrangements, and number of tubes.

5.5.1 Flow Geometry and Flow Parameters

5.5.1.1 Tube-Side Flow Geometry

Tube-side flow area is given as

$$A_{f,t} = n_{t,p} \frac{\pi}{4} d_i^2 \tag{5.29}$$

where
$N_{p,t}$ = Number of tube passes
$n_{t,p}$ = Number of tubes per tube pass

The mass flow rate in the tube side

$$\dot{m}_t = \rho_t A_{f,t} V_t \tag{5.30}$$

where

$$V_t = \text{Average velocity in tube side} = \frac{\dot{m}_t}{\rho_t A_{f,t}}$$

Substituting Eq. (5.29) into Eq. (5.30), we get

$$\dot{m}_t = \rho_t n_{t,p} \frac{\pi}{4} d_i^2 V_t \tag{5.31}$$

Note that the expression for the mass flow rate in the tube side involves two unknowns: the average velocity, V_t, and the number of tubes per tube pass, $n_{t,p}$. In an iterative process, the number of tubes can be varied until the solution is reached. However, no guidance is available for choosing an estimate of the number of tubes. A more effective approach process starts with an assumed average velocity in the tube side and iterate to reach the expected design in terms of size, heat transfer rate, and the pressure drop. Often, the average velocity is restricted by the maximum allowable pressure drop in the tube side.

5.5.1.2 Ratio of Tube-side Free Flow Area to Flow Area of the Tubes

This is an important parameter in the design of heat exchangers, particularly in the computation of pressure drop in the tube side of a shell-and-tube heat exchangers.

$$\sigma = \frac{A_{fr}}{A_t} \tag{5.32}$$

where

A_{fr} = Tube-side free flow area in the tube box = $\frac{\pi}{4}D_s^2$

A_t = Net tube-side flow area = $N_{\text{p,t} \times n_{\text{t,p}}} \times \frac{\pi}{4}d_i^2$

5.5.1.3   Net Surface Area for Heat Transfer

Net surface area for heat transfer in a shell-and-tube heat exchanger is governed by number of variables such as diameter and length of the tubes, number of tubes per tube pass, and number of tube passes given as

$$A = N_{\text{t,p}}n_{\text{t,p}}\pi dL \tag{5.33}$$

where

$N_{\text{p,t}}$ = Number of tube passes

$n_{\text{t,p}}$ = Number of tubes per tube pass

$\quad L$ = Length of the tube

$\quad D$ = diameter of the tube

The length of tube is computed from Eq. (5.33)

$$L = \frac{A}{N_{\text{p,t}}n_{\text{t,p}}\pi d} \tag{5.34}$$

5.5.1.4   Shell-Side Flow Geometry

The flow patterns and characteristics are very complex due to the complex flow geometry formed due to the presence of baffles and tube arrangements. Turbulence is more easily initiated on the shell side because of the typically more complicated flow path. Due to this fact, higher viscosity and low flow rate fluids are used in the shell sides.

Figure 5.21 shows longitudinal and transverse sectional views of a shell-and-tube heat exchangers.

5.5.2   Types and Effects of Baffles

Baffle design plays a significant role in achieving enhanced convective heat transfer in the shell side by making sure that the fluid cannot take short cut; form slow-moving recirculating regions; and can create enhanced mixing and turbulence. Figure 5.22 shows some of the most common baffle types.

5.5.3   Tube Arrangements in Shell Side

The tubes bundle may be arranged in number of different ways depending on the shell design and type of heat-exchanger applications. Some of typical tube patterns are (i) Triangular or staggered, (ii) Rotated triangular, (iii) Square or Inline, and (iv) Rotated square. These patterns are demonstrated in Figure 5.23.

Following these patterns, two most common tube bundle arrangements are (i) an inline arrangement and (ii) a staggered arrangement as demonstrated in Figure 5.24:

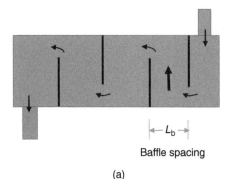

Figure 5.21 Sectional view of the shell showing tube arrangements and flow patterns.
(a) Longitudinal section and (b) transverse section side view.

$\leftarrow L_b \rightarrow$

Baffle spacing

(a)

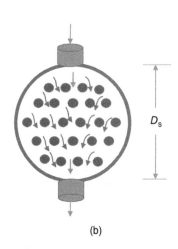

D_s

(b)

While staggered tube arrangement tends to produce increased turbulence/mixing and enhanced heat transfer, it also increases the pressure drop and pumping power requirements. Triangular or rotated triangular tube patterns should not be used where mechanical cleaning of the shell side is required. Tube-to-tube spacing or pitch are termed as **longitudinal pitch**, S_L, and **transverse pitch, S_T**. According to TEMA standard, the minimum spacing is 1.4-in. When mechanical cleaning is desired, higher spacings are recommended. While lower spacing results in enhanced heat transfer because of higher local fluid velocity, it also causes higher pressure to drop.

5.5.4 Shell-Side Flow Area

The major flow and geometrical parameters for shell sides are the baffle type; baffle spacing; shell diameter; clearance spacing between shell wall and baffle; effective shell diameter; and average fluid velocity. Figure 5.25 shows the interior geometry and flow regions within the shell side considering segmental baffles.

Shell-side flow area is defined by a rectangular flow region with length given by the baffle spacing and width given by the effective shell diameter.

$$A_{f,s} = L_b \times D_e s \qquad (5.35)$$

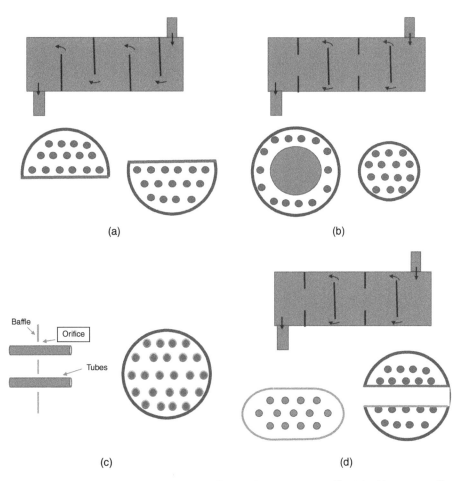

(a)

(b)

(c)

(d)

Figure 5.22 Baffle types. (a) Segmental baffle, (b) disk and donut baffle, (c) orifice-type baffle, and (d) strip baffle.

where

L_b = Baffle spacing = $\frac{L}{(N_b+1)}$

D_e = Effective shell diameter = $D_s - N_{CLD} \times d_o$

D_s = Shell diameter

N_{CLD} = Number of tubes at the centerline of the shell tube bundles

5.5.5 Estimation of Heat-transfer Coefficients in a Shell and Tube

5.5.5.1 Tube Arrangements Inside Shell and Tube Heat Exchanger

Heat is transferred between the fluid flowing inside the tubes and that flowing over the banks of tubes. To determine the overall heat-transfer coefficients, U_i or U_o, for the transport heat balance equation and determine the heat-exchanger size, the convection heat-transfer coefficient for fluid flow inside the tubes and for over the bundle of tubes is

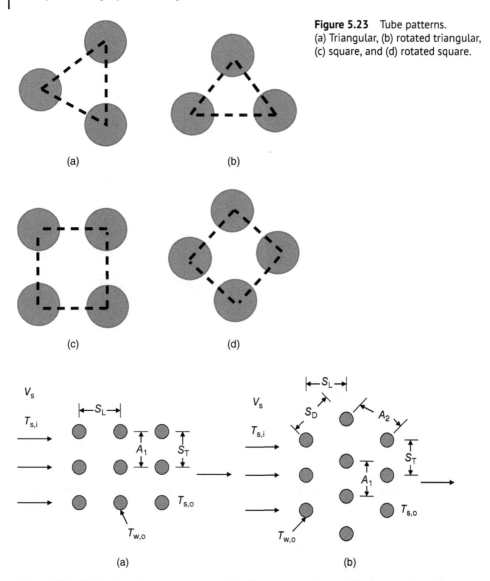

Figure 5.23 Tube patterns.
(a) Triangular, (b) rotated triangular,
(c) square, and (d) rotated square.

Figure 5.24 Shell-side tube arrangements. (a) Inline arrangement and (b) staggered arrangement.

required. Correlations for convection heat-transfer coefficients for different flow geometries, including the fluid flow inside tube and flow over bundle of tubes, are discussed and listed in Chapter 3. Let us identify these correlations in the context of shell-and-tube heat exchanger.

5.5.5.2 Tube-Side Convection Coefficient
For fluid flowing inside the tubes, we will select appropriate correlation based on either fully developed flow or developing flow regions for determining the convection coefficient inside tube flow.

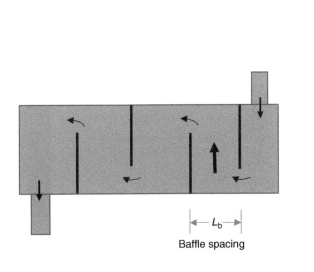
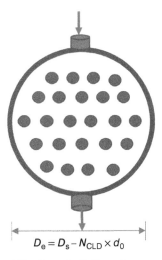

Baffle spacing

$D_e = D_s - N_{CLD} \times d_0$

Effective shell diameter
for shell-side flow

Figure 5.25 Shell-side flow dimensions.

5.5.5.3 Shell-Side Convection Correlation

Let us present here the computation of convection heat-transfer coefficient based on the correlation for external fluid flow across a tube bank given by **Zukauskas** as follows:

$$\overline{Nu_d} = \frac{h_o d_o}{k_{s,m}} = CRe_{d,max}^m Pr^{0.36} \left(\frac{Pr}{Pr_s} \right)^{\frac{1}{4}} \tag{5.36}$$

for $N_L \geq 20$, $0.7 \leq Pr \leq 500$ and $1000 \leq Re_{d,\,max} \leq 2 \times 10^6$

where constant C and **m** are given in Table 3.15.

For $N_L < 20$, a correction factor is used as

$$\overline{Nu_D}\Big|_{N_L < 20} = C_2 \overline{Nu_D}\Big|_{\geq 20} \tag{5.37}$$

where the constant C_2 is given in Table 3.16.

Reynolds number used in this correlation is based on the maximum fluid velocity occurring within the tube banks defined as follows:

$$Re_{d,max} = \frac{\rho V_{max} d_o}{\mu} \tag{5.38}$$

Following these patterns two most common tube bundle arrangements are (i) an in-line arrangement (ii) a staggered arrangement as demonstrated in the Figure 5.24a, b, respectively:

$$V_{max} = \frac{S_T}{S_T - d_o} V_s \tag{5.39}$$

For **staggered alignment**, the maximum velocity may occur at either the normal area A_1 and or the inclined area A_2 as shown in Figure 5.24b based on following conditions:

V_{max} at area A_2 if $2(S_T - d_o) \leq (S_T - d_o)$ and the maximum value:

$$V_{max} = \frac{S_T}{2(S_D - d_o)} V_s \tag{5.40}$$

The external shell-side convection coefficient is then computed as

$$h_o = \frac{\overline{Nu_d} \times k_{s,m}}{d_o} \tag{5.41}$$

This estimate of shell-side convection coefficient is, however, modified using the factors that consider complex flow patterns caused due to the use of baffles as discussed in Section 5.5.7.

5.5.6 Pressure Drops in Tube and Shell Sides

The head loss or pressure drop for the flow of a fluid at an average velocity V through a length L of pipe or tube of diameter D is

$$h_f = f \frac{L}{D} \frac{V^2}{2g_c} \text{ by Darcy} - \text{Weisbach} \tag{5.42}$$

or

$$h_f = 4f_F \frac{L}{D} \frac{V^2}{2g_c} \text{ by Fanning} \tag{5.43}$$

where

f = Darcy–Weisbach friction factor
f_F = Fanning friction factor = $\frac{f}{4}$
$f = f\left(Re, \frac{\varepsilon}{D}\right)$ given by Moody's diagram (See Figure 8.5)
  = for **turbulent** pipe flow
  $\frac{64}{Re}$ for **laminar** pipe flow

Reynolds number for internal pipe flow:

$$Re = \frac{\rho V D_H}{\mu}$$

where

$$D_H = \frac{4A}{P}$$

A curve fit of Moody's data in explicit form is given as

$$f = \frac{0.3086}{\left\{ \log\left[\left(\frac{6.9}{Re_D}\right) + \left(\frac{\varepsilon}{3.7D}\right)^{1.11} \right] \right\}^2} \tag{5.44}$$

Pressure losses in tube and shell sides contribute to the pumping power required and represent the efficiency for heat-transfer process. The objective is to achieve enhanced heat

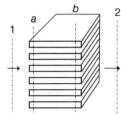

Figure 5.26 Heat-exchanger core along with entrance contraction and exit expansion sections.

transfer with reduced pressure drops. Pressure drop for fluid flowing through the straight section of tube is given by Eq. (5.45) below

$$\Delta P = f\frac{L}{d_i}V_t^2 \tag{5.45}$$

However, the flow geometry in both tube and shell sides of the shell-and-tube heat exchanger as well as many compact heat exchangers is complicated due to the presence of sections that involve contraction of the flow at the inlet sections and expansion of the flow to the exit section of the heat exchangers. Figure 5.26 shows flow through the heat-exchanger core along with expansion and contraction sections.

Kays and London (1984) presented pressure drop expression for flow through the heat-exchanger tube-side core area considering the entrance contraction section and exit expansion section for different tube designs.

The total pressure drop in the heat-exchanger tube core involves multiple components. The entrance pressure drop is composed of two parts: first part is due to the changes in area in contraction section and second part is due to the irreversible expansion involving boundary layer separation and secondary recirculating flow following the contraction section. This entrance pressure drop is expressed as

$$\frac{\Delta P_c}{\rho} = \frac{V^2}{2g_c}(1-\sigma^2) + K_c\frac{V^2}{2g_c} = \frac{V^2}{2g_c}(1-\sigma^2 + K_c) \tag{5.46a}$$

Similarly, the pressure drop at the exit expansion section is composed of two parts: first part is the pressure increase due to the changes in area in expansion section and the second part is due to irreversible expansion involving flow separation. The exit pressure drop is expressed as

$$\frac{\Delta P_e}{\rho} = \frac{V^2}{2g_c}(1-\sigma^2) - K_e\frac{V^2}{2g_c} = \frac{V^2}{2g_c}(1-\sigma^2 - K_e) \tag{5.46b}$$

The tubes' frictional pressure drop in the core length heat exchanger is given as

$$\frac{\Delta P_{core}}{\rho} = f\frac{A}{A_f}\frac{V^2}{2g_c} \tag{5.46c}$$

The last component is the pressure drop due to flow acceleration caused by the changes in flow momentum as density varies between the upstream and downstream sections, and this is expressed as

$$\frac{\Delta P_{acce}}{\rho} = 2(v_2 - v_1)\frac{V^2}{2g_c} = 2v_1\left(\frac{v_2}{v_1} - 1\right)\frac{V^2}{2g_c} \tag{5.46d}$$

Summing up all components given by Eq. (5.46a)–(5.46d), the net heat-exchanger tube core pressure drop is given as

$$\frac{\Delta P_i}{P_1} = \frac{G^2}{2g_c} \frac{v_1}{P_1}\left[\left(K_c + 1 - \sigma^2\right) + 2\left(\frac{v_2}{v_1} - 1\right) + f_F \frac{A}{A_f} \frac{v_m}{v_1} - \left(1 - \sigma^2 - K_e\right)\frac{v_2}{v_1} \right]$$

$$\underbrace{\qquad}_{\substack{\text{Entrance}\\\text{effect}}} \qquad \underbrace{\uparrow}_{\substack{\text{Flow}\\\text{acceleration}}} \qquad \underbrace{\uparrow}_{\substack{\text{Core}\\\text{friction}}} \qquad \underbrace{\uparrow}_{\substack{\text{Exit}\\\text{effect}}}$$

(5.47)

where

$G = \rho_a V_a = \rho_b V_b$ = mass velocity based on free flow area
v_1 = specific volume at entrance
v_2 = specific volume at exit
v_m = mean specific volume
P_1 = inlet pressure
ΔP = Core pressure
K_c = entrance loss coefficient (see Figure 5.27)
K_e = exit loss coefficient (see Figure 5.27)
σ = ratio of free flow area (A_f) to frontal area (A_{fr})
A = total heat-transfer area
A_f = free flow area
f_F = core fanning friction factor = $\frac{f}{4}$
f = Darcy-Weisbach friction factor (given by Moody's Chart)
g_c = 32.174 ft. lb/lbf. s^2

Equation (5.1) is valid for flow inside tube bundle core that involves sudden contraction from frontal area A_{fr} to the flow area A_f and sudden expansion from the free flow area A_f to the exit area A_{fr} as shown in Figure 5.26.

Graphs of the core entrance contraction loss coefficient, K_c, and exit expansion loss coefficient, K_e, for different tube geometries are given by Kays and London (1984) as a function of flow area ratio, σ, and tube flow Reynolds number, $N_R = Re_{D_H}$. Figure 5.27 shows such graphical charts for the case of multiple circular, multiple square, multiple triangular, and multiple tube flat-duct heat-exchanger core.

For circular or simple flow geometries, friction factor can be evaluated from Moody's diagram or from curve-fit Eq. (8.45) **Colebrook's implicit form** or Eq. (8.46) for **Haaland's Explicit form**. Kays and London present many figures for friction factor data for diverse range of core surface geometry.

The pressure drops for flow over the tube bundle, while the entrance and exit loss components are usually considered in the measurement of the friction factor for the specific tube types. The pressure drop in shell side is computed based on the pressure drop for external flow over a bank of tube given as:

$$\Delta P_o = N_L \chi f \left(\frac{\rho V_{max}^2}{2g_c} \right)$$

(5.48)

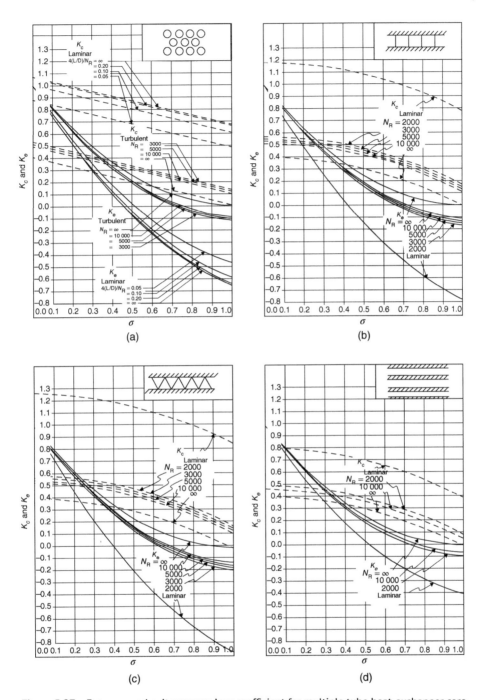

Figure 5.27 Entrance and exit pressure loss coefficient for multiple tube heat-exchanger core. (a) Multiple circular tube core, (b) multiple square tube core, (c) multiple triangular tube core, and (d) multiple core.

where f is the friction factor obtained experimentally for different tube arrangements as shown in Figure 2.17 for inline tube arrangement and Figure 2.18 for staggered tube arrangement; χ is the correction factor introduced for $S_T \neq S_L$ and is given in subplots in Figures 2.17 and 2.18.

5.5.7 Additional Shell-Side Consideration

Heat-transfer correlation and pressure drop data presented for flow over a bank of tubes (by **Zukauskas's correlation**) need to be modified for flow over a bank of tubes bounded by the shell. Shell-side flow is also influenced by the presence of baffles. Number of baffles and baffle spacing strongly influenced the average velocity, flow dynamics and turbulence, and hence the heat-transfer and pressure drop characteristics. A widely accepted technique for shell-side computation is given by Delaware method: Delaware Method (Bell (1981).

In this method, various flow streams are identified in a shell-side design and appropriate correction factors for convection heat-transfer coefficients and pressure drop for each flow stream is proposed. Examples of some of the common flow streams are: (i) Main desired flow stream in shell side, (ii) Flow stream that leaks through the **clearance between the tubes and baffles;** (iii) The bundle bypass stream flowing around the **clearance space between the tube bundle and the shell;** (iv) Shell-to-baffle leakage flowing through the **clearance space between the baffle and the inside shell; and** (v) Flow in the tube bundle caused by the **pass divider** in the header.

5.5.7.1 Corrected Shell-Side Convection Heat-transfer Coefficient

Shell-side convection heat-transfer coefficient is first determined based on the heat-transfer coefficient for external flow over the banks of tubes given by the Zukauskas correlations described before. This correlation value for the ideal flow situation is modified by multiplying it with number of these correction factors:

$$\overline{h}_s = \overline{h}_o F_i F_{ii} F_{iii} F_{iv} F_v \tag{5.49}$$

where

F_i = factor for the desired flow stream in shell side
F_{ii} = correction factor for **baffle cut/spacing**
F_{iii} = correction factor for **baffle leakage**
F_{iv} = Correction factor for the **bundle bypass**
F_{iv} = : correction factor for **baffle spacing**

All these correction factors are presented in graphical forms. To evaluate these factors, it is necessary to know very detailed description of the proposed shell-side geometry, including the **gap size between the tubes** and the **baffle**; and the **spacing between the baffle and the shell.**

As an alternative, Bell (1981) suggested overall correction factor for a first-order approximation for a well-designed heat exchanger as

$$\overline{h}_s = 0.6\overline{h}_o \tag{5.50}$$

where $\overline{h}_o$ is the shell-side heat-transfer coefficient based on Zukauska's correlation for flow over a bundle of tubes with uniform upstream flow.

5.5.7.2 Corrected Shell-Side Pressure Drop

Pressure drop for uniform flow over the banks of tubes is modified to include the effects of (i) pressure drops in the inlet and outlet sections; (ii) pressure drop associated with the interior section defined by the baffles, and pressure drops associated with all by-pass and leakage flows.

Actual pressure drop (ΔP_s) expression suggested by the Delaware method includes number of corrections factors. For simplicity, a first-order approximation for well-designed heat exchanger is given as

$$\Delta P_s = P_f(N_b + 1)\Delta P_o \tag{5.51}$$

where

ΔP_o = Shell-side pressure drop with uniform flow
N_b = Number of baffles
P_f = Correction factor for leakages and usually taken as 0.2–0.3

5.5.8 Temperature-Dependent Fluid Properties and Corrections

Fluid properties such as viscosity, thermal conductivity, specific heat, density, and Prandtl number vary with temperature. Extent to which these properties vary depends on the type of fluid and the phase in which the fluid exists in heat-exchanger design. The fluids in heat-exchanger are often subject to large temperature variations and significantly affect the heat transfer and pressure drop characteristics. These function relationships of fluid properties are often developed using curve fitting discussed in Appendix A. Different approaches are used to consider temperature-dependent property variation in the computations of heat-transfer coefficients and pressure drops. In external fluid flow cases, the fluid properties are computed at some intermediate temperature between solid surface temperature and fluid temperature such as *film temperature* defined as

$$T_f = \frac{1}{2}(T_s + T_\infty) \tag{5.52}$$

All properties are then evaluated at the film temperature using the tabular data or using the functional relations of the property as a function of temperature.

For internal flow cases, fluid properties are evaluated at the mean temperature, called the mixed fluid temperature or mixing cup temperature, T_m. All fluid properties are then evaluated at the mean temperature for use in the correlations for heat-transfer coefficients. Many of the empirical correlations introduce correction factors in form of Prandtl number ratio or viscosity ratios such as $\left(\frac{Pr_m}{Pr_s}\right)^r$ or $\left(\frac{\mu_m}{\mu_s}\right)^r$ as seen in many listed correlations presented in Chapter 3. Here, subscript m refers to mean temperature **and** s refers to surface temperature.

The mean temperature can be defined as either the local mean at a given cross-sections like $T_{m,i}$ at the inlet section and $T_{m,0}$ at the exit section or as an average mean temperature $\overline{T_m}$, i.e. the average of the inlet mean temperature, $T_{m,i}$, and outlet mean temperature, $T_{m,0}$.

An average heat-transfer coefficient is computed based on the properties evaluated at an average mean temperature, $\overline{T_m}$. Effect of variation of temperature on gases and on liquids is different. For gases, properties like thermal conductivity, viscosity, and density all vary considerably, while for liquids, viscosity normally shows considerable variations. For gases, a correction factor based on temperature ratio like $\frac{T_s}{T_m}$ is often used to consider the effect of property variation. For example, the Nusselt number and friction factor values are corrected using this factor as follows:

$$\frac{\overline{Nu}}{\overline{Nu}_m} = \left(\frac{T_s}{\overline{T}_m}\right)^n \tag{5.53a}$$

and

$$\frac{\overline{f}}{\overline{f}_m} = \left(\frac{T_s}{\overline{T}_m}\right)^m \tag{5.53b}$$

For liquids, a factor based on viscosity is used in the following manner

$$\frac{\overline{Nu}}{\overline{Nu}_m} = \left(\frac{\mu_s}{\mu_m}\right)^n \tag{5.54a}$$

and

$$\frac{\overline{f}}{\overline{f}_m} = \left(\frac{\mu_s}{\mu_m}\right)^m \tag{5.54b}$$

Value of m and n varies from type and conditions of heating and cooling.

5.5.9 Classification of Heat-exchanger Design Problems Types

Heat-exchanger design problems can be categorized into different types depending on what variables and parameters are known and what variables are to be determined, and if there are any restrictions. The following is a list of variables in heat-exchanger design problems.

5.5.10 Heat-exchanger Design Analysis: Methodology and Algorithms

5.5.10.1 Design Type-1: Design Methodology
Heat-exchanger type and size are known. Determine heat-exchanger rating in terms of heat-transfer rate, q; exit fluid temperatures; and pressure drops, ΔP.

Strategy: Determine overall heat-transfer coefficient, total surface area for heat transfer, and heat-transfer rate from heat transport equation. Use this heat-transfer rate and determine exit fluid temperatures from energy balance equations for hot and cold fluid. Finally, determine the pressure drop in the tube side and shell side.

Algorithms: Computational algorithm for Design Type-1 problem is demonstrated in the flowchart in Figure 5.28.

Note that for a given heat-exchanger type and size, the calculation procedure is direct in determining the heat-transfer rate, exit fluid temperatures, and pressure drops. However, the procedure can be repeated or iterated while varying some of the geometrical parameters such as the diameter of the tube or the tube pitch and reach the targeted heat-transfer rate or exit fluid temperatures or pressure drops in tube and shell sides.

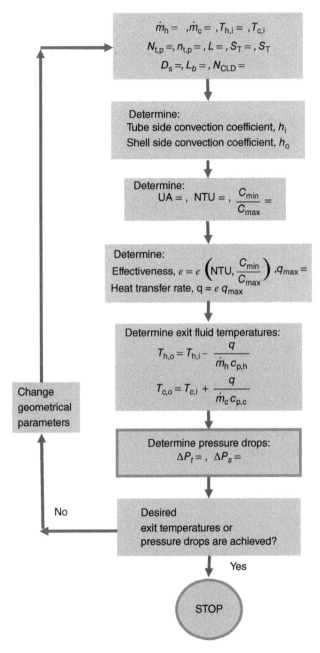

Figure 5.28 Flow chart for design Type-I problems.

5.5.10.1.1 Detailed Computational Steps
Known data:

Cold fluid:	Hot Fluid:
Mass flow rate, $\dot{m}_c$	Mass flow rate, $\dot{m}_h$
Inlet temperature, $T_{c,i}$	Inlet temperature, $T_{h,i}$
Shell-side fluid:	Tube-side fluid:

Assume following data as a first approximation

Number of shell passes, $N_{p,s}$	Number of tube passes, $N_{p,t}$
Tube arrangement	Tube diameter, type, and material
Tube pitch: S_T, S_L	Average tube-side velocity

Determine: Number of tubes per pass, $n_{t,p}$; total number of tubes, n_t; Length of the tube, Number of Baffles, N_b and Shell diameter, D_s

Procedure:

1. Estimate the required fluid properties at the appropriate temperatures:
   Since the exit temperatures are not known, we can either estimate fluid properties based on the inlet temperatures or based on the estimated mean fluid temperature once exit temperature is assumed.
   Estimate the exit or mean fluid temperature to determine fluid properties
   Assume fluid exit temperatures, $T_{h,o}$ = and $T_{c,o}$ =
   Compute mean fluid temperature:

$$T_{hm} = \frac{T_{h,i} + T_{h,o}}{2}$$

$$T_{cm} = \frac{T_{c,i} + T_{c,o}}{2}$$

Get properties of hot and cold fluids based on the mean temperature.

2. Estimate flow areas:

Tube side	Shell Side
$a_t = \dfrac{N_t \times \frac{\pi d_t^2}{4}}{N_{p,t}}$	L_b = Baffle spacing = $\frac{L}{(N_b+1)}$
	D_e = Effective shell diameter = $D_s - N_{CLD} \times d_o$
	$a_s = L_b \times D_e$

Estimate the free flow area and frontal area.

3. Estimate mass velocity

$$V_t = \frac{\dot{m}_t}{\rho_t a_t} \quad V_s = \frac{\dot{m}_s}{\rho_s a_s}$$

4. Compute Reynolds number

$$\text{Re}_t = \frac{\rho_t V_t d_i}{\mu_t} \quad \text{Re}_s = \frac{\rho_s V_s d_o}{\mu_s}$$

$$\text{Re}_{\text{max},s} = \frac{\rho_s V_{\text{max}} d_o}{\mu_s}$$

5. Estimate Nusselt number, heat-transfer coefficient, and friction factor

Tube side	Shell side
(a) Decide fully developed or Entrance region	(a) Use Zakauskus Correlation for flow over a bank of tubes
(b) Select appropriate Correlations	(b) Update shell-side pressure drop to include shell-side flow geometry
$\text{Nu}_t =$	$\text{Nu}_s =$
$h_t = \dfrac{\text{Nu}_t \times k_t}{d_i}$	$h_s = \dfrac{\text{Nu}_s \times k_s}{d_o}$
$f_t =$	$f_s =$

6. Estimate entrance and exit loss coefficient using σ and Re for banks of tubes.
7. Apply the property variation corrections to the Nusselt numbers, heat-transfer coefficients, and friction factors.

$$\frac{\text{Nu}}{\text{Nu}_m} = \left(\frac{\mu_s}{\mu_m}\right)^n \frac{f}{f_m} = \left(\frac{\mu_s}{\mu_m}\right)^n$$

8. Compute fin efficiency, fin surface area, and fin area effectiveness based on fin type and geometry. If there are no fins, then set $\eta_{oi} = 1$ and $\eta_{oo} = 1$
9. Fouling factors for inside and outside tube surfaces: R_{fi} and R_{fo}
10. Compute overall heat-transfer coefficient

$$U_i = \frac{1}{\dfrac{1}{\eta_{oi} h_i} + R_{fi} + \dfrac{\ln \frac{r_o}{r_i}}{2\pi L k} A_i + \dfrac{R_{fo} A_i}{A_o} + \dfrac{1}{\eta_{oo} h_o} \dfrac{A_i}{A_o}}$$

11. Compute inside and outside surface temperature of the tube
    Inner tube wall temperature: $T_{wi} = T_i - \frac{U_i A_i}{h_i}(T_i - T_o)$
    and
    Outer tube wall temperature: $T_{wo} = T_o + \frac{U_i A_i}{h_o A_o}(T_i - T_o)$
12. Check if computed and estimated tube surface temperature agree with some tolerance limit. If there is no agreement, then use the current computed wall temperature and repeat steps 2–12 until convergence.
13. Compute surface area for heat transfer

$$A_i = N_t \pi d_i L$$

14. Compute heat-exchanger rating, q, based on the area obtained in step 14 and U_i obtained in step 11.

Computer Net Transfer Unit:	$NTU = \dfrac{U_i A_i}{C_{min}}$
Compute effectiveness:	$\varepsilon = \varepsilon \left(NTU, \dfrac{C_{min}}{C_{max}} \right)$
Estimate heat-transfer rate:	$q = \varepsilon\, q_{max}$

15. Use the current estimate of the heat-transfer rate to estimate the exit and mean fluid temperatures.

$$T_{h,o} = T_{h,i} - \frac{q}{\dot{m}_h c_{p,h}}$$

$$T_{c,o} = T_{c,i} + \frac{q}{\dot{m}_c c_{p,c}}$$

If the computed exit and mean fluid temperature are not in agreement with the assumed exit or mean fluid temperature, all previous steps 2–16 need to be repeated starting with the computation of mean fluid temperatures and estimating their physical properties. If differences are within the acceptable tolerance limit, then move on to compute the pressure drop values in the tube and shell sides.

16. Estimate pressure drop in tube side, ΔP_t, and pressure drop in shell side, ΔP_s

Pressure drop in tube side:

$$\frac{\Delta P_t}{P_1} = \frac{G_t^2}{2g_c} \frac{v_1}{P_1} \left[(K_c + 1 - \sigma^2) + 2 \left(\frac{v_2}{v_1} \right) + f_F \frac{A}{A_c} \frac{v_m}{v_c} - (1 - \sigma^2 - K_e) \frac{v_2}{v_1} \right]$$

or

We can also use a simplified alternative form considering negligible entrance and exit losses, and negligible effect of variation in density

$$\Delta P_t = v_m \frac{G_t^2}{2g_c} \left[f_F \frac{A}{A_c} \right]$$

where

A = total heat-transfer area = $A_i = N_t \pi d_i L$

$A_t = N_{p,t} \times n_{t,p} \times \frac{\pi}{4} d_i^2$

σ = Ratio of free flow area to shell frontal area = $\frac{A_t}{A_{fr}}$

G_t = Tube – side mass velocity

Re_t = Tube – side Reynolds number

$\frac{1}{v_1} = \rho_i, \frac{1}{v_2} = \rho_o$, and $\frac{1}{v_m} = \rho_m$

k_c and k_e = Entrance and exit loss coefficients

Pressure drop in shell side:

Actual pressure drop (ΔP_s) expression suggested by the Delaware method includes number of corrections factors. For simplicity, a first-order approximation for well-designed heat exchanger is given as

$$\Delta P_s = X_L (N_b + 1)\, \Delta P_0 \tag{5.55}$$

where

N_b = Number of baffles

X_L = Correction factor for leakages and usually taken as 0.2–0.3

ΔP_o = Shell-side pressure drop with uniform flow over bank of tubes:

$$\Delta P_o = N_L \, \chi f \left(\frac{\rho V_{max}^2}{2} \right) \tag{5.56}$$

N_L = Number of tube rows in longitudinal or flow direction = Number of tubes in the center line of the shell= C_{LTD}

f = friction factor for flow over a bundle of tubes arranged either in inline or in staggered manner

χ = Correction factor for nonequal pitch for tube spacing for $S_L \neq S_T$ for inline and $S_D \neq S_T$ for staggered arrangement.

Note that in this Design type-I problem, one can vary number of geometrical parameters like tube diameter or tube pitch in an iterative manner to meet any targeted heat-transfer rate or exit fluid temperatures or pressure drops either in tube side or shell side or both.

Example 5.5 Heat-exchanger Design Type-I

For a shell-and-tube heat exchanger with one-shell pass and one tube pass, the size and characteristics are given as follows:

Cold Fluid: Water in Tube side	Hot Fluid: Water in shell side
55 000 lbm/h	70 000 lbm/h
Inlet Temperature: 60°F	Inlet Temperature: 200°F
Tube Information	**Shell Information**
Tube size: 1 in., 18 BWG(Base case)	Inside Diameter: 1.8 ft
Copper tubes	No. of Baffles: 15
No. of tubes: 85	Inline or square array
Length: 10 ft	$S_L = S_T = 2$-in.
No. of tubes in centerline: 10	

Estimate the ratings of heat exchanger in terms of heat-transfer rate, exit temperature of the tube-side and shell-side fluids, pressure drops in tube side and shell side. Show first iteration by hand calculations.

Solution:

1. **Estimate the required fluid properties at the appropriate temperatures:** Since the exit temperatures are not known, we can either estimate fluid properties based on the inlet temperatures or based on the estimated mean fluid temperature once exit temperatures are assumed.

   Estimate the exit or mean fluid temperature to determine fluid properties

   Assume fluid exit temperatures, $T_{h,o} = 160°F$ and $T_{c,o} = 100°F$

Compute mean fluid temperature:

$$T_{cm} = \frac{T_{c,i} + T_{c,o}}{2} = \frac{60 + 100}{2} = 80°F$$

$$T_{hm} = \frac{T_{h,i} + T_{h,o}}{2} = \frac{200 + 160}{2} = 180°F$$

Get properties of hot and cold fluids based on the mean temperature:

Properties of cold water in tube side at 80°F:

$\rho_{c,i} = 62.3633$ lbm/ft^3, $\rho_{c,o} = 61.9917$ lbm/ft^3,

$\rho_{c,m} = 62.213\ 57$ lbm/ft^3 $\mu_c = 5.761 \times 10^{-4}$ lbm/ft - s, $C_{p,c} = 0.9993$ Btu/lbm ° F, $k_c = 0.3525$ Btu/h ft ° F and

$$Pr_c = 5.879$$

Properties of hot water in shell side at 180°F:

$\rho_{h,i} = 60.12$ lbm/ft^3, $\rho_{h,o} = 60.9981$ lbm/ft^3, $\rho_{h,m} = 60.9981$ lbm/ft^3,

$\mu_h = 2.3165 \times 10^{-4}$ lbm/ft - s, $C_{p,h} = 1.0035$ Btu/lbm °F, $k_h = 0.3879$ Btu/h ft °F and $Pr_h = 2.157$

2. **Estimate flow areas:** *Tube-side flow area:*

   Tube size: For ¾-in 18 BWG gage (from Table C.13):

   $$d_o = 0.75 - \text{in.} = 0.0625 - \text{ft and } d_i = 0.652 - \text{in.} = 0.0543 - \text{ft}$$

   No. of tubes, $N_t = 85$, Tube length: 10 ft.

   Total tube flow area, $a_t = \frac{\pi d_i^2}{4} = \frac{\pi (0.0543)^2}{4} = 0.002\ 316$ ft^2

   $$A_t = \text{Net tube} - \text{side flow area} = N_{p,t} \times n_{t,p} \times \frac{\pi}{4} d_i^2$$

   For $N_{p,t} = 1$ and $n_{t,p} = 85$

   $$A_t = \text{Net tube} - \text{side flow area} = n_{t,p} \times a_t = 85 \times 0.002\ 316$$

   $$A_t = 0.1969 \text{ ft}^2$$

   Shell-side flow areas

   Inside diameter: 1.8 ft., No. of tubes, $N_t = 85$, Number of baffles: 15

   Staggered array: $S_L = S_T = 2$ - in. $= 0.167$ - ft. Tube length, $L = 10$ ft.

   No. of tubes in centerline, $N_{CLD} = 10$

   Baffle spacing, $L_b = \frac{L}{(N_b + 1)} = \frac{10}{15 + 1} = 0.625$ ft

   Effective shell diameter, $D_e = D_s - N_{CLD} \times d_o$, $D_e = 1.8 - 10 \times \frac{0.75}{12} = 1.175$ ft

   Sectional area in between two baffles, $a_s = L_b \times D_e = 0.625 \times 1.175 = 0.734$ ft^2

   Frontal area in tube box, $A_{fr} = \frac{\pi}{4} D_s^2 = \frac{\pi}{4} (1.8)^2 = 2.545$ ft^2

   Net tube-side flow area, $A_t =$ Net tube-side flow area $= n_{t,p} \times \frac{\pi}{4} d_i^2 = 0.1969$ ft^2

   Ratio of free flow area to shell frontal area, $\sigma = \frac{A_t}{A_{fr}} = \frac{0.1969}{2.545} = 0.0774$

3. **Estimate mass velocity** *Tube-side mass velocity*

   $$V_t = \frac{\dot{m}_t}{\rho_t a_t} = \frac{55\ 000}{62.2135 \times 0.1969}, V_t = 1.2471 \text{ ft/s}, G_t = 77.59 \text{ lb\\ft}^2 \text{ s}$$

Shell-side mass velocity

$$V_s = \frac{\dot{m}_s}{\rho_s a_s} = \frac{70\ 000}{60.998 \times 0.734}$$

$$V_s = 0.434 \text{ ft/s}, i_s = 26.491 \text{ lb\\ft}^2 \text{ s}$$

4. **Compute Reynolds number and Heat-transfer Coefficient**

   Tube-side Reynolds number and heat-transfer coefficient

   $Re_t = \frac{\rho_t V_t d_i}{\mu_t} = \frac{62.2135 \times 1.247 \times 0.0543}{0.000\ 576}$, $Re_t = 7312.5$, Flow is turbulent

   $$Nu_i = 0.023 Re_t^{0.8}\ Pr_t^{0.333} = 0.023\ (7312.5)^{0.8}(5.879)^{0.333} = 0.023 \times 1233.82 \times 1.8037$$

   $$Nu_i = 51.185$$

   Tube – side heat transfer coefficient

   $$h_i = \frac{Nu_i \times k_c}{d_i} = \frac{51.185 \times 0.3525}{0.0543}$$

   $$h_i = 332.278 \text{ Btu/h ft}^{20}\text{F}$$

   Shell-side Reynolds number and Heat-transfer Coefficient

   Reynolds number used in the correlation is based on the maximum fluid velocity occurring within the tube banks defined as follows:

   $$Re_{d,max} = \frac{\rho V_{max} d_o}{\mu}$$

   For **inline alignment**, the maximum velocity may occur at either the normal area A_1 and the maximum velocity is computed from

   $$V_{max} = \frac{S_T}{S_T - d_o} V_s = \frac{2}{2 - 0.75} \times 0.434 \text{ ft/s}$$

   $$V_{max} = 0.6944 \text{ ft/s}$$

   Shell-side Reynolds number based on maximum velocity

   $$Re_{max,s} = \frac{\rho_s V_{max} d_o}{\mu_s}$$

   $$Re_{max,s} = \frac{\rho_s V_{max} d_o}{\mu_s} = \frac{60.998 \times 0.6944 \times 0.0625}{0.000\ 231\ 65}$$

   $$Re_{max,s} = 11\ 428$$

   Compute shell-side convection heat-transfer coefficient-based on **Zukauskas correlation** given by Eq. (3.82) as follows:

   $$\overline{Nu_d} = \frac{h_o d_o}{k_{s,m}} = C Re_{d,max}^m\ Pr^{0.36} \left(\frac{Pr}{Pr_s}\right)^{\frac{1}{4}} \tag{3.82}$$

   for $N_L \geq 20$, $0.7 \leq Pr \leq 500$ and $1000 \leq Re_{d,\ max} \leq 2 \times 10^6$

Table 5.4 Classification of heat-exchanger design types.

Design problem type	Known data	To be determined	Restrictions	Approach
Type-1	Heat-exchanger type and size	Heat-transfer rate, exit fluid temperatures, and pressure drops		Direct
Type-1	Heat-transfer rate, exit fluid temperatures, and pressure drops	Heat-exchanger type and size	Length or size	
Type-IIa	Heat-transfer rate, fluid mass flow rates, and inlet temperatures	Size and type: Number of tubes, Length of tubes, number of tube passes, number of shell passes	Length or size	Iterative
Type-IIb	Heat-transfer rate, fluid mass flow rates, and inlet temperatures		Pressure drop either on tube side, or on shell side or both	Iterative

where constant $C = 0.27$ and $m = 0.63$ for aligned tube banks from Table 5.4. Considering $Pr = Pr_s = 2.157$ as the first approximation, Nusselt number is computed from Eq. (3.82) as

$$\overline{Nu_d} = 0.27\, Re_{d,max}^{0.63}\, Pr^{0.36} = 0.27(11\ 428)^{0.63}2.157^{0.36}$$

$$= 0.27 \times 360.1828\, x\, 1.3188$$

$$\overline{Nu_d} = 128.152$$

Considering number tube rows in the longitudinal direction as $N_L = 10$, a correction factor $C_2 = 0.97$ can be used based on data given in Table 5.4, while the shell-side Nusselt number is computed as

$$\overline{Nu_D}\Big|_{N_L-10} = C_2 \overline{Nu_D}\Big|_{\geq 20} = 0.97 \times 128.152$$

$$\overline{Nu_d} = 124.40$$

The external shell-side convection coefficient is then computed as

$$h_o = \frac{\overline{Nu_d} \times k_{s,m}}{d_o} = \frac{124.40 \times 0.3879}{0.0625}$$

$$h_o = 772\ \text{Btu/h ft}^{2\,0}\text{F}$$

This shell-side convection coefficient is now updated taking into account the effect of flow leakages and complex flow pattern caused by the baffles in the shell side

$$h_o = 0.6\, h_o = 0.6 \times 772\ \text{Btu/h ft}^{2\,0}\text{F}$$

$$h_o = 463\ \text{Btu/h ft}^{2\,0}\text{F}$$

5. **Compute overall heat-transfer coefficient**

   Considering no fins in the tubes and neglecting the effects of fouling, the overall heat-transfer coefficient is

   $$U_i = \cfrac{1}{\dfrac{1}{h_i} + \dfrac{\ln \frac{d_o}{d_i}}{2k} d_i + \dfrac{1}{h_o}\dfrac{d_i}{d_o}} = \cfrac{1}{\dfrac{1}{332.278} + \dfrac{\ln \frac{0.0625}{0.0543}}{2\times230} \times 0.0543 + \dfrac{1}{463}\dfrac{0.0543}{0.0625}}$$

   $$= \frac{1}{0.003 + 0.000\,017 + 0.001\,876} = \frac{1}{0.004\,89}$$

   $$U_i = 204.35 \ \frac{\text{Btu}}{\text{hr ft}^{2\circ}\text{F}}$$

6. **Compute inside and outside surface temperature of the tube**

   Inner tube wall temperature: $T_{wi} = T_{cm} + \dfrac{U_i}{h_i}(T_{hm} - T_{cm})$

   $$= 80 + \frac{204.35}{332.278}(180 - 80)$$

   $$T_{wi} = 141.5°\text{F}$$

   and

   Outer tube wall temperature: $T_{wo} = T_{hm} - \dfrac{U_i A_i}{h_o A_o}(T_{hm} - T_{cm})$

   $$= 180 - \frac{204.35}{463}\frac{0.0543}{0.0625}(180 - 80)$$

   $$T_{wo} = 141.65°\text{F}$$

7. **Compute surface area for heat transfer**

   $$A_i = N_t \pi d_i L = 85\pi \times 0.0543 \times 10 = 145 \ \text{ft}^2$$

   $$U_i A_i = 204.35 \ \frac{\text{Btu}}{\text{ft}^2\ \text{hr}°\text{F}} \times 145 \ \text{ft}^2 = 29\,630.8 \ \frac{\text{Btu}}{\text{hr}°\text{F}}$$

8. **Compute heat-exchanger rating, q, using NTU – ε method.**

   $$C_c = \dot{m}_c\, c_{pc} = 55\,000 \ \text{lbm/hr} \times 0.9993 \ \text{Btu/lbm}°\text{F}$$

   $$C_c = 54\,961.5 \ \text{Btu/hr}°\text{F}$$

   $$C_h = \dot{m}_h\, c_{ph} = 70\,000 \ \text{lbm/hr} \times 1.0035 \ \text{Btu/lbm}°\text{F}$$

   $$C_h = 70\,245 \ \text{Btu/hr}°\text{F}$$

   Since $C_c < C_h$, $C_{min} = C_c = 54\,961.5 \ \text{Btu/hr}°\text{F}$, $C_{max} = C_h = 70\,245 \ \text{Btu/hr}°\text{F}$ and
   $C = \dfrac{C_{min}}{C_{max}} = \dfrac{54\,961.5}{70\,245} = 0.7824$

   Computer Net Transfer Unit: $\text{NTU} = \dfrac{U_i A_i}{C_{min}} = \dfrac{29.630.8 \ \text{Btu/hr}°\text{F}}{54{,}961.5 \ \text{Btu/hr}°\text{F}}$

   $$\text{NTU} = 0.539$$

   Compute effectiveness: $\varepsilon = \varepsilon\left(\text{NTU}, \dfrac{C_{min}}{C_{max}}\right)$

For counter flow shell heat exchanger with one-shell pass and one tube

$$\varepsilon = \frac{1 - \exp[-NTU(1 - C))]}{1 - C \exp[-NTU(1 - C)]} \text{ for } C < 1$$

$$\varepsilon = \frac{1 - \exp[-0.539 (1 - 0.7824)]}{1 - 0.7824 \exp[-0.539 (1 - 0.7824)]} = \frac{1 - \exp[-0.11728]}{1 - 0.7824 \exp[-0.11728]}$$

$$\varepsilon = \frac{1 - \exp[-0.11728]}{1 - 0.7824 \exp[-0.11728]} = \frac{1 - 0.8893}{1 - 0.7824 \times 0.8893} = \frac{0.110\,66}{0.3042}$$

Effectiveness, $\varepsilon = 0.3637$

Estimate heat-transfer rate:

$$q = \varepsilon \, q_{max} = \varepsilon \, C_{min} (T_{hi} - T_{ci})$$

$$q = 0.3637 \times 54\,961.5 \text{ Btu/hr}^\circ\text{F} (200 - 60)$$

$$q = 777.369 \tfrac{\text{Btu}}{\text{s}} = 2\,798\,529.66 \text{ Btu/hr}$$

9. **Estimate exit fluid temperature**

   Use the current estimate of the heat-transfer rate to estimate the exit and mean fluid temperatures.

$$T_{h,o} = T_{h,i} - \frac{q}{\dot{m}_h c_{p,h}} = 200 - \frac{2\,798\,529.66 \text{ Btu/hr}}{70\,000 \text{ lbm/hr} \times 1.0035 \text{ Btu/lbm}^\circ\text{F}}$$

$$T_{h,o} = 160.16^\circ\text{F}$$

$$T_{c,o} = T_{c,i} + \frac{q}{\dot{m}_c c_{p,c}} = 60 + \frac{2\,789\,529.66 \text{ Btu/hr}}{55\,000 \times 0.9993 \text{ Btu/lbm}^\circ\text{F}} = 50.75$$

$$T_{c,o} = 110.754^\circ\text{F}$$

Let us compute the mean fluid temperatures and see if results converge.

$$T_{cm} = \frac{T_{c,i} + T_{c,o}}{2} = \frac{60 + 110.754}{2} = 85.377^\circ\text{F}$$

Percentage of relative difference: $\frac{85.377 - 80.00}{85.00} = 6.72\%$

$$T_{hm} = \frac{T_{h,i} + T_{h,o}}{2} = \frac{200 + 160}{2} = 180^\circ\text{F}$$

Percentage of relative difference: $\frac{110.754 - 100.00}{100.00} = 10.754\%$

As we can see, the percent relative errors on estimation of mean temperatures are less than 10% for the cold stream and close to 10% for the hot stream. We can certainly perform one more iteration with these new estimates of the mean fluid temperatures and corresponding fluid properties. For brevity, however, we can proceed on to compute rest of the computations such as the pressure drops in tube and shell sides.

10. **Estimate pressure drop in tube side, ΔP_t, and pressure drop in shell side, ΔP_s**

    Pressure drop in tube side:

$$\Delta P_t = \frac{G_t^2 \, v_1}{2g_c} \left[(K_c + 1 - \sigma^2) + 2 \left(\frac{v_2}{v_1} - 1 \right) + f_F \frac{A}{A_t} \frac{v_m}{v_{c,i}} - (1 - \sigma^2 - K_e) \frac{v_2}{v_1} \right]$$

where

Tube roughness, $\varepsilon = 0.000\,05$, Tube inner diameter, $d_i = 0.652$ in. $= 0.0543$-ft

A = total heat-transfer area $= A_i = N_t \pi d_i L = 85 \, \pi \times 0.0543 \times 10 = 145 \text{ ft}^2$

A_t = Net tube-side flow area $= 0.1969 \text{ ft}^2$

Ratio of free flow area to shell frontal area, $\sigma = \frac{A_L}{A_{fr}} = 0.0774$

G_t = Tube – side mass velocity = 77.59 lb/ft^2 s

Re_t = Tube – side Reynolds number = 7312.5

$\frac{1}{v_1} = \rho_{c,i} = 62.3633$ lbm/ft^3, $\frac{1}{v_2} = \rho_{c,o} = 61.9917$ lbm/ft^3, and $\frac{1}{v_m} = \rho_{c,m} = 61.9917$ lbm/ft^3

From Figure 5.27 $k_c = 0.45$ and $k_e = 0.80$

Let us compute the core Fanning friction factor, f_F, based on **Halland's** explicit form [Eq. (8.46)] of functional representation of Moody's chart (Figure 8.5) for turbulent region and noting that the relationship between the Fanning friction factor and Darcy–Weisbach friction, f, is $f = 4f_F$.

$$f_F = \frac{0.077\,15}{\left\{ \log\left[\left(\frac{\epsilon/D}{3.7}\right)^{1.11} + \frac{6.9}{Re_D} \right] \right\}} = \frac{0.077\,15}{\left\{ \log\left[\left(\frac{0.000\,05/0.0543}{3.7}\right)^{1.11} + \frac{6.9}{7312.5} \right] \right\}}$$

$$f_F = \frac{0.077\,15}{\{\log[0.000\,076\,6 + 0.000\,944]\}} = \frac{0.07715}{\{\log[0.000\,076\,6 + 0.000\,944]\}}$$

$$= \frac{0.077\,15}{8.947} = 0.008\,62$$

$$f_F = 0.008\,62$$

Substituting

$$\Delta P_t = (77.59)^2 \frac{1}{2 \times 62.3633} \left[(0.45 + 1 - (0.0774)^2) + 2\left(\frac{62.3633}{61.9917} - 1\right) \right.$$

$$\left. + f_F \frac{145}{0.1969} \frac{62.3633}{61.9917} - (1 - (0.0774)^2 - 0.80)\frac{62.3633}{61.9917} \right]$$

$$\Delta P_t = 48.267 \left[1.444 + 0.0119 + 0.00862 \times \frac{145}{0.1969} \frac{62.3633}{61.9917} - 0.1952 \right]$$

$$\Delta P_t = 48.267[1.444 + 0.0119 + 6.3859 - 0.1952] =$$

$$\Delta P_t = 367.08/gc = 11.4 \text{ lbf/ft}^2$$

Pressure drop in shell side

Actual pressure drop (ΔP_s) expression suggested by the Delaware method includes number of corrections factors. For simplicity, a first-order approximation for well-designed heat exchanger is given as

From Eq. (5.55)

$$\Delta P_s = P_f(N_b + 1)\Delta P_o \tag{E5.5.1}$$

where

ΔP_o = Shell-side pressure drop with uniform flow over banks of tubes:

From Eq. (5.46)

$$\Delta P_o = N_L \chi f \left(\frac{\rho V_s^2}{2}\right) \tag{E5.5.2}$$

N_b = Number of baffles;

P_f = Correction factor for leakages and usually taken as 0.2–0.3

$$\Delta P_o = N_L \chi f \left(\frac{\rho V_{max}^2}{2}\right) \tag{E5.5.3}$$

$N_L = 10$; $\chi = 1$ for $S_L = S_T$; $V_s = 0.434$ ft/s, $\rho_{h,i} = 60.12$ lbm/ft^3

For shell-side maximum Reynolds number, $Re_{max,s} = 11\,428$, we can read friction factor value from Figure 3.17a for inline tube arrangement as $f = 0.9$.

$$\Delta P_o = N_L \chi f\left(\frac{\rho_{h,i} V_s^2}{2 g_c}\right) = 10 \times 1.0 \times 0.9 \left(\frac{60.12 \times (0.435)^2}{2 \times 32.174}\right) = 102.39 = 3.182 \text{ lbf/ft}^2$$

$$\Delta P_s = P_f(N_b + 1)\Delta P_o = 0.25 \times (15 + 1) \times 3.182 = 12.718 \text{ lbf/ft}^2$$

$$\Delta P_s = 12.718 \text{ lbf/ft}^2$$

5.5.10.2 Heat-exchanger Design Problem Type-IIa

In this type of design type problem with given heat-transfer rating and a restriction given in terms of maximum length of the tube, the procedure is iterative involving determination of the size and type, and pressure drops until the estimated length of the tube is within the assigned value.

5.5.10.2.1 Design Strategy

Several parameters are initially set and computations are then carried out to see if the given heat-transfer rate is met without exceeding the maximum length of the heat-exchanger tube. Computations are repeated with a new value of one of the parameters as the new assumed value until desired heat-transfer rating is met.

Set few parameters in the tube and shell sides and execute the following steps:

I. Assume number of tube passes, $N_{t,p}$
II. Assume tube size tube inside diameter, d_i, and outside diameter, d_o
III. Assume shell-side geometrical parameters:
Tube pitch: $S_L =$ and $S_T =$
Shell diameter, D_s, Number of tubes in centerline, N_{CLD}
Baffle spacing, L_b

I. *Assume and vary* either average velocity in the tube side, V_t, or assume and vary total number of tubes, N_t, until the estimated length is estimated to be less than maximum tube length given as the constraint.

Algorithms: Computational algorithm for Design Type-IIa problem is demonstrated in the flowchart given in Figure 5.29.

Example 5.6 *Heat-exchanger Design problem – Type IIa*

Given heat-transfer rating (q), determine size and type for maximum tube length as a restriction and with following fluid mass flow rates and inlet condition.

Cold fluid in tube side:

Fluid type: Water
Mass flow rate, $\dot{m}_c = 3.8$ kg/s
Cold fluid inlet temperature, $T_{c,i} = 35°C$
Cold fluid outlet temperature, $T_{c,o} = 55°C$

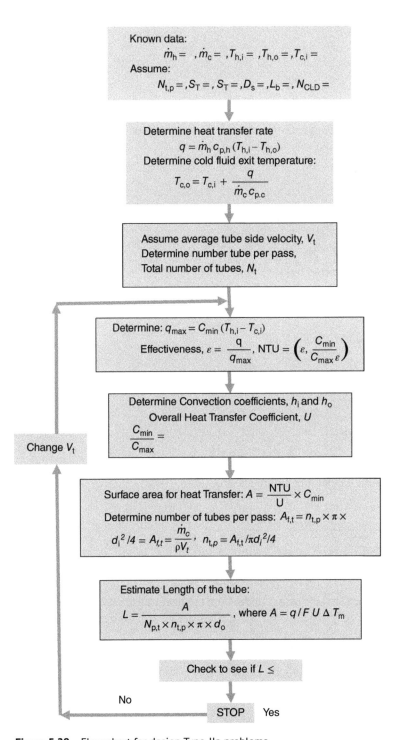

Figure 5.29 Flow chart for design Type-IIa problems.

Hot fluid in shell side:

Fluid type: Water

Mass flow rate, $\dot{m}_h = 1.9\,\text{kg/s}$

Hot fluid inlet temperature, $T_{h,i} = 90°C$

Restriction: Length of tubes, $L_{max} \leq 3.0\,M$

Use initial estimate of the overall heat-transfer coefficient, $U = 1400\,\text{W/m}^2\,C$
Determine number of tubes and tube length.

Initial Assumptions:

1. Number of shell passes, $N_{p,s} = 1$
2. Number of tube passes, $N_{p,t} = 1$,
3. Tube diameter: 3/4-in. (1.905 cm = 0.019 05 m)
4. Average tube-side velocity: $V_{av,t} = 0.35\,\text{m/s}$

From heat-transfer rate for cold fluid:

$q = \dot{m}_c c_{pc}(T_{c,o} - T_{c,i}) = 3.8 \times 4.182\,(55\text{--}35)$
$q = 317.832\,\text{kW}$

From heat-transfer rate for the hot fluid

$$q = \dot{m}_h c_{ph}(T_{h,i} - T_{h,o})$$

$$317.832 = 1.9 \times 4.182(90 - T_{h,o})$$

$$T_{h,o} = 50°C$$

Estimate LMTD:

$$\Delta T_m = Log \text{ mean temperaure difference} = \frac{\Delta T_{max} - \Delta T_{min}}{\ln \frac{\Delta T_{max}}{\Delta T_{min}}}$$

$$\Delta T_m = \frac{(90 - 50) - (55 - 35)}{\ln \frac{(90 - 50)}{(55 - 35)}} = \frac{40 - 20}{\ln \frac{40}{20}}$$

$$\Delta T_m = 28.8539°C$$

Estimate overall surface area for heat transfer from the heat-transfer rate equation

$$q = FUA\Delta T_m$$

For Single-shell pass and Single-tube pass: $N_{p,t} = 1, F = 1$

$$A = \frac{q}{FUA\Delta T_m} = \frac{317\,832\,\text{J/s}}{1 \times 1400\,\frac{j}{s\,m^2\,C} \times 28.8539°C}$$

Total heat-transfer surface area, $A = 7.8680\,\text{m}^2$

Estimate number of tubes in tube side per pass: $n_{t,p}$

For tube-side flow rate: $\dot{m}_c = \rho_t a_{f,t} V_t$

Total area of flow in tube side:

$$a_{f,t} = \frac{\dot{m}_c}{\rho V_t}$$

$$a_{f,t} = \frac{3.8}{1000 \times 0.35} = 0.01086 \text{ m}^2$$

Equating total tube-side flow area to the number of tubes and the flow area of a single tube as

$$a_{f,t} = n_{t,p} \times \pi \times d_i^2/4 = a_{f,t} = \frac{\dot{m}_c}{\rho u_{av}}$$

Number of tubes per pass:

$$n_{t,p} = a_{f,t}/\pi d_i^2/4 = \frac{0.01086}{\pi (0.01905)^2/4}$$

$n_{t,p} = 38.12 \cong 38$

Relating total heat-transfer surface area to the number of tubes and tube length, we get

$$A = q/FU\Delta T_m = N_{p,t} \times n_{t,p} \times \pi dL$$

$$A = 7.8680 \text{ m}^2 = N_{p,t} \times n_{t,p} \times \pi dL$$

Estimate length of the tube:

$$L = \frac{A}{N_{p,t} \times n_{t,p} \times \pi \times d_o}$$

$$L = \frac{7.8680}{1 \times n_{t,p} \times \pi \times d_o} = \frac{7.8680}{1 \times 38 \times \pi \times 0.01905}$$

$$L = 3.4614 \text{ m}$$

Since $L > L_{max}$, we can change our initial assumptions and iterate until the desired length is achieved. One choice is to change the tube size until desired tube length is achieved. Another choice is to the change the number of passes to $N_{p,t} = 2$

For multiple tube passes,

$q = FUA\Delta T_m$

Estimate correction factor for two tube passes:

$F = f(P,Z)$

$$P = \frac{T_{to} - T_{ti}}{T_{si} - T_{ti}} = \frac{55 - 35}{90 - 35} = 0.3636$$

and

$$Z = \frac{T_{si} - T_{so}}{T_{to} - T_{ti}} = \frac{90 - 50}{55 - 35} = 2.0$$

The value of the correction factor can read from Figure 5.14a or can be estimated from Eq. (5.9).

Using Eq. (5.9), we get

$$F = \frac{\sqrt{Z^2 + 1} \ln[(1 - P)/(1 - PZ)]}{(Z - 1)\ln((2 - P(Z + 1 - \sqrt{Z^2 + 1}))/(2 - P(Z + 1 + \sqrt{Z^2 + 1})))}$$

$$F = \frac{\sqrt{2^2 + 1} \ln[(1 - 0.3636)/(1 - 0.3636 \times 2)]}{(2 - 1)\ln((2 - 0.3636(2 + 1 - \sqrt{2^2 + 1}))/(2 - 0.3636(2 + 1 + \sqrt{2^2 + 1})))}$$

$$F = \frac{2.236 \times \ln[0.6364/0.2728]}{\ln(17.932)} = \frac{2.236 \times \ln[0.6364/0.2728]}{2.8866} = \frac{1894}{2.8866} = 0.6561$$

Estimate heat-transfer surface area

For Single-shell pass and two tube passes: $N_{p,t} = 2$, $F = 0.656$

$$A = \frac{q}{FU\Delta T_m} = \frac{317832 \, \text{J/s}}{0.6566 \times 1400 \, \frac{j}{s \, m^2 \, C} \times 28.8539°C}$$

Total heat-transfer surface area, $A = 11.9829 \, \text{m}^2$

Recalculate length of the tube:

$$L = \frac{A}{N_{p,t} \times n_{t,p} \times \pi \times d_o}$$

$$L = \frac{11.9829}{2 \times 38 \times \pi \times 0.01905}$$

$$L = 2.634 \, \text{m} < L_{\text{max}=3,0\text{m}}$$

Summary of the heat-exchanger specification:

Computation now can be carried out to estimate the pressure drops.

5.5.10.3 Heat-exchanger Design Problem Type-IIb

In this category of problems, the heat-transfer rating is given in terms of heat-transfer rate and pressure drop. The objective is to determine size and type that meet the requirements of heat-transfer rating and pressure drop given either in tube side or shell side or both. One way to proceed is to first satisfy the heat-transfer rating and then vary mass velocity, V_t, or G_t in the tube side that leads to the computations of number of tubes and size of the flow geometry in the tube side until the constraints given in terms of pressure drop in the tube side. Another way to proceed is to start with a guess value of the number of tubes for tube-side fluids and vary it until assigned pressure drop is attained.

In a similar manner, for constraint given in terms of pressure drop, we can start the iteration process with a guess value of average velocity, V_s, or mass velocity, G_s, and vary it, which leads to the computations of geometrical parameters such as tube pitch S_L and S_T; baffle spacing, L_b, etc., attains the assigned pressure drop in the shell side.

The strategy is to set few tube-side and shell-side parameters in the following manner:

Assume tube-side parameters:

Number of tube passes, $N_{t,p}$

Assume tube-size tube inside diameter, d_i, and outside diameter, d_o

Assume shell-side geometrical parameters:

Tube pitch: $S_L =$ and $S_T =$, and tube arrangement

Shell diameter, D_s, Number tubes in centerline, N_{CLD}

Baffle spacing, L_b

For a pressure drop constraint given for the tube side, let us assume and vary either average velocity in the tube side, V_t, or vary total number of tubes per tube pass, $a_{t,p}$, until the target pressure drop in the tube side is attained.

Algorithms: Computational algorithm for Design Type-IIb problem is demonstrated in the flowchart in Figure 5.30.

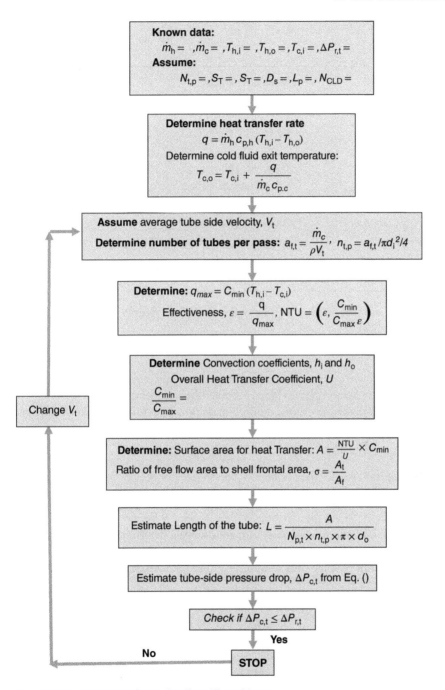

Figure 5.30 Flow chart for design Type-IIb problems.

5.5.10.3.1   Detailed Computational Steps for Design Problem Type-IIb
Known data:

Cold Fluid (Ins Shell Side):	Hot Fluid (Tube side):
Mass flow rate, $\dot{m}_c$	Mass flow rate, $\dot{m}_h$
Inlet temperature, $T_{c,i}$	Inlet temperature, $T_{h,i}$
Outlet temperature, $T_{c,o}$	
Shell-side fluid:	**Tube-side fluid:**
Number of shell passes, $N_{p,s}$	Number of tube passes, $N_{p,t}$
Tube arrangement:	Tube diameter and type (Assume)
Tube pitch: S_T, S_L (**Assume**)	Tube material (**Assume**)
Shell diameter, D_s,	Pressure drop in tube side, ΔP_t
Number of tubes in centerline, N_{CLD}	
Baffle spacing, L_b	

Determine: The number of tubes, $n_{t,p}$; Length of the tube, L; Number of baffles, N_b, and Shell-side pressure drop

Procedure:

1. Determine unknown exit temperature of hot and cold fluids based on the given heat-transfer rating and properties based on inlet fluid temperatures.
   Determine heat-transfer rate from cold fluid mass flow rate, inlet and outlet temperatures

   $$q = \dot{m}_c c_{p,c} \left(T_{c,o} - T_{c,1} \right)$$

   Determine hot fluid exit temperature

   $$T_{h,o} = T_{h,i} - \frac{q}{\dot{m}_h c_{p,h}}$$

2. Estimate mean fluid and tube surface temperatures and determine the required fluid properties.
3. Compute maximum possible heat transfer, q_{max}, and heat-exchanger effectiveness, ε, from heat-transfer rating, q

   $$q_{max} = C_{min} \left(T_{h,i} - T_{c,i} \right)$$

   $$\varepsilon = \frac{q}{q_{max}}$$

4. Assume first guess value **tube-side average velocity V_t (Need to vary this value)**
5. Estimate flow areas:
   Tube side
   Determine tube-side flow area

   $$a_{f,t} = \frac{\dot{m}_c}{\rho V_t},$$

Number of tubes per tube pass:

$$n_{t,p} = a_{f,t}/\pi d_i^2/4$$

Shell Side

$$L_b = \text{Space per baffle} = \frac{L}{(N_b + 1)}$$

$$a_s = L_b \times D_e$$

6. Estimate mass velocity
   Tube side: $V_t = \text{Assumed}$, $G_t = \rho_t V_t$
   Shell side: $V_s = \frac{\dot{m}_s}{\rho_s a_s}$, $G_s = \rho_s V_s$

7. Compute Reynolds number
   $\text{Re}_t = \frac{\rho_t V_t d_i}{\mu_t}$ $\text{Re}_s = \frac{\rho_s V_s d_o}{\mu_s}$

8. Estimate Nusselt number, heat-transfer coefficient, and friction factor. Since the tube length is unknown at the first iteration, consider fully developed correlation as a first initial assumption.

   Tube side

   Select appropriate correlations for Nusselt number:

   For $\text{Re} < 23$ use

   $Nu_t = 3.66$ for laminar flow

   For $\text{Re} > 2300$

   $Nu_t = 0.023 \text{Re}^{0.8} \text{Pr}^{0.333}$

   Convection heat-transfer coefficients:

   $h_t = \frac{Nu_t \times k_t}{d_i}$

   $f_t = \text{Estimate friction factor, f sub t from Moody's diagram (Figure 8.5 or correlation Eq.}$
   (8.45) or (8.46)):

   Shell side

   Compute convection heat-transfer coefficient based on the correlation for external fluid flow across a tube bank given by *Zukauskas* as follows:

$$\overline{Nu_d} = \frac{h_o d_o}{k_{s,m}} = C\text{Re}_{d,max}^m \text{Pr}^{0.36}\left(\frac{\text{Pr}}{\text{Pr}_s}\right)^{\frac{1}{4}} \tag{5.57}$$

$$h_s = \frac{Nu_s \times k_s}{d_o}$$

   $f_s = \text{Estimate friction factor on shell side, fs from Figures 3.17 or 3.18.}$

9. Apply the property variation corrections to the Nusselt numbers, heat-transfer coefficients, and friction factors.

$$\frac{Nu}{Nu_m} = \left(\frac{\mu_s}{\mu_m}\right)^n \frac{f}{f_m} = \left(\frac{\mu_s}{\mu_m}\right)^n$$

10. Compute overall heat-transfer coefficient

$$U_i = \frac{1}{\frac{1}{(\eta_o h)_i} + R_f + \frac{\ln\frac{r_o}{r_i}}{2\pi Lk_w}A_i + \frac{1}{(\eta_o h)_o}\left(\frac{A_i}{A_o}\right)}$$

11. Compute inside and outside surface temperature of the tube

$$T_{w,i} =$$

$$T_{w,o} =$$

12. Check if computed and estimated tube surface temperature agree with some tolerance limit. If there is no agreement, then use the current computed wall temperature and repeat steps 3–13 until convergence.

13. Compute NTU from effectiveness, ε, using relation

$$NTU = NTU\left(\varepsilon, \frac{C_{min}}{C_{max}}\right)$$

14. Compute heat-transfer surface area, A_i, from NTU

$$NTU = \frac{U_i A_i}{C_{min}}$$

$$A_i = \frac{NTU\, C_{min}}{U_i}$$

15. Determine the length of the tubes from surface area for heat transfer

$$A_i = n_{t,p}\pi d_i L$$

$$L = \frac{A_i}{n_{t,p}\pi d_i}$$

16. Estimate the free flow area and frontal area.
Frontal area in tube box, $A_{fr} = \frac{\pi}{4}D_s^2$
Net tube-side flow area, $A_t =$ Net tube-side flow area $= N_{p,txn_{t,p}} \times \frac{\pi}{4}d_i^2$
Ratio of free flow area to shell frontal area, $\sigma = \frac{A_t}{A_{fr}} =$

17. Estimate entrance and exit loss coefficients, K_c and K_e, for the specific tube type and using σ and Re for banks of tubes.

18. Compute tube-side pressure drop, $\Delta P_{c,t}$, using following equation:

$$\Delta P_{t,c} = \frac{G_t^2}{2g_c}v_1\left[(K_c + 1 - \sigma^2) + 2\left(\frac{v_2}{v_1}\right) + f_F\frac{A}{A_c}\frac{v_m}{v_c} - (1 - \sigma^2 - K_e)\frac{v_2}{v_1}\right]$$

19. Repeat steps 6–20 with new estimate of number of tubes, V_t and G_t, until $\Delta P_{t,c} \cong \Delta P_t$

Note that in this approach, average velocity in the tubes is assumed and then the number of tubes in the tube side is computed from the tube-side mass flow rate. The average tube-side velocity is varied until required tube-side pressure drop is reached. Other choice could be to assume number of tubes per tube pass and then estimate tube-side average velocity from mass flow rate. However, no guidance is given for choosing a starting initial guess value of the tube-side velocity or number of tubes per pass.

Example 5.7 *Design Type IIb: With Known Heat Exchanger Ratings Determine Size*
A single-shell-pass and single-tube-pass heat exchanger is to be designed to meet the following specifications:

Light Oil (Tube Side)	Water Side (Shell Side)
Flow rate = 1.30 ft³/s	Flow rate = 4.0 ft³/s
Inlet temperature = 300.0 F	Inlet temperature: 150 F
	Outlet temperature = 200 F
(Use all properties at 250 F)	**(Use all properties at 200 F)**
Tube-side geometrical Information	**Shell-side geometrical variables:**
Material: Copper	Tube arrangement: Inline
Tube: 1- in. 16 BWG	Tube pitch $\frac{S_T}{d}$ = 2.0 and $\frac{S_T}{d}$ = 2.0
	Shell diameter, D_s: 36-in.
	Baffle spacing: L_b = 1.5 ft
	Number of tubes in the centerline, N_{CLD} : 10 (Initial assumption),
Pressure drop	
Tube side, ΔP_t = 5.0 psi (Maximum)	Shell-side pressure drop: no restriction

Estimate the following quantities:

(a) Number of tubes
(b) Length of tubes

Solution:
Let us first set few additional parameters, which can also be modified to meet certain requirements.

Assume number of shell sides as $N_{s,p} = 1$
Assume number of tube passes, $N_{t,p} = 1$

1. Estimate the required fluid properties at the appropriate temperatures:
   Compute mean fluid temperature of the hot fluid (Oil) in the tube side:

$$T_{hm} = \frac{T_{h,i} + T_{h,o}}{2} = \frac{300 + 200}{2} = 250°F$$

Properties of light oil in **tube side** at 250°F:

$$\rho_h = 53 \ \text{lbm/ft}^3, \mu_h = 139 \times 10^{-5} \ \text{lbm/ft-s}, C_{p,h} = 0.52 \ \text{Btu/lbm°F},$$
$$k_h = 0.074 \ \text{Btu/hr ft}^0\text{F and Pr}_h = 35$$

Since the exit temperature of the cold fluid temperature is not known, let us assume the cold fluid properties at inlet temperature as 150°F.
Properties of water in **shell side** at 150°C:

$$\rho_c = 60.2 \ \text{lbm/ft}^3 \ \text{s}, \mu_c = 0.292 \times 10^{-3} \text{lbm/ft}^3 \ \text{s}, C_{p,c} = 1.0 \ \text{Btu/lbm°F},$$
$$k_c = 0.384 \text{Btu/hr ft}^0\text{F and Pr}_c = 1.88$$

2. Determine unknown exit temperature of hot and cold fluids based on the given heat-transfer rating and properties based on inlet fluid temperatures.

Mass flow rates:

Hot fluid – Tube side

$$\dot{m}_h = \rho_h \dot{Q}_h = 53 \text{ lbm/ft}^3 \times 1.30 \text{ ft}^3/s = 68.9 \text{ lbm/s}$$

Cold fluid – Shell side

$$\dot{m}_c = \rho_c \dot{Q}_c = 60.2 \text{ lbm/ft}^3 \times 4.0 \text{ ft}^3/s = 240.8 \text{ lbm/s}$$

Heat-transfer rate

Based on the energy balance on the hot fluid in the tube side, the heat-transfer rate is given as

$$q = \dot{m}_h c_{p,h}(T_{h,i} - T_{h,o}) = 68.9 \times 0.52 \times (300 - 200)$$

$$q = 3582.8 \text{ Btu/s} = 12.898.08 \text{ Btu/h}$$

Based on the energy balance on the cold fluid in the shell side, the exit fluid temperature is given as

$$q = 3582.\dot{8} = m_c c_{p,c}(T_{c,o} - T_{c,i})$$

$$T_{c,o} = T_{c,o} + \frac{q}{\dot{m}_c c_{p,c}}$$

$$T_{c,o} = 150 + \frac{3582.8}{240.8 \text{ lbm/s} \times 1.0 \text{Btu/lbm F}}$$

$$T_{c,o} = 164.87° \text{ F}$$

3. Estimate flow areas and average fluid velocities:

Tube-side flow area

Tube dimensions for 1-in. – 16 BWG from table:
$d_i = 0.9\text{-in.} = 0.0751\text{-ft}$ and $d_o = 1\text{-in.} = 0.0833\text{-ft}$
Area of a single tube

$$a_{t,1} = \frac{\pi}{4}(0.075\ 17)^2 = 0.004\ 438 \text{ ft}^2$$

Let us assume average velocity in the tube side as:

$$V_t = 3.5 \text{ ft/s } (\textbf{\textit{Assumed as first initial guess value}})$$

Mass velocity in tube side, $G_t = \rho_t V_t = 53 \times 3.5 = 185.5 \text{ lbm/ft}^2$.
Determine the number of tubes:

$$\dot{Q}_t = 1.30 \text{ ft}^3/s = n_{t,p} a_{t,1} V_t$$

$$n_{t,p} = \frac{\dot{Q}_t}{a_{t,1} V_t} = \frac{1.30 \text{ ft}^3/s}{0.004\ 438 \times 3.5} = 83.69$$

$$n_{t,p} \approx 84$$

Estimation of the ratio of free flow area to frontal area, σ:

Free flow area:

$$A_{FF} = n_{t,p} a_{t,1} = 84 \times 0.004\ 438$$

$$A_{FF} = 0.3728 \text{ ft}^2$$

Frontal area:

$$A_{fr} = \frac{\pi}{4}D_s^2 = \frac{\pi}{4}(3\text{ ft})^2, A_{fr} = 7.068\text{ ft}^2$$

$$\sigma = \frac{A_{FF}}{A_{fr}} = \frac{0.3728}{7.068} = 0.0527$$

Shell-Side flow area, a_s, and **Mass Velocity**

$$D_e = \text{Effective shell diameter} = D_s - N_{CLD} \times d_o$$

$$D_e = 3.0 - 10 \times 0.0833, D_e = 2.167\text{ ft}$$

Shell-side flow area:

$$a_s = L_b \times D_e = 1.5 \times 2.167$$

$$a_s = 3.25\text{ ft}^2$$

Shell-side fluid average velocity

$$V_s = \frac{\dot{Q}_s}{a_s} = \frac{4.0}{3.25} = 1.23\text{ ft./s}$$

4. Compute convection heat-transfer coefficients

Tube-side convection heat-transfer coefficient

Reynolds number:

$$Re_t = \frac{\rho_t V_t d_i}{\mu_t} = \frac{53 \times 3.5 \times 0.075\,17}{139 \times 10^{-5}} = 10\,031.68$$

Since $Re > 2300$, the flow is turbulent

Let us have a fully developed turbulent convection heat-transfer coefficient based on the heat-transfer correlation given by **Colburn** as

$$Nu_t = 0.023 Re_t^{4/5} Pr_t^{1/3}$$

$$Nu_t = 0.023\,(10\,031.68)^{4/5}\,(35)^{1/3} = 0.023 \times 1588.908 \times 3.27067$$

$$Nu_t = 119.526$$

Convection heat-transfer coefficient for the hot fluid in the tube side:

$$h_i = \frac{Nu \times k_h}{d_i} = \frac{119.526 \times 0.074}{0.007\,517}$$

$$h_i = 117.665\text{Btu/h F}$$

Shell-side convection heat-transfer coefficient

Estimate Nusselt number, heat-transfer coefficient, and friction factor

Shell-side flow Reynolds number

$$Re_s = \frac{\rho_s V_s d_o}{\mu_s} = \frac{60.2 \times 1.23 \times 0.0833}{0.292 \times 10^{-3}}$$

$$Re_s = 21\,123$$

Reynolds number based on the maximum fluid velocity as follows:

For **aligned arrangement**, the maximum velocity is estimated as

$$V_{max} = \frac{S_T}{S_T - d_o}V_s, V_{max} = \frac{S_T/d_o}{S_T/d_o - 1}V_s = \frac{2.0}{2.0 - 1} \times 1.23$$

$$V_{max} = 2.46\text{ ft/s}$$

Shell-side Reynold number based on the maximum velocity:

$$Re_{d,max} = \frac{\rho V_{max} d_o}{\mu}$$

$$Re_{d,max} = \frac{\rho_s V_{max} d_o}{\mu_s} = \frac{60.2 \times 2.46 \times 0.0833}{0.292 \times 10^{-3}}$$

$$Re_{d,max} = 42\,246$$

Compute convection heat-transfer coefficient based on the correlation for external fluid flow across a tube bank is given by **Zukauskas** as follows:

$$\overline{Nu_d} = \frac{h_o d_o}{k_{s,m}} = C\,Re_{d,max}^m\,Pr^{0.36}\left(\frac{Pr}{Pr_s}\right)^{\frac{1}{4}}$$

From Table 5.4. for aligned tube arrangement and for $Re_{d,\,max} = 42\,246$

$$C = 0.27 \text{ and } m = 0.63$$

and assume $Pr = Pr_s$ as a first approximation.

$$Nu_s = 0.27\,(42\,246)^{0.63}(1.88)^{0.25}$$

$$= 0.27 \times 820.814 \times 1.17$$

$$= 259.295$$

Considering shell-side effect based on Delaware's method, the shell-side Nusselt number is given as

$$Nu_s = 0.6\,Nu_s = 0.6 \times 259.295$$

$$Nu_s = 155.577$$

Shell-side convection coefficient

$$h_s = \frac{Nu_s \times k_s}{d_o} = \frac{155.577 \times 0.384}{0.0833}$$

$$h_s = 717.185 \text{ Btu/h F}$$

Estimate Overall Heat-transfer Coefficient

$$U_i = \frac{1}{\frac{1}{\eta_{oi} h_i} + R_{fi} + \frac{\ln\frac{r_o}{r_i}}{2\pi Lk}A_i + \frac{R_{fo}A_i}{A_o} + \frac{1}{\eta_{oo} h_o}\frac{A_i}{A_o}}$$

Use $\eta_{oi} = \eta_{oo} = 1$ for no fins inside and $R_{fi} = R_{fo} = 0$ for negligible fouling.

$$U_i = \frac{1}{\frac{1}{h_i} + \frac{\ln\frac{d_o}{d_i}}{2k_w}d_i + \frac{1}{h_o}\frac{d_i}{d_o}}$$

$$U_i = \frac{1}{\frac{1}{117.665} + \frac{\ln\frac{0.0833}{0.075\,17}}{2\times 226} \times 0.075\,17 + \frac{1}{717.185}\frac{0.075\,17}{0.0833}}$$

$$U_i = \frac{1}{0.008\,49 + 0.000\,017 + 0.001\,258}$$

$$U_i = 102.4 \text{ Btu/h F}$$

5. **Estimate of total surface area and tube length using LMTD method**

$$\Delta T_m = \text{Log mean temperaure difference} = \frac{\Delta T_{max} - \Delta T_{min}}{\ln \frac{\Delta T_{max}}{\Delta T_{min}}}$$

$$\Delta T_m = \frac{(300-169)-(200-150)}{\ln \frac{(300-169)}{(200-150)}} = \frac{131-50}{\ln \frac{131}{50}}$$

$$\Delta T_m = 84.09°F$$

Estimate overall surface area for heat transfer from the heat-transfer rate equation

$$q = FU_i A_i \Delta T_m$$

For single-shell pass and single-tube pass: $N_{p,t} = 1, F = 1$

$$A_i = \frac{q}{FU_i \Delta T_m} = \frac{3582.8 \text{ Btu/s} \times 3600 \text{ s/hr}}{1 \times 102.4 \text{ Btu/s ft}^2 F \times 84.09°F}$$

Total heat-transfer surface area, $A = 1497.89 \text{ ft}^2$

Length of the tubes

$$L = \frac{A_i}{N_{p,t} \times n_{t,p} \times \pi d_i} = \frac{1497.89}{1 \times 84 \times \pi \times 0.07517} = 75.510 \text{ ft.}$$

Estimate pressure drop in tube side and shell side

6. **Estimate pressure drop in tube side, ΔP_t, and pressure drop in shell side ΔP_s**

Pressure drop in tube side:

$$\Delta P_t = \frac{G_t^2 v_1}{2g_c} \left[(K_c + 1 - \sigma^2) + 2\left(\frac{v_2}{v_1} - 1\right) + f_F \frac{A}{A_t} \frac{v_m}{v_{c,i}} - (1 - \sigma^2 - K_e)\frac{v_2}{v_1} \right]$$

where

Tube roughness, $\varepsilon = 0.000\,05$, Tube inner diameter, $d_i = 0.9\text{-in.} = 0.0751\text{-ft}$
$A = \text{total heat-transfer area} = A_i = N_t \pi d_i L = 84\,\pi \times 0.0751 \times 75.519 = 1496.667 \text{ ft}^2$
Free flow area:

$$A_{FF} = A_t = n_{t,p} a_{t,1} = 84 \times 0.004\,438$$

$$A_{FF} = 0.3728 \text{ ft}^2$$

Frontal area:

$$A_{fr} = \frac{\pi}{4} D_s^2 = \frac{\pi}{4}(3 \text{ ft})^2, A_{fr} = 7.068 \text{ ft}^2$$

$$\sigma = \frac{A_{FF}}{A_{fr}} = \frac{0.3728}{7.068} = 0.0527$$

$A_t = \text{Net tube-side flow area} = 0.3728 \text{ ft}^2$
Ratio of free flow area to shell frontal area, $\sigma = \frac{A_t}{A_{fr}} = 0.0527$

$$G_t = \text{Tube} - \text{side mass velocity} = 185.5 \text{ lb/ft}^2 \text{ s}$$

$$Re_t = \text{Tube} - \text{side Reynolds number} = 10\,031.68$$

$$\rho_h = 53 \text{ lbm/ft}^3$$

$$\frac{1}{v_1} = \rho_{h,i} = 53 \text{ lbm/ft}^3, \frac{1}{v_2} = \rho_{h,o} = 53 \text{ lbm/ft}^3, \text{ and } \frac{1}{v_m} = \rho_{c,m} = 53 \text{ lbm/ft}^3$$

From Figure 5.27, for $\sigma - 0.0527$ and tube-side Reynolds number $\text{Re}_t = 10\,031.68$, we can read the loss coefficient values as $k_c \approx 0.5$ and $k_e \approx 1.0$.

Let us compute the core Fanning friction factor, f_F, based on **Halland's** explicit form of functional representation of Moody's chart for turbulent region

$$f_F = \frac{0.077\,15}{\left\{\log\left[\left(\frac{\varepsilon/D}{3.7}\right)^{1.11} + \frac{6.9}{\text{Re}_D}\right]\right\}} = \frac{0.077\,15}{\left\{\log\left[\left(\frac{0.000\,05/0.0751}{3.7}\right)^{1.11} + \frac{6.9}{10\,031.68}\right]\right\}}$$

$$f_F = \frac{0.07715}{\{\log[0.000\,069\,69 + 0.000\,687\,82]\}^2} = \frac{0.077\,15}{9.7382} = 0.007\,922$$

$$f_F = 0.007\,922$$

Substituting

$$\Delta P_{c,t} = \frac{G_t^2 v_1}{2g_c}\left[(K_c + 1 - \sigma^2) + 2\left(\frac{v_2}{v_1} - 1\right) + f_F \frac{A}{A_t}\frac{v_m}{v_{c,i}} - (1 - \sigma^2 - K_e)\frac{v_2}{v_1}\right]$$

$$\Delta P_{c,t} = (185.5)^2 \frac{1}{2 \times 53 \times 32.2}\left[(0.5 + 1 - (0.0527)^2) + 2\left(\frac{53}{53} - 1\right)\right.$$
$$\left. + f_F \frac{1496.667}{0.3728}\frac{53}{53} - (1 - (0.0527)^2 - 1.0)\frac{53}{53}\right]$$

$$\Delta P_{c,t} = 10.081\left[1.4972 + 0.0 + 0.007\,922 \times \frac{1496.667}{0.3728}\frac{53}{53} - (0.002\,77)\right]$$

$$\Delta P_{c,t} = 16.66[1.4972 + 0.0 + 31.804 - (-0.002\,77)]$$

$$\Delta P_{c,t} = 554.847 \text{ lbf/ft}^2 = 3.853 \text{ psi}$$

Since $\Delta P_{c,t} > \Delta P_{\max,t}$, let us go back and reduce the average tube-side velocity and carry out the second iteration.

Iteration # 2: Reduce tube-side average velocity and assume $V_t = 3.2$ ft/s
Mass velocity in tube side, $G_t = \rho_t V_t = 53 \times 3.2 = 169.6$ lhm/ft² s.
Determine the number of tubes:

$$\dot{Q}_t = 1.30 \text{ ft}^3/\text{s} = n_{t,p} a_{t,1} V_t$$

$$n_{t,p} = \frac{\dot{Q}_t}{a_{t,1} V_t} = \frac{1.30 \text{ ft}^3/\text{s}}{0.004\,438 \times 3.2} = 91.538$$

$$n_{t,p} \approx 92$$

Estimation of the ratio of free flow area to frontal area, σ:

Free flow area:

$$A_{FF} = n_{t,p} a_{t,1} = 92 \times 0.004\,438$$

$$A_{FF} = A_t = 0.4083 \text{ ft}^2$$

Frontal area:

$$A_{fr} = \frac{\pi}{4}D_s^2 = \frac{\pi}{4}(3\ \text{ft})^2, A_{fr} = 7.068\ \text{ft}^2$$

$$\sigma = \frac{A_{FF}}{A_{fr}} = \frac{0.4083}{7.068} = 0.0577$$

Compute convection heat-transfer coefficients

Tube-side convection heat-transfer coefficient

Reynolds number:

$$\text{Re}_t = \frac{\rho_t V_t d_i}{\mu_t} = \frac{53 \times 3.2 \times 0.07517}{139 \times 10^{-5}} = 9171.82$$

Since Re > 2300, the flow is turbulent

Let us fully develop turbulent convection heat-transfer coefficient based on the heat-transfer correlation given by **Colburn** as

$$\text{Nu}_t = 0.023\ \text{Re}_t^{4/5}\ \text{Pr}_t^{1/3}$$

$$\text{Nu}_t = 0.023\ (9171.82)^{4/5}\ (35)^{1/3}$$

$$= 0.023 \times 1478.987 \times 3.27067$$

$$\text{Nu}_t = 111.2574$$

Convection heat-transfer coefficient for the hot fluid in the tube side:

$$h_i = \frac{\text{Nu} \times k_h}{d_i} = \frac{111.2574 \times 0.074}{0.07517}$$

$$h_i = 109.53\ \text{Btu/h F}$$

Estimate Overall Heat-transfer Coefficient

$$U_i = \frac{1}{\frac{1}{\eta_{oi}h_i} + R_{fi} + \frac{\ln\frac{r_o}{r_i}}{2\pi Lk}A_i + \frac{R_{fo}A_i}{A_o} + \frac{1}{\eta_{oo}h_o}\frac{A_i}{A_o}}$$

Use $\eta_{oi} = \eta_{oo} = 1$ for no fins inside and $R_{fi} = R_{fo} = 0$ for negligible fouling.

$$U_i = \frac{1}{\frac{1}{h_i} + \frac{\ln\frac{d_o}{d_i}}{2k_w}d_i + \frac{1}{h_o}\frac{d_i}{d_o}}$$

$$U_i = \frac{1}{\frac{1}{109.53} + \frac{\ln\frac{0.0833}{0.07517}}{2\times226} \times 0.07517 + \frac{1}{717.185}\frac{0.07517}{0.0833}}$$

$$U_i = \frac{1}{0.00913 + 0.000017 + 0.001258}$$

$$U_i = 96.108\ \text{Btu/h F}$$

Estimate overall surface area for heat transfer from the heat-transfer rate equation

$$q = FU_i A_i \Delta T_m$$

For single-shell pass and single-tube pass: $N_{p,t} = 1, F = 1$

$$A_i = \frac{q}{FU_i \Delta T_m} = \frac{3582.8 \text{ Btu/s} \times 3600 \text{ s/hr}}{1 \times 96.108 \frac{\text{Btu}}{\text{s ft}^2\text{F}} \times 84.09°\text{F}}$$

Total heat-transfer surface area, $A = 1595.956 \text{ ft}^2$

Length of the tubes

$$L = \frac{A_i}{N_{p,t} \times n_{t,p} \times \pi d_i} = \frac{1595.956}{1 \times 92 \times \pi \times 0.07517} = 73.4579 \text{ ft.}$$

Estimate pressure drop in tube side and shell side

Estimate pressure drop in tube side, ΔP_t, and pressure drop in shell side, ΔP_s

Pressure drop in tube side:

$$\Delta P_t = \frac{G_t^2 v_1}{2g_c} \left[(K_c + 1 - \sigma^2) + 2\left(\frac{v_2}{v_1} - 1\right) + f_F \frac{A}{A_t} \frac{v_m}{v_{c,i}} - (1 - \sigma^2 - K_e)\frac{v_2}{v_1} \right]$$

where

Tube roughness, $\varepsilon = 0.000\,05$, Tube inner diameter, $d_i = 0.9\text{-in.} = 0.0751\text{-ft}$

$A =$ total heat-transfer area $= A_i = 1595.956 \text{ ft}^2$

$A_{FF} = A_t = 0.4083 \text{ ft}^2$,

$\sigma = \frac{A_{FF}}{A_{fr}} = \frac{0.4083}{7.068} = 0.0577$

Mass velocity in tube side, $G_t = \rho_t V_t = 53 \times 3.2 = 169.6 \text{ lbm/ft}^2 \text{ s.}$

$G_t =$ Tube $-$ side mass velocity $= 169.6 \text{ lb/ft}^2 \text{ s}$

$Re_t =$ Tube $-$ side Reynolds number $= 9171.82$

$\rho_h = 53 \text{ lbm/ft}^3$

$$\frac{1}{v_1} = \rho_{h,i} = 53 \text{ lbm/ft}^3, \frac{1}{v_2} = \rho_{h,o} = 53 \text{ lbm/ft}^3 \text{ and } \frac{1}{v_m} = \rho_{c,m} = 53 \text{ lbm/ft}^3$$

From Figure 5.27 for $\sigma - 0.0577$ and tube-side Reynold number $Re_t = 9171.82$, we can read the loss coefficient values as $k_c \approx 0.5$ and $k_e \approx 1.0$.

Let us compute the core Fanning friction factor, f_F, based on ***Halland's*** explicit form of functional representation of Moody's chart [] for turbulent region

$$f_F = \frac{0.077\,15}{\left\{ \log\left[\left(\frac{\varepsilon/D}{3.7}\right)^{1.11} + \frac{6.9}{Re_D} \right] \right\}} = \frac{0.077\,15}{\left\{ \log\left[\left(\frac{0.000\,05/0.0751}{3.7}\right)^{1.11} + \frac{6.9}{9171.82} \right] \right\}}$$

$$f_F = \frac{0.07715}{\{\log[0.000\,069\,69 + 0.000\,752\,3]\}^2} = \frac{0.07715}{9.518} = 0.008\,105$$

$$f_F = 0.008\,105$$

Substituting

$$\Delta P_{c,t} = \frac{G_t^2 v_1}{2g_c} \left[(K_c + 1 - \sigma^2) + 2\left(\frac{v_2}{v_1} - 1\right) + f_F \frac{A}{A_t} \frac{v_m}{v_{c,i}} - (1 - \sigma^2 - K_e)\frac{v_2}{v_1} \right]$$

$$\Delta P_{c,t} = (169.6)^2 \frac{1}{2 \times 53 \times 32.2} \left[(0.5 + 1 - (0.0577)^2) + 2\left(\frac{53}{53} - 1\right) \right.$$
$$\left. + f_F \frac{1595.956}{0.4083} \frac{53}{53} - (1 - (0.0527)^2 - 1.0)\frac{53}{53} \right]$$

$$\Delta P_{c,t} = 8.427 \left[1.4967 + 0.0 + 0.008\,105 \times \frac{1595.956}{0.4083} \frac{53}{53} - (0.00277) \right]$$

$$\Delta P_{c,t} = 8.427[1.496\,77 + 0.0 + 31.68 - (-0.002\,77)]$$

$$\Delta P_{c,t} = 279.60\;\text{lbf/ft}^2 = 1.9416\;\text{psi}$$

Since $\Delta P_{c,t} > \Delta P_{\max,t}$, let us go back and reduce the average tube-side velocity and carry out the third iteration.

Iteration # 3: Reduce tube-side average velocity and assume $V_t = 3.1$ **ft/s**

Mass velocity in tube side, $G_t = \rho_t V_t = 53 \times 3.1 = 164.3\;\text{lbm/ft}^2\;\text{s}$.

Determine the number of tubes:

$$\dot{Q}_t = 1.30\;\text{ft}^3/s = n_{t,p} a_{t,1} V_t$$

$$n_{t,p} = \frac{\dot{Q}_t}{a_{t,1} V_t} = \frac{1.30\;\text{ft}^3/s}{0.004\,438 \times 3.1} = 94.49$$

$$n_{t,p} \approx 95$$

Estimation of the ratio of free flow area to frontal area, σ:

Free flow area:

$$A_{FF} = n_{t,p} a_{t,1} = 95 \times 0.004\,438$$

$$A_{FF} = A_t = 0.42161\;\text{ft}^2$$

Frontal area:

$$A_{fr} = \frac{\pi}{4} D_s^2 = \frac{\pi}{4}(3\;\text{ft})^2, \; A_{fr} = 7.068\;\text{ft}^2$$

$$\sigma = \frac{A_{FF}}{A_{fr}} = \frac{0.42161}{7.068} = 0.0596$$

Compute convection heat-transfer coefficients

Tube-side convection heat-transfer coefficient

Reynolds number:

$$Re_t = \frac{\rho_t V_t d_i}{\mu_t} = \frac{53 \times 3.1 \times 0.07517}{139 \times 10^{-5}} = 8885.2$$

Since $Re > 2300$, the flow is turbulent

Let us fully develop turbulent convection heat-transfer coefficient based on the heat-transfer correlation given by **Colburn** as

$$Nu_t = 0.023 Re_t^{4/5} Pr_t^{1/3}$$

$$Nu_t = 0.023\,(8885.2)^{4/5}\,(35)^{1/3}$$

$$= 0.023 \times 1441.895 \times 3.27067$$

$$Nu_t = 108.467$$

Convection heat-transfer coefficient for the hot fluid in the tube side:

$$h_i = \frac{Nu \times k_h}{d_i} = \frac{108.467 \times 0.074}{0.07517}$$

$$h_i = 106.779 \frac{Btu}{h\ F}$$

Estimate Overall Heat-Transfer Coefficient

$$U_i = \frac{1}{\frac{1}{\eta_{oi} h_i} + R_{fi} + \frac{\ln \frac{r_o}{r_i}}{2\pi L k} A_i + \frac{R_{fo} A_i}{A_o} + \frac{1}{\eta_{oo} h_o} \frac{A_i}{A_o}}$$

Use

$$\eta_{oi} = \eta_{oo} = 1 \text{ for no fins inside and } R_{fi} = R_{fo} = 0 \text{ for negligible fouling.}$$

$$U_i = \frac{1}{\frac{1}{h_i} + \frac{\ln \frac{d_o}{d_i}}{2k_w} d_i + \frac{1}{h_o} \frac{d_i}{d_o}}$$

$$U_i = \frac{1}{\frac{1}{106.779} + \frac{\ln \frac{0.0833}{0.07517}}{2 \times 226} \times 0.075\,17 + \frac{1}{717.185} \frac{0.07517}{0.0833}}$$

$$U_i = \frac{1}{0.009365 + 0.000017 + 0.001258}$$

$$U_i = 93.984 \text{ Btu/h F}$$

Estimate overall surface area for heat transfer from the heat-transfer rate equation

$$q = FU_i A_i \Delta T_m$$

For single-shell pass and single-tube pass: $N_{p,t} = 1$, $F = 1$

$$A_i = \frac{q}{FU_i \Delta T_m} = \frac{3582.8 \text{ Btu/s} \times 3600 \text{ s/hr}}{1 \times 93.984 \text{ Btu/s ft}^2 F \times 84.09° F}$$

Total heat-transfer surface area, $A = 1632.02 \text{ ft}^2$
Length of the tubes

$$L = \frac{A_i}{N_{p,t} \times n_{t,p} \times \pi d_i} = \frac{1632.02}{1 \times 95 \times \pi \times 0.07517} = 72.7457 \text{ ft.}$$

Estimate pressure drop in tube side and shell side
Estimate pressure drop in tube side, ΔP_t, and pressure drop in shell side ΔP_s
Pressure drop in tube side:

$$\Delta P_t = \frac{G_t^2 \, v_1}{2g_c} \left[(K_c + 1 - \sigma^2) + 2 \left(\frac{v_2}{v_1} - 1 \right) + f_F \frac{A}{A_t} \frac{v_m}{v_{c,i}} - (1 - \sigma^2 - K_e) \frac{v_2}{v_1} \right]$$

where

Tube roughness, $\varepsilon = 0.000\,05$, Tube inner diameter, $d_i = 0.9\text{-in.} = 0.0751\text{-ft}$

$$A = \text{total heat} - \text{transfer area} = A_i = 1632.02 \text{ ft}^2$$

$$A_{FF} = A_t = 0.421\,61 \text{ ft}^2$$

$$A_{fr} = 7.068 \text{ ft}^2$$

$$\sigma = \frac{A_{FF}}{A_{fr}} = \frac{0.42161}{7.068} = 0.0596$$

Mass velocity in tube side, $G_t = \rho_t V_t = 53 \times 3.1 = 164.3 \; \text{lbm/ft}^2 \; \text{s}$.

$G_t = \text{Tube} - \text{side mass velocity} = 164.3 \; \text{lb/ft}^2 \; \text{s}$.

$\text{Re}_t = \text{Tube} - \text{side Reynolds number} = 8885.2$

$\rho_h = 53 \; \text{lbm/ft}^3$

$\dfrac{1}{v_1} = \rho_{h,i} = 53 \; \text{lbm/ft}^3, \dfrac{1}{v_2} = \rho_{h,o} = 53 \text{lbm/ft}^3 \; \text{and} \; \dfrac{1}{v_m} = \rho_{c,m} = 53 \; \text{lbm/ft}^3$

From Figure 5.27 for $\sigma - 0.0577$ and tube-side Reynolds number $\text{Re}_t = 9171.82$, we can read the loss coefficient values as $k_c \approx 0.5$ and $k_e \approx 1.0$.

Let us compute the core Fanning friction factor, f_F, based on **Halland's** explicit form of functional representation of Moody's chart [] for turbulent region

$$f_F = \frac{0.07715}{\left\{ \log\left[\left(\frac{\varepsilon/D}{3.7}\right)^{1.11} + \frac{6.9}{\text{Re}_D}\right]\right\}} = \frac{0.077\,15}{\left\{ \log\left[\left(\frac{0.000\,05/0.0751}{3.7}\right)^{1.11} + \frac{6.9}{8885.2}\right]\right\}}$$

$$f_F = \frac{0.077\,15}{\{\log[0.000\,069\,69 + 0.000\,776\,57]\}^2} = \frac{0.077\,15}{9.4403} = 0.008\,17$$

$$f_F = 0.008\,17$$

Substituting

$$\Delta P_{c,t} = \frac{G_t^2 \, v_1}{2g_c}\left[(K_c + 1 - \sigma^2) + 2\left(\frac{v_2}{v_1} - 1\right) + f_F \frac{A}{A_t}\frac{v_m}{v_{c,i}} - (1 - \sigma^2 - K_e)\frac{v_2}{v_1}\right]$$

$$\Delta P_{c,t} = (164.3)^2 \frac{1}{2 \times 53 \times 32.2}\left[(0.5 + 1 - (0.0596)^2) + 2\left(\frac{53}{53} - 1\right)\right.$$
$$\left. + f_F \frac{1632.02}{0.42161}\frac{53}{53} - (1 - (0.0527)^2 - 1.0)\frac{53}{53}\right]$$

$$\Delta P_{c,t} = 7.9088\left[1.4972 + 0.0 + 0.008\,17 \times \frac{1632.02}{0.42161}\frac{53}{53} - (0.00277)\right]$$

$$\Delta P_{c,t} = 7.9088[1.4972 + 0.0 + 31.625 - (-0.00277)]$$

$$\Delta P_{c,t} = 261.9822 \; \text{lbf/ft}^2 = 1.819 \; \text{psi}$$

Iteration # 4: Reduce tube-side average velocity and assume $V_t = 3.0 \; \text{ft/s}$
Mass velocity in tube side, $G_t = \rho_t V_t = 53 \times 3.0 = 159 \; \text{lbm/ft}^2 \; \text{s}$.
Determine the number of tubes:

$$\dot{Q}_t = 1.30 \; \text{ft}^3/\text{s} = n_{t,p} a_{t,1} V_t$$

$$n_{t,p} = \frac{\dot{Q}_t}{a_{t,1} V_t} = \frac{1.30 \text{ft}^3/\text{s}}{0.004\,438 \times 3.0} = 97.64$$

$$n_{t,p} \approx 98$$

Estimation of the ratio of free flow area to frontal area, σ:

Free flow area:

$$A_{FF} = n_{t,p}a_{t,1} = 98 \times 0.004\,438$$

$$A_{FF} = A_t = 0.4349 \text{ ft}^2$$

Frontal area:

$$A_{fr} = \frac{\pi}{4}D_s^2 = \frac{\pi}{4}(3 \text{ ft})^2, A_{fr} = 7.068 \text{ ft}^2$$

$$\sigma = \frac{A_{FF}}{A_{fr}} = \frac{0.4349}{7.068} = 0.0653$$

Compute convection heat-transfer coefficients
Tube-side convection heat-transfer coefficient
Reynolds number:

$$Re_t = \frac{\rho_t V_t d_i}{\mu_t} = \frac{53 \times 3.0 \times 0.07517}{139 \times 10^{-5}} = 8598.5$$

Since Re > 2300, the flow is turbulent

Let us fully develop turbulent convection heat-transfer coefficient based on the heat-transfer correlation given by **Colburn** as

$$Nu_t = 0.023 Re_t^{4/5} Pr_t^{1/3}$$

$$Nu_t = 0.023 \, (8598.5)^{4/5} \, (35)^{1/3}$$

$$= 0.023 \times 1441.895 \times 3.27067$$

$$Nu_t = 105.658$$

Convection heat-transfer coefficient for the hot fluid in the tube side:

$$h_i = \frac{Nu \times k_h}{d_i} = \frac{105.658 \times 0.074}{0.07517}$$

$$h_i = 104.01 \text{ Btu/h F}$$

Estimate Overall Heat-Transfer Coefficient

$$U_i = \cfrac{1}{\cfrac{1}{\eta_{oi}h_i} + R_{fi} + \cfrac{\ln\frac{r_o}{r_i}}{2\pi Lk}A_i + \cfrac{R_{fo}A_i}{A_o} + \cfrac{1}{\eta_{oo}h_o}\cfrac{A_i}{A_o}}$$

Use $\eta_{oi} = \eta_{oo} = 1$ for no fins inside and $R_{fi} = R_{fo} = 0$ for negligible fouling.

$$U_i = \cfrac{1}{\cfrac{1}{h_i} + \cfrac{\ln\frac{d_o}{d_i}}{2k_w}d_i + \cfrac{1}{h_o}\cfrac{d_i}{d_o}}$$

$$U_i = \cfrac{1}{\cfrac{1}{104.01} + \cfrac{\ln\frac{0.0833}{0.07517}}{2\times226} \times 0.07517 + \cfrac{1}{717.185}\cfrac{0.07517}{0.0833}}$$

$$U_i = \cfrac{1}{0.009\,614 + 0.000\,017 + 0.001\,258}$$

$$U_i = 91.834 \text{ Btu/h F}$$

Estimate overall surface area for heat transfer from the heat-transfer rate equation

$$q = FU_i A_i \Delta T_m$$

For single-shell pass and single-tube pass: $N_{p,t} = 1, F = 1$

$$A_i = \frac{q}{FU_i \Delta T_m} = \frac{3582.8 \text{ Btu/s} \times 3600 \text{ s/hr}}{1 \times 91.834 \text{ Btu/s ft}^2 \text{F} \times 84.09°\text{F}}$$

Total heat-transfer surface area, $A_i = 1670.233 \text{ ft}^2$

Length of the tubes

$$L = \frac{A_i}{N_{p,t} \times n_{t,p} \times \pi d_i} = \frac{1670.233}{1 \times 98 \times \pi \times 0.07517} = 72.17 \text{ ft.}$$

Estimate pressure drop in tube side and shell side

Estimate pressure drop in tube side, ΔP_t, and pressure drop in shell side, ΔP_s

Pressure drop in tube side:

$$\Delta P_t = \frac{G_t^2 v_1}{2g_c} \left[(K_c + 1 - \sigma^2) + 2 \left(\frac{v_2}{v_1} - 1 \right) + f_F \frac{A}{A_t} \frac{v_m}{v_{c,i}} - (1 - \sigma^2 - K_e) \frac{v_2}{v_1} \right]$$

where

Tube roughness, $\varepsilon = 0.000\,05$, Tube inner diameter, $d_i = 0.9\text{-in.} = 0.0751\text{-ft}$

$$A = \text{total heat} - \text{transfer area} = A_i = \text{ft}^2$$

$$A_{FF} = A_t = 0.42161 \text{ ft}^2$$

$$A_{fr} = 7.068 \text{ ft}^2$$

$$\sigma = \frac{A_{FF}}{A_{fr}} = \frac{0.42161}{7.068} = 0.0596$$

Mass velocity in tube side, $G_t = \rho_t V_t = 53 \times 3.1 = 164.3 \text{ lbm/ft}^2 \text{ s.}$

$$G_t = \text{Tube} - \text{side mass velocity} = 164.3 \text{ lb/ft}^2 \text{ s}$$

$$Re_t = \text{Tube} - \text{side Reynolds number} = 8885.2$$

$$\rho_h = 53 \text{ lbm/ft}^3$$

$$\frac{1}{v_1} = \rho_{h,i} = 53 \text{ lbm/ft}^3, \frac{1}{v_2} = \rho_{h,o} = 53 \text{ lbm/ft}^3 \text{ and } \frac{1}{v_m} = \rho_{c,m} = 53 \text{ lbm/ft}^3$$

From Figure 5.27, for $\sigma - 0.0577$ and tube-side Reynolds number $Re_t = = 9171.82$, we can read the loss coefficient values as $k_c \approx 0.5$ and $k_e \approx 1.0$.

Let us compute the core Fanning friction factor, f_F, based on **Halland's** explicit form of functional representation of Moody's chart [] for turbulent region

$$f_F = \frac{0.077\,15}{\left\{ \log \left[\left(\frac{\varepsilon/D}{3.7} \right)^{1.11} + \frac{6.9}{Re_D} \right] \right\}} = \frac{0.077\,15}{\left\{ \log \left[\left(\frac{0.000\,05/0.0751}{3.7} \right)^{1.11} + \frac{6.9}{8598.5} \right] \right\}}$$

$$f_F = \frac{0.077\,15}{\{\log[0.000\,069\,69 + 0.001\,162\,9]\}^2} = \frac{0.077\,15}{8.463} = 0.009\,11$$

$$f_F = 0.008\,17$$

Substituting

$$\Delta P_{c,t} = \frac{G_t^2 \, v_1}{2g_c} \left[(K_c + 1 - \sigma^2) + 2\left(\frac{v_2}{v_1} - 1\right) + f_F \frac{A}{A_t} \frac{v_m}{v_{c,i}} - (1 - \sigma^2 - K_e)\frac{v_2}{v_1} \right]$$

$$\Delta P_{c,t} = (164.3)^2 \frac{1}{2 \times 53 \times 32.2} \left[(0.5 + 1 - (0.0596)^2) + 2\left(\frac{53}{53} - 1\right) \right.$$
$$\left. + f_F \frac{1670.233}{0.42161} \frac{53}{53} - (1 - (0.0527)^2 - 1.0)\frac{53}{53} \right]$$

$$\Delta P_{c,t} = 7.9088 \left[1.4972 + 0.0 + 0.00911 \times \frac{1670.233}{0.42161} \frac{53}{53} - (0.00277) \right]$$

$$\Delta P_{c,t} = 7.9088[1.4972 + 0.0 + 36.11 - (-0.00277)]$$

$$\Delta P_{c,t} = 267.838 \ \text{lbf/ft}^2 = 1.859 \ \text{psi}$$

Computed results for different iterations are summarized in the table below:

Iteration No	Tube-side guess velocity	Tube-side mass velocity(G_t), lbm/ft^2 s	h_i	f_F	Number of tubes per tube pass ($n_{t,p}$)	Length of tubes (L), ft	Tube-side pressure drop
1	3.5	185.5.			84	75.510	3.853
2	3.2	169.6			92	73.4579	1.9416
3	3.1	164.3			95	72.746	1.819
4	3.0	159			98	72.17	1.859

As we can notice that the resulting estimate of the length of the tube is about 72-ft, which may be too long considering lengthwise space requirement. We can always reset our assumption of number of tubes passes to 2 or 4 and repeat the preceding computational steps until the recomputed length reaches some acceptable limit.

Pressure drop in shell side:

Actual pressure drop (ΔP_s) expression suggested by the Delaware method includes number of corrections factors. For simplicity, a first-order approximation for well-designed heat exchanger is given as

$$\Delta P_s = P_f(N_b + 1)\Delta P_o \tag{E5.7.1}$$

where

ΔP_o = Shell-side pressure drop with uniform flow over banks of tubes:

$$\Delta P_o = N_L \chi f \left(\frac{\rho V_s^2}{2} \right) \tag{E5.7.2}$$

N_b = Number of baffles;
P_f = Correction factor for leakages and usually taken as 0.2–0.3

$$\Delta P_o = N_L \chi f \left(\frac{\rho V_{max}^2}{2} \right) \tag{E5.7.3}$$

$N_L = 10$; $\chi = 1$ for $S_L = S_T$; $V_s = 0.434$ ft/s, $\rho_{h,i} = 60.12$ lbm/ft^3

For shell-side maximum Reynolds number, $Re_{max,s} = 11\,428$, we can read friction factor value from Figure 3.17a for inline tube arrangement as $f = 0.9$.

$$\Delta P_o = N_L \chi f \left(\frac{\rho_{h,i} V_s^2}{2g_c} \right) = 10 \times 1.0 \times 0.9 \left(\frac{60.12 \times (0.435)^2}{2 \times 32.174} \right) = 102.39 = 3.182 \text{ lbf/ft}^2$$

$$\Delta P_s = P_f (N_b + 1) \Delta P_o = 0.25 \times (15 + 1) \times 3.182 = 12.718 \text{ lbf/ft}^2$$

$$\Delta P_s = 12.718 \text{ lbf/ft}^2$$

5.5.10.4 Shah's Method for Enhanced Convergence in Type-II Design Problems

Note that the procedure outlined in previous section for Type IIb heat-exchanger design problems does not have straightforward guidelines in assuming an initial guess value for the mass velocity to start the iterative process, and vary the average tube-side velocity until tube-side or shell-sides assigned drop values are reached. This often leads to a very time-consuming solution process.

Shah (1984) presented a procedure that involves using an expression for estimating the initial estimate of mass velocity, $G_t = \rho_t V_t$, and hence the number of tubes, N_t

This involves recasting the expression for pressure drop given by Eq. (5.47)

$$\frac{\Delta P_t}{P_1} = \frac{G_t^2}{2g_c} \frac{v_1}{P_1} \left[(K_c + 1 - \sigma^2) + 2 \left(\frac{v_2}{v_1} \right) + f_F \frac{A}{A_c} \frac{v_m}{v_c} - (1 - \sigma^2 - K_e) \frac{v_2}{v_1} \right] \tag{5.58}$$

With some simplification and using empirical data in the form of **j/f factor** or **Colburn j factor** for a given tube type and tube banks, the expression for an estimate of the mass velocity is given as

$$G = \left[\frac{2g_c \eta_o}{v_m \text{Pr}^{2/3} \text{NTU}} \Delta P \frac{J}{f} \right]^{1/2}$$

where

η_o = total area effectiveness for fins,

NTU = NTU for one side

j = Colburn j-factor = St Pr$^{2/3}$

The j/f factor or St Pr$^{2/3}/f$ factor represents a relative ratio of heat transfer to friction pressure drop and is obtained experimentally for different tube surfaces. A good source of this database is illustrated in the book by Kays and London (1984) in numerous figures. One such illustration is Figure 5.31.

5.5.11 Design Procedure for Type-II Heat Exchanger Based on Shah's Method

Strategy: Enforce pressure drop requirement and vary geometry. The computational algorithm for Design Type-IIa problem is demonstrated in the flowchart shown in Figure 5.32.

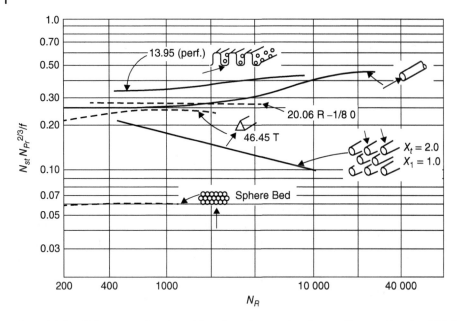

Figure 5.31 *J/f* factor as a function of flow Reynolds number. Source: Kays and London (1984). © 1984, McGraw-Hill.

Algorithm 5.1

Detailed Computational Steps

1. Estimate the exit, mean fluid, and surface temperatures.
2. Evaluate the required fluid properties at the appropriate temperatures.
3. Calculate $\frac{C_{min}}{C_{max}}$ and effectiveness, ε and determine NTU of the heat exchanger for the selected flow arrangement.
4. For an initial guess values of the NTU for one side, use the following approximation:
   $NTU_{1\text{-side}} = 2$ NTU for both fluids with same phase.
   For gas-liquid heat exchangers
   $NTU_{gas\text{-side}} = 1.1$ NTU assuming heat-transfer coefficient in the gas side is one-tenth or order of magnitude less than that of the of the liquid-side
   and
   $NTU_{liq\text{-side}} = 11$ NTU assuming liquid-side heat-transfer coefficient order of magnitude higher that n gas-side for liquid–gas arrangement
5. Obtain a ballpark value of j/f for the selected surface from the j/f vs. Re plots
   If fins are used on this side, estimate fin area effectiveness η_o based on selected fin type and size. Use $\eta_o = 0.8$ as the first approximation.
6. Compute mass velocity from

$$G = \left[\frac{2g_c\eta_o}{v_m \text{Pr}^{2/3}\text{NTU}} \Delta P \frac{j}{f} \right]^{1/2}$$

as the first estimate.

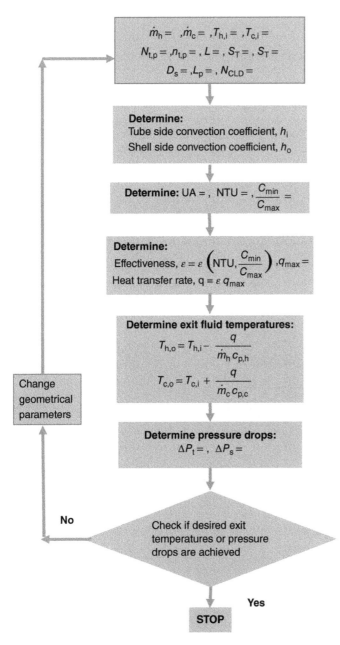

Figure 5.32 Flowchart for design Type-II problems based on Shah's method.

7. Compute Reynolds number

$$\mathrm{Re}_t = \frac{G_t d_i}{\mu_t} \quad \mathrm{Re}_s = \frac{G_s d_o}{\mu_s}$$

8. Estimate Nusselt number, heat-transfer coefficient, and friction factor

Tube side	Shell side
(a) Decide fully developed or entrance region	(a) Use Zakauskus Correlation for flow over a bank of tubes
(b) Select appropriate correlations	(b) Update shell side pressure drop to include shell side flow geometry
$\mathrm{Nu}_t =$	$\mathrm{Nu}_s =$
$h_t = \dfrac{\mathrm{Nu}_t \times k_t}{d_i}$	$h_s = \dfrac{\mathrm{Nu}_s \times k_s}{d_o}$
$f_t =$	$f_s =$

9. Estimate entrance and exit loss coefficient using σ and Re for banks of tubes.
10. Apply the property variation corrections to the Nusselt numbers, heat-transfer coefficients, and friction factors.

$$\frac{\mathrm{Nu}}{\mathrm{Nu}_m} = \left(\frac{\mu_s}{\mu_m}\right)^n \quad \frac{f}{f_m} = \left(\frac{\mu_s}{\mu_m}\right)^n$$

11. Compute overall heat-transfer coefficient

$$U_i = \cfrac{1}{\dfrac{1}{(\eta_o h)_i} + R_f + \dfrac{\ln \frac{r_o}{r_i}}{2\pi L k_w} A_i + \dfrac{1}{(\eta_o h)_o}\left(\dfrac{A_i}{A_o}\right)}$$

12. Compute inside and outside surface temperature of the tube

$$T_{w,i} =$$

$$T_{w,o} =$$

13. Check if computed and estimated tube surface temperatures agree with some tolerance limit. If there is no agreement, then use the current computed wall temperature and repeat steps 2–12 until convergence.
14. Compute surface area for heat transfer

$$A = \mathrm{NTU} \times \frac{C_{min}}{U_i}$$

15. Compute area for tube-side flow

$$A_f = \frac{\dot{m}}{G}$$

16. Estimate number of tubes and frontal area:

$$N_t = \frac{A_f}{\pi \frac{d_i^2}{4}}$$

$$A_{fr} = \frac{A_f}{\sigma}$$

17. Compute tube length

$$L = \frac{A}{N_t \pi \, d_i} = \frac{d_i A}{4A_f}$$

18. Calculate pressure drop from

$$\Delta P_t = \frac{G_t^2}{2g_c} \frac{1}{\rho_1} \left[(K_c + 1 - \sigma^2) + 2 \left(\frac{\rho_1}{\rho_2} \right) + f_F \frac{A}{A_c} \frac{\rho_1}{\rho_2} - (1 - \sigma^2 - K_e) \frac{\rho_1}{\rho_2} \right]$$

19. Compare computed pressure drop ΔP_c with given allowed pressure drop. If pressure drop does not converge, then recomputed mass velocity based on the current estimate of the pressure drop.
20. Repeat steps 7–19 until convergence.

5.6 Compact Heat Exchangers

As discussed in Section 5.1, a compact heat exchanger provides large heat-transfer surface area per unit volume by using denser array of finned tubes or lower flow passage area and/or plates of different flow geometries as compared to the shell-and-tube heat exchanger. The flow configurations are generally like the crossflow types. Both fluid-side passages are made of an array of microchannels, stacked as a matrix. The size of the individual channels, and the number of subchannels as well as the matrix size are scaled to meet the desired heat exchange value desired by the application. Computational fluid dynamics (CFD) analyses can play a significant role in selecting optimal sizes. Figure 5.33 illustrates examples of such compact heat exchangers in cross-flow configures.

Basically, a compact heat exchanger uses high-performance heat-transfer surface that has high heat flux relative to friction pressure drop or friction power usage. A compact heat exchanger provides high-performance surface and hence compactness by using small flow passages that result in high convection heat-transfer coefficient. Additionally, compactness is also achieved by increasing the net heat-transfer surface area with use of densely populated array fins. While use of fins may result in increase in pressure drop, the overall gain in heat transfer offsets the increased friction power usages. Design analysis of fins in terms of *fin efficiency*, η_f, and *fin area effectiveness*, η_t, for a wide range of types of fins is discussed in Chapter 4.

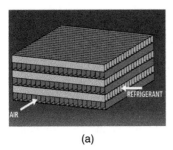
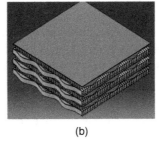

(a)   (b)

Figure 5.33 Compact heat exchangers with flow configurations in cross flow. (a) Compact heat exchanges in cross with dimpled surfaces and fins and (b) compact heat exchanges in cross with wavy surfaces and fins.

High-performance heat-transfer surface also uses additional features such as boundary-layer separation or interrupting boundary layers, vortex, and turbulence generators using inserts, and protruded surface such as delta wing and hemispherical/ellipsoidal bumps; boundary-layer bleed strategies on the vortex generators; and wavy fins with bellowed or sinusoidal configurations, with cut out of louvers.

Some key features of the compact heat exchangers are (i) very small flow passages typically of the **order of millimeters with hydraulic diameter in the range of 5 mm–0.125 mm (0.2 in. to 0.005 in.)**, (ii) high surface area per unit volume or heat-transfer surface area density of the **order of 1000 m^2/m^3 and in the range of m^2/m^3 (200–10 000 ft^2/ft^3)**; (iii) Fins thickness in the **range of 0.05–0.25 mm**; fin density in the range of **100–1000 fins/m** and even higher.

Examples of the compact heat-exchanger surfaces are **plate-fin surfaces, externally finned surfaces, internally finned surfaces, surface roughness, insert devices,** and uses variety of fin geometries such as rectangular fins, triangular fins, wavy fins, louvered fins, circumferential and longitudinal types, continuous or cut, and offset strip fins to compensate for the high thermal resistance by increasing the heat-transfer area.

Wavy fins can be bellowed, serpentine, or corrugated with triangular waves. Studies have demonstrated the enhanced heat rate in wavy fins with less than proportionate increases in pressure drop. *Boundary-layer interruption devices* can be used on the wavy or nonwavy surfaces. Include louvers cut on the high-pressure regions of the bellowed waves to bleed and exchange the boundary layer fluid with adjacent microchannels. The influx and efflux of boundary-layer fluid bring wall layer fluid to the core and increase mixing of the two regions, thereby thinning the thermal boundary layers. Use of roughened surface with macroscale roughness using structures and protrusions such as hemispherical bumps, elliptical elements, interrupted baffles promote turbulence and disrupt the formation hydrodynamic boundary layer. The periodicity, size, and shape of the roughness element play a key role on the degree of changes in turbulence and mixed-flow patterns, and hence increase heat-transfer rates.

All these features result in compact heat exchangers involving wide varieties of complex geometries. Because of the complex nature of the flow geometries, heat transfer and pressure drop characteristics are established through experimental correlations.

The compact heat-exchanger book by Kays and London (1984) includes database of such experimental correlations for different surface geometries, tube or flow passage geometries, and fin types.

As outlined, high-performance heat-transfer surfaces are characterized through use of the heat-transfer-friction power usages plots, obtained based on experimental testing. These high-performance surfaces are subjected to some specific parameters related to the tube geometries, fin geometries, and surface types as outlined below:

Flow and tube geometries:
   Tube types; Tube dimensions; Flow Passage Hydraulic diameter, $D_h = 4 \, r_h$; Plate spacing; Total area/volume between plates, β; Tube spacings in parallel and perpendicular direction; and Free flow area to frontal area, σ; Transfer area/total volume, α
   Fin geometries: Fin types; Fin spacing or Pitch, Fins/in.; Fin metal thickness, Fin size in terms length and width or diameter; Fin area/ total area;
High-performance heat-transfer surface:
   Figure 5.34 shows some examples of high-performance heat-transfer surfaces [Reproduced from Kay and London (1984) in such plots with *x*-axis variable.

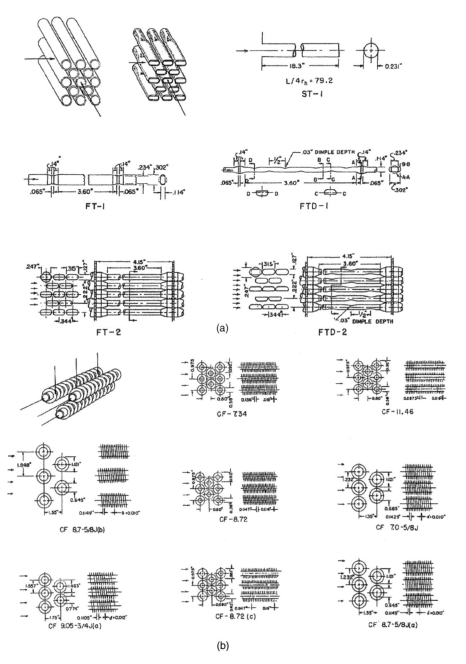

Figure 5.34 Some examples of high-performance heat-transfer surfaces from database. (a) Tubular Surfaces and (b) fin-tube surfaces, circular tubes, circular fins. Source: Kay and London (1984).

This database for high-performance heat-transfer surfaces and the corresponding experimental correlations are widely being used by compact heat-exchanger designers. A few examples of such correlation plots are shown in Figure 5.35. In this figure, the heat-transfer data in terms of $St\, Pr^{2/3} = \dfrac{h}{Gc_p} Pr^{2/3}$ and friction factor, f data are plotted as function of flow Reynolds number, $N_R = \dfrac{G4r_h}{\mu}$, where $N_P = Pr = \dfrac{\mu c_p}{k}$ is the Prandtl number.

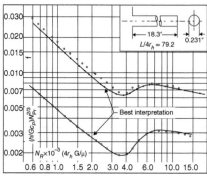

Tube inside diameter = 0.231 in.
Hydraulic diameter = 0.231 in., 0.01925 ft
Flow length/hydraulic diameter, $L/4r_h = 79.2$
Free-flow area per tube = 0.0002908 ft^2

(a)

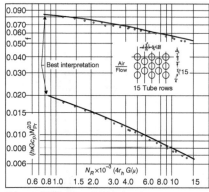

Tube outside diameter = 0.250 in.
Hydraulic diameter, $4r_h = 0.0166$ ft
Free-flow area/frontal area, $\sigma = 0.333$
Heat transfer area/total volume, $\alpha = 80.3$ ft^2/ft^3
Note: Minimum free-flow area is in spaces transverse to flow.

(b)

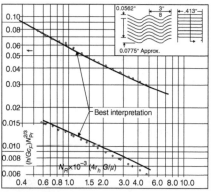

Fin pitch = 17.8 per in.
Plate spacing, $b = 0.413$ in.
Flow passage hydralic diameter, $4r_h = 0.00696$ ft
Fin metal thickness = 0.006 in., aluminum
Total heat transfer area/volume between plate, $\beta = 514$ ft^2/ft^3
Fin area/total area = 0.892
Note: Hydraulic diameter based on free-flow area normal to mean flow direction

(c)

Figure 5.35 Examples of surface geometries and correlation data. (a) Flow inside circular tube – Surface ST-1, (b) Flow normal to staggered tube bank – Surfaces S 1.50-1.25, (c) Wavy plate-fine surface – Surfaces 17.8-3/8W, (d) Flow normal to dimpled flattened tubes – Surfaces FTD-2, and (e) Flow normal to flattened tubes – Surfaces FT-2. Source: Kay and London (1984).

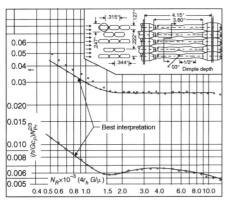

Tube OD before flattening = 0.247 in.
Distance between centers parallel to flow = 0.344 in.
Distance between centers perpendicular to flow = 0.222 in.
Tube dimension parallel to flow = 0.315 in.
Tube dimension perpendicular to flow = 0.127 in.
Distance between spacing plates = 4.15 in.
Length of flat along tube = 3.60 in.
Dimple depth = 0.03 in.
Minimum distance between dimples along tube = 0.5 in.
Flow passage hydraulic diameter, $4r_h$ = 0.016 ft
Total transfer area/total volume, α = 108 ft^2/ft^3
Free-flow area/frontal area, σ = 0.423

(d)

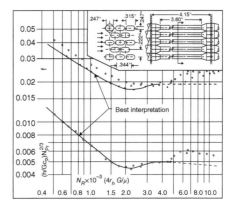

Tube OD before flattening = 0.247 in.
Distance between centers parallel to flow = 0.344 in.
Distance between centers perpendicular to flow = 0.222 in.
Tube dimension parallel to flow = 0.315 in.
Tube dimension perpendicular to flow = 0.127 in.
Distance between spacing plates = 4.15 in.
Length of flat along tube = 3.60 in.
Flow passage hydraulic diameter, $4r_h$ = 0.01433 ft
Total transfer area/total volume, α = 108 ft^2/ft^3
Free-flow area/frontal area, σ = 0.386

(e)

Figure 5.35 (*Continued*)

In these figures, the lower plot is for heat-transfer coefficient in terms of St Pr $^{2/3}$ = $\frac{h}{Gc_p}$ Pr $^{2/3}$ as a function of as function of flow Reynolds number, $N_R = \frac{G4r_h}{\mu}$. The second upper plot is for friction factor, **f**, data as function of flow Reynolds number, $N_R = \frac{G4r_h}{\mu}$. Major parameters for the specific heat-transfer surface are also included below the plot. For a single-point computation with known flow rates and geometry, one can simply read off the data from the chart. However, for design problems involving iterations with flow rates and geometrical parameters, it will be more convenient to develop and use a curve-fitted expression of the correlation using the techniques described in the Appendix A: Parametric Representation.

5.6.1 Algorithm for Compact heat-exchanger Design and Analysis

The computational algorithms for compact heat exchangers are like the algorithms described for the design of shell-and-tube heat exchangers, with the exception of the computation of the convection heat-transfer coefficients for both fluids. Figure 5.36 demonstrates a flowchart for a crossflow compact heat exchanger where air flows normal to flattened tubes with surface FT-2 shown in Figure 5.35e and water flows through the flattened tubes. This is a design Type-1 problem where mass flow rates and the inlet temperatures of both fluids along with the geometrical dimensions, surface types, and parameters on both fluid sides are known. It is required to determine the heat-exchanger ratings in terms of heat-transfer rate, exit fluid temperatures, and pressure drops.

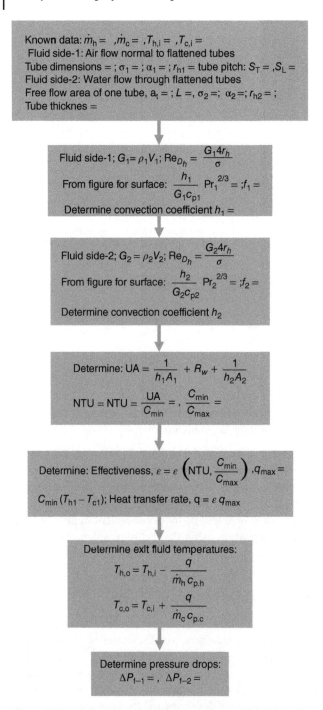

Known data: $\dot{m}_h =$, $\dot{m}_c =$, $T_{h,i} =$, $T_{c,i} =$
Fluid side-1: Air flow normal to flattened tubes
Tube dimensions = ; $\sigma_1 =$; $\alpha_1 =$; $r_{h1} =$ tube pitch: $S_T =$, $S_L =$
Fluid side-2: Water flow through flattened tubes
Free flow area of one tube, $a_t =$; $L =$, $\sigma_2 =$; $\alpha_2 =$; $r_{h2} =$;
Tube thicknes =

Fluid side-1; $G_1 = \rho_1 V_1$; $Re_{D_h} = \dfrac{G_1 4 r_h}{\sigma}$
From figure for surface: $\dfrac{h_1}{G_1 c_{p1}} Pr_1^{2/3} =$; $f_1 =$
Determine convection coefficient $h_1 =$

Fluid side-2; $G_2 = \rho_2 V_2$; $Re_{D_h} = \dfrac{G_2 4 r_h}{\sigma}$
From figure for surface: $\dfrac{h_2}{G_2 c_{p2}} Pr_2^{2/3} =$; $f_2 =$
Determine convection coefficient h_2

Determine: $UA = \dfrac{1}{h_1 A_1} + R_w + \dfrac{1}{h_2 A_2}$
$NTU = NTU = \dfrac{UA}{C_{min}} =$, $\dfrac{C_{min}}{C_{max}} =$

Determine: Effectiveness, $\varepsilon = \varepsilon \left(NTU, \dfrac{C_{min}}{C_{max}} \right)$, $q_{max} =$
$C_{min} (T_{h1} - T_{c1})$; Heat transfer rate, $q = \varepsilon\, q_{max}$

Determine exit fluid temperatures:
$T_{h,o} = T_{h,i} - \dfrac{q}{\dot{m}_h c_{p.h}}$
$T_{c,o} = T_{c,i} + \dfrac{q}{\dot{m}_c c_{p.c}}$

Determine pressure drops:
$\Delta P_{f-1} =$, $\Delta P_{f-2} =$

Figure 5.36 Flow chart compact heat exchanger for design Type-I problems.

5.7 Heat-exchanger Network (HEN) Analysis

In many process and utility plants, several heat exchangers are arranged and connected in a series or in a parallel network while transferring heat from a hot fluid stream to a cold fluid stream for the purpose of heat recovery or for heating and cooling purposes. In many such applications, the key objective is to achieve an optimized design of the heat-exchanger network (HEN) in terms of minimum overall surface area for heat transfer, minimum number of heat exchangers units, and energy savings as well as for minimum total cost – capital and operating costs. The minimum capital cost is achieved based on minimum heat-transfer surface area and minimum number of heat exchangers in the network. The major component of operating cost is the pumping power, which is influenced by the variation in pressures drop based on the flow rates in the network. Figure 5.37 shows a few examples of such HEN in process plants.

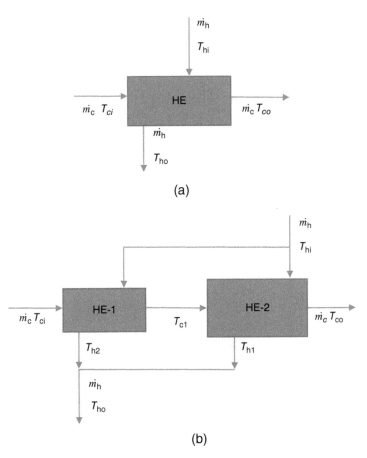

(a)

(b)

Figure 5.37 Single heat exchanger and heat exchangers in a network. (a) Single heat exchanger, (b) Two heat exchangers in the network: Hot stream in series and cold stream is split into two parallel streams.

5.7.1 Basic Analysis Process for HEN

In Figure 5.37, two options are shown: (i) use of one heat exchanger and (ii) use of two heat exchangers where the cold fluid stream is passing through the heat exchanger in series while the hot fluid stream is split into two streams that pass through the two heat exchangers is parallel. The objective is to analyze and confirm if use of two or more heat exchangers could improve the design in terms of performance and cost. One of the key parameters for this HEN design is the net temperature difference between the hot and cold stream, $\Delta T = (T_h - T_c)$, as the two streams pass through the network. From thermodynamics point of view, the best possible heat-transfer process is subject to minimum temperature difference, ΔT_{min}, possible in the heat exchanger. The optimum point for minimum total cost, sum of capital cost, and operating energy cost corresponds to an optimum ΔT_{min}. As the cold stream enters the second heat exchangers at an elevated temperature, the LMTD or the effectiveness for the second heat exchanger is expected to be higher than that for the first heat exchanger, and hence it will result in larger area for the second heat exchanger than the first one. What one needs to see if the sum of the two heat-exchangers areas results in a net increase or decrease in cost compared to the cost of the single heat exchanger. Let us demonstrate this analysis process using an example given below.

Example 5.8 A cold water stream with mass flow rate 1.5 kg/s is to be heated from 25 to 80 °C by using a hot fluid stream with a decrease in temperature from 160 to 90 °C. Estimate the net heat exchanger area using the two options shown in Figure 5.37. As a first approximation, let us assume a constant overall heat-transfer coefficient and consider as $U = 800 \, \text{W/m}^2 \, \text{C}$.

Option-1: One Single Heat exchanger
   Cold Fluid: Water, Specific Heat, $c_{ph} = 4.180 \, \text{kJ/kg C}$

$$C_c = \dot{m}_c \, c_{pc} = 1.5 \times 4.180 = 6.27 \, \text{kJ/s C}$$

Heat-transfer rate, $q = \dot{m}_c \, c_{pc} \, (T_{co} - T_{ci}) = 1.5 \times 4.180 \, (80 - 25) = 344.85 \, \text{kW}$
Hot Fluid: Oil, Specific Heat, $c_{ph} = 2.2 \, \text{kJ/kg C}$
Heat-transfer rate, $q = \dot{m}_h \, c_{ph} \, (T_{hi} - T_{ho})$
Hot fluid mass flow rate,

$$C_h = \dot{m}_h \, c_{ph} = \frac{q}{(T_{hi} - T_{ho})} = \frac{344.85 \, \text{kW}}{(160 - 90)} = 4.9264 \, \text{kJ/s C}$$

$$\dot{m}_h = 4.9264/2.2 = 2.239 \, \text{kg/s}, \; C_h = \dot{m}_h \, c_{ph} = 4.9264 \, \text{kJ/s C}$$

Since $C_h < C_c$, $C_{min} = C_h = 4.9264 \, \text{kJ/s C}$ and $C_{max} = C_c = 6.27 \, \text{kJ/s C}$
Effectiveness based on minimum fluid, $\varepsilon_h = \frac{q}{q_{max}} - \frac{T_{hi} - T_{ho}}{T_{hi} - T_{ci}} = \frac{160 - 90}{160 - 25} = 0.5185$

$$C_r = \frac{C_{min}}{C_{max}} = \frac{4.9264}{6.27} = 0.7857$$

For counter flow double pipe heat exchanger:
From Table 5.1, $\text{NTU} = \frac{1}{C_r - 1} \ln\left(\frac{\varepsilon - 1}{C_r \varepsilon - 1}\right)$

$$\text{NTU} = \frac{1}{0.7857 - 1} \ln\left(\frac{0.5185 - 1}{0.7857 \times 0.5185 - 1}\right) = \frac{1}{-0.2143} \ln\left(\frac{-0.4815}{-0.5926}\right) = 0.9688$$

$$\text{NTU} = 0.9688$$

Heat-exchanger surface area for heat transfer, $A = \text{NTU} \frac{C_{min}}{U} = 0.9688 \times \frac{4926.4 \text{ J/s C}}{800 \text{ W/m}^2 \text{ C}}$

$A = 5.966 \text{ m}^2$

Option-2: Two heat exchangers in the network

For cold fluid stream in series and the hot fluid stream split into two parallel streams, we have

$C_{min} = C_h = \frac{4926.4}{2} = 2.4632 \text{ kJ/s C and } C_{max} = C_c = 6.27 \text{ kJ/s C}$

$C_r = \frac{C_{min}}{C_{max}} = \frac{2.4632}{6.27} = 0.3929$

Assuming an equal value of UA or NTU as the first approximation for the two heat exchangers, we can write the effectives expressions for two heat exchangers as

$\varepsilon_1 = \frac{T_{hi} - T_{h1}}{T_{hi} - T_{cl}} = \frac{160 - T_{h1}}{160 - 25}$

$\varepsilon_2 = \frac{T_{hi} - T_{h2}}{T_{hi} - T_{cl}} = \frac{160 - T_{h2}}{160 - T_{cl}}$

Equating $\varepsilon_1 = \varepsilon_2$, we have

$$\frac{160 - T_{h1}}{160 - 25} = \frac{160 - T_{h2}}{160 - T_{cl}} \tag{E5.8.1}$$

An energy balance for the mixing of two hot streams leads to

$m_{hi}c_h T_{h1} + m_{h2}c_h T_{h2} = \dot{m}_h c_h T_{ho} \text{ Or } T_{h1} + T_{h2} = 2T_{ho} = 2 \times 90$

$$T_{h1} + T_{h2} = 2T_{ho} = 180 \tag{E5.8.2}$$

An energy balance on the second heat exchanger

$6270 (T_{co} - T_{cl}) = 2463.2 (T_{hi} - T_{h2})$

or

$6270 (80 - T_{cl}) = 2463.2 (160 - T_{h2})$

$(80 - T_{cl}) = 0.3929 (160 - T_{h2})$

$80 - T_{cl} = 62.864 - 0.3929 T_{h2}$

$$T_{cl} = 17.136 + 0.3929 T_{h2} \tag{E5.8.3}$$

Equations (E5.8.1), (E5.8.2), and (E5.8.3) are solved for the three unknowns

$T_{h1} =, T_{h2} = \text{and } T_{cl} =$

From the effective expression, we have

$\varepsilon_1 = \frac{T_{hi} - T_{h1}}{T_{hi} - T_{cl}} = \frac{160 - T_{h1}}{160 - 25} =$

From, $NTU = \frac{1}{C_r-1} \ln\left(\frac{\varepsilon-1}{C_r\varepsilon-1}\right)$

$$NTU = \frac{1}{0.3929-1} \ln\left(\frac{0.5185-1}{0.3929 \times 0.5185-1}\right) = \frac{1}{-0.2143} \ln\left(\frac{-0.4815}{-0.5926}\right) = 0.9688$$

$NTU = 0.9688$

Heat-exchanger surface area for heat transfer, $A = NTU\frac{C_{min}}{U} = 0.9688 \times \frac{2463.2 \text{ J/s C}}{800 \text{ W/m}^2 \text{ C}}$

$A_1 = A_2 = \mathbf{5.966 \ m^2}$

This is the area for each of the two heat exchangers.
Total heat-exchangers area is then

$A = 2A_1 = 11.932 \ m^2$

The PINCH Design Method for Heat-exchanger Networks

While a network based with just two heat exchangers is simpler to analyze as demonstrated in the previous example problem, the computational process becomes complicated with more number of heat exchangers in the network. In recent time, more systematic approach and algorithms are developed to simplify the analysis process for more complex and large-scale heat exchange networks. One such method is the PINCH design method Linnhoff and Flower (1978), Linnhoff and Hindmarsh (1983), Umeda et al. (1978, 1979). A PINCH point is defined as the location in the heat exchanger where the temperature difference between the hot and cold streams is the minimum. The PINCH design method for HEN relies on the concept of identifying the presence and location of PINCH in the heat exchanger. More comprehensive description of PINCH analysis method is given in the reference such as Linnhoff and Hindmarsh (1983), Martin and Mato (2008), and Smith (2005). Several software such as HINT and ProSIMPlus include PINCH analysis for heat-exchanger network.

Bibliography

Afgan, N. and Schlunder, E.U. (1974). *Heat Exchangers: Design and Sourcebook*. McGraw-Hill.

ANSI/API Standards, 660-1993, Shell-and-tube Heat Exchangers for General Refinery Service, 5th Ed. American Petroleum Institute.

Standards of the ***Tubular Exchangers Manufacturers Associations*** **(TEMA)**, 9th Ed. 2007.

Bell, K.J. (1981). Delaware method for shell side design. In: *Heat Exchanger: Thermal–Hydraulic Fundamentals and Design* (eds. S. Kakac, A.E. Bergles and F. Mayinger), 581–618. New York: Hemisphere Publishing.

Bergman, T.L., Lavine, A.S., Incropera, F.P., and Dewitt, D.P. (2011). *Fundamentals of Heat and Mass Transfer*, 7e. Wiley.

Cengel, A.Y. and Ghajar, A.J. (2015). *Heat and Mass Transfer: Fundamentals & Applications*, 5e. McGraw-Hill.

Heat Exchanger Institute (2013). *HEI Standards for Shell and TubeHeat Exchangers*, 5e (HEI 129). Heat Exchanger Institute.

Hodge, B.K. and Taylor, R.P. (1999). *Analysis and Design of Energy Systems*, 3e. Prentice-Hall.

Holman, J.P. (2010). *Heat Transfer*, 10e. McGraw-Hill.

Incropera, F.P., Dewitt, D.P., Bergman, T.L., and Lavine, A.S. (2007). *Fundamentals of Heat and Mass Transfer*, 6e. Wiley.

Kays, W.M. and London, A.L. (1984). *Compact Heat Exchangers*, 3e. McGraw-Hill.

Kays, W.M. (n.d.). Loss coefficients from abrupt changes in flow cross section with low Reynolds number flow in single and multiple tube systems. *Trans. ASME* 72: 1067–1074.

Linnhoff, B. and Flower, J.R. (1978). Synthesis of heat exchanger networks II – with various criteria of optimality. *AICHE J* 4: 633–642.

Linnhoff, B. and Hindmarsh, E. (1983). The Pinch design method for heat exchanger networks. *Chem. Eng. Sci.* 38 (5): 745–763.

Martin, A. and Fidel, A.M. (2008). HINT: an educational software for heat exchanger network design and pinch method. *Educ. Chem. Eng.* 3: e6–e14.

Smith, R. (2005). *Chemical Process Design and Integration*. Wiley.

Umeda, T., Itoh, J., and Shiroke, K. (1978). *Chem. Eng. Prog.* 74: 70–76.

Umeda, T., Harada, T., and Shiroke, K. (1979). A thermodynamic approach to the synthesis of heat integration systems in chemical processes. *Comput. Chem. Eng.* 3: 273–282.

Webb, R.L. (1994). *Principles of Enhanced Heat Transfer*. New York: Wiley Inter-science.

Problems

5.1 Following shell-and-tube heat exchanger, size and characteristics are given:

Cold Fluid: Water in Tube side	Hot Fluid: Water in shell side
55 000 lbm/hr	70 000 lbm/h
Inlet Temperature: 60° F	Inlet Temperature: 200° F
Tube Information	**Shell Information**
Tube size: 1 in. , 18 BWG(Base case)	Inside Diameter: 1.8 ft.
Copper tubes	No. of Baffles: 15
No. of tubes: 85	Staggered arrays
Length: 10 ft	$S_L = S_T = 2''$
	No. of tubes in center line: 10

(a) Estimate the ratings of heat exchanger in terms of heat-transfer rate, exit temperature of the tube-side and shell-side fluids, and pressure drops in tube side and shell side. Show first iteration by hand calculations.

(b) Repeat step (a) using computer program to determine the ratings in terms of heat-transfer rate, both fluid exit temperature and pressure drop in tube side and shell.

Repeat step (b) with varying tube sizes ($1/2''$, $3/4''$, 1, and $1\frac{1}{2}$) keeping

A commercial steel pipe with ***pin fins of circular profile*** has the following characteristics:

Pipe dimension:

  ID = 1.75-in. and OD = 2-in.

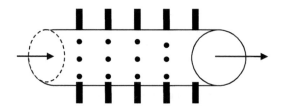

Gas–outside:

Gas temperature: 60 °F, Gas Side Convection Heat-transfer Coefficient: $h_o = 500\,\text{Btu/ft}^2\text{hr}\,°\text{F}$

Gas properties:

$$\rho = 081\,\text{lbm/ft}^3, \mu = 1.165 \times 10^{-5}\,\text{lbm/ft-s}, C_p = 0.24\,\text{Btu/lbm}°\text{F},$$

$$k = 0.0140\,\text{Btu/h ft}°\text{F}, \text{Pr} = 0.70$$

Liquid – (Inside pipe):

Liquid inlet temperature: 150 °F, Liquid outlet temperature: 120 °F

Fins:

Consider **circular pin fins** on outside surface with following dimensions

$$L_f = 0.5 - \text{in.}, d_f = 0.1\text{-in. and } S = 0.1 - \text{in.}$$

Pipe and Fin Material thermal conductivity, $k = 60\,\text{Btu/h ft}°\text{F}$,

(a) Fin efficiency and Fin surface area effectiveness, η_o

(b) Determine the length of the pipe to cool the liquid from 150 to 120 °F.

Show clearly all assumptions and steps for the estimation of the length.

5.2 Design type IIb: With known Heat-exchanger ratings determine size: A single-shell-pass and single-tube-pass heat exchanger is to be designed to meet the following specifications:

Light Oil (Tube Side)	Water Side (Shell Side)
Flow rate = 1.25 ft³/s	Flow rate = 3.0 ft³/s
Inlet temperature = 300.0 °F	Inlet temperature: 150 F
Outlet temperature = 200 °F	
(Use all properties at 250 °F)	**(Use all properties at 200 °F)**
Tube Information	**Shell Information**
Material: Copper	Tube arrangement: Inline
Tube: 1- in. 16 BWG	Tube pitch $\frac{S_T}{d} = 2.0$ and $\frac{S_T}{d} = 2.0$
Average velocity in the tube side:	Baffle spacing,: L_b 10-in.
3.5 ft./s (Assumed)	Shell diameter, D_s: 36-in.
	Number of tubes in the centerline, N_{CLD}:
	(Assume 10 as initial assumption)

Write a computer program implementing the algorithm given for Type IIb class of problems. Show iteration summary in table showing mass velocity, number of tubes, length of tubes, and pressure drops.

5.3 For the given heat-exchanger ratings in terms of heat-transfer rate and pressure drop in the tube side, determine the size in terms of number of tubes, length and number of tube passes using Shah's procedure based on following known data:
Known Data:

Tube Side:	Shell Side:
Water Flow	Steam
Flow rate, Q : 250000 gal/h,	
Mass flow rate, $\dot{m}_t = \dfrac{\rho_{in} \times Q}{3600 \times 7.481}$	
Inlet temperature, $T_{c,i} = 50\,°C$	Condensing Temperature, $T_c = 225\,°C$
Outlet temperature, $T_{c,o} = 100\,°C$	Shell-side heat transfer coefficient, $h_o = 12\,000\ Btu/ft^2 h\,°F$

Tubes: 1 in., Schedule 40
Tube-side pressure drop,

$$\Delta P = 16\ \text{psi} \ (2304\ \text{psf})$$

$$\sigma = \frac{A_f}{A_{fr}} = 0.5$$

5.4 Consider a crossflow compact heat exchanger where air flows normal to flattened tubes with surface FT-2 shown in Figure 5.35e and water flows through the flattened tubes. Air at 25 °C and 1 atm pressure enters at an average velocity of 12 m/s and water at 90 °C and 2 atm pressure at an average velocity of 2 m/s. Determine the UA product per unit volume of the heat exchanger.

6

Analysis and Design of Solar Collector and Solar Thermal System

In this chapter, topics such as solar radiation, radiation optical properties, and analysis and design of solar collector are discussed. For more discussions on solar collector design and solar thermal engineering, reference books such as Duffie and Beckman (2000), Stoecker and Jones (1982), and Kuehn et al. (1998) are recommended.

6.1 Solar Thermal Energy System

The basic purpose of a solar thermal energy system is to collect solar radiation and convert it into useful thermal energy. The system performance depends on several factors, including availability of solar energy, the ambient air temperature, the characteristic of the energy requirement, and especially the thermal characteristics of solar system itself.

As the solar collector intercepts the Sun's energy, a part of this energy is lost as it is absorbed by the cover glass or reflected to the sky. Of the remainder energy absorbed by the collector, *a small portion is lost by convection and re-radiation*, but most is useful thermal energy, which is then transferred via pipes or ducts to a storage mass or directly to the load as required.

An energy storage is usually necessary since the need for energy may not coincide with the time when the solar energy is available. Thermal energy is distributed either directly after collection or from storage to the point of use. The sequence of operation is managed by automatic and/or manual system controls.

6.1.1 Classification of Solar System

The solar collection system for heating and cooling is classified as passive or active.

6.1.1.1 Active System

Active systems consist of components that are to a large extent independent of the building design and often require an auxiliary energy source (pump or fan) for transporting the solar energy collected to its point of use. Active systems are more easily applied to existing buildings.

Design of Thermal Energy Systems, First Edition. Pradip Majumdar.
© 2021 John Wiley & Sons Ltd. Published 2021 by John Wiley & Sons Ltd.
Companion website: www.wiley.com/go/majumdar

6.1.1.2 Passive System

Passive systems collect and distribute solar energy without the use of an auxiliary energy source.

The system critically depends on building design and the thermal characteristics of the material used. In this book, we will primarily discuss active solar system.

6.1.2 Examples of Active Solar Thermal System

6.1.2.1 Solar Water-Heating System

This system uses solar collector mounted on roof top to gather solar radiation involving low temperature in the range of 80–100 °C. Applications involve domestic hot water or swimming-pool-heating systems. Figure 6.1 shows a solar domestic water-heating system in which a flat-plate solar collector is used to heat water. The hot water is then circulated through a coiled heater in a water heater tank for a residential house.

Typical residential water-heating system is either powered by natural gas or by electric power. In a solar water-heating system, an auxiliary heater powered by either natural gas or electric power is also used to sustain hot water supply during morning and evening hours.

6.1.2.2 Solar Space-Heating System

A solar space-heating system can use a liquid collector or an air collector. In a liquid collector, the heat from the collector absorber system is transferred using the liquid circulating in the collector fluid loop. Heated liquid is then circulated through a heat exchanger to transfer the heat to the return air from the space as shown in Figure 6.2a.

In an air collector-based space-heating system, air is the working fluid for collecting heat from the collector absorber plate and transferring heat to circulating space air in the heat exchanger as shown in Figure 6.2b.

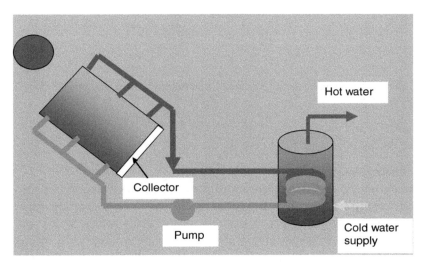

Figure 6.1   Solar water-heating system.

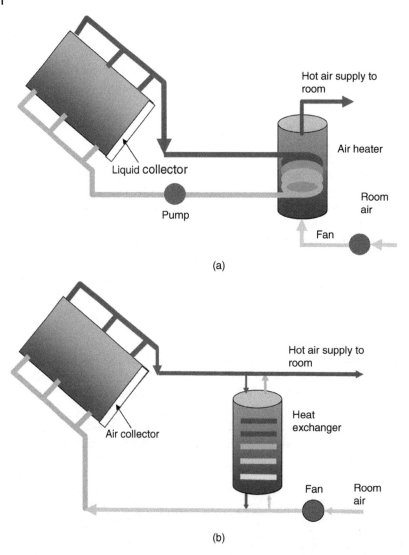

Figure 6.2 (a) Solar system for room air heating using liquid collector. (b) Solar system for room air heating using air collector.

6.1.2.3 Solar-Cooling System

Vapor compression refrigeration systems are normally powered by electric power. A potential use of solar energy is in the solar-driven Rankine power cycle to drive the vapor compression refrigeration systems. Figure 6.3 shows a solar-cooling system in which solar thermal collectors are used to run a Rankine cycle and power a vapor-compression-cooling system. Collector fluid exit temperature in the range of 150–400 °C may be used to run a vapor compression refrigeration chiller with a total cooling load in the range of 5–25 kW for residential air-conditioning and for industrial chillers.

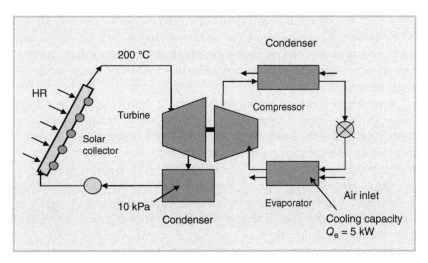

Figure 6.3 Solar-powered cooling system.

In this system, a natural-gas-powered auxiliary heater can also be used in series with solar collectors to sustain the operation during morning and evening hours.

6.1.2.4 A Solar-Driven Irrigation Pump

A solar-energy-driven irrigation pump operating on a solar-driven heat engine can be analyzed and designed for the distribution of water in the irrigation field (Figure 6.4). Solar thermal collector is used to generate high temperature and pressure vapor to produce power output in the turbine and run the irrigation pump.

6.1.2.5 Solar Rankine Cycle Power Generation

A solar-driven Rankine vapor power-generating system is shown in Figure 6.5. Power generation in a relatively low range of 100 kW to 1 MW can be used in concentrating collectors

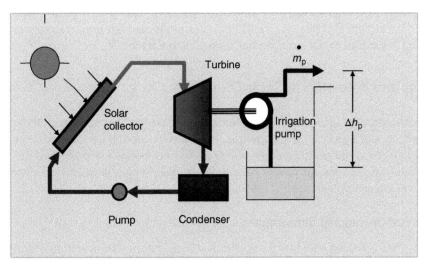

Figure 6.4 Solar-driven irrigation pump.

like concentrating parabolic trough collector (PTC) or compound parabolic concentrator (CPC) collector. Collector fluid is heated to high temperature in range of 300–500 °C and used to produce high temperature and pressure superheated steam to run a steam turbine and produce power. In addition to the condenser, additional feed water heaters could be used for enhanced thermal efficiency.

In Figure 6.5, a **standard Rankine cycle power system** is used to collect heat from a high-temperature parabolic concentrating trough system to run a Rankine cycle power generation system. As we have discussed before, efficiency of a Rankine cycle power generation system increases with the increase in temperature of the vapor at inlet to the turbine. A concentrating-type solar collector can generate high fluid temperature in ranges of 200–400 °C appropriate for generating high-pressure steam or vapor for running turbines. The design of a concentrating solar collector is, however, significantly more complex as compared to a flat-plate collector. Types, design, and characterization of concentrating type collectors will be presented in Section 6.2.5.

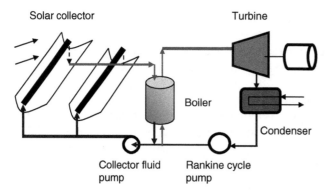

Figure 6.5 Solar Rankine cycle power generation.

The collector side fluid is circulated using a pump and heated in the collector to a high-temperature level. The high-temperature collector fluid is circulated through the heat exchanger/boiler to heat the Rankine cycle fluid and produces high temperature and pressure steam to produce power by expanding through a steam turbine.

6.2 Types and Selection of Solar Collectors

Solar collector absorbs the incoming solar radiation, converts it into thermal energy at the absorbing surface, and transfers the energy to a fluid flowing through the collector. The collector fluid may be liquid or air. Several types of collector are available and are classified based on operational temperature; type of collector fluid: liquid or air; fixed and tracking; and flat plate vs. concentrating.

6.2.1 Collector Operational Temperatures

Solar collectors are classified based on the operational temperature. The operational temperature ranges of solar collectors and operations can be categorized as **low temperature** $(T < 100\,°C)$, **mid temperature** $(100\,°C < T < 400\,°C)$, and **high temperature** $(T > 400\,°C)$.

6.2.2 Fixed vs. Tracking

A *tracking collector* is controlled to follow the Sun throughout the day. A tracking system is rather complicated and generally only used for special high-temperature applications. Fixed collectors are much simpler – their position or orientation, however, may be adjusted on a seasonal basis. They remain fixed over a day's time. Fixed collectors are less efficient than tracking collectors. Nevertheless, they are generally preferred as they are less costly to buy and maintain.

Tracking type is further classified as ***Single-Axis Tracking*** and ***Double-Axis Tracking***. Tracking also can be daily or weekly or sessional or continuous tracking.

6.2.3 Types of Collector Design: Flat Plate vs. Concentrating

A basic flat-plate collector uses plate-and-tube design: flat absorber plate with parallel tubes on the back or top of the plate for collection of solar thermal radiation and transferring it to the circulating collector fluid. Concentrating collectors use mirrored surfaces or lenses to focus the collected solar energy on smaller areas to obtain higher working temperatures and classified as imaging and nonimaging collectors.

Flat-plate collectors may be used for water heating and most space-heating applications. Concentrating collectors are generally required for power generation. Concentrating collector and high-performance flat plate are required for cooling applications since higher temperatures are needed to drive chiller or absorption-type cooling units.

Other variations of flat plate, concentrating collector, and hybrid designs are *(i) the evacuated tube absorber and (ii) compound parabolic concentrator (CPC).* The evacuated tube absorber designs involve mounting the absorbers inside evacuated glass tubes. Evacuation of the tubes eliminates convection and conduction losses in the space between the absorber and the cover. CPC collectors provide the highest possible concentration for any angle of acceptance and can function without the need of any tracking by achieving some level of concentration. CPC systems operate at high- and mid-temperatures.

6.2.4 Flat-Plate Solar Collector

Flat-plate collectors are relatively low-temperature collectors ($T < 100\,°C$) and are widely used in domestic household hot-water heating and for space heating, where the demand of temperature is low. Flat-plate collector are classified as glazed or unglazed and liquid-based or air-based.

The shape of collector highly influences the efficiency of the flat-plate solar thermal collector. Large surface areas can be obtained by various arrangements and shapes of tubes and by making the tubes integral part of the absorber plate by riveting plate segments in regular intervals. If the surface area of the collector is increased, more area of the collector is exposed to the radiation of Sun. The efficiency of the flat-plate collector is increased by optimizing the surface area of the collector and altering its geometry within the same space of traditional flat-plate collector.

Figure 6.6 shows a basic flat-plate collector that uses plate-and-tube design: flat absorber plate with parallel tubes on the back (Figure 6.6a) of the plate or top (Figure 6.6b) of the

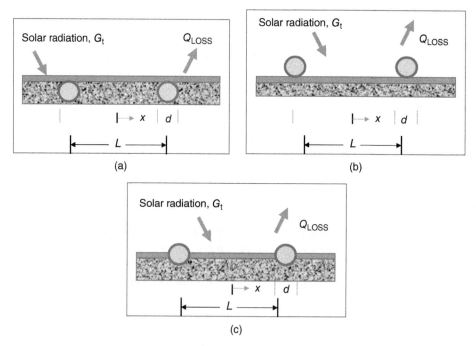

Figure 6.6 Absorber plate with plate-and-tube design in flat-plate collector. (a) Absorber plate with tubes at the bottom. (b) Absorber plate with tube at the top. (c) Absorber plate with tubes as an integral part.

plate for collection of solar thermal radiation and transferring it to the circulating collector fluid.

Figure 6.6c shows an absorber-plate design with the tubes as an integral part of the absorber plate and provides best possible absorber area that is exposed to solar radiation.

The performance of solar flat-plate collector can also be increased by applying the appropriate selective coatings on its surface. Selective coatings increase the energy-absorptive property of the collector absorber plate and at the same time decrease the emissivity of the collector. The radiation emitted by the selective surface is then reflected by using the proper glazing cover.

Figure 6.7 shows a typical constructional design of a flat-plate collector. Major components of a flat-plate design are **an absorber plate, cover glass, insulation**, and **housing.**

The *absorber plate* is usually made of copper and coated to increase the absorption of solar radiation. The *cover glass or glasses* are used to reduce convection and re-radiation losses from the absorber. Insulation is used on the back side and edges of the absorber plate to reduce conduction heat losses. The housing holds the absorber with insulation on the back and edges, and cover plates.

The working fluid (water, ethylene glycol, air, etc.) is circulated in a straight parallel or serpentine fashion through the tubes on the back side of the absorber plate to carry the solar energy to its point of use. The tubes are connected at the top and bottom of the collector to supply and return manifolds to supply and remove the circulating fluid as shown in Figure 6.6a,b.

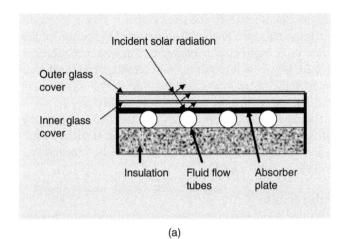

(a)

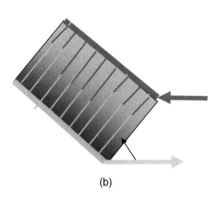

(b)

Figure 6.7 Basic plate-tube flat solar collector design. (a) Sectional view of plate collector. (b) Collector fluid flow in tubes and supply and return manifolds.

Applications of flat-plate collector include water heating, building heating and air-conditioning, and industrial process heating. Flat-plate collectors are usually used for moderate temperatures up to 100 °C. Some of the key features of a flat-plate collector are (i) uses both direct and diffuse radiation; (ii) normally does not need tracking of Sun; (iii) mechanically simple.

6.2.5 Concentrating Collector

Concentrating solar collectors provide high-temperature absorber surface by focusing solar radiation on a smaller surface by either reflecting or refracting incident solar radiation from the primary original surface of surface area, A_o, to a smaller absorber surface of area, A_c. Concentration ratio or concentration factor is one of the major parameters of the concentrating collector, and this is defined as the ratio of the area of the incident surface or the collector aperture area to the area of the absorber surface and written as

$$C_r = \frac{A_o}{A_c} \tag{6.1}$$

The concentration factor defines the extent to which the solar radiation flux is increased over the absorber surface and corresponds to the achievable operating temperature of the collector. Higher the concentration ratio, higher is the temperature at which heat energy can be supplied to the working fluid. However, it brings in more complexity in the design of the optical system of the collector. Achievable concentration ratio can vary from a limiting value for a flat-plate collector to thousands as in a parabolic dish-type collector.

There are different types of concentrating collectors, which are classified based on the type of the concentrating and refracting surface as well as type of the absorber surface used. For example, the concentrating and the reflecting surface could be parabolic or cylindrical and absorber surface can be concave, convex, or flat.

6.2.5.1 Classification Concentrating Collector

Concentrating collectors are classified as focusing and nonfocusing. Focusing types are further classified as line focusing with cylindrical absorber through which the collector fluid flows as in cylindrical parabolic concentrator and point focusing as in parabolic dish. Cylindrical parabolic concentrator is a conventional imaging type of collector and may have designs with concentration ratio in the high range of 10–100 and reaching collector temperature as high as 400 °C. Parabolic dish collectors (PDCs) are designed to achieve very high concentration of the order of thousand and collector temperature as high as 600 °C.

A few examples of concentrating collectors are shown in Figure 6.8.

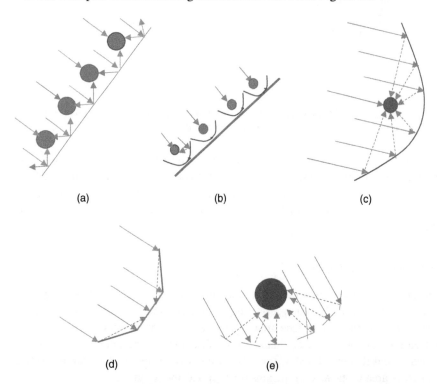

(a) (b) (c)

(d) (e)

Figure 6.8 Concentrating solar collectors. (a) Cylindrical absorber tubes with diffuse back-flat reflector. (b) Cylindrical absorber tubes with individual parabolic reflector. (c) Cylindrical absorber with parabolic concentrator. (d) Plane absorber plate with plane reflector. (e) Cylindrical central receiver with an array of heliostat reflector.

Theoretical upper and lower limits of the concentration ratio for both circular and linear concentrators have been derived based on second law analysis and radiation exchange between the Sun and the absorber (Duffie and Beckman 2000). The lower limits correspond to the condition when absorbed solar radiation equals the thermal losses from the collector. The upper limit corresponds to the condition when absorber temperature approaches the Sun's temperature and represents the useful solar energy gain. Potential solar collector efficiency range is achievable based on range of concentration ratio in between these two limits. The concentration ratio may vary in the range of 10–100 for a target collector absorber temperature in the range of 200–400 °C.

In addition to the optical parameters such as the transmittance (τ) and absorptance (α), another important parameter of concentrating collector is the **intercept factor (γ)**, which is defined as the fraction of the reflected radiation that is incident on absorbing receiver surface. While one objective is to reduce the heat losses by using an absorber with low surface area absorber, a large enough receiver is used to intercept large fraction of the reflected radiation. An interceptor factor (γ) value of 0.9 or higher is preferred in a high-performance collector.

All three parameters τ, α, and γ are a function of the angle of incidence of solar radiation. While individual functional dependence of each one of these parameters on the angle of incidence can be taken into account, a simpler approach involves using a parameter known as **incidence angle modifier** ($\kappa_{\gamma\tau\alpha}$) that uses a combined effect.

The absorbed radiation per unit area of the unshaded aperture of the concentrating collectors is given as

$$q_s'' = \kappa_{\gamma\tau\alpha}(\gamma\tau\alpha)\rho G_t \tag{6.2}$$

where

τ = Transmittance
α = Absorptance
γ = Intercept factor
$\kappa_{\gamma\tau\alpha}$ = Incidence angle modifier

6.2.6 Compound Parabolic Concentrator (CPC) Collector

CPC collector is a nonimaging high-performance collector designed based on optical principles (Winston 1974; Rabl 1976a,b). This is a more advanced collector that is designed to avoid the need for any tracking of the Sun. Two parabolas are combined to form cofocal line over the absorber surface from which the concentrated solar energy is transferred to the circulating collector fluid through a tube or tubes. Such collector just requires seasonal adjustment of collector orientation to keep the collector aperture normal to the Sun at noon, and does not need daily or continuous tracking.

CPC collector can provide high level of concentration ratio by reflecting all the incident radiation over the aperture to receiver surface area for a wide range of incidence angles, defined as the acceptance angle, θ_a. Typical concentration ratio for a CPC collector is around 2–10, which is significantly less than that for PTC and PDC. Some advantages of CPC collectors are the capability to collect diffuse radiation, satisfactory performance for a cloudy day, and no need for continuous tracking.

The basic construction of a CPC collector is shown in Figure 6.9. Geometrically, it is composed of two truncated parabolas on each side and a receiver plate in between two parabolas.

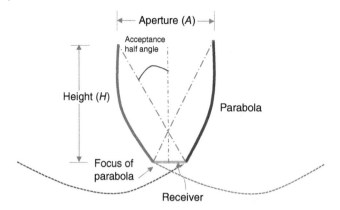

Figure 6.9 Compound parabolic concentrator (CPC).

The two end points of the receiver correspond to the focal points of the two adjacent parabolas. For a perfect specular-reflecting parabola, all solar radiation that enters through the aperture area at an acceptance angle of $2\theta_a$ will be reflected to the receiver plate.

For an ideal two-dimensional CPC collector, the functional relationship between the concentration ratio and the acceptance angle is given as

$$C_r = \frac{1}{\sin \theta_a} \tag{6.3}$$

The acceptance angle varies with the selection of the height-to-aperture ratio, and hence causes variation in the concentration ratio. Table 6.1 shows a variation concentration ratio as a function of the acceptance angle.

The fraction of solar radiation (f_{ar}) incident on an aperture area that reached the absorber surface varies with the angle of incidence, θ. For an ideal CPC with perfect reflector surface, $f_{ar} = 1$ for $\theta = 0$ to $\theta = \theta_a$ and falls to $f_{ar} = 0$ for $\theta > \theta_a$ as demonstrated in Figure 6.10. For a real reflector surface, the function variation of the fraction deviates from the step change

Table 6.1 Concentration ratio as a function of the acceptance angle.

Acceptance angle, θ_a (°C)	Concentration ratio, C_r
5	11.474
10	5.759
15	3.864
20	2.924
25	2.366
30	2.000
45	1.414
60	1.155
75	1.035
90	1.000

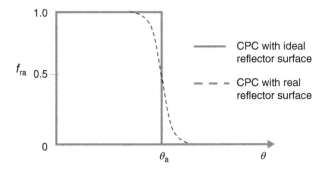

Figure 6.10 Variation of the fraction of radiation that reaches receiver as a function of angle of incident of solar radiation on aperture.

in value around $\theta = \theta_a$, i.e. a fraction of radiation may reach the absorber surface for $\theta > \theta_a$ and a fraction of radiation may not reach the absorber surface for $\theta > \theta_a$.

6.2.6.1 Truncated CPC Collector

One alternative design of the basic CPC collector is the **truncated CPC collector** in which the height of the reflector is reduced to eliminate top vertical part or the less effective area of the reflector without significant reduction in performance. Figure 6.11 shows a typical CPC collector with linear flat absorber plate.

The concentration ratio and the average number of reflections that the incident radiation undergoes before reaching the absorber surface for both full CPC and truncated CPC is affected by the geometrical parameters such as the **Height–Aperture Ratio** ($\frac{H}{A}$ for full CPC and $\frac{h}{a}$ for truncated CPC) and **Reflector Area–Aperture Area Ratio** ($\frac{A_r}{A_a}$), **and the Acceptance Angle** (θ_a).

Ray-tracing technique (Wagmare and Gulhane 2016; Santos-Gonzalez et al. 2014) is often used in designing CPC collector by estimating the optical losses considering the behavior

Figure 6.11 Truncated CPC collector.

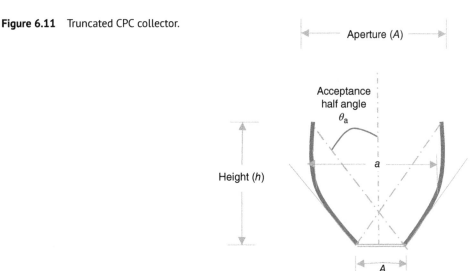

of radiation that falls on collector reflector surface and undergoes several reflections before reaching the absorber surface. Ray-tracing technique is essentially used to estimate the optical losses for different reflector–receiver shapes and orientations.

Basic equations and procedure for the calculation of the concentration ratio and the average number of reflections are presented by Rabl (1976a,b). A number of charts are presented in Rabl (1976a,b) and in Duffie and Beckman (2000) for convenient use by the designer of full and truncated CPC collectors.

Rabl (1976a,b) calculated convective and radiative heat transfer in a CPC collector and presented formulas for evaluating the performance of CPC collector. The procedure included an estimation of optical losses based on calculating the average number of reflections of radiation collected by the collector.

Typical concentration ratio (C_r) may vary in the range of 1–11 as the geometrical parameters vary in the range: $\theta_a = 5\text{--}35°$, $\frac{H}{A} = 0\text{--}6$, and $\frac{A_r}{A_a} = 0\text{--}12$. The typical number of reflections is shown to be in the range of 0–1.5 for these ranges of parameters. For example, for a concentration ratio of $C_r = 4$, a truncated CPC collector can be designed with $\theta_a = $ for $12°$, $\frac{h}{a} = 2.5$, and $\frac{A_r}{A_a} = 1.25$. Higher concentration ratio and higher receiver

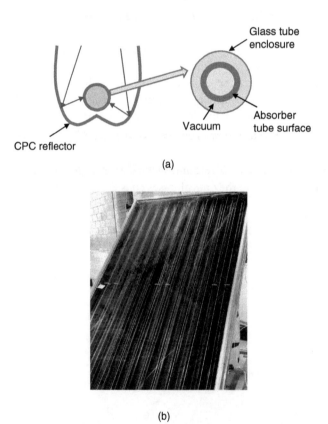

(a)

(b)

Figure 6.12 Truncated CPC collector with evacuated tubular receiver. (a) CPC reflector with evacuated tubular receiver tube design. (b) Solar collector with an array of truncated CPC reflector with evacuated tubular receivers.

temperature can be achieved with the selection of lower **acceptance angle** and higher values of **height–aperture ratio** and **reflector area–aperture area ratio.**

Other variations of the truncated CPC collector with flat receiving surface are the use of truncated CPC collector with tubular receiver as shown in Figure 6.12a and the use of an array of truncated CPC reflectors with evacuated tubular receivers with glass cover over the array as shown in Figure 6.12b. This design variation originated from Winston's CPC design (Rabl et al. 1977, 1979).

6.3 Solar Radiation Characteristics and Estimation

6.3.1 Solar Radiation

Intensity of solar radiation incident on a surface is important in the design of solar collectors, photovoltaic cells, solar-heating and -cooling systems, and thermal management of building. This effect depends on both the location of the Sun in the sky and *the clearness of the atmosphere* as well as on the nature and orientation of the building. We need to know *characteristics of Sun's energy* outside the Earth's atmosphere, its intensity and its spectral distribution, and the variation with Sun's location in the sky during the day and with seasons for various locations on the Earth's surface. The Sun's structure and characteristics determine the nature of the energy it radiates into space. Energy is released due to continuous fusion reaction with interior at a temperature of the order of million degrees. Sun is composed of multiple regions: (i) **Central Region** (Region-I); (ii) Middle region (Region-II) or the **Photosphere** representing the upper layer of the convective zone, composed of strongly ionized gas and capable of absorbing and emitting continuous spectrum of radiation. This region is the source of most solar radiation; (iii) **Chromosphere** – the outer gaseous layer with temperature somewhat higher than the Photosphere and spread over 10 000 km; (iv) **Corona – the** layer consisting of highly rarified gases at temperature as high as 1 000 000 °C, representing the extremity of Sun; (v) **Convection Region (Region-III)** – the outermost region where convection process is involved and the temperature drops to 5000 K. The Sun is represented by the effective black body temperature of 5777°K, which is designated as the representative temperature of a black body radiating the same amount of energy as emitted by the Sun.

6.3.2 Thermal Radiation

Thermal radiation is the intermediate portion (0.1–100 μm) of the electromagnetic radiation spectrum emitted by a substance as a result of its temperature. Thermal radiation heat transfer involves transmission and exchange of electromagnetic waves or photon particles as a result of temperature difference.

6.3.3 Solar Intensity Distribution

Figure 6.13 shows spectral distribution of solar radiation over the bandwidth at outer atmosphere in the wavelength range of 0.25–3.0 μm.

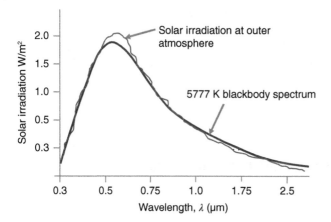

Figure 6.13 Spectral distribution of solar radiation at outer atmosphere.

6.3.4 Extraterrestrial Radiation

Extraterrestrial radiation is the solar radiation that would be received in the absence of the Earth atmosphere. Extraterrestrial solar radiation exhibits a spectral distribution over a range of wavelengths: 0.1–2.5 µm that includes part of ultraviolet, visible, and part of near-infrared radiation.

6.3.5 Solar Constant (G_{sc})

Solar constant is defined as the solar radiation intensity upon a surface normal to Sun ray and at outer atmosphere when the Earth is at its mean distance from the Sun.

$$G_{sc} = 1367 \text{ W/m}^2 = 433 \text{ Btu/ft}^2 \text{ h}$$

This solar radiation amount is simply the total area under the spectral solar radiation distribution shown in Figure 6.13. Extraterrestrial solar radiation varies with the day of the year as the Sun Earth distance varies. It is also referred to as the direct radiation as it strikes surface. An empirical fit of the measured direct solar radiation data is given

$$G_D = G_{sc} \left(1 + 0.033 \ \cos \frac{360n}{365} \right) \tag{6.4}$$

where n represents the day of the year.

An alternate form is given as

$$G_D = G_{sc} \left(\begin{array}{l} 1.000\,110 + 0.034\,221 \ \cos \ B + 0.001\,280 \ \sin \ B \\ + 0.000\,719 \ \cos \ 2B + 0.000\,077 \ \sin \ 2B \end{array} \right) \tag{6.5a}$$

where

$$B = (n-1)\frac{360}{365} \tag{6.5b}$$

As solar energy enters the Earth's atmosphere, the intensity changes due to three different effects: transmission, reflection, and absorption by different elements of the atmosphere. The solar radius spectrum reduces and varies at different depths of the Earth atmosphere due to the effects of absorption and scattering along the length of travel defined by **air mass**. The unit of air mass is defined as the ratio of the length of travel of Sun's ray through the Earth's atmosphere to the length of travel when Sun's position is directly of the position. For example, **air mass one** spectrum curve corresponds to the position of the Sun directly above or at zero zenith angle and the Sun's ray travels a distance of exactly the one-depth of the atmosphere. Extraterrestrial solar radiation spectrum corresponds to **air mass zero** that represents absence of the Earth's atmosphere.

Such spectral reduction of solar radiation is derived based on the **monochromatic extinction coefficient**, K_λ, and the **monochromatic transmittance**, τ_λ, is derived as

$$\tau_\lambda = \exp\left[-\frac{1}{\sin\beta}\int_0^Z K_\lambda(z)dz\right] \tag{6.6}$$

A simple functional relation of air mass with zenith angle (ψ) is given applicable in the range of 0–70° as

$$m = \frac{1}{\psi} \tag{6.7}$$

The solar radiation spectrum shows considerable fluctuations due to this variation in absorption by different element gases of the atmosphere such as H_2O, CO_2, O_3, and O_2 at different wavelength values of the radiation spectrum and by different amounts.

6.3.6 Total Incident Radiation

The total radiation that strikes a surface has three components: (i) direct radiation, (ii) diffuse radiation, and (iii) reflected radiation as demonstrated in Figure 6.14. As solar radiation passes through the Earth atmosphere, a fraction of the radiation passes through the atmosphere and strikes the Earth surface as unaltered and this is referred to as **the direct radiation (G_D)**. Another fraction of the radiation is absorbed and scattered by the atmospheric air and clouds in all directions and also strikes the Earth surface in a reduced amount, and this is referred to as the **diffuse radiation (G_d)**. In addition, there exists a third component that originates from the reflection of total solar radiation from nearby surfaces. This is referred to as the **reflected radiation (G_r)**.

Total irradiation of a surface is given as

$$G_t = G_D + G_d + G_r \tag{6.8}$$

The direct radiation on a surface of any arbitrary orientation is computed from the direct normal radiation on a surface oriented directly normal to the Sun as demonstrated in Figure 6.14.

Figure 6.14 Components of solar radiation.

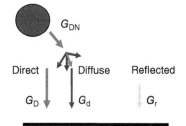

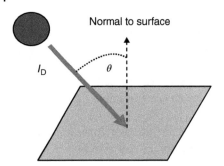

Figure 6.15 Direct solar radiation on a surface of any arbitrary orientation.

The relationship between the direct radiation and direct normal radiation is given as

$$G_{D\theta} = G_{DN} \cos \theta \qquad (6.9)$$

where θ is the incidence angle of Sun's ray as it strikes the Earth surface. It is the defined as the angle between the normal to the surface and the direction of Sun's ray as demonstrated in Figure 6.15.

Incidence angle, θ, or orientation of a surface on Earth with respect to the Sun or normal to Sun's ray can be determined in terms of basic Earth and Sun angles. Before we proceed to give a brief description of the basic Earth and Sun angles, it is also important to understand the relationship between the actual clock time and the solar time as all solar radiation computations are done based on solar time.

6.3.7 Computation of Solar Time

Clock or solar time is considered to begin at zero longitudinal passing through Greenwich, England. The Earth is divided into 360° circular arc by longitudinal lines passing through poles. The zero longitudinal line passes through Greenwich, England. Since the Earth takes 24 hours to complete rotation, 1 hour of clock time corresponds to 15° of longitude. This means that a point on Earth surface exactly 15° west of another point will see the Sun in exactly the same position after one hour.

6.3.8 Greenwich Civil Time (LCT)

Greenwich Civil Time: GCT time or universal time.

This is time along the zero longitudinal line that passes through Greenwich, England. Time starts from midnight at the Greenwich.

6.3.9 Local Civil Time (LCT)

Local Civil Time (LCT) is determined by the longitudinal position of the observer and considering that difference is being four minutes of time for each degree or one hour for 15° difference in longitudinal position from zero longitudinal line.

Example 6.1   *Computation of Local Civil Time*
Compute the LCT at 75° west longitude corresponding to 12:00 noon at GCT.

Solution:
Since 75° longitudinal position corresponds to 75°/15° = 5 hours of time difference from Greenwich, the zero longitudinal position or time (GCT). The LCT at 75° west longitude is then five hours less advanced in clock-time from GCT and is 7 a.m. as local time.

6.3.10   Local Standard Time

The local standard time is the LCT for a selected meridian near the center of the zone. Clocks are usually set for the same time throughout a time zone, covering approximately 15° of longitudinal range. For example, in United States, different standard times are set over the different time zones based on the meridian of the zone. Following is a list of meridian line in United States: Eastern Standard Time (EST): 75°; Central Standard Time (CST): 90°; Mountain Standard Time (MST): 105°; and Pacific Standard Time (PST): 120°. Also, there is Day Light Savings Time (DST) in the United States, which corresponds to the time that is set by forwarding the clock time by an hour in each spring and setting it back again in each fall.

6.3.11   Local Solar Time (LST)

Local Solar Time (LST) is estimated by adjusting LCT by taking into account the apparent daily motion of the Sun across the sky with solar noon being the time the Sun crosses the meridian of the observer. The solar time does not coincide with the local clock time. It is necessary to convert standard time to solar time by applying two corrections. *First*, there is a constant correction for the difference in longitude between the observer's meridian and the meridian on which the local time is based. The Sun takes four minutes to transverse 1° of longitude. The ***second correction*** is from the ***equation of time***, which takes into account of the variation in the Earth's rate of rotation, which affects the time the Sun crosses the observer's meridian. The relationship between the difference in the solar time and standard time is then expressed as

$$\text{Local Solar Time, LST} = \text{LCT} + \text{Equation of time } (E) \tag{6.10}$$

where ***Equation of time*** (E) gives the difference between the LST and the LCT by considering nonsymmetry of the earthly orbit, irregularity of earthly rotational speed, and other factors. While values of Equation of time are for all days of the year, Table 6.2 lists only values of Equation of Time for the 1st, 8th, 15th, and 22nd days of each month. An empirical formula for equation of time as a function of the day of the year is given as follows:

$$E = 229.2 \, (0.000\,075 + 0.001\,868 \cos B - 0.032\,077 \sin B$$
$$- 0.014\,615 \cos 2B - 0.040\,89 \sin 2B) \tag{6.11}$$

Table 6.2 Equation of time € and Sun's declination angle (*d*).

Month	Day-1 Dec. (*d*) Deg:Min	Day-1 Eq. of Time Min:Sec	Day-8 Dec. (*d*) Deg:Min	Day-8 Eq. of Time Min:Sec	Day-15 Dec. (*d*) Deg:Min	Day-15 Eq. of Time Min:Sec	Day-22 Dec. (*d*) Deg:Min	Day-22 Eq. of Time Min:Sec
January	−23:08	−3:16	−22:20	−6:26	−21:15	−9:12	−19:50	−11:27
February	−17:18	−13:34	−15:13	−14:14	−12:55	−14:15	−10:27	−13:41
March	−7:51	−12:36	−5:10	−11:04	−2:25	−9:14	0:21	−7:12
April	4:16	−4:11	6:56	−2:07	9:30	−0:15	11:57	1:19
May	14:51	2:50	16:53	3:31	18:41	3:44	20:14	3:30
June	21:57	2:25	22:47	1:15	23:17	−0:09	23:27	−1:40
July	23:10	−3:33	22:34	−4:48	21:39	−5:45	20:25	−6:19
August	18:12	−6:17	16:21	−5:40	14:17	−4:35	12:02	−3:04
September	8:33	0:15	5:58	2:03	3:19	4:29	0:36	6:58
October	−2:54	10:02	−5:36	12:11	−8:15	13:59	−10:48	15:20
'November	−14:12	16:20	−16:22	16:16	−18:18	15:29	−19:59	14:02
December	−21:41	11:14	−22:38	8:26	−23:14	5:13	−23:27	1:47

where

$B = (n - 1) 360/335$

n = Day of the year

Example 6.2 *Computation of Local Civil Time*

Determine LST corresponding to 11:00 a.m. Central day light saving time (CDST) on 8 February in the United States at 95° west longitude.

Solution:

CST (Central Standard Time) = Central Day Light Saving Time (CDST) − 1 hour

CST = 11:00 − 1 = 10:00 a.m.

This time is for 90° west longitudinal line, the meridian of the central time zone. LCT at 95° west longitude is $5 \times 4 = 20$ minutes less advanced and given as

LCT = CST − 20 minutes = 10:00 a.m. − 20 minutes = 9:40 a.m.

LST is then given as

LST = LCT + Equation of Time (*E*)

From Table 6.2 for 8 February, the Equation of time = −14:14

LST = 9:40 − 14:14 = 9:26 a.m.

Figure 6.16 Orientation of the Earth with respect to the direction of the Sun's radiation.

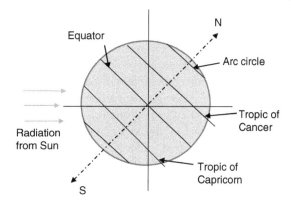

6.3.12 Basic Earth and Sun Angles

Figure 6.16 shows the orientation of Earth with respect to the direction of the Sun's radiation along with basic arc lines equator that separates the northern and the southern hemispheres; the Tropic of Cancer; and the Tropic of Capricorn.

6.3.13 Solar and Wall Angles

Following solar and wall angles are needed for solar radiation calculations: *Latitude angle*, *l*; *Declination angle*, *d*; and *hour angle, h*; *Sun's zenith angle*, Ψ; *Altitude angle*, β; *Azimuth angle*, γ or ϕ; *Sun's incidence angle*, θ; *Wall-solar Azimuth angle*, γ'; and *Wall-Azimuth angle*, ψ. Definitions of these angles are explained here as refresher to the readers.

6.3.13.1 Solar Angles

The position of a point P on Earth's surface with respect to Sun's ray is known at any instant if following angles are known: *Latitude line or Latitude (l) angle, Hour angle (h),* and *Sun's Declination angle (d)*. These solar angles are shown in Figure 6.17 with the help of the position of a point. P with respect to the center, O, of the Earth.

Figure 6.17 Solar angles.

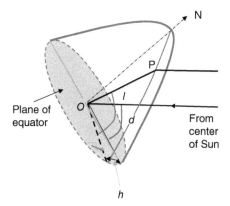

Latitude (l): Latitude angle is defined as the angular distance of the point P north (or south) of the equator. l = Angle between the line **OP** and the projection of the line **OP** on the equatorial plane. For sign conventions, the north latitude angles are considered as positive for any locations on the northern hemisphere and south latitude angles as negative for any position on the southern hemisphere.

Declination (d): Angle between a line extending from the center of the Sun to the center of the Earth and the projection of this line upon the Earth equatorial plane. It is the angular distance of the Sun's rays north (or south) of the equator. The declination angle is considered as positive for Sun's rays strike falling on the northern hemisphere and as negative for Sun's rays striking the southern hemisphere.

Figure 6.18 shows the variation Sun's angle of declination along with the variation of Sun's position along its path.

We can see that at winter solstice $d = -23.5°$, i.e. Sun's rays would be 23.5° south of the Earth's equator. Similarly, at summer solstice $d = +23.5°$, i.e. Sun's rays would be 23.5° north of the Earth's equator. Declination angle $d = 0$ at the equinoxes. An empirical expression of declination angle as a function of the day is

$$d = 23.45 \sin [360(284 + n)/365] \tag{6.12}$$

The Sun's angle of declination is also given for some selected days of the month in a year in Table 6.2.

Hour Angle (h): Hour angle is defined as the angle measured in the Earth's equatorial plane between the projection of the line **OP** and the projection of a line from center of the Sun to the center of Earth. The hour angle expresses the time of the day with respect to solar noon considering the **hour angle (h) as zero at the solar noon**. In the morning hour, angle is considered as negative and in the afternoon as positive. One-hour time is represented by 360/24 or 15° of hour angle.

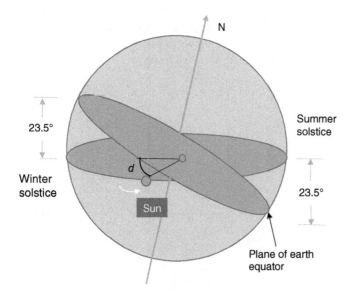

Figure 6.18 Display Sun's path and declination angle.

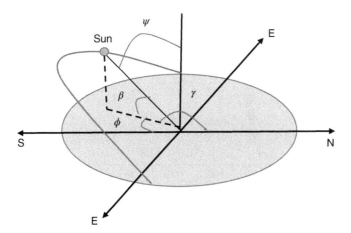

Figure 6.19 Demonstration of solar angles: Sun's zenith, altitude angle, and azimuth angles.

Additional solar angles such as **Sun's Zenith Angle**, Ψ; **Altitude angle**, β; and **Azimuth** angle, γ or ϕ, are shown in Figure 6.19.

Zenith angle (ψ): Zenith angle is the angle between Sun's ray and a line perpendicular to the horizontal plane at P.

Altitude angle (β): Altitude angle is the angle in vertical plane between the Sun's rays and projection of the Sun's ray on a horizontal plane. It follows that both zenith angle and the altitude angle are measured on the same vertical plan and are related by

$$\beta + \psi = \pi/2$$

Azimuth Angle (γ) or (ϕ): Azimuth angle, (γ), is the angle measured from north to the horizontal projection of the Sun's ray and the azimuth angle, (ϕ), is the angle measured from the south to the horizontal projection of the Sun's ray. These two angles are related by the following equation:

$$\gamma + \phi = 180°$$

Based on analytical geometry, all solar angles are computed. Some of these expressions for solar angle are summarized here:

Sun's zenith angle is given as

$$\cos(\psi) = \cos(l)\cos(h)\cos(d) + \sin(l)\sin(d) \tag{6.13a}$$

Sun's altitude angle (β):

$$\sin(\beta) = \cos(l)\cos(h)\cos(d) + \sin(l)\sin(d) \tag{6.13b}$$

Also note that the altitude angle is given as $\beta = \pi/2 - \psi$.

Sun's azimuth, (γ) or (ϕ):

$$\cos(\gamma) = \sec(\beta)\{\cos(l)\sin(d) - \cos(d)\sin(l)\cos(h)\} \tag{6.13c}$$

or

$$\cos(\phi) = (\sin \beta \sin l - \sin d)/(\cos \beta \cos l) \tag{6.13d}$$

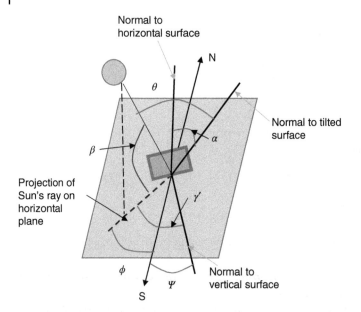

Normal to
horizontal surface

N

θ

Normal to tilted
surface

α

β

Projection of
Sun's ray on
horizontal
plane

γ

ϕ

Normal to
vertical surface

Ψ

S

Figure 6.20 Demonstration of a different wall angle for a tilted wall surface.

6.3.13.2 Wall Angles

A number of wall angles are also used to define the incidence angle between Sun's ray and surface normal to a wall. Figure 6.20 shows angles associated with a tilted wall.

Angle of tilt (α): It is the angle between the normal to wall surface and normal to horizontal surface.

Wall azimuth angle (ψ): It is the angle between the normal to vertical surface and south direction.

Wall-solar azimuth angle (γ'): For a vertical surface, it is the angle measured in horizontal plane between the projection of the Sun's ray on that horizontal plane and a normal to that vertical surface.

The incidence angle is derived as

$$\cos\theta = \cos\beta\,\cos\gamma\,\sin\alpha + \sin\beta\,\cos\alpha \tag{6.14}$$

6.3.13.3 ASHRAE Clear-Day Model for Estimation of Solar Radiation Flux

This is one of the simplest models for solar load computation. ASHRAE Clear-day model (ASHRAE Handbook, Fundamentals 1993) gives an estimation of average clear-day direct normal solar radiation flux and average clear-day diffuse solar radiation flux. A brief description of this model is given here.

6.3.13.3.1 Direct Normal Solar Radiation Flux: G_{ND}

The value of the direct normal solar radiation flux at the Earth surface on an average clear day is given by the empirical formula:

$$G_{ND} = Ae^{-\frac{B}{\sin\beta}} \tag{6.15}$$

A = Apparent direct normal solar radiation flux at outer atmosphere (W/m²)

B = Apparent atmosphere extinction coefficient

β = Solar altitude angle

In this expression, G_{ND} represents the monthly average direct normal radiation at Earth's surface on average clear days. Constants A and B are empirically determined data measured on typical clear days, and t varies throughout year due to the variation of the distance between the Earth and the Sun, and due to the variation in atmospheric conditions. Table 6.3 shows typical values of these constants on 21st day of each month.

6.3.13.3.2 Direct Radiation Flux

Direct radiation on the surface of an arbitrary orientation

$$G_D = G_{ND} \cos \theta \tag{6.16}$$

θ = Angle of incident of Sun's ray to the surface

6.3.13.3.3 Diffuse Radiation Flux: G_d

Average clear-day diffuse radiation flux on a ***horizontal surface*** is given as

$$G_d = CG_{ND} \tag{6.17}$$

where C = ratio of diffuse to direct normal radiation on a horizontal surface and is assumed to be constant for an average clear day for a particular month. Representative 21st-day monthly values of the const, C, are also given in Table 6.3.

6.3.13.4 Diffuse Radiation on Nonhorizontal Surface

$$G_{d\theta} = CG_{ND}F_{WS} \tag{6.18}$$

F_{WS} = Configuration factor between the wall and the sky = $(1 + \cos \varepsilon)/2$

Table 6.3 Extraterrestrial monthly average direct normal solar radiation.

Month	A (W/m²) (Btu/h ft²)	B	C
January	1230 (390)	0.142	0.058
February	1215 (385)	0.144	0.060
March	1186 (376)	0.156	0.071
April	1136 (360)	0.180	0.097
May	1104 (350)	0.196	0.121
June	1088 (345)	0.205	0.134
July	1085 (344)	0.207	0.136
August	1107 (351)	0.201	0.122
September	1151 (365)	0.177	0.092
October	1192 (378)	0.160	0.073
November	1221 (387)	0.149	0.063
December	1233 (391)	0.142	0.057

Source: From ASHRAE (1993). © 1993 ASHRAE.

where

ε = Tilt angle of the wall from horizontal = $(90 - \alpha)$

6.3.13.5 Reflected Radiation (GR)

Reflection of solar radiation from ground to a tilted surface or vertical wall is given as

$$G_r = G_{th}\rho_g F_{wg} \tag{6.19}$$

where

G_{th} = Total radiation (direct + diffuse) on horizontal or ground in front of the wall
ρ_g = Reflectance of ground or horizontal surface
F_{wg} = Angles or configuration factor from wall to ground = $(1 - \cos \varepsilon)/2$
   ε = Wall at a tilt angle *to* the horizontal

Example 6.3 *Horizontal Surface Radiation*

Calculate the clear-day direct, diffuse, and total solar radiation on horizontal surface at 36° north latitude and 84° west longitude on 1 June at 12:00 noon CST.

Solution:

Local Solar Time: LST = LCT + E_{qu} of time

$$LCT = LCT + (90 - 84)/15 * 60 = 12:00 + (90 - 84)/15 * 60$$

At Mid 90°

$$LST = 12:00 + (90 - 84)/15 * 60 + 0:02:25 = 12:26$$

Hour angle: $h = (12:00–12:26)*15/60 = 65°$
Declination angle: $d = 21°$ 57 minutes
Sun's altitude angle:

$$\sin \beta = \cos(l)\cos(d)\cos(h) + \sin(l)\sin(d)$$

$$= \cos(36) \times \cos(21°57\text{ minutes}) + \sin(36)\sin(21°57)$$

$$= (0.994)(0.928)(0.809) + 0.588 \times 0.376$$

$$\sin \beta = 0.965$$

Incidence angle of solar radiation for a horizontal surface:

$$\cos \theta = \sin \beta = 0.965$$

Direct Normal Radiation Flux:

$$G_{ND} = A/[\exp(B/\sin \beta)] = 345/[\exp(0.205/0.965)]$$
$$G_{ND} = 279 \text{ Btu/h ft}^2$$

The direct radiation

$$G_D = G_{ND} \cos \theta = 279 \times 0.965 = 269 \text{ Btu/h ft}^2$$

The diffuse radiation

$$G_d = CG_{ND} = 0.136 \times 279 = 37.4 \text{ Btu/h ft}^2$$

Total Irradiation

$$G = G_D + G_d = 269 + 37.6 = 300 \text{ Btu/h ft}^2$$

6.4 Optical Properties of Absorber Plate and Glazing Materials

6.4.1 Solar Radiation – Material Interaction

As solar radiation strikes a surface, a fraction of the radiation is reflected from the surface, a fraction is transmitted through the material and the rest is absorbed by the material (Figure 6.21).

An overall balance of energy at the absorber surface leads to the following equation:

$$G_t = G_{re} + G_{tr} + G_{ab} \tag{6.20}$$

where

G_{re} = Reflected radiation
G_{tr} = Transmitted radiation
G_{ab} = Absorbed radiation

Equation (6.20) can also be written in terms of the optical properties of the materials as

$$\rho + \tau + \alpha = 1 \tag{6.21}$$

where

ρ = Reflectivity of the material = $\frac{G_{re}}{G_t}$
τ = Transmissivity of the material = $\frac{G_{tr}}{G_t}$
α = Absorptivity of the material = $\frac{G_{ab}}{G_t}$

Materials have additional optical property named ***emissivity*** that characterizes a material with its ability to emit thermal radiation energy as compared to the ideal black body emissive power.

Figure 6.21 Interaction solar radiation material.

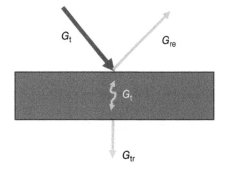

Optical properties considered in Eq. (6.21) are the total optical properties. In general, all these optical properties are spectral, i.e. function of the wavelength, and also depend on the direction of incident radiation:

$$\rho_{\lambda,\theta} = \rho(\lambda, \theta), \quad \tau_{\lambda,\theta} = \tau(\lambda, \theta), \quad \alpha_{\lambda,\theta} = \alpha(\lambda, \theta), \quad \text{and } \varepsilon_{\lambda,\theta} = \varepsilon(\lambda, \theta) \tag{6.22}$$

The total properties are the integrated average over the wavelength and over the hemispherical angular directions.

For an opaque material, $\tau = 0$ and Eq. (6.21), based on total optical properties, reduces to

$$\rho = 1 - \alpha \tag{6.23}$$

Amount of solar energy absorbed in the material is given as

$$G_{ab} = \alpha G \psi \tag{6.24}$$

For an ideal transparent surface, $\alpha = 0$ and Eq. (6.21) reduces

$$\rho = 1 - \tau \tag{6.25}$$

6.4.2 Optical Property of Absorber Plate

The main purpose of the absorber plate is to absorb as much energy as possible, lose as little heat as possible to environment by conduction, convection, and radiation through collector glaze covers, insulations, and structure, and transfer the heat to the collector-circulating fluids. To achieve high performance, the collector absorber plate is designed to achieve high absorptivity (α) for the incident solar radiation or wavelength and low emissivity (ε) for the radiation emitted at the operational temperature of the absorber plate.

The absorber plate is usually made of copper, aluminum, and steel considering several factors such as high thermal conductivity, weight, cost, ease of handling and fabrication, and durability. The optical properties depend on the type and quality of the surface such as roughness as well as the spectrum of radiation over the range of wavelengths and the temperature of the emitting surface. While absorptivity of a collector absorber plate depends on the distribution of solar radiation based on the Sun's temperature of 5500 K, emissivity of the surface depends on the temperature of the collector plate.

To achieve higher optical properties, the absorber plate is coated with paints with high absorptivity. A flat-black paint is one such coating that exhibits high absorptivity of the order of 95% for incident solar radiation. However, such common paint also exhibits high emissivity and so reradiate significant amount heat based on the temperature of the absorber plate.

6.4.3 Selective Coating

The ideal surface is one that exhibits high absorptivity based on solar radiation and a low emissivity based on the absorber-plate temperature. Such conditions are achievable through the use of specially designed coat referred to as the *selective coatings*, which exhibit high absorptivity over the incident solar radiation wavelength spectrum and a low emissivity for the long-wavelength radiation emitted at the absorber-plate temperature. This combination of surface characteristics is possible because 98% of the energy in incoming solar radiation is contained within wavelengths below 3 μm, whereas 99% of the radiation emitted by black or gray surfaces at 400 K is at wavelengths longer than 3 μm. A surface is selective if it absorbs all the solar wavelengths and emits none of the heat wavelengths, so that more heat could be transferred to the working fluid.

Almost all black selective surfaces are generally applied on the metal base, which provides low emittance for thermal radiation and simultaneously good heat transfer characteristics.

Such selective surfaces are formed by depositing multiple layers with each layer made of different elements. For example, a selective absorber surface is formed by depositing or coating very thin black metallic oxide over a bright metal base. The metallic oxide layer provides good absorptivity over 95% for the incident solar radiation spectrum of wavelengths (0.3 to <3.0 μm) and is transparent to longer wavelength heat radiation (>3.0 μm), i.e. neither absorbing nor emitting.

Some examples of these types of selective coatings are *black chrome, copper-on-nickel,* and *black nickel-on-galvanized iron.* Table 6.4 lists a number of such selective coatings along with the absorptivity and emissivity values.

A selective coating is formed by using number of different coating and deposition techniques such as spraying, electroplating, electrodepositing, chemical vapor deposition (CVD), and chemical oxidation.

One of the most popular selective coatings is black chrome, which is fabricated by electroplating a thin layer of nickel over the absorber plate, and then depositing another thin layer of chromium oxide, known as black chrome, on top of the nickel substrate layer. Another common selective coating is the black copper oxide coating on copper or aluminum absorber plate.

To create a selective solar-absorbing surface, two types of special surfaces play a major role – selective and reflective surfaces. Selective surfaces combine a high absorptance for

Table 6.4 Optical properties of selective coatings for absorber plate.

Surface type	α	ε	$\dfrac{\alpha}{\varepsilon}$
Black chrome	0.93	0.10	9.3
Black chrome on Ni-plated steel	0.95	0.09	10.56
Nickel black on galvanized steel			
Black nickel on polished nickel	0.81	0.17	4.76
Black nickel on galvanized iron			
Cu black on cu	0.92	0.11	8.36
Ebonol C on Cu			
CuO on nickel	0.89	0.12	7.4
Co_3O_4 on silver	0.89	0.17	5.24
CuO on aluminum	0.90	0.16	5.63
CuO on anodized aluminum	0.81	0.17	4.76
Solchrome	0.90	0.27	3.3
Black paint	0.93	0.11	8.45

a radiation with a low emittance for temperature range in which it emits radiation. This combination of surface characteristics is possible because 98% of the energy in incoming solar radiation is contained within wavelengths below 3 μm, whereas 99% of the radiation emitted by black or gray surfaces at 400 K is at wavelengths longer than 3 μm. Almost all black selective surfaces are generally applied on the metal base, which provides low emittance for thermal radiation and simultaneously good heat transfer characteristics for photo thermal applications.

A good absorber of heat is always a good radiator of heat. A surface is selective if it absorbs all radiation over the solar wavelength range and emits none at heat wavelengths, so that more heat could be transferred to the working fluid.

6.4.4 Optical Properties of Glazing Materials

Optical properties, refractive index, and extinction coefficients are major factors in determining the fraction solar radiation that passes through glazing covers and reaches the absorbing surface. As we have discussed, all these material properties are functions of wavelengths and directions of incident solar radiation. However, for most solar collector glazing cover materials, use of total properties is good approximation. Let us consider the direct solar radiation as it strikes and transmits through a single sheet of glass as shown in Figure 6.22.

Based on the multiple internal reflections, the reflected, absorbed, and transmitted radiation quantities are given by the sum of infinite series. The total optical properties are given as follows:

Transmissivity:

$$\tau = \frac{(1-r)^2 a}{1 - r^2 a^2} \tag{6.26a}$$

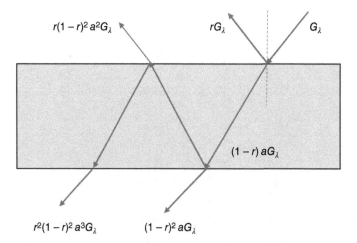

Figure 6.22 Incident solar radiation over single transparent glass involving multiple internal reflections.

Reflectivity:

$$\rho = r + \frac{r(1-r)^2 a^2}{1 - r^2 a^2} \tag{6.26b}$$

and

Absorptivity:

From Eq. (6.21)

$$\alpha = 1 - r - \frac{(1-r)^2 a}{1 - ra} \tag{6.26c}$$

where

$r = $ **Reflectance coefficient** or fraction of each component reflected
$a = $ **Absorption coefficient** or fraction of each component remains for absorption

6.4.4.1 Absorption Coefficient

The *absorption coefficient* is given in terms of the thickness of the glass (t_g), angle of incidence of the solar radiation (θ), and extinction coefficient (K_{extin}) as

$$a = e^{-K_{\text{extin}} t_g'} \tag{6.27a}$$

where

$$t_g' = \frac{t_g}{\sqrt{1 - \frac{\sin^2 \theta}{n^2}}} \quad \text{or} \quad t_g' = \frac{t_g}{\cos \theta'} \tag{6.27b}$$

and

$$n = \text{index of refraction and } \theta' = \text{angle of refraction}$$

Note that for normal incidence, $t_g = t_g'$.

As we have mentioned before, extinction coefficient represents the absorption and scattering as solar radiation transmits through a medium such as in a glazing glass medium. Table 6.5 includes extinction coefficient values for three classes of glass. The typical values vary depending on the actual composition of glass materials. High extinction coefficient values in a heat-absorbing type glass are achieved by using iron oxide to increase absorption of heat from the solar radiation spectrum. Most common types of oxides are ferrous oxide or ferric oxide or a combination of both. Such heat-absorbing glasses are considered as window glasses as in for passive solar applications of a building design.

Table 6.5 Typical extinction coefficient values.

Glass type	Extinction coefficient, K (K_{extin})	
	(m^{-1})	(in.$^{-1}$)
Water-white	4–10	0.1016–0.254
Double strength	20–40	0.508–1.016
Heat absorbing	60–130	1.424–3.3

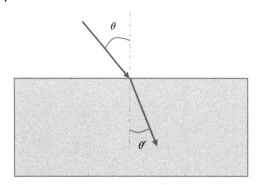

Figure 6.23 Reflection of unpolarized radiation as it passes through a medium.

6.4.4.2 Reflectance Coefficient

Let us consider reflection of unpolarized radiation as it passes from a medium-1 with index of refraction, n_1, to a medium-2 with index of refraction, n_2, as shown in Figure 6.23.

The angle of incidence, θ, and the angle of refraction, θ', are related by Snell's law:

$$n_1 \sin \theta = n_2 \sin \theta' \tag{6.28}$$

or

$$\sin \theta' = \frac{n_1}{n_2} \sin \theta \tag{6.29}$$

Reflection of unpolarized radiation consists of two components representing two vibrating components of the radiation – one vibrating perpendicular to the plane and second vibrating parallel to the plane. This is expressed as

$$r(\theta) = \frac{1}{2}[r_\perp + r_\|] \tag{6.30}$$

Each component of reflection is written based on Fresnel's equation, and the **reflectance coefficient** is derived as

$$r(\theta) = \frac{1}{2}\left[\frac{\sin^2(\theta - \theta')}{\sin^2(\theta + \theta')} + \frac{\tan^2(\theta - \theta')}{\tan^2(\theta + \theta')}\right] \tag{6.31a}$$

At normal incidence of radiation, $\theta = \theta' = 0$ and the expression for *reflectance coefficient* can be simplified to

$$r(0) = \left(\frac{n_1 - n_2}{n_1 + n_2}\right)^2 \tag{6.31b}$$

At normal incidence of solar radiation with medium-1 as **air** ($n_1 = 1$) and medium-2 as the glazing medium ($n_2 = n$). The expression for *reflectance coefficient* reduces to

$$r(0) = \left(\frac{n - 1}{n + 1}\right)^2 \tag{6.31c}$$

Typical values of index of refraction for different glazing materials are given in Table 6.6.

6.4.5 Transmittance Through Glass Cover

As solar radiation is incident on a glass cover, it undergoes multiple reflections between the two surfaces shown in Figure 6.24.

Table 6.6 Index of refraction of glazing materials.

Cover glazing materials	Index of refraction (n)
Glass	1.526
Polycarbonate	1.60
Polytetrafluoroethylene	1.37
Polyvinyl fluoride	1.45

Figure 6.24 Transmission through a nonabsorbing glass layer.

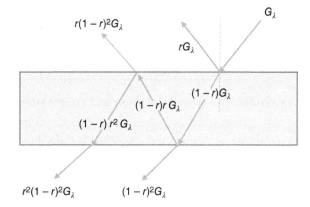

Transmittance of the **nonabsorbing** glass cover is given as the average of the two infinite sums of the parallel and perpendicular components of polarized radiation and given as

$$\tau_r = \frac{1}{2}\left[(1+r_\parallel)^2\sum_{n=0}^{\infty} r_\parallel^{2n} + (1+r_\perp)^2\sum_{n=0}^{\infty} r_\perp^{2n}\right] \tag{6.32a}$$

or

$$\tau_r = \frac{1}{2}\left[\frac{1-r_\parallel}{1+r_\parallel} + \frac{1-r_\perp}{1+r_\perp}\right] \tag{6.32b}$$

Equations (6.32a) and (6.32b) can be generalized to **n-glazing or n-cover glasses** as follows:

$$\tau_{rN} = \frac{1}{2}\left[\frac{1-r_\parallel}{1+(2N-1)r_\parallel} + \frac{1-r_\perp}{1+(2N-1)r_\perp}\right] \tag{6.33}$$

In Eqs. (6.32a), (6.32b), and (6.33), the subscripts r indicate that only reflection is considered for the nonabsorbent cover.

6.4.6 Optical Properties for Absorbing Glazing Cover

The optical properties of an absorbing glazing cover can be written in a simplified form as follows:

Transmittance: $\tau = a\tau_r$ (6.34a)

Absorptance: $\alpha = 1 - a$ (6.34b)

and

$$\text{Reflectance: } \rho = 1 - \alpha - \tau = a - \tau \tag{6.34c}$$

A more detailed computation of the optical properties of absorbing glazing cover is given by Duffie and Beckman (2000).

For a multicover system with identical glazing materials, the procedure outlined for a single glass cover can be used. For a two-cover system with dissimilar gazing materials, following equations are suggested:

$$\tau = \frac{1}{2}(\tau_\perp + \tau_\|) = \frac{1}{2}\left\{\left(\frac{\tau_1\tau_2}{1+\rho_1\rho_2}\right)_\perp + \left(\frac{\tau_1\tau_2}{1+\rho_1\rho_2}\right)_\|\right\} \tag{6.35a}$$

and

$$\rho = \frac{1}{2}(\rho_\perp + \rho_\|) = \frac{1}{2}\left\{\left(\rho_1 + \frac{\tau\rho_2\tau_1}{\tau_2}\right)_\perp + \left(\rho_1 + \frac{\tau\rho_2\tau_1}{\tau_2}\right)_\|\right\} \tag{6.35b}$$

where subscript 1 refers to top cover and 2 refers to the bottom cover.

Example 6.4 Calculate the transmittance of one cover of a solar collector at normal incidence and at 45° angle of incidence. Assume the glazing as standard nonabsorbing glass.

Solution:
For a standard nonabsorbing glass, let us consider the refractive index as $n = 1.5526$ and use Snell's law given by Eq. (6.31c) to calculate the angle of refraction as

$$\text{At normal incidence}: \theta'(0°) = \sin^{-1}\left(\frac{\sin 0}{1.526}\right) = 0$$

Use Eq. (6.31c) to calculate the normal reflectance

$$r(0) = \left(\frac{1.526 - 1}{1.526 + 1}\right)^2 = 0.0434$$

At 45° angle of incidence:

$$\theta'(45°) = \sin^{-1}\left(\frac{\sin 45}{1.526}\right) = 27.604°$$

and use Eq. (6.31a) to calculate reflectance at 45° angle of incidence

$$r(45°) = \frac{1}{2}\left[\frac{\sin^2(45 - \theta')}{\sin^2(45 + \theta')} + \frac{\tan^2(45 - \theta')}{\tan^2(45 + \theta')}\right]$$

or

$$r(45°) = \frac{1}{2}\left[\frac{\sin^2(45 - 27.604)}{\sin^2(45 + 27.604)} + \frac{\tan^2(45 - 27.604)}{\tan^2(45 + 27.604)}\right]$$

$$r(45°) = \frac{1}{2}\left[\frac{0.089\,38}{0.9106} + \frac{0.098\,15}{10.1874}\right] = \frac{1}{2}[0.0981 + 0.009\,63]$$

$$r(45°) = 0.053\,86$$

Example 6.5   Consider a single water-white glass sheet and determine the optical properties of this glass sheet for solar radiation at normal incidence, at (a) $\theta = 0°$ and (b) at $\theta = 45°$ for a glass of thickness, $t_g = 5$-mm.

Solution:

(a) Normal incidence: $\theta = 0°$ and $t_g = t'_g = 0.005$-m.

$$K_{extin} = 5 \text{ m}^{-1}$$

$$a = e^{-K_{extin} t'_g}, a = e^{-5x(0.005)} = 0.9753$$

$$r(0) = \left(\frac{n-1}{n+1}\right)^2 = \left(\frac{1.526-1}{1.526+1}\right)^2 = \frac{0.526}{2.526} = 0.0434$$

Transmissivity:

$$\tau = \frac{(1-r)^2 a}{1-r^2 a^2} = \frac{(1-0.0434)^2 \times 0.9753}{1-(0.0434)^2 (0.9753)^2} = \frac{0.892\,48}{0.9982} = 0.8940$$

Absorptivity:

$$\alpha = 1 - r - \frac{(1-r)^2 a}{1-ra} = 1 - 0.0434 - \frac{(1-0.0434)^2 \times 0.9753}{1-0.0434 \times 0.9753}$$

$$= 1 - 0.0434 - \frac{0.892\,48}{0.9576}$$

$$\alpha = 0.0875$$

Reflectivity:

$$\rho = 1 - \tau - \alpha = 1 - 0.8940 - 0.0875 = 0.0185$$

(b) For 45° incidence: $\theta = 45°$

$$\theta'(45°) = \sin^{-1}\left(\frac{\sin 45}{1.526}\right) = 27.604°$$

$$t'_g = \frac{t_g}{\cos \theta'} = \frac{0.005}{\cos(27.604)} = 0.005\,64 \text{ m}$$

$$a = e^{-K_{extin} t'_g}, a = e^{-5x(0.005\,64)} = 0.9722$$

$$r(45°) = 0.053\,86$$

Transmissivity:

$$\tau = \frac{(1-r)^2 a}{1-r^2 a^2} = \frac{(1-0.053\,86)^2 \times 0.9722}{1-(0.053\,86)^2 (0.9722)^2} = \frac{0.870\,29}{0.997\,25} = 0.872\,68$$

Absorptivity:

$$\alpha = 1 - r - \frac{(1-r)^2 a}{1-ra} = 1 - 0.053\,86 - \frac{(1-0.053\,86)^2 \times 0.9722}{1-0.053\,86 \times 0.9722}$$

$$= 1 - 0.053\,86 - \frac{0.872\,68}{0.944\,76}$$

$$\alpha = 0.0252$$

Reflectivity:

$$\rho = 1 - \tau - \alpha = 1 - 0.872\,68 - 0.0252$$

$$\rho = 0.102\,08$$

6.4.7 Transmittance–Absorptance Product of Collector ($\tau\alpha$)

Transmittance–Absorptance product ($\tau\alpha$) is an important parameter of a solar collator that includes effect of glazing or cover glasses and the absorber plate. As solar radiation transmits through a glass cover and is incident on absorber plate, fraction of the energy is absorbed by the plate and the rest is reflected toward the glass cover as diffused radiation. The reflected radiation undergoes multiple internal reflections and absorptions at the absorber plate as demonstrated in Figure 6.25.

Multiple reflections contribute to the energy absorption in the plate and

$$(\tau\alpha) = \tau\alpha + \tau\alpha(1 - \alpha)\rho_{g,d} + \tau\alpha(1 - \alpha)^2\rho_{g,d}^2 + \cdots \tag{6.36a}$$

or

$$(\tau\alpha) = \tau\alpha\sum_{0}^{\infty}[(1 - \alpha)\rho_{g,d}]^n = \frac{\tau\alpha}{1 - (1 - \alpha)\rho_{g,d}} \tag{6.36b}$$

where $\rho_{g,d}$ represents the reflectivity of glass cover for diffused radiation.

In the derivation of this equation, the reflection from the absorber plate is considered as diffused radiation and so the reflection from the cover system is also based on diffused radiation.

Example 6.6   Estimation of Transmittance – Absorptance Product

Estimate the transmittance–absorptance product ($\tau\alpha$) of a single glass-glazing cover of thickness 2.0 mm and at an angle of incidence of 45°. Assume the glass as water-white with an extinction coefficient value of 20 m^{-1} and the absorptivity of the absorber plate as $\alpha = 0.95$.

Incident solar
radiation

Glass cover

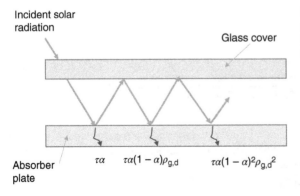

Absorber
plate

$\tau\alpha$ $\tau\alpha(1 - \alpha)\rho_{g,d}$ $\tau\alpha(1 - \alpha)^2\rho_{g,d}^2$

Figure 6.25   Solar radiation transmission through a glass cover with multiple internal reflections and absorptions at the absorber plate.

Solution:

Let us first compute the optical properties of glass cover as follows:

For 45° incidence: $\theta = 45°$

$$\theta'(45°) = \sin^{-1}\left(\frac{\sin 45}{1.526}\right) = 27.605°$$

$$t_g' = \frac{t_g}{\cos \theta'} = \frac{0.002}{\cos(27.605)} = 0.002\,26\ \text{m}$$

$$a = e^{-K_{extin}t_g'}, a = e^{-20\times(0.002\,26)} = 0.9493$$

From Example 6.5, $r(45°) = 0.053\,86$

Cover glass transmissivity:

$$\tau = \frac{(1-r)^2 a}{1 - r^2 a^2} = \frac{(1 - 0.053\,86)^2 \times 0.9493}{1 - (0.053\,86)^2(0.9493)^2} = \frac{0.8498}{0.997\,38} = 0.852$$

Cover glass absorptivity:

$$\alpha = 1 - r - \frac{(1-r)^2 a}{1 - ra} = 1 - 0.053\,86 - \frac{(1 - 0.053\,86)^2 \times 0.9493}{1 - 0.053\,86 \times 0.9493}$$

$$= 1 - 0.053\,86 - \frac{0.8498}{0.9489}$$

$$\alpha = 0.0505$$

Diffuse reflectivity bottom side of the cover glass

$$\rho_d = a - \tau = 0.9493 - 0.852 = 0.0973$$

From Eq. (6.26b), the transmittance–absorptance product

$$(\tau\alpha) = \frac{\tau\alpha}{1 - (1 - \alpha)\rho_{g,d}} = \frac{0.852 \times 0.95}{1 - (1 - 0.95)0.0973} = 0.8134$$

6.4.8 Absorbed Solar Radiation on a Collector Surface

Absorbed solar radiation on solar collector absorber plate consists of contribution from three components: direct radiation, G_D; diffuse radiation, G_d; and reflected radiation, G_r. Once the absorptance–transmittance product is computed for all three components of radiation, total absorbed radiation per unit area is then given as

$$q_s'' = G_D(\tau\alpha)_D + G_d(\tau\alpha)_d \left(\frac{1 + \cos \varepsilon}{2}\right) + \rho_g G_{TH}(\tau\alpha)_g \left(\frac{1 - \cos \varepsilon}{2}\right) \qquad (6.37)$$

where

$$G_D = \text{Direct radiation normal to the surface}$$
$$G_d = \text{Diffuse radiation}$$
$$G_{TH} = \text{Total radiation on a horizontal ground surface}$$
$$\rho_g = \text{Ground reflectivity}$$
$$F_{WS} = \left(\frac{1 + \cos \varepsilon}{2}\right) = \text{View factor from the surface to the sky}$$
$$F_{WG} = \left(\frac{1 - \cos \varepsilon}{2}\right) = \text{View factor from the surface to the ground}$$

6.4.9 Types and Selection of Glazing

As we have mentioned, one way to reduce remitted radiation and convection losses from collector absorber surface is by using glazing cover over the absorber surface. Figure 6.26 shows different mechanisms involved in using glazing cover to reduce thermal energy loss from collector.

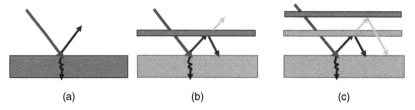

(a)   (b)   (c)

Figure 6.26 Solar radiation interactions with absorber plate without and with glazing covers. (a) No glazing, (b) one glazing cover, and (c) two glazing covers.

Diathermanous materials are the materials, which can transmit radiant energy, including solar energy. From the standpoint of the utilization of solar energy, the important characteristics are reflection (ρ), absorption (α), and transmission (τ). α and ρ values should be as small as possible and the transmission should be as high as possible for maximum efficiency.

Glass is a typical choice for a glazing layer of a flat-plate collector because of its high transmittance value of over 90% for the short-wavelength solar radiation and very low transmittance value for the long-wavelength radiation emitted by the absorber plate. Optical properties of glass vary with the composition of the glass, especially the iron content. A low iron content glass typically possesses transmittance value in the range of 0.85–0.90 at normal incidence and for the solar radiation spectrum of 0.3–3.0 μm. Absorption capacity of glasses increases by maintaining a certain level of iron content within glass composition. For a perfect glazing material, absorptivity, $\alpha = 0$, and emissivity, $\varepsilon = 0$, transmissivity, $\tau = 1$ based on solar radiation wavelength corresponding to 5000 K, and transmissivity, $\tau = 0$ based on radiation emitted by the absorbing plate at the collector temperature. ***Polyvinyl Fluoride Glass*** is one such glazing material due to its high transmissivity and low absorptance and emissivity. A list of various glazing materials with their respective transmittance and conductivity values is given in Table 6.7.

Table 6.7 Transmittance and thermal conductivity values for some glazing materials.

Material	Solar transmittance	Thermal conductivity (W/m K)
Window glass	0.5	0.96
Polycarbonate	0.84	0.19
Polyethylene terephthalate	0.84	0.15
Polyvinyl fluoride	0.93	0.12
Polyamide	0.80	0.24
Fiberglass-reinforced polyester	0.87	0.336

6.4.10   Thermal Insulation

Thermal insulation is used at back side of flat-plate collectors to reduce the downward heat losses from the back side of the plate and the tubes attached to absorber plate. Besides thermophysical properties, there are other factors that need to be considered. These are the dimensional and chemical stability of the insulating materials at high temperatures and resistance moisture and condensation. Typical insulating materials are glass wool, mineral wool, polystyrene, polyurethane, and fiber glass. Table 6.8 lists the thermophysical properties of these insulating materials.

Table 6.8   Thermophysical properties of insulating materials.

Insulating materials	Temperature range	Thermal conductivity (W/m K)
Polystyrene	−60 to 175	0.03
Polyurethane	−350 to 250	0.03
Mineral wool (glass)	32–480	0.04
Mineral wool (stone)	32–1400	0.04
Fiberglass	−20 to 1000	0.04

6.5   Solar Thermal Collector Analysis and Performance

6.5.1   Flat-Plate Collector

Thermal analysis of solar collector is essential for evaluating collector performance and optimizing the collector. Let us consider a section of a flat-plate collector in Figure 6.27 to demonstrate the balance of absorbed solar radiation at absorber surface with useful heat transferred to circulating collector fluid and heat loss from the collector.

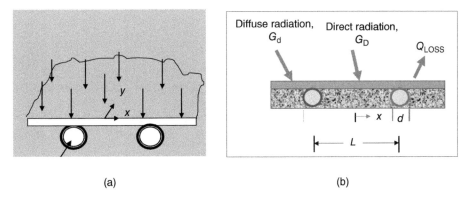

(a)   (b)

Figure 6.27   Flat-plate solar collector design with no glass cover or glazing. (a) Section of a flat-plate collector irradiated absorber plate and (b) Energy balance over the section of the collector.

An overall energy balance over the entire solar collector can be written as follows:

$$[(\tau\alpha)_D A_c G_D]_D + [(\tau\alpha)_d A_c G_d]_d = Q_u + Q_{Loss} \tag{6.38}$$

where

G_D = Direct solar radiation
G_d = Diffuse solar radiation
A_c = Collector area
$(\tau\alpha)_D$ = Transmittance–absorptance product for direct radiation
$(\tau\alpha)_d$ = Transmittance–absorptance product for diffuse radiation
Q_u = Useful heat energy transferred to circulating fluid in tubes
Q_{Loss} = Energy loss from the collector to the surrounding by radiation, convection, and conduction

The total heat energy loss from the collector is given by the sum of heat loss from the top and bottom of the collector.

$$Q_{Loss} = Q_{Loss,top} + Q_{Loss,bottom} = U_a A_c (T_p - T_a) + U_b A_c (T_p - T_a) \tag{6.39}$$

Solar collector performance or efficiency is low due to the significant amount heat loss from the top of the absorber plate. As mentioned before, glass covers as glazing layers are used to reduce component of heat loss from the top.

6.5.1.1 Solar Collector Heat Loss and Overall Heat Transfer

Let us consider the thermal resistance diagram for a two-glass-cover solar collector as shown in Figure 6.28.

The **overall bottom heat loss coefficient** is given as

$$U_b = \frac{1}{\sum R_b} = \frac{1}{R_2 + R_1} = \frac{1}{\frac{L}{K} + \frac{1}{h_b}} \tag{6.40a}$$

where

$$R_1 = \frac{1}{h_b} \quad \text{and} \quad R_2 = \frac{l}{k_{ins}} \tag{6.40b}$$

The overall **top heat loss coefficient** is given as

$$U_t = \frac{1}{\sum R_t} = \frac{1}{R_3 + R_4 + R_5} \tag{6.41a}$$

where

$$R_3 = \frac{1}{h_{c1} + h_{r1}}, R_4 = \frac{1}{h_{c2} + h_{r2}} \quad \text{and } R_5 = \frac{1}{h_{ct} + h_{rt}} \tag{6.41b}$$

Under steady state, the heat loss from the top can be estimated based on heat transfer from the absorber plate to the first glass cover as follows:

$$q_{Loss,top} = h_{p-c1}(T_p - T_{c1}) + \frac{\sigma(T_p^4 - T_{c1}^4)}{\frac{1}{\epsilon_p} + \frac{1}{\epsilon_{c1}} - 1} \tag{6.42}$$

where

h_{p-c1} = Convection coefficient between two inclined parallel plates
ϵ_p = Emissivity of the absorber plate
ϵ_{c1} = Emissivity of the glass cover-1

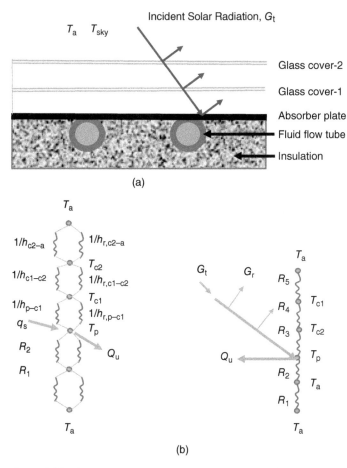

Figure 6.28 Flat-plate collector with two glass covers. (a) Flat-plate collector. (b) Thermal resistance diagram for a two-glass-cover solar collector.

Equation (6.42) can be simplified as

$$q_{\text{Loss,top}} = h_{\text{p-c1}}(T_{\text{p}} - T_{\text{c1}}) + h_{\text{r1}}(T_{\text{p}} - T_{\text{c1}}) \tag{6.43a}$$

or

$$q_{\text{Loss,top}} = (h_{\text{p-c1}} + h_{\text{r1}})(T_{\text{p}} - T_{\text{c1}}) \tag{6.43b}$$

where the radiation film coefficient

$$h_{\text{r1}} = \sigma(T_{\text{p}} - T_{\text{c1}})(T_{\text{p}}^2 + T_{\text{c1}}^2)\frac{1}{\dfrac{1}{\epsilon_{\text{p}}} + \dfrac{1}{\epsilon_{\text{c1}}} - 1} \tag{6.43c}$$

The thermal heat resistance between the absorber plate and the first glass cover is given as

$$R_3 = \frac{1}{h_{\text{p-c1}} + h_{\text{r3}}} \tag{6.44}$$

In a similar manner, the heat loss from the cover glass-1 to cover glass-2 can be

$$q_{Loss,c1-c2} = h_{p-c1}(T_p - T_{c1}) + h_{r1}(T_p - T_{c1})$$ (6.45a)

or

$$q_{Loss,c1-c2} = (h_{c1-c2} + h_{r1})(T_{c1} - T_{c2})$$ (6.45b)

where

h_{c1-c2} = Convection coefficient between two inclined parallel glass cover plates

$h_{r2} = \sigma(T_{c1} + T_{c2})(T_{c1}^2 + T_{c2}^2) \dfrac{1}{\frac{1}{\varepsilon_{c1}} + \frac{1}{\varepsilon_{c2}} - 1}$

ε_{c1} = Emissivity of the glass cover-1

ε_{c2} = Emissivity of the glass cover-2

The thermal heat resistance between the glass cover-1 and the glass cover-2 is given as

$$R_4 = \frac{1}{h_{c1-c2} + h_{r2}}$$ (6.46)

For top glass cover, heat loss is by convection due to the wind and radiation heat transfer from the glass cover to the sky. The heat loss from the top glass cover is given as

$$q_{Loss,top} = h_{c2-air}(T_{c2} - T_a) + h_{r5}(T_{c2} - T_{sky})$$ (6.47a)

where the radiation film coefficient between the top glass cover to surrounding is

$$h_{r5} = \varepsilon_{c2}\sigma(T_{c2} + T_{sky})(T_{c2}^2 + T_{sky}^2)$$ (6.47b)

For simplicity, we can consider $T_a = T_{sky}$ and write the top glass cover resistance as

$$R_5 = \frac{1}{h_{c2-air} + h_{r5}}$$ (6.48)

The top overall heat transfer coefficient is given as

$$U_t = \frac{1}{R_3 + R_4 + R_5}$$ (6.49)

6.5.1.2 Temperature Distribution in Absorbing Plate

Let us consider a section of the absorber plate in between two adjacent tubes as shown in Figure 6.29.

The temperature distribution in the absorber plate is given by the following differential heat equation derived based on two-dimensional heat equation assuming a thin absorber plate and solar heat absorption at the top surface

$$\frac{d^2T}{dx^2} - \frac{U_t}{k_p t_p}(T - T_a) + \frac{q_s''}{k_p t_p} = 0$$ (6.50a)

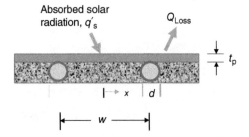

Absorbed solar radiation, q'_s

Q_{Loss}

Figure 6.29 Temperature distribution along absorber plate of flat-plate collector.

or

$$\frac{d^2\theta}{dx^2} - \lambda^2\theta + S = 0 \tag{6.50b}$$

where

$$\theta = (T - T_a), \quad \lambda^2 = \frac{U_t}{k_p t_p}, \text{ and } S = \frac{q_s''}{k_p t_p} \tag{6.50c}$$

Boundary Conditions:

$$\text{At } X = 0, T(w/2) = T_w \quad \text{and} \quad \text{at } x = 0, \ \frac{\partial T}{\partial x} = 0 \tag{6.51}$$

Solution to this differential equation leads to the expression for temperature distribution in the absorber plate.

$$T(x) = T_a + \left(T_w - T_a - S/\lambda^2\right)\left[\frac{e^{\lambda x}}{1 + e^{\lambda w/2}} + \frac{e^{-\lambda x}}{1 + e^{-\lambda w/2}}\right] + S/\lambda^2 \tag{6.52a}$$

or

$$\frac{T(x) - T_a - S/\lambda^2}{\left(T_w - T_a - S/\lambda^2\right)} = \left[\frac{e^{\lambda(x-L)}}{e^{\lambda L} + e^{-\lambda L}} + \frac{e^{-\lambda(x-L)}}{e^{\lambda L} + e^{-\lambda L}}\right] \tag{6.52b}$$

or

$$\frac{T(x) - T_a - q_s'/U_t}{\left(T_w - T_a - q_s'/U_t\right)} = \frac{\cosh \lambda x}{\cosh \lambda L} \tag{6.52c}$$

The typical temperature distribution $T(x)$ in the absorber at any y location is shown in Figure 6.30.

Heat transfer rate at the plate and tube contact area is given as

$$q_w' = -k_p t_p \frac{dT}{dx}\bigg|_{x=w} \tag{6.53}$$

The temperature distribution in the absorber plate depends strongly on the absorbed solar radiation, thermal conductivity of the plate as well as the overall heat transfer coefficient for heat loss from the top surface plate and about air, and the tube wall temperature.

6.5.1.3 Collector Performance
An energy balance for the absorber plate is used to determine the useful energy available from the collector and is expressed as

$$Q_u = A_c [q_s'' - U_c (T_p - T_a)] \tag{6.54a}$$

Figure 6.30 Typical temperature distribution along absorber plate of flat-plate collector.

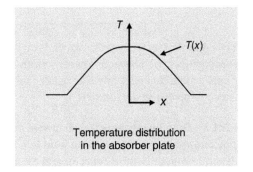

Temperature distribution in the absorber plate

or

$$Q_u = A_c[G_t(\tau\alpha) - U_c(T_p - T_a)]$$ (6.54b)

where

Q_u = Useful energy from the collector
A_c = Collector surface area
q_s'' = Heat absorbed at the collector absorber plate per unit area
G_t = Total incident solar radiation over the collector surface
$(\tau\alpha)$ = Collector transmittance–absorptance product
U_c = Overall collector heat loss coefficient
T_p = Absorber-plate temperature
T_a = Ambient air temperature

6.5.1.4 Collector Efficiency Factor (F')

Useful heat energy gained by the fluid in collector is given by

$$q_u' = F'[q_s' - U_c(T_f - T_a)]$$ (6.55)

where F' is the **collector efficiency factor** defined as the ratio of useful heat energy gained by the collector fluid to the net energy available based on absorbed solar energy and heat loss through the collector system. In other words, it can be expressed as the ratio of thermal resistance based on temperature difference between the absorber plate and the ambient air and the resistance based on temperature difference between the fluid inside the collector tube and the ambient air

$$F' = \frac{\frac{1}{U_c}}{\frac{1}{U_o}} = \frac{U_o}{U_c}$$ (6.56)

The resistance based on temperature difference between the fluid inside the collector tube and the ambient air is the sum of three resistances: (i) resistance between the plate and ambient air, (ii) the bond and contact resistance between the plate and the tube, and (iii) the convective fluid resistance inside the tube.

$$\frac{1}{U_o} = w\left[\frac{1}{U_t\{d_o + (w - d_o)\}} + R_b' + \frac{1}{\pi d_i h_{fi}}\right]$$ (6.57)

Substituting Eq. (6.57) into Eq. (6.56), we have the expression for the collector efficiency factor as

$$F' = \frac{1/U_t}{w\left[\frac{1}{U_c\{d_o + (w - d_o)\}} + R_b' + \frac{1}{\pi d_i h_{fi}}\right]}$$ (6.58)

The collector efficiency factor depends on a number of factors such as the size of tube (d_o), spacing tubes (w), convective heat transfer for flow inside the collector tubes (h_{fi}), and additional conduction resistances associated with the contact and bond resistance between the collector plate and the tubes (R_b').

6.5.1.5 Fluid Temperature Distribution in the Collector Tube

The temperature of the working fluid in a flat-plate collector may range from 30 to 90 °C, depending on the type of collector and the applications. Let us consider constant solar heat flux at the top surface of a solar collector tube array. An energy balance over a differential section of the tube leads to the following differential equation for the collector fluid as

$$\dot{m}_t c_p \frac{dT_f}{dy} = q'_u \tag{6.59a}$$

Substituting Eq. (6.55) for q'_u

$$\dot{m}_t c_p \frac{dT_f}{dx} = wF'[q'_s - U_c(T_f - T_a)] \tag{6.59b}$$

Considering inlet fluid temperature as $T = T_{fi}$ at $x = 0$, the solution to this first-order equation is given as

$$\frac{T_f - T_a - q'_s/U_c}{T_{fi} - T_a - q'_s/U_c} = \exp\left(\frac{U_c w F' x}{\dot{m}_t}\right) \tag{6.60a}$$

Let us consider the collector with n – number of tubes in parallel and assuming uniform flow in all tubes because of the assumption of one-dimensional analysis, we have

Total fluid flow rate: $\dot{m}_f = n\dot{m}_t$ and Total solar collector area: $A_c = nwL_c$

Equation (6.60a) now can be written for the collector fluid exit temperature in terms of solar collector area and total fluid flow rate as

$$\frac{T_{fo} - T_a - q'_s/U_t}{T_{fi} - T_a - q'_s/U_t} = \exp\left(\frac{F'U_c A_c}{\dot{m}_f}\right) \tag{6.60b}$$

or

$$\frac{T_{fo} - T_a - \left[\frac{G_t(\tau\alpha)}{U_t}\right]}{T_{fi} - T_a - \left[\frac{G_t(\tau\alpha)}{U_t}\right]} = \exp\left(\frac{F'U_c A_c}{\dot{m}_f}\right) \tag{6.60c}$$

where

A_c=Solar collector area = nwL_c
$\dot{m}$=Net mass flow rate through the solar collector
n=Number parallel tubes in the solar collector
w=Width spacing between tubes
L_c=Length of tubes or collector

6.5.1.6 Collector Heat Removal Factor (F_R)

Collector heat removal factor, F_R, is defined as the ratio of actual useful heat gained by the collector fluid to the maximum possible heat gained by the fluid if the entire solar collector surface is maintained at the fluid inlet temperature. This is expressed as

$$F_R = \frac{Q_u}{A_c[G_t(\tau\alpha) - U_c(T_{fi} - T_a)]} \tag{6.61}$$

The most common way of determining the useful heat gained by the collector in a single-phase fluid flow is by measuring the mass flow rate and the temperature rise. This is given as

$$Q_u = \dot{m}_f c_p(T_{fo} - T_{fi}) \tag{6.62}$$

Substituting Eq. (6.62) into Eq. (6.61), we get

$$F_R = \frac{\dot{m}_f c_p (T_{fo} - T_{fi})}{A_c [G_t(\tau\alpha) - U_c(T_{fi} - T_a)]} \tag{6.63a}$$

Rearranging, we can express the collector heat removal factor in the following form:

$$F_R = \frac{\dot{m}_f c_p}{A_c U_c} \left[\frac{(T_{fo} - T_{fi})}{\left[\frac{G_t(\tau\alpha)}{U_c} - (T_{fi} - T_a) \right]} \right] \tag{6.63b}$$

Equation (6.63b) can be rearranged to the following form

$$F_R = \frac{\dot{m}_f c_p}{A_c U_c} \left[\frac{\left[\frac{G_t(\tau\alpha)}{U_c} - (T_{fi} - T_a) \right] - \left[\frac{G_t(\tau\alpha)}{U_c} - (T_{fo} - T_a) \right]}{\left[\frac{G_t(\tau\alpha)}{U_c} - (T_{fi} - T_a) \right]} \right] \tag{6.63c}$$

and finally, by combining with Eq. (6.60c), we can express collector heat removal factor as

$$F_R = \frac{\dot{m}_f c_p}{A_c U_c} \left[1 - \exp \left(-\frac{A_c U_c F'}{\dot{m}_f c_p} \right) \right] \tag{6.63d}$$

Collector heat removal factor, F_R, is a very useful parameter that characterizes the performance of a solar collector design. For a known solar collector heat removal parameter, F_R, the heat energy gain of collector can be estimated from Eq. (6.61) as

$$Q_u = F_R A_c [G_t(\tau\alpha) - U_c(T_{fi} - T_a)] \tag{6.64}$$

The analysis presented here is primarily based on assuming several straight parallel tubes over the surface of the collector for recovering heat absorbed by the collector absorber plate. A variety of other tube flow networks, including series–parallel straight or serpentine layout designs, could be considered for enhanced performance of the solar collector. However, flow and heat transfer in such flow networks are complex and require three-dimensional computational fluid dynamic analysis. One such computational analysis for a flat-plate solar collector is presented in the Section 6.5.1.7.

6.5.1.7 Three-Dimensional Analysis

While one-dimensional analysis is good for qualitative understanding performance of solar collector, a more comprehensive and quantitative understanding through a three-dimensional analysis and experimental testing of the collector is essential. Figure 6.31 shows three-dimensional CAD model of a solar thermal collector.

The computational thermal solution of this model is shown in Figure 6.32.

Fluid is significantly hotter around the outer layer of the fluid adjacent to the tube, specially at the top section, which is in direct contact with the absorber plate heated by the solar irradiation. Table 6.9 presents typical design specifications for flat-plate solar collector.

6.5.2 Concentrating Collector

For demonstrating the thermal analysis of a concentrating solar collector, let us consider a typical section of a cylindrical receiver tube covered with cylindrical glass tube as shown in Figure 6.33.

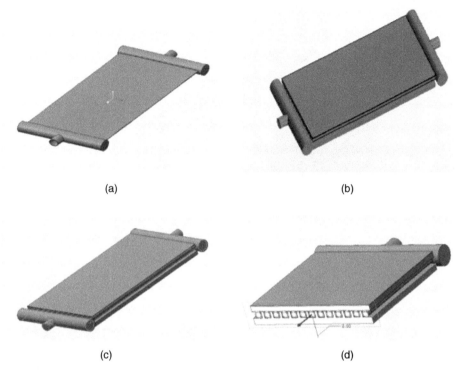

(a)

(b)

(c)

(d)

Figure 6.31   Three-dimensional CAD model of a flat-plate solar collector. (a) Absorber plate with riser and pipes. (b) Solar thermal collector with glass glazing. (c) Absorber plate and flow channels. (d) Sectional view of a collector showing.

An overall energy balance over the entire solar collector can be written as follows:

$$(\tau\alpha)A_c G_t = Q_u + Q_{\text{Loss}} \tag{6.65}$$

where

G_t = Total concentrated solar radiation

A_c = Collector area

$(\tau\alpha)$ = Transmittance–absorptance product for solar radiation

Q_u = Useful heat energy transferred to circulating fluid in tubes

Q_{Loss} = Energy loss from the collector to the surrounding by radiation, convection, and conduction

Heat loss term, Q_{Loss}, in Eq. (6.65) can be expressed considering the resistance diagram shown in Figure 6.33c. As we can see from the resistance diagram, the heat loss from outer surface of the receiver tube involves number of mechanisms: (i) combined conduction and radiation heat transfer conduction through the medium in the space between the receiver surface and the cover glass; (ii) heat conduction through the cover glass; and (iii) combined radiation heat transfer from cover glass surface to sky and wind-driven forced convection heat loss to ambient air through the media gas in the space between the receiver surface and the cover glass. The expressions for these loss components can be derived assuming

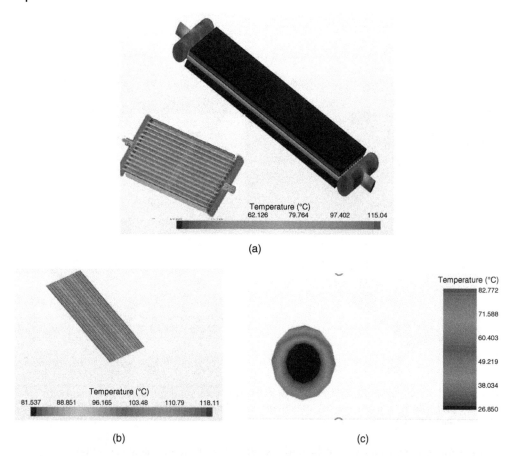

(a)

(b)   (c)

Figure 6.32 Computational solution for temperature distribution a flat-plate collector. (a) Temperature variation within the collector. (b) Absorber-plate temperature distribution. (c) Fluid temperature at outlet section of the tube.

negligible temperature variation around the cylindrical receiver tubes surface and assuming steady-state condition as described below.

Heat loss from receiver outer surface to glass cover inner surface by combined conduction and radiation

$$Q_{\text{Loss}} = \frac{(T_{\text{ro}} - T_{\text{ci}})}{\frac{\ln \frac{r_{\text{ci}}}{r_{\text{ro}}}}{2\pi L K}} + \frac{2\pi r_{\text{ro}}(T_{\text{ro}}^4 - T_{\text{ci}}^4)}{\frac{1}{\varepsilon_r} - \frac{1-\varepsilon_c}{\varepsilon_c}\left(\frac{r_{\text{ro}}}{r_{\text{ci}}}\right)} \tag{6.66a}$$

Heat loss by conduction through glass cover

$$Q_{\text{u}} = \frac{(T_{\text{co}} - T_{\text{ci}})}{\frac{\ln \frac{r_{\text{co}}}{r_{\text{ci}}}}{2\pi L k_{\text{g}}}} \tag{6.66b}$$

Heal loss by convection and radiation from glass cover outer surface

$$Q_{\text{Loss}} = 2\pi r_{\text{co}} h_{\text{co}}(T_{\text{ro}} - T_{\text{a}}) + \varepsilon_{\text{g}} 2\pi r_{\text{co}} L\sigma (T_{\text{co}}^4 - T_{\text{sky}}^4) \tag{6.66c}$$

Table 6.9 Typical design specifications for flat-plate solar collector.

Collector dimensions	62-in. × 26-in. × 4-in.
Collector type	Flat plate
Number of glazing covers	1
Glazing cover material	Polyvinyl fluoride
Glazing cover transmissivity	0.93
Cover thickness	0.125-in.
Absorber-plate dimensions	48-in. × 24-in. × 0.33-in.
Absorber-plate material	Aluminum
Construction type	Fin and tube
Insulation type	Polystyrene
Insulation thickness	1.3-in.
Working fluid	Glycol water
Riser diameter	3.5 in.

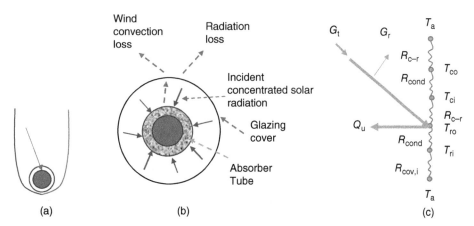

Figure 6.33 Energy transfer in a concentrating solar collector. (a) Concentrating solar collector. (b) Receiver tube section. (c) Resistance diagram showing heat transmission paths.

Useful heat energy term, Q_u, in Eq. (6.65) represents the heat transferred to circulating fluid in the receiver tube and is influenced by the conduction resistance through the receiver tube wall thickness and convection heat transfer resistance to circulating collector fluid. Conduction heat transfer through the receiver wall thickness:

$$Q_u = \frac{(T_{ro} - T_{ri})}{R_{Cond}} = \frac{(T_{ro} - T_{ri})}{\frac{\ln \frac{r_i}{r_o}}{2\pi L k_r}} \tag{6.67a}$$

Convection heat transfer from the inner receiver tube surface to the circulating fluid:

$$Q_u = \frac{(T_{ri} - T_f)}{R_{Conv,i}} = \frac{(T_{ri} - T_f)}{\frac{1}{2\pi r_i h_{conv,i}}} \tag{6.67b}$$

Combining Eqs. (6.67a) and (6.67b), we have

$$Q_u = \frac{(T_{ro} - T_f)}{R_{cond} + R_{conv,i}} \tag{6.68a}$$

or

$$Q_u = \frac{(T_{ro} - T_f)}{\frac{\ln \frac{r_i}{r_o}}{2\pi LK} + \frac{1}{2\pi r_i h_{conv,i}}} \tag{6.68b}$$

Useful heat energy transferred to circulating fluid can also be written based on an energy balance over the circulating fluid as

$$Q_u = \dot{m}_f c_p (T_{fo} - T_{fi}) \tag{6.69}$$

where

$\dot{m}_f$ = Mass flow rate of the collector fluid
T_{fi} = Collector fluid inlet temperature
T_{fo} = Collector fluid outlet temperature

6.5.3 Collector Performance Characterization

The collector efficiency of flat-plate collectors varies with design orientation, time of day, and the temperature of the working fluid. The amount of useful energy collected will also depend on the optical properties (transmissivity and reflectivity) of cover glasses or glazing, the properties of the absorber plate (absorptivity and emissivity) and losses by conduction, convection, and radiation.

The amount of solar irradiation reaching the top of the outside glazing will depend on the location, orientation, and the tilt of the collector. Depending on the transmittance value of the glass cover plate (τ), a fraction of the solar radiation falls on absorber plate and depending on the absorptivity value (α) of the absorber plate, a fraction energy is absorbed and raises the temperature of the absorber plate. Absorbed heat diffuses along the length toward the tube and is transferred to the circulating fluid. Temperature of the absorber plate varies along the plate with peak at the midsection. A fraction of the absorbed energy is lost by convection from the top by radiation due to energy emitted by the absorber plate and convection due to moving air over the collector surface, and a fraction of energy is lost from the bottom through conduction and convection.

It can be reiterated here again that the thermal heat energy absorbed at the collector absorber plate is given as

$$q_s'' = G_t(\tau \alpha) \tag{6.70}$$

6.5.3.1 Solar Collector Efficiency (η_c)
Solar collector efficiency is defined as the ratio of the useful heat gained by the collector fluid to the incident solar radiation at the collector surface, G_t. and expressed as

$$\eta_c = \frac{Q_u}{A_c G_t} \tag{6.71a}$$

or

$$\eta_c = \frac{\dot{m}_f c_p (T_{fo} - T_{fi})}{A_c G_t} \tag{6.71b}$$

Substituting Eq. (6.64) into Eq. (6.71a), we get the alternate expression for solar collector efficiency as

$$\eta_c = F_R \left[(\tau\alpha) - \frac{U_c(T_{fi} - T_a)}{G_t} \right] \tag{6.71c}$$

Equation (6.71c) is most popularly used to characterize the solar collector performance using two parameters: $F_R(\tau\alpha)_n$ and $F_R U_c$, which are usually determined from experimental testing of the collector following a test procedure such as that outlined in ASHRAE/ANSI 93-2003 standard (2003). The test procedure involves measuring collector fluid mass flow rate, $\dot{m}_f$, fluid temperature, T_{fi}, and outlet temperature, T_{fo}, and incident solar radiation, G_t. The collector efficiency is then calculated from $\eta_c = \frac{\dot{m}_f c_p (T_{fo} - T_{fi})}{A_c G_t}$ given by Eq. (6.71b) and plotted against $\frac{U_c(T_{fo} - T_{fi})}{G_t}$. A simple data-fitted form of this plot is straight line, while the quantities such as U_c, F_R, and $(\tau\alpha)$ are assumed as constant.

Figure 6.34 shows a typical theoretical representation of collector performance based on experimental correlated data and following collector efficiency equation (6.71b). In this plot, collector efficiency is plotted as a function of the collector heat loss variable $\frac{U_c(T_{fo} - T_{fi})}{G_t}$. For constant values of F_R, $(\tau\alpha)$, and U_c, the plot shows a straight-line variation where intercept on the y-axis gives the value of $F_R(\tau\alpha)$ and the **negative slope** of the line gives the value of $F_R U_c$. As we can see, the collector fluid inlet temperature equals the ambient temperature, and there is no thermal heat loss and the collector efficiency reaches the maximum value, $F_R(\tau\alpha)$, given by the intercept with y-axis. As the line intersects the x-axis, the collector efficiency drops to zero. This is a condition that corresponds to low-incident solar radiation or high collector fluid inlet temperature making the heal loss equal to the absorbed solar radiation.

It can be reiterated here again that the overall heat transfer coefficient, U_c, is a function of wind speed, ambient air temperature, and number of glazing covers. Also, it is expected

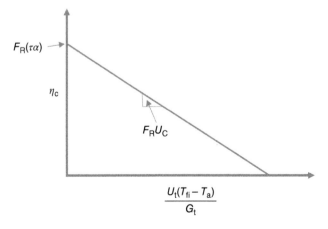

Figure 6.34 Graphical representation of solar collector performance characteristics.

that the collector heat removal factor, F_R, depends to some extent on the temperature, and total solar radiation, G_t, varies with the time of the day and year. This simplified model can be extended to include the dependence of the parameter F_R and $\tau\alpha$ on time of the day or the angle of incidence of solar radiation, and temperature dependence value of F_R and U_c.

Example 6.7 *Flat-Plate Solar Collector Efficiency*

A 1-ft by 3-ft flat-plate double-glazed collector is available for a solar-heating application. The transmittance of each of the two non-heat-absorbing glass cover plates is 0.87 and the aluminum absorber plate has an absorptivity of $\alpha = 0.95$. Determine the collector efficiency when $G_t = 900$ W/m^2, $T_{fi} = 50\,^\circ$ C, and $T_\infty = 10\,^\circ$ C. Use $U_t = 8$ W/m^2 K and $F_R = 0.92$.

Solution:

Compute the solar collector efficiency using Eq. (6.71c)

$$\eta_{col} = F_R \left(\tau_{c1}\tau_{c2}\alpha_a - \frac{U_c(T_{fi} - T_\infty)}{G_t} \right)$$

$$= 0.92 \left(0.97 \times 0.97 \times 0.95 - \frac{8(50 - 10)}{900} \right)$$

$$\eta_{col} = 33.4\%$$

Bibliography

Abhas, A. (2000). Solchrome solar selective coatings – an effective way for solar water heater globally. *Renewable Energy* 19 (1): 145–154.

ASHRAE (1993). *ASHRAE Handbook 1993: Fundamentals*. Atlanta, GA: American Society of Heating, Refrigeration, and Air Conditioning Engineers.

ASHRAE/ANSI Standard 93-2003 2003. *Methods of testing to determine the thermal performance of solar collectors*. Atlanta, GA: American Society of Heating, Refrigeration, and Air Conditioning Engineers.

Duffie, J.A. and Beckman, W.A. (2000). *Solar Engineering of Thermal Processes*, 3e. Wiley.

Harrington, R.V. and Karstetter, B.R. (1967). Heat absorbing glass. Corning, New York. US Patent 3,326,703. Awarded to Corning, New York.

Kuehn, T.H., Ramsey, J.W., and Threlkeld, J.L. (1998). *Thermal Environmental Engineering*, 3e. Prentice Hall Publisher.

Lowery, J.R. (1977). Solar Absorption Characteristics of Several Coatings and Surface Finishes. NASA Technical Memorandum. *NASA TM-X-3509*. National Aeronautics and Space Administration, Washington.

Mooch, G.C. and Ricker, R.W. (1946). Heat absorbing glass. US Patent 2,397,195. Application date: January 9, 1941, Award date: March 26, 1946, Awarded to Libbey-Owens-Ford Glass Company, Toledo, Ohio, USA.

Parmelee, G.V. (1945). Transmission of solar radiation through flat glass. *ASHVE Trans.* 51: 317–350.

Qunwu, H., Yiping, W., and Jinhua, L. (2007). Preparation of solar selective absorbing CuO coating for medium temperature applications. *Front. Chem. Eng. China* 1 (3): 256–260.

Rabl, A. (1976a). Comparisons of solar collectors. *Sol. Energy* 18 (2): 93–111.

Rabl, A. (1976b). Optical and thermal properties of compound parabolic concentrators. *Sol. Energy* 18 (6): 497–511.

Rabl, A., O'Gallagher, J., and Winston, R. (1977). Design and Test of Non-Evacuated Solar Collector with Compound Parabolic Concentrators. University of Chicago. *DOE Report IY-77-S-02-2446*.

Rabl, A., Goodman, N.B., and Winston, R. (1979). Practical design of considerations for CPC solar collectors. *Sol. Energy* 22: 373–381.

Santos-Gonzalez, I., Sandoval-Reyese, M., Garcia-Valladares, O. et al. (2014). Design and evaluation of a compound parabolic concentrator for heat generation of thermal processes. *Energy Procedia* 57: 2956–2965.

Scherer, A.M., Inal, O.T., and Singh, A.J. (1983). Investigation of copper oxide coatings for solar selective applications. *Sol. Energy Mater.* 9 (2): 139–158.

Stoecker, W.F. and Jones, J.W. (1982). *Refrigeration and Air Conditioning*, 2e. McGraw-Hill.

Wagmare, S.A. and Gulhane, N. (2016). Design and ray tracing of a compound parabolic collector with tubular receiver. *Sol. Energy* 137: 165–172.

Welford, W.T. and Winston, R. (1978). *The Optics of Non-imaging Concentrators*. Academic.

Winston, R. (1974). Solar concentrations of novel design. *Sol. Energy* 16: 89.

Problems

6.1 Calculate the transmittance of one cover of a solar collector at normal incidence and at 70° angle of incidence. Assume the glazing as standard nonabsorbing glass.

6.2 Consider a single water-white glass sheet and determine the optical properties of this glass sheet for solar radiation at normal incidence, at $\theta = 30°$ for a glass of thickness, $t_g = 4$-mm.

6.3 Consider a single water-white glass sheet and determine the optical properties of this glass sheet for solar radiation at normal incidence, at (a) $\theta = 0°$, (b) $\theta = 30°$, and $\theta = 60°$ for three different glass thicknesses: (i) 0.0625, (ii) 0.125, and (iii) 0.25. Give results in a summary table.

6.4 Estimate the transmittance–absorptance product $(\tau\alpha)$ of a single cover glass-glazing cover of thickness 3.0 mm and at an angle of incidence of 50°. Assume the glass as water-white with an extinction coefficient value of 10 m^{-1} and the absorptivity of the absorber plate as $\alpha = 0.98$.

6.5 Consider 1-ft by 3-ft flat-plate collector with following specifications: absorber plate: copper with thermal conductivity: 360 W/m C; plate thickness: 1.2-mm; tube spacing: 100-mm; tube diameter: 12.7-mm; collector loss coefficient: 12.0 W/m² C. Calculate the calculator efficiency factor, F' and heat removal factor, F_R considering water flow rate of 0.45 ft/s, water inlet temperature and ambient temperature as 42 and 15 °C.

6.6 For the flat-plate collector.

6.7 A 1-ft by 3-ft flat-plate double-glazed collector is available for a solar-heating application. The transmittance of each of the two non-heat-absorbing glass cover plates is 0.87 and the aluminum absorber plate has an absorptivity of $\alpha = 0.90$. Determine the collector efficiency when $G_t = 500$ W/m^2, $T_{fi} = 50°$ C, and $T_\infty = 25°$ C. Use $U_t = 12$ W/m^2 K and $F_R = 0.91$.

7

Rotary Components in Thermal Systems

One of the major classes of rotary machines are the turbomachines, which are fluid-handling devices that direct the flow with the aid of vanes attached to a rotating shaft for the purpose of either producing power or increasing pressure. Examples of such turbomachines are **turbines** and **prime movers** like pumps, fans, blowers, and compressors. Other types of power-producing devices like reciprocating piston-cylinder engine as well as pressure-increasing devices like gear pumps, are generally not included in the classification of turbomachines. Major distinction is that the turbomachines are rotating machines used to produce power or increase pressure or head of the working fluid. The pressure change is obtained by the combination of centrifugal force and Coriolis force caused due to rotation and change in radius from the inlet to the outlet.

7.1 Turbomachine Types

Turbomachines are primarily classified into two groups based on the direction of energy transfer between the fluid and the machine. **Group one** includes machines in which power is the input as fluid receives energy from the rotating impeller shaft and increases head or pressure. This group includes **pumps, fans, blowers, and compressors**. Group two includes machines that produce power output, as fluid moves through the machines. This group includes **turbines** such as **steam turbines**, **gas turbines**, **wind turbines,** and **hydraulic turbines**.

Turbines are used to extract energy from fluids and produce power in two steps: convert fluid energy into mechanical energy through rotating blades in an impeller and shaft, and then convert the mechanical energy into electrical energy by connecting the rotating shaft to an electrical generator. Turbines are categorized based on the type of fluid and the major energy forms available for conversion. For example, steam turbines are used for converting thermal energy from steam; gas turbines for converting thermal energy from high temperature and pressure gas; water turbines for converting potential energy of water; and wind turbines for converting kinetic energy of wind. Among the prime movers, **pumps** are used for moving liquids and slurries and creating increase in pressure; fans and blowers are used for moving gases and vapors while creating small pressure rise; and **compressors** are used for raising large pressure rise in air and gases. Centrifugal fans are used in applications where the pressure rise across the channel is high for a lower mass flow rate.

Design of Thermal Energy Systems, First Edition. Pradip Majumdar.
© 2021 John Wiley & Sons Ltd. Published 2021 by John Wiley & Sons Ltd.
Companion website: www.wiley.com/go/majumdar

Turbomachines are also categorized as **axial-flow**, **mixed-flow,** and **radial-flow** devices based on the flow configuration such as the meridional path of the working fluid. In an axial-flow machine, the fluid path from inlet to outlet is mainly in the axial direction. On the other hand, the fluid may enter the machine axially, but exit radially in a radial-flow machine, and somewhere in between directions in a mixed-flow machine. **Axial pumps** are designed to handle large flows and for lower-pressure rise at high efficiency. **Radial machines** are designed for large-pressure rise and lower flow rates at high efficiency. Mixed-flow machines are designed to produce medium-pressure rise with medium-flow rates. For example, **cooling water pumps** in a secondary cooling water system in a power plant are **axial-flow pumps** that circulate large amount of cooling water through the condensers and cooling towers. **Boiler water pumps** and **feed water heater pumps** on the other hand are **radial-flow pumps** to produce large-pressure rise.

The flow fields in these devices are three dimensional, and often are unsteady and turbulent depending on the type of fluids and flow conditions. The flow passages in between the vanes are narrow and long with varying cross-sectional shape, and are dominated by viscous effects, rotation, Coriolis forces, separation, and secondary flows.

This chapter primarily discusses the designs, operation, and performance characteristics of centrifugal pumps.

7.2 Basic Equations of Turbomachines

All turbomachines are composed of two major components: a **rotor** with attached blades or vanes forming fluid-flow passages and connected to rotating shaft; and a **stator or volute** an enclosure for the rotor and forms the volute exit passage of the fluid out of the blade passages. The energy transfer in the form of work energy takes place between the rotor and the fluid as it travels through the rotor-blade passages as shown in Figure 7.1.

7.2.1 Conservation of Angular Momentum

The design analysis of all turbomachines was carried out based on the application of a form of Newton's laws of motion applied to the fluid moving through the blade passages of the

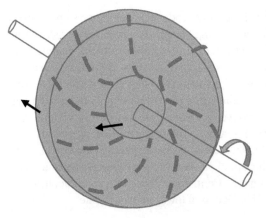

Figure 7.1 A rotor or impeller with attached number of blades mounted on a rotating shaft.

rotor. Basic theoretical equations to analyze turbomachines are derived from conservation of mass and conservation moment of momentum equations.

The **conservation of mass**:

$$\frac{\partial}{\partial t}\left(\int_{CV} \rho d\forall\right) + \int_{CS} \rho\vec{V} \cdot d\vec{A} = 0 \tag{7.1a}$$

For steady flow

$$\int_{CS} \rho\vec{V} \cdot d\vec{A} = 0 \tag{7.1b}$$

For incompressible flow in a nondeformable control volume

$$\int_{CS} \vec{V} \cdot d\vec{A} = 0 \tag{7.1c}$$

The conservation of angular momentum or **moment of momentum equation** for a system is given as

$$\vec{T} = \left.\frac{d\vec{H}}{dt}\right)_{Sys} \tag{7.2}$$

where

$\vec{T}$ = Total torque exerted on the system
$\vec{H}$ = Angular momentum

Angular momentum is defined as

$$\vec{H} = \int_m \left(\vec{r} \times \vec{V} \, dm\right) = \int_m \left(r \times \vec{V}\rho \, d\forall\right) \tag{7.3}$$

Total torque acting over a system can be produced by a number of forces like surface forces, body forces, and by shafts that pass through the system boundary. This is expressed as

$$\vec{T} = \vec{r} \times \vec{F}_S + \int_m \vec{r} \times \vec{g} dm + \vec{T}_{\text{Shaft}} \tag{7.4}$$

Substituting Eq. (7.4) into Eq. (7.2), we get

$$\vec{r} \times \vec{F}_S + \int_m \vec{r} \times \vec{g} dm + \vec{T}_{\text{Shaft}} = \left.\frac{d\vec{H}}{dt}\right)_{Syst} \tag{7.5}$$

Now with the use of **Reynolds transport theorem** for the absolute derivative for the system, we can express the moment of momentum equation for a control volume as

$$\vec{r} \times \vec{F}_S + \int_m \vec{r} \times \vec{g}\rho d\forall + \vec{T}_{\text{Shaft}} = \frac{\partial}{\partial t}\int_{CV} \left(\vec{r} \times \vec{V}\right)\rho\forall + \int_{CS} \left(\vec{r} \times \vec{V}\right)\rho\vec{V} \cdot d\vec{A} \tag{7.6}$$

Three terms on left-hand side of the Eq. (7.6) are torques corresponding to forces like surface forces, body forces, and shafts. The first term on the right-hand side of Eq. (7.6) represents the rate of change of angular momentum within the control volume. The second term represents the net outflow of angular momentum through the control volume surfaces.

7.2.2 The Euler Equation of Energy Transfer in Turbomachines

For the analysis of the torque in a rotary turbomachine, let us consider a fixed control volume enclosing the rotor or the impeller with the number of blades attached radially as shown in Figure 7.2.

The axis of rotation is considered along the z-axis and the shaft rotates at an angular velocity, ω. So, blades or vanes attached to the rotor also rotate in a circular path with the angular velocity, ω. Fluid enters the impeller at the suction eye and inlets the blades pages through the inner radius, r_1, and exits at the outer radius, r_2. The blade width is assumed to vary from $\mathbf{b_1}$ from inlet section to $\mathbf{b_2}$ outlet section. The velocity vectors may be resolved into three mutually perpendicular directions: (i) Axial component velocity, V_a, along the axis of rotation; (ii) Radial component velocity, V_r; and (iii) Tangential component velocity, V_t.

Assume following assumptions to derive a simplified conservation of torque equation for the rotor: (**i**) steady flow; (**ii**) Flow through the impeller is assumed as one dimension for a simplified theory and assumed as uniform flow at each flow section; (**iii**) State of the fluid at any point in the flow passage is constant; (**iv**) rate of heat transfer and work transfer across the rotor control volume are constant; (**v**) neglect torque due to surface forces such as viscous force and pressure force as a first approximation; (**vi**) neglect torque generated by the body forces due to the symmetry of the flow geometry; (**vii**) no leakage flows. With these assumptions, Eq. (7.6) reduces to a simplified form for the ***shaft torque*** given as

$$\vec{T}_{\text{Shaft}} = \int_{\text{CS}} \left(\vec{r} \times \vec{V} \right) \rho \vec{V} \cdot d\vec{A} \tag{7.7a}$$

For a turbomachine with input work, this shaft torque causes a change in angular momentum of the fluid, and for a turbomachine, which produces work output; this is the torque produced by the changes in fluid angular momentum.

This equation can be written in scalar form by considering the z-axis aligned with the axis of rotation and considering inflow at inner radius and outflow at the outer radius of the rotor as

$$T_{\text{Shaft}}\hat{k} = \left[\left(\vec{r} \times \vec{V} \right)_{\text{inlet}} - \left(\vec{r} \times \vec{V} \right)_{\text{outlet}} \right] \dot{m} \tag{7.7b}$$

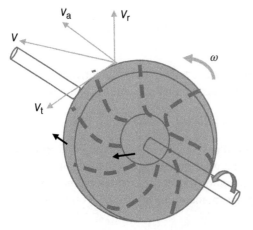

Figure 7.2 Impeller blades with velocity components.

Flow through the impeller is quite complex and three dimensional in nature. One-dimensional mean line analysis is often used to estimate the conceptual flow path, blading, pressure loads, and performance characteristics of turbomachines.

For simplicity, we assume uniform average velocity at inlet and out sections to transform Eq. (7.7b) to the following form:

$$T_{\text{Shaft}} = \left(r_2 V_{t2} - r_1 V_{t1} \right) \dot{m} \tag{7.8}$$

where

r_1 = radius of the impeller at the blade passage inlet
r_2 = radius of the impeller at the blade passage outlet
V_{t1} = tangential component of the fluid velocity at blade inlet
V_{t2} = tangential component of the fluid velocity at blade outletNet

Equation (7.8) is referred to as **Euler's equation** for turbomachines. Changes in the radius of the rotor and in the magnitude of the tangential component of the fluid velocity from inlet to outlet sections of blade contribute to the torque or change in angular momentum of the fluid.

Network done on the turbomachine rotor is given as the dot product of the angular velocity, $\vec{\omega}$, and the applied torque, $\vec{T}$.

$$W_{\text{shaft}} = \vec{\omega} \cdot \vec{T}_{\text{Shaft}} = \omega \hat{k} \cdot T_{\text{Shaft}} \hat{k} \tag{7.9a}$$

$$W_{\text{shaft}} = \omega T_{\text{Shaft}} \tag{7.9b}$$

Substituting Eq. (7.8) into Eq. (7.9b), we have the work energy

$$W_{\text{shaft}} = \omega \left(r_2 V_{t2} - r_1 V_{t1} \right) \dot{m} \tag{7.10}$$

where

V_{t1} = tangential component of the absolute velocity, V_1, at the inlet of the blade
V_{t2} = tangential component of the absolute velocity, V_2, at the outlet of the blade

Noting that the blade velocity at any radius is $U = \omega r$, the shaft work energy can be written as

$$W_{\text{shaft}} = \left(U_2 V_{t2} - U_1 V_{t1} \right) \dot{m} \tag{7.11a}$$

or

$$W_{\text{shaft}} = \left(U_2 V_{t2} - U_1 V_{t1} \right) \rho Q \tag{7.11b}$$

and the shaft work energy per unit mass flow rate is written

$$W_{\text{shaft}} = \left(U_2 V_{t2} - U_1 V_{t1} \right) \tag{7.11c}$$

The shaft work energy or the change in angular momentum of fluid, given by Eq. (7.11c), may be positive or negative depending on the type of turbomachines. Based on a general thermodynamic sign convention, work done by the fluid is considered as positive, i.e. $U_1 V_{t1} > U_2 V_{t2}$ like in a turbine in which work is produced. Work done on a fluid by the machine is negative, i.e. $U_2 V_{t2} > U_1 V_{t1}$ for pumps and compressors in which work energy is added. As we can see, it is the product term UV_t that is the controlling factor in designating or designing a turbomachine as turbine or as pump.

The shaft work energy transfer per unit mass flow of the fluid is the head increased or reduced in the fluid by the turbomachines. This change of the fluid head can be written as

$$h_{shaft} = \frac{W_{in}}{\dot{m}g} = \frac{1}{g}\left(U_2 V_{t2} - U_1 V_{t1}\right) \tag{7.12}$$

The shaft head represents the change in total head of the fluid between inlet and outlet section across the machine in a fluid system and can also be written from energy equation as

$$h_{shaft} = H_{out} - H_{in} = \left(\frac{P_2}{\rho g} + \frac{V_2^2}{2g} + Z_2\right) - \left(\frac{P_1}{\rho g} + \frac{V_1^2}{2g} + Z_1\right) \tag{7.13}$$

7.2.3 Velocity Diagrams

The equations derived in the previous section show that to evaluate performance in terms of head, work, and torque of an impeller with a given blade or vane profile, we need to evaluate the **velocity diagrams**, also known as **velocity triangles**, at the inlet and outlet sections of vane inlet and at the outlet. Velocity triangles provide us with some insights about the design under consideration as a first-order estimation. Let us define following blade or vane angles and fluid velocities at inlet and outlet sections of the vane:

At inlet:

β_1 = Vane angle at the vane inlet. Fluid enters the vane at this angle with a swirl
$U_1 = \omega r_1$ = Impeller or vane speed at inlet in the peripheral direction
V_1 = Absolute fluid velocity at inlet
α_1 = Flow angle of the absolute velocity measured from the direction normal to flow area at inlet
V_{t1} = Tangential or peripheral component of the absolute velocity, V_1, at the inlet of the vane
V_{n1} = Meridional or normal component of the absolute velocity, V_1, at the inlet of the vane
W_1 = Relative fluid velocity

Outlet:

β_2 = Vane angle exit of the blade inlet. Fluid exits the vane at this angle
$U_2 = \omega r_2$ = Impeller or vane angular speed at inlet in the peripheral direction
V_2 = Absolute fluid velocity measured from the direction normal to the flow area at exit
α_2 = Flow angle of the absolute velocity measured from the direction normal to flow area at exit
V_{t2} = Tangential or peripheral component of the absolute velocity, V_2, at the outlet of the vane
V_{n2} = Meridional or normal component of the absolute velocity, V_2, at the inlet of the vane
W_2 = Relative fluid velocity

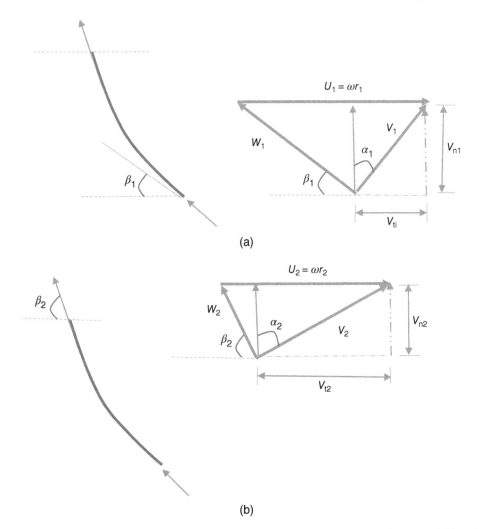

Figure 7.3 Vane or blade geometry and velocity diagrams. (a) Inlet velocity diagram and (b) outlet velocity diagram.

Note that the vane angles β_1 and β_2 are geometrical angles at the inlet and exit ends of the vane, representing its physical shape as shown in Figure 7.3.

The **absolute velocity components,** V_1 and V_2, at the inlet and outlet sections are referred to the velocity in the stationary pipes, guide vanes, and volute parts as measured with respect to the stationary coordinate system. **The relative velocity components,** W_1 and W_2, on the other hand are given with respect to the coordinate system that rotates with the impeller at a speed of $\omega_1 = \frac{U_1}{r_1}$ and $\omega_1 = \frac{U_1}{r_1}$, respectively. Both the absolute and relative velocity components considered here are the ideal velocities if the fluid enters and leaves the vane passages with perfect contact with the vane wall surface. The velocity triangle is completed by drawing the absolute velocity, V, as the vector sum of the relative velocity, W, and the impeller speed, U.

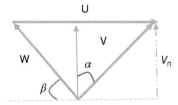

Figure 7.4 A general velocity diagram.

The relative velocity components are given by dividing the flow per vane flow channel by the cross-sectional flow area given as follows:

$$W_1 = \frac{Q}{n_v A_{vc1}} \quad \text{and} \quad W_1 = \frac{Q}{n_v A_{vc1}} \tag{7.14}$$

where

Q = total volume flow rate

n_v = number of vanes or number of vane flow channels

A_{vc1} and A_{vc2} = area of vane flow channel at inlet and outlet, respectively

The absolute velocity can be resolved into a tangential component, V_t, and a meridional or normal component, V_n, as shown in the velocity diagram in Figure 7.4. The meridional or normal velocity components, V_{n1} and V_{n2}, are given on the meridional planes of the absolute velocity and relative velocity at the inlet and outlet of the vanes. The direction of normal component velocity, V_n, for flow through the rotor can vary and can be in the axial, radial, or in a mixed direction depending on the type of turbomachines. For example, in a radial-flow machine, it is the radial component of the absolute velocity and it is the axial component of the absolute velocity in an axial-flow machine.

For a typical blade passage, the absolute velocity, V_1, of the fluid entering the blades passage is the vector sum of the rotor-blade velocity at inlet, $U_1 = \omega r_1$, and the relative velocity, W_1, and so $V_1 = W_1 + U_1$. Similarly, for the outlet section of the blade passage, the absolute velocity, V_2, is the vector sum of rotor-blade velocity, U_2, and the fluid relative velocity relative at the exit of blade passage, i.e. $V_2 = W_2 + U_2$.

Note that in the derivation of the torque or changes in the angular momentum, only the changes in the tangential component of velocities at the inlet and outlet sections of the rotor are considered. Changes in the axial component velocity as well as change in the radial component velocities do not contribute to the torque or the angular momentum. The changes in the axial component velocities from inlet to the outlet section impeller lead to a net axial force, which is required to be withstood by the bearings in the stator section of the turbomachines. Similarly, the changes in the radial component of velocities lead to a net radial force that is withstood by the journal bearing in the rotating shaft.

The absolute velocity will be in the radial direction when the flow at the inlet of the vane is free from any *swirl,* which is introduced by using *inlet guide vanes.* Presence of any such preswirl will force the absolute inlet velocity direction to differ from the radial direction.

To get a better understanding of the energy transfer in the turbomachine, an alternative form of Eq. (7.12) is often used. This is derived based on the velocity diagrams at the inlet and exit sections. Let us consider a typical velocity polygon shown below:

From the right-angle triangle on the right-hand side, we can write

$$V^2 = V_t^2 + V_n^2 \tag{7.15a}$$

or

$$V_n^2 = V^2 - V_t^2 \tag{7.15b}$$

Similarly, from right-angle triangle on the left, we can write the normal component velocity as

$$V_n^2 = W^2 - \left(U - V_t\right)^2 \tag{7.16}$$

Combining Eqs. (7.15b) and (7.16), we get

$$UV_t = \frac{V^2 + U^2 - W^2}{2} \tag{7.17}$$

Let us now use this Eq. (7.17) for the product term UV_t for the inlet and exit sections in Eq. (7.11c) and express the shaft work in alternative form as

$$w_{\text{shaft}} = \frac{\left(V_2^2 - V_1^2\right) + \left(U_2^2 - U_1^2\right) - \left(W_2^2 - W_1^2\right)}{2} \tag{7.18}$$

and substituting in Eq. (7.12) in terms of fluid head change as

$$h_{\text{shaft}} = \frac{\left(V_2^2 - V_1^2\right) + \left(U_2^2 - U_1^2\right) - \left(W_2^2 - W_1^2\right)}{2g} \tag{7.19}$$

This alternative form for Eq. (7.12) shows that the net shaft work energy or shaft head is composed of multiple different energy forms: (i) The first term, $\frac{V_2^2 - V_1^2}{2g}$, is the changes in the absolute kinetic energy and referred to as the change in **dynamic head or dynamic pressure**; (ii) The second term, $\frac{U_2^2 - U_1^2}{2g}$, is the change in the network or fluid energy due to the rotational movement of fluid through the rotor from inlet radius to the outlet radius of the rotor blade, and this is referred to as one of the components of the change in the **static head** or **static pressure** caused by the centrifugal force, and (iii) The third term $\frac{W_2^2 - W_1^2}{2g}$ represents the change in the kinetic energy due to changes in relative velocity and contributes to the change in **static head** or **static pressure** of the fluid.

7.2.4 Slip Consideration

In the presentation of the vane velocity triangles, we have considered the absolute velocity, V, as well as relative velocity, W, as ideal velocities assuming a perfect contact of the fluid with the vane walls and in the vane flow passages. Primary reason for such an assumption is that computation of the ideal velocity components is much simpler than that of the real velocity components. While this makes the analysis much simpler, the results also inherit errors to some extent. In reality, because of the effect of fluid viscosity and boundary layer formation, the fluid flow is subjected to no-slip velocity condition at the vane wall and results in a nonuniform velocity profile in vane flow channels as well as flow separations and turbulence. The error comes from the resulting deviations of the fluid-flow directions from the vane angles at the outlet as well as at the inlet, because it modifies the peripheral or tangential components of the velocities, V_{t1} and V_{t2}, and results in proportionate variations in the estimated values of the power input and total head as given by Eqs. (7.11c) and (7.12), respectively.

Determination of the slip conditions and the deviation in flow direction are very involved and require computational fluid dynamics analysis of flow in the complexly shaped vane flow channel geometries. Usual design practice is to use a slip factor to consider the variations in the peripheral or tangential component of the velocity and this is defined as

$$\mu_{\text{Slip}} = \frac{V_t'}{V_t} \tag{7.20}$$

where

V_t = ideal tangential component velocity
V_t' = real tangential component velocity

Figure 7.5 shows an exit velocity triangle with the effect of slip factor on the velocity components.

The slip factor, μ_{Slip}, defined by Eq. (7.20) is a combined factor that takes into account of two effects: (i) one is the ideal slip factor, μ_e, that takes into account of the no-slip velocity condition at the vane wall and results in the reduction of Euler head to ideal head that takes into account of the slip, and (ii) The velocity distribution factor, C_v, that accounts for the nonuniform velocity profile in vane flow channels. The slip factor can be written as product of these two factors as

$$\mu_{\text{Slip}} = C_v \mu_e \tag{7.21}$$

The **slip factor** depends on a number of variables such as the **number of vanes** (n_v), **vane angles** (β_2) and vane shape, and vane size given by the vane outlet to inlet radius ratio, $\frac{r_2}{r_1}$, in a complex manner.

With the introduction of the **slip factor**, the total head can now be estimated as

$$H_e = \mu_{\text{Slip}} H_i = C_v \mu_e H_i \tag{7.22}$$

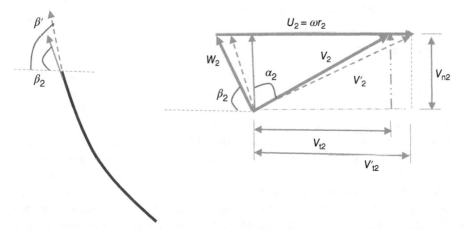

Figure 7.5 Exit velocity triangle with slip factor.

Further, with introduction of the **hydraulic efficiency** as $\eta_h = \frac{H}{H_e}$, defined as the ratio of the ideal head to Euler head, real head expression is given as

$$H = \eta_h C_v \mu_e H_e = \eta_h \mu_{Slip} H_e \tag{7.23}$$

A theoretical analysis of total head shows that the effect of C_v is dominant with the decrease in backward curves of the vane, i.e. at smaller β_2. In fact, the effect disappears for a radial vane, $\beta_2 = 90°$.

One of the simplest analytical expressions for the Euler slip factor was given by Stodola (1927) and Shepherd (1956) as

$$\mu_e = 1 - \left(\frac{U_2}{U_2 - V_{n2} \cot \beta_2} \right) \left(\frac{\pi \sin \beta_2}{n_v} \right) \tag{7.24}$$

Stodola expression shows that slip factor decreases with increase in flow rate, Q, as the normal component velocity, V_{n2}, is directly proportional to the flow rate. Also, slip factor increases and approaches a value of one with the increase in the number of vanes.

7.3 Impeller-Blade Design and Flow Channels

The principles of blade-profile designs in turbomachines include dynamics air flow over airfoil like that in an aircraft wing design as shown in Figure 7.6.

Some geometrical parameters of an airfoil are (i) Length of the airfoil, l_c,(ii) Maximum thickness across the **camberline**, y_c, (iii) The position of maximum thickness, x_c as shown in Figure 7.6a.

The energy transfer between the impeller vanes or blades and the fluid takes place due to the presence of a pressure difference between the two sides of the blade in proportion to the applied torque. As fluid flows over the blade, the pressure over the blade surface changes as the fluid changes direction based on blade profile. For nonsymmetric blade profile, the flow results in a high pressure on the leading side, also known as the **pressure side**, and a low pressure on the trailing side, known as the **suction side**, as demonstrated in Figure 7.7.

This difference in the pressure distributions between the pressure side and the suction sides of the vane results in a net force. The normal component of the force is known as

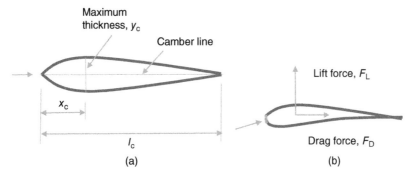

Figure 7.6 Flow over airfoil with lift and drag forces. (a) Symmetric airfoil profile and (b) curved airfoil profile.

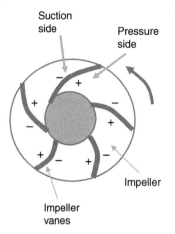

Suction side

Pressure side

Impeller

Impeller vanes

Figure 7.7 Pressure and suction sides of impeller vane.

the *lift force*, F_L. The lift force is a measure of the ability of the blades design to support a weight as it moves through fluid as in the case of airplanes or transfer of energy from the impeller to a fluid as in the case of a pump or fan or compressor, or transfer energy to an impeller as in turbines. The component of the force in the flow directions is referred to as the *drag force*, which contributes to the loss of useful energy.

The shape of the blade varies with the type of turbomachine as the energy transfer interaction varies. All blade types, however, follow some rules like (i) Must have some curvature to create change in flow direction, (ii) minimum thickness from structural strength of view, (iii) Shapes that result in reduced turbulence, flow separation, and recirculation. For ideal performance, one would expect blade shape to be more like an airfoil. However, real blade shapes vary considerably from machine to machine due to the requirement of cost and manufacturing limitations, and so the performance is sacrificed to some extent. Some blade profiles are often fabricated by bending and/or twisting a base airfoil profile out of airfoil section around the mean camber. Some blades are simply composed of a number of smaller circular arcs and straight sections.

In turbomachine, a row of blades, known as the cascades of blades, are mounted on the rotating impeller to produce change in flow directions in the flow channels formed between two adjacent blades. The higher changes in flow direction and the resulting lift force, the higher is the energy transfer.

7.4 Centrifugal Pumps

The general expressions for shaft work and torque described in the previous section now can be applied to the analysis, design, and selection of a centrifugal pump. In a centrifugal pump, a rotating impeller with a number of attached curved vanes enclosed within a casing is used to exert energy on the fluid through centrifugal force as shown in Figure 7.8.

There are two primary components in a pump: (i) A *rotating impeller* with attached blades or vanes, which forces the liquid into a rotary motion through the passages formed by the vanes and the *impeller walls* or *shrouds* by centrifugal force and discharges the liquid at a higher velocity and (ii) A *stationary pump casing*, which directs the liquid

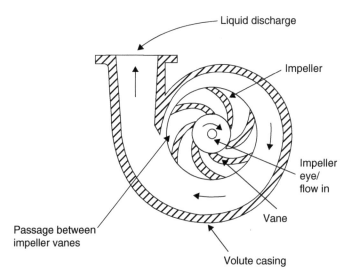

Figure 7.8 Section view of a centrifugal pump with impeller and volute.

toward the eye of the impeller at the inlet section, and forces the liquid coming out of the impeller to move toward the exit. The stationary pump casing encloses the impeller and is connected to the suction and discharge nozzle. A typical construction of rotating impeller is shown in Figure 7.9 showing different components of the pump construction.

In general, liquid exits the impeller vane passages at a high pressure and velocity. The dynamic head associated with the high velocity of the fluid is usually converted into pressure head by using the volute section with expanding flow area as shown in Figure 7.8.

Figure 7.10 shows a single-stage single-suction centrifugal pump mounted at one end of a rotating shaft. The rotating impeller is mounted on a rotating shaft with support bearings. Packings and wearing rings are used around the impeller and the shaft to prevent any leakage of the liquid.

The impeller vanes are attached over the shaft right after the impeller suction eye section where liquid enters after passing through the suction nozzle. The impeller suction eye diameter, $d_1 = 2r_1$, matches with the inside diameter of the impeller shroud. The impeller flow area at the suction eye is the circular area based on the suction eye radius and the subtracted circular area based on the impeller shaft or hub diameter, $d_0 = 2r_0$. Single-suction pumps are generally preferred for the simplicity in design and are easier to maintain. Centrifugal pumps are also available with double-suction inlets. Double-suction pumps are often used because of the additional benefits such as the capability to operate at high volume flow rates and its ability to balance axial impeller forces exerted by fluid flow through the impeller vane passages.

7.4.1 Components of Centrifugal Pumps

Figure 7.11 shows sectional views of the impeller and impeller vanes. Figure 7.11b, in addition, depicts geometrical parameters of the vane flow channel in an impeller.

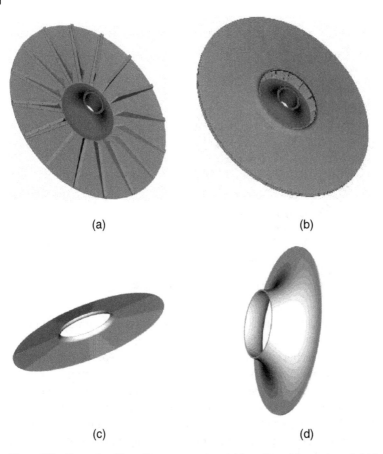

(a)   (b)

(c)   (d)

Figure 7.9 Example of impeller construction. (a) Impeller without shroud. (b) impeller with shroud, (c) impeller shroud, and (d) impeller hub.

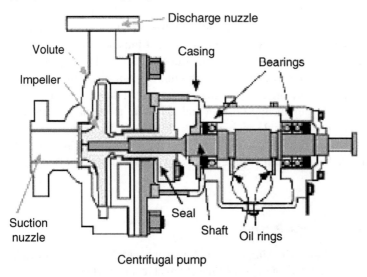

Centrifugal pump

Figure 7.10 Single-stage single-end-suction centrifugal pump construction.

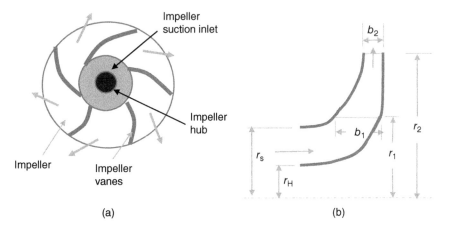

Figure 7.11   Impeller with cylindrical vane. (a) Impeller and (b) section of Impeller vane flow channel.

Because of the impeller rotating action, the fluid exits the impeller vane channel at higher velocity and pressure than it enters at the inlet suction section.

7.4.1.1   Vane or Blade Types

Vanes or blades in a centrifugal pump are classified based on the blade exit angle. Blade can be designed to be turned backward ($\beta_2 < 90°$) or forward ($\beta_2 > 90°$) or simply radial ($\beta_2 = 90°$) as demonstrated in Figure 7.12.

Typically, the inlet blade angle, β_1, for a backward turned blade may be in the range of $15 - 50°$ and exit angle, β_2, in the range of $15 - 35°$. Pumps with forward-curved blades are normally not designed due to the presence of unrealistic and unstable flow.

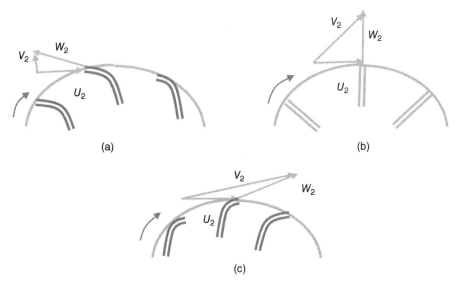

Figure 7.12   Different types of vanes. (a) Backward curved vanes, (b) radial vanes, (c) and forward curved vanes.

7.4.1.2 Impeller Casing

Impeller constructions are often provided with side wall, referred to as the **shroud** that encloses the fluid pathways within the impeller. The impeller can be classified as **shrouded impeller** and **open impeller** with no shrouds.

Impellers are also categorized into three types: radial, axial, and fixed. In a **radial-flow pump**, the blade passages move the liquid in the radial direction or normal to the shaft axis. In **axial-flow pump**, the liquid moves toward the impeller axially and stays directed parallel to the shaft axis. In a fixed-flow pump, the fluid flows out along a direction in between the radial- and axial-flow directions.

As discussed before, impellers are also classified into (i) **Single suction** with one inlet on one side only, (ii) **Double suction** with water flowing into the impeller equally from both sides.

7.4.1.3 Volute and Vane Diffuser Casing

A **volute casing** is the diffusing section at the exit of the impeller casing as shown in Figure 7.9a. Figure 7.9 shows the impeller along with impeller vanes. The volute casing is normally used in a centrifugal pump to convert part of dynamic head of the liquid exiting from vane channels into static head.

The design parameter for the volute is the throat area at start of the diffuser section. The throat area is given as

$$A_{\text{throat, volute}} = \frac{Q}{V_{\text{throat}}} \tag{7.25}$$

where V_{throat} is the liquid velocity at the throat section. The average volute throat velocity is computed from the impeller vane exit tangential component velocity in a proportion to the ratio of the vane exit radius, r_2, to the radius of curvature passing through the center of the throat area, r_{th}, as

$$V_{\text{throat}} = V_{t2} \frac{r_2}{r_{\text{th}}} C_{\text{throat}} \tag{7.26}$$

where C_{throat} is a constant with following recommended values as suggested in Karassik et al. (1976): $C_{\text{throat}} = 1$ for ideal frictionless flow and is a good approximation for volute of large pump with smooth volute of medium-size pump; and $C_{\text{throat}} = 0.9$ for a cast volute of medium and small pumps.

Instead of a volute, often a vane diffuser is used to convert part of dynamic head of the liquid exiting from vane channels into static head as shown in Figure 7.13b. A vane diffuser is composed of several stationary vanes, which are positioned circulatory around the impeller vanes with the purpose of collecting all flows exiting out from the impeller vane channels into a circular casing and directed toward exit pipe connection. The diffuser vane channels provide a greater proportion of the conversion of velocity into pressure or conversion of dynamic head into pressure head. Vane diffuser is geometrically more complex than a volute casing. However, the design procedure is like the volute casing, except the equation for throat velocity, which is applied multiple times for each of the diffuser vane channels. The throat velocity is defined based on the radius of curvature at the diffuser vane channels as follows:

$$V_{\text{throat}} = V_{t2} \frac{r_2}{r_{\text{vd}}} C_{\text{throat}} \tag{7.27}$$

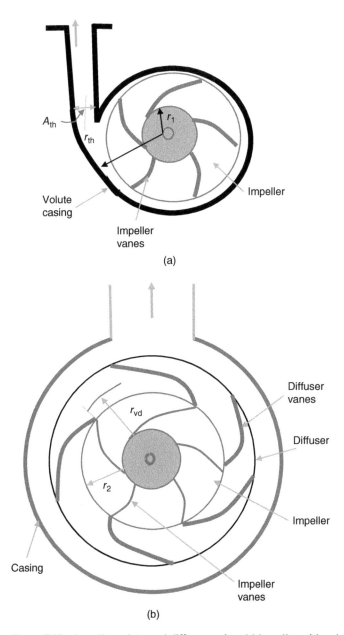

Figure 7.13 Impeller volute and diffuser casing. (a) Impeller with volute casing. (b) Vane Diffuser.

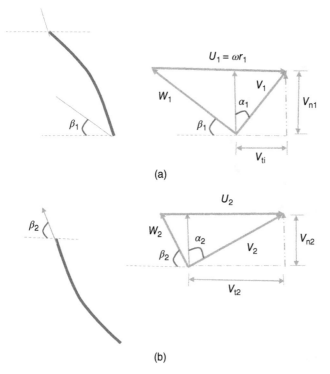

Figure 7.14 Velocity triangles at the entrance and exit of a rotating vane. (a) Velocity diagram at the entrance and (b) velocity triangle at the exit.

where

V_{throat} = velocity at the throat area of vane diffuse channel

r_{vd} = radius of curvature of the vane diffuse channel as shown in Figure 7.13b.

7.4.2 Velocity Triangles and Basic Equations for Pump Performance

To demonstrate all equations necessary for the analysis and design of centrifugal pumps, let us consider the velocity triangles associated with the inlet and outlet of the impeller vanes as shown in Figure 7.14.

From the inlet velocity diagram, we can derive following relationships:

From blade velocity triangle at the inlet

$$U = V_{n1} \left(\tan \alpha_1 + \cot \beta_1 \right) \tag{7.28}$$

and the normal component velocity is

$$V_{n1} = \frac{U}{\tan \alpha_1 + \cot \beta_1} \tag{7.29a}$$

Rearranging,

$$\tan \alpha_1 = \frac{U_1 - V_{n1} \cot \beta_1}{V_{n1}}$$

$$\alpha_1 = \tan^{-1}\left[\frac{U_1 - V_{n1}\cot\beta_1}{V_{n1}}\right] \tag{7.29b}$$

From the outlet velocity triangle, we can derive the

Tangential component velocity

$$V_{t2} = U_2 - V_{n2}\cot\beta_2 = U_2 - \frac{V_{n2}}{\tan\beta_2} \tag{7.30}$$

Normal component velocity at the outlet

$$V_{n2} = \frac{V_{t2}}{\tan\alpha_2} \tag{7.31}$$

where

$$\tan\alpha_2 = \frac{V_{t2}}{V_{n2}} \tag{7.32}$$

7.4.2.1 Volume Flow Rates

$$Q = V_{n1}A_1 = V_{n2}A_2 \tag{7.33}$$

Vane flow area:

$$A_1 = 2\pi r_1 b_1 = \text{Flow area of the vane flow passage at inlet} \tag{7.34a}$$

and

$$A_2 = 2\pi r_2 b_2 = \text{Flow area of the vane flow passage at outlet} \tag{7.34b}$$

where

r_1 = radius of the impeller at vane inlet
r_2 = radius of the impeller at vane outlet
b_1 = width of the vane passage at inlet
b_2 = width of the vane passage at outlet

7.4.2.2 Pump Performance Output

Euler's equations for pump shaft work and pump head are given as

Pump torque:

$$T_{\text{Shaft}} = \dot{m}\left(r_2 V_{t2} - r_1 V_{t1}\right) \tag{7.35a}$$

or

$$T_{\text{Shaft}} = \rho Q\left(r_2 V_{t2} - r_1 V_{t1}\right) \tag{7.35b}$$

or

$$T_{\text{Shaft}} = \frac{\gamma Q\left(r_2 V_{t2} - r_1 V_{t1}\right)}{g} \tag{7.35c}$$

Pump work or power

$$W_{\text{Shaft}} = \omega T = \rho Q\left(U_2 V_{t2} - U_1 V_{t1}\right) = \rho Q h_P \tag{7.36a}$$

or

$$W_{Shaft} = \frac{\gamma Q \left(U_2 V_{t2} - U_1 V_{t1} \right)}{g} = \frac{\gamma Q h_P}{g}$$ (7.36b)

Pump head

$$h_P = \frac{U_2 V_{t2} - U_1 V_{t1}}{g}$$ (7.37a)

or

$$h_P = \left[\frac{\left(V_2^2 - V_1^2 \right) + \left(U_2^2 - U_1^2 \right) - \left(W_2^2 - W_1^2 \right)}{2g} \right]$$ (7.37b)

Equations (7.36b) and (7.37a) represent the ideal pump work and ideal pump head when the tangential component velocities are ideal. These expressions also can be used to represent a real pump so long as the tangential components are real. The real pump head is always less than the ideal pump head due to the head loss within the pump.

Euler Eqs. (7.36a) and (7.37b) for pump power and pump head neglect number losses like leakage losses, disk, and flow friction losses.

Example 7.1 *Estimation of Theoretical Pump Head and Power*

Consider a centrifugal pump with a mixed-flow impeller and running at 3000 rpm and a rated flow rate of 500 gpm. The impeller vane radius at inlet is 2.00-in. and at outlet is 6-in. The vane angles at inlet and outlet of the vanes are 40° and 65°, respectively. The vane passage widths at inlet and outlet are $b_1 = 0.5$-in. and $b_2 = 0.4$-in., respectively. The inlet guide vanes are used to introduce a swirl for the flow at the inlet with absolute velocity making an angle $\alpha_1 = 30°$ from normal.

Determine the theoretical pump head and pump power.

Solution:

Volume flow rate, $Q = 500$ gpm $= 500$ gpm $\times \frac{1}{7.48}$ ft^3/s $\frac{1}{60}$ min /s $= 1.14$ ft^3/s

Peripheral speed:

$$U_1 = \omega r_1 = 2\pi N r_1 = 2\pi \times 3000 \times \frac{1}{60} \times \frac{2}{12} = 52.333 \text{ ft/s}$$

$$U_2 = \omega r_2 = 2\pi N r_2 = 2\pi \times 3000 \times \frac{1}{60} \times \frac{6}{12} = 157 \text{ ft/s}$$

From conservation of mass and incompressible flow:

$$Q = V_{n2} A_2 = V_{n1} A_1$$

Compute the normal components velocities

$$V_{n1} = \frac{Q}{2\pi r_1 b_1} = \frac{1.14}{2\pi \times 2/12 \times 0.5/12} = 26.126 \text{ ft/s}$$

$$V_{n2} = \frac{Q}{2\pi r_2 b_2} = \frac{1.14}{2\pi \times 6/12 \times 0.4/12} = 10.89 \text{ ft/s}$$

From inlet velocity triangle:

$$\tan \alpha_1 = \frac{U_1 - V_{n1} \cot \beta_1}{V_{n1}}$$

$$\alpha_1 = \tan^{-1}\left[\frac{U_1 - V_{n1}\cot\beta_1}{V_{n1}}\right]$$

$$\alpha_1 = \tan^{-1}\left[\frac{52.333 - 26.126\cot 40°}{26.126}\right] = 39.054°$$

From inlet velocity triangle

$$V_{n1} = \frac{U_1}{\tan\alpha_1 + \cot\beta_1} = \frac{52.33}{\tan 39.054° + \cot 40°} = 26.13 \text{ ft/s}$$

Absolute velocity at the inlet,

$$V_1 = \frac{V_{n1}}{\cos\alpha_1} = \frac{26.13}{\cos 40°} = 34.11 \text{ ft/s}$$

Tangential component velocity,

$$V_{t1} = V_1 \sin\alpha_1 = 34.11 \sin 40°$$

$$V_{t1} = 21.925 \text{ ft/s}$$

Similarly, from exit velocity triangle:

$$\tan\alpha_2 = \frac{U_2 - V_{n2}\cot\beta_2}{V_{n2}}$$

$$V_{n2} = \frac{Q}{2\pi r_2 b_2} = \frac{1.14}{2\pi \times 6/12 \times 0.4/12} = 10.89 \text{ ft/s}$$

$$\alpha_2 = \tan^{-1}\left[\frac{U_2 - V_{n2}\cot\beta_2}{V_{n2}}\right]$$

$$\alpha_2 = \tan^{-1}\left[\frac{157 - 10.89\cot 65°}{10.89}\right] = 85.9°$$

Outlet absolute angle, $\alpha_2 = 85.9°$
Absolute velocity at the outlet,

$$V_2 = \frac{V_{n2}}{\cos\alpha_2} = \frac{10.89}{\cos 85.9°} = 152.31 \text{ ft/s}$$

Tangential component velocity,

$$V_{t2} = V_2 \sin\alpha_2 = 152.31 \times \sin 85.9°$$

$$V_{t2} = 151.92 \text{ ft/s}$$

Pump head from Eq. (7.37a)

$$h_P = \frac{U_2 V_{t2} - U_1 V_{t1}}{g} = \frac{157 \times 151.92 - 52.333 \times 21.925}{32.2}$$

$$h_P = \frac{23851.4 - 1147.40}{32.2} = 705 \text{ ft}$$

Pump work from Eq. (7.36a):

$$W_{\text{Shaft}} = \rho Q \left(U_2 V_{t2} - U_1 V_{t1}\right) = \rho Q h_P)$$

or

$$W_{\text{Shaft}} = \gamma Q \, h_{\text{P}} = 62.4 \times 1.14 \times 705$$

$$W_{\text{Shaft}} = 50\,150 \text{ ft lb f} = 91.18 \text{ hp}$$

7.4.2.3 Major Pump Parameters
7.4.2.3.1 Capacity, Q
Capacity, Q, is the rate of volume flow of liquid through by the pump.

7.4.2.4 Pump Head, h_p
Pump head is the change in the energy content of the liquid by the pump impeller. This is expressed as the rate of energy transfer rate per unit mass flow rate and given as

$$h_{\text{p}} = \left[\frac{\left(V_2^2 - V_1^2\right) + \left(U_2^2 - U_1^2\right) - \left(W_2^2 - W_1^2\right)}{2g} = \left(h_{\text{d}} - h_{\text{s}}\right) \right] + \frac{V_{\text{d}}^2}{2} - \frac{V_{\text{s}}^2}{2} \qquad (7.38)$$

where

h_{d} = **Discharged head** of liquid measured at the pump exit nozzle
h_{s} = **Suction head** of liquid measured at the pump inlet nozzle
V_{d} = Average velocity of liquid measured at the pump exit nozzle
V_{s} = Average velocity of liquid measured at the pump exit nozzle

7.4.2.5 Pump Efficiency
7.4.2.5.1 Pump Overall Efficiency (η_{Pump})
Overall pump efficiency is defined as the ratio of the pump waterpower to power input to the shaft, and this is expressed as

$$\text{Pump Efficiency, } \eta_{\text{Pump}} = \frac{\text{Pump liquid input power}}{\text{Pump shaft}} = \frac{P_{\text{L}}}{P_{\text{S}}} \qquad (7.39a)$$

or

$$\eta_{\text{Pump}} = \frac{\rho g \, Q h_{\text{P}}}{P_{\text{S}}} \qquad (7.39b)$$

Pump efficiency, given by the Eq. (7.39b), is referred to as the **overall pump efficiency.** **There are other forms of pump efficiencies** that are often used by the designers to compare different pumps. Some of these additional forms of pump efficiencies are also defined here.

7.4.2.5.2 Pump Vane or Blade Efficiency
Pump **vane efficiency** is defined as the ratio of the input head, $h_{\text{P, input}}$, to liquid to the theoretical Euler head, $h_{\text{P, Euler}}$

$$\eta_{\text{P,Vane}} = \frac{h_{\text{P,input}}}{h_{\text{P,Euler}}} \qquad (7.40)$$

7.4.2.5.3 *Pump Hydraulic Efficiency (η_H)*

Ideal pump head generated by the rotating impeller is reduced due to number losses caused by friction, formation of eddies with the blade passages, and changes in area due to sudden expansion and contraction. **Hydraulic efficiency,** η_h, is defined as the ratio of the total increase in liquid head or available head to the input head of the impeller, and written as

$$\eta_H = \frac{h_{P,input} - \text{Hudraulic Losses}}{h_{P,input}} \tag{7.41a}$$

or

$$\eta_H = \frac{h_{P,input} - \Delta H_{Loss}}{h_{P,input}} = \frac{P_L}{P_L + P_H} \tag{7.41b}$$

where

P_H = Power consumed by the hydraulic losses

With a known value of the hydraulic efficiency, the reduced real pump head is given as

$$h_P = \eta_H \frac{(U_2 V_{t2} - U_1 V_{t1})}{g} \tag{7.42}$$

7.4.2.6 Pump Performance Characteristics

7.4.2.6.1 *Ideal Pump Performance*

An ideal pump impeller contains a very large number of vanes of infinitesimal thickness. The fluid particles are also assumed to move parallel to the vane surfaces and without the presence of any losses. A real pump impeller has a finite number of widely spaced vanes with finite thicknesses. The flow in the vane passages also involves slip and a nonuniform velocity profile, and includes several losses such as hydraulic loss, leakage loss, mechanical loss, and disk friction loss.

Let us demonstrate here the derivation of an ideal pump performance characteristic based on Euler head equation, which is given as

$$h_P = \frac{U_2 V_{t2} - U_1 V_{t1}}{g} \tag{7.43}$$

Substituting Eq. (7.30) into (7.43)

$$h_P = \frac{U_2 \left(U_2 - \dfrac{V_{n2}}{\tan \beta_2} \right)}{g} - \frac{U_1 \left(U_1 - \dfrac{V_{n1}}{\tan \beta_1} \right)}{g}$$

or

$$h_P = \left(\frac{U_2^2}{g} - \frac{U_2 V_{n2}}{g \tan \beta_2} \right) - \left(\frac{U_1^2}{g} - \frac{U_1 V_{n1}}{g \tan \beta_1} \right) \tag{7.44}$$

We can express the normal component velocity as

$$V_{n2} = \frac{Q}{A_2} = \frac{Q}{2 \pi r_2 b_2} \tag{7.45a}$$

and

$$V_{n1} = \frac{Q}{A_2} = \frac{Q}{2 \pi r_1 b_1} \tag{7.45b}$$

Substituting Eqs. (7.45a) and (7.45b) into Eq. (7.44), we have the expression for the pump performance characteristic

$$h_P = \left(\frac{U_2^2}{g} - \frac{U_2 Q}{g A_2 \tan \beta_2} \right) - \left(\frac{U_1^2}{g} - \frac{U_1 Q}{g A_1 \tan \beta_1} \right) \tag{7.46a}$$

Rearranging,

$$h_P = \left(\frac{U_2^2}{g} - \frac{U_1^2}{g} \right) - \left(\frac{U_2}{g A_2 \tan \beta_2} - \frac{U_1}{g A_1 \tan \beta_1} \right) Q \tag{7.46b}$$

This pump characteristics expression shows the variation of pump head with major pump impeller operating variables like pump flow capacity, rotating speed, and impeller-blade dimensions and angles. For a given impeller size, blades design with known blades angles, and a set rotational speed, the pump head is linear function of the pump flow capacity, Q. We can rewrite Eq. (7.46b) as

$$h_P = K_1 - K_2 Q \tag{7.47a}$$

where

$$K_1 = \left(\frac{U_2^2}{g} - \frac{U_1^2}{g} \right) \tag{7.47b}$$

and

$$K_2 = \frac{U_2}{g A_2 \tan \beta_2} - \frac{U_1}{g A_1 \tan \beta_1} \tag{7.47c}$$

A simplified Euler ideal pump performance characteristic equation is often used by considering a special case where the fluid enters the impeller without any prerotation *swirl*, i.e. has no tangential component velocity, V_{t1}. Under this situation, the liquid absolute velocity, V_1, makes an angle, $\alpha_1 = 90°C$, from the tangential direction, and we can reduce the Euler pump Eq. (7.42) to

$$h_P = \frac{U_2 V_{t2}}{g} \tag{7.48}$$

Substituting Eq. (7.30) into (7.48)

$$h_P = \frac{U_2^2}{g} - \frac{U_2 V_{n2}}{g \tan \beta_2} \tag{7.49}$$

Further, substituting the normal component velocity from Eq. (7.33), we get

$$h_P = \frac{U_2^2}{g} - \frac{U_2 Q}{g \tan \beta_2 A_2} \tag{7.50a}$$

or

$$h_P = K_1 - K_2 Q \tag{7.50b}$$

where

$$K_1 = \frac{U_2^2}{g} \tag{7.51a}$$

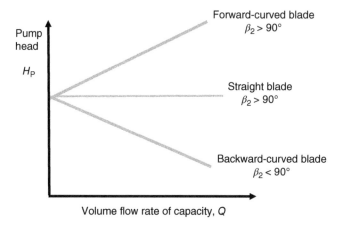

Figure 7.15 Euler ideal pump characteristic curves.

and

$$K_2 = \frac{U_2}{g \tan \beta_2 A_2} = \frac{U_2}{2\pi r_2 b_2 g \tan \beta_2} \tag{7.51b}$$

Equation (7.50a) represents the Euler ideal head (h_p) vs. capacity (Q) pump characteristic curve. This equation shows that for an ideal pump the pump head varies with the capacity in a linear manner with a **shut-off head** or pump head at $Q = 0$ given as $K_1 = \frac{U_2^2}{g}$ and with a slope given by $K_2 = \frac{U_2}{g \tan \beta_2 A_2}$ for a given rotor-blade design and speed. A graphical presentation of the ideal pump characteristic is shown in Figure 7.15 by plotting the pump head with varying flow rate or pump flow capacity.

As demonstrated in Figure 7.15, the slope of the ideal pump characteristic line is a function of the exit blade angle, β_2. For $\beta_2 = 90\,°$, the pump head remains constant with the flow capacity and the pump head-capacity line remains parallel to the horizontal line of capacity axis. For a backward curved vane ($\beta_2 < 90\,°$), the pump characteristic line shows a downward variation of the pump head with increase in flow capacity. On the other hand, for a forward curved vane ($\beta_2 > 90\,°$), the pump head increases with the capacity. However, such pump impeller design with forward curved is not realistic as it gives rise to the situation where the absolute velocity and its tangential component become greater than the blade speed.

We noticed that for all blade angles, pump head reduces to the **shut-off head**, $\frac{U_2^2}{g}$, for zero pump flow capacity. This is because at zero-flow capacity, the absolute velocity of the liquid is equal to the impeller rotating velocity, i.e. $V_2 = U_2$; and the sum total of the dynamic head and static pressure head or the total pump head expression reduces to $H_p = \frac{(V_2^2)+(U_2^2)}{2g} = \frac{U_2^2}{g}$

Euler Ideal power characteristic curve can be derived based on the Euler pump power Eq. (7.36b) and using simplified ideal pump head expression given by Eq. (7.50a) as

$$P_{pump,ideal} = \rho Q \left(\frac{U_2^2}{g} - \frac{U_2}{g \tan \beta_2 A_2} Q \right) \tag{7.52}$$

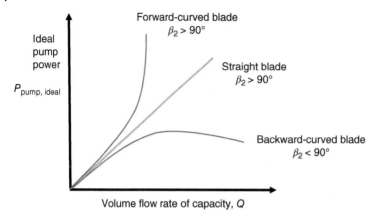

Figure 7.16 Pump power curve with varying flow capacity.

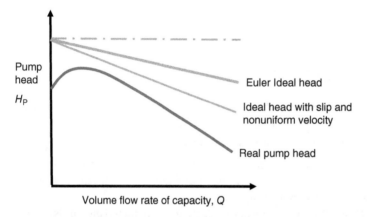

Figure 7.17 Pump characteristics for backward-curved blade with $\beta_2 < 90°$.

As we can see, the pump power varies with flow capacity in a nonlinear quadratic manner. Eq. (7.52) can be used to plot the pump power curve with varying flow capacity for different blade types as shown in Figure 7.16.

A deviation to Euler ideal pump characteristics is given by considering the pump head considering slip and nonuniform velocity distribution, but without considering any other form of losses. Figure 7.17 shows the ideal pump characteristics and their deviation from the Euler ideal characteristics for the impeller with backward curved vanes.

7.4.3 Real Pump Performance

A real pump impeller has a finite number of widely spaced vanes with finite thicknesses. The flow in the vane passages also involves slip and a nonuniform velocity profile, and includes several losses such as hydraulic loss, leakage loss, mechanical loss, and disk friction loss. The hydraulic or flow losses are caused by the flow friction, formation of eddies, and flow separation in flow passages. Hydraulic or flow losses are classified as internal losses and can be divided into three groups:

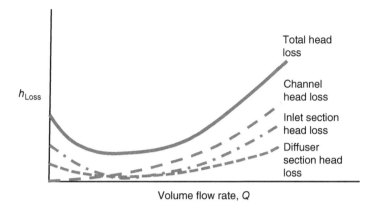

Figure 7.18 Different head losses.

i) *The flow loss in the inlet section of the impeller, including eye and the inducer.* This is in general proportional to the volume flow rate as Q^2 and number flow dynamics factors defined in terms of coefficient of drag. This loss could be high at very low flow rates and leads to parabolic shape for the head loss with volume flow rate as shown in Figure 7.18.

ii) *The flow losses due to friction, flow separation, formation of eddies, and turbulence.* These losses are viscous skin friction losses in the blade flow passages due to the presence of eddies and flow separations, and leakages. The loss is varying in a quadratic manner with volume flow rate as Q^2 for incompressible flow. This is more strongly related to volume flow rate, Q, for compressible flows.

iii) *The flow loss in the discharge section, including the volute and diffuser casing.* These forms of flow losses vary nonlinearly in quadric manner with the flow capacity, Q, leading to a parabolic variation of the loss with volume flow rate as shown in Figure 7.18.

The sum of these losses reduces the real pump head with the flow capacity in a nonlinear manner as shown in Figure 7.17. As we can see, the pump head increases from the shut-off head to maximum head value and then starts decreasing in a nonlinear manner with increase in volume flowrate, Q. More comprehensive description of these losses is given in books by Stepanoff (1948) and Shepherd (1956).

The real pump performance varies from the ideal pump performance due to the reduction in the pump head because of the presence of the slip condition and nonuniform velocity distribution with the flow passages and the hydraulic losses in the pump. Real pump head can be deduced by subtracting hydraulic losses from the ideal pump head. One important feature that can be noticed from the real pump head vs. capacity characteristics is that maximum head may not be the shut-off head at zero-flow capacity. Instead, the pump head may rise to maximum value at some lower flow rate value and then drop with further increase in flow rate. While this feature is seen in certain pump designs such as the one with volute casing, many pump designs exhibit decrease in pump head starting from the shut-off head or closer to the shut-off head.

7.4.3.1 Effect of Slip Factor

As we have discussed in Section 7.2.4, both absolute velocity, V, and relative velocity, W, at the inlet and outlet of the vane wall deviated from the ideal values due to no-slip velocity condition at the vane walls, nonuniform velocity profile in vane flow channels as well as flow separations and turbulence. This results in the deviations of the fluid-flow directions from the vane angles and modifies the peripheral or tangential components of the velocities, V_{t1} and V_{t2}, and hence in the estimated values of the total head and power input. Simplest design practice is to use a slip factor in the expression for real pump head.

Using Eq. (7.23), we can write real pump head from head from Eq. (7.43) as

$$h_P = \mu_{Slip} \eta_{P,H} \frac{(U_2 V_{t2} - U_1 V_{t1})}{g} \tag{7.53a}$$

Or from Eq. (7.46a) as

$$h_P = \mu_{Slip} \eta_{P,H} \left[\left(\frac{U_2^2}{g} - \frac{U_2 V_{n2}}{g \tan \beta_2} \right) - \left(\frac{U_1^2}{g} - \frac{U_1 V_{n1}}{g \tan \beta_1} \right) \right] \tag{7.53b}$$

where

$\eta_{P,H}$ = **Pump hydraulic efficiency**
μ_{Slip} = **Pump slip factor**

Stodola's theoretical expression of slip factor, given by Eq. (7.24), is transformed into a simplified form for vanes of centrifugal impeller as

$$\mu_{Slip} = 1 - \frac{\pi \sin \beta_2}{n_v} \tag{7.54a}$$

Even though this simplified form was originally derived assuming no flow through the impeller vane channels, it has been widely used for design-flow conditions.

For a radial vane with $\beta_2 = 90°$, Stodola's expression reduces to

$$\mu_{Slip} = 1 - \frac{\pi}{n_v} \tag{7.54b}$$

A number of semiempirical and empirical formulas are also derived based on experimental tests to correlate slip factor with variables such as the **number of vanes, vane angles, vane diffuser/volute system,** and **vane size** given by the vane outlet to inlet radius ratio, $\frac{r_2}{r_1}$. One such correlation is given by Karassik et al. (1976).

$$\mu_{Slip} = \frac{1}{1 + \frac{a}{n_v} \left(1 + \frac{\beta_2}{60} \right) \frac{2}{1 - \left(\frac{r_1^2}{r_2^2} \right)}} \tag{7.55}$$

where **a** is an empirical constant that depends on the vane diffuse and volute system with following reported values: **a** = 0.65–0.85 for a volute; **a** = 0.6 for a vane diffuser; and **a** = vanless diffuser.

As discussed, the power loss due hydraulic loss depends strongly on the impeller and impeller vane passages design as well as the pump volume flow capacity, Q. For pumps with

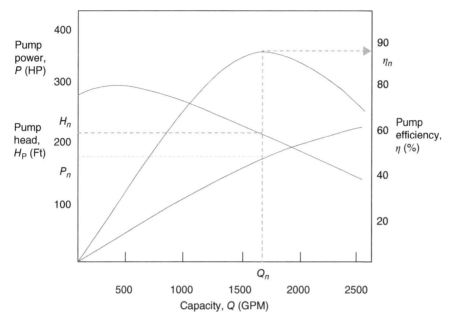

Figure 7.19 Pump performance characteristic at a constant speed.

above average performance, the following approximate correlation of hydraulic efficiency with volume flow rate is given following test data (Karassik et al. 1976):

$$\eta_{\mathrm{H}} \approx 1 - \frac{0.8}{Q^{0.25}} \tag{7.56}$$

7.4.3.2 Pump Performance Characteristics

Figure 7.19 shows typical pump performance characteristic in terms of head, power, and efficiency as a function of flow capacity for a constant speed.

Plots shown in Figure 7.19 are just idealized representation of actual pump characteristic curves, which may include irregularities. As we will see in Chapter 8, operating conditions are set by the intersection of the pump head vs. capacity characteristic performance curve with the piping system head loss characteristic curve. One important criterion for selecting a pump for a piping system is to have the operating head, power, and capacity at the maximum efficiency point. These operating conditions are generally referred to as the **design** or **nominal values**, H_n, P_n, Q_n,and η_n,as demonstrated by the intercepts of the dotted lines with axes values. Often a pump may operate at a capacity, which is lower or higher than the maximum value. This is referred to as the rated value, also known as guaranteed values, given by the manufacturer of the pump. Any operations at values other than the nominal design values are referred to as the **off-design values**, a condition that needs to be avoided as unstable and unsafe situation may result if operated continuously for longer period.

7.4.4 Effect of Operating Impeller Speed

The energy transfer to the fluid in a centrifugal turbomachinery increases with blade or vane tip speed. The required vane tip speed can be achieved either by using a large vane radius in a large impeller size and operating at a lower rotational speed or rpm. Another choice could be using a smaller impeller with smaller vane radius and operating at a higher speed or rpm. So, the choice for increasing the power output from a rotating machine is to increase the rotating speed or increase the impeller-blade-tip diameter. Generally, the option of running the impeller at higher rpm is preferred as it leads to higher efficiency and volume flow while keeping the volume and weight of the machine low. However, higher speed may also lead to higher noise and other external shaft and bearing power losses. Pump manufacturers often offer an impeller design with different impeller size or radius for a given rotating speed.

Figure 7.20 shows a typical variation in pump head-capacity performance curves for different speeds for a fixed impeller size.

As we can see, pump-generated head increases with increase in rotating speed of the pump. Often, it is necessary to develop a functional representation of the pump performance characteristics in terms of pressure rise or pump head as a function of flow rate, Q, for given operating speed, N, as $H_p = f(Q)$. A functional representation of the pump performance characteristics in terms of pump head as a function of two variables: flow rate, Q, and speed, N, can also be developed as $H_p = f(Q, N)$. Procedures for developing parametric representation or curve fitting of performance data are given in Appendix A.

For example, a third-degree polynomial functional representation of the pump performance characteristics is given as

$$H_p = a_0 + a_1 Q + a_2 Q^2 + a_3 Q^3 \tag{7.57a}$$

A functional representation of the pump performance characteristics in terms of two variables $H_p = f(Q, N)$, can be developed in two different forms:

$$H_p = a_0 + a_1 Q + a_2 Q^2 + a_3 N + a_4 N^2 + a_5 QN + a_6 Q^2 N + a_7 QN^2 + a_8 Q^2 N^2 \tag{7.57b}$$

or

$$H_p = \left(A_0 + A_1 Q + A_2 Q^2\right) + \left(B_0 + B_1 Q + B_2 Q^2\right) N + \left(C_0 + C_1 Q + C_2 Q^2\right) N^2 \tag{7.57c}$$

Note that both equations are derived based on using second-degree polynomial.

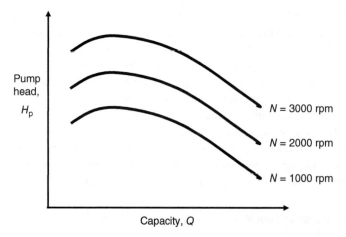

Figure 7.20 Pump head vs. capacity performance curves at different rotating speed for given impeller size.

7.4.5 External Losses

External losses in pump include other losses like leakage loss, mechanical loss, and disk friction loss, and effects on overall pump performance or efficiency.

7.4.5.1 Leakage Loss

Flow leakage takes place in both shrouded impeller and in open impeller. In the case of a shrouded impeller, fluid leaks from the discharge side of the blade passage to the inlet side through the clearance space between the impeller and the shroud. In an open impeller, flow leaks from the pressure side to the suction side of the vane through the edges. Due to this leakage loss, actual delivery capacity from machine is reduced from discharge rate at the exit of the vane passage. This contributes to a loss of the power and results in a reduced efficiency.

The loss in power due to the leakages and recirculation is considered by using the **volumetric efficiency**, η_v, defined as

$$\eta_v = \frac{Q}{Q + Q_L} \tag{7.58a}$$

where

Q = net volume flow rate
Q_L = *Leakage flow rate*

The leakage flow rate is directly related to the clearance space between the impeller and the shroud casing and the operating speed. The volumetric efficiency is normally correlated with the specific speed and the volume flow rate (Karassik et al. 1976; Stepanoff 1948; Shepherd 1956). Typical volumetric efficiency values are in the range of 2% for large pumps and as high as 10% for smaller pumps.

7.4.5.2 Disk Friction Loss

The disk friction loss is due to the circulation flow of the fluid through the space between the rotating shroud or disk surface and the stationary casing. This loss of power due to skin friction and fluid circulation power is also referred to as the **windage loss**. Under the action of the centrifugal force, fluid moves outward over the disk surface and returns through the clearance space bounded by stationary case. The resulting frictional force contributes to the so-called windage power loss or disk friction power loss. In general, the disk friction loss is low for higher rotating speed and smaller impeller size. Stepanoff (1948) suggested that the disc friction or **windage power loss or efficiency**, η_w, roughly can be taken to be inversely proportional to the specific speed.

7.4.5.3 Mechanical Loss

The mechanical loss is the power loss with the bearings and seals. This loss can only be determined by conducting experimental testing with specific bearings and seals. In general, test results show that *the ratio of mechanical power loss to the waterpower* increases with decrease in specific speed and flow capacity. Generally, it may vary from 0.5% to 7% of the pump power in the range volume flow rate 50–10 gpm and specific speed of 500–5000

(Karassik et al. 1976). The mechanical efficiency, η_m, is defined as the ratio of the power transfer to the fluid from the impeller to the power applied to the shaft

$$\eta_m = \frac{\gamma Q h_p}{P_s}$$

(7.58b)

where

$P_s = $ *Shaft power*

The overall machine efficiency, η, is then given as

$$\eta = \eta_v \, \eta_w \, \eta_m$$

(7.59)

7.5 Specific Speed and Pump Selections

Specific speed for a pump is defined as that rotating speed at which a pump impeller geometrically similar one, but different in size, larger or small, will produce a pump head of 1-ft at a flow rate of 1-gpm. Specific speed, N_s, is a dimensionless group used to characterize a group of geometrically similar pumps into a design type. It is derived for the centrifugal pumps using dimensional analysis and principles of similitude. Main objective to derive such a dimensionless group is to group together the key operating variables, representative of the service conditions of the centrifugal pumps so that many pump designs could be lumped together into a single expression. Starting with the main variables like pump flow capacity, Q; diameter of the impeller, D; rotating speed, N; pump head, H_p; energy applied to the shaft, E; dynamic viscosity of liquid, μ; density of liquid, ρ; and acceleration due to gravity, a dimensional analysis study for the centrifugal pump resulted in the following three dimensionless groups (Stepanoff 1948):

$$\pi_1 = \frac{Q}{\left(gH_p\right)^{1/2} D^2}, \pi_2 = \frac{NQ^{1/2}}{\left(gH_p\right)^{3/4}}, \text{ and } \pi_1 = \frac{\rho Q}{\mu D},$$

Among these, the dimensionless group, Π_2, fits into the definition of specific speed while dropping the constant, g; and used to express **the specific speed** for the centrifugal as follows:

$$N_s = \frac{NQ^{1/2}}{H_p^{3/4}}$$

(7.60)

Specific speed simply expresses all values of the operational variables, N, Q, and h_p, that results in that geometrically similar group of pumps into a design class. It is used to characterize the type of the centrifugal pumps as radial pump, mixed-flow pump, and axial-flow pump.

Figure 7.21 shows a classification of centrifugal pumps into different classes like radial, mixed, and axial based on the operation over a range of flow capacities and the corresponding variation in the pump efficiency, including the maximum efficiency. Based on a given set of head – capacity conditions, one can then use this classification criterion to select the best class of pump or the specific speed and the corresponding rotating speed.

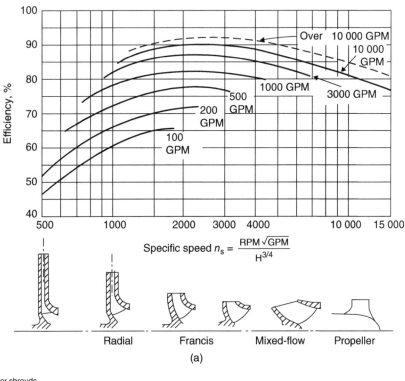

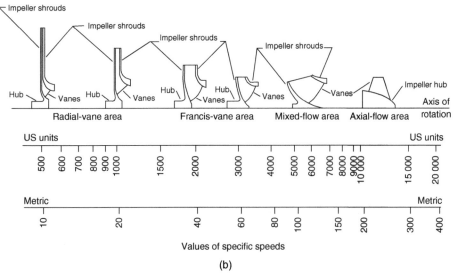

Figure 7.21 Pump efficiency and classification based on specific speed and range of capacity. (a) Selection pump-based specific speed and peak efficiency. Source: From Karassik et al. (1976). © 1976 McGraw-Hill. and (b) Classification of pump based on specific speed. Source: From ANSI/HI 14.1-14.2 American National Standard for Rotodynamic Pumps for Nomenclature and Definitions, Hydraulic Institute, 2019. © 2019 used with permission. Courtesy of Hydraulic Institute, www.pumps.org.

Figure 7.21a shows the classification along with the variation of pump efficiency with specific speed and flow capacity for well-designed pumps in each class of pumps.

We can notice few things as outlined here:

1) For lower specific speed, we can select radial-flow pumps, which may be suitable for pumps with lower flow capacity and large pumps like in the case of boiler feed pump in a power plant cycle.
2) An axial-flow pump with higher specific speed is suitable for flow systems with large flow capacity and lower pumping head like in case of cooling water pumps in the secondary cooling water system.
3) A fixed flow pump is suitable for the midranges of flow capacity and pumping head that operate with a specific speed range of 1500–5000, flow capacity range of 1000–5000 gpm. Pump efficiency may be expected in the range of 80–90%.

Example 7.2 Selection Of Pump Type

Select the pump type and rotating speed for a centrifugal pump that operates at flow capacity of 2500 gpm and pump head of 300 ft. Estimate the following: expected maximum efficiency; classify the type of the pump; estimate the pump head-capacity curve and the pump shaft power input.

Solution:

From Figure 7.21, we can see that for the operating flow capacity of 2000 gpm, a **mixed-flow pump** with a specific speed of $N_s = 2200$ can be selected. The expected maximum efficiency would be around $\eta_p = 84\%$.

Using Eq. (7.60), we can estimate the rotating speed as follows:

$$N_s = \frac{N \, Q^{1/2}}{H_p^{3/4}}$$

$$2200 = \frac{N \, (2500)^{1/2}}{(300)^{3/4}}$$

or

$$N = \frac{2200 \times (300)^{3/4}}{(2500)^{1/2}} = \frac{2200 \times 72.084}{50}$$

Rotating speed of the pump, $N = 3171$ rpm.

7.5.1 Effect of Specific Speed on Pump Performance Characteristics

As we have seen in Figure 7.21a , the peak efficiency of pump changes as the specific speed or the class of the pump impeller is changed. Often during the fluid-flow system, it is necessary to predict the performance characteristics (H_p vs. Q; η vs. Q; and P vs. Q) of a pump when a new type is considered. This can be accomplished from the performance characteristics of existing pump types. There will be variation in the pump performance characteristic curves as pump performs at off-peak conditions or off-nominal flow capacity, and with changes in the specific speed as illustrated in Figure 7.22 (Stepanoff 1948; Karassik et al.

1976). To compare performance characteristics of different class of pumps or different specific speed, the performance parameters are normalized to the condition that corresponds to best or peak efficiency point. Results presented in these figures show variation in the ratio values of $\frac{H}{H_n}$, $\frac{P}{P_n}$, and $\frac{\eta}{\eta_n}$ with flow capacity ratio $\frac{Q}{Q_n}$, where subscript n represents the nominal values corresponding to the peak efficiency for the specific speed considered.

For the type of pump and operating design values of Q_n, H_n, P_n, and η_n as selected and computed in Example 7.1, one can now use data presented in Figure 7.22 to have initial estimate of the pump performance characteristics such as (i) the pump head-capacity curve, (ii) the efficiency-capacity curve, and (iii) the efficiency-capacity curve by computing multiple values of head and power using number off-design points of $\frac{Q}{Q_n}$ (0.2, 0.4, 0.6, 0.8, 1.0, and 1.1) at the specific speed of 2200 and assuming double-suction pump. Real performance characteristics must be obtained from the experimental test data from the manufacturers of the pump.

7.5.2 Affinity Laws for Centrifugal Pumps

Performance curves are normally given for a specific liquid, water for example, and for a pump with an impeller diameter operating at a given speed (rpm). Performance can be extrapolated to other impeller diameter and operating for the same class of pump.

Affinity laws are determined based on dimensional and similitude analysis, and are often used for a class of geometrically similar impellers and operated under similar dynamic conditions, i.e. the class that corresponds a specific speed to estimate or predict the performance of a specific centrifugal impeller for an altered diameter, D, and different rotating speeds. Following affinity laws can be applied to a specific impeller with diameter and rotating speed and assuming constant efficiency

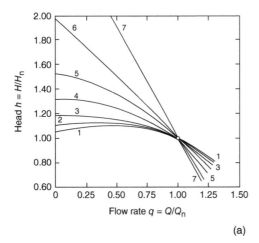

Curve	N_s	Suction type
1	900	Double suction
2	1500	
3	2200	
4	3000	
5	4000	
6	5700	Single suction
7	9200	

(a)

Figure 7.22 Pump head-capacity curves for different specific speed. Source: From Karassik et al. (1976). © 1976 McGraw-Hill. (a) Pump head vs. flow capacity ratio, (b) power vs. flow capacity ratio, and (c) efficiency vs. flow capacity ratio.

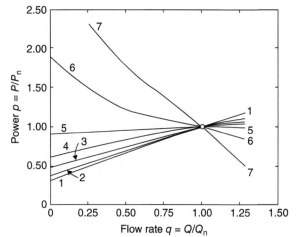

Figure 7.22 (Continued)

(b)

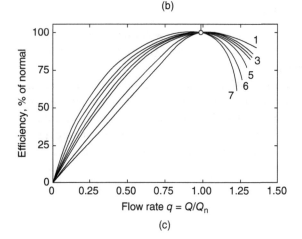

(c)

$$\frac{Q_1}{Q_2} = \frac{N_1}{N_2}\frac{D_1}{D_2} \tag{7.61a}$$

$$\frac{H_1}{H_2} = \frac{(N_1)^2}{(N_2)^2}\frac{(D_1)^2}{(D_2)^2} \tag{7.61b}$$

$$\frac{P_1}{P_2} = \frac{(N_1)^3}{(N_2)^3}\frac{(D_1)^3}{(D_2)^3} \tag{7.61c}$$

Estimation based on the affinity laws should be used as a guidance only, because a number of changes associated with the length of liquid passage, skin friction loss, exit blade angle, and clearance space result as the diameter of the impeller is reduced, and the assumption of a impeller specific speed and constant efficiency may not be a valid one.

For cases where fluids other than water need to be used, the performance characteristics are scaled using correction factors. The performance characteristics for use of a very viscous liquid in a pump whose characteristics are known for water can be estimated using correction factors as follows:

$$Q_{\text{Vis}} = C_Q\, Q_{\text{W}} \tag{7.62a}$$

$$H_{\text{Vis}} = C_{\text{H}}\, H_{\text{W}} \tag{7.62b}$$

$$\eta_{\text{Vis}} = C_\eta\, \eta_{\text{W}} \tag{7.62c}$$

and the pump power is given as

$$P = \frac{\text{S.G} \times Q_{\text{Vis}} \times H_{\text{Vis}}}{3960 \times \eta_{\text{Vis}}} \tag{7.63}$$

where C_Q, C_{H}, and C_η are experimentally determined correction factors (Hydraulic Institute Standard 1994). Figure 7.23 shows such a typical correction factor data chart for a range of viscosity values and in a capacity range of 100–10 000 gpm.

7.6 Cavitation and Net Positive Suction Head (NPSH)

Cavitation is a dynamic phenomenon, which is characterized as a process of formation and subsequent collapse of vapor-filled cavities in liquid due to dynamic actions. The cavities may be bubble, vapor-filled pockets, or a combination of both. The local pressure must be at or below the saturation vapor pressure of the liquid for cavitation to begin and then the cavities must encounter a region of higher pressure than the vapor pressure for the cavities to collapse. Dissolved gases are often released before vaporization starts, and this might be an indication of the onset of cavitation, but real cavitation starts with the vaporization of the liquid. When cavities or bubbles collapse on solid boundaries, it causes severe mechanical damage and cause pitting of the solid body. Cavitation pitting is measured in terms of weight loss of the solid body or weight of boundary material removed per unit time. Table 7.1 shows list of cavitation erosion resistance of some metals.

Cavitation is a serious concern in the design and operation of a centrifugal pump because the pump will start cavitating when the liquid pressure at the suction section or eye of the impeller falls below the saturation pressure of liquid at the local fluid temperature and hence liquid starts boiling forming vapor-filled bubbles or cavities. For example, water at 30 °C (86 °F) will start boiling when water pressure falls below 4.246 kPa (0.616 psi). Selection of pump construction materials and use of any form of cavitation resistance coatings need careful evaluations depending on the applications.

The net positive suction head (NPSH) is a term used in the design and operation of the pump to ensure that the fluid pressure at the suction of the impeller stays above the saturation vapor pressure of the liquid to avoid formation of vapor bubbles, and so that cavitation of the impeller can be avoided. If the pressure of the liquid near the eye of the pump impeller falls below the saturation pressure of the liquid at the local temperature, fluid will start boiling and start forming vapor bubbles. As these bubbles move through the rotating blade passages to the higher-pressure regions, they start collapsing and generate pressure waves. As the pressure waves transmitting it cause erosion of the impeller and pump-housing surfaces, a condition known as cavitation erosion arises. Also, because of the two-phase liquid–vapor nature of the flow, efficiency drops and eventually the impeller fails due to excessive erosion and vibration, causing fatigue of the pump shafts, bearings, and seals.

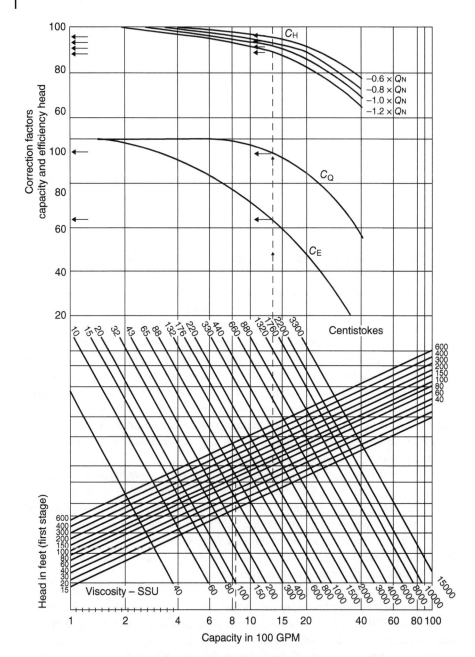

Figure 7.23 Correction factor data chart for a range of viscosity values and in a capacity range of 100–10 000 gpm. Source: Reproduced from Hydraulic Institute Standard (1994). © 1994 used with permission. Courtesy of Hydraulic Institute, Parsippany, NJ, www.pumps.org.

Table 7.1 Cavitation erosion resistance of metals.

Metals and alloys	Magnetostriction weight loss after 2 h (mg)
Rolled stellite	0.6
Welded aluminum bronze	3.2
Cast aluminum bronze	5.8
Welded stainless still (2-layers, 17 Cr-7 Ni)	6.0
Hot-rolled stainless steel (26 Cr-13 Ni)	8.0
Tempered rolled stainless steel (12 Cr)	9.0
Cast stainless steel (18 Cr–8 Ni)	13.00
Cast stainless steel (12 Cr)	20.00
Welded mild steel	97.0
Cast steel	105.0
Aluminum	124.0
Brass	156.0
Cast iron	224.0

Source: From Karassik et al. (1976). © 1976 McGraw-Hill.

While considering NPSH and cavitation possibility during the design phase of the piping system and selection of the pump, designers have to consider two things: *First* the *available net positive suction head* (***NPSH**$_a$*), which depends on the orientation and size of the suction side of the pump, and *second* the required net positive suction head (***NPSH**$_r$*), which depends on the impeller and the interior geometry of the pump. Required ***NPSH**$_r$* is usually determined experimentally by physical testing of the pump and usually provided by the pump manufacturers.

On the other hand, available net positive suction head, ***NPSH**$_a$*, can be computed based on orientation and size of the given piping system and type of fluid considered. To avoid cavitation of the pump, the available net positive suction head should be greater than the required net positive suction head with factor of margin, i.e. $\text{NPSH}_a > \text{NPSH}_r$.

Let us present here the appropriate procedure and equation for computing available net positive suction head or lift based on the two geometrical configurations shown in Figure 7.24.

In the piping system shown in Figure 7.24, a pump draws liquid from reservoir and delivers it to the piping system downstream. Let us first consider the piping system in Figure 7.24a where reservoir locates an elevation, Z_1, from the reference ground and pump is located at elevation, Z_2.

The steady-state energy equation for the suction section of the piping system is written as

$$\left(\frac{P_1}{\rho g} + \frac{V_1^2}{2g} + z_1 \right) = \left(\frac{P_2}{\rho g} + \frac{V_2^2}{2g} + z_2 \right) + \frac{V^2}{2g}\left(\frac{fL}{D} + \sum K \right) \tag{7.64}$$

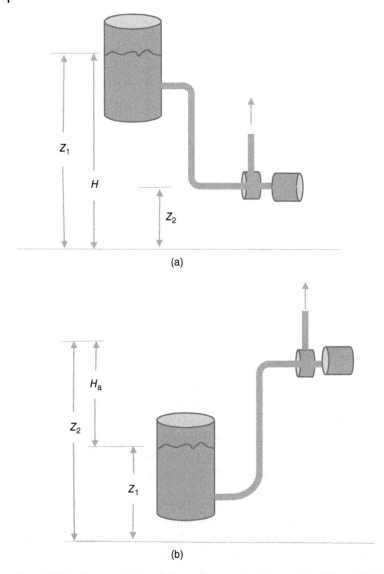

Figure 7.24 Demonstration of net positive suction head suction lift. (a) With suction head. (b) With suction lift.

where location-1 is considered at the free surface of the liquid level in the reservoir and location-2 is at the entry to eye section of the pump impeller. Setting $V_1 \approx 0$ considering a large reservoir and $V_2 = V =$ the average velocity in the pipeline, and we can rearrange Eq. (7.64) in the following form:

$$\frac{P_2}{\rho g} = \frac{P_1}{\rho g} + H_a - \frac{V^2}{2g}\left(\frac{fL}{D} + \sum K + 1\right) \tag{7.65}$$

where

F = friction factor for head loss in the straight section of suction pipe
$\sum K$ = sum of all loss coefficient for valves and fittings in the suction section of pipe

We can now define available net positive suction by subtracting saturation vapor pressure at the local liquid temperature from liquid pressure at location-2 at inlet to the impeller eye as follows:

$$\text{NPSH}_a = H_{sv} = \frac{P_2}{\rho g} - \frac{P_v}{\rho g} = \frac{P_1 - P_v}{\rho g} + H_a - \frac{V^2}{2g}\left(\frac{fL}{D} + \sum K + 1\right) \qquad (7.66a)$$

or

$$\text{NPSH}_a = H_{sv} = \frac{P_1 - P_v}{\rho g} + H_a - h_f \qquad (7.66b)$$

We can notice from Eq. (7.66b) that available NPSH will be positive so long as the sum of the first two terms is greater than the total friction heat loss in the suction line, and this ensures that we can avoid cavitation of the pump. The key factors that influence the onset of cavitation are: (i) Pressure in the reservoir, (ii) Liquid vapor pressure and hence temperature and type of the liquid, (iii) Total frictional head loss in the suction line, which depend on the length and size of the suction pipe, and number and type of valves and fittings in the line.

In the piping system shown in Figure 7.24b, the pump draws liquid from the reservoir that is located below the center line of pump impeller axis, referred to a case of **suction lift**. Following the procedure outlined before, we can write the available NPSH for the case of a suction lift as:

$$\text{NPSH}_a = \frac{P_1 - P_v}{\rho g} - H_a - \frac{V^2}{2g}\left(\frac{fL}{D} + \sum K + 1\right) \text{ with suction lift} \qquad (7.67a)$$

or

$$\text{NPSH}_a = \frac{P_1 - P_v}{\rho g} - H_a - h_f \text{ with suction lift} \qquad (7.67b)$$

We can see from Eq. (7.64b) that the concern for cavitation is high for the case of suction lift as the pressure in the reservoir should be high enough to compensate for the negative values of suction lift, vapor pressure head, and the friction head loss. For a given liquid type and temperature, one must pay attention to the type and number fittings and valves in the suction line, and the maximum allowable suction lift. The criteria should be that available new positive suction head should be greater than the required net positive suction head, i.e. $\text{NPSH}_a > \text{NPSH}_r$ to avoid cavitation.

NPSH requirements change for different pumps handling liquids like hydrocarbons or refrigerants or water at higher temperatures. Readers are recommended to review *Hydraulic Institute Standard and pump Handbook* by Karassik et al. (1976) for more information on cavitation and NPSH requirements for different pump designs, different materials of construction and liquids.

7.6.1 Thoma Cavitation Parameter (σ)

Thoma cavitation parameter is another parameter that is also used as an indicator to predict onset of cavitation. This is the dimensionless form of the total inlet head above the

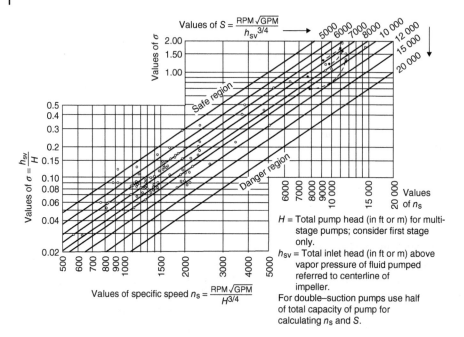

Figure 7.25 Thoma cavitation parameter and cavitation limits for centrifugal pumps. Source: From Karassik et al. (1976). © 1976 McGraw-Hill.

vapor pressure when normalized with pump head, and defined as

$$\sigma = \frac{H_{sv}}{H} \tag{7.68}$$

When the local pressure drops down to the vapor pressure of the liquid, cavitation starts and the Thoma cavitation parameter reaches the **critical sigma value**, σ_c; i.e. $\sigma = \sigma_c$.

For the given specific speed for a pump, the larger the critical sigma value, safer it is for the onset of cavitation. **Thoma cavitation parameter** needs to be determined experimentally. Figure 7.25 shows such experimentally determined values of cavitation parameter to identify the safer and the dangerous regions of cavitation over a range of specific speeds, $N_s = \frac{RPM \sqrt{Q}}{H_p}$, and over a range of the suction specific speed parameter $S = \frac{RPM \sqrt{Q}}{H_{sv}}$, computed based on inlet suction head, H_{sv}.

In such typical cavitation experiments, the suction head, H_{sv}, is decreased while keeping the flow capacity, Q, as constant and observe at what point cavitation starts. The experimental uncertainty in identifying the onset of cavitation is evident through the presence of scatter in the data.

A few things to be remembered while referring to this chart: (i) For multistage pumps, use the pump head based on the single stage; (ii) For double-suction pumps, use half of the total pump flow capacity while computing the specific speed, N_s, and the value of suction speed, S.

7.6.2 Cavitation Resistance Coatings

As we can notice from the data presented in Table 7.1, some of the most common pump materials like cast iron, brass, aluminum, cast steel, and mild steel have very low resistance to cavitation erosion. A large variety of elastomeric coatings such as polyurethane are available and are used to enhance the resistance characteristics of impeller made of those low-cavitation erosion-resistance materials. A list of such recommended elastomer coatings is given in the pump handbook by Karassik et al. (1976).

7.7 Pumps in Series or in Parallel

In many applications, multiple pumps may be added in a line and operated either in series or in parallel as demonstrated in Figure 7.26.

7.7.1 Pumps Connected in Series

Pumps are connected in series to achieve higher head than those of the individual pumps following the two basic rules given as:

Total increase in head in the line is the sum of increase in head from individual pumps

$$\Delta H_{A-B} = \Delta H_{P-1} + \Delta H_{P-2} + \Delta H_{P-3} \tag{7.69a}$$

and

Flow capacity remains same in all pipes in the line:

$$Q_{A-B} = Q_{P-1} = Q_{P-2} = Q_{P-3} \tag{7.69b}$$

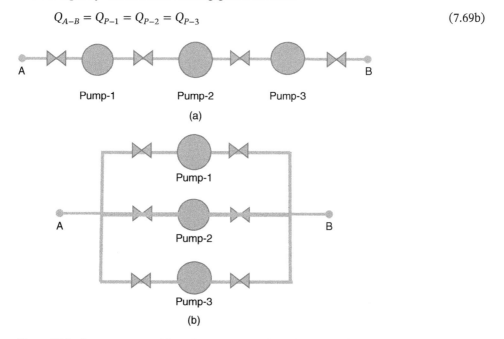

Figure 7.26 Pumps connected in series and parallel in a pipeline. (a) Pumps in series and (b) pumps in parallel.

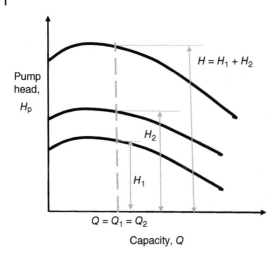

Figure 7.27 Head-capacity curve for pipes connected in series.

The combined head-capacity curve can be derived by simply adding pump heads from individual pumps and keeping capacity same. The combined head-capacity curves as well as two individual pump head-capacity curves are shown in Figure 7.27.

7.7.2 Pumps Connected in Parallel

Pumps are connected in parallel to achieve variable and higher-flow capacity than those of the individual pumps following the two basic rules given as:

If both the pumps are identical, then the total increase in head is same as those given by the individual pumps

$$\Delta H_{A-B} = \Delta H_{P-1} = \Delta H_{P-2} = \Delta H_{P-3} \tag{7.70a}$$

and total flow capacity in the line is sum of the flow capacities given by the individual pumps:

$$Q_{A-B} = Q_{P-1} + Q_{P-2} + Q_{P-3} \tag{7.70b}$$

The combined head-capacity curve can be derived by simply adding flow capacities from all from individual pumps at any value of pump head. The combined head-capacity curves as well as for two individual pump head-capacity curves are shown in Figure 7.28.

7.8 Pump Standards and Codes

The following is a list of most popular and widely used standards for the designs, applications, testing, operations, and maintenance of pumps.

7.8.1 ASME Centrifugal Pumps Standards – PTC 8.2

This standard provides directions for performing tests for centrifugal pumps in different categories of axial- and mixed-flow types and application for liquids or mixture miscible liquids.

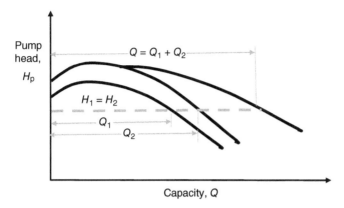

Figure 7.28 Two pumps connected in parallel.

7.8.2 ANSI PUMPS – ASME B73.1 Standards for Chemical/Industrial Process Pumps

This standard brings in dimensional standardization for three groups of pumps in terms of pump sizes. This allows the user to replace pumps with different vendors without affecting piping orientation or location, bedplate dimensions, and motor connections.

7.8.3 ANSI/HI: Hydraulic Institute Standards for Pumps and Pumping Systems

Provides number of different standards for the definitions, industrial terminology, design and applications, testing, installation, operation, and maintenance guides. Standards are available for different types of pumps, including centrifugal pumps, vertical pumps, rotary as well as reciprocating pumps. Some examples of these standards are (i) ANSI/HI – 1.3 for Rotodynamic Centrifugal pumps for Design and Application, (ii) ANSI/HI – 1.4 for Rotodynamic Centrifugal pumps for Installation, Operation, and Maintenance, (iii) ANSI/HI 2.3 for Rotodynamic Centrifugal pumps of Radial-, mixed-, and Axial-flow types for Design and Operation, (iv) ANSI/HI 3.1-3.5, 3.6. For Rotary pumps, (v) ANSI/HI 4.1-5.1-5.6 for Seal-less Rotary Pumps, and (vi) ANSI/H 6.1-6.5, 6.6 for Reciprocating pumps.

Bibliography

American Institute of Chemical Engineers. (1984), *Equipment testing procedure: Centrifugal pumps.* New York: AICHE.

American Society of Mechanical Engineers, (1990). Performance test codes: Centrifugal pumps, ASME PTC 8.2-1990, New York.

ANSI PUMPS – ASME B73.1 Standards (2012). *Specification for Horizontal End Suction Centrifugal Pumps for Chemical Process.* New York: American Society of Mechanical Engineers.

ANSI/HI: Hydraulic Institute Standards. (2014). Hydraulic Institute, New York.

Fox, R.W. (1992). *Introduction to Fluid Mechanics.* Wiley.

Hodge, B.K. and Taylor, R.P. (1999). *Analysis and Design of Energy Systems*, 3e. Hoboken, NJ: Prentice Hall.

Karassik, I.J., Krutzsch, W.C., and Fraser, W.H. (1976). *Pump Handbook*. New York: McGraw-Hill, Inc.

Hydraulic Institute. (1994). Pump Standards, New York.

Shepherd, D.G. (1956). *Principles of Turbomachinery*. New York: The Macmillan Company.

Stepanoff, A.J. (1948). *Centrifugal and Axial Flow Pumps – Theory, Design, and Application*. New York: Wiley.

Stodola, A. (1927). *Steam and Gas Turbines*. New York: McGraw – Hill Book Co.

Problems

7.1 Consider a centrifugal pump with a mixed-flow impeller and running at 3000 rpm and a rated flow rate of 1000 gpm. The impeller vane radius at inlet is 3.00-in. and at outlet is 8-in. The vane angles at inlet and outlet of the vanes are 30° and 70°, respectively. The vane passage widths at inlet and outlet are $b_1 = 0.65$-in. and $b_2 = 0.5$-in., respectively. The inlet guide vanes are used to introduce a swirl for the flow at the inlet with absolute velocity making an angle $\alpha_1 = 30°$ from normal. Determine the theoretical pump head and pump power.

7.2 Consider the Problem 7.1 and repeat computations to show the variations in pump head and power in the rpm range of 2000–5000 rpm.

7.3 Consider the Problem 7.1 and repeat computation with variation in the impeller outlet radius in the range of 6.00, 6.5, 7.0, and 7.5-in.

7.4 Consider a radial centrifugal pump impeller in which water at 600 gpm enters the impeller axially through the impeller inlet section of 1.00-in. radius and exits the impeller with exit width *of* $b_2 = 0.45$-in. Determine pump head, torque, and power.

7.5 Select the pump type and rotating speed (rpm) for a centrifugal pump that operates at flow capacity of 8000 gpm and pump head of 450-ft. Estimate the following: specific speed, expected maximum efficiency and power, and classify the type of the pump.

7.6 Select the pump type and rotating speed for a centrifugal pump that operates at flow capacity of 2500 gpm and pump head of 300-ft. Estimate the following: (i) Specific speed and the expected maximum efficiency and power; and classify the type of the pump. (ii) For the type of pump and operating design values of Q_n, H_n, P_n, and η_n as selected and computed, use data presented in Figure 7.22 to have initial estimate of the pump performance characteristics such as (a) the pump head-capacity curve, (b) the efficiency-capacity curve and (c) the efficiency-capacity curve by computing multiple values of head and power using number of off-design points of $\frac{Q}{Q_n}$ (0.2, 0.4, 0.6, 0.8, 1.0, and 1.1) at the specific speed and assuming double-suction pump.

7.7 For the selected specific speed, use data presented in Figure 7.22 to estimate (a) the pump head-capacity curve, (b) the efficiency-capacity curve, and (c) the efficiency-capacity curve by computing multiple values of head and power using number off-design points of $\frac{Q}{Q_n}$ (0.2, 0.4, 0.6, 0.8, 1.0, and 1.1).

7.8 A piping system application requires a centrifugal pump with the following operating requirements: Flow capacity, $Q = 1200$ gpm; Total pump head, $H_p = 300$ ft and Available $NPSH_a = 30$ ft. Determine the following: (i) Selection speed; (ii) Selection type of pump; (iii) Estimate over pump efficiency and input Power; (iv) Estimate the operating head – capacity characteristics; (v) Impeller discharge velocity triangles; (vi) Impeller discharge dimensions; (vii) Impeller inlet diameter; Impeller inlet vane angle; (viii) Impeller inlet velocity triangle; (ix) Flow areas in between vanes and volute casing; and estimate losses, efficiency, and shaft power.

8

Analysis and Design of Fluid-Flow Systems

Fluid-flow systems essentially serve as an integral part of many thermal energy systems to effectively transport and distribute fluid and energy among many components of the system. Fluid-flow systems include prime fluid movers like pumps, fans, blowers, etc. along with a piping or duct system that may be just a simple single pipeline or it may be a complex series–parallel system. This chapter presents some of the basic principles and concepts for the analysis and the design of fluid-flow systems that integrate with thermal energy systems. Readers are recommended to refer to books by Fox and MacDonald (1973), Munson et al. (2016), ASHRAE Fundamentals (1989), Crane technical paper No 410M (2009), Hodge and Taylor (1999), Janna (1993), Sauer and Howell (1990), and many others as mentioned in Bibliography section.

8.1 Basic Equations of Fluid Flow

The basic equation for analyzing fluid flow through pipes and devices starts with the conservation of mass equation, conservation of momentum equation, and conservation of energy equations for flow through a control volume as shown in Figure 8.1.

8.1.1 Conservation of Mass

$$\frac{\partial}{\partial t}\left(\int_{CV} \rho \, d\forall\right) + \int_{CS} \rho\vec{V} \cdot d\vec{A} = 0 \tag{8.1a}$$

For steady flow

$$\int_{CS} \rho\vec{V} \cdot d\vec{A} = 0 \tag{8.1b}$$

For steady flows, the net mass flow rates into a control volume must be equal to the net mass flow out of the control volume. For one-dimensional case with uniform mass flow and the energy flux across the open control surface, the net mass flux term in Eq. (8.1b) can be expressed as

$$\sum (\rho VA)_{out} = \sum (\rho VA)_{in} \tag{8.1c}$$

Design of Thermal Energy Systems, First Edition. Pradip Majumdar.
© 2021 John Wiley & Sons Ltd. Published 2021 by John Wiley & Sons Ltd.
Companion website: www.wiley.com/go/majumdar

Figure 8.1 Flow and energy transfer across a control volume.

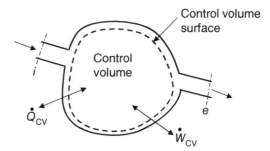

For **steady flow** with one flow in at section 1 and one flow out at section 2, the energy equation reduces to

$$\rho_1 V_1 A_1 = \rho_2 V_2 A_2 \tag{8.1d}$$

For **incompressible flow** through a control volume

$$\int_{CS} \vec{V} \cdot d\vec{A} = 0 \tag{8.2a}$$

For one-dimensional case with uniform volume flow across the open control surface, the net volume flow rate Eq. (8.2a) can be expressed as

$$\sum (VA)_{out} = \sum (VA)_{in} \tag{8.2b}$$

and for uniform flow inlet at section 1 and uniform flow out at section 2, this equation reduces to

$$V_1 A_1 = V_2 A_2 \tag{8.2c}$$

8.1.2 Conservation of Energy

$$\frac{\partial}{\partial t} \left(\int_\forall e\rho \, d\forall \right) + \int_{CS} e\rho \vec{V} \cdot d\vec{A} = \pm \dot{Q} \pm \dot{W} + \dot{Q}_{gen} \tag{8.3}$$

where e is the total energy per unit mass and represents a sum of three energy forms: internal energy, kinetic energy, and potential energy, and is expressed as $e = u + \frac{V^2}{2} + gZ$; $\dot{Q} =$ rate of heat transfer; $\dot{W} =$ rate of work done; and $\dot{Q}_{gen} =$ rate of heat generation within the control volume. The first term on the left-hand side of Eq. (8.3) represents the rate of change of energy within the control volume, and second term on the left-hand side represents net energy transfer across the surfaces of the control volume.

8.1.3 Basic Energy Equation for Analyzing Pipe Flow

The basic equation for analyzing piping system design starts with the conservation mass and conservation of energy equation for the flow through a control volume. Considering a piping system with no heat generation within, the general conservation energy equation for flow through a control volume is given as

$$\dot{Q} = \frac{\partial}{\partial t} \iiint_{CV} e\rho \, dv + \iint_{CS} e\rho \vec{V} \cdot dA + \dot{W} \tag{8.4}$$

The term for rate of work done, $\dot{W}$, consists of three major components: (i) Energy transfer through work done due to any shaft work, $\dot{W}_s$; (ii) Work done due to pressure flow work, $\dot{W}_p$; and (iii) Energy transfer by viscous shear work, $\dot{W}_v$.

Expressing the pressure work term in terms of product of Pv and combining with the internal energy term through use of the thermodynamic relation $h = u + pv$, the energy equation (8.4) is expressed

$$\dot{Q} = \frac{\partial}{\partial t} \iiint_{CV} \left(h + \frac{V^2}{2} + gZ \right) \rho \, dv + \iint_{CS} \left(h + \frac{V^2}{2} + gZ \right) \rho \vec{V} \cdot dA + \dot{W}_s + \dot{W}_v$$

(8.5)

Under steady-state condition, the second term of the right-hand side can be dropped and Eq. (8.5) reduces to the following simplified form:

$$\iint_{CS} \left(h + \frac{V^2}{2} + gZ \right) \rho \vec{V} \cdot dA - \dot{Q} + \dot{W}_s + \dot{W}_v = 0$$

(8.6)

For one-dimensional case with uniform mass flow and the energy flux across the open control surfaces, the net energy flux term can be expressed as

$$\iint_{CS} \left(h + \frac{V^2}{2} + gZ \right) \rho \vec{V} \cdot dA = \sum \left(h + \frac{V^2}{2} + gz \right) \dot{m}_{out} - \sum \left(h + \frac{V^2}{2} + gz \right) \dot{m}_{in}$$

(8.7)

For a control volume with one flow-in at section-1 and one flow-out at section-2 of a pipe section, a mass balance equation (8.1d) results in

$$\dot{m}_{out} = \dot{m}_{in} = \dot{m}$$

(8.8)

For steady flow with one flow-in at section 1 and one flow-out at section 2, the energy equation reduces to

$$\dot{m} \left(h_1 + \frac{V_1^2}{2} + gz_1 \right) = \dot{m} \left(h_2 + \frac{V_2^2}{2} + gz_2 \right) - \dot{Q} + \dot{W}_s + \dot{W}_v$$

(8.9)

Dividing both sides by the mass flow rate, $\dot{m}$

$$\left(h_1 + \frac{V_1^2}{2} + gz_1 \right) = \left(h_2 + \frac{V_2^2}{2} + gz_2 \right) - q + w_s + w_v$$

(8.10)

where

$$q = \frac{\dot{Q}}{\dot{m}}, \quad W_s = \frac{\dot{W}_s}{\dot{m}}, \quad \text{and} \quad W_v = \frac{\dot{W}_v}{\dot{m}}$$

For fluid flow in a pipe, let us substitute enthalpy $h = u + \frac{P}{\rho}$, and write the energy equation in the following form

$$\left(\frac{P_1}{\rho} + \frac{V_1^2}{2} + gz_1 \right) = \left(\frac{P_2}{\rho} + \frac{V_2^2}{2} + gz_2 \right) + W_s + W_v + (u_2 - u_1 - q)$$

(8.11)

Dividing both sides by g and expressing each term as head, we get

$$\left(\frac{P_1}{\rho g} + \frac{V_1^2}{2g} + z_1 \right) = \left(\frac{P_2}{\rho g} + \frac{V_2^2}{2g} + z_2 \right) + h_s + h_v + \frac{(u_2 - u_1 - q)}{g}$$

(8.12)

where

$$h_s = \frac{W_s}{g} = \text{Shaft head gain or loss}$$

$$h_v = \frac{W_v}{g} = \text{Head loss due to viscous shear work}$$

Also, **Bernoulli's equation** for inviscid frictionless flow is given as

$$\left(\frac{P_1}{\rho g} + \frac{V_1^2}{2g} + z_1 \right) = \left(\frac{P_2}{\rho g} + \frac{V_2^2}{2g} + z_2 \right) \tag{8.13}$$

Since Bernoulli's equation represents an ideal fluid flow with no losses, a comparison of Eqs. (8.12) and (8.13) gives the energy or head loss in a pipe section of a fluid flow as

$$\frac{(u_2 - u_1 - q)}{g} + h_v = \Delta h_{\text{rev,loss}} + \Delta h_{\text{irrev,loss}} \tag{8.14}$$

The irreversible head loss, $\Delta h_{\text{irrev, loss}}$, occurs in all real flows and is the result of viscous dissipation, converting mechanical energy into nonreversible internal energy and heat transfer. $\Delta h_{\text{rev, loss}}$ represents the reversible loss due to heat addition or rejection or interchange between mechanical energy and internal energy.

Combining Eqs. (8.12) and (8.14), we get the **energy equation for fluid flow in a pipe** section as

$$\left(\frac{P_1}{\rho g} + \frac{V_1^2}{2g} + z_1 \right) = \left(\frac{P_2}{\rho g} + \frac{V_2^2}{2g} + z_2 \right) + h_s + \Delta h_{\text{rev,loss}} + \Delta h_{\text{irrev,loss}} \tag{8.15}$$

For fluid flow under isothermal condition with no energy transfer through heat transfer, the reversible loss is neglected. Considering the irreversible frictional loss due to viscous shear, we can rewrite **the steady-state energy equation** as

$$\left(\frac{P_1}{\rho g} + \frac{V_1^2}{2g} + z_1 \right) = \left(\frac{P_2}{\rho g} + \frac{V_2^2}{2g} + z_2 \right) + h_s + h_f \tag{8.16a}$$

where $h_f = \Delta h_{\text{irrev, loss}}$ is the **head loss** due to viscous friction in a piping system involving fictional head loss in the straight section of pipe and the frictional head loss in the valves and fittings in the piping system. The head loss in the straight section of the pipe is referred to as the **major loss**, $h_f = h_{f, \text{major}}$, and the head loss in the valves and fittings is referred to as the **minor loss**.

In Eq. (8.16a), the term h_s is the shaft head gain or loss defined as

$$h_s = \frac{W_s}{g} \tag{8.16b}$$

A general form of the piping system energy equation (8.16a) is written as

$$\frac{P_1}{\rho g} + \frac{V_1^2}{2g} + Z_1 = \frac{P_2}{\rho g} + \frac{V_2^2}{2g} + Z_2 - h_{\text{pump}} + h_{\text{turbine}} + h_f \tag{8.16c}$$

where $h_{\text{pump}} = \frac{W_{\text{pump}}}{g}$ is the useful pump delivered to the fluid and $h_{\text{turbine}} = \frac{W_{\text{turbine}}}{g}$ is the turbine head extracted from the fluid. Note that the shaft head is now replaced with the *negative pump head* and a *positive turbine head*. The pump head is zero if the piping system does not include a pump and the turbine head is zero if the system does not include a turbine.

8.1.4 Frictional Head Loss for Flow in Pipes: Major Loss

To obtain a quantitative expression to evaluate the major loss in the straight section of the pipe, let us consider the pressure-driven flow in a circular pipe of length L and diameter D as shown in Figure 8.2.

For fully developed flow with no heat transfer and shaft work in the straight section of a pipe, the application of the energy equation (8.16a) leads to

$$h_f = \frac{\Delta P}{\rho g} = \frac{P_1 - P_2}{\rho g} \tag{8.17}$$

A balance of forces due to pressure difference across sections of the pipe and viscous wall shear stress, as depicted in Figure 8.3, leads to:

$$\Delta P \times \frac{\pi}{4} D^2 = \tau_w \times \pi DL \tag{8.18}$$

where

$\Delta P = P_1 - P_2 = $ Net pressure difference between inlet and outlet sections of the pipe
$\tau_w = $ Wall shear stress

Rearranging

$$\Delta P = \frac{4\tau_w L}{D} \tag{8.19}$$

Equating (8.17) and (8.18)

$$h_f = \frac{\Delta P}{\rho g} = \frac{4\tau_w L}{\rho g D} \tag{8.20}$$

Rearranging

$$\tau_w = \frac{\Delta P}{\rho g} = \frac{h_f \rho g D}{4L} \tag{8.21}$$

Note that Eq. (8.21) is valid for both laminar and turbulent flows. Major task now is to relate the shear stress to all flow parameters. In general, the shear stress for fluid through a pipe

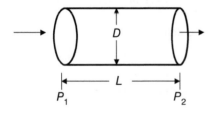

Figure 8.2 Flow in a straight pipe section.

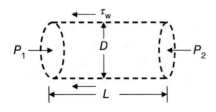

Figure 8.3 Force balance over a straight pipe section.

is a function of several flow variables and parameters, and can be expressed as follows:

$$\tau_w = F(\rho, V, \mu, D, \varepsilon) \tag{8.22}$$

where

V = Average velocity
D = Diameter of the pipe
E = Roughness of the pipe
P = Density of fluid
M = Viscosity of fluid

It can be shown from a **dimensional analysis** that Eq. (8.22) can be expressed in terms of few dimensionless variables:

$$\frac{8\tau_w}{\rho V^2} = F\left(Re_D, \frac{\varepsilon}{D}\right) = f \tag{8.23}$$

where

Re_D = Reynolds number = $\frac{\rho VD}{\mu}$
$\frac{\varepsilon}{D}$ = Relative roughness
ε = Roughness of the pipe wall
D = Diameter of the pipe

The functional relationship on right-hand side of Eq. (8.23) is defined as the **friction factor** for flow in the pipe, and it shows that the friction factor is a function of two parameters: **flow Reynolds number, Re_D**, and the **relative roughness, $\frac{\varepsilon}{D}$**, of pipe wall as defined below:

$$f = F\left(Re_D, \frac{\varepsilon}{D}\right) \tag{8.24}$$

We can now express the shear stress for fluid flow in pipe from Eq. (8.23) as

$$\tau_w = f\frac{\rho V^2}{8} \tag{8.25}$$

Combining Eqs. (8.21) and (8.25), we get

$$\tau_w = f\frac{\rho V^2}{8} = \frac{h_f \rho g D}{4L} \tag{8.26}$$

Solving for friction head loss, we get

$$h_{f,major} = f\frac{L}{D}\frac{V^2}{2g} \tag{8.27}$$

Equation (8.27) represents the **major loss** or the **friction pressure drop in straight section** of the pipe and is known as the **Darcy–Weisbach equation**. This is also valid for duct flows of any cross-sections as well as for both laminar and turbulent flows. For noncircular pipes, diameter D is replaced with the hydraulic diameter, D_H.

Darcy–Weisbach Equation (8.27) for major loss or frictional pressure drop in the straight section of the pipe is also expressed in an alternate form known as **Fanning's equation** given as

Manning's Equation:

$$h_{f,major} = 4f_F \frac{L}{D} \frac{V^2}{2g} \tag{8.28}$$

where

f = Darcy–Weisbach friction factor
f_F = Fanning friction factor $= f/4$

8.1.4.1 Friction Factor: Fully Developed Laminar Flow in Circular Pipe

For piping system designs, the pressure drop in a long pipe is often dominated by that in the fully developed section of the pipe as shown in Figure 8.4. The pressure drop in the fully developed section of the pipe is given by the **Hagen–Poiseuille flow**, which is derived from the approximation of Navier–Stokes equations for fully developed laminar flow in circular pipes.

The simplified governing equations and boundary conditions for fully developed flow in a circular pipe are given as

Governing equation:

$$\frac{1}{r}\frac{d}{dr}\left(\mu r \frac{du}{dr}\right) = \frac{dP}{dx} \tag{8.29}$$

Boundary conditions:

$$r = 0, \text{velocity is finite or } \frac{du}{dr} = 0 \tag{8.30a}$$

$$r = R, u = 0 \tag{8.30b}$$

Analytical solution for the axial component of velocity is derived as

$$u = -\frac{R^2}{4\mu}\left(\frac{dP}{dx}\right)\left[1 - \left(\frac{r}{R}\right)^2\right] \tag{8.31}$$

or

$$u = u_{max}\left[1 - \left(\frac{r}{R}\right)^2\right] \tag{8.32}$$

where

$$u_{max} = \frac{R^2}{4\mu}\left|-\frac{dP}{dx}\right| \tag{8.33}$$

Volume flow rate is then given as

$$Q = \int_A u \, dA = \int_0^R u_{max}\left(1 - \frac{r^2}{R^2}\right) 2\pi r \, dr \tag{8.34}$$

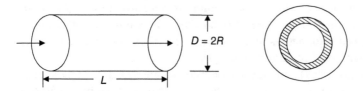

$D = 2R$

L

Figure 8.4 Section of a pipe with fully developed flow.

Solving,

$$Q = \frac{\pi R^4}{8\mu} \left(-\frac{dp}{dx}\right) \tag{8.35}$$

For the pipe of length, L

$$-\frac{dp}{dx} = \frac{\Delta P}{L} \tag{8.36}$$

Substituting Eq. (8.36) into Eq. (8.35), we get

$$Q = \frac{\pi R^4 \Delta P}{8\mu L} \tag{8.37}$$

Rearranging,

$$\Delta P = \frac{8\mu L Q}{\pi R^4} \tag{8.38}$$

and

$$\frac{\Delta P}{\rho g} = \frac{8\mu L Q}{\pi R^4 \rho g} \tag{8.39}$$

or

$$\frac{\Delta P}{\rho g} = \frac{128\mu L Q}{\rho g \pi D^4} \tag{8.40}$$

Now from the application of energy equation (8.16a) for a pipe system, we get

$$\left(\frac{P_1}{\rho g} + \frac{V_1^2}{2g} + z_1\right) = \left(\frac{P_2}{\rho g} + \frac{V_2^2}{2g} + z_2\right) + h_s + h_f \tag{8.16}$$

For $Z_1 = Z_2$, $V_1 = V_2$ and for no rotary device or shaft work, $h_s = 0$, Eq. (8.16a) reduces to

$$\frac{P_1 - P_2}{\rho g} = \frac{\Delta P}{\rho g} = h_f \tag{8.41}$$

Combining Eqs. (8.27), (8.40), and (8.41)

$$h_f = f\frac{L}{D}\frac{V^2}{2g} = \frac{\Delta P}{\rho g} = \frac{128\mu L Q}{\rho g \pi D^4} \tag{8.42}$$

Now substituting $Q = \rho \frac{\pi}{4}D^2 V$, we get

$$f = \frac{64\mu}{\rho V D}$$

or

$$f = \frac{64}{Re} \tag{8.43}$$

where

$Re = \frac{\rho V D}{\mu}$ = Reynolds number for internal flow in a pipe section

Equation (8.43) represents the friction factor for fully developed laminar flow in a circular pipe. Similarly, analytical solution for laminar fully developed flow in a number of different flow channel geometries is derived and a list of such flows is given in Table 8.1.

Table 8.1 Friction factor for laminar fully developed flow.

Flow geometry	Aspect ratio, b/a	Friction factor for fully developed flow, $f Re_D$
⬭ (circle)		64
▢ (square, sides a, b)	1.0	57
▭ (rectangle, a, b)	2.0	62
▭	4.0	73
▭	8.0	82
──	∞	96
△ (triangle)		53
⬡ (hexagon)		60.25
Octagon		61.52

Source: From Incropera et al. (2007). © 2011, John Wiley & Sons.

8.1.4.2 Frictional Pressure Drops for Turbulent Flow

For turbulent flow, roughness of the pipe significantly influences the viscous shear and the pressure drop, and the friction factor data are given as a function of flow Reynolds number and relative roughness of the pipes:

$$f = f \left(Re_D, \frac{\varepsilon}{D} \right) \tag{8.44}$$

where

Re_D = Reynolds number based on pipe diameter
$\frac{\varepsilon}{D}$ = Relative roughness of pipe wall

This functional relationship or friction factor was developed based on experimental analysis of thousands of tests from various sources and presented as a diagram known as **Moody's diagram** (Moody 1944; Fox and McDonald 1992; Incropera et al. 2007; Crane Technical Report 1980) as demonstrated in Figure 8.5. The data represent Darcy–Weisbach friction factor for fully developed flow in a circular pipe.

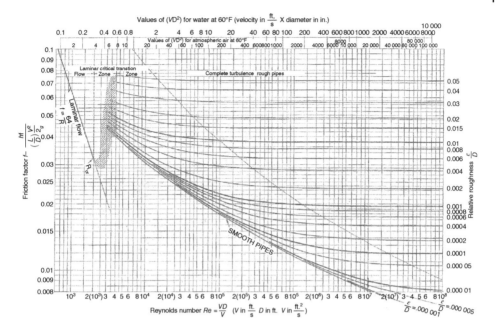

Figure 8.5 Moody's diagram for pipe. Source: From Moody (1944). © 1994 The American Society of Mechanical Engineers.

The chart shows two distinct regimes: Laminar-flow and turbulent-flow regimes. Transition from laminar flow to turbulent flow takes place over a range of Reynolds number 2100–2400 depending on the flow conditions. For analysis purposes, laminar flow is assumed to take place for Reynolds number $Re < 2300$. In Moody's diagram, the laminar flow for flow in circular pipe is plotted based on the expression $f = \frac{64}{Re}$ as a function of Reynolds number Re only. Roughness of the pipe has no effect on the friction factor or friction pressure drop in the Reynolds number range for laminar flows as evident by the analytical solutions presented in Table 8.1.

For pipe flows with flow Reynolds number greater than critical-flow Reynolds number, $Re_c \geq 2300$, the flow is characterized as the turbulent flow. The turbulent flow region has two distinct subregions: **transition region** and **complete turbulence** or **fully rough region**. For turbulent flow, the flow friction fact value increases with higher pipe roughness. Another important observation that can be observed here is that the friction factor value decreases with flow Reynolds number and relative roughness and approaches a constant value in the fully turbulent region for a given relative roughness value. These constant friction factor values are referred to as the **fully rough friction factors**.

Roughness of the pipe or tube depends on the type of pipe materials and fabrication methods used, and it strongly affects the wall shear stress and frictional head loss in turbulent flows. A list of reported measured roughness values for some pipe materials is given in Table 8.2.

Figure 8.6 shows the pipe relative roughness values of some commonly used materials as a function of pipe diameter.

Table 8.2 Roughness values for pipes or tubes.

Materials	Roughness (ε)	
	ft	mm
Cast iron	0.000 85	0.26
Asphalted cast iron	0.0004	0.12
Galvanized iron	0.0005	0.15
Commercial steel or wrought iron	0.000 15	0.046
Riveted steel	0.003–0.03	0.9–9
Concrete	0.001–0.01	0.3–3
Drawn tubing	0.000 005	0.0015

The friction factor data given in Moody's diagram in Figure 8.5 and relative roughness data given in Figure 8.6 are widely used for pressure drop and flow-rate computations in piping system designs.

Many curve-fitted expressions to represent Moody's diagram in turbulent region are developed for computational simplicity using compute programs. A most widely used curve-fit expression for computing friction factor, f, based on Moody's diagram in the turbulent-flow region is given by **Colebrook's implicit form**:

$$\frac{1}{f^{\frac{1}{2}}} = -2.0 \log\left(\frac{\varepsilon/D}{3.7} + \frac{2.51}{Re f^{1/2}}\right) \qquad (8.45)$$

where

$Re_D = \frac{\rho VD}{\mu}$ = Reynolds number

ε = Roughness of pipe

This is considered as the **implicit curve-fit form** of Moody's diagram and is the most accurate form available. However, use of this implicit form of equation requires iterations to determine the friction factor value for a given pipe roughness and Reynolds number. An iterative algorithm based on Colebrook's equation can simply be implemented in a computer program for piping system design analysis.

For direct computations of the friction factor, several **explicit forms** of curve-fit equations are also developed. Each of these equations is associated with certain level of error and accuracy. Among several explicit forms, the most comparable one to Colebrook's equation is the following **Haaland's explicit form**:

$$f = \frac{0.3086}{\left\{\log\left[\left(\frac{\varepsilon/D}{3.7}\right)^{1.11} + \frac{6.9}{Re_D}\right]\right\}^2} \quad \text{for } \frac{\varepsilon}{D} > 10^{-4} \qquad (8.46)$$

Moody's diagram can also be used for noncircular pipes based on hydraulic diameter. However, this is not suitable for eccentric annular flow or pipe cross-sections with high aspect ratios as it results in high secondary recirculation losses at corners.

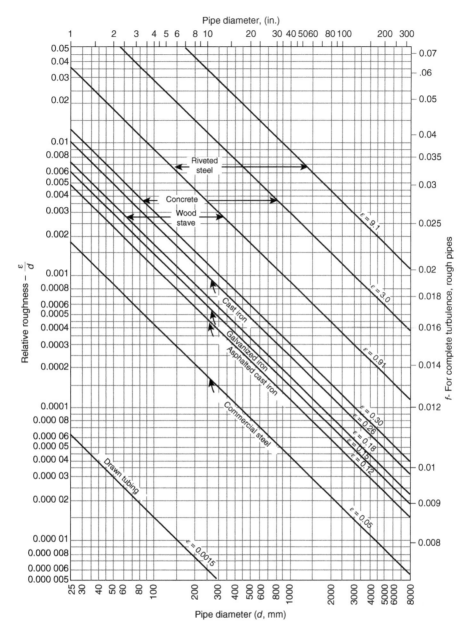

Figure 8.6 Relative roughness of different materials and friction factor of complete turbulence. Source: From Crane Technical Paper 410 (2009).©2009 Crane Co.

8.1.4.3 Minor Losses in Valves and Fittings

Minor losses include additional frictional pressure drops in the piping system components like valves; bends, tee-sections, and branches; expansions and contraction sections; and inlets and exits. Minor losses in fittings and valves are categorized into following forms: (i) Pipe entrance and exit; (ii) Sudden expansion and contraction; (iii) Gradual expansion and contraction; (iv) Bends, elbows, tees, and other fittings; and (iv) Valves – *fully open* and *partially closed*. Flow dynamics in valves and fittings are complex and involve additional pressure drop or head loss due to flow separations and recirculatory secondary flows. Also, due to the complexity of flow and lack of analytical close form of solutions, pressure drop in valves and fittings is estimated based on experimental data given in the form of empirical constants. There are two different approaches in estimating pressure drops in fittings and valves as discussed in Sections 8.1.4.3.1 and 8.1.4.3.2.

8.1.4.3.1 Approach-I: Use of Loss Coefficient

In this approach, pressure drop (ΔP) across a fitting or valves is measured in an experimental setup and presented as a function of average velocity in the following form:

$$\Delta P \sim \frac{V^2}{2}$$

and with the use of the constant of proportionality as

$$\Delta P = K\frac{V^2}{2} \tag{8.47a}$$

and in terms of friction head loss in a piping system as

$$h_m = \frac{\Delta P}{\rho g} = K\frac{V^2}{2g} \tag{8.47b}$$

where K, the constant of proportionality, is defined as the *friction loss coefficient* and is estimated as an empirical constant from measured pressure drop data in the following manner:

$$K = \frac{\Delta P}{\frac{V^2}{2}} \tag{8.48}$$

It can be noted here that minor loss is directly proportional to the square of the average velocity in the pipe as in the case of major loss. The minor loss coefficients depend on the key geometrical parameters of the valves and fittings.

8.1.4.3.2 Approach-II: Use of Equivalent Length

In the alternative approach, the minor loss is expressed in terms of equivalent length, L_{eq}, of the pipe following *Darcy–Weisbach Equation* for major loss as

$$h_m = f\frac{L_{eq}}{D}\frac{V^2}{2g} = K\frac{V^2}{2g} \tag{8.49}$$

Equivalent Length, L_{eq}, is given based on the experimental pressure drop data and related to the loss coefficient as

$$L_{eq} = \frac{KD}{f} \tag{8.50}$$

Sum of all minor losses in a piping system is given as

$$h_m = \sum K\frac{V^2}{2g} = f\frac{\sum L_{eq}}{D}\frac{V^2}{2g} \tag{8.51}$$

Total frictional head loss in a piping system is given as

$$h_f = h_{f,major} + h_{f,minor} = \frac{V^2}{2g}\left(\frac{fL}{D} + \sum K\right) \tag{8.52}$$

8.1.4.4 Minor Loss Coefficient Values

Minor losses contribute significantly and may dominate the overall pipe loss in a piping system, especially in shorter length pipeline. There are many different database sources for the minor loss coefficients such as Hydraulic Institute (1979), Crane Technical Paper 410 (), Chemical Engineering handbook by Perry and Chilton (1963). While most of the data available are obtained based on physical testing performed by many different entities, considerable variation in the data can be noticed due to the variation in the sources or vendors of the valve and fittings, and the associated experimental uncertainties. Some of the very useful loss coefficient values of valves and fitting are summarized in the following section to demonstrate the design and analysis procedure for some fluid-flow systems.

8.1.4.4.1 Entrance and Exit Loss Coefficient

Exit Losses   There are different exit types as shown in Figure 8.7.

Exit loss coefficients are independent of the type of exit, and for all exit types the loss coefficient is given as

$$K_{loss} = 1.0$$

Entrance Losses   Entrance losses depend strongly upon the entrance geometry. Figure 8.8 shows different types of entrance geometries: (i) Re-entrant, (ii) Rounded, and (iii) Beveled.

For a *square or sharp-edged inlet*, the loss coefficient is *given as* $K = 0.5$.

For a *rounded entrance*, the loss coefficient varies with the r/D ratio as given in the table below:

r/D	K
0.02	0.28
0.04	0.24
0.06	0.15
0.10	0.09
>0.15	0.04

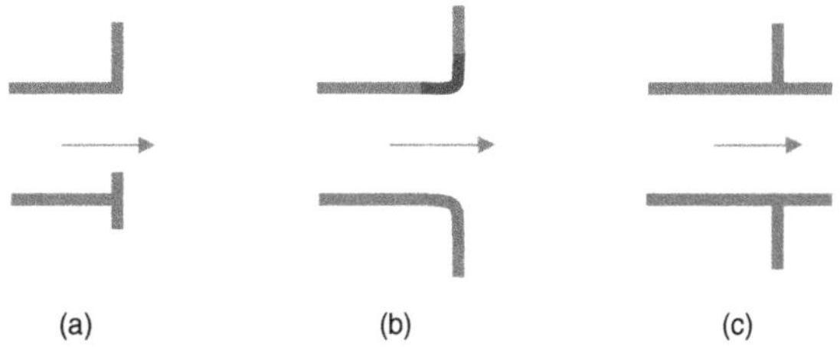

Figure 8.7   Different exit types. (a) Sharp-edge, (b) rounded, and (c) projecting.

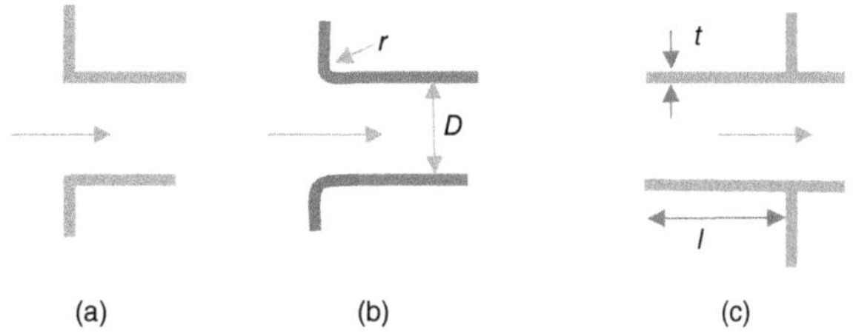

(a)   (b)   (c)

Figure 8.8   Different entrance types. (a) Sharp-edge, (b) rounded, and (c) reentrant.

(a)   (b)   (c)

Re-entrant with **inward projecting entrance:** $K = 0.78$.

8.1.4.4.2 Expansion and Contractions

Piping systems often involve sudden or gradual expansions and contraction flow section to accommodate changes in flow rates or velocity or density. Such flow sections involve higher head losses due to flow separation and corner region recirculatory flows as demonstrated in Figure 8.9.

For the case of sudden expansion, the head loss coefficient is given by the following approximate formula derived based on the equations of mass, momentum, and energy

$$K_{SE} = \alpha \left(1 - \frac{A_{small}^2}{A_{large}^2} \right) = \alpha \left(1 - \frac{d^2}{D^2} \right) \tag{8.53}$$

There is no such formula available for the case of sudden contraction. The loss coefficient values for sudden contract are given in Tables 8.3.

8.1.4.5 Gradual Expansion and Contraction

Figure 8.10 demonstrate the sections with gradual expansion and contraction.

For gradual expansion, the loss coefficient values are given as a function of d/D ratio as in Table 8.4.

For gradual contraction, the loss coefficient values are given as a function of $\frac{d^2}{D^2}$ ratio and angle of contraction theta as in Table 8.5.

8.1.4.6 Valves and Fittings

The loss coefficient values depend on number of geometrical parameters such as size of the pipe and angle of internal components that cause changes in flow directions and hence in

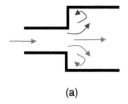

(a)

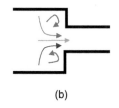

(b)

Figure 8.9 Flow through sudden expansion and contraction sections. (a) Sudden expansion and (b) sudden contraction.

Table 8.3 Loss coefficient for sudden contraction.

d^2/D^2	K_{SC}
~0	0.5
0.2	0.42
0.4	0.3
0.6	0.18
0.8	0.06
1.0	0

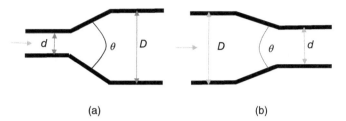

Figure 8.10 Gradual expansion and contraction. (a) Gradual expansion and (b) gradual contraction.

Table 8.4 Loss coefficient for gradual expansion.

$d/_D$	K_{GE}
0.2	0.30
0.4	0.25
0.6	0.15
0.8	0.10
1.0	0

Table 8.5 Loss coefficients for gradual contraction.

$\frac{d^2}{D^2}$	θ			
	10	15–40	50–60	90
0.5	0.05	0.05	0.06	0.12
0.25	0.05	0.04	0.07	0.17
0.10	0.05	0.05	0.08	0.19

pressure drop or head loss. One of the simplest approaches is to neglect the effect of such parameters and use constant values of loss coefficients for the valves and fittings to get an estimation of the minor loss. Tables 8.6 and 8.7 show loss constant values of many valves and fittings that are commonly used in piping systems.

The minor loss for the flow through the valves increases as the valves are partially closed for the purpose of controlling flow rates through the flow system. Table 8.7 shows some typical values of loss coefficient values for partially closed values with respect to the fully open values.

8.1.4.7 Elbows and Bends

Bends and curves induce additional losses due to flow separation and presence of swirling secondary flows caused by the centripetal acceleration. This additional loss is given by the

Table 8.6 Loss coefficients for 100% open valves.

Diameter	Screwed				Flanged			
	1/2	1	2	4	1	2	4	8
Valves								
Gate	0.30	0.24	0.16	0.11	0.80	0.35	0.16	0.07
Globe	14	8.2	6.9	5.7	13	8.5	6.0	5.8
Angle	9.0	4.7	2.0	1.0	4.5	2.4	2.0	2.0
Swing check	5.1	2.9	2.1	2.0	2.0	2.0	2.0	2.0

Table 8.7 Loss coefficients for partially open valves.

Valve closing (%)	Ratio of loss coefficients, K/K_{Open}	
	Gate valve	Globe value
25	3.0–5.0	1.5–2.0
50	12–22	2.0–3.0
75	70–120	6.0–8.0

loss coefficient (K) as

$$h_{bend} = K_{bend} \frac{V^2}{2g} \tag{8.54}$$

Friction head loss due to the axial length of the bend must also be computed and considered, i.e. the bend should be included in the pipe length while computing the major loss (Table 8.8).

Another more comprehensive approach is to specify the loss coefficients in terms of the geometrical parameters and get more accurate estimate of the pipe minor head loss. Comprehensive list of such values is given in Crane Technical Paper No. 410M. A few of such minor loss coefficient values for different types of valves and other fittings are given in Table 8.10 and 8.11. Friction loss coefficient for valves and data are presented in terms of complete turbulence friction factor for clean commercial steel pipe, f_T, and as formulas in terms of geometrical parameters. Table 8.9 presents the complete turbulence friction factor for clean commercial steel pipe.

Some selected valves along with the associated minor loss coefficient values and formulas are summarized in Table 8.10. Note that some of the coefficient loss values are given as multiples of complete turbulence friction factor, f_T. Most of the formulas are given as a function of valve geometrical parameters like $\beta = \frac{d_1}{d_2}$ and angle, θ (Table 8.10 for valves and Table 8.11 for bends and elbows).

Table 8.8 Loss coefficients for elbows and tees.

Diameter (in.)	Screwed				Flanged			
	1/2	1	2	4	1	2	4	8
45 regular	0.39	0.32	0.30	0.29				
45 long radius					0.21	0.20	0.19	0.16
90 regular	2.0	1.5	0.95	0.64	0.5	0.39	0.30	0.26
90 long radius	1.00	0.72	0.41	0.23	0.40	0.30	0.19	0.15
180 regular	2.00	1.5	0.95	0.64	0.41	0.35	0.30	0.25
180 long radius					0.40	0.30	0.21	0.15
			Tees					
Line flows	0.90	0.90	0.90	0.90	0.24	0.19	0.14	0.10
Branch flows	2.4	1.8	1.4	1.1	1.0	0.80	0.64	0.58

Table 8.9 Complete turbulence friction factor for clean commercial steel pipe.

Nominal size (in.)	1/2	3/4	1	1 1/4	1 1/2	2	3	4	5	6	8–10	12–16	18–24
Friction factor, f_T	0.027	0.025	0.023	0.022	0.021	0.019	0.018	0.017	0.016	0.015	0.014	0.013	0.012

8.2 Piping Systems with Rotary Devices

For a piping system that involves a pump or a turbine or both, the steady flow energy is given based on combining Eqs. (8.16c) and (8.52) as follows:

$$\left(\frac{P_1}{\rho g} + \frac{V_1^2}{2g} + z_1\right) = \left(\frac{P_2}{\rho g} + \frac{V_2^2}{2g} + z_2\right) - h_{pump} + h_{turbine} + \frac{V^2}{2g}\left(\frac{fL}{D} + \sum K\right) \quad (8.55)$$

where $h_{pump} = \frac{w_{pump}}{g}$ is the useful pump work delivered to the fluid and $h_{turbine} = \frac{w_{turbine}}{g}$ is the turbine head extracted from the fluid. Note that the shaft head is now replaced with the *negative pump head* and a *positive turbine head*. The pump head is zero if the piping system does not include a pump and the turbine head is zero if the system does not include a turbine.

Figure 8.11 shows a piping system in which a pump delivers liquid from reservoir-1 to reservoir-2. The pump head in the piping system can be estimated as

$$h_{pump} = \left(\frac{P_2}{\rho g} + \frac{V_2^2}{2g} + z_2\right) - \left(\frac{P_1}{\rho g} + \frac{V_1^2}{2g} + z_1\right) + \frac{V^2}{2g}\left(\frac{fL}{D} + \sum K\right) \quad (8.56)$$

Table 8.10 Loss coefficient values for some selected valves and fittings.

Valves and fittings	Loss coefficient values and formula
Gate valve	$\beta = 1$ and $\theta = 0$, $K = 8f_T$ $\beta < 1$ and $\theta \leq 45\,°C$, $K =$ $\dfrac{8f_T + \sin \dfrac{\theta}{2}\,[0.8(1 - \beta^2) + 2.6(1 - \beta^2)^2]}{\beta^4}$ $\beta < 1$ and $45°C < \theta \leq 180°C$, $K =$ $\dfrac{8f_T + 0.5\sqrt{\sin \dfrac{\theta}{2}(1 - \beta^2)} + (1 - \beta^2)^2}{\beta^4}$
Globe valve	$\beta = 1, K = 340f_T$ $\beta < 1$ and $\theta \leq 45\,°$ C, $K = \dfrac{340f_T + \beta[0.5(1 - \beta^2) + (1 - \beta^2)^2]}{\beta^4}$
Ball valve	$\beta = 1$ and $\theta = 0$, $K = 3f_T$ $\beta < 1$ and $\theta \leq 45°C$, $K =$ $\dfrac{3f_T + \sin \dfrac{\theta}{2}[0.8(1 - \beta^2) + 2.6(1 - \beta^2)^2]}{\beta^4}$ $\beta < 1$ and $45°C < \theta \leq 180°C$, $K =$ $\dfrac{3f_T + 0.5\sqrt{\sin \dfrac{\theta}{2}(1 - \beta^2)} + (1 - \beta^2)^2}{\beta^4}$
Butterfly valve	$\beta = 1, d_1 = d_2 = 2\text{-}8$ in., $K = 45f_T$ $\beta = 1, d_1 = d_2 = 10\text{-}14$ in., $K = 35f_T$ $\beta = 1, d_1 = d_2 = 14\text{-}16$ in., $K = 25f_T$
Check valve Swing check	$K = 100f_T$

Table 8.10 (Continued)

Valves and fittings	Loss coefficient values and formula
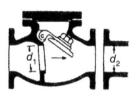	$K = 100f_T$
Lift check valve 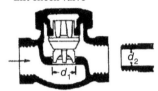	$\beta = 1, K = 600f_T$ $\beta < 1, K =$ $\dfrac{600f_T + \beta[0.5(1 - \beta^2) + (1 - \beta^2)^2]}{\beta^4}$
Plug valve Straight way 	$\beta = 1, K = 18f_T$ $\beta < 1, K =$ $\dfrac{18f_T + \sin\dfrac{\theta}{2}[0.8(1 - \beta^2) + 2.6(1 - \beta^2)^2]}{\beta^4}$
Sudden and gradual contraction 	$\theta \leq 45^\circ \text{C}, K = \dfrac{0.8 \sin\dfrac{\theta}{2}(1 - \beta^2)}{\beta^4}$ $45^\circ\text{C} < \theta \leq 180^\circ\text{C}, K =$ $\dfrac{0.5\sqrt{\sin\dfrac{\theta}{2}(1 - \beta^2)}}{\beta^4}$
Sudden and gradual expansion 	$\theta \leq 45^\circ \text{C}, K = \dfrac{2.6 \sin\dfrac{\theta}{2}(1 - \beta^2)^2}{\beta^4}$ $45^\circ\text{C} < \theta \leq 180^\circ\text{C}, K = \dfrac{(1 - \beta^2)^2}{\beta^4}$

Note: $\beta = \dfrac{d_1}{d_2}$.

Source: All data from Crane Company's Technical Paper 410 (2009).

Table 8.11 Bends and elbows.

Bends	Loss of coefficient values		Loss of coefficient values	
	$\frac{r}{d}$	K	$\frac{r}{d}$	K
Flanged or Butt-welded 90° elbows	1	$20f_T$	8	$24f_T$
	1.5	$14f_T$	10	$30f_T$
	2	$12f_T$	12	$34f_T$
	3	$12f_T$	14	$38f_T$
	4	$14f_T$	16	$42f_T$
	6	$17f_T$	20	$50f_T$

Pipe bends other than 90°

$$K_b = (n-1)\left(0.25 + f_T\frac{r}{d} + 0.5K\right) + K$$

where

n = number of 90°

K = resistance coefficient for the 90° bend as per table

Miter Bends	α	K
	0°	$2f_T$
	15°	$4f_T$
	30°	$8f_T$
	45°	$15f_T$
	60°	$25f_T$
	75°	$40f_T$
	90°	$60f_T$

or

$$h_{\text{pump}} = \frac{P_2 - P_1}{\rho g} + \frac{V_2^2 - V_1^2}{2g} + (z_2 - z_1) + \frac{V^2}{2g}\left(\frac{fL}{D} + \sum K\right) \tag{8.57}$$

The mechanical power input to the fluid is given as

$$\dot{W}_{\text{pump,m}} = \rho Q g h_{\text{pump}} \tag{8.58}$$

The mechanical power input to the shaft is given as

$$\dot{W}_{\text{pump,m}} = \frac{\dot{W}_{\text{pump,m}}}{\eta_{\text{pump}}} = \frac{\rho g Q\, h_{\text{pump}}}{\eta_{\text{pump}}} \tag{8.59}$$

where η_{pump} is the pump efficiency.

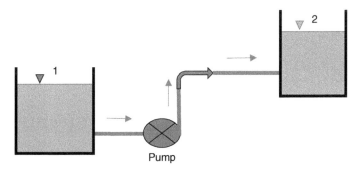

Figure 8.11 Piping system with a circulating fluid pump.

The electrical power input is given as

$$\dot{W}_{\text{pump,e}} = \frac{\rho g Q\, h_{\text{pump}}}{\eta_{\text{pump}} \eta_{\text{motor}}} \tag{8.60}$$

where η_{motor} is the electric motor efficiency.

8.3 Piping System Characteristics

The procedure outlined in Section 8.2 that a pump is required to overcome the total head loss in a piping system to sustain a flow rate. In a true sense, this pump head represents the piping system head loss, Δh_{Sys}, for a given flow rate, Q.

From Eq. (8.60), we can see that a system head loss is the sum total of change in pressure head, change in kinetic energy head, change in potential energy head, and the frictional head loss in the pipe, fittings, and valves. We can rearrange Eq. (8.60) and express the system head as

$$\Delta h_{\text{Sys}} = \frac{P_2 - P_1}{\rho g} + \frac{V_2^2 - V_1^2}{2g} + (z_2 - z_1) + \frac{V^2}{2g}\left(\frac{fL}{D} + \sum K\right) \tag{8.61}$$

For negligible changes in the kinetics head for $V_1 \cong V_2$ and substituting the average velocity as $V = \frac{Q}{A}$, we can rewrite system head loss as

$$\Delta h_{\text{Sys}} = \frac{P_2 - P_1}{\rho g} + (z_2 - z_1) + \frac{Q^2}{2gA}\left(\frac{fL}{D} + \sum K\right) \tag{8.62}$$

A piping system characteristic is the functional variation of the piping system head loss, Δh_{Sys}, with the flow rate, Q, given by Eq. (8.62). A typical piping system curve, which is just a plot of the function $\Delta h_{\text{Sys}} = f(Q)$ given by the solid curve, is shown in Figure 8.12.

Note that in problems where $P_1 \approx P_2$ and $Z_1 \approx Z_2$, the system head is primarily given by the pipe friction drop, which varies with volume flow rate, Q, in a quadratic manner as shown by the dotted curve, starting from $Q = 0$.

Example 8.1 *Estimation of Pump Head and Power for a Pumping System*

Water-pumping system ($\rho = 1.94\,\text{slugs/ft}^3$ and $v = 0.000\,011\,\text{ft}^2/\text{s}$), shown in the figure below, is used to transfer water between two water reservoirs at a volume flow rate of 90 gpm ($Q = 0.2\,\text{ft}^2/\text{s}$) through a pipe of length 400 ft, diameter $D = 2$-in., and relative roughness of

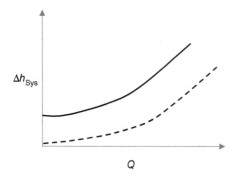

Figure 8.12 Piping system characteristic curve.

$\varepsilon/D = 0.001$. Determine the pump head and pump power assuming a pump efficiency of $\eta_p = 80\%$.

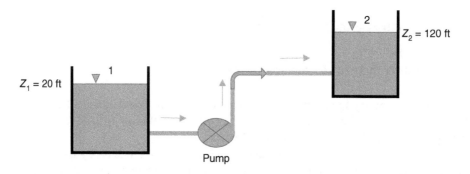

Pump

Solution

The pump head in the piping system can be estimated from the energy equation (8.56) as

$$h_{pump} = \left(\frac{P_2}{\rho g} + \frac{V_2^2}{2g} + z_2 \right) - \left(\frac{P_1}{\rho g} + \frac{V_1^2}{2g} + z_1 \right) + \frac{V^2}{2g} \left(\frac{fL}{D} + \sum K \right)$$

or

$$h_{pump} = \frac{P_2 - P_1}{\rho g} + \frac{V_2^2 - V_1^2}{2g} + (z_2 - z_1) + \frac{V^2}{2g} \left(\frac{fL}{D} + \sum K \right)$$

For pipe-flow system shown in the figure, we can substitute

$$P_1 = P_2 = \text{Atmospheric pressure}$$

$$V_1 = V_2 \approx 0$$

Pump head computed as

$$h_p = (Z_2 - Z_1) + \frac{V^2}{2g} \left(f\frac{L}{D} + \sum K \right)$$

For the given flow rate, we can estimate the average velocity and Reynolds number as follows:

$$\text{Average velocity, } V_{\text{avg}} = \frac{Q}{A} = \frac{0.2 \text{ ft}^3/\text{s}}{\frac{1}{4}\pi\left(\frac{2}{12}\text{ft}\right)^2} = 9.17 \text{ ft/s}$$

and Reynolds number

$$Re = \frac{V_{\text{avg}}D}{v} = \frac{9.17 \times 2/12}{0.000\,01} = 139\,000$$

From Moody's chart for $Re = 139\,000$ and $\varepsilon/D = 0.001$, we have the friction factor $f = 0.0216$.

Let us now calculate the minor loss coefficients as follows:

Type of fittings	Loss coefficients, K
Sharp entrance	0.5
Open globe valve	6.9
12-in. bend	0.15
Elbow (regular 90°)	0.95
Gate valve – half open	0.16 (for 100% open) × 17 for 50% open = 2.7
Sharp exit	1.0
ΣK	12.2

$$\text{Pump head: } h_p = (120 - 20) + \frac{(9.17 \text{ ft/s})^2}{2(32.2)\text{ft} \atop \text{s}^2}\left(0.0216 \times \frac{400 \text{ ft}}{2/12 \text{ ft}} + 12.2\right)$$

$$h_p = 100 \text{ ft} + 84 \text{ ft} = 184 \text{ ft} \quad h_p = 100 \text{ ft} + 84 \text{ ft} = 184 \text{ ft}$$

The mechanical power input to the fluid is given as

$$\dot{W}_{\text{pump,m}} = \rho g Q h_{\text{pump}}$$
$$\dot{W}_{\text{pump,m}} = (1.94 \times 32.2) \text{ lbf/ft}^3 \times 0.2 \text{ ft}^3 \times 184 \text{ ft}$$
$$= 2300 \text{ (ft lbf)/s}$$
$$P = \frac{2300}{550} = 4.3 \text{ HP}$$

The power input to the pump is given as

$$\dot{W}_{\text{pump,m}} = \frac{\dot{W}_{\text{pump,m}}}{\eta_{\text{pump}}} = \frac{4.3}{0.8} = 5.375 \text{ HP}$$

Example 8.2 *Estimation of Pump Head and Power for a Piping System using Explicit Curve-fit Equation for the Friction Factor*

The water transfer system shown in the figure below is used to transfer water from one reservoir to another reservoir at a flow rate of 110 gpm through a pipe of length 1400-ft. The pressure in the reservoirs is 20 and 200 psig, respectively. All pipes are schedule 40, 3-in., and made of commercial steel with relative roughness, $\varepsilon/D = 0.0006$. The water temperature is 60 °F. A list of valves and fitting is given below:

1-check valve, 1-globe valves, 12-standard regular elbows.

Find the pump head and power output required to pump 110 gpm of flow.

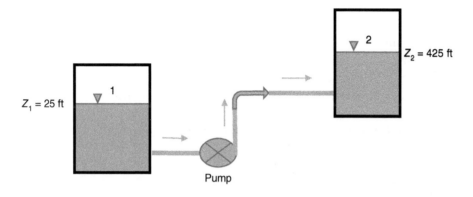

Solution

The pump head in the piping system can be estimated as

$$h_{pump} = \left(\frac{P_2}{\rho g} + \frac{V_2^2}{2g} + z_2\right) - \left(\frac{P_1}{\rho g} + \frac{V_1^2}{2g} + z_1\right) + \frac{V^2}{2g}\left(\frac{fL}{D} + \sum K\right)$$

or

$$h_{pump} = \frac{P_2 - P_1}{\rho g} + \frac{V_2^2 - V_1^2}{2g} + (z_2 - z_1) + \frac{V^2}{2g}\left(\frac{fL}{D} + \sum K\right)$$

For pipe flow system shown in the figure, let us substitute

$$V_1 = V_2 \approx 0$$

Pump head is computed as

$$h_{pump} = \frac{P_2 - P_1}{\rho g} + (Z_2 - Z_1) + \frac{V^2}{2g}\left(f\frac{L}{D} + \sum K\right)$$

For the given flow rate, we can estimate the average velocity and Reynolds number as follows:

$$\text{Volume flow rate, } Q = 110 \text{ gpm} = 110 \text{ gpm} \times \frac{0.002\,675 \text{ ft}^3 \text{ s}}{\text{gpm}} = 0.2943 \text{ ft}^3/\text{s}$$

Pipe inner diameter: for 3-in. and Schedule 40, $d_i = 3.068$-in.

$$\text{Average velocity, } V_{avg} = \frac{Q}{A} = \frac{0.2943 \text{ ft}^3/\text{s}}{\frac{1}{4}\pi\left(\frac{3.068}{12}\text{ ft}\right)^2} = 5.7323 \text{ ft/s}$$

and Reynolds number

$$Re = \frac{V_{avg}D}{\nu} = \frac{5.7323 \times 3.068/12}{0.000\,01} = 146\,555$$

From Haaland's explicit curve-fit equation Moody's chart

$$f = \frac{0.3086}{\left\{ \log \left[\left(\frac{\varepsilon/D}{3.7} \right)^{1.11} + \frac{6.9}{Re_D} \right] \right\}^2} = \frac{0.3086}{\left\{ \log \left[\left(\frac{0.0006}{3.7} \right)^{1.11} + \frac{6.9}{146\ 555} \right] \right\}^2}$$

$$f = \frac{0.3086}{\left\{ \log \left[\left(\frac{0.0006}{3.7} \right)^{1.11} + \frac{6.9}{146\ 555} \right] \right\}^2} = \frac{0.3086}{\left\{ \log \left[0.000\ 062\ 1 + \frac{6.9}{146\ 555} \right] \right\}^2}$$

$$f = 0.019\ 66$$

From Table 8.8, the complete turbulence friction factor for 3-in. nominal clean commercial steel pipe, $f_T = 0.018$.

Let us now calculate the minor loss coefficients from Table 8.9:

Swing check valve: $K_{check} = 100 f_T$

Globe valve: for $\beta = 1$, $K_{globe} = 340 f_T$

Standard regular elbows: for $r/d = 10$, $K_{elbow} = 30 f_T$

Entrance: $K_{entrance} = 0.5$

Exit: $K_{exit} = 0.5$

$$\sum K = K_{entrance} + K_{check} + K_{globe} + 10 K_{elbow} + K_{exit}$$

$$\sum K = 0.5 + 100 f_T + 340 f_T + 10 (30 f_T) + 1.0$$

$$\sum K = 0.5 + 740 f_T + 1.0 = 0.5 + 740 \times 0.018 + 1.0$$

$$\sum K = 14.82$$

Pump head: $h_p = \frac{P_2 - P_1}{\rho g} + (Z_2 - Z_1) + \frac{V^2}{2g} \left(f \frac{L}{d} + \sum K \right)$

$$h_p = \frac{(200 - 20) \times 144}{1.96 \times 32.2} + (425 - 25)$$
$$+ \frac{(5.7323\ \text{ft/s})^2}{2(32.2)\ \text{ft/s}^2} \left(\frac{400\ \text{ft}}{3.068/12\ \text{ft}} + 14.82 \right)$$

$$h_p = \frac{(100 - 20) \times 144}{1.96 \times 32.2} + (425 - 25)$$
$$+ \frac{(5.7323\ \text{ft/s})^2}{2(32.2)\ \text{ft/s}^2} \left(0.019\ 66 \times \frac{400\ \text{ft}}{3.068/12\ \text{ft}} + 14.82 \right)$$

$$= 182.53 + 400 + \frac{(5.7323\ \text{ft/s})^2}{2(32.2)\ \text{ft/s}^2} (30.76 + 14.82)$$

$$= 182.53 + 400 + 23.257$$

$$h_p = 605.79\ \text{ft}$$

The mechanical power input to the fluid is given as

$$\dot{W}_{pump,m} = \rho g h_p$$

$$\dot{W}_{pump,m} = (1.96 \times 32.2) \, lbf/ft^3 \times 0.2943 \, ft^3/s \times 605.79 \, ft$$

$$= 11\,251.85 \, (ft\,lbf)/s$$

$$P = \frac{2300}{550} = 20.46 \, HP$$

8.4 Piping System Design Procedure

The examples of the piping system considered so far involved sequential computations: first estimate the friction head loss and then use the energy equation to determine the required pump head and pump power. However, the procedure becomes quite complex and involves iterations depending on the variation in the design situations. For a single-path series system and a given system configuration, the general functional relationship among all piping system variables is given as

$$\Delta P = f(L, D, Q, \varepsilon, \Delta Z, \rho, \mu) \tag{8.63}$$

and for a given piping system configuration and fluid type, the equation can be reduced to a form involving a fewer number of system variables such as

$$\Delta P = f(L, D, Q) \tag{8.64}$$

Several different design problems can be formulated depending on the known variables and what needs to be determined. A few examples of such design problems and the associated computational sequences are listed here:

Case-1: For known variables L, D, and Q, determine unknown variable ΔH or ΔP

Basic steps:

1) Estimate friction factor from Moody's chart or Colebrook's equation based on Re and $\frac{\varepsilon}{D}$.

2) Estimate major loss: $h_{f,major} = f\frac{L}{D}\frac{V^2}{2g}$

3) Estimate minor loss: $h_{f,minor} = \sum K\frac{V^2}{2g}$ or $f\frac{\sum L_e}{D}$

4) Estimate total head loss: $h_{f,t} = h_{f,major} + h_{f,minor}$

5) Estimate pressure drop from pipe flow energy equation

$$\frac{P_1}{\rho g} + \frac{V_1^2}{2g} + Z_1 = \frac{P_2}{\rho g} + \frac{V_2^2}{2g} + Z_2 + h_f$$

Case-2: For known pressure drop, $\Delta P = P_1 - P_2$, and known variables D and Q, determine the length L.

Basic steps:

1) Estimate friction factor from Moody's chart or Colebrook's equation based on Re and $\frac{\varepsilon}{D}$.

2) Estimate total head loss from energy equation for pipe flow: $h_f = \left(\frac{P_1}{\rho g} + \frac{V_1^2}{2g} + Z_1\right) - \left(\frac{P_2}{\rho g} + \frac{V_2^2}{2g} + Z_2\right)$

3) Estimate length from total head loss: $h_f = \frac{V^2}{2g}\left(f\frac{L}{D} + \Sigma K\right)$

Case-3: For known ΔP, L, and D, determine volume flow rate, Q

Basic steps:

1) Guess friction factor f as the fully rough region value for the known value of and $\frac{\epsilon}{D}$. Estimate friction factor from Moody's chart or Colebrook's equation based on Re

2) Using the guess value of f, estimate average velocity, V, from the pipe flow energy equation

$$\frac{P_1}{\rho g} + \frac{V_1^2}{2g} + Z_1 = \frac{P_2}{\rho g} + \frac{V_2^2}{2g} + Z_2 + \frac{V^2}{2g}\left(f\frac{L}{D} + \Sigma K\right)$$

3) Estimate the volume flow rate, Q, and Re

4) Estimate new guess value of friction factor f from Moody's diagram or Colebrook equation based on the estimated value Re and $\frac{\epsilon}{D}$.

5) Repeat steps 2–4 for newer values of friction factor, f, until convergence.

Case-4: For known pressure drop ΔP and variables L and Q, determine pipe diameter, D.

Basic steps

1) Assume pipe diameter as initial first guess.

2) Estimate average velocity V, Reynolds number, and relative roughness $\frac{\epsilon}{D}$.

3) Estimate friction factor from Moody's chart or Colebrook's equation based on Re and $\frac{\epsilon}{D}$.

4) Estimate total head loss from: $h_f = \frac{V^2}{2g}\left(f\frac{L}{D} + \Sigma K\right)$.

5) Estimate system head $\Delta H_s = (P_1 - P_2)$ from energy equation for pipe flow:

$$\Delta H_s = \left(\frac{P_1 - P_2}{\rho g}\right) + (Z_1 - Z_2) + \left(\frac{V_1^2 - V_2^2}{2g}\right)$$

6) Compare changes in system head ΔH_s with the system friction head loss h_f.
   - If $h_f > \Delta H_s$, then **increase the guess value of the pipe diameter** and repeat steps 2–5.
   - If $h_f < \Delta H_s$, then **decrease the guess value of the pipe diameter** and repeat steps 2–5.

7) Continue iterations until a converged value of pressure drop is reached using a stopping criterion: $\left|\frac{\Delta H_s - h_f}{\Delta H_s}\right| \le \epsilon_s$ (tolerance limit).

Example 8.3   *Determine the Required Pipe Diameter for a Piping System*

Consider a piping system shown in the figure below to deliver water flow from a reservoir at a higher elevation $Z_A = 200$-ft to a reservoir at lower elevation $Z_B = 20$-ft. The pressures in the tanks are $P_A = 70$ psig and $P_B = 15$ psig, respectively. The total length of the pipe is $L = 500$-ft and it includes following list of valves and fittings: two gate valves, one globe valves, and four regular elbows. What diameter of schedule 30 commercial steel pipe would be required to deliver 0.35 ft^3/s of water?

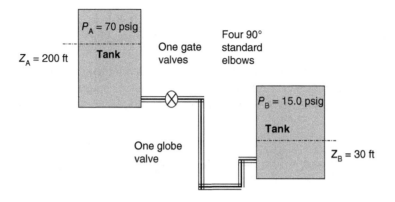

Solution

Let us start from the energy equation for a piping system without any pumps

$$\left(\frac{P_1}{\rho g} + \frac{V_1^2}{2g} + Z_1\right) = \left(\frac{P_2}{\rho g} + \frac{V_2^2}{2g} + Z_2\right) + \frac{V^2}{2g}\left(f\frac{L}{D} + \sum K\right)$$

Rearranging,

$$\frac{P_A - P_B}{\rho g} + (Z_A - Z_B) = \frac{V^2}{2g}\left[f\frac{L}{D} + (K_{entrance} + K_{exit} + 4K_{elbow} + 2K_{gate} + K_{globe})\right]$$

(E8.3.1)

As we can see in this problem, net volume flow in the system is set by the fact that the sum of pressure and potential head differences is balanced by the net pipe friction head loss. For a selected pipe diameter, the available positive head due to the difference pressure and potential head changes due to elevation difference is balanced by the frictional head loss in the pipe length, valves, and fittings. Similar to the procedure outlined for the case-4 design situations, we can determine the pipe diameter using an iterative process.

Consider following loss coefficients from Table 8.9:

Gate valve: for $\beta = 1$, $K_{gate} = 8f_T$
Globe valve: for $\beta = 1$, $K_{globe} = 340f_T$
Standard regular elbows: for $r/D = 10$, $K_{elbow} = 30f_T$
Entrance: $K_{entrance} = 0.5$
Exit: $K_{exit} = 0.5$

$$\sum K = (K_{entrance} + 4K_{elbow} + 2K_{gate} + K_{globe} + K_{exit})$$

$$\sum K = 0.5 + 4 \times 30f_T + 2(8f_T) + 340f_T + 1.0$$

Minor loss coefficient, $\sum K = (1.5 + 476f_T)$.

Substituting in Eq. (E8.3.1)

$$\frac{P_A - P_B}{\rho g} + (Z_A - Z_B) = \frac{V^2}{2g}\left[f\frac{L}{D} + (1.5 + 476f_T)\right]$$

(E8.3.2)

$$\frac{(70-15) \times 144}{1.96 \times 32.2} + (200-30) = \frac{V^2}{2g}\left[f\frac{L}{D} + (1.5 + 476f_T)\right]$$

$$125.49 + 170 = \frac{V^2}{2g}\left[f\frac{L}{D} + (1.5 + 476f_T)\right]$$

or

$$295.9 = \frac{V^2}{2g}\left[f\frac{L}{D} + (1.5 + 476f_T)\right] \tag{E8.3.3}$$

For the correct selection of the pipe size, the total friction pipe loss on the right-hand side must be equal to the available sum of pressure and potential head of

$$\frac{P_A - P_B}{\rho g} + (Z_A - Z_B) = 295.9 - \text{ft}$$

on left-hand side of the Eq. (E8.3.3)

Iteration # 1: Guess diameter $D = 1.9$-in. $= 0.158$-ft

For 1-in. commercial steel (from Figure 8.6),

$$\text{Relative roughness, } \frac{\epsilon}{D} = 0.0009$$

From Table 8.8, the complete turbulence friction factor for 1.9-in. nominal clean commercial steel pipe, $f_T = 0.02$.

For the given volume flow rate $0.35 \text{ ft}^3/\text{s}$, we can estimate the average velocity and Reynolds number as follows:

Average velocity, $V = \frac{Q}{\frac{\pi}{4}D^2} = \frac{0.35}{\frac{\pi}{4}(0.1583)^2} = 17.783 \text{ ft/s}$

Reynolds number: $Re = \frac{V_{avg}D}{\nu} = \frac{17.783 \times 0.158}{0.000\,01} = 280\,971$

From Haaland's explicit curve-fit equation Moody's chart

$$f = \frac{0.3086}{\left\{\log\left[\left(\frac{\epsilon/D}{3.7}\right)^{1.11} + \frac{6.9}{Re_D}\right]\right\}^2} = \frac{0.3086}{\left\{\log\left[\left(\frac{0.0009}{3.7}\right)^{1.11} + \frac{6.9}{280\,971}\right]\right\}^2}$$

$$= \frac{0.3086}{\left\{\log\left[0.000\,097\,4 + \frac{6.9}{280\,971}\right]\right\}^2}$$

$$f = 0.020\,15$$

From (E.8.3.3) for the energy equation

$$\epsilon_s = 295.9 - \frac{(24.45)^2}{2 \times 32.2}\left[0.020\,15 \times \frac{500}{0.158} + (1.5 + 476 \times 0.02)\right]$$

$$\epsilon_s = 295.9 - \frac{(17.783)^2}{2 \times 32.2}[63.7658 + 9.52] = 295.9 - 359.86$$

$$\epsilon_s = -63.968$$

Iteration # 2: Guess diameter $D = 2.2$-in. $= 0.1833$-ft

For 2.2-in. commercial steel (from Figure 8.6),

$$\text{Relative roughness, } \frac{\epsilon}{D} = 0.0009$$

From Table 8.8, the complete turbulence friction factor for 2-in. nominal clean commercial steel pipe, $f_T = 0.019$.

For the given volume flow rate $0.35 \text{ ft}^3/\text{s}$, we can estimate the average velocity and Reynolds number as follows:

Average velocity, $V = \frac{Q}{\frac{\pi}{4}D^2} = \frac{0.35}{\frac{\pi}{4}(0.1833)^2} = 13.263 \text{ ft/s}$

Reynolds number: $Re = \frac{V_{avg}D}{\nu} = \frac{13.263 \times 0.1833}{0.000\,01} = 243\,110$

From Haaland's explicit curve-fit equation Moody's chart

$$f = \frac{0.3086}{\left\{ \log\left[\left(\frac{\epsilon/D}{3.7}\right)^{1.11} + \frac{6.9}{Re_D}\right]\right\}^2} = \frac{0.3086}{\left\{ \log\left[\left(\frac{0.0009}{3.7}\right)^{1.11} + \frac{6.9}{243\,110}\right]\right\}^2}$$

$$= \frac{0.3086}{\left\{ \log\left[0.000\,097\,4 + \frac{6.9}{243\,110}\right]\right\}^2}$$

$f = 0.2028$

From (E8.3.3) for the energy equation

$$\epsilon_s = 295.9 - \frac{(13.263)^2}{2 \times 32.2}\left[0.020\,28 \times \frac{500}{0.1833} + (1.5 + 476 \times 0.019)\right]$$

$$\epsilon_s = 295.9 - \frac{(13.263)^2}{2 \times 32.2}[55.319 + 9.044] = 295.9 - 175.806$$

$\epsilon_s = 120.09$

Iteration # 3: Guess diameter $D = 2.1$-in. $= 0.175$-ft
For 2.1-in. commercial steel (from Figure 8.6),

$$\text{Relative roughness, } \frac{\epsilon}{D} = 0.0009$$

From Table 8.8, the complete turbulence friction factor for 2-in. nominal clean commercial steel pipe, $f_T = 0.019$.

For the given volume flow rate $0.35 \text{ ft}^3/\text{s}$, we can estimate the average velocity and Reynolds number as follows:

Average velocity, $V = \frac{Q}{\frac{\pi}{4}D^2} = \frac{0.35}{\frac{\pi}{4}(0.175)^2} = 14.551 \text{ ft/s}$

Reynolds number: $Re = \frac{V_{avg}D}{\nu} = \frac{14.551 \times 0.175}{0.000\,01} = 254\,642$

From Haaland's explicit curve-fit equation Moody's chart

$$f = \frac{0.3086}{\left\{ \log\left[\left(\frac{\epsilon/D}{3.7}\right)^{1.11} + \frac{6.9}{Re_D}\right]\right\}^2} = \frac{0.3086}{\left\{ \log\left[\left(\frac{0.0009}{3.7}\right)^{1.11} + \frac{6.9}{254\,642}\right]\right\}^2}$$

$$= \frac{0.3086}{\left\{ \log\left[0.000\,097\,4 + \frac{6.9}{267\,320}\right]\right\}^2}$$

$f = 0.2019$

From (E8.3.3) for the energy equation

$$\epsilon_s = 295.9 - \frac{(14.551)^2}{2 \times 32.2}\left[0.020\,192 \times \frac{500}{0.1667} + (1.5 + 476 \times 0.019)\right]$$

$$\epsilon_s = 295.9 - \frac{(14.551)^2}{2 \times 32.2}[60.564 + 9.044] = 295.9 - 228.85$$

$$\epsilon_s = 67.05$$

Iteration # 4: Guess diameter $D = 2$-in. $= 0.1667$-ft
For 2-in. commercial steel (from Figure 8.6),

$$\text{Relative roughness, } \frac{\epsilon}{D} = 0.0009$$

From Table 8.8, the complete turbulence friction factor for 2-in. nominal clean commercial steel pipe, $f_T = 0.019$.

For the given volume flow rate $0.35\,\text{ft}^3/\text{s}$, we can estimate the average velocity and Reynolds number as follows:

$$\text{Average velocity, } V = \frac{Q}{\frac{\pi}{4}D^2} = \frac{0.35}{\frac{\pi}{4}(0.1667)^2} = 16.036\,\text{ft/s}$$

$$\text{Reynolds number: } Re = \frac{V_{\text{avg}}D}{\nu} = \frac{16.036 \times 0.1667}{0.000\,01} = 267\,320$$

From Haaland's explicit curve-fit equation Moody's chart

$$f = \frac{0.3086}{\left\{\log\left[\left(\frac{\epsilon/D}{3.7}\right)^{1.11} + \frac{6.9}{Re_D}\right]\right\}^2} = \frac{0.3086}{\left\{\log\left[\left(\frac{0.0009}{3.7}\right)^{1.11} + \frac{6.9}{267\,320}\right]\right\}^2}$$

$$= \frac{0.3086}{\left\{\log\left[0.000\,097\,4 + \frac{6.9}{267\,320}\right]\right\}^2}$$

$$f = 0.020\,19$$

From (E8.3.3) for the energy equation

$$\epsilon_s = 295.9 - \frac{(16.036)^2}{2 \times 32.2}\left[0.020\,19 \times \frac{500}{0.1667} + (1.5 + 476 \times 0.019)\right]$$

$$\epsilon_s = 295.9 - \frac{(16.036)^2}{2 \times 32.2}[60.565 + 9.044] = 295.9 - 277.95$$

$$\epsilon_s = 17.947$$

Iteration # 5: Guess diameter $D = 1.95$-in. $= 0.1625$-ft
For 1.95-in. commercial steel (from Figure 8.6),

$$\text{Relative roughness, } \frac{\epsilon}{D} = 0.0009$$

From Table 8.8, the complete turbulence friction factor for 2-in. nominal clean commercial steel pipe, $f_T = 0.019$.

For the given volume flow rate $0.35\,\text{ft}^3/\text{s}$, we can estimate the average velocity and Reynolds number as follows:

$$\text{Average velocity, } V = \frac{Q}{\frac{\pi}{4}D^2} = \frac{0.35}{\frac{\pi}{4}(0.1625)^2} = 16.876\,\text{ft/s}$$

Reynolds number: $Re = \dfrac{V_{avg}D}{v} = \dfrac{16.876 \times 0.1625}{0.000\,01} = 274\,236$

From Haaland's explicit curve-fit equation Moody's chart

$$f = \dfrac{0.3086}{\left\{ \log\left[\left(\dfrac{\varepsilon/D}{3.7}\right)^{1.11} + \dfrac{6.9}{Re_D}\right]\right\}^2} = \dfrac{0.3086}{\left\{ \log\left[\left(\dfrac{0.0009}{3.7}\right)^{1.11} + \dfrac{6.9}{274\,236}\right]\right\}^2}$$

$$= \dfrac{0.3086}{\left\{ \log\left[0.000\,097\,4 + \dfrac{6.9}{274\,326}\right]\right\}^2}$$

$$f = 0.020\,17$$

From (E8.3.3) for the energy equation

$$\varepsilon_s = 295.9 - \dfrac{(16.876)^2}{2 \times 32.2}\left[0.020\,17 \times \dfrac{500}{0.1625} + (1.5 + 476 \times 0.019)\right]$$

$$\varepsilon_s = 295.9 - \dfrac{(16.876)^2}{2 \times 32.2}[60.0615 + 9.044] = 295.9 - 314.453$$

$$\varepsilon_s = -18.553$$

Iteration # 6: Guess diameter $D = 1.975$-in. $= 0.164\,58$-ft
For 1.95-in. commercial steel (from Figure 8.6),

$$\text{Relative roughness, } \dfrac{\varepsilon}{D} = 0.0009$$

From Table 8.8, the complete turbulence friction factor for 2-in. nominal clean commercial steel pipe, $f_T = 0.019$.

For the given volume flow rate $0.35\,\text{ft}^3/\text{s}$, we can estimate the average velocity and Reynolds number as follows:

Average velocity, $V = \dfrac{Q}{\frac{\pi}{4}D^2} = \dfrac{0.35}{\frac{\pi}{4}(0.164\,58)^2} = 16.952\,\text{ft/s}$

Reynolds number: $Re = \dfrac{V_{avg}D}{v} = \dfrac{16.452 \times 0.164\,58}{0.000\,01} = 270\,767$

From Haaland's explicit curve-fit equation Moody's chart

$$f = \dfrac{0.3086}{\left\{ \log\left[\left(\dfrac{\varepsilon/D}{3.7}\right)^{1.11} + \dfrac{6.9}{Re_D}\right]\right\}^2} = \dfrac{0.3086}{\left\{ \log\left[\left(\dfrac{0.0009}{3.7}\right)^{1.11} + \dfrac{6.9}{270\,767}\right]\right\}^2}$$

$$= \dfrac{0.3086}{\left\{ \log\left[0.000\,097\,4 + \dfrac{6.9}{270\,767}\right]\right\}^2}$$

$$f = 0.020\,18$$

From (E8.3.3) for the energy equation

$$\varepsilon_s = 295.9 - \dfrac{(16.452)^2}{2 \times 32.2}\left[0.020\,18 \times \dfrac{500}{0.164\,58} + (1.5 + 476 \times 0.019)\right]$$

$$\varepsilon_s = 295.9 - \dfrac{(16.452)^2}{2 \times 32.2}[61.3075 + 9.044] = 295.9 - 295.682$$

$$\varepsilon_s = -0.082$$

A summary of the iterative computations is given in the table below:

Iteration	D (in.)	V (ft/s)	Re	f	ε_s
1	1.9	17.783	280 971	0.020 15	−63.97
2	2.2	13.263	243 110	0.2028	120.09
3	2.1	14.551	254 642	0.2019	67.05
4	2.0	16.036	267 320	0.020 19	17.947
5	1.95	16.876	274 236	0.020 17	−18.55
6	1.975	16.452	270 767	0.020 18	−0.082

As we can see from the iteration process, the old guess values of the diameter are updated with new values in a manual manner either by increasing or decreasing the diameter values based on the net values computed from Eq. (E8.3.3), and this process is continued for six-iteration steps until the absolute error, ε_s, is reduced to a value of 0.082-ft. In reality, the iterative scheme could use an automated process until the absolute error or a relative percentage error is less than an assigned error tolerance value, ε_a, i.e. $\varepsilon_s \leq \varepsilon_a$.

8.5 Piping Network Classifications

Piping networks are classified into three basic categories: series network, parallel network, and series–parallel network as demonstrated in Figure 8.13. Each one of these classified piping systems can be solved using the general energy equation for pipe flow given by Eq. (8.55) and following two basic rules: (i) Conservation mass or volume flow and (ii) Uniqueness of pressure at all junction points in the network.

8.5.1 Pipes in Series

A series piping system consists of a number of pipes connected in series with each other as shown in Figure 8.13a. For the three pipes connected in series, we can apply the following two basic rules:

Rule-1: Conservation of volume flow rate for compressible flow of fluid through the series network:

$$Q_1 = Q_2 = Q_3 = \text{Constant} \tag{8.65a}$$

or

$$\frac{\pi}{4}(D_1)^2 V_1 = \frac{\pi}{4}(D_2)^2 V_2 = \frac{\pi}{4}(D_3)^2 V_3 \tag{8.65b}$$

We can express V_2 and V_3 in terms of V_1 as follows:

$$V_2 = V_1\left(\frac{D_1}{D_2}\right)^2 \quad \text{or} \quad \frac{V_2}{V_1} = \left(\frac{D_1}{D_2}\right)^2 \tag{8.66a}$$

and

$$V_3 = V_1\left(\frac{D_1}{D_3}\right)^2 \quad \text{or} \quad \frac{V_3}{V_1} = \left(\frac{D_1}{D_3}\right)^2 \tag{8.66b}$$

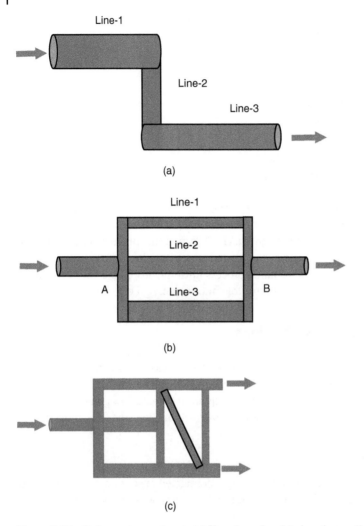

Figure 8.13 Piping system network. (a) Pipes in series, (b) pipes in parallel, and (c) pipes in series–parallel.

Rule-2: Total head loss between nodes A and B is the sum of head losses in all pipes connected in series:

$$\Delta h_{A-B} = \Delta h_1 + \Delta h_2 + \Delta h_3 \tag{8.67a}$$

or

$$\Delta h_{A-B} = \frac{V_1^2}{2g}\left[f_1\frac{L_1}{D_1} + \sum K_1\right] + \frac{V_2^2}{2g}\left[f_2\frac{L_2}{D_2} + \sum K_2\right] + \frac{V_3^2}{2g}\left[f_3\frac{L_3}{D_3} + \sum K_3\right] \tag{8.67b}$$

or

$$\Delta h_{A-B} = \frac{V_1^2}{2g} \left\{ \left[f_1 \frac{L_1}{D_1} + \sum K_1 \right] + \left(\frac{D_2}{D_1} \right)^2 \right.$$

$$\left. \times \left[f_2 \frac{L_2}{D_2} + \sum K_2 \right] + \left(\frac{D_1}{D_3} \right)^2 \left[f_3 \frac{L_3}{D_3} + \sum K_3 \right] \right\} \qquad (8.67c)$$

Let us now consider a few design cases to demonstrate the use of these two rules for solving a series-pipe system.

Case-1: For a series-pipe system with a given volume flow rate Q, evaluate the net head loss. For the given flow rate, the average velocity can be computed from Eq. (8.65b) using the Rule-1 for conservation of volume flow rate in the series piping system. In the next step, the total head loss can be evaluated from Eq. (8.67b) or (8.67c) using the Rule-2.

Case-2: For the known head loss ΔH and pipe sizes, evaluate volume flow rate, Q. The iteration process starts with the guess friction factor values in each line with the fully rough region values for the known relative roughness values. In the next step, Eq. (8.67c), derived based on the Rule-2, can be used to determine the first estimate of the velocity in the line-1. Volume flow rate is then computed using Eq. (8.65a) or (8.65b) based on Rule-1. In the next iteration step, new estimates of the friction factors from Moody's chart or by using Colebrook's equation for all pipes are obtained by using the relative roughness values and by computing the velocities and Reynolds number. Iteration process is continued until convergence is reached for the volume flow rate.

Example 8.4 *Determine Volume Flow Rate in a System with Pipes in Series*

Consider a series of piping system consisting of three pipes with diameters and lengths given as below:

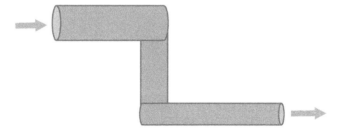

Pipe	Length, L (m)	Diameter, D (cm)	Roughness, ε (mm)	$\frac{\varepsilon}{d}$
1	80	10	0.30	0.003
2	100	8	0.16	0.002
3	10	4	0.20	0.005

Inlet and outlet pressure of the system are 500 and 100 kPa, respectively. Change in the elevation from inlet to outlet section is 30 m. Estimate the volume flow rate of water in the piping system. Calculate the volume flow rate water through the piping system. Assume following water properties: density, $\rho = 1000$ kg/m³, and kinematic viscosity, $v = 1.02 \times 10^{-6}$ m³/s.

Solution

Total changes in the total energy head between the inlet and outlet sections of the piping system are given from the energy equation as

$$\Delta H_{A-B} = \frac{P_1 - P_2}{\rho g} + \frac{V_1^2 - V_2^2}{2g} + (Z_1 - Z_2)$$

Neglecting the changes in the kinetic energy head as a first approximation, we get the simplified form as

$$\Delta H_{A-B} = \frac{P_1 - P_2}{\rho g} + (Z_1 - Z_2)$$

$$= \frac{500\,000 - 100\,000}{1000 \times 9.81} + 30$$

$$\Delta H_{A-B} = 40.77 + 30 = 70.77 \text{ m}$$

From Eq. (8.66a) based on Rule-1 for series pipe:

$$V_2 = V_1 \left(\frac{D_1}{D_2}\right)^2 = \frac{25}{16} V_1 \tag{E.8.4.1a}$$

and

$$V_3 = V_1 \left(\frac{D_1}{D_3}\right)^2 = \frac{25}{4} V_1 \tag{E.8.4.1b}$$

Reynolds numbers in the lines are:

$$Re_2 = \frac{V_2 \, D_2}{V_1 \, D_1} Re_1 = \frac{5}{3} Re_1 \quad \text{and} \quad Re_3 = \frac{V_3 \, D_3}{V_1 \, D_1} Re_1 = \frac{5}{2} Re_1$$

Let us now equate the total changes in energy head with the friction head loss based on the Rule-2 given by Eq. (8.67c)

$$\Delta h_{A-B} = \frac{V_1^2}{2g} \left\{ \left[f_1 \frac{L_1}{D_1} + \sum K_1 \right] + \left(\frac{D_1}{D_2}\right)^2 \left[f_2 \frac{L_2}{D_2} + \sum K_2 \right] + \left(\frac{D_1}{D_3}\right)^2 \left[f_3 \frac{L_3}{D_3} + \sum K_3 \right] \right\}$$

Neglecting the minor losses

$$\Delta h_{A-B} = \frac{V_1^2}{2g} \left\{ \left[f_1 \frac{L_1}{D_1} \right] + \left(\frac{D_1}{D_2}\right)^2 \left[f_2 \frac{L_2}{D_2} \right] + \left(\frac{D_1}{D_3}\right)^2 \left[f_3 \frac{L_3}{D_3} \right] \right\}$$

Substituting the numerical values

$$70.77 = \frac{V_1^2}{2 \times 9.81} \left\{ \left[f_1 \frac{80}{0.1} \right] + \left(\frac{25}{9}\right)^2 \left[f_2 \frac{100}{0.08} \right] + \left(\frac{25}{4}\right)^2 \left[f_3 \frac{10}{0.04} \right] \right\}$$

$$70.77 = \frac{V_1^2}{2 \times 9.81} \{[800 f_1] + [9645 f_2] + [9765 f_3]\} \tag{E.8.4.2}$$

Note that the friction head losses in pipes 2 and 3 dominate the total head loss.

Let us now initiate the iteration process by estimating the friction factor values in the pipes based on fully rough or fully turbulent values:

Iteration # 1

For $\frac{\varepsilon_1}{D_1} = 0.003$, fully rough value $f_1 = 0.0262$

For $\frac{\varepsilon_2}{D_2} = 0.002$, fully rough value $f_2 = 0.0234$

For $\frac{\varepsilon_3}{D_3} = 0.005$, fully rough value $f_3 = 0.0304$

Substituting in Eq. (E.8.4.2)

$$70.77 = \frac{V_1^2}{2 \times 9.81}\{[800 \times 0.0262] + [9645 \times 0.0234] + [9765 \times 0.0304]\}$$

Solving for velocity V_1

$$V_1^2 = \frac{70.77 \times 2 \times 9.81}{\{[800 \times 0.0262] + [9645 \times 0.0234] + [9765 \times 0.0304]\}}$$

$$V_1^2 = \frac{1388.5}{20.96 + 225.69 + 296.86} = \frac{1388.5}{543.51} = 2.555$$

$V_1 = 1.598$ m/s and Volume flow rate, $Q_1^{(1)} = \frac{\pi}{4}(d_1)^2 V_1$

From Eq. (E.8.4.1a)

$$V_2 = \frac{25}{16}V_1 = 2.496 \text{ m/s}$$

and from Eq. (E.8.4.1b)

$$V_3 = \frac{25}{4}V_1 = 9.988 \text{ m/s}$$

Iteration #2

$$Re_1 = \frac{V_1 D_1}{\nu} = \frac{1.598 \times 0.1}{1.02 \times 10^{-6}} = 156\ 666$$

$$f_1 = \frac{0.3086}{\left\{\log\left[\left(\frac{\varepsilon_1/D_1}{3.7}\right)^{1.11} + \frac{6.9}{Re_1}\right]\right\}^2} = \frac{0.3086}{\left\{\log\left[(0.003/3.7)^{1.11} + \frac{6.9}{156\ 666}\right]\right\}^2}$$

$$= \frac{0.3086}{\left\{\log\left[0.000\ 37 + \frac{6.9}{156\ 666}\right]\right\}^2}$$

$$f_1 = 0.0269$$

$$Re_2 = \frac{5}{3}Re_1 = 261\ 110$$

$$f_2 = \frac{0.3086}{\left\{\log\left[\left(\frac{\varepsilon_2/D_2}{3.7}\right)^{1.11} + \frac{6.9}{Re_2}\right]\right\}^2} = \frac{0.3086}{\left\{\log\left[(0.002/3.7)^{1.11} + \frac{6.9}{261\ 110}\right]\right\}^2}$$

$$= \frac{0.3086}{\left\{\log\left[0.000\ 236 + \frac{6.9}{261\ 110}\right]\right\}^2}$$

$$f_2 = 0.0240$$

and

$$Re_3 = \frac{5}{2}Re_1 = 391\ 665$$

$$f_3 = \frac{0.3086}{\left\{\log\left[\left(\frac{\varepsilon_3/D_3}{3.7}\right)^{1.11} + \frac{6.9}{Re_3}\right]\right\}^2} = \frac{0.3086}{\left\{\log\left[(0.005/3.7)^{1.11} + \frac{6.9}{391\ 665}\right]\right\}^2}$$

$$= \frac{0.3086}{\left\{ \log \left[0.000\,653 + \frac{6.9}{391\,665} \right] \right\}^2}$$

$$f_3 = 0.030\,64$$

Substituting in Eq. (E.8.4.2)

$$70.77 = \frac{V_1^2}{2 \times 9.81} \{ [800 \times 0.0269] + [9645 \times 0.0240] + [9765 \times 0.030\,64] \}$$

$$V_1^2 = \frac{70.77 \times 2 \times 9.81}{\{ [800 \times 0.0269] + [9645 \times 0.0240] + [9765 \times 0.030\,64] \}}$$

$$V_1^2 = \frac{1388.5}{21.52 + 231.48 + 299.19} = \frac{1388.5}{552.1996} = 2.514$$

$$V_1^{(2)} = 1.586\,\text{m/s}$$

and

Volume flow rate, $Q_1^{(2)} = \frac{\pi}{4}(D_1)^2 V_1^{(2)} = 0.012\,456\,\text{m}^3/\text{s}$

From Eq. (E.8.4.1a)

$$V_2^{(2)} = \frac{25}{16}V_1^{(2)} = 2.478\,\text{m/s}, \quad Q_2^{(2)} = \frac{\pi}{4}(D_2)^2 V_2^{(2)} = 0.012\,455\,\text{m}^3/\text{s}$$

and from Eq. (E.8.4.1b)

$$V_3^{(2)} = \frac{25}{4}V_1^{(2)} = 9.9915\,\text{m/s}, \quad Q_3^{(2)} = \frac{\pi}{4}(D_3)^2 V_1^{(2)} = 0.012\,555\,\text{m}^3/\text{s}$$

Iteration # 3

$$Re_1 = \frac{V_1^{(2)} D_1}{\nu} = \frac{1.586 \times 0.1}{1.02 \times 10^{-6}} = 155\,490$$

$$f = \frac{0.3086}{\left\{ \log \left[\left(\frac{\varepsilon/d}{3.7} \right)^{1.11} + \frac{6.9}{Re_D} \right] \right\}^2}$$

$$f_1 = \frac{0.3086}{\left\{ \log \left[\left(\frac{\varepsilon_1/D_1}{3.7} \right)^{1.11} + \frac{6.9}{Re_1} \right] \right\}^2} = \frac{0.3086}{\left\{ \log \left[(0.003/3.7)^{1.11} + \frac{6.9}{155\,490} \right] \right\}^2}$$

$$= \frac{0.3086}{\left\{ \log \left[0.000\,37 + \frac{6.9}{155\,490} \right] \right\}^2}$$

$$f_1 = 0.026\,97$$

$$Re_2 = \frac{5}{3}Re_1 = 259\,150$$

$$f_2 = \frac{0.3086}{\left\{ \log \left[\left(\frac{\varepsilon_2/D_2}{3.7} \right)^{1.11} + \frac{6.9}{Re_2} \right] \right\}^2} = \frac{0.3086}{\left\{ \log \left[(0.002/3.7)^{1.11} + \frac{6.9}{259\,150} \right] \right\}^2}$$

$$= \frac{0.3086}{\left\{ \log \left[0.000\,236 + \frac{6.9}{259\,150} \right] \right\}^2}$$

$$f_2 = 0.0235$$

and

$$Re_3 = \frac{5}{2}Re_1 = 388\,725$$

$$f_3 = \frac{0.3086}{\left\{\log\left[\left(\frac{\varepsilon_3/D_3}{3.7}\right)^{1.11} + \frac{6.9}{Re_3}\right]\right\}^2} = \frac{0.3086}{\left\{\log\left[(0.005/3.7)^{1.11} + \frac{6.9}{388\,725}\right]\right\}^2}$$

$$= \frac{0.3086}{\left\{\log\left[0.000\,653 + \frac{6.9}{388\,725}\right]\right\}^2}$$

$$f_3 = 0.031\,08$$

Substituting in Eq. (E.8.4.2)

$$70.77 = \frac{V_1^2}{2\times9.81}\{[800\times0.026\,97] + [9645\times0.0235] + [9765\times0.031\,08]\}$$

$$V_1^2 = \frac{70.77\times2\times9.81}{\{[800\times0.026\,97] + [9645\times0.0235] + [9765\times0.031\,08]\}}$$

$$V_1^2 = \frac{1388.5}{21.576 + 226.6575 + 303.4962} = \frac{1388.5}{551.7297} = 2.5166$$

$$V_1^{(3)} = 1.5863\,\text{m/s}$$

and

$$\text{Volume flow rate, } Q_1^{(3)} = \frac{\pi}{4}(D_1)^2 V_1^{(3)} = 0.012\,458\,\text{m}^3/\text{s}$$

From Eq. (E.8.4.1a)

$$V_2^{(3)} = \frac{25}{16}V_1^{(3)} = 2.4785\,\text{m/s},\ Q_2^{(3)} = \frac{\pi}{4}(D)^2 V_2^{(3)} = 0.012\,458\,\text{m}^3/\text{s}$$

and from Eq. (E.8.4.1b)

$$V_3^{(3)} = \frac{25}{4}V_1^{(3)} = 9.9143\,\text{m/s},\ Q_3^{(3)} = \frac{\pi}{4}(D_3)^2 V_1^{(2)} = 0.012\,45\,\text{m}^3/\text{s}$$

Summary of iteration results.

	Pipe	Length, L (m)	$\frac{\varepsilon}{D}$	f	V (m/s)	Q (m³/s)
Iteration-1	1	80	0.003	0.0262	2.598	
	2	200	0.002	0.0234	2.496	
	3	50	0.005	0.0304	9.988	
Iteration-2	1	80	0.003	0.0269	1.586	0.012 456
	2	200	0.002	0.0240	2.478	0.021 245 5
	3	50	0.005	0.0306	9.992	0.012 555
Iteration-3	1	80	0.003	0.0269	1.5863	0.012 458
	2	200	0.002	0.0235	2.4785	0.012 458
	3	50	0.005	0.0310	9.9143	0.012 45

We can see very good convergence in results for the three pipes achieved in just three iterations. The speed of convergence is good because of the initial good guess values based on fully rough or complete turbulence region values.

8.5.2 Pipes in Parallel

A parallel piping system consists of a number of pipes connected in parallel with each other as shown in Figure 8.13b. For the three pipes connected in parallel, we can apply the following two basic rules:

Rule-1: Conservation of volume flow rate for compressible flow of fluid through the series network:

$$Q_1 + Q_2 + Q_3 = Q \tag{8.68a}$$

or

$$V_1 \frac{\pi}{4} D_1^2 + V_2 \frac{\pi}{4} D_2^2 + V_3 \frac{\pi}{4} D_3^2 = Q \tag{8.68b}$$

Rule-2: The head loss between nodes A and B is same in all three parallel pipes to satisfy the uniqueness of pressure at all nodes. The sum of head losses in all pipes connected in series:

$$\Delta h_{A-B} = \Delta h_1 = \Delta h_2 = \Delta h_3 \tag{8.69a}$$

or

$$\Delta h_{A-B} = \frac{V_1^2}{2g} \left[f_1 \frac{L_1}{D_1} + \sum K_1 \right] = \frac{V_2^2}{2g} \left[f_2 \frac{L_2}{D_2} + \sum K_2 \right] = \frac{V_3^2}{2g} \left[f_3 \frac{L_3}{D_3} + \sum K_3 \right] \tag{8.69b}$$

Case-1: For known head loss ΔH, L, and D, determine volume flow rate, Q

Since the total head loss is same in all lines in a parallel system given, it is simple to find Q_i, $i = 1$, 2, and 3 in line separately and then take the sum $Q = Q_1 + Q_2 + Q_3$ as the total flow rate as per the **Rule-1** given by Eq. (8.68b). In this process, since the total head loss, we can use the **Rule-2** given by Eq. (8.69b) and determine velocity and volume flow rate in each line separately using an iterative process.

1) Guess friction factor f as the fully rough region value for the known value of and $\frac{\epsilon}{D}$ as an initial guess value in each line and determine the velocity V_i, $i = 1, 2, ...N$ using Eq. (8.69b) and the volume flow rate, $Q_i^{(1)}$, for each line of the parallel system.
2) Use these new estimated values of the velocity and volume flow rates and estimate Reynolds numbers in all parallel lines.
3) Estimate new guess value of friction factor f from Moody's diagram or curve-fit equation by Colebrook's equation or Haaland's equation based on the estimated value Re and $\frac{\epsilon}{D}$.
4) Use these newer values of friction factors to recompute the velocities and flow rates for each line.
5) Repeat steps 1–4 until convergence is reached.
6) Compute total flow from the Rule-1 $Q = \sum_i^N Q_i$.

Case-2: For known total flow rate, Q; the pipe sizes in terms of pipe sizes L and D; pipe roughness values, $\frac{\epsilon}{D}$; and presence of valves and fitting in the pipes; determine the head loss ΔH and the flow distributions in all the pipes in a parallel pipe system.

In the **first step**, the iterative process in this class problems starts with guessing the flow rate in one of the pipes. For example, one can start iteration assuming the flow rate in the first pipe as $Q_1 = \frac{Q}{N}$, where N is the number of parallel pipes. In the next step, the head loss in this pipe can be computed using the head loss expression given by Eq. (8.69b)

$$\Delta h_1 = \Delta h_{A-B} = \frac{V_1^2}{2g}\left[f_1\frac{L_1}{D_1} + \sum K_1\right]$$

where $Q_1 = V_1\frac{\pi}{4}D_1^2$

In the next step, we can use **Rule-2** given by Eq. (8.69b) by equating the estimated head loss in line-1 as equal to other head losses in other lines and then solving for velocities and flow rates.

The iterative process is summarized as follows:

1) Start initial guess value of the flow rate in the first pipe as $Q_1 = \frac{Q}{N}$.
2) Estimate the head loss in this pipe using the head loss expression given by Eq. (8.69b).

$$\Delta h_1 = \frac{V_1^2}{2g}\left[f_1\frac{L_1}{D_1} + \sum K_1\right]$$

where $V_1 = \frac{Q_1}{\frac{\pi}{4}D_1^2}$.

3) Compute Reynolds number, Re_2, and then estimate friction factor, f_2, from Moody's diagram or curve-fit equation by Colebrook's equation or Haaland's equation based on the estimated value Re_2 and $\frac{\varepsilon_2}{D_2}$.
4) Set $\Delta h_2 = \Delta h_1$ and determine velocity, V_2, in line-2 from

$$\Delta h_2 = \frac{V_2^2}{2g}\left[f_2\frac{L_2}{D_2} + \sum K_2\right]$$

and volume flow rate, $Q_2 = V_2\frac{\pi}{4}D_2^2$.

5) Repeat steps 3 and 4 for rest of the pipes:
   Following Rule-2 given by Eq. (8.69b), set $\Delta h_i = \Delta h_1$ for $i = 3, 4, ...N$ and determine velocity, V_i for $i = 3, 4, ...N$ for rest of the pipes from

$$\Delta h_i = \frac{V_i^2}{2g}\left[f_i\frac{L_i}{D_i} + \sum K_i\right]$$

   Estimate friction factor, f_i, from Moody's diagram or curve-fit equation by Colebrook's equation or Haaland's equation based on the estimated value Re_i and $\frac{\varepsilon_i}{d_i}$ and compute volume flow rate, $Q_i = V_i\frac{\pi}{4}D_i^2$.

6) Compute total flow rate from

$$Q = \sum_i^N Q_i.$$

7) Repeat steps 1–6 with new estimate of flow in pipe-1 as

$$Q_1^{(k+1)} = \phi Q_1^{(k)}$$

   where
   Superscripts $k + 1$ and k are the present and previous iterative values.
   ϕ = relaxation factor that could be greater or lower than 1 depending on the rate of convergence.

8) Repeat steps 1–7 until required convergence for the total flow rate, Q, is reached.

Example 8.5 *Determine Volume Flow Rate in a System with Pipes in Parallel*

Consider a parallel piping system consisting of three pipes with diameters and lengths given as below:

Pipe	Length, L (m)	Diameter, D (cm)	Roughness, ε (mm)	$\frac{\varepsilon}{D}$
1	80	10	0.30	0.003
2	200	8	0.16	0.002
3	50	4	0.20	0.005

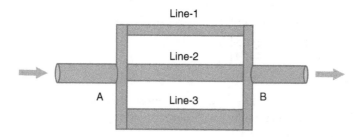

The inlet and outlet pressure conditions, elevation changes, and the geometrical informa-tion of the piping system are same as in the series piping system considered in 8.4. Calculate the volume flow rate water through the parallel piping system.

Solution

Equate the total change in energy head with friction head loss in each line based on Rule-2 as follows:

$$\Delta h_{A-B} = \frac{V_1^2}{2g}\left[f_1\frac{L_1}{D_1} + \sum K_1\right] = \frac{V_2^2}{2g}\left[f_2\frac{L_2}{D_2} + \sum K_2\right] = \frac{V_3^2}{2g}\left[f_3\frac{L_3}{D_3} + \sum K_3\right]$$

(E8.5.1)

Let us neglect the minor losses in the lines and the changes in the kinetic energy and equate with the net head loss computed in Example 8.4 and using Eq. (E.8.5.1)

$$70.77 = \frac{V_1^2}{2g}\left[f_1\frac{80}{0.1}\right] = \frac{V_2^2}{2g}\left[f_2\frac{200}{0.08}\right] = \frac{V_3^2}{2g}\left[f_3\frac{50}{0.04}\right]$$

or

$$70.77 = \frac{V_1^2}{2g}[800f_1] = \frac{V_2^2}{2g}[2500f_2] = \frac{V_3^2}{2g}[1250f_3]$$

(E8.5.2)

Let us now use above equation and use an iterative refinement process to determine the velocity and volume flow rate in each line. Let us start with the initial guess values of the friction factor values as the fully rough values:

Initial guess values

For $\frac{\varepsilon_1}{D_1} = 0.003$, fully rough value $f_1 = 0.0262$
For $\frac{\varepsilon_2}{D_2} = 0.002$, fully rough value $f_2 = 0.0234$
For $\frac{\varepsilon_3}{D_3} = 0.005$, fully rough value $f_3 = 0.0304$

Determine the flow rates in three pipes one at time iterative as follows:

Pipeline-1

Iteration #1

Let us assume the initial guess value for the friction factor as the fully rough value from Moody's chart. For pipe-1 with the relative roughness of $\frac{\varepsilon}{D} = 0.003$, we get the initial guess value as $f_1 = 0.0262$. Substitute this value in Eq. (E8.5.2) and solve for the velocity in the line as

$$70.77 = \frac{V_1^2}{2g}\left[f_1 \frac{80}{0.1}\right] = \frac{V_2^2}{2g}\left[f_2 \frac{200}{0.08}\right] = \frac{V_3^2}{2g}\left[f_3 \frac{50}{0.04}\right]$$

or

$$70.77 = \frac{V_1^2}{2g}[800f_1] = \frac{V_2^2}{2g}[2500f_2] = \frac{V_3^2}{2g}[1250f_3]$$

$$70.77 = \frac{V_1^2}{2g}[800f_1]$$

$$V_1^2 = \frac{70.77 \times 2 \times 9.8}{800 \times 0.0262} = 66.178$$

or

$$V_1^{(1)} = 8.135 \text{ m/s}$$

Volume flow rate, $Q_1^{(1)} = \frac{\pi}{4}D_1^2 V_1^{(1)} = \frac{\pi}{4}(0.1)^2 \times 8.135$

$$Q_1^{(1)} = 0.0638 \text{ m}^3/\text{s}$$

Iteration #2

$$Re_1 = \frac{V_1^{(1)} D_1}{\nu} = \frac{8.135 \times 0.1}{1.02 \times 10^{-6}} = 797\,549$$

Estimate friction factor using curve-fit equation of Moody's chart:

$$f = \frac{0.3086}{\left\{\log\left[\left(\frac{\varepsilon/D}{3.7}\right)^{1.11} + \frac{6.9}{Re_D}\right]\right\}^2}$$

$$f_1 = \frac{0.3086}{\left\{\log\left[\left(\frac{\varepsilon_1/D_1}{3.7}\right)^{1.11} + \frac{6.9}{Re_1}\right]\right\}^2} = \frac{0.3086}{\left\{\log\left[\left(\frac{0.003}{3.7}\right)^{1.11} + \frac{6.9}{797\,549}\right]\right\}^2}$$

$$= \frac{0.3086}{\left\{\log\left[0.000\,370\,5 + \frac{6.9}{797\,549}\right]\right\}^2}$$

$$f_1 = 0.026\,22$$

From Eq. (E8.5.2)

$$70.77 = \frac{V_1^2}{2g}\left[f_1\frac{80}{0.1}\right]$$

$$V_1^2 = \frac{70.77 \times 2 \times 9.8}{0.026\,22 \times 800} = 66.1275$$

$$V_1^{(2)} = 8.1318 \text{ m/s}$$

Volume flow rate, $Q_1^{(2)} = \frac{\pi}{4}D_1^2 V_1^{(2)} = \frac{\pi}{4}(0.1)^2 \times 8.1318$

$$Q_1^{(2)} = 0.063\,88/s$$

In a similar manner, we should compute the flow rates in Lines-2 and -3.

Line-2

Iteration # 1

Let us assume the initial guess value for the friction factor as the fully rough value from Moody's chart. For pipe-2 with the relative roughness of $\frac{\varepsilon}{D} = 0.002$, we get the initial guess value as $f_2 = 0.0234$. Substitute this value in Eq. (E8.5.2) and solve for the velocity in the line as

$$70.77 = \frac{V_2^2}{2g}\left[f_2\frac{200}{0.08}\right] = \frac{V_2^2}{2g}[2500f_2]$$

$$V_2^2 = \frac{70.77 \times 2 \times 9.8}{2500 \times 0.0234} = 23.71$$

or

$$V_2^{(1)} = 4.869 \text{ m/s}$$

Volume flow rate, $Q_2^{(1)} = \frac{\pi}{4}D_2^2 V_2^{(1)} = \frac{\pi}{4}(0.08)^2 \times 4.869$

$$Q_2^{(1)} = 0.024\,47 \text{ m}^3/s$$

Iteration # 2

$$Re_2 = \frac{V_2^{(1)}D_2}{v} = \frac{4.869 \times 0.08}{1.02 \times 10^{-6}} = 381\,882$$

Estimate friction factor using curve-fit equation of Moody's chart:

$$f = \frac{0.3086}{\left\{\log\left[\left(\frac{\varepsilon/D}{3.7}\right)^{1.11} + \frac{6.9}{Re_D}\right]\right\}^2}$$

$$f_2 = \frac{0.3086}{\left\{\log\left[\left(\frac{\varepsilon_2/D_2}{3.7}\right)^{1.11} + \frac{6.9}{Re_2}\right]\right\}^2} = \frac{0.3086}{\left\{\log\left[\left(\frac{0.002}{3.7}\right)^{1.11} + \frac{6.9}{381\,882.35}\right]\right\}^2}$$

$$= \frac{0.3086}{\left\{\log\left[0.000\,236\,2 + \frac{6.9}{381\,882}\right]\right\}^2}$$

$$f_2 = 0.023\,88$$

From Eq. (E8.5.2)

$$70.77 = \frac{V_2^2}{2g}\left[f_2\frac{200}{0.08}\right] = \frac{V_2^2}{2g}[2500f_2]$$

$$V_2^2 = \frac{70.77 \times 2 \times 9.8}{0.023\,88 \times 2500} = 23.234$$

$$V_2^{(2)} = 4.820\,\text{m/s}$$

Volume flow rate, $Q_2^{(2)} = \frac{\pi}{4}D_2^2V_2^{(2)} = \frac{\pi}{4}(0.08)^2 \times 4.820$

$$Q_2^{(2)} = 0.024\,22/\text{s}$$

Line-3

Iteration #1

Let us assume the initial guess value for the friction factor as the fully rough value from Moody's chart. For pipe-3 with the relative roughness of $\frac{\varepsilon}{D} = 0.005$, we get the initial guess value as $f_3 = 0.0304$. Substitute this value in Eq. (E.8.5.2) and solve for the velocity in the line as

$$70.77 = \frac{V_3^2}{2g}\left[f_3\frac{50}{0.04}\right]$$

or

$$70.77 = \frac{V_3^2}{2g}[1250f_3]$$

$$V_3^2 = \frac{70.77 \times 2 \times 9.8}{1250 \times 0.0304} = 36.502$$

or

$$V_3^{(1)} = 6.0417\,\text{m/s}$$

Volume flow rate, $Q_3^{(1)} = \frac{\pi}{4}D_3^2V_3^{(1)} = \frac{\pi}{4}(0.04)^2 \times 6.0417$

$$Q_3^{(1)} = 0.007\,592\,\text{m}^3/\text{s}$$

Iteration #2

$$Re_3 = \frac{V_3^{(1)}D_3}{\nu} = \frac{6.0417 \times 0.04}{1.02 \times 10^{-6}} = 236\,929$$

Estimate friction factor using curve-fit equation of Moody's chart:

$$f_1 = \frac{0.3086}{\left\{\log\left[\left(\frac{\varepsilon_3/D_3}{3.7}\right)^{1.11} + \frac{6.9}{Re_3}\right]\right\}^2} = \frac{0.3086}{\left\{\log\left[\left(\frac{0.005}{3.7}\right)^{1.11} + \frac{6.9}{236\,929}\right]\right\}^2}$$

$$= \frac{0.3086}{\left\{\log\left[0.000\,653\,3 + \frac{6.9}{236\,929}\right]\right\}^2}$$

$$f_1 = 0.030\,78$$

From Eq. (E8.5.2)

$$70.77 = \frac{V_3^2}{2g}\left[f_3\frac{50}{0.04}\right]$$

$$V_3^2 = \frac{70.77 \times 2 \times 9.8}{0.030\,78 \times 1250} = 36.051\,77$$

$$V_3^{(2)} = 6.004 \text{ m/s}$$

Volume flow rate, $Q_3^{(2)} = \frac{\pi}{4}D_3^2V_3^{(2)} = \frac{\pi}{4}(0.04)^2 \times 6.004$

$$Q_3^{(2)} = 0.0075 \text{ m}^3/\text{s}$$

The table below summarizes iterative solutions in the three pipes based on first two iterations.

Summary of iteration results.

Pipe		Length, L (m)	Diameter, D (cm)	$\frac{\varepsilon}{D}$	f	V (m/s)	Q (m³/s)
Pipe-1	Iteration-1	80	10	0.003	0.0262	8.135	0.0638
	Iteration-2	80	10	0.003	0.026 22	8.1318	0.063 88
Pipe-2	Iteration-1	200	8	0.002	0.0240	4.869	0.024 47
	Iteration-2	200	8	0.002	0.0239	4.820	0.024 22
Pipe-3	Iteration-1	50	4	0.005	0.0304	6.014 17	0.007 592
	Iteration-2	50	4	0.005	0.0308	6.004	0.007 50

We can see that the estimates of the flow rates in three pipes approach the converged values even in the first two iterations. Iterations could further be continued depending on accuracy required in a problem. Net flow in the piping system now can be computed based on Rule-1 given by Eq. (8.68a) of the parallel-flow piping system as

$$Q = Q_1 + Q_2 + Q_3 = 0.063\,88 + 0.024\,22 + 0.0075$$

$$Q = 0.0956 \text{ m}^3/\text{s}$$

8.6 Piping System in Series–Parallel Network

A series–parallel piping system consists of a number of pipes connected in series as well as in parallel as shown in Figure 8.13c. From analysis point of view, a series–parallel system is computationally more complex and requires special attention to the algorithm used for a stable iterative solution. However, the basis for the development of the solution algorithm for a series–parallel system is also based on the same two basic rules: (i) Conservation of mass and (ii) Uniqueness of pressure.

Basic methodology developed for series and parallel piping systems can be extended to the more complex series–parallel network. Before proceeding with the presentation of the methodology, let us introduce a few terms and sign conventions with the following series–parallel piping system shown in Figure 8.14.

A *loop* is a closed path formed by connecting series of pipes. Several such loops are identified in the series–parallel system as shown in Figure 8.14. A *node* is a point where two or more lines are connected.

Figure 8.14 Series–parallel piping network.

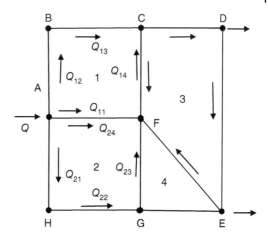

Regarding *sign convention* for flow around a loop, we can consider **head loss**, h_f, as positive for the *counterclockwise direction* and as negative for the *clockwise direction* around a loop. Flow *toward* a node is considered as *positive* and as negative for flow *outward* from a node.

For the pipes connected in a series–parallel network, we can apply the following two basic conditions:

Rule-1: Conservation of mass at all nodes.

For any *node* with a number of connected lines, the sum of all flow rates is zero due to the absence of any source or sink at the node and ensuring that there is no accumulation of mass, i.e. meeting conservation mass. Mathematically, we can state that for a *node* α with N number of connected lines, the conservation of mass is satisfied in the following manner:

$$\sum_{\beta=1}^{N} Q_{\alpha\beta} = 0 \qquad (8.70)$$

where

α = Index for node number
β = Index for connected lines at anode
N = Number of connected lines at the node
Q = Volume flow rate reaching or leaving a node.

For example, for the node F shown in the figure below, the sum of all volume flow rates meeting at the node should equate to zero and this should establish the correct flow directions.

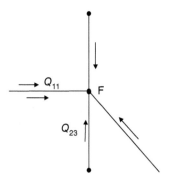

Rule-2: Uniqueness of pressure at any node or total head loss in a loop is zero.

The notation for flow rate in pipes is given as Q_{ij} with index j used to represent the loop number and index i used to represent the pipe number in the j-th loop.

As a consequence of this rule, we can state that the sum of all head losses in all pipes in a loop is zero. Mathematically for any loop j in a network system, this translates to the following statement

$$\sum_{i=1}^{M} h_{ij} = 0 \quad \text{for the } j\text{-th loop} \tag{8.71a}$$

where

j = Index for loop number in piping network
i = Index for lines in the j-th loop
M = Number of lines in the j-th loop
h = Head loss in a line

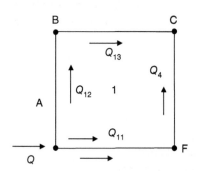

As a consequence of this rule, we can state that head losses between any two nodes must be same for all pathways between these two nodes. For example, for the loop-1 shown above, we can state

$$h_{f_{12}} + h_{f_{13}} = h_{f_{11}} + h_{f_{14}}$$

For pipes that are common to two loops, the flow rates are identical in two pipes, but with opposite signs. For example, for the pipe A–F, $Q_{11} = -Q_{24}$ according to sign convention.

An approach with these two rules to solve a complex series–parallel network involves iterations due to nonlinear nature of the resulting system of equations, and often may lead to an **ill-posed** system of equations. These computational difficulties are more predominant for class II-design problems where all pipe geometries are known as well as the inputs and outputs of the piping network, but the flow rates in each individual pipe or the flow distributions within the network are unknown and need to be determined. From our basic understanding of the flow dynamics, we know that higher amount of mass or volume of fluid tends to flow through the pipe that possess lower friction head loss and least amount through the pipe with highest friction head loss. Flow distribution within the network is established by this balance of pipe friction drops and can be determined by using an iterative process.

A systematic iterative algorithm that leads to stable and enhanced convergence was first proposed ***by Hardy Cross (1936)***, and this is referred to as the ***Hardy Cross method***. A brief description of the method is presented here.

8.6.1 Hardy Cross Method – Based on Darcy–Weisbach Friction Factor

Hardy Cross computational algorithm is based on the use of Eqs. (8.70) and (8.71a) for Rules-1 and -2, respectively, along with ***Darcy–Weisbach Equation*** Eq. (8.27) for the friction factor.

The iterative procedure starts with satisfying conservation of mass at each node as given by the **Rule-1** and determining the initial estimated values of all flow rates in all pipes in the system. However, in many situations the system may have a number of lines, N_l, greater than the number of nodes, N_n; hence, it may lead to a situations where the number of equations derived from mass balances remains less than the number unknowns, i.e. the number of unknown flow rates, Q_{ij}. This is because the mass balances at N_n number of nodes leads to N_n linear algebraic equations with N_l unknowns. For such linearly dependent systems, it is necessary to guess $N_l - N_n + 1$ values of unknowns and solve the system of equations for rest of the unknowns. Completion of this step results in the first set of guess values of flow rates, $Q_{ij}^{(0)}$, in all pipes (i) in a loop (j).

In the next step, these estimated values are used to satisfy the uniqueness of pressure at any given point in a loop as given by the **Rule-2** and determine to **calculate correction factors for mass or volume flow** rates for each loop. These correction values are then used to determine the improved estimates of the flow rates for the next iteration step. This procedure is repeated until convergence is reached for all flow rates.

Since the flow-rate values are inaccurate, with the substitution, this initial set of guess flow values $Q_{ij}^{(0)}$ in Eq. (8.71a) results in some residue given as

$$\sum_{i=1}^{M} h_{fij}^0 \neq 0 \quad \text{for loop } j \tag{8.71b}$$

and the residue term helps us to estimate **the *correction factor values, ΔQ_j,*** for flows in each loop.

The correction value is then used to update the guess values in all pipes (i) in the loop, j, as follows:

$$Q_{ij}^{(1)} = Q_{ij}^{(0)} + \Delta Q_j \tag{8.72}$$

The iteration process is then repeated until convergence in all values is reached. At convergence, all correction values ΔQ_j also approach zero values. This is then repeated for all other piping loops in the network.

The correction values for the flow rates are obtained based on **Newton–Raphson scheme** (Chapra and Canale 2006; Rao 2002), a numerical technique for solving a nonlinear equation expressed as $f(x) = 0$ by using a truncated Taylor series expansion. The basic first-order **Newton–Raphson scheme** is outlined as follows:

For a correct solution or root value, we can equate the function to zero as follows:

$$f(x + \Delta x) = f(x) + \frac{df(x)}{dx} \Delta x + \frac{1}{2!} \frac{d^2 f(x)}{dx^2} \Delta x^2 + \cdots = 0 \tag{8.73a}$$

Truncating the function by dropping the higher-order terms, we get

$$f(x + \Delta x) \approx f(x) + \frac{df(x)}{dx} \Delta x = 0 \tag{8.73b}$$

and the correction values is then solved as

$$\Delta x = -\frac{f(x)}{\frac{df(x)}{dx}} \tag{8.73c}$$

The iterative scheme is

$$x^{(k+1)} = x^{(k)} + \Delta x \tag{8.73d}$$

where

$x^{(k+1)}$ = Present iterative values

$x^{(k)}$ = Previous iterative value

The stopping criteria for a converged solution are given as

$$\left| \frac{x^{(k+1)} - x^{(k)}}{x^{(k+1)}} \right| \leq \varepsilon_s \tag{8.73e}$$

where ε_s is the assigned tolerance limit.

The **Newton–Raphson scheme** outlined through (8.74a)–(8.74d) can be applied to determine the correction values for the flow rates in the loops of the pipe network by using a truncated Taylor series expansion of the head loss function

$$h_f(Q_{ij}) = 0 \tag{8.74a}$$

For a correct solution, we can equate the function to zero as follows:

$$h_f(Q_{ij} + \Delta Q_j) = h_f(Q_{ij}) + \frac{dh_{f_{ij}}}{dQ_{ij}} \Delta Q_j + \frac{1}{2!} \frac{d^2 h_{f_{ij}}}{dQ_{ij}^2} \Delta Q_j^2 + \cdots = 0 \tag{8.74b}$$

Truncating the function by dropping the higher-order terms, we get

$$h_{f_{ij}}(Q_{ij} + \Delta Q_j) \approx h_f(Q_{ij}) + \frac{dh_{f_{ij}}}{dQ_{ij}} \Delta Q_j = 0 \tag{8.74c}$$

and the correction values are solved as

$$\Delta Q_j = -\frac{h_{f_{ij}}(Q_{ij})}{\dfrac{dh_{f_{ij}}}{dQ_{ij}}}$$

(8.74d)

For the solution of series–parallel network problem, the procedure starts with the **Darcy–Weisbach expression**, given by Eq. (8.27), for the friction head loss in a pipe as

$$h_m = f\frac{L}{D}\frac{V^2}{2g}$$

(8.27)

A general form of the friction head loss for all pipes in a loop, based on **Darcy–Weisbach expression** (8.27), is given as

$$h_{f_{ij}} = f_{ij}\frac{L_{ij}}{D_{ij}^{5}}\frac{V_{ij}^2}{2g}$$

(8.75)

We can now substitute Eq. (8.75) into Eq. (8.71a) to satisfy Rule-2 for the uniqueness of pressure at any node or total head loss in a loop.

$$\sum_{i=1}^{M}\frac{V_{ij}^2}{2g}\left[f_{ij}\frac{L_{ij}}{D_{ij}}\right] = 0 \quad \text{for the } j\text{-th loop}$$

(8.76)

where the friction factor, f_{ij}, is computed for each pipe from Moody's diagram or using a curve-fit expression such as Colebrook's implicit form of Eq. (8.45) or Haaland's explicit form of Eq. (8.46). In this procedure, the friction factor f is computed based on flow velocity, V; diameter, D; and pipe roughness, ε.

As a first step, this friction head loss in a pipe is recast by substituting velocity in terms of volume flow rate ($V = \frac{Q}{A}$) to express friction head loss in a pipe in terms of geometrical parameters L and D, and flow variable Q as

$$h_{f_{ij}} = \frac{8}{\pi^2 g}f_{ij}\frac{L_{ij}}{D_{ij}^{5}}Q_{ij}^2$$

(8.77)

or

$$h_{f_{ij}} = K_1 f_{ij}\frac{L_{ij}}{D_{ij}^{5}}Q_{ij}^2$$

(8.78a)

where K_1 is a constant given as

$$K_1 = \frac{8}{\pi^2 g}$$

(8.78b)

Taking derivative of Eq. (8.78a), we get

$$\frac{dh_{f_{ij}}}{dQ} = 2k_1 f_{ij}\frac{L_{ij}}{D_{ij}^{5}}Q_{ij}$$

(8.79)

Use of Eq. (8.71a) to satisfy Rule-2 for the uniqueness of pressure at any node or total head loss in a loop is expressed as

$$\sum_{i=1}^{M}K_1 f_{ij}\frac{L_{ij}}{D_{ij}^{5}}Q_{ij}^2 = 0 \quad \text{for the } j\text{-th loop}$$

(8.80)

Substituting Eqs. (8.78a) and (8.79) into Eq. (8.74d) while considering the sum of head losses in all pipes of a loop to satisfy the Rule-2 for the uniqueness of the pressure at any point of the loop, we get the volume flow correction factor as follows:

$$\Delta Q_j = -\frac{\sum_{i=1}^{M} K_1 f_{ij} \frac{L_{ij}}{D_{ij}^5} Q_{ij}^2}{2\sum_{i=1}^{M} k_1 f_{ij} \frac{L_{ij}}{D_{ij}^5} Q_{ij}} \tag{8.81}$$

In the use of the **Rule-2** given by Eq. (8.71a), the sign convention for the head loss $h_{f_{ij}}$ follows the signs of the flow rate, i.e. head loss is written as negative for a negative flow rate Q_{ij} in the following manner.

To implement the **sign convention** for flow around a loop, i.e. **head loss, $h_f > 0$**, for the **counterclockwise direction** and **head loss, $h_f < 0$**, for the **clockwise direction** around a loop, we can set following

$$h_{f_{ij}} = K_1 f_{ij} \frac{L_{ij}}{D_{ij}^5} Q_{ij}^2 \quad \text{for} \quad Q_{ij} > 0 \tag{8.82a}$$

$$h_{f_{ij}} = -K_1 f_{ij} \frac{L_{ij}}{D_{ij}^5}(-Q_{ij})^2 \quad \text{for} \quad Q_{ij} < 0 \tag{8.82b}$$

and the derivatives are

$$\frac{dh_{f_{ij}}}{dQ} = 2k_1 f_{ij} \frac{L_{ij}}{D_{ij}^5} Q_{ij} \quad \text{for} \quad Q_{ij} > 0 \tag{8.83a}$$

$$\frac{dh_{f_{ij}}}{dQ} = 2k_1 f_{ij} \frac{L_{ij}}{D_{ij}^5}(-Q_{ij}) \quad \text{for} \quad Q_{ij} < 0 \tag{8.83b}$$

Equation (8.81) for the estimation of the **volume flow rate correction** can now be expressed in a simplified form using the absolute-value notation as

$$\Delta Q_j = -\frac{\sum_{i=1}^{M} K_1 f_{ij} \frac{L_{ij}}{D_{ij}^5} Q_{ij}|Q_{ij}|}{2\sum_{i=1}^{M} k_1 f_{ij} \frac{L_{ij}}{D_{ij}^5} Q_{ij}} \tag{8.84}$$

Another simplified form of Eq. (8.84) is written

$$\Delta Q_j = -\frac{\sum_{i=1}^{M} K_{ij} Q_{ij}|Q_{ij}|}{2\sum_{i=1}^{M} K_{ij} Q_{ij}} \tag{8.85a}$$

where

$$K_{ij} = K_1 f_{ij} \frac{L_{ij}}{D_{ij}^5} = \frac{8}{\pi^2 g} f_{ij} \frac{L_{ij}}{D_{ij}^5} \tag{8.85b}$$

For pipes with considerable amount of **minor losses due to valve and fittings**, the total head loss equation is written as

$$h_{f_{ij}} = f_{ij} \frac{L_{ij}}{D_{ij}^5} \frac{V_{ij}^2}{2g} + \sum K \frac{V_{ij}^2}{2g} \tag{8.86a}$$

or

$$h_{f_{ij}} = \frac{V_{ij}^2}{2g}\left[f_{ij}\frac{L_{ij}}{D_{ij}^5} + \sum K\right]$$

(8.86b)

and in terms of volume flow rate as

$$h_{f_{ij}} = \frac{8}{\pi^2 g}f_{ij}\frac{L_{ij}}{D_{ij}^5}Q_{ij}^2 + \frac{8}{\pi^2 g}\sum K\frac{1}{D_{ij}^4}Q_{ij}^2$$

(8.87a)

or

$$h_{f_{ij}} = K_1 f_{ij}\frac{L_{ij}}{D_{ij}^5}Q_{ij}^2 + K_1\sum K\frac{1}{D_{ij}^4}Q_{ij}^2$$

(8.87b)

where

$\sum K$ = Sum of all minor loss coefficients for valves and fittings

$K_1 = \frac{8}{\pi^2 g}$

Use of Eq. (8.72) to satisfy Rule-2 for the uniqueness of pressure at any node or total head loss in a loop leads to

$$\sum_{j=1}^{M}\left[K_1 f_{ij}\frac{L_{ij}}{D_{ij}^5}Q_{ij}^2 + K_1\sum K\frac{1}{D_{ij}^4}Q_{ij}^2\right] = 0 \quad \text{for the} \quad i\text{-th loop}$$

(8.88)

and the volume flow correction factor as follows:

$$\Delta Q_j = -\frac{\sum_{i=1}^{M}\left[K_1 f_{ij}\frac{L_{ij}}{D_{ij}^5}Q_{ij}^2 + K_1\sum K\frac{1}{D_{ij}^4}Q_{ij}^2\right]}{2\sum_{i=1}^{M}\left[k_1 f_{ij}\frac{L_{ij}}{D_{ij}^5}Q_{ij} + K_1\sum K\frac{1}{D_{ij}^4}Q_{ij}\right]}$$

(8.89a)

and in terms of using proper sign convention for flow directions as

$$\Delta Q_j = -\frac{\sum_{i=1}^{M}\left[K_1 f_{ij}\frac{L_{ij}}{D_{ij}^5}Q_{ij}|Q_{ij}| + K_1\sum K\frac{1}{D_{ij}^4}Q_{ij}|Q_{ij}|\right]}{2\sum_{i=1}^{M}\left[k_1 f_{ij}\frac{L_{ij}}{D_{ij}^5}Q_{ij} + K_1\sum K\frac{1}{D_{ij}^4}Q_{ij}\right]}$$

(8.89b)

or

$$\Delta Q_j = -\frac{\sum_{i=1}^{M}\left[K_{ij}Q_{ij}|Q_{ij}| + K_1\sum KK_{ij,m}Q_{ij}^2\right]}{2\sum_{i=1}^{M}K_{ij}Q_{ij}}$$

(8.89c)

where

$$K_{ij} = K_1 f_{ij}\frac{L_{ij}}{D_{ij}^5} = \frac{8}{\pi^2 g}f_{ij}\frac{L_{ij}}{D_{ij}^5}$$

(8.90a)

and

$$K_{ij,m} = K_1\sum K\frac{1}{D_{ij}^4}$$

(8.90b)

The correction value is then used to update the guess values in all pipes, i, in the loop, j, using Eq. (8.72):

$$Q_{ij}^{(1)} = Q_{ij}^{(0)} + \Delta Q_j \qquad (8.72)$$

The iteration process is then repeated until convergence in all values is reached following the convergence criteria:

$$\left| \frac{Q_{ij}^k - Q_{ij}^{k-1}}{Q_{ij}^k} \right| \leq \varepsilon_s \qquad (8.91a)$$

where

$k = $ Index for the iteration number

$\varepsilon_s = $ Assigned tolerance limit for convergence

Alternatively, we can also check if flow correction factors approach a zero value and we can use following criterion for convergence:

$$|\Delta Q_j| \leq \varepsilon_s \qquad (8.91b)$$

At convergence, all correction values ΔQ_j also approach zero values.

This is then repeated for all other piping loops in the network.

The Hardy Cross method described so far and referred to is the **Darcy–Weisbach friction factor – based Hardy Cross Method** using friction factor. This method is considered to be an accurate way of determining the flow distributions in a series–parallel pipe network.

8.6.2 Hazen Williams – Based Hardy Cross Method

This alternate approach was developed with the purpose of having a much simpler way to include the effect of friction pipe head loss, h_f, using an empirical friction coefficient, known as Hazen–Williams coefficient, C. The development of this approach and procedure is described in Section 8.6.2.1.

8.6.2.1 Hazen Williams Expression and Coefficients

In this method, the major head loss in a pipe given by Darcy–Weisbach equation Eq. (8.27) is replaced by a generalized expression given as

$$h_f = KQ^n \qquad (8.92a)$$

where K and n can be determined for any flow geometry and pipe types through experiments. Such an expression can be used for any fluid types. However, prior determination of these values is required before proceeding with the solution of a problem using Hardy Cross method.

A modified version based on Darcy–Weisbach expression is developed primarily for using water as the working fluid. This version is known as **Hazen–Williams expression.** Since friction factor is also a function of volume flow, Q, and diameter, D, through its dependence on Reynolds number, an empirical relationship is used to transform Eq. (8.92a) into the **Hazen–Williams expression** for water flow through pipes given as

$$h_f = KQ^{1.852} = KQ^n \qquad (8.92b)$$

where $n = 1.852$ for water flow in a pipe and

$$K = \frac{K_1 L}{C^{1.852} D^{4.8704}}$$
(8.92c)

where

K_1 = Depending on the units of Q. ($K_1 = 1.318$ for English unit and 0.85 for SI unit).
C = Hazen–Williams coefficient, representative of a dimensionless number for the roughness of the pipe.

It can be noted here that the higher the value of C, the smoother is the pipe surface. Table 8.12 presents a list of Hazen–Williams coefficients for commonly used pipe materials.

Note that K and n can similarly be solved for any fluid types and for any problem with given pipe cross-section or geometrical parameters like length, L; diameter, D for circular pipe or hydraulic diameter for noncircular pipes; and with known pipe roughness using curve fitting of experimental data.

Following this procedure, we can now satisfy the **Rule-2** for a loop j with i lines given by Eq. (8.75) as $\sum_{i=1}^{M} h_{fij}^0 = 0$ and write a general expression for correction value for the loop $- j$:

$$\Delta Q_j = -\frac{\sum_{i=1}^{M} [K_{ij} Q_{ij}^0 (|Q_{ij}^0|)^{n-1}]}{n \sum_{i=1}^{M} K_{ij} (Q_{ij}^0)^{n-1}} \quad \text{for loop} - j$$
(8.93a)

Table 8.12 Some selected Hazen–William coefficients.

Hazen–Williams coefficients	C
Aluminum	130–150
Cast iron – new unlined	130
Cast iron – 10 yr old	107–113
Cast iron – 20 yr old	89–110
Cast iron – 30 yr old	75–90
Cast iron – 40 yr old	64–83
Concrete	100–140
Copper	130–140
Ductile iron	140
Fiber	140
Fiber glass	130
Galvanized iron	120
Glass	130
Plastic	130–150
Polyethylene	130–150
Polyvinyl chloride (PVC)	150
Smooth pipe	140
Steel – corrugated	60
Steel welded or seem less	110

where

$$K_{ij} = \frac{K_1 L_{ij}}{C^{1.852}(D_{ij})^{4.8704}} \tag{8.93b}$$

For the series–parallel network shown in Figure 8.14, the index j varies from 1 to 4 number of loops.

8.6.3 Hardy Cross Method Algorithm

The Hardy Cross method and the procedure discussed so far is now summarized:

1) Subdivide the network into several loops and make sure that all pipes are included in at least one loop.
2) Make initial guess values for $(N_l - N_n + 1)$ number of flow rates $Q^0_{\alpha\beta}$ and then use mass flow balance given by Rule-1 to determine first estimate of rest of the flow rates

$$Q^0_{\alpha\beta}$$

3) Relabel the set of values $\{Q^0_{\alpha\beta}\}$ obtained in Step-2 according to the loop rule and sign conventions and finalize the initial guess values of the unknown set of flow rates, $\{Q^0_{ij}\}$.
4) Determine the constant K_{ij} for all lines in the network using Eq. (8.74b):
   For **Darcy–Weisbach friction factor – based Hardy Cross Method**

$$K_{ij} = K_1 f_{ij} \frac{L_{ij}}{D^5_{ij}} = \frac{8}{\pi^2 g} f_{ij} \frac{L_{ij}}{D^5_{ij}} \tag{8.89a}$$

and

$$K_{ij,m} = K_1 \sum K \frac{1}{D^4_{ij}} \tag{8.89b}$$

   For **Hazen Williams – based Hardy Cross Method**

$$K_{ij} = \frac{K_1 L_{ij}}{C^{1.852}_2 (D_{ij})^{4.8704}}. \tag{8.93b}$$

5) Determine the **correction factor** ΔQ_j for all loops in network using:
   For **Darcy–Weisbach friction factor – based Hardy Cross Method**

$$\Delta Q_j = -\frac{\sum_{i=1}^{M}\left[K_{ij} Q^0_{ij}|Q^0_{ij}| + K_1 \sum KK_{ij,m}(Q^0_{ij})^2\right]}{2\sum_{i=1}^{M} K_{ij}(Q^0_{ij})^2} \quad \text{for loop} - j \tag{8.90}$$

   For **Hazen Williams – based Hardy Cross Method**

$$\Delta Q_j = -\frac{\sum_{i=1}^{M}[K_{ij} Q^0_{ij}(|Q^0_{ij}|)^{n-1}]}{n\sum_{i=1}^{M} K_{ij}(Q^0_{ij})^{n-1}} \quad \text{for loop} - j \tag{8.93a}$$

6) Use correction factors ΔQ_j and obtain new corrected values of the flow rates in each line

$$Q^1_{ij} = Q^0_{ij} + \Delta Q_j \tag{8.72}$$

For any line, which is common in two loops, we must update the flow rate using the two correction values, one from each loop.

7) Check for convergence:

$$\left| \frac{Q_{ij}^k - Q_{ij}^{k-1}}{Q_{ij}^k} \right| \le \varepsilon_s \tag{8.91a}$$

where

k = Index for the iteration number

ε_s = Assigned tolerance limit for convergence

Alternatively, we can also check if flow correction factors approach a zero value and we can use following criterion for convergence:

$$|\Delta Q_i| \le \varepsilon_s \tag{8.91b}$$

Example 8.6   Hardy-Cross Method to Determine Flow Rate in Series-Parallel Pipe Network

Use the **Hardy Cross** method to determine the flow rates in each of the lines of the series–parallel pipe network shown in the figure below. (a) Show an initial estimate of mass for all lines, (b) estimate the correction factor for flow rate in each loop, and (c) correct flow rate in each pipe. Use Hazen–Williams coefficient as $C = 130$. Pipe lengths and diameters are given in the table below

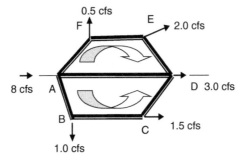

Line	Length (ft)	Diameter (in.)
AB (11)	800	8
BC (12)	1500	6
CD (13)	1000	4
AD (14) (25)	2000	10
AF (26)	1200	6
FE (27)	1000	4
ED (28)	1000	6

Show all calculations for all steps as outlined in the Hardy Cross algorithm and show the corrected flow rates for all lines.

Solution

The piping network is divided into two loops and the initial flow direction is assumed as shown. For this problem, we have the number of lines as $N_l = 7$ and number of nodes as $N_n = 6$.

Use Rule-1 to satisfy conservation of flow rates at each node and estimate initial estimated values of flow rates in all pipes. Number required for guess values for linearly dependent systems: $N_l - N_n + 1 = 7 - 6 + 1 = 2$, and so we need to guess flow values in two pipes. Let us now apply mass balance for each node at a time as per Rule-1 and determine initial guess values of flow in lines connected to each node.

Node-B

$$\sum Q = 0, Q_1 - Q_2 - 1.0 = 0$$

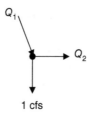

1 cfs

Use **first guess:** $Q_1 = 2.0$ cfs and determine rest of the flow in all lines connected to the node as follows

$$2.0 - Q_2 - 1.0 = 0$$
$$Q_2 = 1.0 \text{ cfs outward and counterclockwise in loop}$$

Node-C

$$\sum Q = 0, Q_2 - Q_3 - 1.5 = 0, 1.0 - Q_3 - 1.5 = 0$$

$Q_3 = -0.5$ cfs inward to the node and clockwise in loop

Node-D

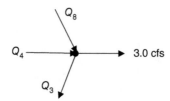

3.0 cfs

$$\sum Q = 0, Q_4 + Q_8 - Q_3 - 3.0 = 0, Q_4 + Q_8 - 0.5 - 3.0 = 0$$

Use second guess: $Q_4 = -2.0$ cfs

$$2.0 + Q_8 - 0.5 - 3.0 = 0$$

$Q_8 = 1.5$ toward the node-D and **clockwise in loop-2 and so set as**

$$Q_8 = -1.5$$

Node-E

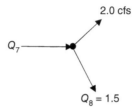

2.0 cfs

Q_7

$Q_8 = 1.5$

$$\sum Q = 0, Q_7 - Q_8 - 2.0 = 0, Q_7 - 1.5 - 2.0 = 0$$

$Q_7 = 3.5$ cfs toward the node E and **clockwise in loop-2** and so set as negative

$$Q_7 = -3.5 \text{ cfs}$$

Node-F

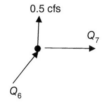

0.5 cfs

Q_7

Q_6

$$\sum Q = 0, Q_6 - Q_7 - 0.5 = 0, Q_6 - 3.5 - 0.5 = 0$$

$Q_6 = 4.0$ cfs toward the node F and **clockwise in loop-2 so set as negative**

$$Q_6 = -4.0 \text{ cfs}$$

Node-A

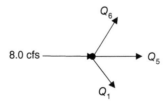

Q_6

8.0 cfs

Q_5

Q_1

$$\sum Q = 0, 8.0 - Q_1 - Q_6 - Q_5 = 0, 8.0 - 2.0 - 4.0 + Q_5 = 0$$

$$Q_5 = 2 \text{ cfs}$$

Initial estimate of all flow rates in the pipes is summarized as follows:

Loop-1

$$Q = \begin{array}{ll} 2.0 & \text{Line 1} \\ 1.0 & \text{Line 2} \\ -0.5 & \text{Line 3} \\ -2.0 & \text{Line 4} \end{array}$$

Loop-2

$$Q = \begin{array}{ll} 2.0 & \text{Line 5} \\ -4.0 & \text{Line 6} \\ -3.5 & \text{Line 7} \\ -1.5 & \text{Line 4} \end{array}$$

Note that we have use loop index as i and pipe no index as j

Iteration-1

Use **Rule-2** to satisfy the uniqueness of pressure value at any point satisfying Eq. (8.75) in all pipes. Let us show the calculation steps for the line-1 of the loop-1.

Constant values: Hazen–Williams coefficient, $C = 130$; conversion unit factor, $K_1 = 4.727$; and friction head constant, $n = 1.852$.

Loop-1

Determine the constant K_{11} for all lines in the network using Eq. (8.74b):

Line-1: $L_{11} = 800$ ft and $D_{11} = 8$ in. $= 0.666$ ft

$$K_{11} = \frac{K_1 L_{11}}{C_2^{1.852} D_{11}^{4.8704}}, K_{11} = \frac{4.727 \times (800)}{(130)^{1.852} \times (0.666)^{4.8704}}, K_{11} = 3.3135$$

Line-2: $L_{12} = 1500$ ft and $D_{12} = 6$ in. $= 0.5$ ft

$$K_{12} = \frac{K_{12}}{C_2^{1.852} D_{12}^{4.8704}}, K_{12} = \frac{4.727 \times (1500)}{(130)^{1.852} \times (0.5)^{4.8704}}, K_{12} = 25.226$$

Line-3: $L_{11} = 1000$ ft and $D_{13} = 4$ in. $= 0.333$ ft

$$K_{13} = \frac{K_1 L_{13}}{C_2^{1.852} D_{13}^{4.8704}}, K_{13} = \frac{4.727 \times (1000)}{(130)^{1.852} \times (0.333)^{4.8704}}, K_{13} = 121.1529$$

Line-4: $L_{14} = 2000$ ft and $D_{14} = 10$ in. $= 0.8333$ ft

$$K_{14} = \frac{K_1 L_{14}}{C_2^{1.852} D_{14}^{4.8704}}, K_{14} = \frac{4.727 \times (2000)}{(130)^{1.852} \times (0.8333)^{4.8704}}, K_{14} = 2.7941$$

Loop-2

Line-5: common to line-4: $L_{25} = 2000$ ft and $D_{25} = 10$ in. $= 0.8333$ ft

$$K_{25} = \frac{K_1 L_{25}}{C_2^{1.852} D_{25}^{4.8704}}, K_{25} = \frac{4.727 \times (1000)}{(130)^{1.852} \times (0.8333)^{4.8704}}, K_{25} = 2.7941$$

Line-6: $L_{26} = 1200$ ft and $D_{25} = 6$ in. $= 0.5$ ft

$$K_{26} = \frac{K_1 L_{26}}{C_2^{1.852} D_{26}^{4.8704}}, K_{26} = \frac{4.727 \times (1200)}{(130)^{1.852} \times (0.5)^{4.8704}}, K_{26} = 20.1781$$

Line-7: $L_{27} = 1000$ ft and $D_{27} = 4$ in. $= 0.3333$ ft

$$K_{27} = \frac{K_1 L_{27}}{C_2^{1.852} D_{27}^{4.8704}}, K_{27} = \frac{4.727 \times (1000)}{(130)^{1.852} \times (0.5)^{4.8704}}, K_{27} = 121.1529$$

Line-8: $L_{28} = 1000$ ft and $D_{28} = 6$ in. $= 0.5$ ft

$$K_{28} = \frac{K_1 L_{28}}{C_2^{1.852} D_{28}^{4.8704}}, K_{28} = \frac{4.727 \times (1000)}{(130)^{1.852} \times (0.5)^{4.8704}}, K_{28} = 16.8151$$

Let us now compute the numerator and denominator terms of Eq. (8.93a) for the flow correction values:

$$\Delta Q = -\frac{KQ|Q|^{n-1}}{nK|Q|^{n-1}}$$

$K_{11}|Q_{11}^0|^{n-1} = (3.3135) \times |2|^{0.852} = 5.980$

and $K_{11}Q_{11}^0|Q_{11}^0|^{n-1} = (3.3135) \times 2 \times |2|^{0.852} = 11.96$

$K_{12}|Q_{12}^0|^{n-1} = (25.2226) \times |1|^{0.852} = 25.2226$

and $K_{12}Q_{12}^0|Q_{12}^0|^{n-1} = (25.2226) \times 1 \times |1|^{0.852} = 25.2226$

$K_{13}|Q_{13}^0|^{n-1} = (121.1529) \times |-0.5|^{0.852} = 67.1206$

and $K_{13}Q_{13}^0|Q_{13}^0|^{n-1} = (121.1529) \times (-0.5) \times |-0.5|^{0.852} = -33.5603$

$K_{14}|Q_{14}^0|^{n-1} = (2.7941) \times |-2|^{0.852} = 5.0433$

and $K_{14}Q_{14}^0|Q_{14}^0|^{n-1} = (2.7941) \times (-2) \times |-2|^{0.852} = -33.5603$

$K_{25}|Q_{25}^0|^{n-1} = (2.7941) \times |2|^{0.852} = 5.0433$

and $K_{25}Q_{25}^0|Q_{25}^0|^{n-1} = (2.7941) \times (2) \times |2|^{0.852} = -10.0866$

$K_{26}|Q_{26}^0|^{n-1} = (20.1781) \times |-4|^{0.852} = 65.7408$

and $K_{26}Q_{26}^0|Q_{26}^0|^{n-1} = (2.7941) \times (-4) \times |-4|^{0.852} = -262.9634$

$K_{27}|Q_{27}^0|^{n-1} = (121.1529) \times |-3.5|^{0.852} = 352.2733$

and $K_{27}Q_{27}^0|Q_{27}^0|^{n-1} = (2.7941) \times (-3.5) \times (-3.5)^{0.852} = -1232.9567$

$K_{28}|Q_{28}^0|^{n-1} = (16.8151) \times |-1.5|^{0.852} = 23.7535$

and $K_{28}Q_{28}^0|Q_{28}^0|^{n-1} = (2.7941) \times (-1.5) \times (-1.5)^{0.852} = -35.6303$

Table below shows a summary of all data computed in iteration 1.

Preceding computations are repeated for all pipes in loops 1 and 2, and summarized in the following table for the iteration step-1:

						Iteration-1		
Loop #	Pipe no	D	L	Q	K	$K\|Q\|^{n-1}$	$KQ\|Q\|^{n-1}$	
1	1	0.6666	800	2	3.3135	5.98	11.96	$\Delta Q_1 = 0.0340$
	2	0.50	1500	1	25.2226	25.2226	25.2226	
	3	0.3333	1000	−0.5	121.1529	67.1206	−33.5603	
	4	0.8333	2000	−2	2.7941	5.0433	−10.0866	
				Sum	103.3665	− 6.4637		
2	5	0.8333	2000	2	2.7941	5.0433	10.0866	$\Delta Q_2 = 1.840$
	6	0.50	1200	−4	20.1781	65.7408	−262.9634	
	7	0.3333	1000	−3.5	121.1529	352.2733	−1232.9567	
	8	0.50	1000	−1.5	16.8151	23.7535	−35.6303	
				Sum	446.8109	− 1521.4638		

We can now compute the correction values for flow rates in the two loops using the summation quantities computed and presented in the table.

Correction value for *loop-1* from Eq. (8.93a) using $j = 1$

$$\Delta Q_1 = -\frac{\sum_{j=1}^{M=4}[K_{1j}Q_{1j}^0|Q_{1j}^0|^{n-1}]}{n\sum_{j=1}^{M}K_{1j}(Q_{1j}^0)^{n-1}}$$

$$\Delta Q_1 - \frac{-6.4637}{1.852 \times 103.3665} = 0.0340$$

Correction value for *loop-2* from Eq. (8.93a) **using** $j = 2$:

$$\Delta Q_2 = -\frac{\sum_{j=1}^{M=4}[K_{2j}Q_{2j}^0|Q_{2j}^0|^{n-1}]}{n\sum_{j=1}^{M}K_{2j}(Q_{2j}^0)^{n-1}}$$

$$\Delta Q_2 = -\frac{-1521.4638}{1.852 \times 446.8109} = 1.8340$$

Update flow rates with these correction values using Eq. (8.72):

Loop-1:

$$Q_{11}^{(1)} = Q_{11}^{(0)} + \Delta Q_1 = 2 + 0.0340 = 2.034$$

$$Q_{12}^{(1)} = Q_{12}^{(0)} + \Delta Q_1 = 1 + 0.0340 = 1.034$$

$$Q_{13}^{(1)} = Q_{13}^{(0)} + \Delta Q_1 = -0.5 + 0.0340 = -0.466$$

$$Q_{14}^{(1)} = Q_{14}^{(0)} + \Delta Q_1 - \Delta Q_2 = -2 + 0.0340 - (1.840) = -3.806 \ (\textbf{Common Line})$$

Loop-2:

$$Q_{25}^{(1)} = Q_{25}^{(0)} + \Delta Q_1 - \Delta Q_2 = 2 - 0.031\,04 + (1.840) = 3.806 \ (\boldsymbol{Common\ Line})$$

$$Q_{26}^{(1)} = Q_{26}^{(0)} + \Delta Q_2 = -4 + (1.840) = -2.16$$

$$Q_{27}^{(1)} = Q_{27}^{(0)} + \Delta Q_2 = -3.5 + (1.840) = -1.6$$

$$Q_{28}^{(1)} = Q_{28}^{(0)} + \Delta Q_2 = -1.5 + (1.840) = -0.340$$

This completes the iteration-1.

Iteration-2

Summary of new estimate of flow rates.

Loop-1

$$Q_{11}^{(1)} = 2.034$$

$$Q_{12}^{(1)} = 1.034$$

$$Q_{13}^{(1)} = -0.466$$

$$Q_{14}^{(1)} = -3.806 \ (\boldsymbol{Common\ Line})$$

Loop-2

$$Q_{25}^{(1)} = 3.806 \ (\boldsymbol{Common\ Line})$$

$$Q_{26}^{(1)} = -2.16$$

$$Q_{27}^{(1)} = -1.6$$

$$Q_{28}^{(1)} = -0.340$$

Computations are now repeated with these new estimates of the flow rates for all pipes in loops 1 and 2, and summarized in the following table for the iteration-2:

					Iteration-2				
Loop #	Pipe no	D	L	Q	K	$K\|Q\|^{n-1}$	$KQ\|Q\|^{n-1}$		
1	1	0.67	800	2.031 04	3.3135	6.063	12.331	$\Delta Q_1 = 0.1219$
	2	0.50	1500	1.031 04	25.226	25.966	26.849	
	3	0.33	1000	−0.466	121.1529	65.350	−29.522	
	4	0.83	2000	−3.806	2.7941	8.709	−33.150	
			Sum	−9.579		106.088	−23.492	
2	5	0.83	2000	3.806	2.7941	8.709	33.150	$\Delta Q_2 = -0.8035$
	6	0.50	1200	−2.160	20.1781	38.847	−83.891	
	7	0.33	1000	−1.660	121.1529	186.463	−309.438	
	8	0.50	1000	0.340	16.8151	6.734	2.293	
			Sum			240.753	−362.472	

This iteration process is continued until the convergence is reached within tolerance limit of less than 0.0001 for the correction values of flow in each flow loop. A summary of these iterations is shown here. Notice that once this convergence limit is reached for the correction values, we can further continue the iterations until we attain converged value of flow rates in all pipes in flow network.

Iteration-3

Loop #	Pipe no	D	L	Q	K	$K|Q|^{n-1}$	$KQ|Q|^{n-1}$	
1	1	0.67	800	2.156	3.3135	6.370	13.733	$\Delta Q_1 = 0.0867$
	2	0.50	1500	1.156	25.226	28.545	32.994	
	3	0.33	1000	−0.344	121.1529	48.960	−16.849	
	4	0.83	2000	−4.488	2.7941	10.018	−44.961	
			Sum			93.893	−15.083	
2	5	0.83	2000	4.488	2.7941	10.018	44.961	$\Delta Q_2 = 0.2006$
	6	0.50	1200	−1.356	20.1781	26.156	−35.467	
	7	0.33	1000	−0.856	121.1529	106.220	−90.923	
	8	0.50	1000	1.144	16.8151	18.864	21.581	
			Sum			161.258	−59.848	

Iteration-4

Loop #	Pipe no	D	L	Q	K	$K|Q|^{n-1}$	$KQ|Q|^{n-1}$	
1	1	0.67	800	2.243	3.3135	6.587	14.772	$\Delta Q_1 = 0.0281$
	2	0.50	1500	1.243	25.226	30.356	37.720	
	3	0.33	1000	−0.257	121.1529	38.250	−9.846	
	4	0.83	2000	−4.602	2.7941	10.233	−47.094	
			Sum			85.426	−4.448	
2	5	0.83	2000	4.602	2.7941	10.233	47.094	$\Delta Q_2 = 0.0219$
	6	0.50	1200	−1.155	20.1781	22.828	−26.375	
	7	0.33	1000	−0.655	121.1529	84.650	−55.477	
	8	0.50	1000	1.345	16.8151	21.641	29.100	
			Sum			139.352	−5.658	

				Iteration-5			
Loop #	Pipe no	D	L	Q	K	$K\|Q\|^{n-1}$	$KQ\|Q\|^{n-1}$
1	1	0.67	800	2.271	3.3135	6.557	15.117	$\Delta Q_1 = 0.0005$
	2	0.50	1500	1.271	25.226	30.339	39.314
	3	0.33	1000	−0.229	121.1529	34.669	−7.949
	4	0.83	2000	−4.574	2.7941	10.180	−46.564
			Sum			81.745	0.082
2	5	0.83	2000	4.574	2.7941	10.180	46.564	$\Delta Q_2 = 0.0039$
	6	0.50	1200	−1.133	20.1781	22.459	−25.456
	7	0.33	1000	−0.633	121.1529	82.234	−52.089
	8	0.50	1000	1.367	16.8151	21.941	29.984
			Sum			136.814	0.997

				Iteration-6			
Loop #	Pipe no	D	L	Q	K	$K\|Q\|^{n-1}$	$KQ\|Q\|^{n-1}$
1	1	0.67	800	2.271	3.3135	6.559	15.123	$\Delta Q_1 = 0.0000$
	2	0.50	1500	1.271	25.226	30.950	39.345
	3	0.33	1000	−0.229	121.1529	34.600	−7.915
	4	0.83	2000	−4.573	2.7941	10.179	−46.553
			Sum			82.288	0.000
2	5	0.83	2000	4.573	2.7941	10.179	46.553	$\Delta Q_2 = 0.000\,26$
	6	0.50	1200	−1.129	20.1781	22.393	−25.292
	7	0.33	1000	−0.629	121.1529	81.799	−51.492
	8	0.50	1000	1.371	16.8151	21.995	30.144
			Sum			179.117	0.087

					Iteration-7			
Loop #	Pipe no	D	L	Q	K	$K\|Q\|^{n-1}$	$KQ\|Q\|^{n-1}$	
1	1 (AB)	0.67	800	2.271	3.3135	6.559	15.123	$\Delta Q_1 = 0.0000$
	2 (BC)	0.50	1500	1.271	25.226	30.950	39.345	
	3 (CD)	0.33	1000	−0.229	121.1529	34.600	−7.915	
	4 (AD)	0.83	2000	−4.573	2.7941	10.179	−46.553	
			Sum			82.288	0.000	
2	5 (AD)	0.83	2000	4.573	2.7941	10.179	46.553	$\Delta Q_2 = 0.0000$
	6 (AF)	0.50	1200	−1.129	20.1781	22.387	−25.292	
	7 (FE)	0.33	1000	−0.629	121.1529	81.762	−51.440	
	8 (ED)	0.50	1000	1.371	16.8151	22.000	30.158	
			Sum			136.328	0.021	

We noticed that flow rates in all pipes in the network have reached close to converged values and indicated correct flow directions in all pipes as demonstrated in the figure below.

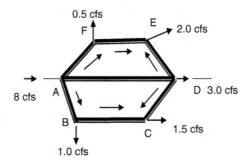

8.6.4 Generalized Hardy Cross

The Hardy Cross method and procedure presented in Sections 8.5.1 and 8.5.2 included only the major losses in lines and to some extent minor losses. Let us now extend this method to a generalized manner to include not only the major and minor losses due to valves and fittings but also the presence of any major devices like heat exchanger and pumps. Let us express these components of head loss or head gain in a generalized form in terms of volume flow rate, Q, as described below:

8.6.4.1 Minor Losses
Head loss in fittings and valves is expressed as

$$h_{f,\min\,ij} = \sum K_{ij}\frac{V_{ij}^2}{2g} \tag{8.94a}$$

Let us express substitute $V_{ij} = \frac{Q_{ij}}{\frac{\pi}{4}D_{ij}^2}$ in Eq. (8.94a) and express as a function of volume flow rate, Q, as

$$h_{f,min\ ij} = \sum K_{ij}\frac{8}{\pi g_c D_{ij}^2}Q_{ij}^2 \tag{8.94b}$$

where

$\sum K_{ij}$ = Sum of all loss coefficients in valves and fittings
D_{ij} = Diameter or hydraulic diameter of the line-j in loop-i
Q_{ij} = Volume flow rate in line-j in loop-i

8.6.4.2 Devices

Pressure drop, ΔP_{HE}, or head loss, Δh_{HE}, in a heat exchanger can be estimated over a range of flow rates and a functional relationship can be derived using the curve-fitting technique discussed in **Appendix A** in the following form

$$\Delta h_{HE} = a_1 Q_{ij} + a_2 Q_{ij}^2 + \cdots + a_m Q_{ij}^m = \sum_{n=1}^{N} a_m Q_{ij}^m$$

or

$$\Delta h_{HE} = \sum_{m=1}^{M} a_m Q_{ij}^m \tag{8.95}$$

A similar expression can be derived to express pressure drop in other devices in a line. This relationship can either be determined by theoretical computation or by conducting experiments and using curve fitting of the available pressure drop data set.

8.6.4.3 Pumps

Since pumps are used in pipelines to create pressure rise, which is equivalent to negative head loss, we can express the polynomial curve-fitted from the pump characteristic curve as

$$h_P = -[a_0 + a_1 Q_{ij} + a_2 Q_{ij}^2 + \cdots + Q_{ij}^m] \tag{8.96a}$$

or

$$h_P = -\left[a_0 + \sum_{m=1}^{M} a_m Q_{ij}^m\right] \tag{8.96b}$$

8.6.4.4 Generalized Expression

Equations (8.94a), (8.95) and (8.96b) can now be combined to expresses in a generalized form given as

$$h_{D_{ij}} = \left[a_{ij0} + \sum_{m=1}^{M} a_{ijm} Q_{ij}^m\right] \tag{8.97}$$

where

m = Order of the polynomial

Incorporating the general representation of the devices in the Hardy Cross method, the expression for the flow correction equation is given as

$$\Delta Q_j = -\frac{\sum_{j=1}^{m} K_{ij} Q_{ij}^0 |Q_{ij}^0|^{n-1} + \text{SIGN\_}Q_{ij} a_{ij0} + \sum_{m=1}^{M} a_{ijm} Q_{ij} |Q_{ij}|^{m-1}}{\sum_{j=1}^{m} n K_{ij} |Q_{ij}^0|^{n-1} + \sum_{m=1}^{M} m a_{ijm} |Q_{ij}|^{m-1}} \qquad (8.91)$$

where

SIGN$\_Q_{ij} = 1$ for $Q_{ij} > 0$ and Sign$\_Q_{ij} = -1$ for $Q_{ij} < 0$

The order of the polynomial m and the coefficient values, a_{ijm} are assigned based on Eqs. (8.94a), (8.95) and (8.96b) for minor losses, devices and pumps, respectively.

Bibliography

ASHRAE (1989). *Handbook – Fundamentals*. Atlanta: American Society of Heating, Refrigeration and Air Conditioning Engineers, Inc.

Chapra, S.C. and Canale, R.P. (2006). *Numerical Methods for Engineers*. New York: McGraw Hill.

Colebrook, C.F. (1939). Turbulent flow in pipes, with particular reference to the transition between the smooth and rough pipe laws. *J. Inst. Civ. Eng.* 11: 133–156.

Cross, H. (November 1936). Analysis of Flow in Networks of Conduits or Conductors. University of Illinois. *Bulletin No. 286*.

Brkic, D.B. and Praks, P. (2019). Short overview of early developments of the hardy cross type methods for computation of flow distribution in pipe networks. *Appl. Sci.* 9: 1–15.

Crane Co. Flow of fluids through valves, fitting and pipes, Technical paper No. 410M, Crane Co, 2009.

Fox, R.W. and McDonald, A. (1992). *Introduction to Fluid Mechanics*. New York: Wiley.

Gerhart, P.M., Gerhart, A.L., Hochstein, J. et al. (2016). *Munson, Young and Okiishi's Fundamentals of Fluid Mechanics*, 8e. New York: Wiley.

Sauer, H.J. and Howell, R.H. (1990). *Principles of Heating Ventilation and Air Conditioning Engineers, a Textbook Supplement to the 1989 ASHRAE Handbook – Fundamentals*. Atlanta: American Society of Heating, Refrigeration and Air Conditioning, Inc.

Hodge, B.K. and Taylor, R.P. (1999). *Analysis and Design of Energy Systems*, 3e. New Jersey: Prentice Hall.

Hydraulic Institute Standard (1969). *Engineering Data Book*. Cleveland, OH: Hydraulic Institute Standard.

Incropera, F.P., Dewitt, D.P., Bergman, T.L., and Lavine, A.S. (2007). *Fundamentals of Heat and Mass Transfer*, 6e. Wiley.

Janna, S.W. (1993). *Design of Fluid Thermal Systems*. PWS-KENT Publishing Co.

Larock, B.E., Jeppson, R.W., and Watters, G.Z. (1999). *Hydraulics of Pipeline Systems*. Boca Raton, FL: CRC Press.

Moody, L.F. (1944). Fiction factors for pipe flow. *Trans. Am. Soc. Mech. Eng.* 66: 671–681.

(1961). *Pipe Friction Manual*, 3e. New York: The Hydraulic Institute.

Pritchard, P.J. and John, W. (2015). *Mitchell, Fox and McDonald's Introduction to Fluid Mechanics*, 9e. Wiley.

Rao, S.S. (2002). *Applied Numerical Methods for Engineers and Scientists*. Prentice Hall.

Problems

8.1 The figure below shows secondary water transfer system for delivering cooling water at a flow rate of 8000 gpm from a cooling tower to the condenser. The potential head due to height difference between the cooling tower water reservoir and the cooling tower spray header is $\Delta Z_{A-B} = 30$ ft and the pressure drop across the condenser is $\Delta P_{HE} = 20$ psi. The system consists of a 10-in., schedule 40, 12 000-ft commercial steel pipe and following set of valves and fittings: 1 check valve, 2 gate valves, 1 ball globe valves, and 12 std regular elbows.

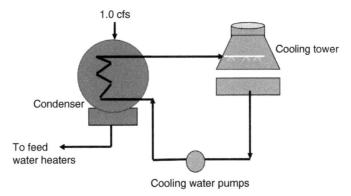

Consider three different options **Option-1**: consider a single line with 8-in., schedule 40, commercial steel pipe, and a larger pump; **Option-2:** Consider two parallel 6-in., schedule 40, commercial steel pipes with two smaller pumps; **Option-3:** Consider a single line with a 10-in. plastic pipe and larger pump. The water temperature is 50 °F.
(a) Compute the pump head and power output for all three options.
(b) If the pump is 75% efficient and electricity costs 6 cents per kWh, estimate the electrical cost of the three options for 8000 hours of usage per year.

8.2 An emergency flooding system for a nuclear reactor core is shown. Using schedule 80 pipe, specify the pipe size, increase in pump head, and power output of the pump if the 40 000 gal of water must be delivered in five minutes.

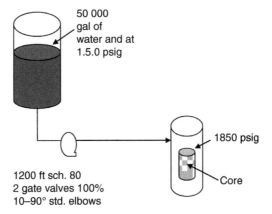

8.3   A pump is used to pump water through a 3-in.-ID, 1200-ft long commercial steel pipe from a feed-water heater maintained at a pressure 150 psig and to a boiler situated at 500 ft higher and maintained at a higher pressure of 350 psig. Minor losses have a combined loss coefficient of 800. The pump characteristics (1 × 2, 8-in. rotor) are given in the figure below as an example. The water-transfer system shown in the figure is used to move water from one reservoir to another as shown in the figure. All pipes are schedule 40, 3-in., and made of commercial steel. The water temperature is 60 °F.

(a) Find the pump head and power output required to pump 90 gpm of flow.

(b) Determine the system curve by considering flow rates over the range from 20 to 130 gpm (20, 45, 90, and 130 gpm). Assume the friction factor obtained in step-a as constant at increased flow rate.

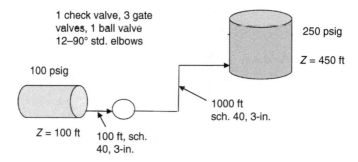

8.4   A pump is used to deliver 100 gpm of water at 70 °F through a 5-in. ID, 1000-ft-long commercial steel pipe from one reservoir to another as shown in the figure below. The valve and fittings are also shown

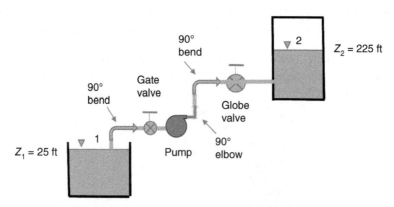

Determine: (a) Required pump head and power (HP); (b) Compute specific speed type of pump and identify type of pump; and (c) Available net positive suction head and Thoma cavitation parameter.

8.5   Use the Hazen–Williams based Hardy Cross method to obtain the flow rates in each of the lines of the network shown in the figure. (a) Show initial estimate of mass at each node, (b) estimate the correction flow factor for flow rate in each loop, and (c) correct flow rate in each pipe. Use Hazen–Williams coefficient as $C = 130$. Pipe lengths and diameters are given in the table below:

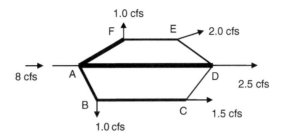

Line	Length (ft)	Diameter (in.)
AB	800	8
BC	1500	6
CD	1000	4
AD	2000	10
DE	1000	6
AF	1200	6
FE	1000	4

a) Show all hand calculations for the first iteration and show the corrected flow rates in each line.

b) Write a computer code implementing the Hazen–Williams Hardy Cross algorithm and show the summary of results in all iterations until a convergence close to $\varepsilon_s \leq 0.001$ for all flow rates.

8.6 Consider problem in Example 8.5 and solve the problem using Darcy–Weisbach friction factor – based Hardy Cross algorithm.

(a) Show all hand calculations for the first iteration and show the corrected flow rates in each line.

(b) Write a computer code implementing the Hazen–Williams Hardy Cross algorithm and show the summary of results in all iterations until a convergence close to $\varepsilon_s \leq 0.001$ for all flow rates.

9

Simulation of Thermal Systems

As mentioned earlier, a thermal system is defined as a collection of a number thermal, fluid, and power-generating components whose parameters are interrelated. A system simulation is like observing a synthetic system that imitates the performance of a real system.

The major purpose of carrying out a system simulation is to determine the expected operating conditions and performance at design as well as off-load or part-load conditions at all operating points of the system. This helps determine the limiting conditions of the operating variables. While this is generally used in the design stage to explore alternatives and/or an improved design, it can also be applied to an existing design to explore the possibility of modifications to improve performance.

A system simulation involves calculation of all operating variables such as mass flow rates, temperature, pressure, density, and all other variables throughout the system. It presumes knowledge of the performance characteristics of all components, constitutive relations as well as equations for thermodynamic properties of the working substances, and all mass and energy balance equations. The equations for performance characteristics of all components and thermodynamic properties along with the conservation of mass and energy balances form a set of simultaneous equations.

9.1 Basic Principles, Types, and Classes of Simulations

A system simulation analysis is usually performed in the design stage for multiple purposes: (i) To develop an improved design, (ii) to explore the possibility of modifying an existing design for enhanced performance and economics, and (iii) to determine the operating conditions or performance at off-load or part-load conditions and the limiting conditions of the operating variables.

There are different classifications of simulation:

(I) *Continuous system vs. discrete system*:
   Continuous simulations are those in which the quantities and associated states change continuously with time such as in reservoirs in which the amount of fluid and thermodynamic state vary with time. Continuous simulation can be carried out through solutions of differential equations. Discrete simulation in contrast is carried out in systems in which the quantities and the states vary only at discrete points in time.

Design of Thermal Energy Systems, First Edition. Pradip Majumdar.
© 2021 John Wiley & Sons Ltd. Published 2021 by John Wiley & Sons Ltd.
Companion website: www.wiley.com/go/majumdar

(II) *Steady-state simulation vs. dynamic simulation*:

In steady-state thermal simulations, the mass flow rate, heat and work transfer rates, and thermodynamic states remain constant. Most of the design analysis of thermal systems and components is performed based on steady-state simulation. Dynamics simulations are time dependent and are performed for systems where one of the quantities such as mass flow rate, or temperature or heat transfer rate, or work transfer rate varies with time. Such simulations are important for initial start-up or shutdown operations of the systems or components. One example of such analysis is the transient thermal simulation of room conditions that are influenced by the transfer of thermal solar load throughout the day. Design of an air-conditioning and cooling system for such environment is carried out based on transient thermal analysis. Most of the control systems for thermal and fluid flow systems are based on dynamic simulations.

(III) *Deterministic vs. stochastic*:

In a deterministic system simulation, all input variables are precisely specified. This contrasts with the stochastic system simulation in which the variables either vary randomly or follow some probabilistic distribution.

9.2 Simulation Procedure and Methodology

9.2.1 Information Flow Diagram

An information flow diagram of a system consists of block diagrams for all major components, showing the flow of all input and output variables. Each component in the system is represented in the form of a **block diagram** along with a **transfer function** that represents the performance characteristics. An example of such a block diagram and transfer function for a centrifugal pump is shown in Figure 9.1. For a centrifugal pump operating at a specified speed, the major variables are mass flow rate, $\dot{m}$; inlet pressure, P_i; and outlet pressure, P_o. The transfer function that shows the relationship among the variables is given as $f(P_i, P_o, \dot{m}) = 0$

9.2.2 Development of the Information Flow Diagram

Let us demonstrate the development of the functional relations of all major components of the system and ways to form an information flow diagram using two example problems.

Example 9.1 *Pump-Piping Flow Distribution System*

Consider a pump that draws water from a reservoir and delivers through a length of pipeline shown in Figure 9.2. Water is delivered at two outlets A and B with flow rates $\dot{m}_A$ and $\dot{m}_B$ to supply water to some process devices or applications through two nozzles. Flow rates through the piping system and through the two exit nozzles depend on the performance characteristics of the pump as well the friction head loss characteristics of the piping system. A steady-state system simulation can be performed to determine the flow rates through the two exits to supply water for two applications.

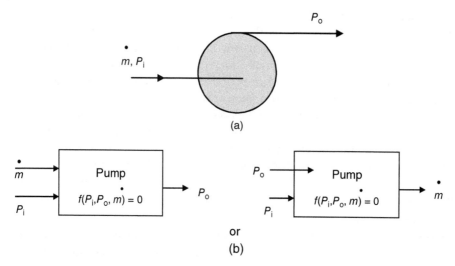

Figure 9.1 Representation of a centrifugal pump in an information flow diagram: (a) schematic diagram and (b) block diagram in an information flow diagram.

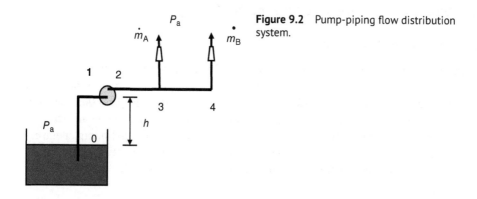

Figure 9.2 Pump-piping flow distribution system.

As a first step to develop the simulation model, it is necessary to derive the transfer functions or the functional relations that represent the performance characteristics of all components including pipe-section frictional pressure drops. The following set of equations has been derived for different sections of the pipe and for different devices.

Pipe section 0–1:

Let us consider the pipe flow energy equation:

$$\frac{P_1}{\rho g} + \frac{V_1^2}{2g} + Z_1 = \frac{P_2}{\rho g} + \frac{V_2^2}{2g} + Z_2 + \frac{V^2}{2g}\left(f\frac{L}{D} + \sum K\right)$$

Noting that for the pipe section 0–1, $P_1 = P_0 = P_a$, $P_2 = P_1$, $Z_2 - Z_1 = h$, and $V_1 = V_2$, application of the pipe flow energy equation to the pipe section 0–1 leads to

$$\frac{P_a}{\rho g} - \frac{P_1}{\rho g} = \frac{V^2}{2g}\left(f\frac{L}{D} + \sum K\right) + (Z_1 - Z_0) \qquad\qquad \text{(E9.1.1a)}$$

Figure 9.3 Graphical representation of the pump performance characteristics in terms of pump pressure rises vs. mass flow rate.

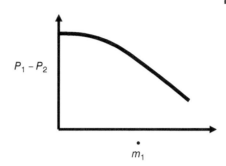

$P_1 - P_2$

m_1

or

$$P_a - P_1 = \rho \frac{V^2}{2}\left(f\frac{L_{0-1}}{D} + \sum K \right) + \rho g h \qquad (E9.1.1b)$$

Now substituting $V = \frac{\dot{m}}{\rho A}$, Eq. (E9.1.1b) can be expressed as

$$P_a - P_1 = C_{0-1}\,\dot{m}^2 + \rho g \qquad (E9.1.1c)$$

where

$$C_{0-1} = \frac{1}{2\rho A^2}\left(f\frac{L_{0-1}}{D} + \sum K \right) \qquad (E9.1.1d)$$

The transfer function for this inlet pipe section 0–1 is then given as

$$f_1(\dot{m}_1, P_1) = 0 \qquad (E9.1.1e)$$

Pump:

Performance characteristics of a typical pump operating at a given speed are shown in Figure 9.3.

A parametric representation of this pump performance is given by polynomial curve fit as

$$P_1 - P_2 = a_0 + a_1 Q + a_2 Q^2 + a_3 Q^3 \qquad (E9.1.2a)$$

Substituting $Q = \frac{\dot{m}_1}{\rho}$, we can express the transfer function for the pumps as

$$f_2(\dot{m}_1, P_1, P_2) = 0 \qquad (E9.1.2b)$$

Pipe section 2–3:

Following the procedure for the pipe section 0–1, we can also express the transfer function for the pipe section 2–3 as

$$P_2 - P_3 = C_{2-3}\,\dot{m}^2 + \rho g h \qquad (E9.1.3a)$$

where

$$C_{2-3} = \frac{1}{2\rho A^2}\left(f\frac{L_{2-3}}{D} + \sum K \right) \qquad (E9.1.3b)$$

and the transfer function for the pipe section 2–3 is given as

$$f_3(\dot{m}_1, P_2, P_3) = 0 \qquad (E9.1.3c)$$

Pipe section 3–4:

Following a similar procedure, we can derive the transfer function as

$$P_3 - P_4 = C_{3-4}\, \dot{m}_2^2 \tag{E9.1.4a}$$

where

$$C_{3-4} = \frac{1}{2\rho A^2}\left(f\frac{L_{3-4}}{D} + \sum K\right) \tag{E9.1.4b}$$

and the transfer function for the pipe section 3–4 is given as

$$f_4(\dot{m}_2, P_3, P_4) = 0 \tag{E9.1.4c}$$

Flow exit A:

$$\dot{m}_A = C_A\sqrt{P_3 - P_a} \tag{E9.1.5a}$$

where C_A= Coefficient of discharge for the nozzle A and the corresponding transfer function is given as

$$f_5(\dot{m}_A, P_3, P_a) = 0 \tag{E9.1.5b}$$

Similarly, for flow exit B:

$$\dot{m}_B = C_B\sqrt{P_4 - P_a} \tag{E9.1.6a}$$

where C_B= Coefficient of discharge for the nozzle B and the corresponding transfer function is given as

$$f_6(\dot{m}_B, P_4, P_a) = 0 \tag{E9.1.6b}$$

Mass balance at node 3:

$$\dot{m}_1 = \dot{m}_2 + \dot{m}_A \tag{E9.1.7a}$$

and the transfer function for the mass balance at node 3 is given as

$$f_7(\dot{m}_1, \dot{m}_2, \dot{m}_A) = 0 \tag{E9.1.7b}$$

Mass balance at node 4:

$$\dot{m}_2 = \dot{m}_B \tag{E9.1.8a}$$

and the transfer function for the mass balance at node 4 is given as

$$f_8(\dot{m}_2, \dot{m}_B) = 0 \tag{E9.1.8b}$$

Equations (E9.1.1c or E9.1.1e), (E9.1.2a or E9.1.2b), (E9.1.3a or E9.1.3c), (E9.1.4a or E9.1.4c), (E9.1.5a or E9.1.5b), (E9.1.6a or E9.1.6b), (E9.1.7a or E9.1.7b) and E9.1.8a or E9.1.8b form a system of eight algebraic equations with eight unknown variables: $P_1, P_2, P_3, P_4, \dot{m}_1, \dot{m}_2, \dot{m}_A, \dot{m}_B$

The transfer function equations (E9.1.1c or E9.1.1e), (E9.1.2a or E9.1.2b), (E9.1.3c), (E9.1.4c), (E9.1.5b)–(E9.1.6b), (E9.1.7b), and (E9.1.8b) for all components in the system can be used to develop an information flow diagram for this system. One important aspect about forming the information flow diagram is to ensure that only one variable is computed as output from each of the transfer function equations, and so there should be only one

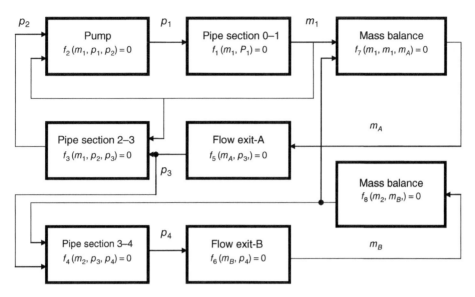

Figure 9.4 Information flow diagram for pump-piping flow distribution.

output from each of the component information block and all remaining variables should be used as inputs. Many different information flow diagrams can be formed depending on how one decides to compute the transfer functions and transfer the information from one block to the other. One such possible information flow diagram for the example system is shown in Figure 9.4.

Example 9.2   *Solar Water Heating System*

A solar collector mounted on the roof top of a single-family house is used to run a solar domestic water heating system as shown in Figure 9.5. Flat plate solar collectors are used to heat the water, which is circulated through a pump-piping system.

Solar collector:

Solar collector performance characteristics:

Flat collector efficiency:

$$\eta_c = F_R(\tau\alpha) - \frac{F_R U_t (T_{c,i} - T_a)}{G_t} \qquad (E9.2.1a)$$

where $F_R(\tau\alpha)$ and $F_R U_t$ are two collector performance parameters as defined in Chapter 6.

Heat removal through collector heat transfer:

$$Q_u = A_c G_t \eta_c = A_c G_t \left[F_R(\tau\alpha) - \frac{U_t(T_{c,i} - T_a)}{G_t} \right] \qquad (E9.2.1b)$$

*Transfer function for **solar collector heat transfer**:*

$$f_1(Q_u, G_t, F_R, \tau\alpha, T_a, T_{c,i}) = Q_u - A_c G_t \left[F_R(\tau\alpha) - \frac{U_t(T_{c,i} - T_a)}{G_t} \right] = 0$$

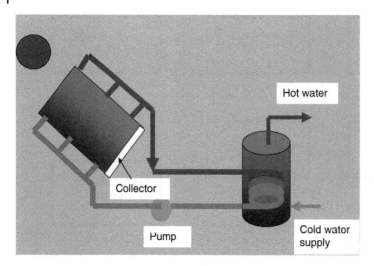

Figure 9.5 Solar water heating system.

Setting the following parameters for the collector's performance characteristics – solar incident radiation and ambient outdoor temperature: A_c, G_t, η_c, F_R, $(\tau\alpha)$, U_t, T_a, we can write the transfer function equation for the solar collector as

$$f_1(Q_u,\ T_{c,i}) = Q_u - A_c G_t \left[F_R(\tau\alpha) - \frac{U_t(T_{c,i} - T_a)}{G_t} \right] = 0 \tag{E9.2.1c}$$

Heat transfer to solar collector fluid:

$$Q_u = \dot{m}_c C_{pc} (T_{c,o} - T_{c,i}) \tag{E9.2.2a}$$

Transfer function equation for **solar collector fluid**:

$$f_2(Q_u,\ \dot{m}_c, T_{c,i}, T_{c,o}) = Q_u - \dot{m}_c C_{pc} (T_{c,o} - T_{c,i}) = 0 \tag{E9.2.2b}$$

Water heater-collector fluid:

Heat transfer from the collector fluid:

$$Q_H = \dot{m}_c\, C_{pc}(T_{c,i} - T_{c,o}) \tag{E9.2.3a}$$

Equating this heat transfer to the supply water,

$$\dot{m}_w\, C_{pc}(T_{w,o} - T_{w,i}) = Q_H$$

Transfer function equation for heat transfer from the **collector fluid**:

$$f_3(Q_H, \dot{m}_c, T_{c,i}, T_{c,o}) = Q_H - \dot{m}_c\, C_{pc}(T_{c,i} - T_{c,o}) = 0 \tag{E9.2.3b}$$

The heat transfer in the water heater can be computed using effectiveness – net transfer unit (ε-NTU) relationship as

$$Q_H = \varepsilon Q_{max} = \varepsilon\, C_{min}\, (T_{c,i} - T_{w,i}) \tag{E9.2.4a}$$

or

$$Q_H = \varepsilon \left(NTU, \frac{C_{min}}{C_{max}} \right) C_{min}\, (T_{c,i} - T_{w,i}) \tag{E9.2.4b}$$

We can write the transfer function for the heat transport across the heater tubes based on the heat exchanger effectiveness value as

$$f_4\, (Q_H,\ \varepsilon,\ C_{min}, T_{c,i}, T_{w,i}) = 0 \tag{E9.2.4c}$$

where $\varepsilon = \varepsilon \left(NTU, \frac{C_{min}}{C_{max}} \right)$

For known values of inlet water temperature, $T_{w,\,i}$, we can write the transfer function equation for **heat transport**:

$$f_4\, (Q_H,\ \varepsilon,\ C_{min}, T_{c,i}) = 0 \tag{E9.2.4d}$$

Water Heater – Overall heat balance between the collector fluid and the hot water supply:

$$Q_H = \dot{m}_w\, C_{pc}(T_{w,o} - T_{w,i}) = \dot{m}_c\, C_{pc}(T_{c,i} - T_{c,o}) \tag{E9.2.5a}$$

Transfer function equation for the **overall heat energy balance in the water heater:**

$$f_5(Q_H, \dot{m}_c\, T_{c,i}, T_{c,o}, \dot{m}_w\, T_{w,i}, T_{w,o}) = 0$$

For the given water supply inlet temperature, $T_{w,\,i}$, and the required supply water outlet temperature, $T_{w,\,o}$, the transfer equation for the overall energy balance can be reduced to

$$f_5(Q_H, \dot{m}_c\, T_{c,i}, T_{c,o}, \dot{m}_w) = 0 \tag{E9.2.5b}$$

The number of parameters and variables may be set so that rest of the unknown operating variables can be solved by performing system simulation analysis using the transfer function equations. Figure 9.6 shows the information flow diagram for the solar water heating system.

9.3 Solution Methods for System Simulation

The system simulation is performed by solving the system of simultaneous algebraic equations, representing the collected set of transfer functions for different devices and processes in the system. Since for this set of equations, some or all of which may be non-linear, iterative solution methods such as single-point iteration, **successive substitution,** and the **Newton–Raphson** method can be used. The method of successive substitution is straightforward and may simply be initiated by assuming initial guess values of a few unknown variables. In the next step, transfer function algebraic equations are solved in sequence following the information flow diagram until all unknown variables are computed including the initially assumed variables. The iterative procedure is continued with the newly found estimated values and the procedure is repeated until the desired convergence limit is reached.

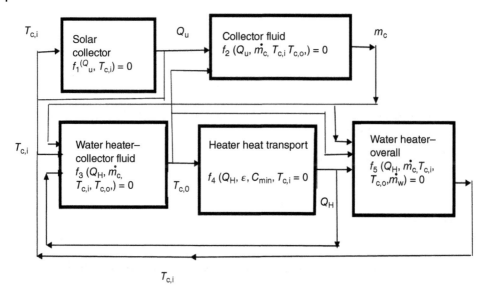

Figure 9.6 Information flow diagram for a solar water heating system.

The basic algorithm for the method of **successive substitution** is summarized below:

1) Assume one or more variables
2) Begin Calculation
3) Continue until originally assumed variables have been recalculated
4) Check for Convergence
5) Continue iterations until convergence is reached.

The method of successive substitution is straightforward but sensitive to the information flow diagram assumed. The method may be subjected to instability in the computations and the solution may be divergent depending on the computation algorithm set by the assumed information flow diagram. This may happen even with the availability of good initial guess values.

The Newton–Raphson method, on the other hand, provides stable and faster convergence for solving the system of algebraic equations (James et al. 1985; Rao 2002).

9.4 Newton–Raphson Method for the Solution of Nonlinear Equations

The Newton–Raphson method is one of the most widely used root finding solution methods for nonlinear equations of both types: algebraic and transcendental. This is an iterative method that starts with an initial guess value of the unknown variable and continues iterative computations using the new estimates until a convergence criterion is met. The basic concept can be demonstrated for the solution of a function $f(x) = 0$ graphically as depicted in Figure 9.7.

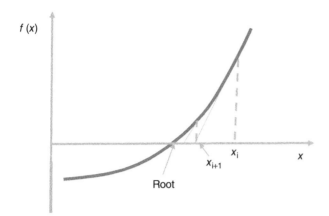

Figure 9.7 Root finding scheme based on the Newton–Raphson method.

The iterative process starts with the initial guess value, x_i. A tangential line is then drawn on the function curve at $x = x_i$. The intersection of this tangent with the x-axis is the next improved approximation for the root value. The process is repeated until the approximate value gets closer to the root value.

Before presenting the solution method for system simulation based on the Newton–Raphson method, let us introduce the basic computational methodology of the Newton–Raphson method for solving a single variable function followed by the solution of a multivariable system of equations.

The Newton–Raphson iterative scheme for determining the roots of a single variable equation, $f(x) = 0$, is derived from the truncated Taylor series expansion of the function about the point $x = x_i$ as demonstrated below:

$$f(x) = f(x_i) + \frac{df}{dx}(x_{i+1} - x_i) + \frac{d^2f}{dx^2}(x_{i+1} - x_i)^2 + \ldots \tag{9.1a}$$

or

$$f(x) = f(x_i) + f'(x_i)(x_{i+1} - x_i) + f''(x_i)(x_{i+1} - x_i)^2 + \ldots \tag{9.1b}$$

Neglecting the higher order terms, we get

$$f(x) \approx f(x_i) + f'(x_i)(x_{i+1} - x_i) \tag{9.1c}$$

Equation (9.1c) is used as an iterative scheme to solve the root of the equation $f(x) = 0$ as

$$f(x) \approx f(x_i) + f'(x_i)(x_{i+1} - x_i) \approx 0 \tag{9.2}$$

Solving,

$$x_{i+1} = x_i - \frac{f(x_i)}{f'(x_i)} \tag{9.3}$$

where

x_i = Previous iterative value
x_{i+1} = Present iterative value

For a function with two independent variables $f(x_1, x_2)$, the Taylor series expansion is taken about the point $x_{1,i}$ and $x_{2,i}$ as

$$f(x_1, x_2) = f(x_{1,i}, x_{2,i}) + \left[\frac{\partial f(x_1, x_2)}{\partial x_1}(x_{1,i+1} - x_{1,i}) + \frac{\partial f(x_1, x_2)}{\partial x_2}(x_{2,i+1} - x_{2,i}) \right]$$

$$+ \left[\frac{1}{2} \frac{\partial^2 f(x_1, x_2)}{\partial x_1^2}(x_{1,i+1} - x_{1,i})^2 + \frac{1}{2} \frac{\partial^2 f(x_1, x_2)}{\partial x_1 \partial x_2}(x_{1,i+1} - x_i)(x_{2,i+1} - x_2) \right.$$

$$\left. + \frac{1}{2} \frac{\partial^2 f(x_1, x_2)}{\partial x_2^2}(x_{2,i+1} - x_{2,i})^2 \right] + \cdots \qquad (9.4)$$

Neglecting the second-order terms, we get

$$f(x_1, x_2) \approx f(x_{1,i}, x_{2,i}) + \left[\frac{\partial f(x_1, x_2)}{\partial x_1}(x_{1,i+1} - x_{1,i}) + \frac{\partial f(x_1, x_2)}{\partial x_2}(x_{2,i+1} - x_{2,i}) \right] = 0 \qquad (9.5)$$

In a similar manner, we can write the Taylor series expansion of a function with **n** variables: as follows:

$$f(x_1, x_2, \ldots, x_n) = \sum_{j=1}^{j=n} \frac{\partial f(a_1, a_2, \ldots, a_n)}{\partial x_j}(x_j - a_j) + \sum_{i=1}^{i=n} \sum_{j=1}^{j=n} \frac{\partial^2 f(a_1, a_2, \ldots, a_n)}{\partial x_i \partial x_j} + \cdots \qquad (9.6)$$

9.5 Newton–Raphson Method for the Solution of a System of Equations

Let us now consider the following general system of equations with multiple variables $(x_1, x_2, \ldots, x_n)$:

$$f_1(x_1, x_2, x_3, \ldots, x_n) = 0 \qquad (1)$$
$$f_2(x_1, x_2, x_3, \ldots, x_n) = 0 \qquad (2)$$
$$f_3(x_1, x_2, x_3, \ldots, x_n) = 0 \qquad (3)$$
$$\vdots$$
$$f_m(x_1, x_2, x_3, \ldots, x_n) = 0 \qquad (m) \qquad (9.7)$$

where index i represents the variable numbers in the range of 1–n. The index j represents the equation numbers in the range of 1–m.

Following the procedure outlined for deriving the iterative scheme for a single function equation while neglecting the higher order terms, we can derive the Newton–Raphson iterative scheme for the system of equations.

The truncated Taylor series expansions of the system of equations are given as

$$f_1(x_1, x_2, \ldots, x_n) \approx f_1(x_{1,k}, x_{2,k}, \ldots, x_{n,k})$$
$$+ \left[\frac{\partial f_1}{\partial x_1}(x_{1,k+1} - x_{1,k}) + \frac{\partial f_1}{\partial x_2}(x_{2,k+1} - x_{2,k}) + \cdots \frac{\partial f_1}{\partial x_n}(x_{n,k+1} - x_{n,k}) \right] = 0$$

$$f_2(x_1, x_2, \ldots, x_n) \approx f_2(x_{1,k}, x_{2,k}, \ldots, x_{n,k})$$
$$+ \left[\frac{\partial f_2}{\partial x_1}(x_{1,k+1} - x_{1,k}) + \frac{\partial f_2}{\partial x_2}(x_{2,k+1} - x_{2,k}) + \cdots \frac{\partial f_2}{\partial x_n}(x_{n,k+1} - x_{n,k}) \right] = 0$$

$$\vdots$$

$$f_m(x_1, x_2, \ldots, x_n) \approx f_m(x_{1,k}, x_{2,k}, \ldots, x_{n,k})$$

$$+ \left[\frac{\partial f_m}{\partial x_1}(x_{1,k+1} - x_{1,k}) + \frac{\partial f_m}{\partial x_2}(x_{2,k+1} - x_{2,k}) + \cdots \frac{\partial m}{\partial x_n}(x_{n,k+1} - x_{n,k}) \right] = 0 \qquad (9.8)$$

where

$x_{1,k}, x_{2,k}, \ldots, x_{n,k}$ = estimated values of the unknown variable in the **previous iteration**

$x_{1,k+1}, x_{2,k+1}, \ldots, x_{n,k+1}$ = estimated values of the unknown variable in the **present iteration**

The system of linear equations, Eq. (9.8), can now be written in a matrix form as

$$\begin{bmatrix} \frac{\partial f_1}{\partial x_1} & \cdots & \frac{\partial f_1}{\partial x_n} \\ \vdots & \ddots & \vdots \\ \frac{\partial f_m}{\partial x_1} & \cdots & \frac{\partial f_m}{x_n} \end{bmatrix} \begin{Bmatrix} x_{1,k+1} - x_{1,k} \\ \vdots \\ x_{n,k+1} - x_{n,k} \end{Bmatrix} = \begin{Bmatrix} f_1(x_{1,k}, x_{2,k}, \ldots, x_{n,k}) \\ \vdots \\ f_m(x_{1,k}, x_{2,k}, \ldots, x_{n,i}) \end{Bmatrix} \qquad (9.9a)$$

or

$$A\mathbf{x} = c \qquad (9.9b)$$

where A = the coefficient matrix with all elements given by the derivatives of the functions

$$= \begin{bmatrix} \frac{\partial f_1}{\partial x_1} & \cdots & \frac{\partial f_1}{\partial x_n} \\ \vdots & \ddots & \vdots \\ \frac{\partial f_m}{\partial x_1} & \cdots & \frac{\partial f_m}{x_n} \end{bmatrix} \qquad (9.9c)$$

C = the column vector with all elements given by the set of function at the previous approximate values, $f(x_{1,k}, x_{2,k}, \ldots, x_{n,k})$

$$= \begin{Bmatrix} x_{1,k+1} - x_{1,k} \\ \vdots \\ x_{n,k+1} - x_{n,k} \end{Bmatrix} \qquad (9.9d)$$

$\mathbf{x}$ = the column vector with elements given by the correction value $\Delta x = x_{k+1} - x_k$, the difference between the present approximate value and previous approximate value:

$$= \begin{Bmatrix} f_1(x_{1,k}, x_{2,k} \ldots x_{n,k}) \\ \vdots \\ f_m(x_{1,k}, x_{2,k} \ldots x_{n,k}) \end{Bmatrix} \qquad (9.9e)$$

Note that the elements of the coefficient matrix are the partial derivatives of the transfer functions. The elements of the coefficient matrix and the functional values are estimated based on the previous estimates of the unknown variables, $(x_{1,k}, x_{2,k}, \ldots, x_{n,k})$. As we can see, Eq. (9.9a) represents a system of linear equations.

The partial derivative of all functions can be derived analytically for many function types depending on the type of problems. In many situations, however, the functions are not differentiable using an exact analytical way. In such problems, the derivates are derived numerically as will be demonstrated in the later sections.

The linear system of equations is solved for Δx_j, the difference between the present iteration value, x_{k+1}, and the previous iteration value, x_k:

$$\begin{Bmatrix} \Delta x_1 \\ \vdots \\ \Delta x_n \end{Bmatrix} = \begin{Bmatrix} x_{1,k+1} - x_{1,k} \\ \vdots \\ x_{n,k+1} - x_{n,k} \end{Bmatrix} \tag{9.10}$$

The new estimate of the variables can be computed from

$$x_{1,k+1} = x_{1,k} + \Delta x_1$$
$$x_{2,k+1} = x_{2,k} + \Delta x_2$$
$$\vdots$$
$$x_{n,k+1} = x_{n,k} + \Delta x_n \tag{9.11}$$

The procedure is repeated with the present or new estimate of the variables until the successive iterated values of the approximate solution differ by less than an assigned tolerance limit, ε_s. The Newton–Raphson solution procedure for the system of equations is summarized using an algorithm.

9.6 Newton–Raphson Solution Algorithm

Step 1: Rewrite the transfer function equations in the following form:

$$f_j(x_i, x_2, \dots, x_n) = 0$$

Step 2: Evaluate the partial derivative of each function with respect to the independent variables:

$$j = 1: \quad \frac{\partial f_1}{\partial x_1} = \frac{\partial f_1}{\partial x_2} = \frac{\partial f_1}{\partial x_n} =$$

$$j = 2: \quad \frac{\partial f_2}{\partial x_1} = \frac{\partial f_2}{\partial x_2} = \frac{\partial f_2}{\partial x_n} =$$

$$\vdots$$

$$j = m: \quad \frac{\partial f_2}{\partial x_1} = \frac{\partial f_2}{\partial x_2} = \frac{\partial f_2}{\partial x_n} =$$

Step 3: Write the set of transfer function equations for system simulation in a matrix form:

$$\begin{bmatrix} \frac{\partial f_1}{\partial x_1} & \cdots & \frac{\partial f_1}{\partial x_n} \\ \vdots & \ddots & \vdots \\ \frac{\partial f_m}{\partial x_1} & \cdots & \frac{\partial f_m}{x_n} \end{bmatrix} \begin{Bmatrix} x_{1,k+1} - x_{1,k} \\ \vdots \\ x_{n,k+1} - x_{n,k} \end{Bmatrix} = \begin{Bmatrix} f_1(x_{1,k}, \dots, x_{n,k}) \\ \vdots \\ f_n(x_{1,k}, \dots, x_{n,k}) \end{Bmatrix}$$

Step 4: Assigned convergence tolerance limit value, $\varepsilon_s =$
Step 5: Assume initial guess values: $x_1^0, x_2^0, \dots, x_n^0$
Step 6: Set present iterated values to initial guess values or present iterated values:

$$x_{1,k} = x_1^0, x_{2,k} = x_2^0, \dots, x_{n,k} = x_n^0$$

Step 7: Compute function values using the present iterated values

$$f_1(x_{1,k}, x_{2,k}, \ldots, x_{n,k}) = 0$$
$$f_2(x_{1,k}, x_{2,k}, \ldots, x_{n,k}) = 0$$
$$\vdots$$
$$f_n(x_{1,k}, x_{2,k}, \ldots, x_{n,k}) = 0$$

Step 8: Compute the partial derivatives $\frac{\partial f_j}{\partial x_i}$ to represent the elements of the coefficient matrix and the function values using the present iterated values:

$$j = 1: \quad \frac{\partial f_1}{\partial x_1} =; \frac{\partial f_1}{\partial x_2} =; \frac{\partial f_1}{\partial x_n} =$$

$$j = 2: \quad \frac{\partial f_2}{\partial x_1} =; \frac{\partial f_2}{\partial x_2} =; \frac{\partial f_2}{\partial x_n} =$$

$$\vdots$$

$$j = m: \quad \frac{\partial f_m}{\partial x_1} =; \frac{\partial f_m}{\partial x_2} =; \frac{\partial f_m}{\partial x_n} =$$

Step 9: Solve the system of equations (9.9) using the Newton–Raphson method.

$$\begin{bmatrix} \frac{\partial f_1}{\partial x_1} & \cdots & \frac{\partial f_1}{\partial x_n} \\ \vdots & \ddots & \vdots \\ \frac{\partial f_m}{\partial x_1} & \cdots & \frac{\partial f_m}{x_n} \end{bmatrix} \begin{Bmatrix} x_{1,k+1} - x_{1,k} \\ \vdots \\ x_{n,k+1} - x_{n,k} \end{Bmatrix} = \begin{Bmatrix} f_1(x_{1,k}, \ldots, x_{n,k}) \\ \vdots \\ f_m(x_{1,k}, \ldots, x_{n,k}) \end{Bmatrix}$$

$$\begin{Bmatrix} \Delta x_1 \\ \vdots \\ \Delta x_n \end{Bmatrix} = \begin{Bmatrix} x_{1,k+1} - x_{1,k} \\ \vdots \\ x_{n,k+1} - x_{n,k} \end{Bmatrix}$$

Step 10: Get a new estimate of the iterated values:

$$x_{1,k+1} = x_{1,k} + \Delta x_1$$
$$x_{2,k+1} = x_{2,k} + \Delta x_2$$
$$\vdots$$
$$x_{n,k+1} = x_{n,k} + \Delta x_n$$

where

K = Previous iteration values
$k + 1$ = Present iterative values

Step 11: Check for convergence:

$$\epsilon_{1,a} = \left| \frac{x_{1,k+1} - x_{1,k}}{x_{1,k+1}} \right| \leq \epsilon_{1,s}$$

$$\epsilon_{2,a} = \left| \frac{x_{2,k+1} - x_{2,k}}{x_{2,k}} \right| \leq \epsilon_{2,s}$$

$$\vdots$$

$$\epsilon_{n,a} = \left| \frac{x_{n,k+1} - x_{n,k}}{x_{n,k}} \right| \le \epsilon_{s,n}$$

Step 12: If convergence is reached, then stop computing. Otherwise, reset the present values with the newly computed values:

$$x_{1,k} = x_{1,k+1}$$

$$x_{2,k} = x_{2,k+1}$$

$$\vdots$$

$$x_{n,k} = x_{n,k+1}$$

Step 13: Repeat steps 6 to 12 until convergence.

Example 9.3 *System Simulation of a Flow System With Pump and By-pass Line*

A centrifugal pump operates with a bypass as shown in the figure below. The flow rate in the main pipe is controlled by controlling the flow rate in the bypass line that includes a flow control valve.

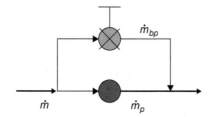

The pressure drop through the bypass line is computed based on the total friction head loss in the bypass line and the pressure drop in the bypass flow control valve, and is given by the curve-fit equation

$$\Delta P = 1.5 \dot{m}_{bp}^2$$

The pump characteristics in terms of pump pressure as a function of flow rate is given by a second-order curve-fit equation

$$\Delta P = 60 + 5\dot{m}_p - 0.1\dot{m}_p^2$$

The pressure drop in the main piping line is computed based on the total friction head loss and is given as

$$\Delta P = 0.03 \dot{m}_p^2$$

where the flow rates are in kg/s and the pressure drop in kPa.

(a) Develop the system of equations for the correction values of the variables m, m_p, m_{bp}, and ΔP using the **Newton–Raphson technique**.
(b) Use the initial guess values as $m_p = 50.0$ kg/s, $m = 55.0$ kg/s, $m_{bp} = 5.00$ kg/s, and $\Delta P = 40.0$ kPa and show iterations to determine the corrections and update the values.

Solution

Notice that the unknown variable is the total mass flow rate in the system. Following the algorithm outlined for the Newton–Raphson method for system simulation, let us show the iterated values for the first iteration as follows:

Step 1: Write all the transfer functions of the components of the system as follows:

Bypass line:

$$f_1 = -\Delta P + 1.5\dot{m}_{bp}^2 = 0 \qquad\qquad (E9.3.1)$$

Pump characteristics: (E9.1.1)

$$f_2 = \Delta P = -60 + 5\dot{m}_p - 0.1\dot{m}_p^2 = 0 \qquad\qquad (E9.3.2)$$

Friction Pressure drop in the main line:

$$f_3 = \Delta P = 60 + 5\dot{m}_p - 0.1\dot{m}_p^2 = 0 \qquad\qquad (E9.3.3)$$

Conservation of mass flow rate:

$$f_4 = -\dot{m} + \dot{m}_p + \dot{m}_{bp} = 0 \qquad\qquad (E9.3.4)$$

Step 2: Evaluate the partial derivative of each function with respective to the independent variables:

$$\frac{\partial f_1}{\partial(\Delta P)} =; \quad \frac{\partial f_1}{\partial \dot{m}} =; \quad \frac{\partial f_1}{\partial \dot{m}_p}; \quad \frac{\partial f_1}{\partial \dot{m}_{bp}}$$

$$\frac{\partial f_2}{\partial(\Delta P)} =; \quad \frac{\partial f_2}{\partial \dot{m}} =; \quad \frac{\partial f_2}{\partial \dot{m}_p} = \frac{\partial f_2}{\partial \dot{m}_{bp}}$$

$$\frac{\partial f_3}{\partial(\Delta P)} =; \quad \frac{\partial f_3}{\partial \dot{m}} =; \quad \frac{\partial f_3}{\partial \dot{m}_p} =; \quad \frac{\partial f_3}{\partial \dot{m}_{bp}}$$

$$\frac{\partial f_4}{\partial(\Delta P)} =; \quad \frac{\partial f_4}{\partial \dot{m}} =; \quad \frac{\partial f_4}{\partial \dot{m}_p} =; \quad \frac{\partial f_4}{\partial \dot{m}_{bp}}$$

Step 3: Write the system of equations in the matrix form:

$$\begin{bmatrix} \dfrac{\partial f_1}{\partial(\Delta P)} & \dfrac{\partial f_1}{\partial \dot{m}} & \dfrac{\partial f_1}{\partial \dot{m}_p} & \dfrac{\partial f_1}{\partial \dot{m}_{bp}} \\[2mm] \dfrac{\partial f_2}{\partial(\Delta P)} & \dfrac{\partial f_2}{\partial \dot{m}} & \dfrac{\partial f_2}{\partial \dot{m}_p} & \dfrac{\partial f_2}{\partial \dot{m}_{bp}} \\[2mm] \dfrac{\partial f_3}{\partial(\Delta P)} & \dfrac{\partial f_3}{\partial \dot{m}} & \dfrac{\partial f_3}{\partial \dot{m}_p} & \dfrac{\partial f_3}{\partial \dot{m}_{bp}} \\[2mm] \dfrac{\partial f_4}{\partial(\Delta P)} & \dfrac{\partial f_4}{\partial \dot{m}} & \dfrac{\partial f_4}{\partial \dot{m}_p} & \dfrac{\partial f_4}{\partial \dot{m}_{bp}} \end{bmatrix} \begin{Bmatrix} \Delta(\Delta P) \\ \Delta \dot{m} \\ \Delta \dot{m}_p \\ \Delta \dot{m}_{bp} \end{Bmatrix} = \begin{Bmatrix} f_1 \\ f_2 \\ f_3 \\ f_4 \end{Bmatrix} \qquad (E9.3.5)$$

Derive the partial derivatives of the functions from Eqs. (E9.3.1) to (E9.3.4) and substitute into the elements of the coefficient matrix:

$$\begin{bmatrix} -1 & 0 & 0 & 3\dot{m}_{bp} \\ -1 & 0 & 5 - 0.2\dot{m}_p & 0 \\ -1 & 0 & 0.06\dot{m}_p & 0 \\ 0 & -1 & 1 & 1 \end{bmatrix} \begin{Bmatrix} \Delta(\Delta P) \\ \Delta \dot{m} \\ \Delta \dot{m}_p \\ \Delta \dot{m}_{bp} \end{Bmatrix} = \begin{Bmatrix} f_1 \\ f_2 \\ f_3 \\ f_4 \end{Bmatrix} \qquad (E9.3.6)$$

Step 4: Assume the initial guess values:

$$\dot{m}_p^{(0)} = 50.0 \text{ kg/s}, \dot{m}^{(0)} = 55.0 \text{ kg/s}, \dot{m}_{bp}^{(0)} = 5.0 \text{ kg/s}, \Delta P^{(0)} = 40.0 \text{ kPa}$$

Step 5: Compute the partial derivatives to represent the elements of the coefficient matrix and the function values in Eq. (E9.3.6) using the present iterated values:

$$f_1 = -40 + 1.5 \, (5)^2 = -2.5$$

$$f_2 = -40 + 60 + 5 \, (50) - 0.1 \, (50)^2 = 20$$

$$f_3 = -40 + 0.030 \, (50)^2 = 35$$

$$f_4 = -55 + 50 + 5 = 0$$

We get the system of equations:

$$\begin{bmatrix} -1 & 0 & 0 & 3(15) = 15 \\ -1 & 0 & 5 - 0.2\,(50) = -5 & 0 \\ -1 & 0 & 0.06(50) = 3 & 0 \\ 0 & -1 & 1 & 1 \end{bmatrix} \begin{bmatrix} \Delta(\Delta P) \\ \Delta \dot{m} \\ \Delta \dot{m}_p \\ \Delta \dot{m}_{bp} \end{bmatrix} = \begin{Bmatrix} -2.5 \\ 2.0 \\ 35 \\ 0 \end{Bmatrix}$$

or

$$\begin{bmatrix} -1 & 0 & 0 & 15 \\ -1 & 0 & -5 & 0 \\ -1 & 0 & 3 & 0 \\ 0 & -1 & 1 & 1 \end{bmatrix} \begin{Bmatrix} \Delta(\Delta P) \\ \Delta \dot{m} \\ \Delta \dot{m}_p \\ \Delta \dot{m}_{bp} \end{Bmatrix} = \begin{Bmatrix} -2.5 \\ 2.0 \\ 35 \\ 0 \end{Bmatrix}$$

Step 6: Solving, we get the correction values:

$$\begin{Bmatrix} \Delta(\Delta P) \\ \Delta \dot{m} \\ \Delta \dot{m}_p \\ \Delta \dot{m}_{bp} \end{Bmatrix} = \begin{Bmatrix} -29.375 \\ -0.25 \\ 1.875 \\ -2.135 \end{Bmatrix}$$

Step 7: Compute the improved iterated values:

$$\Delta P^1 = \Delta P^0 - \Delta(\Delta P) = 40 - (-29.375) = 69.375 \text{ kPa}$$

$$\dot{m}^1 = \dot{m}^0 - \Delta \dot{m} = 55 - (-0.25) = 55.25 \text{ kg/s}$$

$$\dot{m}_p^{(1)} = \dot{m}_p^{(0)} - \Delta \dot{m}_p = 50 - (1.875) = 48.125 \text{ kg/s}$$

$$\dot{m}_{bp}^{(1)} = \dot{m}_{bp}^{(0)} - \Delta \dot{m}_{bp} = 5 - (-2.125) = 7.125$$

Step 8: Update variables with the new iterated values.

$$\left\{\begin{matrix} \Delta P^{(1)} \\ \dot{m}^{(1)} \\ \dot{m}_p^{(1)} \\ \dot{m}_{bp}^{(1)} \end{matrix}\right\} = \left\{\begin{matrix} 69.375 \text{ kPa} \\ 55.25 \text{ kg/s} \\ 48.125 \text{ kg/s} \\ 7.125 \text{ kg/s} \end{matrix}\right\}$$

Repeat the steps to recompute the function values, partial derivatives, and solve the system of equations (E9.3.6) using the new iterated values. Repeat the steps until the convergence limit is reached.

Example 9.4   Chilled Water Cooling Coil

The outside air at $T_{a,i} = 30°$ C is cooled in a cooling coil using chilled water as shown in the figure below. The chilled water at a temperature of $T_{w,i} = 10°$ C is supplied through a chilled water circulating pump and a flow control valve to the cooling coil. The flow rate through the control valve is regulated based on the feedback from the desired outlet air temperature from the cooling coil.

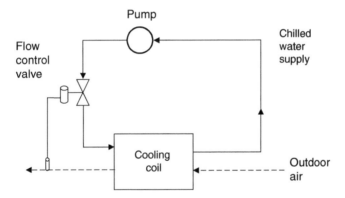

The cooling coil heat transfer characteristics and the size are established by the product of the overall heat transfer characteristics and the surface area for heat transfer, i.e. UA. The circulating pump pressure head characteristics are equal to the net pressure drop through the control valve and the cooling coil. The outlet air temperature is used to regulate the opening of the control valve and hence the chilled water flow rate. It is desired to keep the outlet air temperature within a range of 12–15°C. The linear relationship of the control valve flow coefficient with the cooled air outlet temperature is given as

$$K_{CV} = K_{CV,0}\left(\frac{T_{a,0} - 12}{\Delta T}\right)$$

where $K_{CV,0}$ = the flow coefficient value for the fully open valve at an air outlet temperature of 12° C.

The performance characteristics of the major devices are summarized below:

Pump performance characteristics:

$$\Delta P_{Pump} = P_2 - P_1 = 100\,000 - 16\,000\,\dot{m}_w^2$$

Control valve flow pressure drop characteristics:

$$\Delta P_{CV} = K_{cv}\dot{m}_w^2 \text{ or } \dot{m}_w = K_v\,\sqrt{P_2 - P_1}$$

Pressure drop through the cooling coil:

$$\Delta P_{CC} = K_{CC}\,\dot{m}_w^2 \text{ or } P_3 - P_4 = K_{CC}\dot{m}_w^2$$

Cooling coil heat transfer characteristics:

$$UA = 5000\,W/K$$

Use the Newton–Raphson method to derive the system of equations for system simulation analysis for unknown variables: P_2, $\dot{m}_w$, K_V, P_3, $T_{a,o}$, q, and $T_{w,o}$. Solve the system based on the following data: $T_{a,i} = 30°\,C$, $T_{w,i} = 10°\,C$, $K_{CC} = 9000$, $K_{CV,0} = 0.015$, $m_a = 3.5\,kg/s$, $C_{pa} = 1.0\,kJ/kg\,K$, $C_{pw} = 4.2\,kJ/kg\,K$, $UA = 10\,kW/K$, $P_4 = P_1 = 0$.

Solution
The transfer function equations are as follows:

Chilled water pump:

$$\Delta P_{Pump} = P_2 - P_1 = 100\,000 - 16\,000\,\dot{m}_w^2$$

Transfer function:

$$f_1 = 100\,000 - 16\,000\,\dot{m}_w^2 - P_2 + P_1 = 0 \qquad\qquad\text{(E9.4.1a)}$$

Partial derivatives:

$$\frac{\partial f_1}{\partial P_2} = -1, \frac{\partial f_1}{\partial \dot{m}_w} = 32\,000\dot{m}_w \qquad\qquad\text{(E9.4.1b)}$$

Control valve:

$$\Delta P_{CV} = K_{cv}\dot{m}_w^2 \text{ or } \dot{m}_w = K_v\,\sqrt{P_2 - P_3}$$

Transfer function:

$$f_2 = (P_2 - P_3)K_{cv}^2 - \dot{m}_w^2 = 0 \qquad\qquad\text{(E9.4.2a)}$$

Partial derivatives:

$$\frac{\partial f_2}{\partial P_2} = K_{cv}^2, \frac{\partial f_2}{\partial \dot{m}_w} = -2\,\dot{m}_w, \frac{\partial f_2}{\partial K_{CV}} = 2K_{cv}(P_2 - P_3), \text{ and } \frac{\partial f_2}{\partial P_3} = -K_{cv}^2 \quad\text{(E9.4.2b)}$$

Cooling coil:

$$\Delta P_{CC} = K_{CC}\,\dot{m}_w^2 \text{ or } P_3 - P_4 = K_{CC}\dot{m}_w^2$$

Transfer function:

$$f_3 = K_{CC}\dot{m}_w^2 - P_3 + P_4 = 0 \qquad (E9.4.3a)$$

Partial derivatives:

$$\frac{\partial f_3}{\partial P_2} = 0, \ \frac{\partial f_3}{\partial \dot{m}_w} = 2 K_{CC}\dot{m}_w, \ \frac{\partial f_3}{\partial K_{CV}} = 0, \ \frac{\partial f_3}{\partial P_3} = -1, \ \frac{\partial f_3}{\partial T_{a,0}} = 0, \ \frac{\partial f_3}{\partial q} = 0, \ \text{and} \ \frac{\partial f_3}{\partial T_{w,0}} = 0 \qquad (E9.4.3b)$$

Control valve flow coefficient:

$$K_{CV} = K_{CV,0}\left(\frac{T_{a,0} - 12}{\Delta T}\right)$$

Transfer function:

$$f_4 = K_{CV,0}(T_{a,0} - 12) - \Delta T\, K_{CV} \qquad (E9.4.4a)$$

Partial derivatives:

$$\frac{\partial 4}{\partial P_2} = 0, \ \frac{\partial f_4}{\partial \dot{m}_w} = 0, \ \frac{\partial f_4}{\partial K_{CV}} = -\Delta T, \ \frac{\partial f_4}{\partial P_3} = 0, \ \frac{\partial f_4}{\partial T_{a,0}} = K_{CV,0}, \ \frac{\partial f_4}{\partial q} = 0, \text{and} \ \frac{\partial f_4}{\partial T_{w,0}} = 0 \qquad (E9.4.4b)$$

Additional transfer function equations are derived from the heat transfer rate equations for the cooling coil as follows:

Overall heat transfer rate equation for the cooling coil heat exchanger:

$$q = UA\,\Delta T_m = UA\,\frac{(T_{a,i} - T_{w,0}) - (T_{a,0} - T_{w,i})}{\ln\left(\frac{(T_{a,i}-T_{w,0})}{(T_{a,0}-T_{w,i})}\right)}$$

Transfer function:

$$f_5 = q\,\ln\left(\frac{(T_{a,i} - T_{w,0})}{(T_{a,0} - T_{w,i})}\right) - UA\,[(T_{a,i} - T_{w,0}) - (T_{a,0} - T_{w,i})] = 0 \qquad (E9.4.5a)$$

Partial derivatives:

$$\frac{\partial f_5}{\partial P_2} = 0, \ \frac{\partial f_3}{\partial \dot{m}_w} = 0, \ \frac{\partial f_3}{\partial K_{CV}} = 0, \ \frac{\partial f_3}{\partial P_3} = 0$$

$$\frac{\partial f_5}{\partial T_{a,0}} = q\left(\frac{T_{a,0} - T_{w,i}}{T_{a,i} - T_{w,0}}\right)\left(-\frac{T_{a,i} - T_{w,0}}{(T_{a,0} - T_{w,i})^2}\right) + UA = -q\left(\frac{1}{T_{a,0} - T_{w,i}}\right) + UA$$

$$\frac{\partial f_5}{\partial q} = \ln\left(\frac{T_{a,i} - T_{w,0}}{T_{a,0} - T_{w,i}}\right)$$

and

$$\frac{\partial f_5}{\partial T_{w,0}} = q\left(\frac{T_{a,0} - T_{w,i}}{T_{a,i} - T_{w,0}}\right)\left(\frac{-1}{T_{a,0} - T_{w,i}}\right) + UA = -\left(\frac{q}{T_{a,i} - T_{w,0}}\right) + UA \qquad (E9.4.5b)$$

Heat transfer rate equation for the air side:

$$q = \dot{m}_a \, C_{pa} \, (T_{a,i} - T_{a,o})$$

Transfer function:

$$f_6 = \dot{m}_a \, C_{pa} \, (T_{a,i} - T_{a,o}) - q \qquad \text{(E9.4.6a)}$$

Partial derivatives:

$$\frac{\partial f_6}{\partial P_2} = 0, \quad \frac{\partial f_6}{\partial \dot{m}_w} = 0, \quad \frac{\partial f_6}{\partial K_{CV}} = 0, \quad \frac{\partial f_6}{\partial P_3} = 0, \quad \frac{\partial f_6}{\partial T_{a,o}} = -\dot{m}_a \, C_{pa}, \quad \frac{\partial f_6}{\partial q} = -1,$$

$$\text{and} \quad \frac{\partial f_6}{\partial T_{w,o}} = 0 \qquad \text{(E9.4.6b)}$$

Heat transfer rate equation for the chilled water side:

$$q = \dot{m}_w \, C_{pw} \, (T_{w,o} - T_{w,i})$$

Transfer function:

$$f_7 = \dot{m}_w \, C_{pw} \, (T_{w,o} - T_{w,i}) - q \qquad \text{(E9.4.7a)}$$

Partial derivatives:

$$\frac{\partial f_7}{\partial P_2} = 0, \quad \frac{\partial f_7}{\partial \dot{m}_w} = C_{pw} \, (T_{w,o} - T_{w,i}), \quad \frac{\partial f_7}{\partial K_{CV}} = 0, \quad \frac{\partial f_7}{\partial P_3} = 0, \quad \frac{\partial f_6}{\partial T_{a,o}} = 0, \quad \frac{\partial f_6}{\partial q} = -1,$$

$$\text{and} \quad \frac{\partial f_6}{\partial T_{w,o}} = \dot{m}_w \, C_{pw} \qquad \text{(E9.4.7b)}$$

System equations:

$$
\begin{bmatrix}
\dfrac{\partial f_1}{\partial P_2} & \dfrac{\partial f_1}{\partial \dot{m}_w} & \dfrac{\partial f_1}{\partial K_V} & \dfrac{\partial f_1}{\partial P_3} & \dfrac{\partial f_1}{\partial T_{a,o}} & \dfrac{\partial f_1}{\partial q} & \dfrac{\partial f_1}{\partial T_{w,0}} \\[2mm]
\dfrac{\partial f_2}{\partial P_2} & \dfrac{\partial f_2}{\partial \dot{m}_w} & \dfrac{\partial f_2}{\partial K_V} & \dfrac{\partial f_2}{\partial P_3} & \dfrac{\partial f_2}{\partial T_{a,o}} & \dfrac{\partial f_2}{\partial q} & \dfrac{\partial f_2}{\partial T_{w,0}} \\[2mm]
\dfrac{\partial f_3}{\partial P_2} & \dfrac{\partial f_3}{\partial \dot{m}_w} & \dfrac{\partial f_3}{\partial K_V} & \dfrac{\partial f_3}{\partial P_3} & \dfrac{\partial f_2}{\partial T_{a,o}} & \dfrac{\partial f_3}{\partial q} & \dfrac{\partial f_3}{\partial T_{w,0}} \\[2mm]
\dfrac{\partial f_4}{\partial P_2} & \dfrac{\partial f_4}{\partial \dot{m}_w} & \dfrac{\partial f_4}{\partial K_V} & \dfrac{\partial f_4}{\partial P_3} & \dfrac{\partial f_4}{\partial T_{a,o}} & \dfrac{\partial f_4}{\partial q} & \dfrac{\partial f_4}{\partial T_{w,0}} \\[2mm]
\dfrac{\partial f_5}{\partial P_2} & \dfrac{\partial f_5}{\partial \dot{m}_w} & \dfrac{\partial f_5}{\partial K_V} & \dfrac{\partial f_5}{\partial P_3} & \dfrac{\partial f_5}{\partial T_{a,o}} & \dfrac{\partial f_5}{\partial q} & \dfrac{\partial f_5}{\partial T_{w,0}} \\[2mm]
\dfrac{\partial f_6}{\partial P_2} & \dfrac{\partial f_6}{\partial \dot{m}_w} & \dfrac{\partial f_6}{\partial K_V} & \dfrac{\partial f_6}{\partial P_3} & \dfrac{\partial f_6}{\partial T_{a,o}} & \dfrac{\partial f_6}{\partial q} & \dfrac{\partial f_6}{\partial T_{w,0}} \\[2mm]
\dfrac{\partial f_7}{\partial P_2} & \dfrac{\partial f_7}{\partial \dot{m}_w} & \dfrac{\partial f_7}{\partial K_V} & \dfrac{\partial f_7}{\partial P_3} & \dfrac{\partial f_7}{\partial T_{a,o}} & \dfrac{\partial f_7}{\partial q} & \dfrac{\partial f_7}{\partial T_{w,0}} \\[2mm]
\dfrac{\partial f}{\partial P_2} & \dfrac{\partial f}{\partial \dot{m}_w} & \dfrac{\partial f}{\partial K_V} & \dfrac{\partial f}{\partial P_3} & \dfrac{\partial f}{\partial T_{a,o}} & \dfrac{\partial f}{\partial q} & \dfrac{\partial f}{\partial T_{w,0}}
\end{bmatrix}
\begin{Bmatrix}
\Delta P_2 \\
\Delta \dot{m}_w \\
\Delta K_{CV} \\
\Delta P_3 \\
\Delta T_{a,o} \\
\Delta q \\
\Delta T_{w,0}
\end{Bmatrix}
=
\begin{Bmatrix}
f_1 \\
f_2 \\
f_3 \\
f_4 \\
f_5 \\
f_6 \\
f_7
\end{Bmatrix}
$$

$$\text{(E9.4.8)}$$

Let us compute the partial derivatives of the transfer function equations, Eqs. (E9.4.1b)–(E9.4.7b).

Substituting the values and expressions of the partial derivatives as elements of the coefficient matrix in Eq. (E9.4.8), we get the system of equations:

$$
\begin{bmatrix}
-1 & -32\,000\dot{m}_\mathrm{w} & 0 & 0 & 0 & 0 & 0 \\
K_\mathrm{cv}^2 & -2\dot{m}_\mathrm{w} & 2K_\mathrm{cv}(P_2-P_3) & -K_\mathrm{cv}^2 & 0 & 0 & 0 \\
0 & 18\,000\dot{m}_\mathrm{w}, & 0 & -1 & 0 & 0 & 0 \\
0 & 0 & \Delta T & 0 & 0.015 & 0 & 0 \\
0 & 0 & 0 & 0 & -q\left(\frac{1}{(T_{\mathrm{a,o}}-10)}\right)+10 & \ln\left(\frac{30-T_\mathrm{w,o}}{T_{\mathrm{a,o}}-10}\right) & -\left(\frac{q}{30-T_\mathrm{w,o}}\right)+10 \\
0 & 0 & 0 & 0 & -3.0 & -1 & 0 \\
0 & 4.2\left(T_\mathrm{w,o}-T_\mathrm{w,i}\right) & 0 & 0 & 0 & -1 & 4.2\dot{m}_\mathrm{w}
\end{bmatrix}
$$

$$
\begin{Bmatrix}
\Delta P_2 \\
\Delta \dot{m}_\mathrm{w} \\
\Delta K_\mathrm{CV} \\
\Delta P_3 \\
\Delta T_\mathrm{a,o} \\
\Delta q \\
\Delta T_\mathrm{w,0}
\end{Bmatrix}
=
\begin{Bmatrix}
f_1 \\
f_2 \\
f_3 \\
f_4 \\
f_5 \\
f_6 \\
f_7
\end{Bmatrix}
$$

$$\text{(E9.4.9)}$$

Let us substitute the given data for the system parameters and express the system as follows:

$$
\begin{bmatrix}
-1 & -32\,000\dot{m}_\mathrm{w} & 0 & 0 & 0 & 0 & 0 \\
K_\mathrm{cv}^2 & -2\dot{m}_\mathrm{w} & 2K_\mathrm{cv}(P_2-P_3) & -K_\mathrm{cv}^2 & 0 & 0 & 0 \\
0 & 18\,000\dot{m}_\mathrm{w}, & 0 & -1 & 0 & 0 & 0 \\
0 & 0 & -2 & 0 & 0.015 & 0 & 0 \\
0 & 0 & 0 & 0 & -q\left(\frac{1}{T_{\mathrm{a,o}}-10}\right)+10 & \ln\left(\frac{30-T_\mathrm{w,o}}{T_{\mathrm{a,o}}-10}\right) & -\left(\frac{q}{30-T_\mathrm{w,o}}\right)+10 \\
0 & 0 & 0 & 0 & -3.0 & -1 & 0 \\
0 & 4.2\left(T_\mathrm{w,o}-T_\mathrm{w,i}\right) & 0 & 0 & 0 & -1 & 4.2\dot{m}_\mathrm{w}
\end{bmatrix}
$$

$$
\begin{Bmatrix}
\Delta P_2 \\
\Delta \dot{m}_\mathrm{w} \\
\Delta K_\mathrm{CV} \\
\Delta P_3 \\
\Delta T_\mathrm{a,o} \\
\Delta q \\
\Delta T_\mathrm{w,0}
\end{Bmatrix}
=
\begin{Bmatrix}
f_1 \\
f_2 \\
f_3 \\
f_4 \\
f_5 \\
f_6 \\
f_7
\end{Bmatrix}
$$

$$\text{(E9.4.10)}$$

Let us apply the Newton–Raphson solution algorithm for solving this system of equations (E9.4.10):

Step 1:

Let us assume the initial guess values of the unknown variables as

$$P_2^0 = 30\,000\ \mathrm{Pa},\ \dot{m}_\mathrm{w}^0 = 2.0\ \mathrm{kg/s},\ K_\mathrm{CV}^0 = 0.01,\ P_3^0 = 20\,000\ \mathrm{Pa},\ T_\mathrm{a,o}^0 = 14,$$
$$q^0 = 50.0\ \mathrm{kW\ and}\ T_\mathrm{w,0}^0 = 18.0^\circ\mathrm{C}$$

Step 2:

Based on these guess values compute the elements of the coefficient matrix and all seven functional values.

From Eq. (E9.4.1a),

$$f_1 = 100\,000 - 16\,000\, \dot{m}_{\mathrm{w}}^{0\,2} - P_2^0 + P_1 \approx 0$$

or

$$f_1 = 100\,000 - 16\,000 \times 4.0 - 30\,000 + 0 = -6000$$

From Eq. (E9.4.2a),

$$f_2 = (P_2^0 - P_2^0)(K_{\mathrm{cv}})^2 - (\dot{m}_{\mathrm{w}}^0)^2 \approx 0$$

or

$$f_2 = (30\,000 - 20\,000)(0.01)^2 - (2)^2 = -3.0$$

From Eq. (E9.4.3a),

$$f_3 = K_{\mathrm{CC}}\,(\dot{m}_{\mathrm{w}}^0)^2 - P_3 + P_4 \approx 0$$

or

$$f_3 = 9\,000 - 20\,000 + 0 = 11\,000$$

From Eq. (E9.4.4a),

$$f_4 = K_{\mathrm{CV},0}(T_{\mathrm{a,o}}^0 - 12) - \Delta T\, K_{\mathrm{CV}}^0 \approx 0$$

or

$$f_4 = 0.015\,(14 - 12) - 3.0\,(0.01) = 0$$

From Eq. (E9.4.5a),

$$f_5 = q^0\ \ln\left(\frac{(T_{\mathrm{a,i}} - T_{\mathrm{w,0}}^0)}{(T_{\mathrm{a,o}}^0 - T_{\mathrm{w,i}})}\right) - \mathrm{UA}\ [(T_{\mathrm{a,i}} - T_{\mathrm{w,0}}^0) - (T_{\mathrm{a,o}}^0 - T_{\mathrm{w,i}})] \approx 0$$

or

$$f_5 = 50.0\ \ln\left(\frac{(30 - 18)}{(14 - 10)}\right) - 10\ [(30 - 18) - (14 - 10)] = 25.07$$

From Eq. (E9.4.6a),

$$f_6 = \dot{m}_{\mathrm{a}}\, C_{\mathrm{pa}}\,(T_{\mathrm{a,i}} - T_{\mathrm{a,o}}^0) - q^0 \approx 0$$

or

$$f_6 = 3.5 \times 1.0\,(30 - 14) - 50 = 6$$

From Eq. (E9.4.7a),

$$f_7 = \dot{m}_{\mathrm{w}}^0\, C_{\mathrm{pw}}\,(T_{\mathrm{w,0}}^0 - T_{\mathrm{w,i}}) - q^0$$

or

$$f_7 = 2.0\,(4.2)\,(18 - 10) - 50 = 17.2$$

Let us also compute all elements of the coefficient matrix of the system using the initial guess values. Based on the initial guess values of the unknown variables, the system of equations can be written as follows:

$$
\begin{bmatrix}
-1 & -64\,000 & 0 & 0 & 0 & 0 & 0 \\
0.0001 & -4 & 200 & -0.0001 & 0 & 0 & 0 \\
0 & 36\,000 & 0 & -1 & 0 & 0 & 0 \\
0 & 0 & -2 & 0 & 0.015 & 0 & 0 \\
0 & 0 & 0 & 0 & -2.5 & 1.098 & 5.833 \\
0 & 0 & 0 & 0 & -3.0 & -1 & 0 \\
0 & 33.6 & 0 & 0 & 0 & -1 & 8.4
\end{bmatrix}
\begin{Bmatrix}
\Delta P_2 \\
\Delta \dot{m}_w \\
\Delta K_{CV} \\
\Delta P_3 \\
\Delta T_{a,0} \\
\Delta q \\
\Delta T_{w,0}
\end{Bmatrix}
=
\begin{Bmatrix}
-6000 \\
-3.0 \\
11\,000 \\
0 \\
25.07 \\
6 \\
17.2
\end{Bmatrix}
$$

$$\text{(E9.4.11)}$$

Solving the system given by Eq. (E9.4.11), we get the correction values for the unknown variables:

$$
\begin{Bmatrix}
\Delta P_2 \\
\Delta \dot{m}_w \\
\Delta K_{CV} \\
\Delta P_3 \\
\Delta T_{a,0} \\
\Delta q \\
\Delta T_{w,0}
\end{Bmatrix}
=
\begin{Bmatrix}
5470.5 \\
0.000\,00 \\
-0.000\,00 \\
-10\,702.16 \\
-3.056\,11 \\
3.168\,35 \\
2.391\,711
\end{Bmatrix}
$$

$$\text{(E9.4.12)}$$

A MATLAB® solution program is given below:

MATLAB Solution: Ax = C

```
format long
A = [-1 -64000 0 0 0 0 0; 0.0001 -4 200 -0.0001 0 0 0;
0 36000 0 -1 0 0 0; 0 0 -2 0 0.015 0 0; 0 0 0 0 -2.5
1.098 5.833; 0 0 0 0 -3 -1 0; 0 33.6 0 0 0 -1 8.4];
C = [-6000; -3; 11000; 0; 25.07; 6; 17.2];
X = linsolve(A,C);

X =

   1.0e+04 *

   0.547052184319236
   0.000000827309620
  -0.000002292088327
  -1.070216853679570
  -0.000305611776880
   0.000316835330640
   0.000239171158025
```

Corrected values after first iteration:

$$P_2^{(1)} = P_2^{(0)} + \Delta P_2 = 30\,000 + 5470.52 = 35\,470.52$$

$$\dot{m}_w^{(1)} = \dot{m}_w^{(0)} + \Delta \dot{m}_w = 2.0 + 0.0000 = 2.0$$

$$K_{CV}^{(1)} = K_{CV}^{(0)} + \Delta K_{CV} = 0.01 - 0.000\,002\,29 = 0.009\,99$$

$$P_3^{(1)} = P_3^{(0)} + \Delta P_3 = 20\,000 - 10\,702.1 = 9297.9$$

$$T_{a,0}^{(1)} = T_{a,0}^{(0)} + \Delta T_{a,0} = 14 - 3.0561 = 10.9439$$

$$q^{(1)} = q^{(0)} + \Delta q = 50.0 + 3.16835 = 53.168$$

$$T_{w,0}^{(1)} = T_{w,0}^{(0)} + \Delta T_{w,0} = 18.0 + 2.3917 = 20.39$$

Repeat the steps to recompute the function values and partial derivatives and solve the system of equations (E9.4.10) using the new iterated values. Repeat the steps until the convergence limit is reached.

9.7 Some Facts About the Newton–Raphson Method

One of the major advantages of the Newton–Raphson method is that it is not sensitive to the selection of the calculation sequence of the function equations or the selection of the information flow diagram. Other major characteristics of the Newton–Raphson method is that the method ensures speedy convergence in most cases when the initial guess value is close to the true solution. This is demonstrated graphically in Figure 9.7. However, the method may not converge in many situations such as in the cases where the derivative or the slope of the function is close to zero or involves sharp variation, or where the initial guess values cannot be guessed close to the true solution value and the method converges as demonstrated in Figure 9.8. These difficulties can be mitigated by starting with a new guess value.

Because of this reason, the Newton–Raphson method is often preceded by one of the searching or bracketing root finding algorithms such the Bisection method (James et al. 1985; Rao 2002) or watching the graphical representations of the functions associated with slow converging variables. To enhance the convergence rate, often an under-relaxation scheme is used within the Newton–Raphson algorithm for some variables that may vary by a large amount during the few initial iteration steps. The following is a typical *successive*

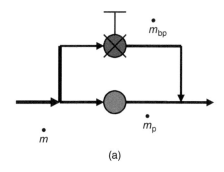

(a)

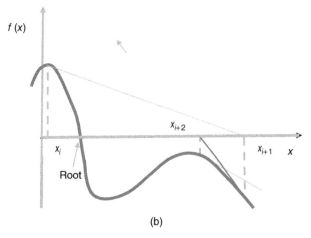

(b)

Figure 9.8 Example case of no convergence.

under-relaxation scheme that can be implemented in the Newton–Raphson algorithm for the solution of a system of nonlinear equations (Majumdar 2005):

$$x_{1,k+1} = \lambda x_{1,k+1} + (1 - \lambda) x_{1,k}$$
$$x_{1,k+1} = \lambda x_{1,k+1} + (1 - \lambda) x_{1,k}$$
$$\vdots$$
$$x_{1,k+1} = \lambda x_{1,k+1} + (1 - \lambda) x_{1,k} \tag{9.12}$$

where λ = the **successive under-relaxation parameter**

Typical values of the under-relaxation parameter lie in the range of $0 < \lambda < 1$. An optimum value of the under-relaxation parameter is selected based on several numerical experimentations conducted.

9.8 Numerical Evaluations of Partial Derivatives in System Simulation

The general Newton–Raphson method described so far requires evaluation of derivates of the functions. While the derivatives can be easily evaluated for polynomial functions, differentiation of some of the functions may be quite complex for those involving transcendental functions and those not expressible in explicit forms. In such cases, the derivatives can be evaluated using numerical methods such as finite difference approximations (Majumdar 2005). Such an approach may also be quite advantageous for developing a generalized computer program implementing the Newton–Raphson method for solving a nonlinear system of equations. A first-order forward-difference finite difference approximation scheme (Majumdar 2005) can be used to numerically evaluate the derivatives in the following manner:

$$\frac{\partial f_j}{\partial x_i} = \frac{f_j(x_1, \ x_2, \ldots, x_i + \Delta x_i, \ldots, x_n) - f_j(x_1, \ x_2, \ldots, x_i, \ldots, x_n)}{\Delta x_i} \tag{9.13}$$

A subroutine, ***parderiv*** could be utilized to calculate the partial derivatives of all functions.

9.9 Different Solution Options for a Linear System of Equations

We have noticed that one of the essential steps in the solution of a nonlinear system of equations using the Newton–Raphson method requires solving a linear system of equations Eq. (E9.9a), given as

$$\begin{bmatrix} \dfrac{\partial f_1}{\partial x_1} & \cdots & \dfrac{\partial f_1}{\partial x_n} \\ \vdots & \ddots & \vdots \\ \dfrac{\partial f_n}{\partial x_1} & \cdots & \dfrac{\partial f_n}{x_n} \end{bmatrix} \begin{Bmatrix} x_{1,i+1} - x_{1,i} \\ \vdots \\ x_{n,i+1} - x_{n,i} \end{Bmatrix} = \begin{Bmatrix} f_1(x_{1,i} \ldots x_{n,i}) \\ \vdots \\ f_n(x_{1,i} \ldots x_{n,i}) \end{Bmatrix} \tag{9.9a}$$

or

$$A \, \Delta \mathbf{x} = \mathbf{c} \tag{9.9b}$$

where

$$A = \begin{bmatrix} \dfrac{\partial f_1}{\partial x_1} & \cdots & \dfrac{\partial f_1}{\partial x_n} \\ \vdots & \ddots & \vdots \\ \dfrac{\partial f_n}{\partial x_1} & \cdots & \dfrac{\partial f_n}{x_n} \end{bmatrix}$$

$$\Delta \mathbf{x} = \begin{Bmatrix} x_{1,i+1} - x_{1,i} \\ \vdots \\ x_{n,i+1} - x_{n,i} \end{Bmatrix}$$

$$\mathbf{c} = \begin{Bmatrix} f_1(x_{1,i} \ldots x_{n,i}) \\ \vdots \\ f_n(x_{1,i} \ldots x_{n,i}) \end{Bmatrix}$$

There are a number of solution methods that can be used for solving such linear systems of equations. A direct solver such as the Gaussian elimination method or an iterative solver such as the Gauss–Seidel method can be used for solving Eq. (9.9). Although the Gauss elimination solver is the most popular linear solver it may be limited to relatively smaller sizes of the matrix due to limitations of computer storage and time. For larger systems of equations, the Gauss–Seidel method or other iterative solvers are widely used. A Gauss elimination solver subroutine, **GAUSS** could be used for developing computer programs for thermal system simulation using the Newton–Raphson method.

9.10 A Generalized Newton–Raphson Algorithm for System Simulation

A generalized Newton–Raphson algorithm for developing a thermal system simulation program is summarized in an algorithm flow diagram shown in Figure 9.9.

A generalized computer program implementing the Newton–Raphson algorithm for steady-state simulation of thermal systems can easily be developed following the flow chart outlined in Figure 9.9.

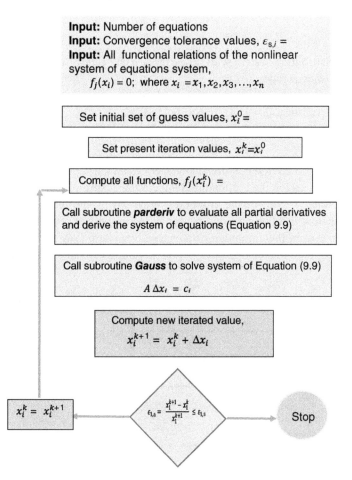

Figure 9.9 A flow chart for the generalized Newton–Raphson algorithm for a system simulation.

Bibliography

Bathie, W.W. (1996). *Fundamentals of Gas Turbines*, 2e. New York, NY: Wiley.

Chapra, S.C. and Canale, R.P. (1996). *Numerical Methods for Engineers*, 5e. New York, NY: McGraw-Hill.

Ferguson, C.R. and Kirkpatrick, A.T. (2016). *Internal Combustion Engines*, 3e. Wiley.

Harmon, R.T.C. (1982). *Gas Turbine Engineering*. New York, NY: Macmillan.

Hodge, B.K. and Taylor, R.P. (1999). *Analysis and Design of Energy Systems*, 3e. Prentice-Hill.

Jaluria, Y. (1997). *Design and Optimization of Thermal Systems*. New York, NY: McGraw-Hill.

James, M.L., Smith, G.M., and Wolford, J.C. (1985). *Applied Numerical Methods for Digital Computation*, 3e. New York, NY: Harper Collins Publishers.

Majumdar, P. (2005). *Computational Methods for Heat and Mass Transfer*. New York, NY: Taylor & Francis.

Rao, S. (2002). *Applied Numerical Methods for Engineers and Scientists*. New Jersey, NJ: Prentice Hall.

Stoecker, W.F. (1971). A generalized program for steady-state system simulation. *ASHRAE Transactions* 77: 140–148.

Stoecker, W.F. (1989). *Design of Thermal Systems*, 3e. New York: Mc-Graw Hill.

Problems

9.1 Consider an air flow distribution system with duct pressure characteristics and fan performance characteristics as follows:

Duct flow pressure drop characteristics: $\Delta P = 92.0 + 15.2\, Q^{1-92}$

Fan performance characteristics: $Q = 23.4 - (0.6 \times 10^{-4})(\Delta P)^2$

Use the Newton–Raphson method to determine the rated operating point in terms of flow rate Q and pressure ΔP of the system.

9.2 The chilled water system shown in the figure below operates on the vapor compression refrigeration system between the evaporation temperature T_e and condensation temperature T_c. The water stream in the evaporator is cooled by the refrigerant, undergoing two-phase flow and boiling heat transfer in the evaporator. The superheated refrigerant vapor from the compressor undergoes two-phase flow and heat transfer in the condensation, transferring heat to the cooling water stream returning from the cooling tower.

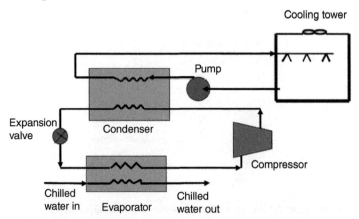

Data on individual component performances are as follows:

The overall heat transfer coefficient value and size of the two heat exchangers are given as follows, for the condenser UA = 25 000, and for evaporation UA = 18 000. The refrigeration or cooling capacity of a vapor compression refrigeration system is presented in the table below:

Evaporator cooling capacity, q_e (kW).

Evaporative temperature, T_e (°C)	Condensing temperature, T_c (°C)		
	25	35	45
0	150.5	115.3	82.5
5	181.8	140.6	101.1
10	214.6	169.1	126.2

Determine the parametric representation (see Appendix A) of the refrigeration capacity, q_e, as a function of the evaporation temperature, T_e, and the condenser temperature, T_c, using the data given in the table using the following form:

$$q_e = a_0 + a_1 T_e + a_2 T_e^2 + a_3 T_c + a_4 T_c^2 + a_5 T_e T_c + a_6 T_e^2 T_c + a_7 T_e T_c^2 + a_8 T_e^2 T_c^2$$

Determine the parametric representation (see Appendix A) of the refrigeration capacity, q_e, as a function of the evaporation temperature, T_e, and the condenser temperature, T_c, using the data given in the table below using the following form:

$$w_c = a_0 + a_1 T_e + a_2 T_e^2 + a_3 T_c + a_4 T_c^2 + a_5 T_e T_c + a_6 T_e^2 T_c + a_7 T_e T_c^2 + a_8 T_e^2 T_c^2$$

Compressor power, w_c (kW)

Evaporative temperature, T_e (°C)	Condensing temperature, T_c (°C)		
	25	35	45
0	36.7	28.12	20.12
5	44.56	34.46	24.78
10	52.86	41.65	31.08

Determine the parametric representation (see Appendix A) of the refrigeration capacity, q_e, as a function of the evaporation temperature, T_e, and the condenser temperature, T_c, using the data given in the table using the following form:

$$w_c = a_0 + a_1 T_e + a_2 T_e^2 + a_3 T_c + a_4 T_c^2 + a_5 T_e T_c + a_6 T_e^2 T_c + a_7 T_e T_c^2 + a_8 T_e^2 T_c^2$$

Determine the evaporator temperature, T_e, and the condenser temperature, T_c, by conducting a steady-state simulation of the refrigeration chiller system assuming the water inlet temperature to the evaporator as $T_{ew,i} = 8°C$ and cooing water temperature at the inlet to the condenser as $T_{cw,i} = 20°C$.

9.3 A centrifugal pump operates with two bypass lines as shown in the figure below. The pressure drop through the bypass line with the control valve is given by the equation

$$\Delta P = 1.2 \, (m_{bp})^2$$

The pump characteristics are given by the equation

$$\Delta P = 60 + 5 \, m_p - 0.1 \, m_p^2$$

and the equation for the pressure drop in this piping system is

$$\Delta P = 0.018 \, m_p^2$$

The pressure drop in the line with the heat exchanger is given by

$$\Delta P = 0.5 \, m_{hx}^2$$

where the flow rates are in kg/s and the pressure drop in kPa.

(a) Develop the system of equations for the correction values of the variables $\dot{m}, \dot{m}_p, \dot{m}_{bp}, \dot{m}_{hx}$ and ΔP using the Newton–Raphson technique.

(b) Use initial guess values as and iterate to determine these values with a tolerance limit of 0.01. (Show at least two iterations based on hand calculation).

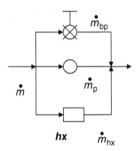

9.4 In a steam power plant condenser, shown in the figure below, steam undergoes two-phase condensation heat transfer and heats an incoming water stream entering at 28°C. Condenser heat transfer rate and the size are given as 100 kW and 2.2 m², respectively. The overall heat transfer coefficient U in the condenser as a function of water mass flow rate $\dot{m}$ is expressed as $U = 18.6 \, \dot{m}^2$, where U is in kW/m² K and $\dot{m}$ is in kg/s.

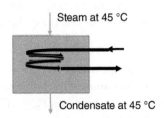

Steam at 45 °C

Condensate at 45 °C

9.5 Consider the simulation of a standard ideal gas turbine cycle as shown below:

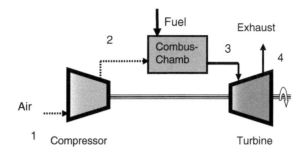

Let us consider the following data set:

Temperature and pressure data set at the inlet to the turbine – state 1:

Ambient pressure at the exit of the compressor exhaust – state 4:

Heat input in the combustion chamber, $Q_H = 10,000$ kW

Turbine performance characteristics: $\dot{m} = f(P_2)$ and $w_c = f(P_2)$

Turbine discharge pressure vs. flow rate		Discharge pressure vs. compressor power	
m (kg/s)	Discharge pressure (kPa)	Discharge pressure (kPa)	Compressor power (kW)
8	450	500	1900
12	300	350	1400
16	120	200	1175

Turbine characteristics can be represented in the following quadratic curve-fitted form:

$$P_2 = a_0 + a_1 \dot{m} + a_2 \dot{m}^2 \text{ and } w_c = b_0 + b_1 P_2 + b_2 P_2^2$$

Compressor performance characteristics:

Turbine mass flow rate: $\dot{m} = f(P_2, T_3)$

1200°C		1000°C		800°C	
Inlet pressure, P	Flow rate, kg/s	Inlet pressure, P	Flow rate, kg/s	Inlet pressure, P	Flow rate, kg/s
200	6.2	200	6.5	200	6.8
300	9.5	300	9.8	300	10.2
400	10.2	400	11.0	400	12.1

Parametric representation can be obtained in the following form:

$$\dot{m} = c_0 + c_1 P_2 + c_2 P_2^2 + c_3 P_2 + c_4 P_2^2 + c_5 P_2 T_3 + c_6 P_2^2 T_3 + c_7 P_2 T_3^2 + c_8 P_2^2 T_3^2$$

Turbine Power, $w_t = f(P_2, T_3)$

1200°C		1000°C		800°C	
Inlet pressure, P	Turbine power (kW)	Inlet pressure, P	Turbine power (kW)	Inlet pressure, P	Turbine power (kW)
200	1200	200	2210	200	3400
300	1240	300	2255	300	3470
400	1290	400	2215	400	3540

Parametric representation can be obtained in the following form:

$$w_t = d_0 + d_1 P_2 + d_2 P_2^2 + d_3 P_2 + d_4 P_2^2 + d_5 P_2 T_3 + d_6 P_2^2 T_3 + d_7 P_2 T_3^2 + d_8 P_2^2 T_3^2$$

Perform a steady system simulation of the gas turbine power system to determine the following variables in the system: air-gas flow rate, $\dot{m}$; compressor exhaust pressure and temperature, P_2 and T_2; compressor power input, w_c; combustion chamber exit temperature, T_3; turbine power output, w_t; and net power out of the system, w_{net}. Assume air inlet conditions as $T_1 = 20°C$ and $P_1 = 98\,kPa$, and turbine exhaust, $P_4 = 102\,kPa$. Follow the following steps to perform the simulation: (i) Determine all transfer function equations and the system of equations needed to determine all the simulation variables $\dot{m}$, P_2, T_2, w_c, T_3, w_t, and w_{net}; (ii) perform the steady-state simulation and show all the iterated values of all variables until a close tolerance. *Note*: Assume the fluid as an ideal gas and mass flow rate as constant with negligible variation due to the addition of fuel in the combustion chamber.

9.6 Repeat Problem 9.5, assuming a regenerative gas turbine cycle as shown below:

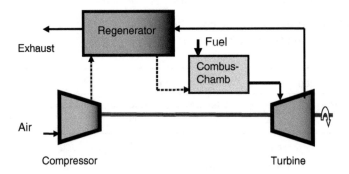

Consider the regenerator as a gas–gas heat exchanger with rated UA = 600 W/m² K. Perform the steady-state simulation and show unknown values including the state points at all inlets and outlets of the preheated air stream and the regenerative exhaust stream for all iterations until convergence. Comment on the improvement made in using the regenerative heat exchanger.

9.7 Develop a computer program implementing the Newton–Raphson algorithm and show an iterative solution of Example 9.4 considering a tolerance limit of 0.1% for all variables. Present a summary table showing the iterative values of all variables along with the changes in the percent relative errors for all variables.

9.8 Consider the solar water heating system problem outlined in Example 9.2. A solar collector mounted on the rooftop of a single-family house is used to run a solar domestic water heating system as shown in Figure 9.5. Flat plate solar collectors are used to heat water that is circulated through a pump-piping system. Assume the following parameters for the system to perform a steady-state simulation of the system:

Solar collector: 3-ft × 5-ft single cover glass-glazing cover flat plate solar collector.
Transmittance–absorptance product: $(\tau\alpha) = 0.82$
Collector heat removal factor: $F_R = 0.92$.
Total incident solar irradiation: $G_t = 900 \text{ W/m}^2$
Collector working fluid: Glycol water
Heater water inlet and outlet temperatures: $T_{wi} = 20°\text{ C}$ and $T_{wo} = 40°\text{ C}$
Collector overall heat transfer coefficient: $U_t = 15 \text{ W/m}^2 \text{ K}$
Ambient temperature: $T_a = 15°\text{ C}$

10

Optimization of Thermal Components and Systems

Optimization process is a critical step in the design and development of a process or a device or a system. It is the process of determining a set of variables or parameters that leads to an optimum design, i.e. the maximum or minimum of a function that characterizes the performance of the system. The major parameters and variables are varied until a combination is reached that optimizes the design. The maximum and minimum values of a representative analysis function, often a cost function or a performance function, are normally sought subject to some constraints such as the minimum weight for a space vehicle or an aircraft; size or minimum volume of the system such as in automobiles or electronics devices. Component design and system simulations are often performed before engaging into the process of optimizing a thermal system. For small projects, the cost may often be difficult to justify for engaging into a comprehensive optimization process. Often, the optimization step is considered as an integral part of the iterative design cycle subject to some given specifications. Through an iterative process, designers can evaluate many potential solutions of the system based on cost effectiveness, efficiency, reliability, and durability of the system.

Based on our discussion on thermal devices and thermal systems in Chapters 4–8, a few examples of optimization problems can be given here: (i) Optimum size and shape of fins based on optimum fin efficiency or fin effectiveness subject to a constraint such as the minimum weight for aerospace applications or subject to the minimum volume for use in electronics cooling.

Cost estimation, both capital and operating, and economic analysis play a critical role in the evaluation of cost-effective design. Optimization studies often require some form of functional relationships of cost as a function of some key component or system parameters. Readers are suggested to study the brief review of economic evaluation and cost estimation given in Appendix B.

10.1   Optimization Analysis Models

An optimization analysis model uses *input variables* and computes the *analysis functions* as outputs. A subset of the *analysis input variables* is considered as the *design variables* that are subjected to change to achieve an optimum function value. The rest of the analysis variables are assigned with some constant values. The analysis functions are nothing but

Design of Thermal Energy Systems, First Edition. Pradip Majumdar.
© 2021 John Wiley & Sons Ltd. Published 2021 by John Wiley & Sons Ltd.
Companion website: www.wiley.com/go/majumdar

Figure 10.1 Straight fin of rectangular profile for optimization.

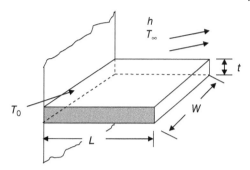

the predictive engineering models for a design problem such as the predictive models for heat transfer, pressure drop, and thermal efficiency.

As an example, let us assume all the major components of the optimization analysis model for a fin design problem as follows.

For a straight fin of rectangular profile, shown in Figure 10.1, some of the **predictive analysis models** are the fin heat loss, fin efficiency, and fin weight defined as follows:

Fin heat loss:

$$q_f = \sqrt{hPkA_c} \frac{\sinh mL + (h/mk)\cosh mL}{\cosh mL + (h/mk)\sinh mL}(T_0 - T_\infty) \tag{10.1}$$

Fin efficiency:

$$\eta_f = \frac{\dfrac{\sinh mL + (h/mk)\cosh mL}{\cosh mL + (h/mk)\sinh mL}}{mL} \tag{10.2}$$

and

Fin weight:

$$w_f = \rho_f LWt \tag{10.3}$$

where

$$m = \sqrt{\frac{hP}{kA_c}}, A_c = Wt \text{ and } P = 2(W + t) \tag{10.4}$$

Analysis functions: Fin heat transfer rate, fin efficiency, and fin weight.

Analysis variables: Fin length, L; fin width, W; thickness, t; fin material conductivity, k; convection heat transfer coefficient, h; surrounding ambient temperature, T_∞; and fin base temperature, T_0.

Design variables: Fin length, L, and fin width, W. Other variables such as material type or thermal conductivity, convection coefficient, and the ambient temperature can be assumed.

10.2 Formulation and Mathematical Representation of Optimization Problems in Thermal Systems

Formulation of an optimum design problem always involves describing the problem in the form of a mathematical statement, which includes an **objective function** to judge the design; a set of **constraints** within which the system must perform; and a set of identified **design variables**. A typical mathematical statement of an optimization problem is comprised of an **objective function** that characterizes the cost or the performance, and the **constraints** that arise from the physical conditions or restrictions. An **objective function** is a dependent variable stated as a function of one or several independent variables as follows:

$$f = \mathbf{f}(x_1, x_2, \ldots x_n) \tag{10.5}$$

where

        f=objective function representing a criterion such as cost or performance characteristics of the system

$x_1, x_2, \ldots x_n$=Independent variables representing operating variables and/or parameters

In order to start a design optimization process, a designer needs to identify threes items: (i) an objective function in the form of a performance function to be maximized or in the form of an cost function to be minimized, (ii) a set of design variables, and (iii) the constraint functions.

10.2.1 Analysis and Design Variables

Analysis variables are the parameters that characterize the behavior of a system and define the design of the system. As mentioned, often a few parameters are assigned with constant values to reduce the number of parameters that are varied in the optimization process. **Design variables** are a subset of the analysis variables, which are defined as the variables that are free to vary during the optimization process, and each combination of these design variable values results in a specific design. Not all such designs are, however, feasible or workable. Some combinations of these parameters that satisfy all assigned constraints result in feasible or workable designs. One such workable design that results in a minimum or maximum value of the objective function is defined as the optimum design.

10.2.2 Objective Function

An objective function or a criterion is used for comparing various designs to search for an optimum design. It is expressed in terms of the design variables and attains a numerical value when the final optimum design is achieved. Often, the objective function is referred to as the cost function, which is used as the criterion to identify a design with minimum total cost of the system. For an optimum design, the objective function can either be minimized or maximized depending on the nature of the problems. If one is looking for an optimum design in terms of enhanced performance such as efficiency or thermal effectiveness of a thermal system or device, the objective function is written in the form of a performance function, and the function is maximized to reach for an optimum design. On the other hand, for problems where we are looking for a design with minimum cost of the system or minimum weight or minimum pressure drop, the objective function or the cost function is minimized to reach an optimum design. In some situations, we may have more than one objective function. For example, in the case of a solar thermal system design, we may like to achieve both maximum thermal efficiency and minimum cost of the system. One common way to treat such a problem is to maximize the performance function and use the cost function as a constraint. For the design of an optimized active cooling system, we can attempt to maximize the convective heat transfer coefficient subject to minimum pressure drop as a constraint. Another way to address such optimization problems with multiple objective functions is to compose a ***composite objective function*** that may involve the weighted sum of all objective functions. Selection of the weighting factors is critical as it influences the resulting optimum design.

Another important aspect of the design optimization process is that the system may be complex involving many subsystems or devices. A general approach is to analyze and optimize the design of each one of the subsystems independently and then analyze and optimize the entire system integrating all subsystems. For example, in the optimum design of a compact heat exchanger, one needs to analyze different types of fins and optimize the fin design in terms of the shape, length, width, and thickness. The compact heat exchanger is then optimized in terms of heat transfer and pressure drop considering finned tubes or channels and considering channel shape, size, length, and fin spacing, and the number of fins and their orientations.

To define an objective function in the form of cost of a thermal system, it is also essential to have a comprehensive knowledge of the relationships of the cost of different components with its major parameters. Readers are requested to refer to Appendix B on "Economic Analysis and Cost Estimation of the Thermal Components and Systems."

10.2.3 Design Constraints

Constrains are the restrictions placed on a design while searching for an optimum design.

There are different categories of the design restrictions: (i) Linear and nonlinear constraints and (ii) equality and inequality constraints.

10.2.3.1 Equality and Inequality Constraints

Design problems may involve equality as well inequality constraints as given by the following examples:

Equality constraints:

$$cx_1 = Q \tag{10.6}$$

Inequality constraints

$$cx_1 \leq Q \tag{10.7a}$$

or

$$cx_1 \geq Q \tag{10.7b}$$

where

$x_1 = $ design variable
$c = $ constant
$Q = $ limiting value

One example of such an inequality constraint in thermal system design is that the calculated exit temperature of the cooling stream should be less than some target temperature value such as the freezing temperature to prevent frosting or less than the dew point temperature to prevent condensation.

The inequality constraints can be a less-than inequality or a greater-than inequality as defined by Eqs. (10.7a) or (10.7b) respectively. The sign notation can be changed from the less-than constraint to a greater-than constraint and vice versa by just multiplying both sides of the constraint function equation by a factor of −**1**.

10.2.3.2 Linear and Nonlinear Constraints

Linear constraints only involve linear relationships among dependent variables, and the mathematical expression contains only the first-order terms. An example of such a linear constraint can be given as follows:

$$ax_2 + bx_3 < M \tag{10.8}$$

where

$x_2, x_3 = $ design variables
  $a, b = $ weighting factors or constants
    $M = $ Constant representing a limiting value

10.2.3.3 Nonlinear Constraints

Nonlinear constraints consist of second-order or higher order terms representing nonlinear relationships among design variables. An example of a nonlinear constraint can be given as follows:

$$cx_4^d x_5 > N \tag{10.9}$$

where

$x_4, x_5 = $ design variables
  $c, d = $ weighting factors or constants
    $N = $ Constant representing a limiting value

10.2.4 Implicit Constraints

Implicit constraints involve a complex and indirect relationship with some design variables, $\mathbf{x}$, as well as some dependent variable, $\mathbf{Y(x)}$, which itself depends on the design variable. A mathematical statement of an implicit constraint for an optimization problem is given as follows:

$$g_j(\mathbf{x}, \mathbf{Y}) \le 0 \tag{10.10}$$

where j is the index number for several implicit constraints.

For example, in the optimum design of a fin in terms of maximum fin efficiency, η_i, the fin heat loss or heat transfer rate at the base of the fin, q_f, and fin efficiency, η, is computed based on the temperature solution $T(\mathbf{x})$, which itself depends on some design variable, $\mathbf{x}$, such as the fin shape or profile and fin materials.

10.2.5 Formulation of the Optimization Problem

A basic outline for the formulation of the optimization problem is given here: (i) Description of the design problem and design alternative or options; (ii) definition of an objective function: This will require establishing a relationship or correlation between the major performance quantities and the controlling design variables – for example, for the objective function in terms of cost, a relationship between the cost of the components or systems with the major performance quantities such as the ratings of the heat transfer, pressure drop, sizes of the components, and operating design variables; (iii) identification of all the analysis variables and a set of design variables; (iv) definition of design constraints and identification of the type of constraints; (v) use of an optimization method to analyze the selected designs. Figure 10.2 shows a flow chart for the design optimization process.

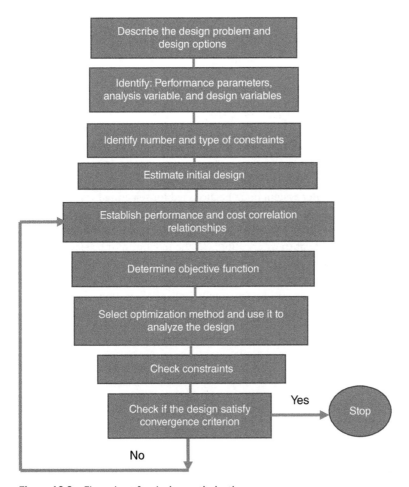

Figure 10.2 Flow chart for design optimization process.

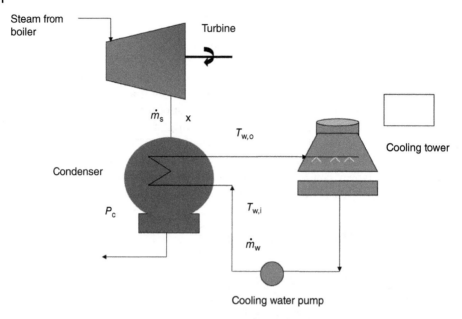

Figure 10.3 Optimum condenser design for a secondary water system.

Example 10.1 *Example of Optimization process for Heat Exchanger Design*
Let us now consider the design optimization process for the optimum design of a condenser in a secondary water system in a power plant as shown in Figure 10.3.

A condenser in a secondary water system must be designed for rejecting heat in the condenser for a steam power plant. The steam flow rate, $\dot{m}_s$, and the quality of the steam, x, at the inlet to the condenser are known. It is desired to design the condenser for a minimum first cost subject to the assigned heat transfer rating, Q_c, the maximum tube side pressure drops, $\Delta P_t < \Delta P_{t, max}$, and the maximum tube length, $L < L_{max}$.

In the analysis of the condenser a number of analysis functions could be computed after selecting several analysis variables representing geometrical variables, flow variables, and thermo-physical properties. A list of the analysis functions and analysis variables is given below:

Analysis Variables
Shell diameter, D_s
Tube size, d_o and d_i
Tube length, L
Tube spacing, S_L and S_T
Number of tube passes, $N_{p, t}$
Number of shell pass, $N_{p, t}$
Number of tubes per tube pass, $n_{t, p}$
Tube material construction and properties

Baffle spacing, S_b

Number of baffles, N_b

Steam flow rate, $\dot{m}_s$

Steam quality, x

Condenser pressure, P_c

Cooling water flow rate, $\dot{m}_w$

Cooling water inlet and outlet temperatures, $T_{w,i}$ and $T_{w,o}$

Cooling tower range, R and Approach, A

Design outdoor dry bulb temperature, T_{db}, and wet bulb temperature, T_{wb}

Analysis Function

Condenser cost, C_{cond}

Condenser heat transfer rate, Q_c

Condenser tube side pressure drop, ΔP_t

In a design analysis, some of the analysis variables such fluid properties, material properties, a few of the geometrical variables, and a few of the flow variables are assumed and set as constant, and some others are considered as design variables, which are varied during design analysis computations. A list of design functions and analysis variables is given below:

Design Variables

Shell diameter, D_s

Number of tube passes, $N_{p,t}$

Tube size, d_o and d_i

Tube length, L

Baffle spacing, S_b

Design Function

Objective function:
    Minimize – Condenser cost, C_{cond}

Constraints

Heat transfer rating, Q_c

Tube length, $L < L_{max}$

Condenser tube side pressure drop, $\Delta P_t < \Delta P_{t,max}$

10.2.6 General Mathematical Statement of Optimization Problems

A general statement of an optimization is given in terms of objective function, and design constrains as a function of design variables.

Objective Function:

$$f = f(x_1, x_2, x_3, \ldots, x_n) \tag{10.11}$$

Design Constraints:

Equality constraints:

$$g_1 = g_1(x_1, x_2, x_3, \dots, x_n)$$
$$g_2 = g_2(x_1, x_2, x_3, \dots, x_n)$$
$$\vdots$$
$$g_m = g_m(x_1, x_2, x_3, \dots, x_n) \tag{10.12}$$

Inequality constraints:

$$h_1 = h_1(x_1, x_2, x_3, \dots, x_n)$$
$$h_2 = h_2(x_1, x_2, x_3, \dots, x_n)$$
$$\vdots$$
$$h_0 = h_0(x_1, x_2, x_3, \dots, x_n) \tag{10.13}$$

10.2.7 Examples of Design Optimization Problems

Let us consider a few examples to illustrate the mathematical statement of a thermal system optimization problem.

Example 10.2 *Solar Thermal Power Generation*
A solar-driven power generating system, shown in Figure 10.4, is to be used to generate 5 MW of mechanical power to run an electric generator. The collector side fluid is circulated using a pump and heated in the collector to a high temperature level. The high temperature collector fluid is circulated through the heat exchanger/boiler to heat the Rankine cycle fluid and produces high temperature and pressure steam to produce power by expanding through a steam turbine. Collector fluid is heated to high temperature, but not exceeding 400 ° C, and used to produce high temperature and pressure steam to run a steam turbine and produce power. The condenser pressure is assumed to be 20 kPa.

The cost of the solar collector is expressed as function of the area, A_c, as

$$C_{Col} = C_{r,c} A_c^{m,c}$$

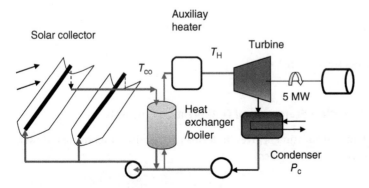

Figure 10.4 Solar thermal power generation.

The cost of the boiler/heat exchange is expressed as a function of heat transfer from the collector fluid to the water side as

$$C_{\text{Boiler}} = C_{\text{r,b}} Q_{\text{B}}^{\text{m,b}}$$

Cost of the auxiliary heater:

$$C_{\text{Aux}} = C_{\text{r,a}} Q_{\text{Aux}}^{\text{m,a}}$$

Cost of the turbine:

$$C_{\text{Turbine}} = C_{\text{r},T-1} \dot{m}_{\text{w}T}^{\text{m},T} + C_{\text{r},T-2} \left(\frac{T_{\text{H}}}{P_{\text{c}}} \right)^{\text{m},T}$$

Cost of the collector fluid pump:

$$C_{\text{Pump-C}} = C_{\text{r,cp}} \dot{m}_{\text{c}}^{\text{m,c}}$$

Cost of the feed water pump:

$$C_{\text{Pump-w}} = C_{\text{r,wp}} \dot{m}_{\text{w}}^{\text{m,w}}$$

Total cost is the sum of all the major components in the system:

$$C(A_{\text{c}}, Q_{\text{B}}, Q_{\text{Aux}}, \dot{m}_{\text{w}T}, T_{\text{H}}, \dot{m}_{\text{c}}, \dot{m}_{\text{w}}) = C_{\text{r,c}} A_{\text{c}}^{\text{m,c}} + C_{\text{r,b}} Q_{\text{B}}^{\text{m,b}} + C_{\text{r,a}} Q_{\text{Aux}}^{\text{m,a}}$$
$$+ C_{\text{r},T-1} \dot{m}_{\text{w}T}^{\text{m},T} + C_{\text{r},T-2} \left(\frac{T_{\text{H}}}{P_{\text{c}}} \right)^{\text{m},T}$$
$$+ C_{\text{r,cp}} \dot{m}_{\text{c}}^{\text{m,c}} + C_{\text{r,wp}} \dot{m}_{\text{w}}^{\text{m,w}} \qquad \text{(E10.2.1)}$$

This total cost is defined as the ***objective function,*** which is required to be minimized.

The ***first constraint function*** is given based on the minimum required power out of the turbine, i.e.

$$W_{\text{T}}(A_{\text{c}}, Q_{\text{B}}, Q_{\text{Aux}}, \dot{m}_{\text{w}T}, T_{\text{H}}, \dot{m}_{\text{c}}, \dot{m}_{\text{w}}) \geq 5 \text{ MW} \qquad \text{(E10.2.2)}$$

The second constraint function is given based on the requirement that the collector fluid exit temperature should not exceed $400\,^\circ$C, i.e.

$$T_{\text{Col}} = T_{\text{Col}}(A_{\text{c}}, Q_{\text{B}}, Q_{\text{Aux}}, \dot{m}_{\text{w}T}, T_{\text{H}}, \dot{m}_{\text{c}}, \dot{m}_{\text{w}}) \leq 400\,^\circ\text{C} \qquad \text{(E10.2.3)}$$

Example 10.3   *Solar Thermal Power Generation*

The objective function and the constraint function for the optimization study of the solar heater system shown below:

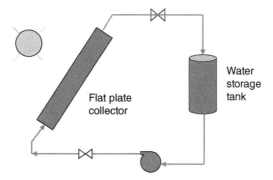

Flat plate collector

Water storage tank

The flat plate solar collector receives an average total solar irradiation of 260.0 $G_t =$ 260 W/m² and heats the circulating cooling water from 25° C to a high temperature, T_H, and stores it in the water storage tank. The collector optical parameter and the overall heat transfer coefficient are given as $\tau\alpha = 0.90$ and $U = 2.0$ W/m² K respectively. The thermal energy stored in the tank during the 10 hours daylight time is 200 MJ. The cost function for the solar collector as a function of the area is given as $C_{collector} = 20\,A_C$, where $A_C =$ area of the collector. The cost function of the storage tank is given as $C_{Storage} = 101.5\,V_T$, where V_T is the volume of the water storage tank. Derive the objective function and the constraint function, considering the area of the collector and the volume of the storage tank as the two primary variables.

Solution
The objective function is the sum of the collector cost and the water storage tank, and is given as

$$C(A_C, V_T) = C_{Collector} + C_{Storage} = 20\,A_C + 101.5\,V_T \tag{E10.3.1}$$

$$\text{Objective function}: C(A_C, V_T) = 20\,A_C + 101.5\,V_T \tag{E10.3.2}$$

The total cost needs to be minimized with the selection of the solar collector area and the water storage tank volume subject to the constraint that the total energy to be collected and stored as water thermal energy in the storage tank is 200 MJ. Notice that the objective function given by the Eq. (E10.3.1) is expressed as a function of two primary variables: collector area A_c and storage volume V_T.

The constraint function for this problem can now be derived and expressed in terms of the two primary variables in the following manner:
Net solar energy collection rate is

$$Q_u = G_t\,A_C - Q_{loss} \tag{E10.3.3}$$

where the net heat energy loss from the collector is given as

$$Q_{loss} = UA_C \left[\frac{(25 + T_H)}{2} - T_a \right] \tag{E10.3.4}$$

Substituting Eq. (E10.3.3) into Eq. (E10.3.2), we get

$$Q_u = G_t\,A_C - \left[\frac{(25 + T_H)}{2} - T_a \right] \tag{E10.3.5}$$

Considering an overall energy balance, we can state that the net energy collected is equal to the total thermal energy stored over the daylight time, i.e.

$$Q_u \times 3600 = Q_{Stored}$$

Substituting Eq. (E10.3.5) for the solar energy collected and $Q_{Stored} = 200\,000$ kJ

$$\left\{ G_t\,A_C - UA_C \left[\frac{(25 + T_H)}{2} - T_a \right] \right\} \times 36\,000 = 200\,000 \text{ kJ} \tag{E10.3.6}$$

Notice that the constraint function given by Eq. (E10.3.6) includes multiple variables such as G_t, A_C, U, T_H, and T_a. Substituting the given data for total incident solar irradiation as $G_t = 200 \text{ W/m}^2$, the overall heat transfer loss coefficient, $U = 2.0 \text{ W/m}^2 \text{ K}$, and the outside ambient temperature as $T_a = 10° \text{ C}$, we get

$$\left\{ 260 A_C - 2.0 A_C \left[\frac{(25 + T_H)}{2} - 10 \right] \right\} \times \frac{36\,000 \text{s}}{1000} = 200\,000 \text{ kJ} \qquad (E10.3.7)$$

or

$$\left\{ 260 A_C - 2.0 A_C \left[\frac{(T_H)}{2} + 2.5 \right] \right\} = 5555.56 \text{ kJ} \qquad (E10.3.8)$$

We can express the collector exit temperature or the water temperature at the inlet to the storage tank by considering the sensible energy storage in the mass of water in the storage tank in the following:

$$Q_{\text{Stored}} = m_W C_{pw}(T_H - 25) = \rho_w C_{pw} V_T (T_H - 25)$$

Solving for T_H,

$$T_H = 25 + \frac{Q_{\text{Stored}}}{\rho_w C_{pw} V_T} = 25 + \frac{200\,000}{1000 \times 4.19 \, V_T}$$

or

$$T_H = 25 + \frac{47.7}{V_T} \qquad (E10.3.9)$$

Let us now substitute the collector exit temperature given by Eq. (E10.3.9) into Eq. (E10.3.8) to express the constraint function in terms of the two primary variables as

$$\left[260 A_C - 2.0 A_C \left[\frac{(25)}{2} + \frac{47.7}{2V_T} + 2.5 \right] \right] = 5555.56 \text{ kJ} \qquad (E10.3.10)$$

or

$$\left[260 A_C - 2.0 A_C \left[15 + \frac{47.7}{2V_T} \right] \right] = 5555.56 \text{ kJ}$$

or

$$\textbf{Constraint function} : g(A_C, V_T) = \left(230 A_C - \frac{47.7 A_C}{V_T} = 5555.56 \right) \text{ kJ} \quad (E10.3.11)$$

10.3 Optimization Methods

There are a number of different optimization methods that are applicable to the design optimization of thermal systems and components. A few examples of these methods are as follows:

1. Graphical optimization method
2. Optimizing method of differential calculus
3. Method of Lagrange multipliers
4. Search methods

5. Dynamic programming
6. Geometric programming
7. Linear programming

Among these optimization methods, several are widely used in the optimization study of many thermal energy system components. Some of these methods are discussed in the following section with applications to thermal energy system designs.

10.3.1 Graphical Optimization Method

This is one of the simplest ways to search for an optimum solution to a problem that involves a few variables. The problem involves plotting the objective function against two or three variables on a two-dimensional space or three-dimensional space, respectively. All constraint functions are also plotted against the independent variables on the same two-dimensional and three-dimensional space. Such plots give a visual image and inspection of the region inside which all points will satisfy the constraint functions. The plots of the objective function inside this feasible space show how the optimum value of the objective function is reached and with what combination of the independent variable while satisfying all constraint functions.

Example 10.4   *Design and Optimization of a Straight Fin of Rectangular Profile*
In the design of a straight fin of rectangular profile, one needs to optimize the fin geometrical parameters such length L and width W to enhance the convection heat loss while keeping reduced conduction resistance by providing increased surface area and maximizing the fin performance parameter such as fin efficiency. As we try to increase the surface area by varying the length and width, the weight of the fin also increases. As a restriction for the allowable weight of the fin, one can maximize the efficiency while keeping the weight less than the allowable weight.

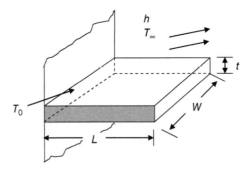

Solution
The fin efficiency is given as

$$\eta_f = \frac{\sqrt{hpKA_c}\left(\dfrac{\sinh\ mL + (h/mk)\cosh\ mL}{\cosh\ mL + (h/mK)\sinh\ mL}\right)}{hPL} \tag{E10.4.1}$$

where

$$m^2 = \frac{hP}{kA_c} \tag{E10.4.2}$$

h = Convection heat transfer coefficient

k = Thermal conductivity of the fin material of construction

$A_c = Wt$ = fin base area for heat conduction

$P = 2(W + t)$ = Perimeter of the fin surface

As we can see, the fin efficiency is a function of a number of parameters:

$$\eta_f = f(L, W, t, k, h)$$

By setting some parameter values such as h, k, and t, we can write the objective function of the fin efficiency as a function of the two primary variables, the length and width of the fin:

Objective function: $\eta_f = \eta_f(L, W)$

Constraint function: $W_{wt}(L, W) = \rho LWt \leq W_{wt}$ $\qquad$ (E10.4.3)

$$L \geq 0 \tag{E10.4.4a}$$

and

$$W \geq 0 \tag{E10.4.4b}$$

In order to search for the optimal value of the fin efficiency, we can compute the fin efficiency given by the Eq. (E10.4.1) over a range of values for the length, L, and width, W, and show these computed values in a two-dimensional plot with L and W as the two coordinates as shown in Figure 10.5.

The inequality constraints given by Eqs. (E10.4.4a) and (E10.4.4b) indicate that the solution should lie in the top right quadrant of the plot. The dark line represents the constraint function plot given by Eq. (E10.4.3). The hatched area is enclosed by the constraint function line, and the L and W coordinate lines represent the region where all acceptable fin efficiency values lie, and the highest value represents the optimum maximum efficiency value.

10.3.2 Optimization Method of Differential Calculus

This is one of most popular optimization methods and is often applied to find the maximum or minimum of many optimization problems that involve continuous and differentiable functions.

In this method of differential calculus, a **necessary condition** for the existence of the local maximum and minimum of a continuous and differentiable function $f(x)$ at a point $x = x^*$ within the range (a, b) is given by setting the first derivative of the function to zero as follows:

$$f'(x) = \frac{\partial f}{\partial x} = 0 \tag{10.14}$$

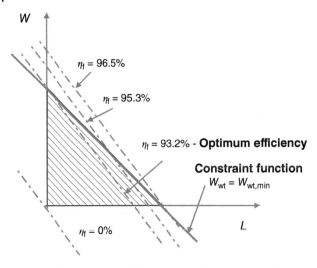

Figure 10.5 Graphical optimization method.

Additionally, the **sufficient condition** is that the point x^* is guaranteed to be the **local minimum** if $f'(x^*) = \frac{\partial^2 f(x^*)}{\partial x^2} > 0$ and to be the **local maximum** if $f'(x^*) = \frac{\partial^2 y}{\partial x^2} < 0$.

For a function $\mathbf{f(x, y)}$ with two independent variables (x, y) and with the existence of the continuous partial derivatives, the necessary conditions for the function $f(x, y)$ to possess a relative or local maximum or minimum at point (x^*, y^*) within the domain $\mathbf{R}$ are obtained by setting the exact differential to zero as

$$df = \frac{\partial f}{\partial x}dx + \frac{\partial f}{\partial y}dy = 0 \tag{10.15}$$

For independent variables x and y with arbitrary values of dx and dy, Eq. (10.15) is equivalent to the following two statements:

$$f_x = \frac{\partial f}{\partial x} = 0 \tag{10.16a}$$

$$f_y = \frac{\partial f}{\partial y} = 0 \tag{10.16b}$$

Solving Eqs. (10.16a) and (10.16b) gives the critical point (x^*, y^*) of the function.

The **second derivative test** is required to confirm if $f(x^*, y^*)$ is a local maximum or local minimum. This sufficient condition is derived and stated thus: this solution vector is guaranteed to be a local minimum solution if the matrix $[\mathbf{A}]$ of the second partial derivatives of the function $f(\vec{x})$ is a **positive derivative**. This is evaluated by checking the determinant of the second derivative matrix given as

$$\mathbf{D} = \begin{vmatrix} f_{xx} & f_{xy} \\ f_{xy} & f_{yy} \end{vmatrix} = \begin{vmatrix} \dfrac{\partial^2 f}{\partial x^2} & \dfrac{\partial^2 f}{\partial x \partial y} \\ \dfrac{\partial^2 f}{\partial x \partial y} & \dfrac{\partial^2 f}{\partial y^2} \end{vmatrix} \tag{10.17a}$$

or

$$\mathbf{D}(x^*, y^*) = f_{xx}f_{yy} - (f_{xy})^2 \tag{10.17b}$$

The **sufficient condition** is then stated as:

If $D > 0$, and $f_{xx}(x^*, y^*) > 0$ then the objective function $f(x^*, y^*)$ is a local **minimum**.
If $D > 0$, and $f_{xx}(x^*, y^*) < 0$ then the objective function $f(x^*, y^*)$ is a local **maximum**.
If $D < 0$, then the objective function $f(x^*, y^*)$ is not a local **maximum** or **minimum** and the point (x^*, y^*) is referred to as the **saddle point**.

Example 10.5 *Determine Local Maximum and Minimum of Function*
Determine the local maximum and minimum of the following function:

$$f(x, y) = x^4 + y^4 - 4xy + 2$$

Solution
Let us first apply the first derivative necessary condition to determine the critical points:

$$f_x = 4x^3 - 4y = 0$$

or

$$x^3 - y = 0 \qquad\qquad (E10.5.1)$$

and

$$f_y = 4y^3 - 4x = 0$$

or

$$y^3 - x = 0 \qquad\qquad (E10.5.2)$$

Equations (E10.5.1) and (E10.5.2) now can be solved for the critical points. Substituting Eq. (E10.5.1) for y into Eq. (E10.5.2), we obtain

$$x^9 - x = 0 \qquad\qquad (E10.5.3)$$

Solving Eq. (E10.5.3), we obtain three real roots: $x = 0$, 1, and -1, and the three critical points (x^*, y^*): $(-1, -1)$, $(1, 1)$, and (0.0).

We now apply the sufficient condition in terms of the second-order partial derivative matrix determinant:

$$D(x^*, y^*) = f_{xx}f_{yy} - (f_{xy})^2$$

Second derivatives are evaluated as

$$f_{xx} = 12x^2, f_{yy} = 12y^2, \text{and } f_{xy} = -4$$

Substituting,

$$D(x^*, y^*) = 144x^2y^2 - 16$$

Let us now substitute the critical points and check:

$D(-1, -1) = 128 > 0$ and $f_{xx} = 12 > 0$ and so $f(-1, -1)$ is a local minimum.
$D(1,1) = 128 > 0$ and $f_{xx} = 12 > 0$ and so $f(1, 1)$ is a local minimum.
$D(0, 0) = -16 < 0$ so the point $(0, 0)$ is a saddle point and has no local maximum or minimum.

10.3.2.1 Functions with Many Variables

Let us now consider a general function with n independent variables of the following form:

$$f(\vec{x}) = f(x_1, x_2, \ldots x_n) \tag{10.18}$$

The necessary condition is then given as

$$\frac{\partial f}{\partial x_1} = \frac{\partial f}{\partial x_1} = \cdots = \frac{\partial f}{\partial x_n} = 0 \tag{10.19a}$$

or

$$\frac{\partial f}{\partial x_i}(x_1, x_2, \ldots x_n) = 0 \text{ for } i = 1, 2, \ldots n \tag{10.19b}$$

For a nonlinear function $f(\vec{x})$, the necessary condition given by Eqs. (10.19a) and (10.19b) leads to a set of n simultaneous nonlinear equations. The solution of this set gives the point $\vec{x}^*$ at which the local maximum or minimum function value is obtained.

$$\vec{x}^* = \begin{Bmatrix} x_1^* \\ x_2^* \\ \vdots \\ x_n^* \end{Bmatrix} \tag{10.20}$$

A **sufficient condition** is derived and stated thus: this solution vector is guaranteed to be a local minimum solution if the matrix $[\mathbf{A}]$ of the second partial derivatives of the function $f(\vec{x})$ is a **positive derivative**. The matrix $[\mathbf{A}]$ is written as

$$\begin{bmatrix} \dfrac{\partial^2 f}{\partial x_1^2} & \dfrac{\partial^2 f}{\partial x_1 \partial x_2} & \cdots & \dfrac{\partial^2 f}{\partial x_1 \partial x_n} \\ \dfrac{\partial^2 f}{\partial x_1 \partial x_2} & \dfrac{\partial^2 f}{\partial x_2^2} & \cdots & \dfrac{\partial^2 f}{\partial x_2 \partial x_n} \\ \vdots & \vdots & \vdots & \vdots \\ \vdots & \vdots & \vdots & \vdots \\ \dfrac{\partial^2 f}{\partial x_1 \partial x_n} & \dfrac{\partial^2 f}{\partial x_1 \partial x_n} & \cdots & \dfrac{\partial^2 f}{\partial x_n^2} \end{bmatrix} \tag{10.21}$$

The matrix $[\mathbf{A}] = [a_j]$ is a symmetric matrix and is referred to as the **Hessian matrix**. The condition for a matrix $[\mathbf{A}]$ to be positive is that all its eigenvalues are positive.

Several numerical methods are available to determine the eigenvalues of a symmetric matrix. However, some of the most popular methods such as **Jacobi method, Given's method,** and **Householder's method** (Rao 2002, James et al. 1985) are limited to problems where the order of the matrix is low. One straightforward method of determining that the matrix $[\mathbf{A}]$ is positive definite is to check if all the following determinants of the sub-matrices including the n-order matrix are positive.

$$\mathbf{A}_1 = [a_{11}], \mathbf{A}_2 = \begin{bmatrix} a_{11} & a_{12} \\ a_{21} & a_{22} \end{bmatrix}$$

$$A_3 = \begin{bmatrix} a_{11} & a_{12} & a_{13} \\ a_{21} & a_{22} & a_{23} \\ a_{31} & a_{32} & a_{33} \end{bmatrix} \cdots A_n = \begin{bmatrix} a_{11} & a_{12} & \cdots & a_{1n} \\ a_{21} & a_{22} & \cdots & a_{2n} \\ \vdots & \vdots & \vdots & \vdots \\ \vdots & \vdots & \vdots & \vdots \\ a_{n1} & a_{n1} & \cdots & a_{nn} \end{bmatrix} \tag{10.22}$$

However, determining the determinant of a matrix becomes increasingly difficult using a direct method such as the Cramer rule because of excessive computation time for a matrix as the order of the matrix is four or higher.

10.3.3 Method of Lagrange Multiplier

This method is applicable for problems where the objective functions are differentiable and all constraints are equality constraints. The method is based on the principles of differential calculus and determines the optimum values based on the nature of the derivatives of the function. The method of Lagrange multiplier can be classified as constrained optimization problems and unconstrained optimization problems. A constraints optimization problem can be transformed into an unconstrained optimization problem by combining objective functions and constraint functions by substitutions.

In order to demonstrate the procedure for the method of Lagrange multiplier, let us consider the problem of finding the maximum or minimum values or, in other words, the extreme values of the following function with three independent variables x, y, and z:

$$f(x, y, z) = 0 \tag{10.23}$$

and subject to two equality constraints:

$$g_1 = g_1(x, y, z) = C_1 \tag{10.24a}$$

$$g_2 = g_2(x, y, z) = C_2 \tag{10.24b}$$

As the function $f(x, y, z)$ possesses a relative maximum or minimum at point $(x^*, y^*,)$ within the domain R, *the necessary condition* is obtained by setting the exact differential to zero as

$$df = \frac{\partial f}{\partial x} dx + \frac{\partial f}{\partial y} dy + \frac{\partial f}{\partial y} dy = 0 \tag{10.25}$$

In order to ensure that the three independent variables satisfy the two equality conditions given by Eqs. (10.20a) and (10.20b), we can consider the exact differential dg_1 and dg_2 of the two equality constraints as zero and obtain following two equations:

$$dg_1 = \frac{\partial g_1}{\partial x} dx + \frac{\partial g_1}{\partial y} dy + \frac{\partial g_1}{\partial z} dz = 0 \tag{10.26}$$

and

$$dg_1 = \frac{\partial g_2}{\partial x} dx + \frac{\partial g_2}{\partial y} dy + \frac{\partial g_2}{\partial z} dz = 0 \tag{10.27}$$

Let us multiply Eqs. (10.26) and (10.27) by arbitrary factors λ_1 and λ_2, and add to Eq. (10.25) to obtain the following:

$$\left(\frac{\partial f}{\partial x} + \lambda_1 \frac{\partial g_1}{\partial x} + \lambda_2 \frac{\partial 2}{\partial x} \right) dx + \left(\frac{\partial f}{\partial y} + \lambda_1 \frac{\partial g_1}{\partial y} + \lambda_2 \frac{\partial 2}{\partial y} \right) dy$$
$$+ \left(\frac{\partial f}{\partial z} + \lambda_1 \frac{\partial g_2}{\partial z} + \lambda_2 \frac{\partial g_2}{\partial z} \right) dz = 0 \tag{10.28}$$

As we can assign arbitrary values to differentials dx, dy, and dz, we can set each term of the Eq. (10.28) to zero and obtain the following equations:

$$\frac{\partial f}{\partial x} + \lambda_1 \frac{\partial g_1}{\partial x} + \lambda_2 \frac{\partial g_2}{\partial x} = 0 \qquad\qquad (10.29a)$$

$$\frac{\partial f}{\partial y} + \lambda_1 \frac{\partial g_1}{\partial y} + \lambda_2 \frac{\partial g_2}{\partial y} = 0 \qquad\qquad (10.29b)$$

$$\frac{\partial f}{\partial z} + \lambda_1 \frac{\partial g_2}{\partial z} + \lambda_2 \frac{\partial g_2}{\partial z} = 0 \qquad\qquad (10.29c)$$

Now we have a system of five equations that includes three Eqs. (10.29a)–(10.29c) and two constraints Eqs (10.24a) and (10.24b), to determine the three unknown variables x, y, z and the two unknown factors λ_1 and λ_2. The arbitrary factors λ_1 and λ_2 are referred to as the **Lagrange multipliers, which are introduced to** simplifying this optimization procedure. Often it is not necessary to determine the explicit values for Lagrange multipliers. In some applications, the Lagrange multiplier may represent some physical characteristics of the problem.

Example 10.6   *Optimization Using Method of Lagrange Multiplier*
Determine the maximum volume of a rectangular storage tank that needs to have a total surface area of 10 m².

Solution
Let us represent the three geometrical variables length, width, and height as l, w, and h, respectively. We now have to maximize the volume of the tank given as

$$V = lwh \qquad\qquad (E10.6.1)$$

subject to the constraint function given for the total surface area as

$$g(x, y, z) = 2lw + 2lh + 2hw = 10 \qquad\qquad (E10.6.2)$$

Using the procedure outlined for the Lagrange multiplier for the objective function given by Eq. (E10.6.1) and one equality constraint given by Eq. (E10.6.2), we get

$$\frac{\partial V}{\partial l} + \lambda \frac{\partial g}{\partial l} = 0 \qquad\qquad (E10.6.3a)$$

$$\frac{\partial V}{\partial w} + \lambda \frac{\partial g}{\partial w} = 0 \qquad\qquad (E10.6.3b)$$

$$\frac{\partial V}{\partial h} + \lambda \frac{\partial g}{\partial h} = 0 \qquad\qquad (E10.6.3c)$$

$$g(l, w, h) = 2lw + 2lh + 2lw = 10 \qquad\qquad (E10.6.3d)$$

Evaluating and then substituting the derivatives of the volume as the objective function and the surface area as the constraint function, Eqs. (E10.6.3a)–(E10.6.3d) reduce to

$$wh + \lambda(2w + 2h) \qquad\qquad (E10.6.4a)$$

$$lh + \lambda(2l + 2h) \qquad\qquad (E10.6.4b)$$

$$lw + \lambda(2l + 2w) \qquad\qquad (E10.6.4c)$$

$$2lw + 2lh + 2lw = 10 \qquad\qquad (E10.6.4d)$$

This system of four equations can be solved through elimination and substitutions.

10.4 A General Procedure for Lagrange Multiplier

In order to demonstrate a general procedure for the method of Lagrange multiplier, let us consider the mathematical statement of the optimization problem outlined by the Eqs. (10.11) and (10.12) for the objective function and the equality constraints respectively as follows:

Optimize the **objective function**:

$$f(\vec{x}) = f(x_1, x_2, x_3, \ldots, x_n) \tag{10.30}$$

where n is the number of unknown variables.

$$\vec{x} = \{x_1, x_2 \cdots x_n\}^T = \text{Unknown vector}$$

and **subject** to design **equality constraints**:

$$g_j(\vec{x}) = \text{Constraint functions with } j = 1, 2, \ldots m \tag{10.31}$$

where

$$g_1 = g_1(x_1, x_2, x_3, \ldots, x_n)$$
$$g_2 = g_2(x_1, x_2, x_3, \ldots, x_n)$$
$$\vdots$$
$$g_m = g_m(x_1, x_2, x_3, \ldots, x_n) \tag{10.32}$$

where m is the number of constraints.

According to the method of Lagrange multiplier, we can define the Lagrange function as

$$I(\vec{x}, \vec{\lambda}) = f(\vec{x}) + \sum_{j=1}^{j-m} \lambda_j g_j(\vec{x}) \tag{10.33}$$

where $\vec{\lambda} = \{\lambda_1, \lambda_2, \ldots \lambda_m\}^T = $ constant, known as the Lagrange multiplier vector.

We can state that the necessary condition for the stationary or extremum value of the Lagrange function $I(\vec{x}, \vec{\lambda})$ is obtained by

$$\frac{\partial I}{\partial x_i} = \frac{\partial f}{\partial x_i} + \sum_{j=1}^{j=m} \lambda_j \frac{\partial g_j}{\partial x_i} = 0, I = 1, 2 \ldots n \tag{10.34a}$$

or written as n-equation:

$$\frac{\partial f}{\partial x_1} + \lambda_1 \frac{\partial g_1}{\partial x_1} + \lambda_2 \frac{\partial g_2}{\partial x_1} - \cdots + \lambda_m \frac{\partial g_m}{\partial x_1} = 0$$

$$\frac{\partial f}{\partial x_2} + \lambda_1 \frac{\partial g_1}{\partial x_2} + \lambda_2 \frac{\partial g_2}{\partial x_2} - \cdots + \lambda_m \frac{\partial g_m}{\partial x_2} = 0$$

$$\vdots$$

$$\frac{\partial f}{\partial x_n} + \lambda_1 \frac{\partial g_1}{\partial x_n} + \lambda_2 \frac{\partial g_2}{\partial x_n} - \cdots + \lambda_m \frac{\partial g_m}{\partial x_n} = 0 \tag{10.34b}$$

and

$$\frac{\partial I}{\partial \lambda_j} = g_j = 0, j = 1, 2, \ldots m \tag{10.35}$$

or as *m*-constraints equation

$$g_1 = g_1(x_1, x_2, x_3, \ldots, x_n)$$
$$g_2 = g_2(x_1, x_2, x_3, \ldots, x_n)$$
$$\vdots$$
$$g_m = g_m(x_1, x_2, x_3, \ldots, x_n) \qquad (10.36)$$
$$g_m = g_m(x_1, x_2, x_3, \ldots, x_n) = 0$$

Equations (10.34a) and (10.34b) and Eq. (10.36) are solved for n unknown values $x_1^*, x_2^*, \ldots x_n^*$ of **independent variables** and **m-Lagrange multiplier** values. The procedure outlined is applicable to cases where the number of constraints **m** is less than the number of unknown variables, **n.**, i.e. **m < n**.

For **m = n**, i.e. the number of available equations is equal to the number of unknowns. Such a condition leads to a unique solution to the problem, but not an optimized solution. For the case of $m > n$, the available number of equations is greater than the number of unknowns and this results in a situation where no solution exists.

One important thing to notice about the method of Lagrange multipliers is that it leads to the requirement of solving a system of **simultaneous nonlinear equations**.

Example 10.7 Piping System for a Secondary Water Cooling System
In a secondary water piping, a piping system with two parallel pipes and two pumps are used to deliver cold water from a cooling tower base to a power plant condenser as shown in the figure below:

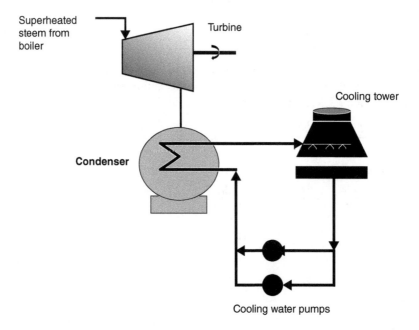

The total cold water flow rate, Q, required at the condenser is 0.01 m³/s. The pressure drops in the two pipelines are given as $\Delta P_1 = 4.0 \times 10^7\, Q_1^2$ and $\Delta P_2 = 5.75 \times 10^7\, Q_2^2$ with

Q_1 and Q_2 as the flow rate in the two lines respectively. Consider identical pump and electric motor efficiencies. State the objective function and the constraints to minimize the net power for the pumps in the system. Use the Lagrange multiplier method to determine the flow rates in the two pipes that result in minimum net pump power.

Solution

Power requirements for the two pumps are

Pump – 1:

$$P_1 = \rho Q_1 \Delta P_1 = 1.00 \times Q_1 \Delta P_1 = Q_1 \times 4.0 \times 10^7 \; Q_1^2 = 4.0 \times 10^7 \; Q_1^3$$

Pump – 2:

$$P_2 = \rho Q_2 \Delta P_2 = 1.00 \times Q_2 \Delta P_2 = Q_2 \times 5.75 \times 10^7 \; Q_2^2 = 5.75 \times 10^7 \; Q_2^3$$

Objective function : New power, $P = 4.0 \times 10^7 \; Q_1^3 + 5.75 \times 10^7 \; Q_2^3$

Constraint : $g = Q = Q_1 + Q_2 = 0.01 \; \text{m}^3/\text{s}$.

Following the method of Lagrange multiplier, we can define the Lagrange function as

$$I(\vec{x}, \vec{\lambda}) = P(Q_1, Q_2) + \lambda g (Q_1, Q_2)$$

We can state the necessary condition for the minimum value of the Lagrange function $I(\vec{x}, \vec{\lambda})$ as

$$\frac{\partial I}{\partial Q_1} = \frac{\partial P}{\partial Q_i} + \lambda \frac{\partial g}{\partial Q_1} = 0$$

or

$$3(4.0 \times 10^7) \, Q_1^2 + \lambda = 0$$

and

$$\frac{\partial I}{\partial Q_2} = \frac{\partial P}{\partial Q_2} + \lambda \frac{\partial g}{\partial Q_2} = 0$$

or

$$3 \, (5.75 \times 10^7) \, Q_2^2 + \lambda = 0$$

We now have a system of the following **three simultaneous nonlinear equations** for three unknowns, Q_1, Q_2, and λ:

$$(12.0 \times 10^7) \, Q_1^2 - \lambda = 0$$
$$(17.25 \times 10^7) \, Q_2^2 - \lambda = 0$$
$$Q_1 + Q_2 = 0.01 \; \text{m}^3/\text{s}$$

Solving this set through elimination and substitution, we obtain

$$Q_1 = 0.005 \, 470 \, 2 \; \text{m}^2/\text{s}, Q_2 = 0.004 \, 529 \, 8 \; \text{m}^2/\text{s}$$

New power, $P = 4.0 \times 10^7 \; Q_1^3 + 5.75 \times 10^7 \; Q_2^3$

$$= 4.0 \times 10^7 \times (0.005 \, 470 \, 2)^3 + 5.75 \times 10^7 (0.004 \, 529 \, 8)^3$$
$$= 4.0 \times 10^7 \times (0.005 \, 470 \, 2)^3 + 5.75 \times 10^7 (0.004 \, 529 \, 8)^3$$

$$P = 11.5147 \; \text{kW}$$

Example 10.8   Solar Water Storage System

Consider the objective function and the constraint function derived for the solar water heater system given in Example 10.2 and solve for the optimum values for the collector area and the storage volume using the Lagrange multiplier method.

Solution

$$\text{Objective function: } C(A_C, V_T) = 20 A_C + 101.5 V_T \qquad (E10.8.1)$$

$$\text{Constraint function: } g(A_C, V_T) = \left(230 A_C - \frac{47.7 A_C}{V_T} = 5555.56 \text{ kJ} \right) \qquad (E10.8.2)$$

Using the procedure outlined for the Lagrange multiplier for the objective function and equality constraint, we get

$$\frac{\partial C}{\partial A_C} + \lambda \frac{\partial g}{\partial A_C} = 0 \qquad (E10.8.3a)$$

$$\frac{\partial C}{\partial V_T} + \lambda \frac{\partial g}{\partial V_T} = 0 \qquad (E10.8.3b)$$

$$g(A_C, V_T) = (230 A_C - \frac{47.7 A_C}{V_T} = 5555.56 \text{ kJ} \qquad (E10.8.3c)$$

Evaluating the derivative terms in Eq. (E10.8.3a) using the objective function (E10.8.1) and constraint function (E10.8.2), we get

$$20 + \lambda \left(230 - \frac{47.7}{V_T} \right) = 0 \qquad (E10.8.4)$$

Similarly, Eq. (E10.8.3b) can be evaluated as

$$101.5 - \lambda \frac{47.7 A_C}{V_T^2} = 0 \qquad (E10.8.5)$$

Equations (E10.8.3c), (E10.3.4), and (E10.3.5) form a system of **three simultaneous non-linear equations** with three variables A_C, V_T, λ.

$$20 + \lambda \left(230 - \frac{47.7}{V_T} \right) = 0 \qquad (E10.8.6a)$$

$$101.5 - \lambda \frac{47.7 A_C}{V_T^2} = 0 \qquad (E10.8.6b)$$

$$230 A_C - \frac{47.7 A_C}{V_T} = 5555.56 \text{ kJ} \qquad (E10.8.6c)$$

Solving this set, we get the following optimum values for the collector area and the storage volume and the value for the Lagrange multiplier:

Collector area, $A_C = 29.2 \text{ m}^2$, storage volume,
$V_T = 1.2 \text{ m}^3$ and Lagrange multiplier, $\lambda = 0.105$.

10.4.1 Geometric Programming

This method is particularly suitable for problems where both the objective functions and the constraints equations are expressed as a sum of polynomial functions. A few examples of such expressions are given as follows:

Minimize: $f(x) = a_1 x + a_2 x^{-0.4}$

Minimize: $f(x_1, x_2) = a_1 + a_2 x_1 + a_3 x_1^{-0.5} x_2^2$

Subject to constraint: $g(x_1, x_2) = x_1 - \frac{b_1}{x_2} = 0$

A major feature of the geometric programming is to first determine the optimum value of the function rather than to first determine the values of the independent variables, which leads to the optimum objective function value. The method is also very efficient for solving large-scale problems.

10.4.1.1 Degree of Difficulty

Degree of Difficulty (dod) is used as a measure of complexity involved in the use of geometrical programming in an optimization problem. This is defined as a function of the number of polynomial terms in the objective function as well in constraint functions, and the number of independent variables as follows:

$$\boldsymbol{dod} = N_{pt} - (N_v + 1) \tag{10.37}$$

where

N_{pt} = Total number of polynomial terms in the objective function and in the constraint functions

N_v = Total number of independent variables

For a **zero-dod**, the geometrical programming is the simplest method of choice. However, for a dod-number greater than zero, the geometrical programming method becomes increasingly difficult because of the time-consuming solution of the involved nonlinear equations.

10.4.1.2 General Optimization Procedure by Geometric Programming

In this section, we first present a general procedure of geometric programing involving an objective function with **one independent variable**. The procedure is then extended to problems involving multiple independent variables.

Let us consider the problem of finding the optimum value of the following objective function that constitutes the sum of two polynomial function terms, f_1 and f_2:

$$f(x) = a_1 x^{n_1} + a_2 x^{n_2} = f_1 + f_2 \tag{10.38a}$$

where

$$f_1(x) = a_1 x^{n_1} \text{ and } f_2(x) = a_2 x^{n_2} \tag{10.38b}$$

As per the geometric programming method, the optimum value f^* of the objective function can also be given in the following product form:

$$f = \left(\frac{a_1 x^{n_1}}{w_1}\right)^{w_1} \left(\frac{a_2 x^{n_2}}{w_2}\right)^{w_2} = \left(\frac{f_1}{w_1}\right)^{w_1} \left(\frac{f_2}{w_2}\right)^{w_2} \tag{10.39}$$

subject to the criteria that

$$w_1 + w_2 = 1 \tag{10.40}$$

Selection of weighting factor values are derived by applying the Lagrange multipliers method to minimize the objective function $\ln f(x) = \ln [f_1 + f_2] = w_1(\ln f_1 - \ln w_1) + w_2(\ln f_2 - \ln w_2)$ and the constraint function $g = w_1 + w_2 - 1 = 0$ in the following manner:

$$\nabla(\ln f) - \lambda \nabla g = 0 \text{ and } g = 0,$$

which lead to the following set of equations:

$$\ln f_1 - 1 - \ln w_1 - \lambda = 0 \tag{10.41a}$$

$$\ln f_2 - 1 - \ln w_2 - \lambda = 0 \tag{10.41b}$$

$$w_1 + w_2 - 1 = 0 \tag{10.41c}$$

Solving the set of three equations results in

$$w_1 = \frac{f_1}{f_1 + f_2} \tag{10.42a}$$

$$w_1 = \frac{f_2}{f_1 + f_2} \tag{10.42b}$$

The objective is to select variables $x_1, x_2, \ldots, x_n$ to minimize the objective function $f(x)$. These values of $x_1^*, x_2^* \ldots, x_3^*$ are found by setting the derivatives of Eqs. (10.38a) and (10.38b) to zero.

$$n_1 a_1 x^{n_1-1} + n_2 a_2 x^{n_2-1} = 0$$

Multiplying both sides by x, we can express

$$n_1 f_1^* + n_2 f_2^* = 0 \tag{10.43}$$

where f_1^* and f_2^* are the values of f_1 and f_2 at the optimum values of **f**.

Substituting $f_1^* = -\frac{n_2 f_2^*}{n_1}$ from Eq. (10.43) into Eqs. (10.42a) and (10.42b) for the weighting factors,

$$w_1 = \frac{f_1}{f_1 + f_2} = \frac{-\dfrac{n_2 f_2^*}{n_1}}{-\dfrac{n_2 f_2^*}{n_1} + f_2^*} = \frac{-n_2}{n_1 - n_2} \tag{10.44a}$$

and

$$w_1 = \frac{f_2}{f_1 + f_2} = \frac{f_2^*}{-\dfrac{n_2 f_2^*}{n_1} + f_2^*} = \frac{n_1}{n_1 - n_2} \tag{10.44b}$$

Substitution of these values of weighting factors into the objective function Eq. (10.39) leads to the optimum value of the objective function in the following form:

$$f^* = \left(\frac{a_1}{w_1}\right)^{w_1} \left(\frac{a_2}{w_2}\right)^{w_2} \tag{10.45}$$

subject to

$$w_1 + w_2 = 1 \tag{10.46}$$

$$n_1 w_1 + n_2 w_2 = 0 \tag{10.47}$$

where w_1 and w_2 are two weighting factors, which are defined as

$$w_1 = \frac{f_1^*}{f_1^* + f_2^*} = \frac{f_1^*}{f^*} \tag{10.48a}$$

and

$$w_2 = \frac{f_2^*}{f_1^* + f_2^*} = \frac{f_2^*}{f^*} \tag{10.48b}$$

One of the important characteristics of geometric programming is that the procedure leads to a system of simultaneous linear equations.

Example 10.9   Selection of Pipe Diameter in a Secondary Water System

The total cost of a secondary water piping system constitutes the capital cost of the pipe, cost of the pump, and the operating cost of the pump for the net pressure drop in the system. The capital cost of the piping is given as $C_{pipe} = 15\,D^2$. Let us assume that the operating cost is controlled by the major friction pressure drop in the pipe of length L. The pump operating head and operating power are given as

$$\Delta H_{pump} = \frac{\Delta P}{\rho} = f\frac{L}{D}\frac{V^2}{2} \text{ and } P_{pump} = \gamma Q \Delta H$$

Substituting $V = \frac{Q}{\frac{\pi}{4}D^2}$, the pump operating power can be expressed as a function of diameter as

$$P_{pump} = a_2 D^{-5}$$

Considering the electrical operating cost, let us express the functional relationship of operating cost as a function of diameter as

$$C_{pump} = 60 \times 10^{10} D^{-5}$$

Determine the optimum pipe diameter, D, for the minimum total cost of the piping system.

Solution
The objective function is given by the total cost as

$$f = C_{pipe} + C_{pump}$$

$$f = 15\,D^2 + 60 \times 10^{10} D^{-5} \tag{E10.9.1a}$$

or

$$f = f_1 + f_2 \tag{E10.9.1b}$$

where

$$f_1 = 15\,D^2 \text{ and } f_2 = 60 \times 10^{10} D^{-5}$$

Following the procedure for geometric programing, the total minimum cost function in the product form is

$$f^* = \left(\frac{f_1}{w_1}\right)^{w_1} \left(\frac{f_2}{w_2}\right)^{w_2}$$

or

$$f^* = \left(\frac{15D^2}{w_1}\right)^{w_1} \left(\frac{60 \times 10^{10} D^{-5}}{w_2}\right)^{w_2} \tag{E10.9.2}$$

subject to the conditions that

$$w_1 + w_2 = 1 \tag{E10.9.3}$$

and

$$n_1 w_1 + n_2 w_2 = 2w_1 - 5w_2 = 0 \tag{E10.9.4}$$

Solving Eqs. (E10.9.3), (10.29a)–(10.29c), and (E10.9.4) for the weighting factors, we get

$$w_1 = \frac{5}{7} \text{ and } w_2 = \frac{2}{7}$$

Substituting these weighting factor values into the optimum function for the minimum cost of the system as given by Eq. (E10.9.2),

$$f^* = \left(\frac{15}{5/7}\right)^{5/7} \left(\frac{60 \times 10^{10}}{2/7}\right)^{2/7}$$

Minimum cost : $f^* = 1.2429 + 15\,272 = US\,\$15\,273$

The optimum pipe diameter is given by Eq. (10.48a) or Eq. (10.48b) at the optimum value. Using Eq. (10.48a) at the optimum value, we get the optimum pipe diameter as

$$w_1 = \frac{f_1^*}{f_1^* + f_2^*} = \frac{f_1^*}{f^*} = \frac{15D_*^2}{f^*}$$

$$\frac{5}{7} = \frac{15D_*^2}{15\,273}$$

or

$$D^* = 26.96 \text{ mm}$$

Example 10.10 A water heating system uses hot gas stream from a combustion chamber to process the water stream in a water heater as shown in the figure below:

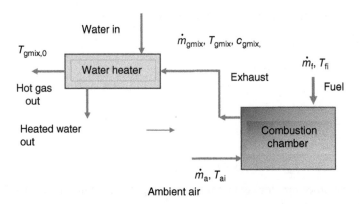

Gasoline fuel at a flow rate of $\dot{m}_f = 0.003$ kg/sand heat of combustion of $Q_{HV} = 44\,500$ kJ/kg is injected into the combustion chamber and undergoes chemical combustion with oxygen, which is supplied in the form of air at a flow rate of $\dot{m}_a$ at ambient condition with the temperature $T_{ai} = 25\,°C$. A combustion analysis show dependence of the combustion efficiency, η_{comb} on the mass flow rate of air, supply in the following empirical form $\eta_{comb} = 1.0 - \frac{C_1}{(\dot{m}_a + \dot{m}_f)^2}$. Derive the objective function for maximizing heat transfer in the water heater, Q_{WH}. Use geometric programming to determine the mass flow rate of water, which results in a maximum heat transfer in the water heater, using the following assumed set of known data: $Q_{HV} = 44\,500\,kJ/kg$, mixture exit temperature from water heater,, $T_{gmax,o} = 145\,°C$, Estimated mixture gas specific heat, ambient air temperature, $C_{p,gmix} = 1.03\,kJ/kg\,K$, and the empirical constant for combustion efficiency as $C_1 = 0.00001$.

Solution

Heat release in the combustion chamber is given as

$$\dot{Q}_{Comb} = \eta_{Comb} \times \dot{m}_f \times Q_{HV} \tag{E10.10.1}$$

Energy balance on the combustion chamber:
Heat release in the combustion chamber = Heat transferred to the product of combustion gas

$$\dot{Q}_{Comb} = (\dot{m}_a + \dot{m}_f)C_{p,mix}(T_{gmix} - T_{a,i}) \tag{E10.10.2}$$

Rearranging, we can write the expression for the product gas temperature as

$$T_{gmix} = T_{a,i} + \frac{\dot{Q}_{Comb}}{(\dot{m}_a + \dot{m}_f)C_{p,mix}} \tag{E10.10.3}$$

Substituting Eq. (E10.10.1) into (E10.10.3),

$$T_{gmix} = T_{a,i} + \frac{\eta_{Comb} \times \dot{m}_f \times Q_{f,HV}}{(\dot{m}_a + \dot{m}_f)C_{p,mix}} \tag{E10.10.4}$$

From the energy balance over the water heater

$$Q_{WH} = (\dot{m}_a + \dot{m}_f)\,C_{p,mix}(T_{gmix} - T_{gmax,o}) \tag{E10.10.5}$$

Substituting the expression given by Eq. (E10.10.4) for T_{gmix}, we get

$$Q_{WH} = (\dot{m}_a + \dot{m}_f)C_{p,mix}\left(T_{a,i} + \frac{\eta_{Comb} \times \dot{m}_f \times Q_{f,HV}}{(\dot{m}_a + \dot{m}_f)C_{p,mix}} - T_{gmax,o}\right) \tag{E10.10.6}$$

For simplification, let us assume $X = (\dot{m}_a + \dot{m}_f)$ and rewrite Eq. (E10.10.6) as

$$Q_{WH} = X\,C_{p,mix}\left(T_{a,i} + \frac{\eta_{Comb} \times \dot{m}_f \times Q_{f,HV}}{X\,C_{p,mix}} - T_{gmax,o}\right) \tag{E10.10.7a}$$

or

$$Q_{WH} = X\,C_{p,mix}(T_{a,i} - T_{gmax,o}) + \eta_{Comb} \times \dot{m}_f \times Q_{f,HV} \tag{E10.10.7b}$$

Let us now substitute the expression for the combustion efficiency

$$\eta_{comb} = 1.0 - \frac{C_1}{(\dot{m}_a + \dot{m}_f)^2} = 1.0 - \frac{C_1}{X^2}$$

into Eq. (E10.10.7b)

$$Q_{WH} = X\,C_{p,mix}(T_{a,i} - T_{gmax,o}) + \left(1.0 - \frac{C_1}{X^2}\right) \times \dot{m}_f \times Q_{f,HV}$$

Rearranging,

$$Q_{WH} = \dot{m}_f \times Q_{f,HV} + X\,C_{p,mix}(T_{a,i} - T_{gmax,o}) - \frac{\dot{m}_f \times Q_{f,HV} \times C_1}{X^2} \tag{E10.10.8}$$

Let us now substitute all known data: $\dot{m}_f = 0.003$ kg/s, $Q_{HV} = 44\,500$ kJ/kg, $T_{ai} = 25\,°C$, $T_{gmax,\,o} = 145\,°C$, $T_{a,\,i} = 25\,°C$, and

$$Q_{WH} = 0.003 \times 44\,500 + 1.03 \times (25 - 145)\,X - 0.003 \times 44\,500 \times 0.0002\,X^{-2}$$

or

$$Q_{WH} = 133.5 - 123.6\,X - 0.0267\,X^{\;2} \tag{E10.10.9}$$

Since the first term of the Eq. (E10.10.9) is constant, we can define the rest of the two terms as the objective function to determine the air mass flow rate. A minimum value of the sum of these two terms will result in maximum heat transfer in the water heater.

Minimize the **Objective function:**

$$f = 123.6\,X + 0.0267\,X^{-2} = f_1 + f_2 \tag{E10.10.10a}$$

where

$$f_1 = 123.6\,X \text{ and } f_2 = 0.0267\,X^{-2} \tag{E10.10.10b}$$

Following the procedure for geometric programing, the minimum value of the function in the product form is

$$f^* = \left(\frac{123.6}{w_1}\right)^{w_1} \left(\frac{0.0267}{w_2}\right)^{w_2} \tag{E10.10.11}$$

subject to the conditions that

$$w_1 + w_2 = 1 \tag{E10.10.12a}$$

and

$$n_1 w_1 + n_2 w_2 = w_1 - 2w_2 = 0 \tag{E10.10.12b}$$

Solving Eqs. (E10.9.3), (10.29a)–(10.29c), and (E10.9.4) for the weighting factors, we get

$$w_1 = \frac{2}{3} \text{ and } w_2 = \frac{1}{3}$$

Substituting these weighting factor values into the minimum value of the objective function given by Eq. (E10.10.11),

$$f^* = \left(\frac{123.6}{2/3}\right)^{2/3} \left(\frac{0.0267}{1/3}\right)^{1/3}$$

$$= (32.51)\,(0.43107)$$

$$f^* = 14.01$$

Now the values X can be found from the weighting factor given by Eq. (10.48a) at the optimum value as

$$w_1 = \frac{f_1^*}{f_1^* + f_2^*} = \frac{f_1^*}{f^*} \text{ or } 2/3 = \frac{f_1^*}{14.01} = \frac{123.6\,X^*}{f^*}$$

$$123.6\,X^* = 2/3 \times 14.01$$

Solving, $X^* = 0.0756$, and the air mass flow rate is

$$\dot{m}_a = X^* - \dot{m}_f = 0.0756\text{–}0.003$$

Mass flow rate of air, $\dot{m}_a = 0.0726 \text{ kg/s}$
Heat transfer rate

$$Q_{WH} = 133.5 - 123.6 \times 0.0756 - 0.0267\,(0.07256)^{-2}$$

$$= 133.5\text{–}9.344\text{–}5.071$$

$$= 119.08 \text{ KJ}$$

10.4.1.3 Multivariable Geometric Programming

The procedure outlined for one independent variable can be extended to problems with two or more independent variables. Let us consider the objective function involving two independent variables, x_1 and x_2, and with multiple polynomial functions given as

$$f(x) = a_1 x_1^{n_1} + a_2 x_2^{n_2} + a_3 (x_1 x_2)^{n_3} = f_1 + f_2 + f_3 \tag{10.49}$$

where

$$f_1 = a_1 x_1^{n_1}, f_2 = a_2 x_2^{n_2}, \text{ and } f_3 = a_3 (x_1 x_2)^{n_3}$$

As per the geometric programming method, the optimum value f^* of the objective function can also be given in the following product form:

$$f^* = \left(\frac{a_1}{w_1}\right)^{w_1} \left(\frac{a_2}{w_2}\right)^{w_2} \left(\frac{a_3}{w_3}\right)^{w_3} \tag{10.50}$$

and subject to the criteria that

$$w_1 + w_2 + w_3 = 1 \tag{10.51}$$

and to satisfy the following two equations derived based on two variables:

For variable x_1 : $n_1 w_1 + n_3 w_3 = 0$ \hfill (10.52a)

For variable x_2 : $n_2 w_2 + n_3 w_3 = 0$ \hfill (10.52b)

where w_1, w_2, and w_3 are three weighting factors, which are defined as

$$w_1 = \frac{f_1^*}{f_1^* + f_2^* + f_3^*} = \frac{f_1^*}{f^*} \tag{10.53a}$$

$$w_2 = \frac{f_2^*}{f_1^* + f_2^* + f_3^*} = \frac{f_2^*}{f^*} \tag{10.53b}$$

$$w_3 = \frac{f_3^*}{f_1^* + f_2^* + f_3^*} = \frac{f_3^*}{f^*} \tag{10.53c}$$

After solving the weighting factor values from Eqs. (10.51), (10.52a), and (10.52b), we can now compute the optimum function f^* from Eq. (10.50)

$$f^* = \left(\frac{a_1}{w_1}\right)^{w_1} \left(\frac{a_2}{w_2}\right)^{w_2} \left(\frac{a_3}{w_3}\right)^{w_3} \tag{10.54}$$

Equations (10.53a)–(10.53c) can then be used for solving the optimum values of the variables x_1^* and x_2^* using the computed values of the weighting factors.

10.4.2 Procedure for Solving Multivariable Problem Using Geometric Programing

The general procedural steps are given as follows:

1. Select an optimization function, **f**, following Eq. (10.49).
2. Set the sum of all weighting factors given by Eq. (10.51).
3. Solve the system of simultaneous linear equations Eqs. (10.51), (10.52a), and (10.52b) for the weighting factors w_1, w_2, and w_3.
4. Use these weighting factor values and solve for the **optimum function** value f^* using Eq. (10.50).
5. Solve for optimum values of the variables from Eqs. (10.53a)–(10.53c).

In a similar manner, a general procedure can be written involving multiple independent variables as follows:

Let us consider the objective function involving **n number of independent variables** $x_1, x_2, \ldots, x_n$ and expressed as a sum of multiple polynomials given as

$$f(x) = a_1 x_1^{n_1} + a_2 x_2^{n_2} + \cdots + a_n (x_n)^{n_n} = f_1 + f_2 + \cdots + f_n \tag{10.55a}$$

or

$$f(x) = \sum_{i=1}^{n} f_i \tag{10.55b}$$

where n = total number of terms in the objective function.

As per the geometric programming method, the objective function can also be given by the following product form:

$$f = \left(\frac{a_1 x_1^{n_1}}{w_1}\right)^{w_1} \left(\frac{a_2 x_2^{n_2}}{w_2}\right)^{w_2} \cdots \left(\frac{a_n x_n^{n_n}}{w_n}\right)^{w_n} \tag{10.56}$$

With the selection of the optimum values and by the selection of the weighting factors the function given by Eq. (10.45) is made equivalent to the expression of objective function given by Eq. (10.50).

The objective is to select variables $x_1, x_2, \ldots, x_n$ to minimize the objective function $f(x)$. These values of $x_1^*, x_2^* \ldots, x_3^*$ are found by setting the derivatives of the Eq. (10.50) to zero. This leads to the optimum value of the objective function in following form:

$$f^* = \left(\frac{a_1}{w_1}\right)^{w_1} \left(\frac{a_2}{w_2}\right)^{w_2} \cdots \left(\frac{a_n}{w_n}\right)^{w_n}$$

subject to the criteria that

$$w_1 + w_2 + \cdots + w_n = 1 \tag{10.57a}$$

and for each independent variable from the system of simultaneous linear equations

$$\sum_{i=1}^{n} a_{in} w_i = 0 \tag{10.57b}$$

where

$i =$ index for polynomial term number in the objective function

$n =$ total number of polynomial terms in the objective function

$j =$ variable ranging from 1 to the total number of variables, M

where $w_1, w_2, \ldots w_n$ are weighting factors, which are defined as

$$w_1 = \frac{f_1^*}{\sum_{i=1}^{n} f_n^*} \tag{10.58a}$$

$$w_2 = \frac{f_2^*}{\sum_{i=1}^{n} f_n^*} \tag{10.58b}$$

$$w_n = \frac{f_n^*}{\sum_{i=1}^{n} f_n^*} \tag{10.58c}$$

10.4.2.1 Multivariable Geometric Programing with Constraints

The procedure outlined now can be extended to a multivariable objective function subject to a constraint considering the following statement of the objective function and constraint function:

Minimize: Objective function:

$$f(x) = f_1 + f_2 + f_2 \tag{10.59}$$

subject to the equality constraint function

$$g(x) = g_1 + g_2 = 1 \tag{10.60}$$

where f_1, f_2, f_3, g_1, and g_2 are in polynomial forms involving three independent variables. Let us assume that n_1, n_2, and n_3 are the power index of the variables in the polynomial functions associated with the objective function $f(x)$, and m_1, and m_2 are the power index of the variables in the polynomial functions associated with the constraint function $g(x)$.

Note here that the polynomial functions on the left-hand side of the constraint function are expressed in such a way that the sum of these functions is equated to unity on the right-hand side.

As per the geometric programming method, the objective function can be given by the following product form:

$$f = \left(\frac{a_1 x_1^{n_1}}{w_1}\right)^{w_1} \left(\frac{a_2 x_2^{n_2}}{w_2}\right)^{w_2} \left(\frac{a_3 x_1 x_2^{n_3}}{w_3}\right)^{w_3} \tag{10.61a}$$

or

$$f = \left(\frac{f_1}{w_1}\right)^{w_1} \left(\frac{f_2}{w_2}\right)^{w_2} \left(\frac{f_3}{w_3}\right)^{w_3} \tag{10.61b}$$

subject to the criteria that the sum of the weighting factors is unity and given as

$$w_1 + w_2 + w_3 = 1 \tag{10.62a}$$

where

$$w_1 = \frac{f_1}{f_1 + f_2 + f_3}, w_2 = \frac{f_2}{f_1 + f_2 + f_3}, \text{ and } w_3 = \frac{f_2}{f_1 + f_2 + f_3} \tag{10.62b}$$

In a similar manner, the constraint function is also expressed in the product form

$$g = g_1 + g_2 = \left(\frac{g_1}{w_4}\right)^{w_4} \left(\frac{g_2}{w_5}\right)^{w_5} = 1 \tag{10.63a}$$

and in the following:

$$\left(\frac{g_1}{w_4}\right)^{Mw_4} \left(\frac{g_2}{w_5}\right)^{Mw_5} = 1 \tag{10.63b}$$

subject to the criteria that

$$Mw_4 + Mw_5 = M \tag{10.64}$$

where the weighting factors for the constraint function are defined as

$$w_4 = \frac{g_1}{g_1 + g_2} \tag{10.65a}$$

$$w_5 = \frac{g_2}{g_1 + g_2} \tag{10.65b}$$

Notice that the arbitrary constant M is used in the constraint function and along with its sum function weighting factor.

The objective function and the constraint functions are now combined into a single product form by direct multiplication of the constraint function given by Eq. (10.63a) as follows:

$$f = \left(\frac{a_1 x_1^{n_1}}{w_1}\right)^{w_1} \left(\frac{a_2 x_2^{n_2}}{w_2}\right)^{w_2} \left(\frac{a_3 x_1 x_2^{n_3}}{w_3}\right)^{w_3} \left[\left(\frac{g_1}{w_4}\right)^{Mw_4} \left(\frac{g_2}{w_5}\right)^{Mw_5}\right] \tag{10.66}$$

Following the procedure outline in Section 10.4.2 for unconstrained geometric programming, we can derive the optimum function in the following form:

$$f^* = \left(\frac{a_1}{w_1}\right)^{w_1} \left(\frac{a_2}{w_2}\right)^{w_2} \left(\frac{a_3}{w_3}\right)^{w_3} \left(\frac{g_1}{w_4}\right)^{Mw_4} \left(\frac{g_2}{w_5}\right)^{Mw_5} \tag{10.67}$$

subject to the following conditions:

$$w_1 + w_2 + w_3 = 1 \tag{10.68a}$$

$$Mw_4 + Mw_5 = M \tag{10.68b}$$

For variable x_1 : $n_1 w_1 + w_3 + m_1 Mw_4 + m_2 Mw_5 = 0$ (10.69a)

For variable x_2 : $n_2 w_2 + n_3 w_3 + m_1 Mw_4 + m_2 Mw_5 = 0$ (10.69b)

The five linear equations given by Eqs. (10.68a), (10.68b), (10.69a), (10.69b), (10.69c) can now be solved for the unknowns w_1, w_2, Mw_4, Mw_5, and M. These unknown values can now be substituted into the optimum function f^* defined by Eq. (10.67), followed by solving for optimum values of variables x_1^* and x_2^*, and using either f_1^* or f_2^* or f_2^*, which are computed from the weighting factor equations given by (10.64b).

Bibliography

Arora, J.S. (1989). *Introduction to Optimum Design*. McGraw-Hill.

Beightler, C. and Philips, D.T. (1976). *Applied Geometric Programing*. New York: Wiley.

Boem, R.F. (1987). *Design Analysis of Thermal Systems*. New York: Wiley.

Chapra, S.C. and Canale, R.P. (2006). *Numerical Methods for Engineers*, 5e. New York: McGraw-Hill.

Diffie, R.J., Peterson, E.L., and Zener, C. (1967). *Geometric Programming: Theory and Application*. New York: Wiley.

Huang, C.-H. (2013). *Engineering Design by Geometric Programming, Mathematical Problems in Engineering*, 568998. Hindawi Publishing Company.

James, M.L., Smith, G.M., and Wolford, J.C. (1985). *Applied Numerical Methods for Digital Computation*, 3e. New York: Harper Collins Publishers.

Majumdar, P. (2005). *Computational Methods for Heat and Mass Transfer*. New York: Taylor & Francis.

Rao, S.S. (2002). *Applied Numerical Methods for Engineers and Scientists*. Hoboken, NJ: Prentice Hall.

Reklattis, G.V., Ravindran, A., and Ragsdell, K.M. (2006). *Engineering Optimization Methods and Applications*, 2e. New York: Wiley.

Stoecker, W.F. (1989). *Design of Thermal Systems*, 3e. McGraw-Hill.

Wilde, D.J. and Beightler, C.S. (1967). *Foundations of Optimization*. Hoboken, NJ: Prentice-Hall.

Zener, C. (1971). *Engineering Design by Geometric Programming*. New York: Wiley-Interscalene.

Problems

10.1 Determine the local maximum and minimum of the following function:

$$f(x, y) = x^3 + y^3 - 6xy + 3$$

10.2 A design analysis of a shell and tube heat exchanger for a certain application resulted in the requirement of the total tube length of $L_{total} = 130$-m for the computed total surface area needed to meet the heat transfer rating. The computed number of tubes per unit area is restricted to only 190 tubes/m² of shell cross-sectional area based on the assumed tube pitch.

The major components of the cost are the following:

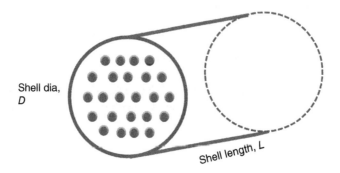

Shell dia, D

Shell length, L

(i) Cost of shell, $C_{Shell} = C_1 D^{C_2} L$, where D = diameter of the shell, L = length of the shell, and C_1 and C_2 are the empirical constants for the shell cost function.

(ii) Cost of tubes is constant and equal to C_3.

(iii) Cost of the floor space for installation of heat exchanger is given as $C_{space} = C_4 \, DL$.

Consider the following empirical constants for the cost function: $C_1 = 1000$, $C_2 = 3.0$, $C_3 = \$1200$, and $C_4 = 500$ DL. Use the calculus method to determine the minimum total cost of the heat exchanger and the corresponding optimum values of the diameter of the shell, D, and the length of the shell, L.

10.3 Solve Problem 10.2 using the Lagrange multiplier method. Repeat the computation of optimum cost and the optimum variables varying the parameter for the restriction of number of tubes per unit area of shell cross-sectional area as 180, 190, and 200 tubes/m² of shell cross-section based on the assumed tube pitch.

10.4 The total cost of a system constitutes the capital cost of the pipe, cost of the pump, the operating cost of the pump for the net pressure drop in the system, and the cost of the device. The capital cost of the pipe is given as $C_{Pipe} = 25D^2$. Let us assume that the operating cost is controlled by the major friction pressure drop in the pipe of length L. The functional relationship of pumping cost as a function of diameter, D, is $C_{pump} = 120 \times 10^{12} D^{-5}$. The cost of the device is given as $C_{device} = 92\,D^{-1}$. Determine the optimum pipe diameter D (mm) and flow rate Q (m³/s) for the minimum total cost of the system.

10.5 A secondary water system, shown in the figure below, is used to transfer heat from the condensing steam in the condenser by circulating water though a piping to the cooling tower where heat is dissipating to reject heat though a cooling tower. The cooling system uses a circulation pump to transfer 8 MW of heat from the condenser to the cooling tower.

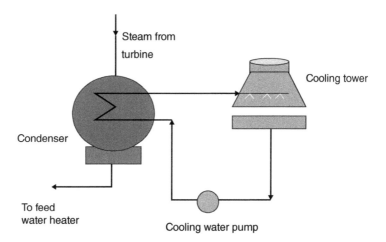

Some of the major cost components of this system are the following:

(i) Capital cost and life cycle maintenance cost of the cooling tower. One of the cost functions for the cooling tower can be expressed as a function of the surface area for heat transfer of the cooling tower for heat and mass transfer, A_{ct}, while the cost function for the cooling tower can be developed based on key parameters such as Approach, A, and Range, R. For simplicity, let us consider cost function correlation based on cooling tower transfer area for heat and mass transfer, as $C_{CT} = C_1 A_{ct}^{0.5}$.

(ii) Pumping cost for the circulating water including capital; operation and maintenance in terms of mass flow rate of cooling water as $C_{pmp} = C_2 \dot{m}_w^3$.

(iii) Cost related to the size of the condenser depends on the required heat transfer surface, which in turn depends on the number of parameters such as the mass flow rate of cooling water, cooling water inlet temperature, and the cooling tower surface area for heat and mass transfer. Let us assume this cost function correlation as

$$C_{Cond} = \frac{C_3 \dot{Q}_c}{\dot{m}_w^{1.4} A_{ct}}$$

Considering constant correlation constant values as $C_1 = 1000$, $C_2 = 003$, and $C_3 = 100$,

(a) derive an unconstrained objective function based on total cost as a function of the two variables A_{ct} and $\dot{m}_w$;

(b) determine the minimum cost and the corresponding optimum values of A_{ct} and $\dot{m}_w$.

10.6 An oil pipeline over a distance of $L = 50$ km uses a series of pumps to deliver oil at a flow rate of 0.75 m^3/s from a seaport to a refinery. The major cost functions are (i) Cost of pipeline C_{pipe} = US\$3 500 000 $D^{1.3}$ and (ii) Cost of pump = $N_{pump}(C_2 + C_3 \, \Delta p^{0.6})$, where $N_{pump} = \frac{L_{pipe}}{L}$ and L = pump location length frequency.

10.7 Determine the constrained objective function for the total cost as a function of the variables, pipe diameter, D, length frequency for pump location, and the pressure drop in each pipe section. (i) Determine the constrained function for the pressure drop in each section Δp as a function of pipe diameter, D, and pipe location length, L, assuming the pipe friction factor as 0.0025; and (ii) use geometric programming to determine the minimum total cost and the optimum value of the variables, L, D, and Δp.

Appendix A

Parametric Representation of Thermal Parameters and Properties

In this appendix, procedures for developing equations for parametric representations of thermal parameters and properties are discussed. The primary objective is to show how to represent the performance characteristics of various thermal components such as turbines, compressors, pumps, fans, cooling towers, fins, or heat sinks; performance characteristics of processes or systems; and thermodynamic properties of substances in terms of saturation and superheat data tables.

There are a number of reasons for developing such equations: (i) To facilitate the process of system simulation; (ii) to develop a mathematical statement of the problem for optimization; (iii) to develop computer codes for the analysis and design as it is more efficient to operate with equations rather than with tabular data or charts; (iv) to store equipment performance data to facilitate computations and simulations; and (v) to automate equipment selection process. The processes in devices such as turbines and compressors are very complex and modeling of such processes involves extensive computational fluid dynamics (CFD) methods. To accelerate the design analysis process, common practices involve the use of experimental or catalog performance data in the form of curve-fitted equations. Heat exchangers, on the other hand, follow certain laws that can be used as a model following a sequence of equations and algorithms.

A.1 Examples of Data for Parametric Representations

A few of the cases where parametric representations are very useful and often used in the design and simulation of thermal systems are shown below.

Data may exist in different formats such as the discrete data in a table or in the form of a performance chart. Performance characteristics of a vapor compression refrigeration system are given in terms of cooling capacity and coefficient of performance (COP) as a function of evaporation temperature, T_e, and condensation temperature, T_c. Figure A.1a shows the cooling capacity values as a function of evaporation temperature and condensation temperature. Figure A.1b shows the performance chart of a centrifugal pump displaying the variation of performance parameters such as pump head with volume flow rate and for different pump speeds. Figure A.1c shows tabular thermodynamics data for saturated water over a range of temperatures. Curve-fitting methods are used to develop a function equation to represent the data. Additionally, thermophysical and transport properties in

Design of Thermal Energy Systems, First Edition. Pradip Majumdar.
© 2021 John Wiley & Sons Ltd. Published 2021 by John Wiley & Sons Ltd.
Companion website: www.wiley.com/go/majumdar

Cooling capacity, Q_c (kW)		
	Condensing Temperature, T_c (C)		
Evaporating Temperature T_e(C)	20	30	40
10	150	130	82
20	192	148	105
25	228	168	120

(a)

Figure A.1 Example of data for parametric representation of tablular data for thermodynamic properties. (a) Cooling capacity of a vapor compression system, (b) performance characteristics of a centrifugal pump, and (c) tabular data of thermodynamic properties of water – saturation data.

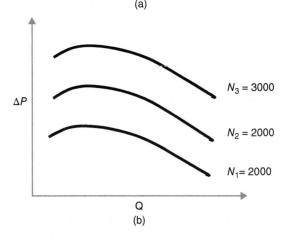

$N_3 = 3000$

$N_2 = 2000$

$N_1 = 2000$

ΔP

Q

(b)

tabular form and charts can also be used to develop the functional form as a function of temperature.

A.2 Basic Approaches for Equation Development

Curve-fitting approaches involve development of equations for the purpose of purely number processing operations. Many devices such as turbine, compressor, and valve involve complex flow processes and can be represented by a mathematical model or a CFD model. However, sometimes the model is so complex that it is more advantageous to use experimental or catalog data to develop the equation using curve-fitting techniques for use in design and simulation analysis of thermal systems. Alternatively, we can develop a *performance chart* using the mathematical model over a range of operating conditions and parameters. Equations can then be developed using curve-fitting techniques based on these charts. In many cases, use of some physical laws helps develop equations. For example, *heat exchangers* follow certain laws that suggest a form for the equation and this insight can be used in the design and simulation process. For a tabular set of data, a functional relation can be developed to use it as an interpolation function to determine data at some intermediate points.

Curve-fitting to a set of data can be achieved using several approaches. Two most common approaches are (i) **polynomial curve fit** and (ii) **least square regression curve fit**. In the first approach, usually polynomial or transcendental functions are used to approximately

Temp (°C)	Pressure (kPa)	Specific Volume (m³/kg)		Enthalpy (kJ/kg)		Entropy (kj/kg.K)	
		Sat. liquid	Sat. vapor	Sat. liquid	Sat. vapor	Sat. liquid	Sat. vapor
0.01							
10	1.2276	0.001 000	106.377	41.99	2519.74	0.1510	8.9007
20	2.339	0.001 002	57.7897	83.94	2538.06	0.2966	8.6671
30	4.246	0.001 004	32.8932	125.77	2556.25	0.4369	8.4533
40	7.384	0.001 008	19.5229	167.54	2574.26	0.5724	8.2569
50	12.350	0.001 012	12.0318	209.31	2592.06	0.7037	8.0762
60	19.941	0.001 017	7.67071	251.11	2609.59	0.8311	7.9095
70	31.19	0.001 023	5.04217	292.96	2626.80	0.9548	7.7552
80	47.39	0.001 029	3.40715	334.88	2643.66	1.0752	7.6121
90	70.14	0.001 036	2.36036	376.90	2660.09	1.1924	7.4790
100	101.3	0.001 044	1.67290	419.02	2676.05	1.3068	7.3548

(c)

Figure A.1 *(Continued)*

represent the data set. In this approach, the derived fitting function passes exactly through all data points. In the second most popular approach, the representing function equation of the curve does not necessarily pass through all data set points, but rather passes close enough to all data points. Such an approach is referred to as the **Least Square Regression** curve fit.

A.3 Parametric Representation Techniques

A.3.1 Polynomial Curve-Fitting

A most common and useful form of equation representation is a polynomial function [$y(x)$] that involves one dependent variable (y) as a function of an independent variable (x).

$$y = a_0 + a_1 x + a_2 x^2 \cdots + a_n x^n \tag{A.1}$$

where n represents the degree of the polynomial. The $(n+1)$ polynomial coefficients $a_0, a_1, \ldots, a_n$ can be determined by fitting the polynomial function to pass exactly through $(n+1)$ data points.

In many situations, the function $[f(x, y)]$ may involve two or more independent variables. Example: Flow rate in an axial compressor is a function of inlet pressure, inlet temperature, compressor speed, and outlet pressure.

A.3.1.1 Polynomial Curve-Fitting – Single Variable

One example of a polynomial curve-fitting is the pressure drop characteristics of a device represented as a function of volume flow rate, $\Delta P = f(Q)$. Depending on the degree of the polynomial required, a number $(\Delta P, Q)$ pair of data points could be read off from the pressure drop characteristics plot shown in Figure A.2.

Similarly, thermophysical properties of fluids and solids vary with temperature and representation of these functional variations can easily be developed using polynomial curve-fitting methods. Another example is the equation representation of thermodynamic data such as the saturation data for enthalpy and entropy values of saturated vapor and liquid corresponding to a range of temperatures as depicted in the Figure A.1c.

Based on these data, the following functional form of saturation water data could be developed to facilitate the design analysis through the use of computer program or for performing simulation analysis of a thermodynamic steam power system: $h_f(T)$, $h_g(T)$, $h_{fg}(T)$, $s_f(T)$, $s_g(T)$, $s_{fg}(T)$, $v_f(T)$, $v_g(T)$, and $v_{fg}(T)$.

When the number of data points available is precisely the same as the degree of the polynomial equation plus 1, i.e. $(n+1)$, a polynomial equation can be derived that expresses these data points exactly. Let us consider a set of n data points (x_1, y_1), (x_2, y_2), ... (x_n, y_n) obtained either from a table or from a performance curve or from a performance map. Substituting this data successively into the n-degree polynomial, we obtain the following system of equations.

$$y_1 = a_0 + a_1 x_1 + a_2 x_1^2 \cdots + a_n x_1^n$$
$$y_2 = a_0 + a_1 x_2 + a_2 x_2^2 \cdots + a_n x_2^n$$
$$\vdots$$
$$y_{n+1} = a_0 + a_1 x_{n+1} + a_2 x_{n+1}^n \cdots + a_n x_{n+1}^n \tag{A.2a}$$

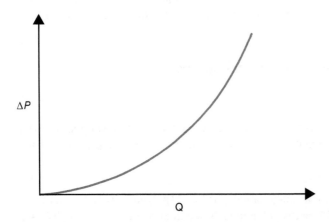

Figure A.2 Typical pressure drop characteristics of a device.

This equation can be written in the following matrix form as

$$
\begin{bmatrix}
1 & x_1 & x_1^2 & \cdots & x_1^n \\
1 & x_2 & x_2^2 & \cdots & x_2^n \\
\vdots & \vdots & \vdots & \cdots & \vdots \\
\vdots & \vdots & \vdots & \cdots & \vdots \\
1 & x_{n+1} & x_{n+1}^2 & \cdots & x_{n+1}^n
\end{bmatrix}
\begin{Bmatrix}
a_0 \\
a_1 \\
\vdots \\
\vdots \\
a_n
\end{Bmatrix}
=
\begin{Bmatrix}
y_0 \\
y_1 \\
\vdots \\
\vdots \\
y_{n+1}
\end{Bmatrix}
\qquad\text{(A.2b)}
$$

The system of linear equations given by Eq. (A.2b) can be solved for the unknown coefficients $a_0, a_1, \ldots a_n$ of the polynomial representation form given by Eq. (A.1).

Example A.1   Determine a curve-fit equation for the enthalpy of saturated liquid water as a function of temperature using ninth degree polynomial equations and using data given in the steam table in the range of 10–100 °C.

Solution
Let us use the following set of tabular thermodynamic data from the saturation data Table A.1c. Let us demonstrate the development of this curve-fitted polynomial equation considering the data for the enthalpy of saturated liquid water property. We can use the ninth degree polynomial for the saturated liquid enthalpy values as follows:

$$
h_f = a_0 + a_1 T + a_2 T^2 + a_3 T^3 + a_4 T^4 + a_5 T^5 + a_6 T^6 + a_7 T^7 + a_8 T^8 + a_9 T^9 \qquad\text{(EA.1)}
$$

Let us select 10 data points and substitute them successively in Eq. (EA.1) and derive the following system of linear equations written in a matrix form as

$$
\begin{bmatrix}
1 & 273.16 & 273.16^2 & 273.16^3 & 273.16^4 & 273.16^5 & 273.16^6 & 273.16^7 & 273.16^8 & 273.16^9 \\
1 & 283.15 & 283.15^2 & 283.15^3 & 283.15^4 & 283.15^5 & 283.15^6 & 283.15^7 & 283.15^8 & 283.15^9 \\
1 & 293.15 & 293.15^2 & 293.15^3 & 293.15^4 & 293.15^5 & 293.15^6 & 293.15^7 & 293.15^8 & 293.15^9 \\
1 & 303.15 & 303.15^2 & 303.15^3 & 303.15^4 & 303.15^5 & 303.15^6 & 303.15^7 & 303.15^8 & 303.15^9 \\
1 & 313.15 & 313.15^2 & 313.15^3 & 313.15^4 & 313.15^5 & 313.15^6 & 313.15^7 & 313.15^8 & 313.15^9 \\
1 & 333.15 & 333.15^2 & 333.15^3 & 333.15^4 & 333.15^5 & 333.15^6 & 333.15^7 & 333.15^8 & 333.15^9 \\
1 & 343.15 & 343.15^2 & 343.15^3 & 343.15^4 & 343.15^5 & 343.15^6 & 343.15^7 & 343.15^8 & 343.15^9 \\
1 & 353.15 & 353.15^2 & 353.15^3 & 353.15^4 & 353.15^5 & 353.15^6 & 353.15^7 & 353.15^8 & 353.15^9 \\
1 & 363.15 & 363.15^2 & 363.15^3 & 363.15^4 & 363.15^5 & 363.15^6 & 363.15^7 & 363.15^8 & 343.15^9 \\
1 & 373.15 & 373.15^2 & 373.15^3 & 373.15^4 & 373.15^5 & 373.15^6 & 373.15^7 & 343.15^8 & 343.15^8
\end{bmatrix}
$$

$$
\times
\begin{Bmatrix}
a_0 \\
a_1 \\
a_2 \\
a_3 \\
a_4 \\
a_5 \\
a_6 \\
a_7 \\
a_8 \\
a_9
\end{Bmatrix}
=
\begin{Bmatrix}
0.01 \\
42.01 \\
83.96 \\
125.79 \\
167.57 \\
251.13 \\
292.98 \\
334.91 \\
376.92 \\
419.04
\end{Bmatrix}
\qquad\text{(EA.2)}
$$

Solving the matrix equation, we get the following set of coefficient constants for the curve-fitted equation of the saturated liquid water:

$a_0 = 9\,959\,642 \times 10^6$, $a_1 = -2.739\,99 \times 10^5$, $a_2 = 3344.126$, $a_3 = -23.7669$,

$a_4 = 0.108\,398$, $a_5 = -0.33 \times 10^{-3}$, $a_6 = 6.65 \times 10^{-7}$, $a_7 = 8.6 \times 10^{-10}$,

$a_8 = 6.5 \times 10^{-13}$, and $a_9 = -2.2 \times 10^{-16}$

The curve-fitted equation for the saturated liquid water is given as

$$h_f = 9959642 \times 10^6 - 2.73999 \times 10^5\,T + 3.344.126\,T^2$$

$$-23.7669\,T^3 + 0.108398\,T^4$$

$$-0.33 \times 10^{-3}\,T^5 + 6.65 \times 10^{-7}\,T^6 + 8.6 \times 10^{-10}\,T^7 + 6.5 \times 10^{-13}\,T^8$$

$$-2.2 \times 10^{-16}\,T^9$$

A.3.1.2 Polynomial Curve-Fitting – Two Variables or More Variables

In many situations, the representative function may involve two or more variables such as $z = f(x, y)$. For example, the pressure rise in a pump is expressed as a function of flow rate Q and RPM (N) as $\Delta P = f(Q, N)$; or the flow rate in an axial compressor as a function of multiple variables such as the inlet pressure, inlet temperature, compressor speed, and outlet pressure. There are multiple approaches to deriving such functional representations.

A.3.1.2.1 Approach-1

One of the simplest approaches to developing such a multivariable representative function can be developed based on the multiple use of second degree or third degree polynomial function. The methodology is described here considering the performance of a pump. Performance characteristics of a pump are presented in terms of pressure rise as a function of speed N and flow rate Q, as demonstrated in the Figure A.1b.

In this approach, a curve-fitted equation is obtained for each curve first. Then the coefficients of these individual equations are combined to form the parametric representation of the complete performance map.

$$z = \left(A_0 + A_1 x + A_2 x^2\right)\left(B_0 + B_1 x + B_2 x^2\right) y + \left(C_0 + C_1 x + C_2 x^2\right) y^2 \tag{A.3}$$

Three points $(\Delta P_{11}, Q_1)$, $(\Delta P_{12}, Q_2)$, and $(\Delta P_{13}, Q_3)$ on the curve for $N_1 = 1000$ would provide the constants in the equation

$$\Delta P = a_1 + b_1 Q + c_1 Q^2 \tag{A.4a}$$

based on the solution of the matrix equation given as

$$\begin{bmatrix} 1 & Q_1 & Q_1^2 \\ 1 & Q_1 & Q_2^2 \\ 1 & Q_1 & Q_3^2 \end{bmatrix} \begin{Bmatrix} a_1 \\ b_1 \\ c_1 \end{Bmatrix} = \begin{Bmatrix} \Delta P_{11} \\ \Delta P_{12} \\ \Delta P_{13} \end{Bmatrix} \tag{A.4b}$$

Similarly, three points $(\Delta P_{21}, Q_1)$, $(\Delta P_{22}, Q_2)$, and $(\Delta P_{23}, Q_3)$ on the curve for $N_2 = 2000$ would provide the constants in the equation

$$\Delta P = a_2 + b_2 Q + c_2 Q^2 \tag{A.5a}$$

based on the solution of the matrix equation given as

$$\begin{bmatrix} 1 & Q_1 & Q_1^2 \\ 1 & Q_1 & Q_2^2 \\ 1 & Q_1 & Q_{3.}^2 \end{bmatrix} \begin{Bmatrix} a_2 \\ b_2 \\ c_2 \end{Bmatrix} = \begin{Bmatrix} \Delta P_{21} \\ \Delta P_{22} \\ \Delta P_{23} \end{Bmatrix} \tag{A.5b}$$

Similarly, three points $(\Delta P_{31}, Q_1)$, $(\Delta P_{32}, Q_2)$, and $(\Delta P_{33}, Q_3)$ on the curve for $N_3 = 3000$ would provide the constants in the equation

$$\Delta P = a_3 + b_3 Q + c_3 Q^2 \tag{A.6a}$$

based on the solution of the matrix equation given as

$$\begin{bmatrix} 1 & Q_1 & Q_1^2 \\ 1 & Q_1 & Q_2^2 \\ 1 & Q_1 & Q_{3.}^2 \end{bmatrix} \begin{Bmatrix} a_3 \\ b_3 \\ c_3 \end{Bmatrix} = \begin{Bmatrix} \Delta P_{31} \\ \Delta P_{32} \\ \Delta P_{33} \end{Bmatrix} \tag{A.6b}$$

In the next step, the constant vector **a** can be determined using the three data points (a_1, N_1), (a_2, N_2), and (a_3, N_3) in fitting the following quadratic form

$$\mathbf{a} = A_0 + A_1 N + A_2 N^2 \tag{A.7a}$$

and based on solving the following matrix form

$$\begin{bmatrix} 1 & N_1 & N_1^2 \\ 1 & N_2 & N_2^2 \\ 1 & N_3 & N_{3.}^2 \end{bmatrix} \begin{Bmatrix} A_0 \\ A_1 \\ A_2 \end{Bmatrix} = \begin{Bmatrix} a_1 \\ a_2 \\ a_3 \end{Bmatrix} \tag{A.7b}$$

Similarly, the constant vector **b** can be determined using the three data points (b_1, N_1), (b_2, N_2), and (b_3, N_3) in fitting the following quadratic form

$$\mathbf{b} = B_0 + B_1 N + B_2 N^2 \tag{A.8a}$$

and based on solving the following matrix form

$$\begin{bmatrix} 1 & N_1 & N_1^2 \\ 1 & N_2 & N_2^2 \\ 1 & N_3 & N_{3.}^2 \end{bmatrix} \begin{Bmatrix} B_0 \\ B_1 \\ B_2 \end{Bmatrix} = \begin{Bmatrix} b_1 \\ b_2 \\ b_3 \end{Bmatrix} \tag{A.8b}$$

Similarly, the constant vector **c** can be determined using the three data points (b_1, N_1), (b_2, N_2), and (b_3, N_3) in fitting the following quadratic form

$$\mathbf{c} = C_0 + C_1 N + C_2 N^2 \tag{A.9a}$$

and based on solving the following matrix form

$$\begin{bmatrix} 1 & N_1 & N_1^2 \\ 1 & N_2 & N_2^2 \\ 1 & N_3 & N_{3.}^2 \end{bmatrix} \begin{Bmatrix} C_0 \\ C_1 \\ C_2 \end{Bmatrix} = \begin{Bmatrix} c_1 \\ c_2 \\ c_3 \end{Bmatrix} \tag{A.9b}$$

Finally, the two-dimensional parametric representation of pump performance characteristics is given by

$$\Delta P = \mathbf{a} + \mathbf{b}\, Q + \mathbf{c}\, Q_2 \tag{A.10}$$

Substituting the expressions for the vectors **a**, **b**, and **c** from Eqs. (A.7a), (A.8a), and (A.9a) respectively into Eq. (A.10), we get the final curve-fitted expression

$$\Delta P = \left(A_0 + A_1 N + A_2 N^2\right) + \left(B_0 + B_1 N + B_2 N^2\right) Q + \left(C_0 + C_1 N + C_2 N\right) Q^2 \tag{A.11}$$

A.3.1.2.2 Approach-2

As an alternate approach, a single parametric equation can be derived by selecting an eighth degree $(n = 8)$ polynomial and selecting a total of $n + 1 = 9$ data points from the performance map with three data points set from each curve. The procedure starts with assuming the eighth degree polynomial of the following form:

$$\Delta P = a_0 + a_1 Q + a_2 Q^2 + a_3 N + a_4 N^2 + a_5 QN + a_6 Q^2 N + a_7 QN^2 + a_8\, Q^2 N^2 \tag{A.12}$$

In the next step, the nine data points are substituted directly into Eq. (A.12) successively and form the system of nine linear simultaneous equations. The solution of the systems of equations results in the values of the unknown coefficients set $(a_0, a_1, a_3, \ldots, a_8)$.

Example A.2   The cooling capacity of a vapor compression refrigeration system is a function of the evaporating and condensing temperature. The performance data as a function of three different evaporating and condensing temperatures are shown in the table.

Determine a parametric representation of the cooling capacity as a function of evaporation and condensation temperatures: $Q_c = f(T_c, T_e)$.

Solution

For $T_c = 20°$, the three pairs of points (10, 150), (20, 192), and (25, 228) can be used to obtain a curve-fit equation of the form

$$Q_c = a_1 + b_1 T_e + c_1 T_e^2$$

Substituting three data points, we get the system of equations as

$$\begin{bmatrix} 1 & 10 & 10^2 \\ 1 & 20 & 20^2 \\ 1 & 25 & 25^2 \end{bmatrix} \begin{bmatrix} a_1 \\ b_1 \\ c_1 \end{bmatrix} = \begin{bmatrix} 150 \\ 192 \\ 228 \end{bmatrix}$$

After solving the above matrix, we get $a_1 = 148$, $b_1 = -1.8$, $c_1 = 0.2$.
The curve fit equation for $T_c = 20°$ can be written as

$$Q_c = 148 - 1.8 T_e + 0.2 T_e^2$$

Similarly, for $T_c = 30°$ the three pairs of points (10, 130), (20, 148), and (25, 168) can be used to obtain a curve-fit equation of the form

$$Q_c = a_2 + b_2 T_e + c_2 T_e^2$$

Substituting the three data points, we get the system of equations as

$$\begin{bmatrix} 1 & 10 & 10^2 \\ 1 & 20 & 20^2 \\ 1 & 25 & 25^2 \end{bmatrix} \begin{bmatrix} a_2 \\ b_2 \\ c_2 \end{bmatrix} = \begin{bmatrix} 130 \\ 148 \\ 168 \end{bmatrix}$$

After solving the above matrix, we get $a_2 = 141.333, b_2 = -2.6, c_2 = 0.1466$
The curve fit equation for $T_c = 30°$ can be written as

$$Q_c = 141.333 - 2.6T_e + 0.1466T_e^2$$

Similarly, for $T_c = 40°$ the three pairs of points (10, 82), (20, 105), and (25, 120) can be used to obtain a curve-fit equation of the form

$$Q_c = a_3 + b_3T_e + c_3T_e^2$$

Substituting the three data points, we get the system of equations as

$$\begin{bmatrix} 1 & 10 & 10^2 \\ 1 & 20 & 20^2 \\ 1 & 25 & 25^2 \end{bmatrix} \begin{bmatrix} a_3 \\ b_3 \\ c_3 \end{bmatrix} = \begin{bmatrix} 82 \\ 105 \\ 120 \end{bmatrix}$$

After solving the above matrix, we get $a_3 = 68.333, b_3 = 0.9, c_3 = 0.04666$.
The curve fit equation for $T_c = 40°$ can be written as

$$Q_c = 68.333 + 0.9T_e + 0.046\,66T_e^2$$

In the next step, we will consider $a_1 = 148$, $a_2 = 141.333$, $a_3 = 68.333$ and use the three data points $(a_1, 25°C), (a_2, 30°C)$, and $(a_3, 40°C)$ to derive the second degree equation of T_c in the following manner:

$$a = A_0 + A_1T_c + A_2T_c^2$$

$$\begin{bmatrix} 1 & 25 & 25^2 \\ 1 & 30 & 30^2 \\ 1 & 40 & 40^2 \end{bmatrix} \begin{bmatrix} A_0 \\ A_1 \\ A_2 \end{bmatrix} = \begin{bmatrix} 148 \\ 141.333 \\ 68.333 \end{bmatrix}$$

By solving, we get
$A_0 = -117, A_1 = 20.544, A_2 = -0.3977$
Similarly, we consider $b_1 = -1.8, b_2 = -2.6, b_3 = 0.9$ and use the three data points $(b_1, 25°C), (b_2, 30°C)$, and $(b_3, 40°C)$ to derive the second degree equation of T_c in the following manner:

$$b = B_0 + B_1T_c + B_2T_c^2$$

$$\begin{bmatrix} 1 & 25 & 25^2 \\ 1 & 30 & 30^2 \\ 1 & 40 & 40^2 \end{bmatrix} \begin{bmatrix} B_0 \\ B_1 \\ B_2 \end{bmatrix} = \begin{bmatrix} -1.8 \\ -2.6 \\ 0.9 \end{bmatrix}$$

Solving, we get
$B_0 = 27.7, B_1 = -2.03, B_2 = 0.034$
Similarly, we consider $c_1 = 0.2, c_2 = 0.1466, c_3 = 0.0466$ and use the three data points $(c_1, 25°C), (c_2, 30°C)$, and $(c_3, 40°C)$ to derive the second degree equation of T_c in the following manner:

$$c = C_0 + C_1T_c + C_2T_c^2$$

$$\begin{bmatrix} 1 & 25 & 25^2 \\ 1 & 30 & 30^2 \\ 1 & 40 & 40^2 \end{bmatrix} \begin{bmatrix} B_0 \\ B_1 \\ B_2 \end{bmatrix} = \begin{bmatrix} 0.2 \\ 0.1466 \\ 0.0466 \end{bmatrix}$$

By solving the matrix form, we get

$C_0 = 0.5$, $C_1 = -0.01311$, $C_2 = 0.000\,044$

Finally, the parametric representation of the cooling capacity as a function of evaporation and condensation temperatures is given as

$$Q_c = \left(-117 + 20.544 T_c - 0.3977 T_c^2\right) + \left(27.7 - 2.03 T_c + 0.034 T_c^2\right) T_e$$
$$+ \left(0.5 - 0.013\,11 T_c + 0.000\,044 T_c^2\right) T_e^2$$

A.3.2 Least Square Regression Curve-Fitting

This is a method of determining a best curve fit to a given set of data. The curve fit function that represents the data set is derived assuming that the curves do not necessarily pass through all data points, but rather pass near all data points following the general trend of the data with minimum overall error. In linear regression, a linear line form is used to fit a set of data points. Consider a set of n data points: (x_1, y_1), (x_2, y_2), (x_3, y_3) ... (x_n, y_n), where y represents the dependent variable and x is the independent variable. Let us assume that this data set can be fitted assuming a linear approximate function of the form

$$Y = C_1 + C_2 x \tag{A.13}$$

where Y represents the estimated value of the dependent variable given by this linear function equation.

Substituting the data points successively into Eq. (A.13),

$$Y_1 = C_1 + C_2 x_1$$
$$Y_2 = C_1 + C_2 x_2$$
$$Y_3 = C_1 + C_2 x_3$$
$$\vdots$$
$$Y_n = C_1 + C_2 x_n \tag{A.14}$$

Next the errors, also known as the **residuals** (r), are computed by comparing the values of Y_i, $i = 1, 2, 2, ...n$ given by the approximate Eq. (A.14) to the original data set values, y_i, as follows:

$$r_1 = C_1 + C_2 x_1 - y_1$$
$$r_2 = C_1 + C_2 x_1 - y_2$$
$$r_3 = C_1 + C_2 x_3 - y_3$$
$$\vdots$$
$$r_n = C_1 + C_2 x_n - y_n \tag{A.15}$$

In the least square method, the objective is to select the correlation coefficient values of C_1 and C_2 in such a way that the sum of the squares of the residuals for all data points is sought to be minimized:

$$S = \sum_{i=1}^{n} r_i^2 = \sum_{i=1}^{n} \left(Y_i - y_i\right)^2 = \sum_{i=1}^{n} \left(C_1 x_i + C_2 - y_i\right)^2 \tag{A.16}$$

For minimizing the residual function $S = S(C_1, C_2)$, let us set the partial derivatives to zero and solve for the constants:

$$\frac{\partial S}{\partial C_2} = 0 \quad \text{or} \quad 2\sum_{i=1}^{n} (C_1 x_i + C_2 - y_i) = 0 \tag{A.17a}$$

$$\frac{\partial S}{\partial C_1} = 0 \quad \text{or} \quad 2\left(\sum_{i=1}^{n} x_i\right) \sum_{i=1}^{n} (C_1 x_i + C_2 - y_i) = 0 \tag{A.17b}$$

These two equations lead to a system of equations:

$$nC_2 + \left(\sum_{i=1}^{n} x_i\right) C_1 = \sum_{i=1}^{n} y_i \tag{A.18a}$$

$$\left(\sum_{i=1}^{n} x_i\right) C_2 + \left(\sum_{i=1}^{n} x_i^2\right) C_1 = \sum_{i=1}^{n} x_i y_i \tag{A.18b}$$

In a matrix form as

$$\begin{bmatrix} n & \left(\sum_{i=1}^{n} x_i\right) \\ \left(\sum_{i=1}^{n} x_i\right) & +\left(\sum_{i=1}^{n} x_i^2\right) \end{bmatrix} \begin{Bmatrix} C_2 \\ C_1 \end{Bmatrix} = \begin{Bmatrix} \sum_{i=1}^{n} y_i \\ \sum_{i=1}^{n} y_i \end{Bmatrix} \tag{A.19}$$

The linear system of equations can simply be solved by using the **Cramer rule**:

$$C_1 = \frac{\begin{vmatrix} \left(\sum_{i=1}^{n} y_i\right) & \left(\sum_{i=1}^{n} x_i\right) \\ \sum_{i=1}^{n} x_i y_i & \left(\sum_{i=1}^{n} x_i^2\right) \end{vmatrix}}{\begin{vmatrix} n & \left(\sum_{i=1}^{n} x_i\right) \\ \left(\sum_{i=1}^{n} x_i\right) & \left(\sum_{i=1}^{n} x_i^2\right) \end{vmatrix}} \tag{A.20a}$$

and

$$C_2 = \frac{\begin{vmatrix} n & \sum_{i=1}^{n} y_i \\ \left(\sum_{i=1}^{n} x_i\right) & \sum_{i=1}^{n} x_i y_i \end{vmatrix}}{\begin{vmatrix} n & \left(\sum_{i=1}^{n} x_i\right) \\ \left(\sum_{i=1}^{n} x_i\right) & \left(\sum_{i=1}^{n} x_i^2\right) \end{vmatrix}} \tag{A.20b}$$

or

$$C_1 = \frac{\left(\sum_{i=1}^{n} y_i\right) \left(\sum_{i=1}^{n} x_i^2\right) - \left(\sum_{i=1}^{n} x_i\right) \sum_{i=1}^{n} x_i y_i}{n\left(\sum_{i=1}^{n} x_i^2\right) - \left(\sum_{i=1}^{n} x_i\right)^2} \tag{A.21a}$$

$$C_2 = \frac{n\left(\sum_{i=1}^{n} x_i y_i\right) - \left(\sum_{i=1}^{n} y_i\right) \left(\sum_{i=1}^{n} x_i\right)}{n\left(\sum_{i=1}^{n} x_i^2\right) - \left(\sum_{i=1}^{n} x_i\right)^2} \tag{A.21b}$$

A.3.2.1 Accuracy of the Least Square Curve Fit

The accuracy of the linear least square regression is defined in terms of standard error as given by the **standard deviation** for the regression line fitting the data set as

$$S_{y/x} = \sqrt{\frac{S}{n-2}} \tag{A.22}$$

where S is the sum of the square of all residuals of the data and n is the number of data points.

Another way of defining the standard error of the least square curve fit is expressed in terms of the correlation coefficient defined as

$$\text{Correlation Coefficient, } r = \sqrt{\left(\frac{S_0 - S}{S_0}\right)} \tag{A.23}$$

where S is the sum of the squares of all residuals of the data around the linear curve fit equation:

$$S = \sum_{i=1}^{n} (Y_i - y_i)^2 = \sum_{i=1}^{n} (C_1 x_i + C_2 - y_i)^2 \tag{A.24}$$

and S_0 is the sum of the squares of the deviations of the dependent variable values of the data set around the mean values of the dependent variable values:

$$S_0 = \sum_{i=1}^{n} (y_i - \bar{y})^2 \tag{A.25}$$

where the mean value is given as

$$\bar{y} = \sqrt{\frac{1}{n} \sum_{i=1}^{n} y_i} \tag{A.26}$$

A good least square curve fit is given by the higher values of the correlation coefficient, r compared to 1. Closer the values of the correlation coefficient r to 1, better is the representation by the least square curve ft.

Example A.3 Parametric Representation for Saturated Pressure as a function of time Temperature

A saturated temperature and pressure relation is typically fitted by an equation of the form:

$$\ln(T) = A + \frac{B}{T} \tag{A.3.1}$$

$$\ln(P) = A + \frac{B}{T} \tag{A.3.2}$$

where

P = saturated pressure

T = absolute temperature

Based on the linear least square method, the unknowns A and B are given by the following two sets of equations:

$$nA + \left(\sum_{i=1}^{n} \frac{1}{T_i}\right) B = \sum_{i=1}^{n} \ln(P)_i$$

$$\left(\sum_{i=1}^{n} \frac{1}{T_i}\right) A + \left(\sum_{i=1}^{n} \frac{1}{T_i^2}\right) B = \sum_{i=1}^{n} \frac{1}{T_i} \ln(P_i)$$

and in matrix form as

$$\begin{bmatrix} n & \sum_{i=1}^{n} \frac{1}{T_i} \\ \sum_{i=1}^{n} \frac{1}{T_i} & \sum_{i=1}^{n} \frac{1}{T_i^2} \end{bmatrix} \begin{Bmatrix} A \\ B \end{Bmatrix} = \begin{Bmatrix} \sum_{i=1}^{n} \ln(P_i) \\ \sum_{i=1}^{n} \frac{1}{T_i} \ln(p_i) \end{Bmatrix}$$

Constants A and B are obtained either by using the Cramer rule or by using liners solvers such as the Gauss elimination solver.

Let us now demonstrate the development of curve-fitted saturated data using the thermodynamics data. Use the least square method to determine the values of A and B that give the best fit of saturation pressure data for steam in the range of 20–350 °C. Compare the values of P at those points using the curve fit equations.

Number, n	Temperature, T (K)	Pressure, P (kPa)	$1/T$	$1/T^2$	ln (P)	$(1/T) \times$ ln (P)
1	293.15	2.339	0.003 411 22	1.163 64E-05	0.849 723 49	0.002 898 6
2	308.15	5.628	0.003 245 17	1.053 11E-05	1.727 754 14	0.005 606 86
3	323.15	12.35	0.003 094 54	9.576 17E-06	2.513 656 06	0.007 778 6
4	338.15	25.03	0.002 957 27	8.745 43E-06	3.220 075 11	0.009 522 62
5	353.15	47.39	0.002 831 66	8.018 29E-06	3.858 411 24	0.010 925 7
6	368.15	84.55	0.002 716 28	7.378 2E-06	4.437 343 08	0.012 053 08
7	383.15	143.3	0.002 609 94	6.811 81E-06	4.964 940 33	0.012 958 22
8	398.15	232.1	0.002 511 62	6.308 22E-06	5.447 168 31	0.013 681 2
9	413.15	361.3	0.002 420 43	5.858 47E-06	5.889 708 64	0.014 255 62
10	428.15	543.1	0.002 335 63	5.455 17E-06	6.297 293 47	0.014 708 15
11	443.15	791.7	0.002 256 57	5.092 12E-06	6.674 182 53	0.015 060 78
12	458.15	1122.7	0.002 182 69	4.764 14E-06	7.023 491 78	0.015 330 11
13	473.15	1553.8	0.002 113 49	4.466 86E-06	7.348 458 82	0.015 530 93
14	488.15	2104.2	0.002 048 55	4.196 56E-06	7.65169063	0.01567488
15	503.15	2794.9	0.001 987 48	3.950 07E-06	7.935 551 61	0.015 771 74
16	518.15	3648.2	0.001 929 94	3.724 68E-06	8.201 989 17	0.015 829 37
17	533.15	4688.6	0.001 875 64	3.518 04E-06	8.452 889 31	0.015 854 62
18	548.15	5941.8	0.001 824 32	3.328 14E-06	8.689 767 4	0.015 852 9
19	563.15	7436	0.001 775 73	3.153 2E-06	8.914 088 35	0.015 828 98
20	578.15	9201.8	0.001 729 65	2.991 71E-06	9.127 154 4	0.015 786 83
21	593.15	11 274	0.001 685 91	2.842 31E-06	9.330 254 47	0.015 730 01
22	608.15	13 694	0.001 644 33	2.703 82E-06	9.524 713 06	0.015 661 78
23	623.15	16 514	0.001 604 75	2.575 22E-06	9.711 963 78	0.015 585 27
$\sum n = 23$		**Sum**	**0.052 792 83**	**0.000 1276 26**	**147.792 269**	**0.307 886 84**

Let us assume the logarithmic form of the function for the saturation pressure as a function of temperature:

$$\ln (P) = A + \frac{B}{T}$$

Noting that this is the same linear form of the Eq. (A.3.2), we can equate the correlation coefficients A and B based on the solution given by Eq. (A.21a and A.21b) are given by the solution of the following linear system of equations:

$$\begin{bmatrix} n & \sum_{i=1}^{23} \frac{1}{T_i} \\ \sum_{i=1}^{23} \frac{1}{T_i} & \sum_{i=1}^{23} \frac{1}{T_i^2} \end{bmatrix} \begin{bmatrix} A \\ B \end{bmatrix} = \begin{bmatrix} \sum_{i=1}^{23} \ln P_i \\ \sum_{i=1}^{23} \frac{1}{T_i} \ln P_i \end{bmatrix}$$

Elements of the matrix equation are computed using the data set and shown in the table. Substituting the sum values from the table

$$\begin{bmatrix} 23 & 0.052\,793 \\ 0.052\,793 & 0.000\,127\,626 \end{bmatrix} \begin{bmatrix} A \\ B \end{bmatrix} = \begin{bmatrix} 147.7923 \\ 0.307\,887 \end{bmatrix}$$

Solving, we get

$$A = 17.585\,37 \text{ and } B = -4861.845\,887$$

The curve fit equation now can be written as

$$\ln(P) = 17.58537 - \frac{4861.845887}{T}$$

The table below shows the comparison of the estimated values of saturation pressure given by the curve-fitted equation to the thermodynamics table data.

Temperature T	Pressure P	Estimated pressure
293.15	2.339	2.719 722 519
308.15	5.628	6.097 262 764
323.15	12.35	12.682 238 5
338.15	25.03	24.719 471 23
353.15	47.39	45.526 087 43
368.15	84.55	79.775 401 66
383.15	143.3	133.783 798 3
398.15	232.1	215.784 232 9
413.15	361.3	336.171 034 5
428.15	543.1	507.703 042 6
443.15	791.7	745.655 283
458.15	1122.7	1067.912 923
473.15	1553.8	1495.004 751
488.15	2104.2	2050.076 618
503.15	2794.9	2758.807 96
518.15	3648.2	3649.276 614
533.15	4688.6	4751.778 617
548.15	5941.8	6098.610 589
563.15	7436	7723.822 686
578.15	9201.8	9662.950 1
593.15	11 274	11 952.730 73
608.15	13 694	14 630.816 06
623.15	16 514	17 735.481 57

A.3.3 Curve-Fitted Correlational Function for Thermophysical Properties

Several polynomial curve-fitted equations are developed by Yaws (1977) for various gases and liquids. A list of such third degree or fourth degree polynomial correlation functions along with correlation coefficients is given in the book by Boem (1987). A brief list of some of the correlation functions for water are summarized in the table below for the convenience of the readers in their computations and in the development of simulation and design analysis.

Water–Vapor:

$k_{wv}(T) = 73.4 \times 10^{-4} - 10.1 \times 10^{-6} T + 18.0 \times 10^{-8} T^2 - 91.0 \times 10^{-12}$ (Range 27 – 1073 ° K)

$C_{p,wv}(T) = 1.88 - 0.167 \times 10^{-3} T + 0.844 \times 10^{-6} - 0.270 \times 10^{-9} (T)$ (Range 273 – 1073 ° K)

$\mu_{wv}(T) = -31.89 + 41.45 \times 10^{-2} T - 8.272 \times 10^{-6} T^2$ (Range 273 – 1073 ° K)

Water–Liquid:

$k_{wl}(T) = -0.3838 - 0.5254 \times 10^{-2} T - 0.0637 \times 10^{-4}$ (Range 273 – 647 ° K)

$C_{p,wl}(T) = 2.823 + 11.83 \times 10^{-3} T - 35.05 \times 10^{-6} T^2 + 36.02 \times 10^{-9} T^3$ (Range 273 – 623 ° K)

$\mu_{wl}(T) = -10.73 + 1828 T - 14.66 \times 10^{-6} T^2$ (Range 273 – 647 ° K)

$\rho_{wl}(T) = 347 + 0.274 T$ (Range 0 – 347.2 ° C)

Bibliography

Boem, R.F. (1987). *Design Analysis of Thermal Systems*. Wiley.

Chapra, C.S. and Canale, R.P. (2015). *Numerical Methods for Engineers*, 7e. New York: Mc-Graw Hill.

Rao, S.S. (2002). *Applied Numerical Methods for Engineers and Scientists*. Prentice Hall.

Stoecker, W.F. (1989). *Design of Thermal Systems*, 3e. McGraw-Hill.

Yaws, C. (1977). *Physical Properties*. New York: Mc-Graw-Hill.

Problems

A.1 Obtain a parametric representation of the saturation temperature as a function of pressure for water using thermodynamic data over a temperature range of 20–350 °C.

A.2 Develop a computer program to determine curve fit equations for the enthalpy and entropy of saturated liquid and saturated vapor of steam as a function of temperature using ninth degree polynomial equation and using the data given in the steam table in the range of 0.01–100 °C.

A.3 Develop a computer program to determine curve fit equations for enthalpy, entropy of saturated liquid and saturated vapor of steam, and specific volume of saturated liquid as a function of temperature using a ninth degree polynomial equation and using data given in the steam table in the range of 20–350 °C.

Appendix B

Economic Analysis and Cost Estimation of Thermal Components and Systems

B.1 Economic Analysis Procedure

As we have discussed, one of the important steps in the selection of thermal components and energy systems is done by evaluating the economics of a project while comparing multiple design options. The purpose of this evaluation is to check the economic feasibility of a new project or to scale up an existing project through projected cost estimation taking into account different cost terms. Some of the major cost items that need to be considered for economic analysis are (i) one-time *capital cost (CC)*; (ii) *regular **Operation and Maintenance cost (O&M)***; (iii) periodic investment cost for the **Replacement** of certain equipment over the life of the project; and (iv) salvage value at the end of the life of the project.

Before proceeding with the economic analysis, let us briefly review some of the basic concepts here.

B.1.1 Some Basic Concepts

B.1.1.1 Interest Rate and Its Effect on Investments

B.1.1.1.1 Future Worth

Future worth is defined as the amount of money that grows in a bank account over a period of time along with an applied interest rate. If X is the amount of money put into a bank account at an applied interest rate that is compounded annually or semiannually, then the *future worth* or the growth of the money depends on the initial amount X, number of years (n) the money is kept in the bank, the interest rate ($i\%$) applied, and the type of interest payments: simple or compounded.

For simple interest, the future accumulated value, X_F, of the invested money, X_P, is

$$X_F = (1 + ni)X_P \tag{B.1}$$

where

X_P	=	Present worth value
X_F	=	Future worth value
i	=	Interest rate
n	=	Number of years invested

Design of Thermal Energy Systems, First Edition. Pradip Majumdar.
© 2021 John Wiley & Sons Ltd. Published 2021 by John Wiley & Sons Ltd.
Companion website: www.wiley.com/go/majumdar

For compound interest, the future accumulated value, X_F, of the invested money, X_P, is

$$X_F = \left(1 + \frac{i}{m}\right)^{mn} X_P \tag{B.2}$$

where

m = Number of compounding periods per year

B.1.1.1.2 Present Worth

Equations (B.1) and (B.2) are often used to determine the present worth (PW) of any future incomes in the following manner.

For simple interest, the **present worth** of any future accumulated values, X_F, is

$$X_P = \frac{X_F}{(1 + ni)} \tag{B.3}$$

where

X_P = Present worth value

X_F = Future worth value

i = Interest rate

n = Number of years invested

For compound interest, the future accumulated value, X_F, of the invested money, X_P, is

$$X_P = \frac{X_F}{\left(1 + \frac{i}{m}\right)^{mn}} \tag{B.4}$$

Present worth estimates are applied for any future incomes such as the salvage value from an equipment or any regular incomes earned based on any services provided.

B.1.1.1.3 Uniform Present Worth

This represents the amount of money estimated at the beginning of a project that is equivalent to **uniform regular savings (X_A)** or **uniform regular earnings** or **uniform regular spending** at regular compounding periods within the life of the project.

For simple interest:

$$X_P = X_A \frac{(1 + i)^n - 1}{i(1 + i)^n} \tag{B.5}$$

For compound interest:

$$X_P = X_A \frac{\left(1 + \frac{1}{m}\right)^{mn} - 1}{\frac{i}{m}\left(1 + \frac{i}{m}\right)^{mn}} \tag{B.6}$$

Example B.1   *Estimate Total Present Worth*

Consider a periodic earning of $4000 per year for 10 years at an annual interest rate of 8% that is compounded semiannually. Find the total present value of the earning at the beginning of the time.

Solution

$$X_P = X_A \frac{\left(1 + \frac{i}{m}\right)^{mn} - 1}{\frac{i}{m}\left(1 + \frac{i}{m}\right)^{mn}}$$

$$X_P = 4000 \times \frac{\left(1 + \frac{0.08}{2}\right)^{2 \times 10} - 1}{\frac{0.08}{2}\left(1 + \frac{0.08}{2}\right)^{2 \times 10}} = 4000 \times \frac{1.191}{0.0876} = \$54\,384$$

Present Value, $X_P = \$54\,384$

B.1.1.1.4 Future Worth of Uniform Series of Amounts

The future worth of a uniform series of payment, X_R, with an interest rate i that is compounded at the same frequency as the payment is computed as

$$X_F = \frac{(1 + i)^n - 1}{i} X_R \tag{B.7a}$$

Equation (B.7a) can also be rearranged to estimate the amount of money required to be deposited periodically in a bank to accumulate a certain amount of money over a period to replace an existing equipment at the end of its life.

$$X_R = \frac{i}{(1 + i)^n - 1} X_F \tag{B.7b}$$

Example B.2   *Estimate Periodic Savings for Future Worth*
A new cooling system has been installed in a power generation plant. It is decided that a certain fixed amount of money will be set aside and invested each year starting from the first year so that $10\,000 will be available in 15 years for the replacement of the cooling system. How much money must be set aside each year in a bank with an interest rate of 6% and compounded annually?

Solution
From Eq. (B.7b)

$$X_R = \frac{i}{(1 + i)^n - 1} X_F$$

Substituting the values

$$X_R = \frac{0.06}{(1 + 0.06)^{15} - 1} \times \$10\,000.00$$

Regular yearly savings, $X_R = \$430.0$

B.1.2   Some Common Methods of Economic Evaluation

Some of the popular methods of economic evaluations are discussed here. Among these, methods such as ***Return on Investment (ROI) Method*** and ***Payback Method*** are quite simple and often used for performing a quick economic analysis to find a comparative ranking of different options for a project. These methods are simple and do not consider all important variables such as the timing of cash flow.

B.1.2.1   Return on Investment (ROI) Method

ROI is defined as the ratio of the Annual Net Benefit (Income – Expenses) to the Original Value of the Project:

$$\text{ROI} = \frac{\text{Net Benefit}}{\text{original value}} = \frac{\text{Net Income} - \text{Expenses}}{\text{Original book value}} \tag{B.8}$$

Example B.3   *Estimate Return on Investment*

Consider two design options for a project with the following estimated data. Determine the economic feasibility of the project based on ROI.

Cost items	Option-1	Option-2
Original capital values	$95 000	$70 000
Annual O&M	$6 000	$3 000
Annual depreciation rate	$4 000	$2 500
Yearly net income	$50 000	$45 000

Option-1:

$$\text{ROI} = \frac{\text{Net Income} - \text{Expenses}}{\text{Original book value}} = \frac{50\,000 - (6000 + 4000)}{95\,000} \times 100$$

$$\text{ROI} = 42.1\%$$

Option-2:

$$\text{ROI} = \frac{\text{Net Income} - \text{Expenses}}{\text{Original book value}} = \frac{45\,000 - (3000 + 2500)}{70\,000} \times 100$$

$$\text{ROI} = 56.4\%$$

As a result of this ROI economic analysis, one may choose Option-2 over Option-1 because the ROI number is higher for Option-2.

B.1.2.2   Payback Method

In this method, comparison is made based on the **number of years to Pay Back** (NPB) the investment or number of years to break even. The NPB is estimated as the ratio of the Capital Cost to the Yearly Benefit minus the yearly expenses:

$$\text{NPB} = \frac{\text{Capital first cost}}{\text{Yearly Benefit} - \text{Yearly expenses}} \tag{B.9}$$

Such methods are applicable for short-term investments and for projects with **uncertain lives.** Some of the major drawbacks of such simple methods are (i) no consideration of the effect of cash flows beyond the payback period, (ii) no consideration of the effectiveness of the investments over the entire project life (PL), and (iii) no consideration of the interest rate and time value of the invested funds.

B.1.3 Life Cycle Cost (LCC) Analysis

There are a number of methods that employ more comprehensive economic analysis by taking into account a number of important economic parameters such as a variety of different cost functions such as capital cost, O&M cost including energy cost, and salvage values; cash flows beyond the payback period; effectiveness of the investments over the life of the project; and interest rate and time value of the invested funds. Additional factors such as inflation and deflation; depreciation; and taxes are also important for economic analysis. For simplicity, these factors are excluded in the life cycle cost (LCC) analysis here. For more information, refer the following books: Stoecker (1989), Boem (1987), and Sullivan et al. (2014).

In the **LCC** analysis method, different options or alternatives of a project are compared based on the total present worth (PW) value, which is a sum of all Present Worth Values (PWVs) for all individual cost functions associated with the project options and assuming a **Project Life (PL)**. As a first step, all cost functions are estimated at the same point in time, and most often at the present time as a Present Worth (PW) value as

$$PW\,(X, PL) = \text{present worth of a cost item } X \text{ over the project life, PL}$$

We assume that the following items are estimated for each of the options considered for the process: Initial Cost (IC); O&M Cost; Discounted Energy Cost (E); Utilizable Life Time of the Process Plant (LP); and Discounted Salvage Value at the end of the life time (S).

The PWV of a project is expressed as

$$PW\,(\text{total cost, PL}) = PW\,(CC, PL) + PW\,(E, PL)$$
$$+ PW\,(O\&M, PL)\,PW\,(\text{total cost, PL}) = PW\,(CC, PL) + PW\,(E, PL)$$
$$+ PW\,(R, PL) - PW\,(S, PL) \tag{B.10}$$

The procedure for LCC economic analysis is summarized here with a list of major steps:

Step-1: Define project LCC analysis period (PL)

If system options have different project lifetimes, then the project life cycle analysis period is assumed to be equal to the life of the longest project option. The option with the shorter life is then replaced at an appropriate time for realistic costs. Also, the significant salvage value of the new system will have to be considered at the end of the analysis. The new system may have a significant salvage value at the end of the LCC analysis period.

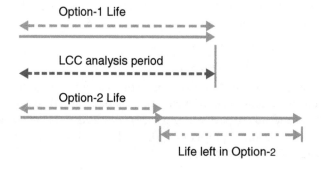

Step-2: Estimate all these costs at the same point in time (Present Worth)
For example, estimate the present worth (PW) of each cost item:

PW (X, PL) = denotes the present worth of cost item X over the life of the project, LP

Step-3: Estimate the total Present Worth cost for each option:

$$\text{PW (total cost, PL)} = \text{PW (CC, PL)} + \text{PW (E, LP)} + \text{PW (O\&M, LP)}$$

$$+ \text{PW (R, LP)} - \text{PW (S, LP)}$$

Step-4: Select the system with the lowest Present Worth cost.

Example B.4 *Comparison of Options using LCC Analysis*

A company must build a new electric power generating plant and can choose between a solar rankine plant and a fuel cell plant. The anticipated comparative data for the two plants are given as follows:

Plant	Capital cost	Generating cost including maintenance, per kWh	Replacement	Equipment life, years	Salvage value, percent of capital cost
Solar rankine	$120 000	$0.0006		20	5
Fuel cell	$80 000	$0.0005	$25 000	15	20

The expected annual consumption of power is 10 000 MWh. If money is borrowed at 5% interest and compounded annually, identify the option with lower LCC.

Solution

Let us estimate the Present Worth cost of each option:

Option-1: Solar Rankine:

Initial Capital cost: CC_1 = $120 000
Generating cost including O&M cost $0.001/kWh

$$OM_1 = 0.0006 \times \$10\,000 \times 10^3 = \$6000$$

Salvage value: 5% of CC_1 $0.05 \times \$120\,000 = \$6000$

$$S_1 = 0.05 \times \$120\,000 = \$6\,000$$

Estimate Present Worth of Option-1
Present Worth of Initial Capital cost, $PW(CC_1, 20)$ = $120 000
Present Worth of O&M:
This represents the amount of money estimated at the beginning of a project that is equivalent to **uniform regular spending** for operation and management (O&M) at regular compounding periods within the life of the project. For simple interest, let us use Eq. (B.5) to compute the Present Worth for O&M:

$$PW\left(OM_1, \text{PL}\right) = OM_1 \frac{(1+i)^n - 1}{i(1+i)^n}$$

$$PW\left(OM_1, PL\right) = OM_1 \frac{(1+0.05)^{20} - 1}{0.05(1+0.05)^{20}} = \$6000 \frac{1.6532}{0.131\,664}$$

$$PW\left(OM_1, PL\right) = \$2929$$

Present Worth of Replacement cost, $PW(R_1, PL) = \$0.0$

Present Worth of Salvage Value:

Salvage values, $PW\left(S_1, PL\right) = S_1 \frac{1}{(1+i)^n}$

$$PW\left(S_1, PL\right) = \$6000 \times \frac{1}{(1+0.05)^{20}}$$

$$PW\left(S_1, PL\right) = \$2261.0$$

Net Present Worth for Option-1:

$$PW\left(\text{total cost}, PL\right) = PW\left(CC_1, PL\right) + PW\left(OM_1, LP\right) + PW\left(R_1, LP\right) PW\left(S_1, LP\right)$$

$$= \$120\,000 + \$2929 - \$2161.0$$

PW (total cost-1, PL) = \$120 768

Option-2: Fuel Cell:

Initial Capital cost: $CC_2 = \$80\,000$

Generating cost including maintenance cost $\$0.001/kWh$

$$OM_2 = 0.0005 \times \$10\,000 \times 10^3 = \$5000$$

Replacement cost: $\$25\,000$ after the 15th year

Salvage value: 20% of $CC_1 = \$80\,000$

$$S_2 = 0.20 \times \$80\,000 = \$16\,000$$

Estimate Present Worth of Option-2

Present Worth of Initial Capital cost, $PW\left(CC_2, 20\right) = \$80\,000$

Present Worth of O&M:

For **uniform regular spending** for operation and management (O&M) at regular compounding periods within the life of the project, use Eq. (B.5) to compute the Present Worth for the O&M.

$$PW\left(OM_2, PL\right) = OM_2 \frac{(1+i)^n - 1}{i(1+i)^n}$$

$$PW\left(OM_2, PL\right) = OM_2 \frac{(1+0.05)^{20} - 1}{0.05\,(1+0.05)^{20}} = \$5\,000 \frac{1.6532}{0.131\,664}$$

$$PW\left(OM_2, PL\right) = \$62\,781$$

Present Worth of Replacement: $\$40,00$ at 15th year: $X_P = X_F \frac{1}{(1+i)^n}$

$$PW\left(R_2, PL\right) = R_2 \frac{1}{(1+i)^n}$$

$$= \$25\,000 \frac{1}{(1+0.05)^{15}}$$

$$PW\left(R_2, PL\right) = \$12\,025$$

Present Worth of Salvage Value:

Salvage values, $\mathrm{PW}\left(S_2, \mathrm{PL}\right) = S_2 \frac{1}{(1+i)^n}$

$$\mathrm{PW}\left(S_2, \mathrm{PL}\right) = \$16\,000 \times \frac{1}{(1+0.05)^{20}}$$

$$\mathrm{PW}\left(S_2, \mathrm{PL}\right) = \$6030$$

Net Present Worth for Option-2:

$$\mathrm{PW}\,(\text{total cost, PL}) = \mathrm{PW}\left(CC_2, \mathrm{PL}\right) + \mathrm{PW}\left(OM_2, \mathrm{LP}\right)$$
$$+ \mathrm{PW}\left(R_2, \mathrm{LP}\right) - \mathrm{PW}\left(S_2, \mathrm{LP}\right)$$
$$= \$80\,000 + \$62781 + \$12\,025 - \$6030$$

PW (total cost – 2, PL) = \$148 776

Summary of Present Worth Options:

Plant	Capital cost	O&M cost	Replacement cost	Salvage value	Total present worth
Solar rankine	$120 000	$2929.0	0	$2161	$120 768
Fuel cell	$80 000	$62.781	$12 025	$	148 776

Based on this LCC analysis, the fuel cell power generation system seems to be the preferred one. However, a number of cost factors are not included in this analysis. This includes production cost of hydrogen fuel and cost of hydrogen storage. Similarly, solar rankine cycle power generation involves additional cost factors such cost of storage and additional energy cost for evening and morning hours.

B.2 Cost Estimation of Thermal Components and Systems

As we have noticed, one of the prior steps to proceed with economic analysis is to estimate the cost of many cost function factors. As we have mentioned, various factors make up the cost of a system, e.g. capital cost, fuel and/or power costs, operating and maintenance costs, taxes, and interest. Estimation of each of these aspects can be quite involved. Even an estimation of the initial capital cost may require major effort. For example, if the purchase of a small turbine is desired, a variety of design approaches could satisfy the end need, and each of these may have its own cost/performance aspects that need to be determined.

B.2.1 Equipment Cost

For economic analysis, a database of cost data over a range of parameters for that product or for alternate designs is required. This database should include not only the costs of materials to be purchased and the costs of product lines, but also the costs of competing products. The most reliable and accurate source of price information for a product is the vendor or manufacturer of the product. Costs of equipment are often given by the manufacturers while quoting the prices of their products at the place of manufacturing whereas the cost of shipping and installation are quoted separately. Such vendors' cost data for a specific piece of equipment represents the highest level of accuracy, but is only valid for a specific period of time. However, such an approach is very time consuming for developing a correlation cost function as a function of a number of variables and parameters.

A most comprehensive and desirable way to present data is to give a detailed multiple variables plot or a multi-variable curve-fitted cost function correlation with important parameters:

$$C = f(X_1, X_2, \ldots X_n) \tag{B.11}$$

This information can be tabulated through contact with many different vendors. This could be quite time consuming and data could be quite difficult to correlate. One such comprehensive cost database was developed by Boem (1987) for thermal energy systems and components based on cost data reported in numerous published articles. Since this collected cost data existed over a period, the historical data was updated to the year 1986–1987 using M&S Index values. A procedure to build a relationship of this historical data to the present day value is also given utilizing an equipment price index.

One of the most common cost functions is often expressed as a power law function of the size parameter as

$$C = C_r \left(\frac{S}{S_r}\right)^m \tag{B.12}$$

where

C	=	The cost of a size of interest, S
S_r	=	One reference size for which the reference cost S_r is known
m	=	Represents the slope of the straight line that represents the power law variation on a log–log plot. Usually the value of m falls in the range of 0.5–1.0, but sometimes m is greater than 1.0. If no information is known, then a value of $m = 0.6$ is often used.

A summary of such data is given in Table B.1.

The prices of products vary over time due to a number of factors. The most important are the inflation/deflation rates, which are indicated by the variation in different indices representing different industrial sections and products. For example, the consumer price index represents a composite of items that consumers purchase. Similarly, the Marshall and Swift (M&S) index provides cost of components belonging to all industries and more specifically to the process industry. This index reflects the costs of major thermal and fluid flow equipment. The values were computed with a base value of 100 in the year 1926. The values of these M&S indices were routinely published in the Chemical Engineering Magazine. Table B.2 shows M&S index values for some selected years. This data table can be extended to additional years as required.

M&S index values for years beyond 1985 can be found in old issues of Chemical Engineering. However, Chemical Engineering Magazine has also started publishing the *Chemical Engineering* Plant Cost Index (CEPCI), which is also used as a resource for plant construction costs. The CEPCI is calculated using various data from the U.S. Bureau of Labor Statistics. This index data is published once a year. For example, the 2018 index value is **603**, and the previous average annual values for the CEPCI are as follows: 568 (2017); 542 (2016); and 555 (2015) (Table B.3).

Table B.1 Some examples of cost database for thermal system components.

Components	m	C_r $1000	S_r	Size range
Centrifugal radial pump,	0.26	2	10	0.2–16 kW
horizontal, cast iron	0.43	5.3	100	16–400 kW
	0.34	3.2	0.5	0.05–30 m^3/min
Centrifugal axial flow, steel w/motor	0.03	47	10	4–40 m^3/min
Compressor - Centrifugal	9	450	10^3	2–4000 kW
Steam turbine	0	1.6	-	5–30 kW
	0.5	1.6	30	30–1000 kW
	0.68	25	1000	$(1–60) \times 10^3$ kW
Gas turbine	0	9	-	5–100 kW
	0.36	9	100	100–5000 kW
	0.54	2300	15E3	$(1 – 2) \times 10^4$ HP
Shell and tube heat exchanger (CS)	0.71	21	100	2–2000 m^3
Shell and tube heat exchanger (SS)	0.67	235	1	400–9000 ft^3
Steam condenser – water cooled	0.55	3	10	$5 – 10^5$ kW
Cooling tower				
Induced or forced draft				
Approach- 5.5 °C;	1.0	70	10	4–60 m^3/min
Wet bulb temp −23.8 °C;	0.64	560	100	60–700 m^3/min
Range −5.5 °C	1.0	72	3.6×10^3	10^3–10^4 kW
Central chiller – vapor compressor				
Reciprocating				
Centrifugal				10–185 tons
	0.5	13.6	50	80–2000 tons
	0.66	92	500	
Solar-driven cooling				
Absorption machine				
Air-cooled Rankine vapor compression		25	25 tons	
		38	25 tons	

Source: Reproduced from Boem (1987).

Table B.2 M&S index values.

Year	M&S index
1970	303
1975	444
1980	660
1985	790

Table B.3 Chemical engineering plant index (CEPCI) from Chemical Engineering magazine.

Year	CEPCI index
2006	500
2007	525
2008	575
2009	521
2010	551
2011	586
2012	585
2013	567
2014	576
2015	555
2016	542
2017	568
2018	603

If the cost of an item is known at any specific year, then the cost of the item at another time can be estimated from the ratio of the index values given below:

$$\text{Cost of the item at the year of interest} = \text{Known cost value at a reference year} \times \frac{\text{Index at the date of interest}}{\text{Index at the reference year}}$$

All costs reported in the database table provided by Boem (1987) were adjusted from their original reported values to an M&S index value of 800.

One important aspect to be remembered is that the price of an item and, hence the index value, change due to several factors. While the most important of them is the inflation/deflation rate, the other important factors are technological innovation, market competition, and market stimulation packages provided by the government through tax incentives or rebates.

Be concerned about whether there may have been any technological or marketing development that might have a major impact on the price index and on the price functions. For example, the price of **Solar PV panels** might have been affected considerably by the major market simulation that took place in the early 1980s and prices have been dropping continuously in years 2000 and beyond because of many technological breakthroughs in PV cells. Also, significant tax incentives were given around 2000–2010 for renewable energy products and services, and most recently the incentives for promoting sales for electric vehicles are given.

Bibliography

Boem, R.F. (1987). *Design Analysis of Thermal Systems*. New York: Wiley.

Stoecker (1989). *Design of Thermal Systems*, 3e. New York: McGraw Hill.

Sullivan, W.G., Wicks, E.M., and Koelling, C.P. (2014). *Engineering Economy*, 16e.

Problems

B.1 Estimate the 1980 prices of the following heat exchangers for the range of areas of 20–900 ft^2: (i) Carbon steel shell and tube; (ii) 304 stainless steel shell and tube; (iii) carbon steel double pipe; and (iv) 304 stainless steel plate. Create a chart using the correlations of cost for each item.

B.2 Estimate the cost of a centrifugal pump with a water flow rate of 1000 gpm and a pressure head of 25 psi.

B.3 Estimate the cost of a forced draft cooling tower that needs to supply 15 000 gpm of cold water to cool the condenser of a thermal power plant. Consider the following cooling tower design parameters and conditions: Approach, $A = 10\,°C$; Range, $R = 15\,°C$, and outside wet bulb temperature, $T_{wb} = 30\,°C$.

Appendix C

Thermodynamic and Thermophysical Properties

Table C.1 Saturation Thermodynamic properties of water–temperature table.

Temperature (°C)	Pressure (kPa)	Specific volume (m³/kg)		Enthalpy (kJ/kg)			Entropy (kJ/kg K)		
		Saturated liquid	Saturated vapor	Saturated liquid	Saturated vapor	Evaporation	Saturated liquid	Saturated vapor	Evaporation
0.01	0.6113	0.001 000	206.131	0.00	2501.3	2501.3	0.0000	9.1562	9.1562
10	1.2276	0.001 000	106.377	41.99	2519.74	2477.7	0.1510	8.9007	8.7498
20	2.339	0.001 002	57.7897	83.94	2538.06	2454.1	0.2966	8.6671	8.3706
30	4.246	0.001 004	32.8932	125.77	2556.25	2430.5	0.4369	8.4533	8.0164
40	7.384	0.001 008	19.5229	167.54	2574.26	2406.7	0.5724	8.2569	7.6845
50	12.350	0.001 012	12.0318	209.31	2592.06	2382.7	0.7037	8.0762	7.3725
60	19.941	0.001 017	7.670 71	251.11	2609.59	2358.5	0.8311	7.9095	7.0784
70	31.19	0.001 023	5.042 17	292.96	2626.80	2333.8	0.9548	7.7552	6.8004
80	47.39	0.001 029	3.407 15	334.88	2643.66	2308.8	1.0752	7.6121	6.5369
90	70.14	0.001 036	2.360 36	376.90	2660.09	2283.2	1.1924	7.4790	6.2866
100	101.3	0.001 044	1.672 90	419.02	2676.05	2257.0	1.3068	7.3548	6.0480
110	143.3	0.001 052	1.210 14	461.27	2691.47	2230.2	1.4184	7.2386	5.8202
120	198.5	0.001 060	0.891 86	503.69	2706.30	2202.6	1.5275	7.1295	5.6020
130	270.1	0.001 070	0.668 50	546.29	2720.46	2174.2	1.6343	7.0269	5.3925
140	361.3	0.001 075	0.508 85	589.11	2733.87	2144.8	1.7390	6.9298	5.1908
150	475.9	0.001 090	0.392 78	632.18	2746.44	2114.3	1.8417	6.8378	4.9960
160	617.8	0.001 102	0.307 06	675.53	2758.09	2082.6	1.9426	6.7501	4.8075
170	791.7	0.001 114	0.242 83	719.20	2768.70	2049.5	2.0418	6.6663	4.6244
180	1002.2	0.001 127	0.194 05	763.21	2778.16	2015.0	2.1395	6.5857	4.4461
190	1254.4	0.001 141	0.156 54	807.61	2786.37	1978.8	2.2358	6.5464	4.2720
200	1553.8	0.001 156	0.127 36	852.43	2793.18	1940.7	2.3308	6.5078	4.1014

(continued)

Design of Thermal Energy Systems, First Edition. Pradip Majumdar.
© 2021 John Wiley & Sons Ltd. Published 2021 by John Wiley & Sons Ltd.
Companion website: www.wiley.com/go/majumdar

Table C.1 (Continued)

Temper-ature (°C)	Pressure (kPa)	Specific volume (m³/kg)		Enthalpy (kJ/kg)			Entropy (kJ/kg K)		
		Saturated liquid	Saturated vapor	Saturated liquid	Saturated vapor	Evapo-ration	Saturated liquid	Saturated vapor	Evapo-ration
210	1906.3	0.001 173	0.104 41	897.75	2798.48	1900.7	2.4247	6.3584	3.9337
220	2317.8	0.001 190	0.086 19	943.61	2802.12	1858.5	2.5177	6.2860	3.7683
230	2794.9	0.001 209	0.071 58	990.10	2803.95	1813.8	2.6099	6.2146	3.6047
240	3344.2	0.001 229	0.059 76	1037.31	2803.81	1766.5	2.7015	6.1436	3.4422
250	3973.0	0.001 251	0.050 13	1085.34	2801.52	1716.2	2.7927	6.0729	3.2802
260	4688.6	0.001 276	0.042 20	1134.35	2796.89	1662.5	2.8837	6.0018	3.1181
270	5498.7	0.001 302	0.035 64	1184.49	2789.65	1605.2	2.9750	5.9301	2.9551
280	6411.7	0.001 332	0.030 17	1235.97	2779.53	1543.6	3.0667	5.8570	2.7903
290	7436.0	0.001 366	0.025 57	1289.04	2766.13	1477.1	3.1593	5.7821	2.6227
300	8581.0	0.001 404	0.021 67	1344.01	2748.94	1404.9	3.2533	5.7044	2.4511
310	9856.6	0.001 447	0.018 35	1401.29	2727.27	1326.0	3.3492	5.6229	2.2737
320	11 274	0.001 499	0.015 49	1461.45	2700.08	1238.6	3.4479	5.5361	2.0882
330	12 845	0.001 561	0.013 00	1525.29	2665.85	1140.6	3.5506	5.4416	1.8909
340	14 586	0.001 638	0.010 80	1594.15	2622.01	1027.9	3.6593	5.3356	1.6763
350	16 514	0.001 740	0.008 81	1670.54	2563.92	893.4	3.7776	5.2111	1.4336
360	18 651	0.001 892	0.006 94	1760.48	2481.00	720.5	3.9146	5.0525	1.1379
370	21 028	0.002 213	0.004 93	1890.37	2332.12	441.8	4.1104	4.7972	0.6868

Source: Reproduced from Sonntag et al. (2003).

Table C.2 Saturation Thermodynamic properties of water–pressure table.

Pressure P (kPa)	Temper-ature T (°C)	Specific volume (m³/kg)		Enthalpy (kJ/kg)			Entropy (kJ/kg K)		
		Saturated liquid	Saturated vapor	Saturated liquid	Saturated vapor	Evapo-ration	Saturated liquid	Saturated vapor	Evapo-ration
kPa									
0.6133	0.01	0.001 000	206.132	0.00	2501.3	2501.3	0	9.1562	9.1562
1.0	6.98	0.001 000	129.208	29.29	2514.18	2484.89	0.1059	8.9756	8.8697
2.0	17.50	0.001 001	67.004	73.47	2533.49	2460.02	0.2607	8.7236	8.4629
3.0	24.08	0.001 003	45.665	101.03	2545.50	2444.47	0.3545	8.5775	8.2231
4.0	28.96	0.001 004	34.800	121.44	2554.37	2432.93	0.4226	8.4746	8.0520
5.0	32.88	0.001 005	28.193	137.79	2561.45	2423.66	0.4763	8.3950	7.9187
10.0	45.81	0.001 010	14.674	191.81	2584.63	2392.82	0.6492	8.1501	7.5010
20.0	60.06	0.001 017	7.649	251.38	2609.70	2358.33	0.8319	7.9085	7.0766
30.0	69.10	0.001 022	5.229	289.21	2625.28	2336.07	0.9439	7.7686	8.8247
40.0	75.87	0.001 026	3.993	317.55	2636.74	2319.19	1.0258	7.6700	6.6441
50.0	81.33	0.001 030	3.240	340.47	2645.87	2305.40	1.0910	7.5939	6.5029

(*continued*)

Table C.2 (Continued)

Pressure P (kPa)	Temperature T (°C)	Specific volume (m³/kg)		Enthalpy (kJ/kg)			Entropy (kJ/kg K)		
		Saturated liquid	Saturated vapor	Saturated liquid	Saturated vapor	Evaporation	Saturated liquid	Saturated vapor	Evaporation
MPa									
0.100	9.62	0.001 043	1.6940	417.44	2675.46	2258.02	1.3025	7.3593	6.0568
0.150	111.37	0.001 053	1.1593	467.08	2693.54	2226.46	1.4335	7.2232	5.7897
0.200	120.23	0.001 061	0.8857	504.68	2706.63	2201.96	1.5300	7.1271	5.5970
0.250	127.43	0.001 067	0.7187	535.34	2716.89	2181.55	1.6072	7.0526	5.4455
0.300	133.55	0.001 073	0.6058	561.45	2725.30	2163.85	1.6717	6.9918	5.3201
0.350	138.88	0.001 079	0.5243	584.31	2732.40	2148.10	1.7274	6.9404	5.2130
0.400	143.63	0.001 084	0.4625	604.73	2738.53	2133.81	1.7766	6.8958	5.1193
0.450	147.93	0.001 088	0.4140	623.24	2743.91	2120.67	1.8606	6.8565	5.0359
0.500	151.86	0.001 093	0.3749	640.21	2748.67	2108.47	1.8972	6.1212	4.9606
0.600	158.85	0.001 101	0.3157	670.54	2756.80	2086.26	1.9627	6.7600	4.8289
0.700	164.97	0.001 108	0.2729	697.20	2763.50	2066.30	2.0199	6.7080	4.7158
0.800	170.43	0.001 115	0.2404	721.10	2769.13	2048.04	2.0464	6.6627	4.6166
0.900	175.38	0.001 121	0.2150	742.82	2773.94	2031.32	2.0946	6.6225	4.5280
1.000	179.91	0.001 127	0.194 44	762.79	2778.08	2015.29	2.1386	6.5864	4.4478
1.100	184.09	0.001 133	0.177 53	781.32	2781.68	2000.36	2.1791	6.5535	4.3744
1.200	187.99	0.001 139	0.163 33	798.64	2784.82	1986.19	2.2165	6.5233	4.3067
1.300	191.64	0.001 144	0.151 25	814.91	2787.58	1972.67	2.2514	6.4953	4.2438
1.400	195.07	0.001 149	0.140 84	830.29	2790.00	1959.72	2.2842	6.4692	4.1850
1.500	198.32	0.001 154	0.131 77	844.87	2792.15	1947.28	2.3150	6.4448	4.1298
1.750	205.76	0.001 166	0.113 49	878.48	2796.43	1917.95	2.3851	6.3895	4.0044
2.000	212.42	0.001 177	0.099 63	908.77	2799.51	1890.74	2.4473	6.3408	3.8935
2.250	218.45	0.001 187	0.088 75	936.48	2801.67	1865.19	2.5034	6.2971	3.7938
2.500	223.99	0.001 197	0.079 98	962.09	2803.07	1840.98	2.5546	6.2574	3.7028
2.750	229.12	0.001 207	0.072 75	985.97	2803.86	1817.89	2.6018	6.2208	3.6190
3.000	233.90	0.001 216	0.066 68	1008.41	2804.14	1795.73	2.6456	6.1869	3.5412
3.250	238.38	0.001 226	0.061 52	1029.60	2803.97	1774.37	2.6866	6.1551	3.4685
3.500	242.60	0.001 235	0.057 07	1049.73	2803.43	1753.70	2.7252	6.1252	3.4000
4.0	250.40	0.001 252	0.049 77	1087.29	2801.38	1714.09	2.7963	6.0700	3.2737
5.0	263.99	0.001 286	0.039 44	1154.21	2794.33	1640.12	2.9201	5.9733	3.0532
6.0	275.64	0.001 319	0.032 44	1213.32	2784.33	1571.00	3.0266	5.8891	2.8625
7.0	285.88	0.001 351	0.027 37	1266.97	2772.07	1505.10	3.1210	5.8132	2.6922
8.0	295.06	0.001 384	0.023 52	1316.61	2757.94	1441.33	3.2067	5.7431	2.5365

(continued)

Table C.2 (Continued)

Pressure P (kPa)	Temperature T (°C)	Specific volume (m³/kg)		Enthalpy (kJ/kg)			Entropy (kJ/kg K)		
		Saturated liquid	Saturated vapor	Saturated liquid	Saturated vapor	Evaporation	Saturated liquid	Saturated vapor	Evaporation
9.0	303.40	0.001 418	0.020 48	1363.23	2742.11	1378.88	3.2857	5.6771	2.3915
10.0	311.06	0.001 452	0.018 02	1407.53	2724.67	1317.14	3.3595	5.6140	2.2545
11.0	318.15	0.001 489	0.015 98	1450.05	2705.60	1255.55	3.4294	5.5527	2.1233
12.0	324.75	0.001 527	0.014 26	1491.24	2684.83	1193.59	3.4961	5.4923	1.9962
13.0	330.93	0.001 567	0.012 78	1531.46	2662.22	1130.76	3.5604	5.4323	1.8719
14.0	336.75	0.001 611	0.011 48	1571.08	2637.55	1066.47	3.6231	5.3716	1.7485
15.0	342.24	0.001 658	0.010 33	1610.45	2610.49	1000.04	3.6847	5.3097	1.6250
16.0	347.43	0.001 711	0.009 31	1650.00	2580.59	930.59	3.7460	5.2454	1.4995
17.0	352.37	0.001 770	0.008 37	1690.25	2547.15	856.90	3.8078	5.1776	1.3698
18.0	357.06	0.001 840	0.007 49	1731.97	2509.09	777.13	3.8713	5.1044	1.2330
19.0	361.54	0.001 924	0.006 66	1776.43	2464.54	688.11	3.9387	5.0227	1.0841
20.0	365.81	0.002 035	0.005 83	1826.18	2409.74	583.56	4.0137	4.9269	0.9132
21.0	369.89	0.002 206	0.004 95	1888.30	2334.72	446.42	4.1073	4.8015	0.6942
22.0	373.80	0.002 808	0.003 53	2034.92	2158.97	124.04	4.3307	4.5224	0.1917
22.09	374.14	0.003 155	0.003 16	2099.26	2099.26	0.0	4.4297	4.4297	0

Table C.3 Superheated Thermodynamic properties of vapor water.

Temperature (°C)	Specific volume (m³/kg)	Enthalpy (kJ/kg)	Entropy (kJ/kg K)	Temperature (°C)	Specific volume (m³/kg)	Enthalpy (kJ/kg)	Entropy (kJ/kg K)
10 kPa (45.81 °C)				50 kPa (81.33 °C)			
Sat	14.673 55	2584.63	8.1501	Sat	3.240 34	2645.87	7.5939
50	14.869 20	2592.56	8.1749	—	—	—	—
100	17.195 61	2687.46	8.4479	100	3.418 33	2682.52	7.6947
150	19.512 51	2782.99	8.6881	150	3.889 37	2780.08	7.9400
200	21.825 07	2879.52	8.9037	200	4.355 95	2877.64	8.1579
250	24.135 59	2977.31	9.1002	250	4.820 45	2975.99	8.3555
300	26.445 08	3076.51	9.2812	300	5.283 91	3075.52	8.5372
400	31.062 52	3279.51	9.6076	400	6.209 29	3278.89	8.8641
500	35.678 96	3489.05	9.8977	500	7.133 64	3488.62	9.1545
600	40.294 88	3705.40	10.1608	600	8.057 48	3705.10	9.4177
700	44.910 52	3928.73	10.4028	700	8.981 04	3928.51	9.6599
800	49.525 99	4159.10	10.6281	800	9.904 44	4158.92	9.8852
900	54.141 37	4396.44	10.8395	900	10.827 73	4396.30	10.0967
1000	58.756 69	4640.58	11.0392	1000	11.750 97	4640.46	10.2964

(*continued*)

Table C.3 (Continued)

Temperature (°C)	Specific volume (m³/kg)	Enthalpy (kJ/kg)	Entropy (kJ/kg K)	Temperature (°C)	Specific volume (m³/kg)	Enthalpy (kJ/kg)	Entropy (kJ/kg K)
100 kPa (99.62 °C)				200 kPa (120.23 °C)			
Sat	1.694 00	2675.46	7.3593	Sat	0.885 73	2706.63	7.1271
150	1.936 36	2776.38	7.6133	150	0.959 64	2768.80	7.2795
200	2.172 26	2875.27	7.8342	200	1.080 34	2870.46	7.5066
250	2.406 04	2974.33	8.0332	250	1.198 80	2970.98	7.7085
300	2.638 79	3074.28	8.2157	300	1.316 16	3071.79	7.8926
400	3.102 63	3278.11	8.5434	400	1.549 30	3276.55	8.2217
500	3.565 47	3488.09	8.8341	500	1.781 39	3487.03	8.5132
600	4.027 81	3704.72	9.0975	600	2.012 97	3703.96	8.7769
700	4.489 86	3928.23	9.3398	700	2.244 26	3927.66	9.0194
800	4.951 74	4158.71	9.5652	800	2.475 39	4158.27	9.2450
900	5.413 53	4396.12	9.7767	900	2.706 43	4395.77	9.4565
1000	5.875 26	4640.31	9.9764	1000	2.937 40	4640.01	9.6563
1100	6.336 96	4890.95	10.1658	1100	3.168 34	4890.68	9.8458
1200	6.798 63	5147.56	10.3462	1200	3.399 27	5147.32	10.0262
300 kPa (133.55 °C)				400 kPa (143.63 °C)			
Sat	0.605 82	2725.30	6.9918	Sat	0.462 46	2738.53	6.8958
150	0.633 88	2760.95	7.0778	150	0.470 84	2752.82	6.9299
200	0.716 29	2865.54	7.3115	200	0.534 22	2860.5171 706	
250	0.796 36	2967.59	7.5165	250	0.595 12	2947.59	7.3788
300	0.875 29	3069.28	7.7022	300	0.654 84	3069.28	7.5661
400	1.031 51	3274.98	8.0329	400	0.772 62	3274.98	7.8984
500	1.186 69	3485.96	8.3250	500	0.889 34	3485.96	8.1912
600	1.341 36	3703.20	8.5892	600	1.005 55	3703.20	8.4557
700	1.495 73	3927.10	8.8319	700	1.121 47	3927.10	8.6987
800	1.649 94	4157.83	9.0575	800	1.237 22	4157.83	8.9244
900	1.804 06	4395.42	9.2691	900	1.352 88	4395.42	9.1361
1000	1.958 12	4639.71	9.4689	1000	1.468 47	4639.71	9.3360
1100	2.112 14	4890.41	9.6585	1100	1.584 04	4890.41	9.5255
1200	2.266 14	5147.07	9.8389	1200	1.699 58	5147.07	9.7059
1300	2.420 13	5409.03	10.0109	1300	1.815 11	5409.03	9.8780

(*continued*)

Table C.3 (Continued)

Temperature (°C)	Specific volume (m³/kg)	Enthalpy (kJ/kg)	Entropy (kJ/kg K)	Temperature (°C)	Specific volume (m³/kg)	Enthalpy (kJ/kg)	Entropy (kJ/kg K)
500 kPa (151.86 °C)				600 kPa (158.85 °C)			
Sat	0.374 89	2748.67	6.8212	Sat	0.315 67	2756.80	6.7600
200	0.424 92	2855.37	7.0592	200	0.352 02	2850.12	6.9665
250	0.474 36	2960.68	7.2708	250	0.393 83	2957.16	7.1816
300	0.522 56	3064.20	7.4598	300	0.434 37	3061.63	7.3723
350	0.570 12	3167.65	7.6328	350	0.474 24	3165.66	7.5463
400	0.617 28	3271.83	7.7937	400	0.513 72	3270.25	7.7078
500	0.710 93	3483.82	8.0872	500	0.591 99	3482.75	8.0020
600	0.804 06	3701.67	8.3521	600	0.669 74	3700.91	8.2673
700	0.896 91	3925.97	8.5952	700	0.747 20	3925.41	8.5107
800	0.989 59	4156.96	8.8211	800	0.824 50	4156.52	8.7367
900	1.082 17	4394.71	9.0329	900	0.901 69	4394.36	8.9485
1000	1.174 69	4639.11	9.2328	1000	0.978 83	4638.81	9.1484
1100	1.267 18	4889.88	9.4224	1100	1.055 94	4889.61	9.3381
1200	1.359 64	5146.58	9.6028	1200	1.133 02	5146.34	9.5185
800 kPa (170.43 °C)				1000 kPa (179.91 °C)			
Sat	0.240 43	2769.13	6.6627	Sat	0.194 44	2778.08	6.5864
200	0.260 80	2839.25	6.8158	200	0.205 96	2827.86	6.6939
250	0.293 14	2949.97	7.0384	250	0.232 68	2942.59	6.9246
300	0.324 11	3056.43	7.2327	300	0.257 94	3051.15	7.1228
350	0.354 39	3161.68	7.4088	350	0.282 47	3157.65	7.3010
400	0.384 26	3267.07	7.5715	400	0.306 59	3263.88	7.4650
500	0.443 31	3480.60	7.8672	500	0.354 11	3478.44	7.7621
600	0.501 84	3699.38	8.1332	600	0.401 09	3697.85	8.0289
700	0.560 07	3924.27	8.3770	700	0.447 79	3923.14	8.2731
800	0.618 13	4155.65	8.6033	800	0.494 32	4154.78	8.4996
900	0.676 10	4393.65	8.8153	900	0.540 75	4392.94	8.7118
1000	0.734 01	4638.20	9.0153	1000	0.587 12	4637.60	8.9119
1100	0.791 88	4889.08	9.2049	1100	0.633 45	4888.55	9.1016
1200	0.849 74	5145.85	9.3854	1200	0.679 77	5145.36	9.2821
1300	0.907 58	5407.87	9.5375	1300	0.726 08	5407.41	9.4542

(*continued*)

Table C.3 (Continued)

Temperature (°C)	Specific volume (m³/kg)	Enthalpy (kJ/kg)	Entropy (kJ/kg K)	Temperature (°C)	Specific volume (m³/kg)	Enthalpy (kJ/kg)	Entropy (kJ/kg K)
1200 kPa (187.99 °C)				1400 kPa (195.07 °C)			
Sat	0.163 33	2784.82	6.5233	Sat	0.140 84	2790.00	6.4692
200	0.169 30	2815.90	6.5898	200	0.143 02	2903.32	6.4975
250	0.192 35	2935.01	6.8293	250	0.163 50	2927.22	6.7467
300	0.213 82	3045.80	7.0316	300	0.182 28	3040.35	6.9533
350	0.234 52	3153.59	7.2120	350	0.200 26	3149.49	7.1359
400	0.254 80	3260.66	7.3773	400	0.217 80	3257.42	7.3025
500	0.294 63	3476.28	7.6758	500	0.252 15	3474.11	7.6026
600	0.333 93	3696.32	7.9434	600	0.285 96	3694.78	7.8710
700	0.372 94	3922.01	8.1881	700	0.319 47	3920.87	8.1160
800	0.411 17	4153.90	8.4149	800	0.352 81	4153.03	8.3431
900	0.450 51	4392.23	8.6272	900	0.386 06	4391.53	8.5555
1000	0.489 19	4637.00	8.8274	1000	0.419 24	4636.41	8.7558
1100	0.527 83	4888.02	9.0171	1100	0.452 39	4887.49	8.9456
1200	0.566 46	5144.87	9.1977	1200	0.485 52	5144.38	9.1262
1300	0.605 07	5406.95	9.3698	1300	0.518 64	5406.49	9.2983
1600 kPa (201.40 °C)				1800 kPa (207.15 °C)			
Sat	0.123 80	2794.02	6.4217	Sat	0.110 42	2797.13	6.3793
250	0.141 84	2919.20	6.6732	250	0.124 97	2910.96	6.6066
300	0.158 62	3034.83	6.8844	300	0.140 21	3029.21	8.8226
350	0.174 56	3145.35	7.0693	350	0.154 57	3141.18	7.0099
400	0.190 05	3254.17	7.2373	400	0.168 47	3250.90	7.1793
500	0.220 29	3471.93	7.5389	500	0.195 50	3469.75	7.4824
600	0.249 98	3693.23	7.8080	600	0.221 99	3691.69	7.7523
700	0.279 37	3919.73	8.0535	700	0.248 18	3918.59	7.9983
800	0.308 59	4152.15	8.2808	800	0.274 20	4151.27	8.2258
900	0.337 72	4390.82	8.4934	900	0.300 12	4390.11	8.4386
1000	0.366 78	4635.81	8.6938	1000	0.325 98	4635.21	8.6390
1100	0.395 81	4886.95	8.8837	1100	0.351 80	4886.42	8.8290
1200	0.424 82	5143.89	9.0642	1200	0.377 61	5143.40	9.0096
1300	0.453 82	5406.02	9.2364	1300	0.403 40	5405.56	9.1817

(*continued*)

Table C.3 (Continued)

Temperature (°C)	Specific volume (m³/kg)	Enthalpy (kJ/kg)	Entropy (kJ/kg K)	Temperature (°C)	Specific volume (m³/kg)	Enthalpy (kJ/kg)	Entropy (kJ/kg K)
2000 kPa (212.42 °C)				2500 kPa (223.99 °C)			
Sat	0.099 63	2799.51	6.3408	Sat	0.079 98	2803.07	6.2574
250	0.111 44	2902.46	6.5452	250	0.087 00	2880.06	6.4084
300	0.125 47	3023.50	6.7663	300	0.098 90	3008.81	6.6437
350	0.138 57	3136.96	6.9562	350	0.109 76	3126.24	6.8402
400	0.151 20	324 760	7.1270	400	0.120 10	3239.28	7.0147
450	0.163 53	3357.48	7.2844	450	0.130 14	3350.77	7.1745
500	0.175 68	3467.55	7.4316	500	0.139 98	3462.04	7.3233
600	0.199 60	3690.14	7.7023	600	0.159 30	3686.25	7.5960
700	0.223 23	3917.45	7.9487	700	0.178 32	3914.59	7.8435
800	0.246 68	4150.40	8.1766	800	0.197 16	4148.20	8.0720
900	0.270 04	4389.40	8.3895	900	0.215 90	4387.64	8.2853
1000	0.293 33	4634.61	8.5900	1000	0.234 58	4633.12	8.4860
1100	0.316 59	4885.89	8.7800	1100	0.253 22	4884.57	8.6761
1200	0.339 84	5142.92	8.9606	1200	0.271 85	5141.70	8.8569
1300	0.363 06	5405.10	9.1328	1300	0.290 46	5403.95	9.0291
3000 kPa (233.90 °C)				4000 kPa (250.40 °C)			
Sat	0.066 68	2804.14	6.1869	Sat	0.049 78	2801.38	6.0700
250	0.070 58	2855.75	6.2871	250	—	—	—
300	0.081 14	2993.48	6.5389	300	0.058 84	2960.68	6.3614
350	0.090 53	3115.25	6.7427	350	0.066 45	3092.43	6.5820
400	0.099 36	3230.82	6.9211	400	0.073 41	3213.51	6.7689
450	0.107 87	3344.00	7.0833	450	0.080 03	3330.23	6.9362
500	0.116 19	3456.48	7.2337	500	0.086 43	3445.21	7.0900
600	0.132 43	3682.34	7.5084	600	0.098 85	3674.44	7.3688
700	0.148 38	3911.72	7.7571	700	0.110 95	3905.94	7.6198
800	0.164 14	4146.00	7.9862	800	0.122 87	4141.59	7.8502
900	0.179 80	4385.87	8.1999	900	0.134 69	4382.34	8.0647
1000	0.195 41	4631.63	8.4009	1000	0.146 45	4628.65	8.2661
1100	0.210 98	4883.26	8.5911	1100	0.158 17	4880.63	8.4566
1200	0.226 52	5140.49	8.7719	1200	0.169 87	5138.07	8.6376
1300	0.242 06	5402.81	8.9442	1300	0.181 56	5400.52	8.8099

(continued)

Table C.3 (Continued)

Temperature (°C)	Specific volume (m³/kg)	Enthalpy (kJ/kg)	Entropy (kJ/kg K)	Temperature (°C)	Specific volume (m³/kg)	Enthalpy (kJ/kg)	Entropy (kJ/kg K)
5000 kPa (263.99 °C)				6000 kPa (275.64 °C)			
Sat	0.039 44	2794.33	5.9733	Sat	0.032 44	2784.33	5.8891
300	0.045 32	2924.53	6.2083	300	0.036 16	2884.19	6.0673
350	0.051 94	3068.39	6.4492	350	0.042 23	3042.97	6.3334
400	0.057 81	3195.64	6.6458	400	0.047 39	3177.17	6.5407
450	0.063 30	3316.15	6.8185	450	0.052 14	3301.76	6.7192
500	0.068 57	3433.76	6.9758	500	0.056 65	3422.12	6.8802
550	0.073 68	3550.23	7.1217	550	0.061 01	3540.62	7.0287
600	0.078 69	3666.47	7.2588	600	0.065 25	3658.40	7.1676
700	0.088 49	3900.13	7.5122	700	0.073 52	3894.28	7.4234
800	0.098 11	4137.17	7.7440	800	0.081 60	4132.74	7.6566
900	0.107 62	4378.82	7.9593	900	0.089 58	4375.29	7.8727
1000	0.117 07	4625.69	8.1612	1000	0.097 49	4622.74	8.0751
1100	0.126 48	4878.02	8.3519	1100	0.105 36	4875.42	8.2661
1200	0.135 87	5135.67	8.5330	1200	0.113 21	5133.28	8.4473
1300	0.145 26	5398.24	8.7055	1300	0.121 06	5395.97	8.6199

Table C.4 R-134 saturated Thermodynamic properties.

Temper-ature (°C)	Pressure (kPa)	Specific volume (m³/kg)		Enthalpy (kJ/kg)			Entropy (kJ/kg K)		
		Saturated liquid	Saturated vapor	Saturated liquid	Saturated vapor	Evapo-ration	Saturated liquid	Saturated vapor	Evapo-ration
−40	51.8	0.000 708	0.356 96	148.98	373.48	224.50	0.7991	1.7620	0.9629
−35	66.8	0.000 715	0.281 22	154.98	376.64	221.67	0.8245	1.7553	0.9308
−30	85.1	0.000 722	0.224 02	161.12	379.80	218.68	0.8499	1.7493	0.8994
−26.3	101.3	0.000 728	0.190 20	165.80	382.16	216.36	0.8690	1.7453	0.8763
−25	107.2	0.000 730	0.180 30	167.38	382.95	215.57	0.8754	1.7441	0.8687
−20	133.7	0.000 738	0.146 49	173.74	386.08	212.34	0.9007	1.7395	0.8388
−15	165.0	0.000 746	0.120 07	180.19	389.20	209.00	0.9256	1.7354	0.8096
−10	201.7	0.000 755	0.099 21	186.72	392.28	205.56	0.9507	1.7319	0.7812
−5	244.5	0.000 764	0.082 57	193.32	395.34	202.02	0.9755	1.7288	0.7534
0	294.0	0.000 773	0.069 19	200.00	398.36	198.36	1.0000	1.7262	0.7262
5	350.0	0.000 783	0.058 33	206.75	401.32	194.57	1.0243	1.7239	0.6995
10	415.8	0.000 794	0.049 45	213.58	404.23	190.65	1.0485	1.7218	0.6733

(*continued*)

Table C.4 (Continued)

Temperature (°C)	Pressure (kPa)	Specific volume (m³/kg)		Enthalpy (kJ/kg)			Entropy (kJ/kg K)		
		Saturated liquid	Saturated vapor	Saturated liquid	Saturated vapor	Evaporation	Saturated liquid	Saturated vapor	Evaporation
15	489.5	0.000 805	0.042 13	220.49	407.07	186.58	1.0725	1.7200	0.6475
20	572.8	0.000 817	0.036 06	227.49	409.84	182.35	1.0963	1.7183	0.6220
25	666.3	0.000 829	0.030 98	234.59	412.51	177.92	1.1201	1.7168	0.5967
30	771.0	0.000 843	0.026 71	241.79	415.08	173.29	1.1437	1.7153	0.5716
35	887.6	0.000 857	0.023 10	240.10	417.52	168.42	1.1673	1.7139	0.5465
40	1017.0	0.000 873	0.020 02	256.54	419.82	163.28	1.1909	1.7123	0.5214
45	1160.2	0.000 890	0.017 39	264.11	421.96	157.85	1.2145	1.7106	0.4962
50	1318.1	0.000 908	0.015 12	271.83	423.91	152.08	1.2381	1.7088	0.4706
55	1491.6	0.000 928	0.013 16	279.72	425.65	145.93	1.2619	1.7066	0.4447
60	1681.8	0.000 951	0.011 46	287.79	427.13	139.33	1.2857	1.7040	0.4182
65	1889.9	0.000 976	0.009 97	296.09	428.30	132.21	1.3099	1.7008	0.3910
70	2117.0	0.001 005	0.008 66	304.64	429.11	124.47	1.3343	1.6970	0.3627
75	2364.4	0.001 038	0.007 49	313.51	429.45	115.94	1.3592	1.6923	0.3330
80	2633.6	0.001 078	0.006 45	322.79	429.19	106.40	1.3849	1.6862	0.3013
85	2926.2	0.001 128	0.005 50	332.65	428.10	95.45	1.4117	1.6782	0.2665
90	3244.5	0.001 195	0.004 61	343.38	425.70	82.31	1.4404	1.6671	0.2267
95	3591.5	0.001 297	0.003 73	355.83	420.81	64.98	1.4733	1.6498	0.1765
100	3973.2	0.001 557	0.002 64	374.74	407.21	32.47	1.5228	1.6098	0.0870

Source: Reproduced from Sonntag et al. (2003).

Table C.5 R-134 superheated Thermodynamic properties.

Temperature (°C)	Specific volume (m³/kg)	Enthalpy (kJ/kg)	Entropy (kJ/kg K)	Temperature (°C)	Specific volume (m³/kg)	Enthalpy (kJ/kg)	Entropy (kJ/kg K)
Pressure – 100 kPa (−26.54 °C)				Pressure – 200 kPa (−10.22 °C)			
Sat	0.368 89	373.06	1.7629	Sat	0.100 02	392.15	1.7320
−20	0.405 07	388.82	1.82.79	−10	0.100 13	392.34	1.7328
−10	0.422 22	396.64	1.85.82	0	0.105 01	400.91	1.7647
0	0.439 21	404.59	1.8878	10	0.109 74	409.50	1.7956
10	0.456 08	412.70	1.9170	20	0.114 36	418.15	1.8256
20	0.472 87	420.96	1.9456	30	0.118 89	426.87	1.8549
30	0.489 58	429.38	1.9739	40	0.123 35	435.71	1.8836
40	0.506 23	437.96	2.0017	50	0.127 76	444.66	1.9117
50	0.522 84	446.70	2.0292	60	0.132 13	453.74	1.9394

(continued)

Table C.5 (Continued)

Temperature (°C)	Specific volume (m³/kg)	Enthalpy (kJ/kg)	Entropy (kJ/kg K)	Temperature (°C)	Specific volume (m³/kg)	Enthalpy (kJ/kg)	Entropy (kJ/kg K)
60	0.539 41	455.60	2.0563	70	0.136 46	462.95	1.9666
70	0.555 95	464.66	2.0834	80	0.140 76	472.30	1.9935
80	0.572 47	473.88	2.1096	90	0.145 04	481.79	2.0200
90	0.588 96	484.26	2.1358	100	0.149 30	491.42	2.0461
100	0.605 44	492.81	2.1617	110	0.153 55	501.21	2.0720
Pressure – 400 kPa (8.84 °C)				Pressure – 600 kPa (21.52 °C)			
Sat	0.051 36	403.56	1.7223	Sat	0.034 42	410.66	1.7179
10	0.051 68	404.65	1.7261	30	0.036 09	419.09	1.7461
20	0.054 36	413.97	1.7584	40	0.037 96	428.88	1.7779
30	0.056 93	423.22	1.7895	50	0.039 74	438.59	1.8084
40	0.059 40	432.46	1.8195	60	0.041 45	448.28	1.8379
50	0.061 81	441.75	1.8487	70	0.043 11	457.99	1.8666
60	0.064 17	451.10	1.8772	80	0.044 73	467.76	1.8947
70	0.066 48	460.55	1.9051	90	0.046 32	477.61	1.9222
80	0.068 77	470.09	1.9325	100	0.047 88	487.55	1.9492
90	0.071 02	479.75	1.9595	110	0.049 43	497.59	1.9758
100	0.073 25	489.52	1.9860	120	0.050 95	507.75	2.0019
110	0.075 47	499.43	2.0122	130	0.052 46	518.03	2.0277
120	0.077 67	509.46	2.0381	140	0.053 96	528.43	2.0532
130	0.079 85	519.63	2.0636	150	0.055 44	538.95	2.0784
800 kPa (31.30 °C)				1000 kPa (39.37 °C)			
Sat	0.025 71	415.72	1.7150	Sat	0.020 38	419.54	1.7125
40	0.027 11	424.86	1.7446	40	0.020 47	420.25	1.7148
50	0.028 61	435.11	1.7768	50	0.021 85	431.24	1.7494
60	0.030 02	445.22	1.8076	60	0.023 11	441.89	1.7818
70	0.031 37	455.27	1.8373	70	0.024 29	452.34	1.8127
80	0.032 68	465.31	1.8662	80	0.025 42	462.70	1.8425
90	0.033 94	475.38	1.8943	90	0.026 50	473.03	1.8713
100	0.035 18	485.50	1.9218	100	0.027 54	483.36	1.8994
110	0.036 39	495.70	1.9487	110	0.028 56	493.74	1.9268
120	0.037 58	505.99	1.9753	120	0.029 56	504.17	1.9537
130	0.038 76	516.38	2.0014	130	0.030 53	514.69	1.9801
140	0.039 92	526.88	2.0271	140	0.031 50	525.30	2.0061
150	0.041 07	537.50	2.0525	150	0.032 44	536.02	2.0318
160	0.042 21	548.23	2.0775	160	0.033 38	546.84	2.0570

(*continued*)

Table C.5 (Continued)

Temperature (°C)	Specific volume (m³/kg)	Enthalpy (kJ/kg)	Entropy (kJ/kg K)	Temperature (°C)	Specific volume (m³/kg)	Enthalpy (kJ/kg)	Entropy (kJ/kg K)
1200 kPa (46.31 °C)				1400 kPa (52.42 °C)			
Sat	0.016 76	422.49	1.7102	Sat	0.014 14	424.78	1.7077
50	0.017 24	426.84	1.7237	60	0.015 03	434.08	1.7360
60	0.018 44	438.21	1.7584	70	0.016 08	445.72	1.7704
70	0.019 53	449.18	1.7908	80	0.017 04	456.94	1.8026
80	0.020 55	459.92	1.8217	90	0.017 93	467.93	1.8333
90	0.021 51	470.55	1.8514	100	0.018 78	478.79	1.8628
100	0.022 44	481.13	1.8801	110	0.019 58	489.59	1.8914
110	0.023 33	491.70	1.9081	120	0.020 36	500.38	1.9192
120	0.024 20	502.31	1.9354	130	0.021 12	511.19	1.9463
130	0.025 04	512.97	1.9621	140	0.021 86	522.05	1.9730
140	0.025 87	523.70	1.9884	150	0.022 58	532.98	1.9991
150	0.026 69	534.51	2.0143	160	0.023 29	543.99	2.0248
160	0.027 50	545.43	2.0398	170	0.023 99	555.10	2.0502
170	0.028 29	556.44	2.0649	180	0.024 68	566.30	2.0752
1600 kPa (57.90 °C)				2000 kPa (67.48 °C)			
Sat	0.012 15	426.54	1.7051	Sat	0.009 30	428.75	**1.6991**
60	0.012 39	429.32	1.7135		—	—	—
70	0.013 45	441.89	1.7507	70	0.009 58	432.53	**1.7101**
80	0.014 38	453.72	1.7847	80	0.010 55	446.30	**1.7497**
90	0.015 22	465.15	1.8166	90	0.011 37	458.95	**1.7850**
100	0.016 01	476.33	1.8469	100	0.012 11	471.00	**1.8177**
110	0.016 76	487.39	1.8762	110	0.012 79	482.69	**1.8487**
120	0.017 48	498.39	1.9045	120	0.013 42	494.19	**1.8783**
130	0.018 17	509.37	1.9321	130	0.014 03	505.57	**1.9069**
140	0.018 84	520.38	1.9591	140	0.014 61	516.90	**1.9346**
150	0.019 49	531.43	1.9855	150	0.015 17	528.22	**1.9617**
160	0.020 13	542.54	2.0115	160	0.015 71	539.57	**1.9882**
170	0.020 76	553.73	2.0370	170	0.016 24	550.96	**2.0142**
180	0.021 38	565.02	2.0622	180	0.016 76	562.42	**2.0398**

Table C.6 Properties of various ideal gases at 25°C, 100 kPa

Substance	Formula	Molecular weight	R (kJ/kg K)	ρ (kg/m³)	C_{p0} (kJ/kg K)	C_{v0} (kJ/kg K)
Ammonia	NH_3	17.031	0.4882	0.694	2.130	1.642
Acetylene	C_3H_2	26.038	0.3193	1.05	1.699	1.380
Argon	Ar	39.948	0.2081	1.613	0.520	0.312
Butane	C_4H_{10}	58.124	0.1430	2.407	1.716	1.573
Carbon Dioxide	CO_2	44.010	0.1889	1.775	0.842	0.653
Carbon monoxide	CO	28.011	0.2968	1.13	1.041	0.744
Ethane	C_2H_6	30.070	0.2765	1.222	1.766	1.490
Ethanol	C_2H_5OH	46.069	0.1805	1.883	1.427	1.246
Ethylene	C_2H_4	29.054	0.2964	1.138	1.548	1.252
Hydrogen	H_2	2.016	4.1243	0.0813	14.209	10.085
Methane	CH_4	16.043	0.5183	0.648	2.254	1.736
Methanol	C_2H_3OH	32.042	0.2595	1.31	1.405	1.146
Neon	Ne	20.183	0.4120	0.814	1.03	0.618
n-Octane	C_8H_{18}	114.232	0.0727	0.092	1.711	1.38
Oxygen	O_2	31.999	0.2598	1.292	0.922	0.662
Propane	C_3H_{18}	44.094	0.1886	1.808	1.679	1.490
R-134a	$CF_{3C}H_2F$	102.03	0.081 49	4.2	0.852	0.771
Sulfur dioxide	SO_2	64.059	0.1298	2.618	0.624	0.494
Sulfur trioxide	SO_3	80.058	0.1038	3.272	0.635	0.531
Nitrogen	N_2	28.013	0.2968	1.13	1.042	0.745
Nitric oxide	NO	30.006	0.2771	1.21	0.993	0.716
Nitrous oxide	N_2O	44.013	0.1889	1.775	0.879	0.690
Steam	H_2O	18.050	0.4615	0.0231	1.872	1.410

Source: Reproduced from Sonntag et al. (2003).

Table C.7 Properties of air.

Temperature (K)	ρ (kg/m³)	C_p (J/kg K)	$\mu \times 10^7$ (N s/m³)	$v \times 10^7$ (m²/s)	$k \times 10^3$ (W/m K)	$\alpha \times 10^6$ (m²/s)	Pr
100	3.5562	1.032	71.1	2.00	9.34	2.54	0.786
150	2.3364	1.012	103.4	4.426	13.8	5.84	0.758
200	1.7458	1.007	132.5	7.590	18.1	10.3	0.737
250	1.3947	1.006	159.6	11.44	22.3	15.9	0.720
300	1.1614	1.007	184.6	15.89	26.3	22.5	0.707
350	0.9950	1.009	208.2	20.92	30.0	29.9	0.700
400	0.8711	1.014	230.1	26.41	33.8	38.3	0.690
450	0.7740	1.021	250.7	32.39	37.3	47.2	0.686
500	0.6964	1.030	270.1	38.79	40.7	56.7	0.684
550	0.6329	1.040	288.4	45.57	43.9	66.7	0.683
600	0.5804	1.051	305.8	52.69	46.9	76.9	0.685
650	0.5356	1.063	322.5	60.21	49.7	87.3	0.690
700	0.4975	1.075	338.8	68.10	52.4	98.0	0.695
750	0.4643	1.087	354.6	76.37	54.9	109.0	0.702
800	0.4354	1.099	369.9	84.93	57.3	120.0	0.709
850	0.4097	1.110	384.3	93.80	59.6	131.0	0.716
900	0.3868	1.121	398.1	102.9	62.0	143.0	0.720
950	0.3666	1.131	411.3	112.2	64.3	155.0	0.723
1000	0.3482	1.141	424.4	121.9	66.7	168.0	0.726
1100	0.3166	1.159	449.0	141.8	71.5	195.0	0.728
1200	0.2902	1.175	473.0	162.9	76.3	224.0	0.728
1300	0.2679	1.189	496.0	185.1	82.0	238.0	0.719
1400	0.2488	1.207	530.0	213.0	91.0	303.0	0.703
1500	0.2322	1.230	557.0	240.0	100.0	350.0	0.685
1600	0.2177	1.248	584.0	268.0	106.0	390.0	0.688
1700	0.2049	1.267	611.0	298.0	113.0	435.0	0.685
1800	0.1935	1.286	637.0	329.0	120.0	482.0	0.683
1900	0.1833	1.307	663.0	362.0	128.0	534.0	0.677
2000	0.1741	1.337	689.0	396.0	137.0	589.0	0.672
2100	0.1658	1.372	715.0	431.0	147.0	646.0	0.667
2200	0.1582	1.417	740.0	468.0	160.0	714.0	0.655
2300	0.1513	1.478	766.0	506.0	175.0	783.0	0.647
2400	0.1448	1.558	792.0	547.0	196.0	869.0	0.630
2500	0.1389	1.665	818.0	589.0	222.0	960.0	0.613
3000	0.1135	2.726	955.0	841.0	486.0	1570.0	0.536

Table C.8 Functional relationships of specific heat with temperature.

Gas	Chemical formula	Constant pressure-specific heat C_p (kJ/kg K)
Air		$C_p = 1.05 - 0.365\theta + 0.85\theta^2 - 0.39\theta^3$
Carbon dioxide	CO_2	$C_p = 0.45 + 1.67\theta - 1.27\theta^2 + 0.39\theta^3$
Carbon monoxide	CO	$C_p = 1.10 - 0.46\theta + 1.0\theta^2 - 0.454\theta^3$
Ethane	C_2H_6	$C_p = 0.18 + 5.92\theta - 2.31\theta^2 + 0.29\theta^3$
Ethanol	C_2H_5OH	$C_p = 0.2 - 4.65\theta - 1.82\theta^2 - 0.03\theta^3$
Hydrogen	H_2	$C_p = 13.46 + 4.6\theta - 6.85\theta^2 + 3.79\theta^3$
Methane	CH_4	$C_p = 1.2 + 3.25\theta + 0.75\theta^2 - 0.71\theta^3$
Methanol	CH_3OH	$C_p = 0.66 + 2.21\theta + 0.81\theta^2 - 0.89\theta^3$
Nitrogen	N_2	$C_p = 1.10 - 0.46\theta + 1.0\theta^2 - 0.454\theta^3$
Nitric oxide	NO	$C_p = 0.98 - 0.031\theta + 0.325\theta^2 - 0.14\theta^3$
Nitrous oxide	N_2O	$C_p = 0.49 + 1.65\theta - 1.31\theta^2 + 0.42\theta^3$
Oxygen	O_2	$C_p = 0.88 - 0.0001\theta + 5.4\theta^2 - 0.33\theta^3$
Sulfur dioxide	SO_2	$C_p = 0.37 + 1.05\theta - 0.77\theta^2 + 0.21\theta^3$
Steam	H_2O	$C_p = 1.79 + 0.107\theta + 0.586\theta^2 - 0.20\theta^3$

$\theta = \frac{T\,(K)}{100}$

Source: Reproduced from Sonntag et al. (2003).

Table C.9 Enthalpy of formation and absolute entropy of various substances at 25°C, 0.1 MPa

Substance	Formula	Molecular weight	Enthalpy of formation, $\overline{h}_f^0$, kJ/kmol	Absolute entropy, $\overline{s}_f^0$, kJ/kmol K
Ammonia	NH_3	17.031	45 720	192.572
Benzene	C_6H_6	78.114	+82 050	219.957
Butane	C_4H_{10}	58.124	−126 200	306.647
Carbon – graphite (solid)	C	12.011	0	5.740
Carbon dioxide	CO_2	44.010	−393 522	213.795
Carbon monoxide	CO	28.011	−110 527	197.653
Ethene	C_2H_4	28.054	+52 467	219.330
Ethane	C_2H_6	30.070	−84 740	229.597
Ethanol	C_2H_5OH	46.069	−235 000	282.444
Ethanol	C_2H_5OH	46.069	−277 380	160.554
Hydrogen	H_2	2.016	0	130.678
Methane	CH_4	16.043	−74 873	186.251
Methanol	C_2H_3OH	32.042	−201 300	239.709
Methanol (liquid)	C_2H_3OH	32.042	−239 220	126.809

(*continued*)

Table C.9   (Continued)

Substance	Formula	Molecular weight	Enthalpy of formation, $\overline{h}_f^0$, kJ/kmol	Absolute entropy, $\overline{s}_f^0$, kJ/kmol K
n-Octane	C_8H_{18}	114.232	−208 600	466.514
n-Octane (liquid)	C_8H_{18}	114.232	−250 105	360.575
Oxygen	O_2	31.999	0	205.148
Sulfur (solid)	S	32.06	0	32.056
Sulfur dioxide	SO_2	64.059	−296 842	248.212
Sulfur trioxide	SO_3	80.058	−395 765	256.769
Nitrogen	N_2	28.013	0	191.609
Nitric oxide	NO	30.006	+90 291	210.759
Nitrogen oxide	N_2O	44.013	+82 050	219.957
Nitrogen dioxide	NO_2	46.005	+33 100	240.034
Water (vapor)	H_2O	18.015	−241 826	188.834
Water (liquid)	H_2O	18.015	−285 830	69.950

Source: Reproduced from Sonntag et al. (2003).

Table C.10   Ideal gas enthalpy and entropy as a function of temperature of various substances.

Temperature, T (K)	$\overline{h}_T = \left(\overline{h} - \overline{h}_{298}^0\right)$ (kJ/kmol)	$\overline{s}_T^0$ (kJ/kmol K)
Carbon dioxide (CO_2)		
0	−9 364	0
100	−6 457	179.010
200	−3 413	199.976
298	0	213.794
300	69	214.024
400	4 003	225.314
500	8 305	234.902
600	12 906	243.284
700	17 754	250.752
800	22 806	257.496
900	28 030	263.646
1 000	33 397	269.299
1 100	38 885	274.528
1 200	44 473	279.390
1 300	50 148	283.931
1 400	55 895	288.190
1 500	61 705	292.199

(continued)

Table C.10 (Continued)

Temperature, T (K)	$\bar{h}_T = \left(\bar{h} - \bar{h}_{298}^0\right)$ (kJ/kmol)	$\bar{s}_T^0$ (kJ/kmol K)
Carbon monoxide (CO)		
0	−8 671	0
100	−5 772	165.852
200	−2 860	186.024
298	0	197.651
300	54	197.831
400	2 977	206.240
500	5 932	212.833
600	8 942	218.321
700	12 021	223.067
800	15 174	227.277
900	18 397	231.074
1 000	21 686	234.538
1 100	25 031	237.726
1 200	28 427	240.679
1 300	31 867	243.431
1 400	35 343	246.006
1 500	38 852	248.426
Hydrogen (H_2)		
0	−8 467	0
100	−5 467	100.727
200	−2 774	119.410
298	0	130.678
300	53	130.856
400	2 961	139.219
500	5 883	145.738
600	8 799	151.078
700	11 730	155.609
800	14 681	159.554
900	17 657	163.060
1 000	20 663	166.225
1 100	23 704	169.121
1 200	26 785	171.798
1 300	29 907	174.294
1 400	33 073	176.637
1 500	36 281	178.849

(continued)

Table C.10 (Continued)

Temperature, T (K)	$\bar{h}_T = \left(\bar{h} - \bar{h}_{298}^0\right)$ (kJ/kmol)	$\bar{s}_T^0$ (kJ/kmol K)
Oxygen (O_2)		
0	−8 683	0
100	−5 777	173.308
200	−2 868	193.483
298	0	205.148
300	54	205.329
400	3 027	213.873
500	6 086	220.693
600	9 245	226.450
700	12 499	231.465
800	15 836	235.920
900	19 241	239.931
1 000	22 703	243.579
1 100	26 212	246.923
1 200	29 761	250.011
1 300	33 345	252.878
1 400	36 958	255.556
1 500	40 600	258.068
Nitrogen (N_2)		
0	−8 670	0
100	−5 768	159.812
200	−2 857	179.985
298	0	191.609
300	54	191.789
400	2 971	200.181
500	5 911	206.740
600	8 894	212.177
700	11 937	216.865
800	15 046	221.016
900	18 223	224.757
1 000	21 463	228.171
1 100	24 760	231.314
1 200	28 109	234.227
1 300	31 503	236.943
1 400	34 936	239.487
1 500	38 405	241.881

(continued)

Table C.10   (Continued)

Temperature, T (K)	$\bar{h}_T = \left(\bar{h} - \bar{h}_{298}^0\right)$ (kJ/kmol)	$\bar{s}_T^0$ (kJ/kmol K)
Nitrogen oxide (NO)		
0	−9 192	0
100	−6 073	177.031
200	−2 951	198.747
298	0	210.759
300	55	210.943
400	3 040	219.529
500	6 059	226.263
600	9 144	231.886
700	12 308	236.762
800	15 548	241.088
900	18 858	244.985
1 000	22 229	248.536
1 100	25 653	251.799
1 200	29 120	254.816
1 300	32 626	257.621
1 400	36 164	260.243
1 500	39 729	262.703
Nitrogen dioxide(NO$_2$)		
0	−10 186	0
100	−6 861	202.563
200	−3 495	225.852
298	0	240.034
300	68	240.263
400	3 927	251.342
500	8 099	260.638
600	12 555	268.755
700	17 250	275.988
800	22 138	282.513
900	27 180	288.450
1 000	32 344	293.889
1 100	37 606	298.904
1 200	42 946	303.551
1 300	48 351	307.876
1 400	53 808	311.920
1 500	59 309	315.715

Source: Reproduced from Sonntag et al. (2003).

Table C.11 Enthalpy of combustion of hydrocarbon fuels at 25°C, 0.1 MPa.

Substance	Formula	Liquid water in the products (kJ/kmol)	Gas water in the products (kJ/kmol)
Ammonia	NH_3	−45 720	192.572
Benzene	C_6H_6	+82 050	219.957
Butane	C_4H_{10}	−126 200	306.647
Carbon – graphite (solid)	C	0	5.740
Carbon dioxide	CO_2	−393 522	213.795
Carbon monoxide	CO	−110 527	197.653
Ethene	C_2H_4	+52 467	219 330
Ethane	C_2H_6	−84 740	229.597
Ethanol	C_2H_5OH	−235 000	282.444
Ethanol	C_2H_5OH	−277 380	160.554
Hydrogen	H_2	0	130.678
Methane	CH_4	−74 873	186.251
Methanol	C_2H_3OH	−201 300	239.709
Methanol (liquid)	C_2H_3OH	−239 220	126.809
n-Octane	C_8H_{18}	−208 600	466.514
n-Octane (liquid)	C_8H_{18}	−250 105	360.575
Oxygen	O_2	0	205.148
Sulfur (solid)	S	0	32.056
Sulfur dioxide	SO_2	−296 842	248.212
Sulfur trioxide	SO_3	−395 765	256.769
Nitrogen	N_2	0	191.609
Nitric oxide	NO	+90 291	210.759
Nitrogen oxide	N_2O	+82 050	219.957
Nitrogen dioxide	NO_2	+33 100	240.034
Water (vapor)	H_2O	−241 826	188.834
Water (liquid)	H_2O	−285 830	69.950

Source: Reproduced from Sonntag et al. (2003).

Table C.12 Properties of saturated liquid.

Temperature (K)	ρ (kg/m³)	C_p (kJ/kg K)	$\mu \times 10^2$ (N s/m³)	$v \times 10^6$ (m²/s)	$k \times 10^3$ (W/m K)	$\alpha \times 10^7$ (m²/s)	Pr
Engine oil							
273	899.1	1.796	385	4 280	147	0.910	47 000
280	895.3	1.827	217	2 430	144	0.880	27 500
290	890.0	1.868	99.9	1 120	145	0.872	12 900
300	884.1	1.909	48.6	550	145	0.859	6 400
310	877.9	1.951	25.3	288	145	0.847	3 400
320	871.8	1.993	14.3	161	143	0.823	1 965
330	865.8	2.035	8.36	96.6	141	0.800	1 205
340	859.9	2.076	5.31	61.7	139	0.779	793
350	853.9	2.118	3.56	41.7	138	0.763	546
360	847.8	2.161	2.52	29.7	138	0.753	395
370	841.8	2.206	1.86	22.0	137	0.738	300
380	836.0	2.250	1.41	16.9	136	0.723	233
390	830.6	2.294	1.10	13.3	135	0.709	187
400	825.1	2.337	0.874	10.6	134	0.695	152
410	818.9	2.381	0.698	8.52	133	0.682	125
420	812.1	2.427	0.564	6.94	133	0.675	103
430	806.5	2.471	0.470	5.83	132	0.662	88
Ethylene glycol							
273	1 130.8	2.294	6.51	57.6	242	0.933	617
280	1 125.8	2.323	4.20	37.3	244	0.933	400
290	1 118.8	2.368	2.47	22.1	248	0.936	236
300	1 114.4	2.415	1.57	14.1	252	0.939	151
310	1 103.7	2.460	1.07	9.65	255	0.939	103
320	1 096.2	2.505	0.757	6.91	258	0.940	73.5
330	1 089.5	2.549	0.561	5.15	260	0.936	55.0
340	1 083.8	2.592	0.431	3.98	261	0.929	42.8
350	1 079.0	2.637	0.342	3.17	261	0.917	34.6
360	1 074.0	2.682	0.278	2.59	261	0.906	28.6
370	1 066.7	2.728	0.228	2.14	262	0.900	23.7
373	1 058.5	2.742	0.215	2.03	263	0.906	22.4

(*continued*)

Table C.12 (Continued)

Temperature (K)	ρ (kg/m^3)	C_p (kJ/kg K)	$\mu \times 10^2$ (N s/m^3)	$v \times 10^6$ (m^2/s)	$k \times 10^3$ (W/m K)	$\alpha \times 10^7$ (m^2/s)	Pr
Refrigerant 134							
230	1 426.8	1.249	0.049 12	0.3443	112.1	0.629	5.5
240	1 397.7	1.267	0.042 02	0.3006	107.3	0.606	5.0
250	1 367.9	1.287	0.036 33	0.2656	102.5	0.583	4.6
260	1 337.1	1.308	0.031 66	0.2368	97.9	0.560	4.2
270	1 305.1	1.333	0.027 75	0.2127	93.4	0.537	4.0
280	1 271.8	1.361	0.024 43	0.1921	89.0	0.514	3.7
290	1 236.8	1.393	0.021 56	0.1744	84.6	0.491	3.5
300	1 199.7	1.432	0.019 05	0.1588	80.3	0.468	3.4
310	1 159.9	1.481	0.016 80	0.1449	76.1	0.443	3.3
320	1 116.8	1.543	0.014 78	0.1323	71.8	0.417	3.2
330	1 069.1	1.627	0.012 92	0.1209	67.5	0.388	3.1
340	1 015.0	1.751	0.011 18	0.1102	63.1	0.355	3.1
350	9 51.3	1.961	0.009 51	0.1000	58.6	0.314	3.2
360	870.1	2.437	0.007 81	0.0898	54.1	0.255	3.5
370	740.3	5.105	0.005 80	0.0783	51.8	0.137	5.7
Water – saturated liquid							
273.15	1 000	4.217	1750		569		12.99
280	1 000	4.198	1422		582		10.26
290	1 001	4.184	1080		598		7.56
300	1 003	4.179	855		613		5.83
310	1 007	4.178	695		628		4.62
320	1 011	4.180	577		640		3.77
330	1 016	4.184	489		650		3.15
340	1 021	4.188	420		660		2.66
350	1 027	4.195	365		668		2.29
360	1 034	4.203	324		674		2.02
370	1 041	4.214	289		679		1.80
373.15	1 044	4.217	279		680		1.76
380	1 049	4.226	260		683		1.61
390	1 058	4.239	237		686		1.47
400	1 067	4.256	217		688		1.34

Table C.13 Properties of water vapor.

Temperature (K)	ρ (kg/m³)	C_p (kJ/kg K)	$\mu \times 10^7$ (N s/m³)	$v \times 10^6$ (m²/s)	$k \times 10^3$ (W/m K)	$\alpha \times 10^7$ (m²/s)	Pr
Water – vapor (steam)							
380	0.5863	2.006	127.1	21.68	24.6	20.4	1.06
400	0.5542	2.014	134.4	24.25	26.1	23.4	1.04
450	0.4902	1.980	152.5	31.11	29.9	30.8	1.01
500	0.4405	1.985	170.4	38.68	33.9	38.8	0.998
550	0.4005	1.997	188.4	47.04	37.9	47.4	0.993
600	0.3652	2.026	206.7	56.60	42.2	57.0	0.993
650	0.3380	2.056	224.7	66.48	46.4	66.8	0.996
700	0.3140	2.085	242.6	77.26	50.5	77.1	1.00
750	0.2931	2.119	260.4	88.84	54.6	88.4	1.00
800	0.2739	2.152	278.6	101.7	59.2	100	1.01
850	0.2579	2.186	296.9	115.1	63.7	113	1.02

Table C.14 Thermophysical properties of selected metallic solids at 300 K.

Substance	ρ (kg/m³)	C_p (J/kg K)	K (W/m K)	$\alpha \times 10^6$ (m²/s)
Aluminum	2 702	903	237	97.1
Boron	2 500	1 107	27.0	9.76
Cobalt	8 862	421	99.2	26.6
Copper-pure	8 933	385	401	117
Commercial bronze	8 800	420	52	14
Cartridge brass				
Gold	19 300	129	317	127
Iron	7 870	447	80.2	23.1
Lithium	534	3.57	84.8	44.6
Carbon steel				
Plain carbon	7 854	434	60.5	17.7
Stainless steel				
AISI 302	8 055	480	15.1	3.91
AISI 304	7 900	477	14.9	3.95
AISI 316	8 238	468	13.4	3.48
Lead	11 340	129	35.3	24.1
Magnesium	1 740	1 024	156	87.6
Nickel-pure	8 900	444	90.7	23.0
Nichrome	8 400	420	12	3.4
Inconel X-750	8 510	439	11.7	3.1
Palladium	12 020	244	71.8	24.5

(*continued*)

Table C.14 (Continued)

Substance	ρ (kg/m³)	C_p (J/kg K)	K (W/m K)	$\alpha \times 10^6$ (m²/s)
Platinum	21 450	133	71.6	25.1
Silicon	2 330	712	148	89.2
Silver	10 500	235	429	174
Thorium	11 700	118	54.0	39.1
Tin	7 300	227	66.6	40.1
Titanium	4 500	522	21.9	9.32
Tungsten	19 300	132	174	68.3
Uranium	19 070	116	27.6	12.5
Vanadium	6 100	489	30.7	10.3
Zinc	7 140	389	116	41..8
Zirconium	6 570	278	22.7	12.4

Table C.15 Thermophysical properties of selected nonmetallic solids.

Substance	ρ (kg/m³)	C_p (J/kg K)	K (W/m K)	$\alpha \times 10^6$ (m²/s)
Aluminum oxide sapphire	3970	765	46	15.1
Aluminum oxide polycrystalline	3970	765	36.0	11.9
Boron	2500	1105	27.6	9.99
Boron fiber epoxy (30% vol.)	2080	1122		
Parallel			2.29	
Perpendicular			0.59	
Carbon				
Amorphous	1950			
Diamond type IIa	3500	509		
k parallel to c-axis			1.60	
k perpendicular to c-axis			2300	
Graphite fiber				
Epoxy composite (25% vol.)	1400	935		
k parallel to c-axis			11.1	
k perpendicular to c-axis			0.87	
Silicon carbide	3160	675	490	230
Silicon dioxide	2650	745		
Crystalline quartz				
k parallel to c-axis			10.4	
k perpendicular to c-axis			6.21	
Silicon dioxide Polycrystalline – fused silica	2220	745	1.38	0.834
Silicon nitride	2400	691	16.0	9.65
Sulfur	2070	708	0.206	0.141
Titanium dioxide Polycrystalline	4157	710	2.8	2.8

Table C.16 Thermophysical properties of selected liquid metals.

Liquid metal	Atomic number	Boiling point (°C)	Melting point (°C)	Viscosity (cP)	Thermal conductivity (Calculated/cm/s/K)	Heat capacity (Calculated/g/K)	Density (g/cm³) × 10⁻³
Bismuth	83	1477	271	0.996	0.037	0.0376	9.66
Lead	82	1737	327	1.85	0.037	0.037	10.39
Potassium	19	760	63.8	0.191	0.0956	0.1826	0.747
Sodium	11	883	97.8	0.269	0.1701	0.3065	0.854

Table C.17 Total and normal emissivity of selected surfaces.

Surface type	Temperature range (K)	Emissivity (ε)
Metallic		
Aluminum		
Highly polished film	100–600	0.02–0.06
Foil – bright	100–300	0.06–0.07
Anodized	300–400	0.82–0.76
Chromium		
Polished or plated	100–600	0.05–0.14
Copper		
Highly polished	300–1000	0.03–0.04
Stably oxidized	600–1000	0.50–0.80
Gold		
Highly polished	100–1000	0.01–0.06
Foil – bright	100–300	0.06–0.07
Molybdenum		
Polished	600–2500	0.06–0.26
Shot blasted – rough	600–1500	0.25–0.40
Stably oxidized	600–800	0.80–0.82
Nickel		
Polished	600–1200	0.09–0.17
Stably oxidized	600–1000	0.40–0.57
Platinum		
Polished	800–1500	0.10–0.18
Silver		
Polished	300–1000	0.02–0.08

(continued)

Table C.17 (Continued)

Surface type	Temperature range (K)	Emissivity (ε)
Stainless steels		
Tungsten		
Polished	1000–2500	0.01–0.29
Filament	3588	0.39
Zinc		
Galvanized sheet iron	300	0.23
Nonmetallic		
Aluminum oxide	600–1500	0.69–0.41
Asphalt pavement	300	0.85–0.93
Building materials		
Brick red	300	0.93–0.96
Wood	300	0.82–9.92
Concrete	300	0.88–0.93
Ice	273	0.95–0.98
Paints		
Black	300	0.98
White acrylic	300	0.90
White – zinc oxide	300	0.92
Sand	300	0.90
Silicon carbide	600–1500	0.87–0.87
Skin	300	0.95
Snow	273	0.82–0.90
Soil	300	0.93–0.96
Rocks	300	0.88–0.95
Teflon	300–500	0.85–0.92
Vegetation	300	0.92–0.96
Water	300	0.96

Table C.18 Solar radiative properties.

Material	Absorptivity	Transmissivity
Aluminum		
Polished	0.09	
Anodized	0.14	
Quartz overcoated	0.11	
Foil	0.15	
Brick – red	0.63	
Concrete	0.60	
Galvanized sheet metal		
Clean, new	0.65	
Oxidized, weathered	0.80	
Glass – 3.2-mm thickness		
Float or tempered		0.79
Low iron oxide		0.88
Metal, plated		
Black sulfide	0.92	
Black cobalt oxide	0.93	
Black nickel oxide	0.92	
Paints		
Black	0.98	
White acrylic	0.26	
White zinc oxide	0.16	
Plexi glass, 3.2-mm thickness		0.90
Snow		
Fine particles-fresh	0.13	
Ice granules	0.33	
Teflon – 0.13-mm thickness		0.92

Table C.19 Steel-pipe dimensions.

Nominal pipe size (in.)	OD (in.)	Schedule no.	Wall thickness (in.)	ID (in.)
$\frac{1}{8}$	0.405	40	0.068	0.269
		80	0.095	0.215
$\frac{1}{4}$	0.540	40	0.088	0.364
		80	0.119	0.302
$\frac{3}{8}$	0.675	40	0.091	0.493
		80	0.126	0.423
$\frac{1}{2}$	0.840	40	0.109	0.622
		80	0.147	0.546
$\frac{3}{4}$	1.050	40	0.113	0.824
		80	0.154	0.742
1	1.315	40	0.133	1.049
		80	0.179	0.957
		160	0.250	0.815
11/4	1.660	40	0.140	1.380
		140	0.250	1.160
$1\frac{1}{2}$	1.900	40	0.145	1.610
		80	0.200	1.500
		160	0.281	1.338
2	2.375	40	0.154	2.067
		80	0.218	1.939
		160	0.343	1.689
2.1/2	2.875	40	0.200	2.469
		80	0.276	2.323
		160	0.375	2.125
3	3.500	40	0.216	3.068
		80	0.300	2.900
		160	0.438	2.624
4	4.500	40	0.237	4.026
		80	0.337	3.826
		120	0.438	3.624
		160	0.531	3.438
5	5.563	40	0.258	5.047
		80	0.375	4.813
		120	0.500	4.563
		160	0.625	4.313
6	6.625	40	0.280	6.065
		80	0.432	5.761
		120	0.562	5.501
		160	0.718	5.189
8	8.625	40	0.322	7.981
		80	0.500	7.625
		120	0.718	7.189
		160	0.906	6.813

(*continued*)

Table C.19 (Continued)

Nominal pipe size (in.)	OD (in.)	Schedule no.	Wall thickness (in.)	ID (in.)
10	10.75	40	0.365	10.020
		80	0.500	9.750
		120	0.843	9.064
		160	1.125	8.500
12	12.75	40	0.406	11.938
		80	0.687	11.376
		120	1.000	10.750
		160	1.312	10.126
14	14.00	40	0.438	13.124
		80	0.750	12.500
		120	0.937	12.126
		160	1.406	11.168
16	16.00	40	0.500	15.000
		80	0.843	14.314
		120	1.218	13.564
		160	1.593	12.814
18	18.00	40	0.562	16.876
		80	0.937	16.126
		120	1.375	15.250
		160	1.781	14.438
20	20.0	40	0.593	18.814
		80	1.031	17.938
		120	1.500	17.000
		160	1.968	16.064
24	24.00	40	0.687	22.626
		80	1.218	21.564
		120	1.812	20.376
		160	2.343	19.314

Table C.20 Tube dimensions.

Diameter		Thickness	
External (in.)	Internal (in.)	BWG gage	Nominal wall (in.)
$\frac{5}{8}$	0.527	18	0.049
	0.495	16	0.065
	0.459	14	0.083
$\frac{3}{4}$	0.652	18	0.049
	0.620	16	0.065
	0.584	14	0.083
	0.560	13	0.095
1	0.902	18	0.049
	0.870	16	0.065
	0.834	14	0.083
	0.810	13	0.095
$1\frac{1}{4}$	1.152	18	0.049
	1.120	16	0.065
	1.084	14	0.083
	1.060	13	0.095
	1.032	12	0.109
$1\frac{1}{2}$	1.402	18	0.049
	1.370	16	0.065
	1.334	14	0.083
	1.310	13	0.095
	1.282	12	0.109
$1\frac{3}{4}$	1.620	16	0.065
	1.584	14	0.083
	1.560	13	0.095
	1.532	12	0.109
	1.490	11	0.120
2	1.870	16	0.065
	1.834	14	0.083
	1.810	13	0.095
	1.782	12	0.109
	1.760	11	0.120

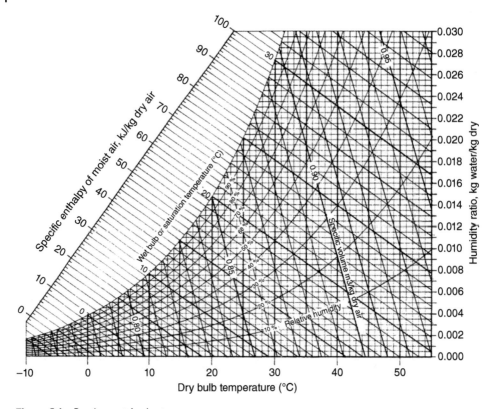

Figure C.1   Psychrometric chart.

Bibliography

Sonntag, R.E., Borgnake, C., and van Wylen, G.J. (2003). *Fundamentals of Thermodynamics*, 6e. Wiley.

Appendix D

Modified Bessel Function of the First and the Second Kinds

Table D.1 Modified Bessel function of the first and second kind.

x	$I_0(x)$	$K_0(x)$	$I_1(x)$	$K_1(x)$
0.0	1.0000		0	
0.1	1.0025	2.4271	0.0501	9.8538
0.2	1.0100	1.7527	0.1005	4.7760
0.3	1.0226	1.3725	0.1517	3.0560
0.4	1.0404	1.1145	0.2040	2.1844
0.5	1.0635	0.9244	0.2579	1.6564
0.6	1.0920	0.7775	0.3137	1.3028
0.7	1.1263	0.6605	0.3719	1.0503
0.8	1.1665	0.5653	0.4329	0.8618
0.9	1.2130	0.4867	0.4371	0.7165
1.0	1.2661	0.4210	0.5652	0.6019
1.1	1.3262	0.3656	0.6375	0.5098
1.2	1.3917	0.3185	0.7147	0.4346
1.3	1.4693	0.2782	0.7973	0.3725
1.4	1.5534	0.2437	0.8861	0.3206
1.5	1.6467	0.2138	0.9817	0.2774
1.6	1.7500	0.1880	1.0848	0.2406
1.7	1.8640	0.1655	1.1963	0.2094
1.8	1.9896	0.1459	1.3172	0.1826
1.9	2.1277	0.1288	1.4482	0.1597
2.0	2.2796	0.113 89	1.5906	0.139 87
2.1	2.4463	0.100 78	1.7455	0.122 75
2.2	2.6291	0.089 26	1.9141	0.107 90
2.3	2.8296	0.079 14	2.0978	0.094 98
2.4	3.0493	0.070 22	2.2981	0.083 72
2.5	3.2898	0.062 35	2.5167	0.073 89

(continued)

Design of Thermal Energy Systems, First Edition. Pradip Majumdar.
© 2021 John Wiley & Sons Ltd. Published 2021 by John Wiley & Sons Ltd.
Companion website: www.wiley.com/go/majumdar

Table D.1 (Continued)

x	$I_0(x)$	$K_0(x)$	$I_1(x)$	$K_1(x)$
2.6	3.5533	0.055 40	2.7554	0.065 28
2.7	3.8417	0.049 26	3.0161	0.057 74
2.8	4.1573	0.043 82	3.3011	0.051 11
2.9	4.5027	0.039 01	3.6126	0.045 29
3.0	4.8808	0.034 74	3.9534	0.0416
3.1	5.2945	0.030 95	4.3262	0.035 63
3.2	5.7472	0.027 59	4.7343	0.031 64
3.3	6.2426	0.024 61	5.1810	0.028 12
3.4	6.7848	0.021 96	5.6701	0.025 00
3.5	7.3782	0.019 60	6.2058	0.022 24
3.6	8.0277	0.017 50	6.7927	0.019 79
3.7	8.7386	0.015 63	7.4357	0.017 63
3.8	9.5169	0.013 97	8.1404	0.015 71
3.9	10.369	0.012 48	8.9128	0.014 00
4.0	11.302	0.011 160	9.7595	0.012 484
4.1	12.324	0.009 980	10.688	0.011 136
4.2	13.442	0.008 927	11.706	0.009 938
4.3	14.668	0.007 988	12.822	0.008 872
4.4	16.010	0.007 149	14.046	0.007 923
4.5	17.481	0.006 400	15.389	0.007 078
4.6	19.093	0.005 730	16.863	0.006 325
4.7	20.858	0.005 132	18.479	0.005 654
4.8	22.794	0.004 597	20.253	0.005 055
4.9	24.915	0.004 119	22.199	0.004 521
5.0	27.240	0.003 691	24.336	0.004 045

Appendix E

Constants and Conversion Units

Table E.1 Fundamental physical constants.

Gravitational acceleration:	$g = 9.806\,65$ m/s^2
	$= 32.174\,05$ ft/s^2
Faraday constant: $F = 96\,486$ C/mol	
Stefan Boltzmann: $\sigma = 5.670\,51 \times 10^{-8}$ W/m^2 K^4	
Universal gas constant:	$\overline{R} = 8.314\ 51\ \frac{\text{kJ}}{\text{kmol}}$ K
	$= 1.985\ 89$ Btu/lbmol R

Table E.2 Conversion factors.

Area (A)

1 mm$^2 = 1.0 \times 10^{-6}$ m^2	1 ft$^2 = 144$ in.^{2}
1 cm$^2 = 1.0 \times 10^{-4}$ m$^2 = 0.1550$ in.^{2}	1 in$^2 = 6.4516$ cm$^2 = 6.4516 \times 10^{-4}$ m^2
1 m$^2 = 10.7639$ ft^2	1 ft$^2 = 0.092\,903$ m^2

Conductivity (k)

1 W/m K $= 0.577\,789$ Btu/h R	1 W/m K $= 0.577\,789$ Btu/h R

Distance

1 cm $= 0.3970$ in.	1 in. $= 2.54$ cm
1 cm $= 0.0328$ ft	1 ft $= 0.3048$ m
1 m $= 3.280\,84$ ft	

Density (ρ)

1 kg/m$^3 = 0.0624$ lbm/ft^3	1 lbm/ft$^3 = 16.018$ kg/m^3
1 g/cm$^3 = 1$ kg/l	

(continued)

Design of Thermal Energy Systems, First Edition. Pradip Majumdar.
© 2021 John Wiley & Sons Ltd. Published 2021 by John Wiley & Sons Ltd.
Companion website: www.wiley.com/go/majumdar

Table E.2   (Continued)

Energy (E)

$1\,J = 1\,N\,m = 1\,kg\,m^2/s^2$

$1\,J = 0.7375\,lbf\,ft$

$1\,Cal = 4.1868\,J$

$1\,J = 2.777\,778 \times 10^{-7}\,kWh$

$1\,J = 9.478\,134 \times 10^{-4}\,Btu$

$1\,J = 6.241\,506 \times 10^{18}\,eV$

$1\,lbf\,ft = 1.3558\,J = 1.285\,07 \times 10^{-3}\,Btu$

$1\,Btu = 1.055\,kJ = 778.1693\,lbf\,ft$

Specific Heat (C)

$1\,kJ/kg\,K = 0.2388\,Btu/lbm\,R$

$1\,Btu/lbm\,R = 4.1868\,kJ/kg\,K$

Heat flux (per unit area)

$1\,W/m^2 = 0.316\,998\,Btu/h\,ft^2$

$1\,Btu/h\,ft^2 = \mathbf{3.154\,59}\,W/m^2$

Heat transfer coefficient

$1\,W/m^2\,K = 0.176\,11\,Btu/h\,ft^2\,R$

$1\,Btu/h\,ft^2\,R = 5.678\,28\,W/m^2\,K$

Length

$1\,cm = 0.3970\,in.$

$1\,m = 3.28\,084\,ft$

$1\,in. = 2.54\,cm$

$1\,ft = 0.3048\,m$

Mass

$1\,kg = 2.204\,623\,lbm$

$1\,lbm = 0.453\,592\,kg$

Moment (torque, T)

$1\,N\,m = 0.737\,562\,lbf\,ft$

$1\,lbf\,ft = 1.355\,818\,N\,m$

Power

$1\,W = 1\,J/s = 1\,N\,m/s$

$1\,kW = 3412.14\,Btu/hr$

$1\,hp\,(metric) = 0.735\,499\,kW$

$1\,kW = 1.34\,hp$

$1\,lbf\,ft/s = 1.355\,818\,W$

$1\,Btu/s = 1.0550\,56\,kW$

$1\,hp\,(UK) = 0.7457\,kW$

$1\,hp\,(UK) = 550\,lbf\,ft/s$

Pressure

$1\,Pa = 1\,N/m^2 = 1\,kg/m\,s^2$

$1\,bar = 100\,kPa$

$1\,atm = 1.013\,25 \times 10^5\,Pa = 101.325\,kPa$

$1\,mmHg\,[0\,°C] = 0.133\,322\,kPa$

$1\,mmH_2O\,[4\,°C] = 9.806\,38\,kPa$

$1\,lbf/in.^2 = 6.894\,757\,kPa$

$1\,atm = 14.695\,94\,lbf/in.^2(psi)$

$1\,in.\,Hg\,[0\,°C] = 0.491\,15\,lbf/in.^2$

$1\,in.\,H_2O\,[4\,°C] = 0.036\,126\,lbf/in.^2$

Specific energy

$1\,kJ/kg = 0.429\,92\,Btu/lbm$

$1\,Btu/lbm = 2.326\,kJ/kg$

(continued)

Table E.2 (Continued)

Specific kinetic energy

$1\,m^2/s^2 = 0.001\,kJ/kg$ $\qquad\qquad$ $1\,ft^2/s^2 = 3.9941 \times 10^{-5}\,Btu/lbm$

$1\,kg/kg = 1000\,m^2/s^2$ $\qquad\qquad$ $1\,Btu/lbm = 25\,037\,ft^2/s^2$

Velocity

$1\,m/s = 3.280\ 84\,ft/s$ $\qquad\qquad$ $1\,ft/s = 0.681\ 818\,mi/h$

$1\,km/h = 0.277\ 78\,m/s$ $\qquad\qquad$ $= 0.3048\,m/s$

$= 0.911\ 34\,ft/s$ $\qquad\qquad$ $1\,mi/h = 1.466\ 67\,ft/s$

$= 0.621\ 37\,mi/h$ $\qquad\qquad$ $= 0.447\,m/s$

$\qquad\qquad$ $= 1.609\ 344\,km/h$

Volume

$1000\,cm^3 = 1\,l$

$1\,l = 0.264\,gal$

$1\,l = 3.53 \times 10^{-2}\,ft^3$

Weight

$1\,kg = 2.2\,lb$

Index

Design of Thermal Energy Systems, First Edition. Pradip Majumdar.
© 2021 John Wiley & Sons Ltd. Published 2021 by John Wiley & Sons Ltd.
Companion website: www.wiley.com/go/majumdar